Lecture Notes in Computer Science

Lecture Notes in Computer Science

Edited by G. Goos and J. Hartmanis

230

8th International Conference on Automated Deduction

Oxford, England, July 27–August 1, 1986
Proceedings

Edited by Jörg H. Siekmann

Springer-Verlag
Berlin Heidelberg New York London Paris Tokyo

Editor

Jörg H. Siekmann
Universität Kaiserslautern, FB Informatik
Postfach 3049, 6750 Kaiserslautern, Federal Republic of Germany

CR Subject Classifications (1985): F.3.1, F.4.1, F.4.2, I.2.3, I.2.4

ISBN 3-540-16780-3 Springer-Verlag Berlin Heidelberg New York
ISBN 0-387-16780-3 Springer-Verlag New York Berlin Heidelberg

Printing and binding: Beltz Offsetdruck, Hemsbach/Bergstr.
2145/3140-543210

FOREWORD

The CADE conferences are the main international forum for the presentation of research on all aspects of automated deduction, computational logic and theoretical foundations of Artificial Intelligence. The main areas being:

- Design and Implementation of Deduction Systems
- Application of Deduction Systems
- Logical Calculi suitable for Computer Applications
- Term Rewriting Systems
- Logical Methods for Knowledge Representation
- Deductive Components of Expert Systems
- Inference Machines
- Logic Programming
- Deductive Databases
- Formal Specification and Verification of Programs
- Program Synthesis
- Unification Theory
- Mechanization of Mathematical Induction
- Nonmonotonic and Default Reasoning
- Nonclassical Logics and Proof Theory
- Computer Algebra

The EIGHTH INTERNATIONAL CONFERENCE ON AUTOMATED DEDUCTION, held in July 1986 at St.Catherine's College in Oxford, England, included the presentation of 25 long papers, 24 short papers and 4 invited talks by P.Andrews (Carnegie Mellon University, USA), G.Huet (INRIA, France), R.Reiter (University of Toronto, Canada) and D.Warren (University of Manchester, England).

The 49 presented papers were selected by the programme committee out of a total of 92 submitted papers – almost twice as many papers as at previous CADE conferences.

Apart from the explosion in the size of the conference there is also a noticable trend towards the broadening of the scope of CADE as witnessed by the increasing number of papers on the mechanization of nonclassical logics as well as the many papers on theoretical and logical foundations of Artificial Intelligence.
The most active subareas (if judged by their number of papers) currently are Unification Theory and Term Rewriting Systems.
Interesting is also the sharp increase in European participation:

Number of papers from Europe:	(24 | 12 | 5 | 10)
Number of papers from USA:	(20 | 17 | 17 | 19)
Other countries:	(5 | 0 | 2 | 1)

In parentheses: (CADE-8 | CADE-7 | CADE-6 | CADE-5)

However, as many subfields of Automated Deduction, whose classical papers were initially presented at CADE conferences (such as Logic Programming, Term Rewriting Systems, Deductive Databases, Nonmonotonic Reasoning), now have flourishing international conferences of their own , CADE does not represent the entire work on the mechanization of deductive reasoning any more and general trends in this area can no longer be assessed solely on the base of this conference.

A BEST PAPER AWARD was donated by SPRINGER Verlag and the programme committee selected three candidate papers for the award:

N.Eisinger: "What you always wanted to know about Clause Graph Resolution"
L.Bachmair, N.Dershowitz: "Commutation, Transformation and Termination"
M.Stickel: "A PROLOG Technology Theorem Prover".

Out of these three papers the contribution by Norbert Eisinger was selected for the award by a subcommittee consisting of A.Bundy, B.Kowalski, R.Reiter and D.Warren.

As a novel feature the proceedings include extended abstracts of 21 major deduction systems, many of them had a development time of more than a decade.

The seven previous conferences on automated deduction have been held at:

Argonne Nat. Lab., USA, 1974 (Proc. IEEE Transactions on Computers, vol. C-25, no. 8, Aug.1976);
Oberwolfach, West Germany, 1976;
Mass. Inst. of Technology, USA, 1977 (Proc. MIT, Cambridge, Mass.);
University of Texas at Austin, USA, 1979 (Proc. Texas University, Austin);
Les Arcs, France, 1980 (Proc. in SPRINGER LNCS, vol.87);
Courant Institute, New York, USA, 1982 (Proc. in SPRINGER LNCS, vol.138);
Napa, California, USA, 1984 (Proc. in SPRINGER LNCS, vol.170).

The conference was organized by:

PROGRAMME COMMITTEE

J.Barwise (CSLI, USA)
B.Buchberger (Univ. Linz, Austria)
A.Bundy (Univ. of Edinburgh, UK)
B.Boyer (MCC, USA)
J.Goguen (SRI, USA)
P.Hajek (Univ. Prag, CSSR)
J.-P.Jouannoud (CRIN, France)
R.Kowalski (Imperial College, UK)
E.L.Lusk (Argonne National Lab., USA)
M.A.McRobbie (Australian National Univ., Australia)
D.Musser (General Electric, USA)
B.Orevkov (Math. Inst. Leningrad, USSR)
R.Reiter (Univ. of Toronto, Canada)
M.M.Richter (Univ. Kaiserslautern, FRG)
D.Scott (Carnegie Mellon Univ., USA)
R.Shostak (SRI, USA)
J.Siekmann (Univ. Kaiserslautern, FRG)
M.Stickel (SRI, USA)
Ch.Walther (Univ. Karlsruhe, FRG)
D.Warren (Univ. of Manchester, UK)

CHAIRMAN

J.H.Siekmann
Fachbereich Informatik
Postfach 3049
6750 Kaiserslautern
West Germany

LOCAL ARRANGEMENTS

R.J.Cunningham
Dept. of Computing
Imperial College of Science and Technology
Exhibition Road
London SW7
England

EXHIBITION

A.J.J. Dick
Building R1/2.60
Rutherford Appleton Laboratory
Chilton, Didcot
Oxfordshire
England

SPONSORS

Imperial College and the Alvey Directorate
in co-operation with the
Association for Automated Reasoning
British Computer Society
United Kingdom Science and Engineering
Research Council

CONTENTS

Equality Reasoning

Program Verification

Graph Based Deduction

Special Deduction Systems

Invited Talk

Constructive ATP

Unification Theory

Theoretical Issues

Connections and Higher-Order Logic

Peter B. Andrews

Mathematics Department,
Carnegie Mellon University,
Pittsburgh, Pa. 15213, U.S.A.

Abstract

Theorem proving is difficult and deals with complex phenomena. The difficulties seem to be compounded when one works with higher-order logic, but the rich expressive power of Church's formulation [10] [3] of this language makes research on theorem proving in this realm very worthwhile. In order to make significant progress on this problem, we need to try many approaches and ideas, and explore many questions. A highly relevant question is "What makes a logical formula valid?".

One approach to this question is semantic. Theorems are true because they express essential truths and thus are true in all models of the language in which they are expressed. Truth can be perceived from many perspectives, so there may be many essentially different proofs of theorems. This point of view is very appealing, but it does not shed much light on the basic question of what makes certain sentences true in all models, while others are not.

Of course, theorems are formulas which have proofs, and every proof in any logical system may provide some insight. This suggests seeing what one can learn by studying the forms proofs can take. While this may be helpful, many of the most prominent features of proofs seem to be influenced as much by the logical system in which the proof is given as by the theorem that is being proved.

We focus on trying to understand what there is about the syntactic structures of theorems that makes them valid. In the case of formulas of propositional calculus, one can test a formula for being a tautology in an explicit syntactic way. However, simply checking each line of a truth table is not really very enlightening, and we may still find ourselves asking "What is there about the structure of this formula which makes it a tautology?". Clearly the pattern of occurrences of positive and negative literals in such a formula is very important, and this leads to the theory of *connections* or *matings* [4] [1] [6]. A connection is a pair of literal-occurrences, and a mating is a set of connections, i.e., a relation between occurrences of literals. A mating is *acceptable* if its structure guarantees that the formula is a tautology. Perhaps much more can be said about criteria for matings to be acceptable.

This work is supported by NSF grant DCR-8402532.

One view is that in higher-order logic as well as in first-order logic, formulas are valid because they can be expanded into tautologies. (Of course, this is the fundamental idea underlying various forms of Herbrand's Theorem.) Valid formulas may be regarded as tautologies which have been abbreviated by existential generalization, disjunctive simplification, introduction of definitions, and λ-conversion. Their basically tautological structure has been disguised, and to prove them we must unmask the disguises.

Of course, to understand in this sense why a formula is valid one needs more than the tautolgy hidden within it; one needs to know how to expand the formula into the tautology. A very elegant, concise, and nonredundant way of representing all this information is an *expansion tree proof (ET-proof)* [14]. Once an ET-proof has been found, it can be converted without further search into a natural deduction proof [15]. Moreover, proofs of formulas of first- and higher-order logic in virtually any format induce ET-proofs. (See [16] and [17]). An ET-proof embodies the essential combinatorial structure of many proofs, and is a key to translating between them.

Redundancy occurs naturally in human discourse, and it is quite appropriate that proofs in natural deduction style contain a great deal of redundancy. However, when one is searching for a proof, it is desirable to work within a context where one can focus on the essential features of the problem as directly and economically as possible. Thus, ET-proofs provide an excellent context in which to conduct the search.

The basic processes involved in constructing an ET-proof are expanding the formula and searching for an acceptable mating of the literal-occurrences in the formula. The formula can be expanded by duplicating quantifiers and by instantiating quantifiers on higher-order variables. The matingsearch process involves building up an acceptable set of connections which is *compatible*, i.e., a set for which there is a substitution which simultaneously makes the literal-occurrences in each connection complementary. The substitution can be found by applying Huet's unification algorith [12], which may never terminate if the required substitution does not exist.

Let us consider certain aspects of the problem of searching for ET-proofs, and the kind of computer environment which will facilitate the process. Since many difficult mathematical questions can be expressed as formulas of higher-order logic, it is clear that a useful theorem proving system in this realm will allow for human input, so it should be able to operate in a mixture of automatic and interactive modes. The search for an ET-proof in higher-order logic may involve a number of simultaneous unbounded searches, so a search procedure which is in principle complete will achieve early success only on relatively simple problems, and should be augmented by general heuristic search processes. Also, we can anticipate that we will wish to take advantage of multiprocessors as suitable ones become readily available. Thus, the search process should deal with well defined data structures which are modified in systematic ways by processes which can work independently and interact in controlled and fruitful ways. To facilitate human interaction, facilities should be available for displaying information and partial proofs in a variety of formats.

Matingsearch heuristics can be based on the need to establish connections between literal-occurrences in such a way that an acceptable mating can eventually be achieved, and to discover incompatibilities quickly. Unless an acceptable mating is quickly found, most of the useful information acquired in the matigsearch process concerns what will not work, i.e., which sets of connections are incompatible. A number of matingsearch processes can profitably operate simultaneously if each contributes to, and makes use of, a collection of incompatible sets of connections which we shall call the *failure record.* The failure record can be constantly augmented not only by incompatibilities discovered by the matingsearch processes, but also by analyses of symmetries in the formula, deeper analyses of potential matings by the unification algorithm, and other methods. When the formula is expanded, the matingsearch processes may have to reexamine what must be done to achieve an acceptable mating, but the information embodied in the failure record remains useful.

The need to expand the formula by instantiating quantifiers with terms which cannot be generated by unification of existing subformulas is one of the vexing problems which distinguishes higher-order logic from first-order logic. Bledsoe has made some significant contributions here [8] [9], but much more needs to be done. Naturally, one can generate such terms incrementally by using primitive substitutions (projections and substitutions which introduce single connectives and quantifiers in a general way), thus achieving the effects of Huet's splitting rules [11]. The development of heuristics to guide this process, and to determine the types of variables introduced in new quantifiers, is a major area for research. It is important to understand how appropriate instantiations can be used to create the literals needed for the connections of an acceptable mating.

Many problems arise in connection with the automation of higher-order logic. By formulating these precisely and dealing with them systematically, we can hope to eventually make progress in this important realm.

References

1. Peter B. Andrews, *Theorem Proving via General Matings,* Journal of the ACM 28 (1981), 193-214.

2. Peter B. Andrews, Dale A. Miller, Eve Longini Cohen, Frank Pfenning, "Automating Higher-Order Logic," in *Automated Theorem Proving: After 25 Years,* edited by W. W. Bledsoe and D. W. Loveland, Contemporary Mathematics series, vol. 29, American Mathematical Society, 1984, 169-192.

3. Peter B. Andrews, *An Introduction to Mathematical Logic and Type Theory: To Truth Through Proof,* Academic Press, 1986.

4. Wolfgang Bibel, *On Matrices with Connections,* Journal of the ACM 28 (1981), 633-645.

5. Wolfgang Bibel, *Automated Theorem Proving,* Vieweg, Braunschweig, 1982.

6. Wolfgang Bibel, *Matings in Matrices,* Communications of the ACM 26 (1983), 844-852.

7. Wolfgang Bibel and Bruno Buchberger, *Towards a Connection Machine for Logical Inference*, Future Generations Computer Systems **1** (1984-1985).

8. W. W. Bledsoe, "A Maximal Method for Set Variables in Automatic Theorem Proving," in *Machine Intelligence 9*, Ellis Harwood Ltd., Chichester, 1979, pp. 53-100.

9. W. W. Bledsoe. Using Examples to Generate Instantiations for Set Variables, ATP-67, University of Texas at Austin, July 1982, 44 pp

10. Alonzo Church, *A Formulation of the Simple Theory of Types*, Journal of Symbolic Logic **5** (1940), 56-68.

11. Gérard P. Huet, "A Mechanization of Type Theory," in *Proceedings of the Third International Joint Conference on Artificial Intelligence*, IJCAI, 1973, 139-146.

12. Gérard P. Huet, *A Unification Algorithm for Typed λ-Calculus*, Theoretical Computer Science **1** (1975), 27-57.

13. D. C. Jensen and T. Pietrzykowski, *Mechanizing ω-Order Type Theory Through Unification*, Theoretical Computer Science **3** (1976), 123-171.

14. Dale A. Miller. *Proofs in Higher-Order Logic*, Ph.D. Thesis, Carnegie-Mellon University, October, 1983. 81 pp.

15. Dale A. Miller, "Expansion Tree Proofs and Their Conversion to Natural Deduction Proofs," in *7th International Conference on Automated Deduction, Napa, California, USA*, edited by R. E. Shostak, Lecture Notes in Computer Science 170, Springer-Verlag, May 14-16, 1984, 375-393.

16. Frank Pfenning, "Analytic and Non-analytic Proofs," in *7th International Conference on Automated Deduction, Napa, California, USA*, edited by R. E. Shostak, Lecture Notes in Computer Science 170, Springer-Verlag, May 14-16, 1984, 394-413.

17. Frank Pfenning. *Proof Transformations in Higher-Order Logic*, Ph.D. Thesis, Carnegie-Mellon University, 1986.

COMMUTATION, TRANSFORMATION, AND TERMINATION*

Leo Bachmair

Nachum Dershowitz

Department of Computer Science
University of Illinois at Urbana–Champaign
Urbana, Illinois 61801
U.S.A.

Abstract. In this paper we study the use of *commutation* properties for proving *termination* of rewrite systems. Commutation properties may be used to prove termination of a combined system $R \cup S$ by proving termination of R and S separately. We present termination methods for ordinary and for equational rewrite systems. Commutation is also important for *transformation* techniques. We outline the application of transforms—mappings from terms to terms—to termination in general, and describe various specific transforms, including transforms for associative–commutative rewrite systems.

1. Introduction

Rewrite techniques have been applied to various problems, including the word problem in universal algebra (Knuth and Bendix, 1970), theorem proving in first order logic (Hsiang, 1985), proofs of inductive properties of abstract data types (Musser, 1980; Huet & Hullot, 1982), and computing with rewrite programs (O'Donnell, 1985; Dershowitz, 1985a). Many of these applications require a *terminating* rewrite system (see Dershowitz, 1985b). In this paper we study the use of *commutation* properties for proving termination of rewrite systems. We present termination methods for ordinary and for *equational* rewrite systems. In particular, we consider termination of *associative-commutative* rewrite systems.

Commutation was used by Rosen (1973) for establishing Church–Rosser properties of combinations of rewrite systems, and by Raoult and Vuillemin (1980) for proving operational and semantic equivalence between recursive programs. Dershowitz (1981) and Guttag, et al. (1983) apply properties similar to commutation to termination. We use commutation to reduce the problem of proving termination of a combined system $R \cup S$ to the problem of proving termination of the individual systems R and S separately. These commutation properties can often be easily established for certain systems, such as linear rewrite systems.

---

* This research was supported in part by the National Science Foundation under grant DCR 85–13417.

The use of well-founded sets is fundamental for termination arguments. Given a rewrite system R and a well-founded ordering $>$ on a set W, the problem is to find a mapping from terms to W, such that well-foundedness of $>$ implies termination of R. We study *transforms*, that is, mappings T from terms to terms, and derive conditions on R, T and $>$ that are sufficient for termination of R. It turns out that commutation properties play an important role in such transformation techniques. We present methods for both ordinary and equational rewrite system. The transforms we describe may be used, for instance, to prove termination of associative-commutative rewrite systems.

2. Definitions

Let $\mathcal{T}$ be the set of *terms* over some set of operator symbols F and some set of variables V. Terms containing no variables are called *ground terms*. We write $s[t]$ to indicate that a term s contains t as a subterm and denote by $s[t/u]$ or just $s[u]$ the result of replacing a particular occurrence of t by u.

A binary relation $\rightarrow$ on $\mathcal{T}$ is *monotonic* if $s \rightarrow t$ implies $u[s] \rightarrow u[t]$, for all terms u, s, and t. It is *stable (under substitution)* if $s \rightarrow t$ implies $s\sigma \rightarrow t\sigma$, for all terms s and t, and every substitution σ. The symbols $\rightarrow^+$, $\rightarrow^*$ and $\leftrightarrow$ denote the transitive, transitive-reflexive, and symmetric closure of $\rightarrow$, respectively. The inverse of $\rightarrow$ is denoted by $\leftarrow$. A relation $\rightarrow$ is *Noetherian* if there is no infinite sequence $t_1 \rightarrow t_2 \rightarrow t_3 \rightarrow \cdots$. A transitive Noetherian relation is called *well-founded*. A *reduction ordering* is a stable and monotonic well-founded ordering.

An *equation* is a pair (s,t), written $s=t$, where s and t are terms. For any set of equations E, $\leftrightarrow_E$ denotes the smallest symmetric relation that contains E and is monotonic and stable. That is, $s \leftrightarrow_E t$ if and only if $s=c[u\sigma]$ and $t=c[v\sigma]$, where $u=v$ or $v=u$ is in E. A reduction ordering $>$ is *compatible* with E if $s \overset{*}{\leftrightarrow}_E u > v \overset{*}{\leftrightarrow}_E t$ implies $s > t$, for all terms s, t, u, and v. *Directed* equations, in which every variable appearing on the right-hand side also appears on the left-hand side, are called *rewrite rules* and are written $s \rightarrow t$. A *rewrite system* is any set R of rewrite rules. The *reduction relation* $\rightarrow_R$ is the smallest stable and monotonic relation that contains R, i.e. $s \rightarrow_R t$ if and only if $s=c[l\sigma]$ and $t=c[r\sigma]$, for some rewrite rule $l \rightarrow r$ in R. We use R^{-1} to denote the inverse of R, and $R^{\leftrightarrow}$ to denote $(R \cup R^{-1})^*$.

Let E be a set of equations and R be a rewrite system. The *equational rewrite system* R/E (R *mod* E) is the set consisting of all rules $l \rightarrow r$ such that $l \overset{*}{\leftrightarrow}_E u \rightarrow_R v \overset{*}{\leftrightarrow}_E r$, for some terms u and v. Consequently, the reduction relation $\rightarrow_{R/E}$ is the relation $\overset{*}{\leftrightarrow}_E \circ \rightarrow_R \circ \overset{*}{\leftrightarrow}_E$, where $\circ$ denotes composition of relations. Analogously, if S is a rewrite system, we let R/S be the set of all rewrite rules $l \rightarrow r$ such that $l \overset{*}{\rightarrow}_S u \rightarrow_R v \overset{*}{\rightarrow}_S r$, for some terms u and v; the relation $\rightarrow_{R/S}$ is $\overset{*}{\rightarrow}_S \circ \rightarrow_R \circ \overset{*}{\rightarrow}_S$.

Let R/E be an equational rewrite system. We write $s \downarrow_{R/E} t$ to indicate that there exists a term u such that $s \overset{*}{\rightarrow}_{R/E} u \overset{*}{\leftarrow}_{R/E} t$. The system R/E is *Church-Rosser* if $s \overset{*}{\leftrightarrow}_{R/E} t$ implies $s \downarrow_{R/E} t$, for all terms s and t. It is *terminating* if $\rightarrow_{R/E}$ is Noetherian. An equational rewrite system R/E terminates if and only if there exists a reduction ordering $>$ that contains R and is compatible with $\overset{*}{\leftrightarrow}_E$. A terminating Church-Rosser rewrite system is called *canonical*. A term

t is *irreducible* in R/E if there is no term t' such that $t \rightarrow_{R/E} t'$. If $t \overset{*}{\rightarrow}_{R/E} t'$ and t' is irreducible in R/E, then t' is called an *R/E-normal form* of t. In a canonical system R/E any two normal forms t_1 and t_2 of a term t are equivalent in E. An ordinary rewrite system R may be regarded as an equational rewrite system R/E, where E is the empty set. Hence, all the definitions above apply to ordinary rewrite systems.

3. Commutation

Commuting rewrite systems have been investigated by Rosen (1973) and Raoult and Vuillemin (1980), among others. In this paper, we present new termination methods based on commutation that apply to ordinary as well as to equational rewrite systems.

Definition 1. Let R and S be rewrite systems. We say that R and S *commute* if $\leftarrow_R \circ \rightarrow_S$ is contained in $\rightarrow_S \circ \leftarrow_R$ (see Fig. 1).

For termination arguments the following *non-symmetric* commutation properties are also important.

Definition 2. A rewrite system R *commutes over* another system S if $\rightarrow_S \circ \rightarrow_R$ is contained in $\rightarrow_R \circ \rightarrow_S$; R *quasi-commutes over* S if $\rightarrow_S \circ \rightarrow_R$ is contained in $\rightarrow_R \circ \overset{*}{\rightarrow}_{R \cup S}$ (see Fig. 1). We say that R commutes over a set of equations E if $\leftrightarrow_E \circ \rightarrow_R$ is contained in $\rightarrow_R \circ \overset{*}{\leftrightarrow}_E$; R quasi-commutes over E if $\leftrightarrow_E \circ \rightarrow_R$ is contained in $\rightarrow_R \circ \overset{*}{\leftrightarrow}_E \circ \overset{*}{\rightarrow}_{R/E}$.

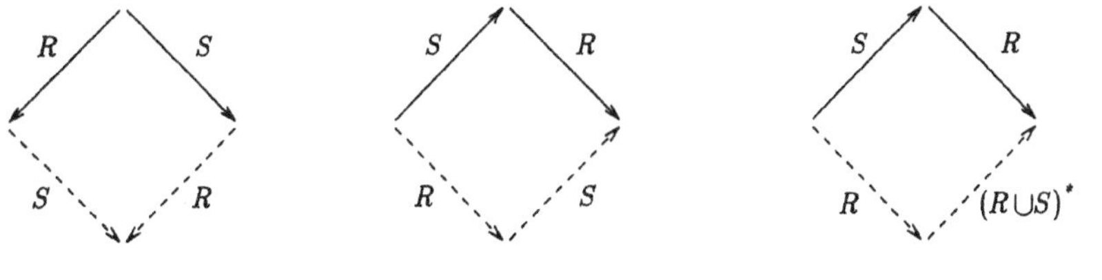

R and S commute   *R commutes over S*   *R quasi-commutes over S*

Fig. 1

Lemma 1. *Let R and S be two rewrite systems. Then the combined system $R \cup S$ is terminating if and only if both R/S and S are.*

Lemma 2. *If a rewrite system R quasi-commutes over another system S, then R/S is terminating if and only if R is.*

Proof. Trivially, if R/S is terminating, so is R. For the other direction, assume that R/S is not terminating. Then there exists an infinite derivation $t_1 \overset{*}{\rightarrow}_S t_2 \rightarrow_R t_3 \overset{*}{\rightarrow}_S t_4 \rightarrow_R \cdots$ containing an infinite number of applications of R. By the fact that R quasi-commutes over S, any application of R (beginning with $t_2 \rightarrow_R t_3$) can be pushed back through all preceding applications of

S. Thus there must also be an infinite derivation for R alone. □

Combining the above two lemmata, we have

THEOREM 1. *If a rewrite system R quasi-commutes over a rewrite system S, then the combined system $R \cup S$ terminates if and only if R and S both do.*

Syntactic properties of rewrite systems, such as linearity, may be helpful for establishing commutation; see, for example, Raoult and Vuillemin (1980). A term in which no variable appears more than once is called *linear.* A rewrite system R is called *left-linear,* if all left-hand sides of rules in R are linear; *right-linear,* if all right-hand sides are linear; and *linear,* if it is both left- and right-linear. A term s *overlaps* a term t if it can be unified with some non-variable subterm of t. We say that there is no overlap between s and t if neither s overlaps t nor t overlaps s.

Lemma 3. (Raoult and Vuillemin, 1980) *Let R be a left-linear and S be a right-linear rewrite system. If there is no overlap between a left-hand side of R and a right-hand side of S, then R quasi-commutes over S.*

Putting Lemma 3 and Theorem 1 together, we obtain

THEOREM 2. (Dershowitz, 1981) *Let R be a left-linear and S be a right-linear rewrite system. If there is no overlap between left-hand sides of R and right-hand sides of S, then the combined system $R \cup S$ terminates, if and only if R and S both do.*

Example 1. The systems

$$(x+y)\cdot z \quad \rightarrow \quad x\cdot z+y\cdot z$$

and

$$\begin{array}{rcl} x\cdot x & \rightarrow & x \\ x+x & \rightarrow & x \end{array}$$

both terminate. The first is left-linear and the second has only variables on the right; therefore their union also terminates.

Similar results hold for equational rewrite systems. Note that the relations $(R \cup S)/E$ and $R/E \cup S/E$ are the same.

PROPOSITION 1. *Let E be a set of equations and R and S be rewrite systems such that R/E quasi-commutes over S/E. Then $(R \cup S)/E$ terminates if and only if R/E and S/E both do.*

The relation R/E quasi-commutes over S/E if and only if $\rightarrow_S \circ \rightarrow_{R/E}$ is contained in $\rightarrow_{R/E} \circ \overset{*}{\rightarrow}_{(R \cup S)/E}$. This condition is slightly weaker than quasi-commutation of R/E over S.

PROPOSITION 2. (Jouannaud and Munoz, 1984) *Let R/E be an equational rewrite system such that R quasi-commutes over E. Then R terminates if and only if R/E does.*

Again, linearity may be used to advantage.

THEOREM 3. *Suppose E is linear, R is left-linear, and S is right-linear. If there is no overlap between a right-hand side of S and a left-hand side of R or either side of an equation in E, then*

$(R\cup S)/E$ terminates if and only if R/E and S both do.

Example 2. *(Distributive lattices)* Let R be

$$(x\cap y)\cup z \quad\rightarrow\quad (x\cup z)\cap(y\cup z)$$

S be

$$\begin{array}{rcl} x\cap(x\cup y) & \rightarrow & x \\ x\cup x & \rightarrow & x \\ x\cap x & \rightarrow & x \end{array}$$

and E be

$$\begin{array}{rcl} x\cup(y\cup z) & = & (x\cup y)\cup z \\ x\cup y & = & y\cup x \\ x\cap(y\cap z) & = & (x\cap y)\cap z \\ x\cap y & = & y\cap x \end{array}$$

E is linear, R is left–linear, and S contains only variables on the right–hand side. By the above theorem, $(R\cup S)/E$ terminates if S and R/E both do. Termination of S is trivial, since every rule in S is length–decreasing. To prove termination of R/E one can, for example, use a polynomial interpretation τ, where $\tau_\cup$ is $\lambda xy.x{*}y$ and $\tau_\cap$ is $\lambda xy.x+y+1$.

4. Transformation

The notion of well–foundedness suggests the following straightforward method of proving termination (Manna and Ness, 1970, and Lankford, 1975). Given a rewrite system R, find a well–founded ordering $>$ on terms, such that

$s\rightarrow_R t$ implies $s>t$, for all terms s and t.

It is frequently convenient to separate the well–founded ordering $>$ into two parts: a *termination function* τ that maps terms in $\mathcal{T}$ to a set $\mathcal{W}$, and a "standard" well–founded ordering $\succ$ on $\mathcal{W}$. We will consider, in this section, mappings τ, called *transforms*, that map terms into terms and can be represented by a canonical rewrite system T. That is, τ maps a term t to its (unique) T–normal form t^*. We denote by $T!$ the rewrite system consisting of all rules $t\rightarrow t^*$. We assume that the ordering $\succ$ is a reduction ordering, and thus may also be characterized by some (possibly infinite) rewrite system S. We will next present termination methods that are based on certain commutation properties of S and T.

Convention. From now on we will use the symbols R, R^* and R^+ to ambiguously denote the relations $\rightarrow_R$, $\overset{*}{\rightarrow}_R$ and $\overset{+}{\rightarrow}_R$, respectively.

Definition 3. A rewrite system R is *reducing relative to S and T* if it is contained in $T^*\circ S\circ(T^*)^{-1}$ (see Fig. 2).

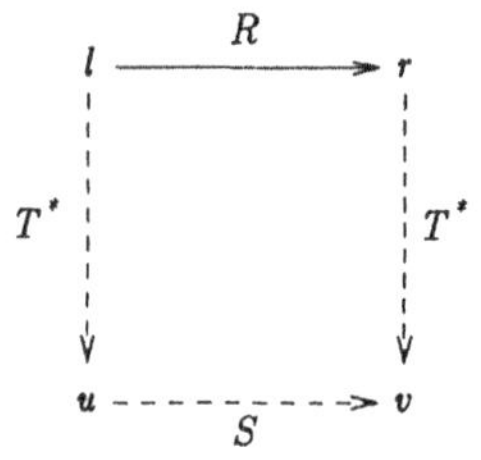

Fig. 2

THEOREM 4. *Let R, S, and T be rewrite systems such that T is canonical, S terminates, and S and T! commute. If R is reducing relative to S and T, then $R/T^{\leftrightarrow}$ terminates.*

Proof. Suppose that $R/T^{\leftrightarrow}$ is not terminating. Then there is an infinite sequence $t_1 \rightarrow_R t_2 \overset{*}{\leftrightarrow}_T t_3 \rightarrow_R t_4 \overset{*}{\leftrightarrow}_T \cdots$. Using the facts that R is reducing, T is canonical, and S and $T!$ commute, we can construct an infinite sequence $u_1 \rightarrow_S u_2 \rightarrow_S u_3 \rightarrow_S \cdots$ as shown in Fig. 3.

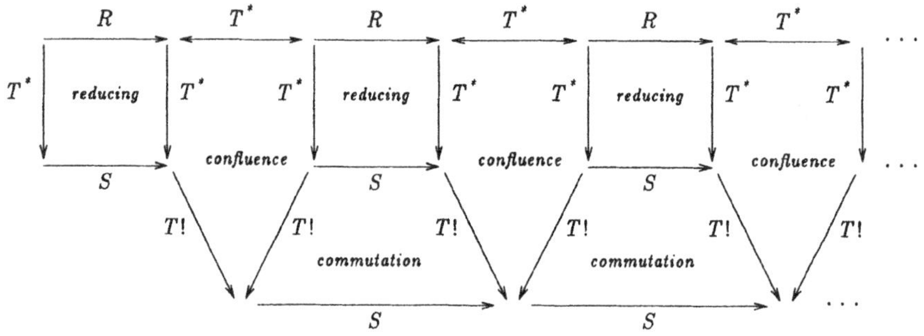

Fig. 3

This contradicts the fact that S is terminating. □

COROLLARY 1. *Let R, S, T, and T′ be rewrite systems such that T is canonical, S terminates, and S and T! commute. If T′ is contained in $T^{\leftrightarrow}$ and R is reducing relative to S and T, then R/T' terminates.*

For termination proofs *symbolic interpretations* of operators are often useful. These consist of a single rewrite rule $f(x_1, \ldots, x_n) \rightarrow t[x_1, \ldots, x_n]$, where t is a term containing all variables $x_1, \ldots, x_n$, but not containing f. Such transforms are obviously canonical. They may be used, for instance, to declare two operators equivalent (for the purpose of proving termination). The *T-normalized version* R_T of R consists of all rewrite rules $l^* \rightarrow r^*$, where l^* and r^* are T-normal forms of l and r, respectively, for some rule $l \rightarrow r$ is in R.

Lemma 4. *Let R be a rewrite system, T be a symbolic interpretation, and R_T be the T–normalized version of R. Then R is reducing relative to R_T and T, and R_T^+ and $T!$ commute.*

Proof. That R is reducing relative to R_T and T follows immediately from the definition of R_T. To prove commutation of R_T^+ and $T!$ we show that, for all rules $l \rightarrow r$ in R_T, $u[l\sigma] \rightarrow_{R_T} u[r\sigma]$ implies $l' \rightarrow_{R_T} r'$, where l' and r' are T–normal forms of $u[l\sigma]$ and $u[r\sigma]$, respectively. These normal forms may be obtained by first applying T in the substitution part of σ, and then applying further reduction steps in the context u. That is, $u[l\sigma] \overset{*}{\rightarrow}_T u[l\rho] \rightarrow_{T!} v[l\rho, \ldots, l\rho] = l'$ and, similarly, $u[r\sigma] \overset{*}{\rightarrow}_T u[r\rho] \rightarrow_{T!} v[r\rho, \ldots, r\rho] = r'$. Obviously, $l' \rightarrow_{R_T}^+ r'$. □

Combining Corollary 1 and Lemma 4, we obtain

PROPOSITION 3. *Let R and T' be rewrite systems and T be a symbolic interpretation and suppose that T' is contained in $T^{\leftrightarrow}$. Then R/T' is terminating if R_T is.*

Example 3. Let R be

$$\begin{array}{rcl} g(x,y) & \rightarrow & h(x,y) \\ h(f(x),y) & \rightarrow & f(g(x,y)) \end{array}$$

We use the first rule as a transform T and let R' be the second rule. The T–normalized version R_T' of R' is

$$h(f(x),y) \quad \rightarrow \quad f(h(x,y))$$

R_T' terminates, since it decreases the summed length of all the terms with outermost operator h. By Proposition 3, this implies termination of R'/T. Since T is terminating, so is $R = R' \cup T$.

"Local" commutation of S and T in general does not imply commutation of S and $T!$, but only commutation of S and T^*. If S/T terminates, then a commutation property may be used that can be established by a local test.

Lemma 5. *Suppose that T is canonical and S/T terminates. Then $(S/T)^+$ and T^* commute, if $\leftarrow_T \circ \rightarrow_S$ is contained in $\rightarrow_{S/T}^+ \circ \overset{*}{\leftarrow}_T$.*

Proof. By Noetherian induction on $S \cup T$. Note that $S \cup T$ is Noetherian, since both T and S/T are. □

Again linearity is useful for establishing commutation. A rewrite rule $l \rightarrow r$ is *non–annihilating* if every variable appearing in l also appears in r.

Lemma 6. *Let S and T be rewrite systems. If T is left–linear and non–annihilating and there is no overlap between left–hand sides of S and T, then $\leftarrow_T \circ \rightarrow_S$ is contained in $\rightarrow_{S/T}^+ \circ \overset{*}{\leftarrow}_T$.*

Proof. Suppose that $c \leftarrow_T t \rightarrow_S d$. We distinguish three cases.

a) If the two reduction steps apply at disjoint positions, i.e. $u[r,l'] \leftarrow_T u[l,r] \rightarrow_S u[l,r']$, then $u[r,l'] \rightarrow_S u[r,r'] \leftarrow_T u[l,r']$.

b) If the S–reduction step applies in the variable part of the T–reduction step, i.e. $v[l, \ldots, l] \leftarrow_T u[l] \rightarrow_S u[r]$, then $v[l, \ldots, l] \rightarrow_S^+ v[r, \ldots, r] \leftarrow_T u[r]$ (there has to be at least one S–reduction step, since T is non–annihilating).

c) If the T–reduction step applies in the variable part of the S–reduction step, i.e.

$u[r] \leftarrow_T u[l] \rightarrow_S v[l, \ldots, l]$, then we have $u[r] \rightarrow_S v[r, \ldots, r] \overset{*}{\leftarrow}_T v[l, \ldots, l]$. □

THEOREM 5. *Let R, S, and T be rewrite systems such that T is canonical, S/T terminates, and $(S/T)^+$ and T^* commute. If R is reducing relative to S/T and T, then $R/T^{\leftrightarrow}$ terminates.*

Proof. The same as the proof of Theorem 4, except that instead of S we have S/T, and instead of $T!$ we have T^*. □

Example 4. Let R be the following rewrite system for computing the factorial function (Kamin and Levy, 1980):

$$\begin{aligned} f(s(x)) &\rightarrow f(p(s(x))) \\ f(0) &\rightarrow s(0) \\ p(s(x)) &\rightarrow x \end{aligned}$$

We use the last rule as a transform T and let R' be $R-T$. T is length-decreasing, hence terminating. The T-normalized version R_T' of R' is

$$\begin{aligned} f(s(x)) &\rightarrow f(x) \\ f(0) &\rightarrow s(0) \end{aligned}$$

$R_T' \cup T$ is terminating, since each rule either decreases the length of a term, or maintains the length and decreases the number of occurrences of f. Also, since T is linear, non-annihilating, and does not overlap with left-hand sides of R_T', the relations $(R_T'/T)^+$ and T^* commute. By Theorem 5, R'/T is terminating, which, together with termination of T, implies termination of R.

The termination methods outlined above may also be applied to equational rewrite systems R/E by using transforms T such that E is contained in $T^{\leftrightarrow}$.

Suppose I consists of the axioms for identity, $f(x,e)=x$ and $f(e,x)=x$. Let T_I be the transform $\{f(x,e) \rightarrow x, f(e,x) \rightarrow x\}$. This transform is canonical. Given a rewrite system R, let R_I' consist of all rules $u \rightarrow v$, where u and v are T_I-normal forms of $l\sigma$ and $r\sigma$, respectively, $l \rightarrow r$ is in R, and σ is a substitution such that $x\sigma$ is either x or e, for all variables x. If $l \rightarrow r$ is in R_I', and $x\sigma$ is either x or e, then $l' \rightarrow r'$ is also in R_I', where l' and r' are T_I-normal forms of $l\sigma$ and $r\sigma$, respectively. Let R_I contain R_I' and, in addition, for every rule $e \rightarrow r$ in R_I', where $r \neq e$, rules $x \rightarrow f(x,r)$ and $x \rightarrow f(r,x)$; for every rule $l \rightarrow e$ in R_I', where $l \neq e$, rules $f(x,l) \rightarrow x$ and $f(l,x) \rightarrow x$; and the rule $x \rightarrow x$, if $e \rightarrow e$ is in R_I' (the additional rules are necessary for commutation of R_I and T_I). If R is finite, so is R_I.

Lemma 7. *Let R be a rewrite system and T_I and R_I be as defined above. Then R is reducing relative to T_I and R_I, and R_I and $T_I!$ commute.*

Proof. That R is reducing relative to T_I and R_I follows from the definition of R_I. For commutation, it suffices to show that, for all rules $l \rightarrow r$ in R_I and all terms c and substitutions σ, $u \rightarrow_{R_I} v$, where u and v are T_I-normal forms of $c[l\sigma]$ and $u[r\sigma]$, respectively. Without loss of generality, we may assume that c and σ are irreducible in T_I. Let σ' be a substitution such that $x\sigma'$ is e, if $x\sigma$ is e, and $x\sigma'$ is x, otherwise. The assertion can be easily shown if $l \rightarrow r$ is not in R_I'. If $l \rightarrow r$ is in R_I', then, by the remark above, $l' \rightarrow r'$ is in R_I', where l' and r' are T-normal forms of $l\sigma'$ and $r\sigma'$, respectively. Since $\sigma = \sigma' \circ \rho$, for some substitution ρ, we obtain $l\sigma \rightarrow_{T_I!} l'\rho$ and $r\sigma \rightarrow_{T_I!} r'\rho$. Since c, $l'\rho$, and $r'\rho$ are irreducible in T_I, are irreducible in T_I, the assertion can be easily established. □

PROPOSITION 4. *An equational system R/I terminates if and only if R_I terminates.*

Proof. The if-direction follows from Lemma 7. The only-if-direction holds because R_I is contained in R/I. □

The requirement—in the theorems above—that the transform T be canonical may be somewhat relaxed. We say that a rewrite system R is *confluent modulo E* if, for all terms s, t, u, and v with $u \overset{*}{\leftarrow}_R s \overset{*}{\leftrightarrow}_E t \overset{*}{\rightarrow}_R v$ there exist terms u' and v' such that $u \overset{*}{\rightarrow}_R u' \overset{*}{\leftrightarrow}_E v' \overset{*}{\leftarrow}_R v$.

Lemma 8. (Huet, 1980) *Let R be a terminating rewrite system. Then R is confluent modulo E if and only if, for all terms s and t, $s \overset{*}{\leftrightarrow}_{E \cup R} t$ implies $s \rightarrow_{R!} \circ \overset{*}{\leftrightarrow}_E \circ \leftarrow_{R!} t$.*

In other words, if R is terminating and confluent modulo E, then two terms are equivalent in $E \cup R$ if and only if their respective R-normal forms are equivalent in E.

THEOREM 6. *Let R, S, and T be rewrite systems and E be an equational theory. Suppose that T is terminating and confluent modulo E, S/E is terminating, and S and $T!$ commute. If R is reducing relative to S and T, then $R/(E \cup T^{\leftrightarrow})$ terminates.*

Proof. Let $t_1 \rightarrow_R t_2 \overset{*}{\leftrightarrow}_{E \cup T} t_3 \rightarrow_R t_4 \overset{*}{\leftrightarrow}_{E \cup T} \cdots$ be an infinite sequence. Under the given assumptions, an infinite sequence of S/E reduction steps can be constructed as follows:

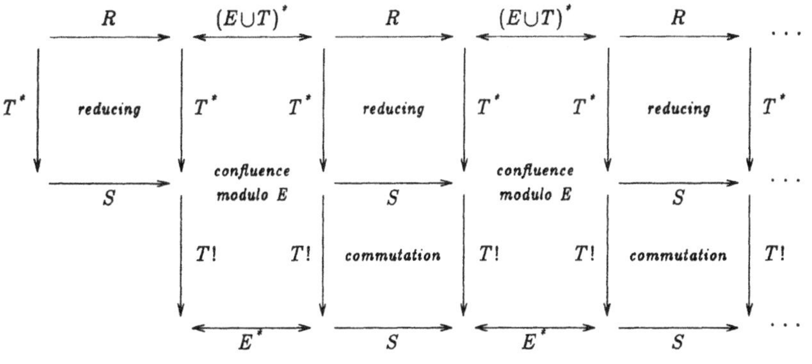

Fig. 4

□

COROLLARY 2. *Let R, S, and T be rewrite systems and E be an equational theory. Suppose that T is terminating and confluent modulo E, $(S/T)/E$ is terminating, and $(S/T)^+$ and T^* commute. If R is reducing relative to S/T and T, then $R/(E \cup T^{\leftrightarrow})$ terminates.*

Recall that Lemma 5 provides a local test for commutation of $(S/T)^+$ and T^*. In the next section we consider particular transforms in depth.

5. Transforms Based on Distributivity and Associativity

Equational rewrite systems R/E, where E is a set of associativity and commutativity axioms, are of particular importance in practice. We will apply the transformation techniques outlined above to the termination problem for such systems (*AC termination*).

Let f be some operator symbol in F. An *associativity axiom* for f is an equation of the form $f(x,f(y,z))=f(f(x,y),z)$ or $f(f(x,y),z)=f(x,f(y,z))$, a *commutativity axiom* is an equation of the form $f(x,y)=f(y,x)$. An equational rewrite system R/E is called *associative-commutative* if E contains only associativity and commutativity axioms. From now on let AC denote a set of associativity and commutativity axioms for which any associative operator is also commutative and vice versa. We say that *f is in AC* to indicate that f is an associative-commutative operator.

Let $>$ be an ordering, called a *precedence ordering*, on the set of operator symbols F. We define the rewrite relation RPO recursively as follows:

a) $f(\cdots s \cdots) \rightarrow_{RPO} s$,

b) $f(\cdots s \cdots) \rightarrow_{RPO} f(\cdots s_1 \cdots s_n \cdots)$, if $s \rightarrow_{RPO} s_i$, for $1 \leq i \leq n$,

c) $s=f(s_1, \ldots, s_n) \rightarrow_{RPO} g(t_1, \ldots, t_k)$, if $f>g$ and $s \rightarrow_{RPO} t_i$, for $1 \leq i \leq k$.

The *recursive path ordering* $>_{rpo}$ associated with $>$ is the transitive closure $\rightarrow^+_{RPO}$ of $\rightarrow_{RPO}$.

Lemma 9. (Dershowitz, 1982) *Let $>$ be a precedence ordering on the set of operator symbols F. Then $>_{rpo}$ is well-founded if and only if $>$ is well-founded.*

Recall that a reduction ordering $>$ is *compatible* with AC if $s \leftrightarrow^*_{AC} u > v \leftrightarrow^*_{AC} t$ implies $s>t$, for all terms s, t, u, and v. A rewrite system R/AC terminates if and only if there is a reduction ordering $>$ that is compatible with AC, such that $l \rightarrow_R r$ implies $l>r$. Unfortunately, many reduction orderings are not compatible with AC. For instance, the recursive path ordering $>_{rpo}$ is not: if f is in AC and $a>_{rpo}b$, then

$$f(a,f(b,b)) \leftrightarrow_{AC} f(f(a,b),b) >_{rpo} f(a,f(b,b)),$$

but $f(a,f(b,b)) >_{rpo} f(a,f(b,b))$ is false.

We will design a *transform* T such that, for some set of equations E, (a) T is terminating and confluent modulo E, (b) AC is contained in $E \cup T^{\leftrightarrow}$, (c) $(S/T)/E$ is terminating, and (d) $(S/T)^+$ and T^* commute. For S we will use the recursive path ordering, restricted to terms irreducible in T. For E we use the *permutation congruence* $\sim$, which is the smallest stable congruence, such that $f(X,u,Y,v,Z) \sim f(X,v,Y,u,Z)$. If property (a) is satisfied then T-irreducible terms are unique up to equivalence in E and may serve as representatives for AC-equivalence classes. A natural choice for such a canonical representation are "flattened" terms. Let L be the rewrite system consisting of all reduction rules (on varyadic terms) of the form $f(X,f(Y),Z) \rightarrow f(X,Y,Z)$, where f is in AC, Y denotes a sequence of variables $y_1, \ldots, y_n$ of length $n \geq 2$, and X and Z are sequence of variables of length k and l, respectively, where $k+l \geq 1$. For example, $f(x,f(y,z)) \rightarrow f(x,y,z)$ is a "flattening rule", but $f(f(x)) \rightarrow f(x)$ is not.

Terms irreducible in L are called *flattened.*

Lemma 10. *The rewrite system L is canonical, $L/\sim$ is terminating, and AC is contained in $\sim \cup L^{\leftrightarrow}$.*

Any recursive path ordering $>_{rpo}$ contains L and is compatible with the permutation congruence $\sim$. Therefore $(RPO/L)/\sim$ is terminating. Unfortunately, the commutation property (d) is not satisfied, as the following example illustrates: if f is in AC and $f>g$, then $f(a,b)\rightarrow_{RPO} g(a,b)$ and

$$f(a,b,c)\leftarrow_L f(f(a,b),c)\rightarrow_{RPO} f(g(a,b),c).$$

Both $f(a,b,c)$ and $f(g(a,b),c)$ are flattened, but $f(g(a,b),c)\rightarrow_{RPO} f(a,b,c)$. However, if the transform T contains, in addition to L, the rewrite rule $f(g(x,y),z)\rightarrow g(f(x,z),f(y,z))$, then

$$f(a,b,c)\rightarrow_{RPO} g(f(a,c),f(b,c))\leftarrow_T g(f(a,b),c).$$

Let $>$ be a well-founded precedence ordering. A *distributivity rule* for f and g is a rewrite rule of the form

$$f(X,g(Y),Z) \rightarrow g(f(X,y_1,Z), \cdots ,f(X,y_n,Z)),$$

where Y is a sequence $y_1, \ldots, y_n$ of length $n\geq 1$, $f>g$, and neither f nor g are constants. For example, $x*(y+z)\rightarrow x*y+x*z$ and $-(x+y)\rightarrow(-x)+(-y)$ are distributivity rules. Such sets of distributivity rules are terminating (they are contained in $>_{rpo}$) but not canonical, in general. For example, if f distributes over both g and h, then the term $f(g(x),h(y))$ can be transformed to two different terms, $g(h(f(x,y)))$ or $h(g(f(x,y)))$. To guarantee that properties (a)–(d) above are satisfied, we have to impose certain restrictions on sets of distributivity rules.

Let F_D be a set of non-constant operator symbols f containing all AC operators. Let D be the set of all distributivity rules for f and g, where f and g are in F_D and $f>g$. The rewrite system $T=L\cup D$, where L consists of all flattening rules for operators in F_D, is called the *A-transform* corresponding to $>$ and F_D. Let F' be $F-\{c\}$, if c is minimal among all constants, or F, if there is no such constant.

Definition 4. A precedence ordering $>$ satisfies the *associative path condition* for F_D, if F_D can be partitioned into two sets $\{f_1, \ldots, f_n\}$ and $\{g_1, \ldots, g_m\}$, such that $n\leq m$ and

a) g_i is minimal in F', for $1\leq i\leq m$,

b) $f_i>g_i$, for $1\leq i\leq n$,

c) f_i is minimal in $F'-\{g_i\}$, for $1\leq i\leq n$.

For example, if f, g, h and i are in F_D, then the precedence orderings shown in Figs. 5(a) and 5(b) do not satisfy the associative path condition, but the ordering in Fig. 5(c) does.

Lemma 11. *Let $>$ be a precedence ordering that satisfies the associative path condition, T be the corresponding A-transform, and S be the corresponding rewrite system consisting of all pairs $l\rightarrow r$ such that $l>_{rpo}r$ and l and r are irreducible in T. Then*

a) T is terminating and confluent modulo $\sim$,

b) AC is contained in $E\cup T^{\leftrightarrow}$,

c) $(S/T)/\sim$ is terminating, and

d) $(S/T)^+$ and T^* commute.

Sketch of proof. Part (b) follows from Lemma 10. The recursive path ordering $>_{rpo}$ contains both S and T and is compatible with the permutation congruence $\sim$. Therefore $(S/T)/\sim$ is terminating. For confluence T modulo $\sim$ it suffices to prove $c \xrightarrow{*}_T s \sim t \xleftarrow{*}_T d$, for all "critical overlaps" $c \leftarrow_T u \rightarrow_T d$ or $c \leftarrow_T u \sim d$. The restrictions on the precedence ordering $>$ are essential for the proof of this confluence property.

By Lemma 5, $(S/T)^+$ and T^* commute if, for all terms s, t, and u with $s \leftarrow_T u \rightarrow_S t$, there exist terms v and w, such that $s \xrightarrow{*}_T v \rightarrow_S w \xleftarrow{*}_T$. This is implied by the following two properties:

(i) *Monotonicity.* If $l \rightarrow r$ is in S, then, for any term c, $l' \rightarrow_S^+ r'$, where l' and r' are T–normal forms of $c[l]$ and $c[r]$, respectively.

(ii) *Stability.* If $l \rightarrow r$ is in S, then, for any substitution σ, $l' \rightarrow_S^+ r'$, where l' and r' are T–normal forms of $l\sigma$ and $r\sigma$, respectively.

Both properties can be proved by induction on the length of l and r. □

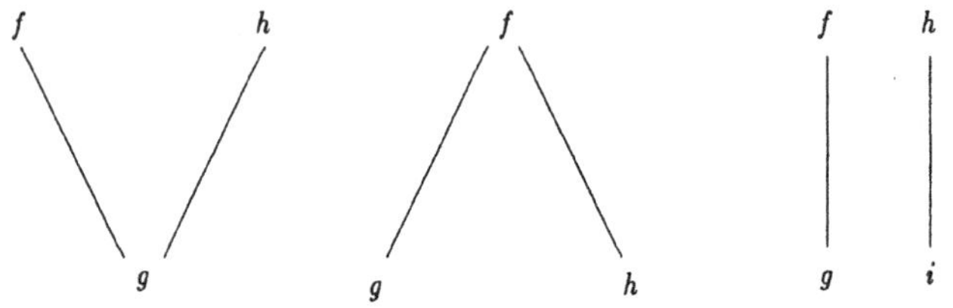

a) b) c)

Fig. 5

Definition 5. Let T be the A–transform corresponding to some precedence ordering $>$. The *associative path ordering* $>_{apo}$ is defined by:

$s >_{apo} t$ *if and only if,* $s^* >_{rpo} t^*$,

where s^* and t^* are T–normal forms of s and t, respectively.

Summarizing the results above we have the following theorems for AC termination.

THEOREM 7. *If $>$ is a well-founded precedence ordering that satisfies the associative path condition, then the corresponding associative path ordering $>_{apo}$ is a reduction ordering and is compatible with AC.*

THEOREM 8. *Let $>$ be a precedence ordering that satisfies the associative path condition and T be the corresponding A–transform. Suppose that T' is contained in $T^{\leftrightarrow}$. If $l >_{apo} r$, for every rule $l \rightarrow r$ in R, then $R/(T' \cup AC)$ terminates.*

Transformation techniques for AC termination were first suggested by Dershowitz, et al. (1983). The associative path ordering described above is simpler than the ordering given by Bachmair and Plaisted (1985). In particular, Theorem 7 implies that

if $s >_{apo} t$, *then* $s\sigma >_{apo} t\sigma$, *for any substitution* σ.

This "lifting lemma" allows efficient implementations of the associative path ordering based on the recursive path ordering. The A-transform may also be used in combination with a lexicographic path ordering. More precisely, operators that are not in AC may be given lexicographic status, i.e. some positions in a term may be given more significance than others (see Kamin & Levy, 1980). A-transforms may be extended to include symbolic interpretations of non-AC operators. That is, the results above also hold for transforms $T=T_1 \cup T_2$, where T_1 is an A-transform corresponding to a precedence ordering $>$ and a set of operator symbols F_D, and T_2 consists of a single rule $f(x_1, \ldots, x_n) \rightarrow t[x_1, \ldots, x_n]$, where f is not in F_D.

Example 5. *(Boolean algebra).* The following canonical rewrite system for boolean algebra is due to Hsiang (1985). We outline a termination proof using an associative path ordering. R consists of the following rules:

$$\begin{array}{rcl}
x \oplus \mathit{false} & \rightarrow & x \\
x \wedge \mathit{false} & \rightarrow & \mathit{false} \\
x \wedge \mathit{true} & \rightarrow & x \\
x \wedge x & \rightarrow & x \\
(x \oplus y) \wedge z & \rightarrow & (x \wedge z) \oplus (y \wedge z) \\
x \oplus x & \rightarrow & \mathit{false} \\
x \vee y & \rightarrow & (x \wedge y) \oplus (x \oplus y) \\
x \supset y & \rightarrow & (x \wedge y) \oplus (x \oplus \mathit{true}) \\
x \equiv y & \rightarrow & (x \oplus y) \oplus \mathit{true} \\
\neg x & \rightarrow & x \oplus \mathit{true}
\end{array}$$

The operators $\oplus$ and $\wedge$ are in AC. Let $>$ be the precedence ordering shown in the Hasse diagram in Fig. 6, and T be the A-transform corresponding to $>$ and $F_D = \{\wedge, \oplus\}$, extended by a symbolic interpretation $\{\mathit{false} \rightarrow \mathit{true}\}$. The fifth rule of R is a distributivity rule and is placed in T'. Let R' be $R - T'$. Since $l >_{apo} r$, for all rules $l \rightarrow r$ in R', we may conclude, by Theorem 8, that $R'/(T' \cup AC)$ terminates. The system T'/AC also terminates (see Example 2), which implies termination of $R/AC = (R' \cup T')/AC$.

Example 6. *(Modules).* Let A be an associative-commutative ring with identity. An A-module M over A is an algebraic structure consisting of operations $\oplus : M \times M \rightarrow M$ and $\cdot : A \times M \rightarrow M$, such that $(M, \oplus)$ is an abelian group (the identity of the group is denoted by Ω, the inverse to $\oplus$ by I) and the following identities hold: $\alpha \cdot (\beta \cdot x) = (\alpha * \beta) \cdot x$, $1 \cdot x = x$, $(\alpha + \beta) \cdot x = (\alpha \cdot x)(\beta \cdot x)$ and $\alpha \cdot (x y) = (\alpha \cdot x)(\alpha \cdot y)$. For the sake of readability we use Greek letters for variables ranging over elements of A, and Roman letters for variables ranging over elements of M. The following rewrite system R was obtained with the rewrite rule laboratory RRL (see Kapur & Sivakumar, 1984):

$$
\begin{array}{rcl}
\alpha+0 & \longrightarrow & \alpha \\
\alpha+(-\alpha) & \longrightarrow & 0 \\
-0 & \longrightarrow & 0 \\
-(-\alpha) & \longrightarrow & \alpha \\
-(\alpha+\beta) & \longrightarrow & (-\alpha)+(-\beta) \\
\alpha*(\beta+\gamma) & \longrightarrow & (\alpha*\beta)+(\alpha*\gamma) \\
\alpha*0 & \longrightarrow & 0 \\
\alpha*(-\beta) & \longrightarrow & -(\alpha*\beta) \\
\alpha*1 & \longrightarrow & \alpha \\
x\oplus\Omega & \longrightarrow & x \\
\alpha\cdot(\beta\cdot x) & \longrightarrow & (\alpha*\beta)\cdot x \\
1\cdot x & \longrightarrow & x \\
(\alpha+\beta)\cdot x & \longrightarrow & (\alpha\cdot x)\oplus(\beta\cdot x) \\
\alpha\cdot(x\oplus y) & \longrightarrow & (\alpha\cdot x)\oplus(\alpha\cdot y) \\
(-\alpha\cdot x)\oplus(\alpha\cdot x) & \longrightarrow & \Omega \\
(-1\cdot x)\oplus x & \longrightarrow & \Omega \\
0\cdot x & \longrightarrow & \Omega \\
\alpha\cdot\Omega & \longrightarrow & \Omega \\
I(x) & \longrightarrow & (-1)\cdot x
\end{array}
$$

The operators $+$, $*$, and $\oplus$ are in AC. To prove termination of R/AC we use the associative path ordering corresponding to the precedence ordering $>$ shown in the Hasse diagram in Fig. 7. The operator $\cdot$ has lexicographic status (right to left). Let T' consist of the sixth and eighth rule, and R' be $R-T'$. Then T' is contained in the A-transform T corresponding to $>$. Since $l >_{apo} r$, for all rules $l \longrightarrow r$ in R', $R'/(T' \cup AC)$ is terminating. Termination of T'/AC can be proved separately, which implies termination of R/AC.

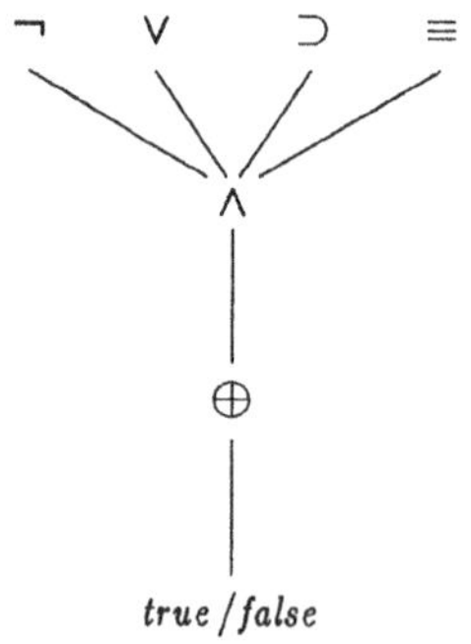

Fig. 6

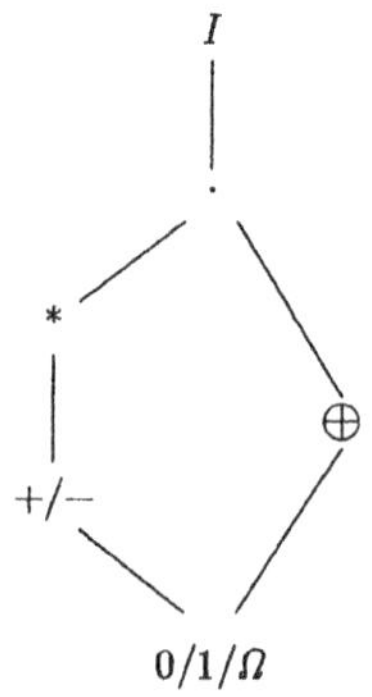

Fig. 7

A-transforms may also be used for proving termination of ordinary rewrite systems.

Example 7. *(Associativity and endomorphism).* Let R be the following rewrite system (Ben Cherifa and Lescanne, 1985):

$$\begin{array}{rcl} (x\cdot y)\cdot z & \rightarrow & x\cdot(y\cdot z) \\ f(x)\cdot f(y) & \rightarrow & f(x\cdot y) \\ f(x)\cdot(f(y)\cdot z) & \rightarrow & f(x\cdot y)\cdot z \end{array}$$

Let T' be the first rule of R and R' be $R-R'$. Since T' is terminating, R terminates if R'/T' terminates. Let T be the A-transform corresponding to a precedence ordering $>$, where f is smaller than $\cdot$ and f and $\cdot$ are in F_D. Then we have $l >_{apo} r$, for both rewrite rules $l \rightarrow r$ in R'. Since T' is contained in $T^{\leftrightarrow}$, R'/T' is terminating.

6. Summary

We have presented termination methods based on commutation properties, and have developed an abstract framework for describing transformation techniques. These general results have led us to the development of various particular transforms, including methods for proving termination of equational rewrite systems R/E, where E contains associativity, commutativity, and identity axioms. It should be possible to automate, to a certain degree, the process of developing transforms for certain classes of rewrite systems by using a "completion-like" procedure as suggested by Jouannaud and Munoz (1984).

References

[1] Bachmair, L., and Plaisted, D.A. (1985). Termination orderings for associative-commutative rewriting systems, *J. of Symbolic Computation 1,* 329–349.

[2] Ben Cherifa, A., and Lescanne, P. (1985). A method for proving termination of rewriting systems based on elementary computations on polynomials, unpublished manuscript.

[3] Dershowitz, N. (1981). Termination of linear rewriting systems, *Proc. 8th EATCS Int. Colloquium on Automata, Languages and Programming,* S. Even and O. Kariv, eds., Lect. Notes in Comp. Science 115, New York, Springer, 448–458.

[4] Dershowitz, N. (1982). Orderings for term-rewriting systems, *Theoretical Computer Science 17,* 279–301.

[5] Dershowitz, N. (1985a). Computing with rewrite systems, *Information and Control 64,* 122–157.

[6] Dershowitz, N. (1985b). Termination. *Proc. 1st Int. Conf. on Rewriting Techniques and Applications,* Dijon, France, Lect. Notes in Comp. Science, Springer, 180–224.

[7] Dershowitz, N., Hsiang, J., Josephson, N.A., and Plaisted, D.A. (1984). Associative-commutative rewriting. *Proc. 8th IJCAI,* Karlsruhe, 940–944.

[8] Guttag, J.V., Kapur, D., and Musser, D.R. (1983). On proving uniform termination and restricted termination of rewriting systems. *SIAM Computing 12,* 189–214.

[9] Hsiang, J. (1985). Refutational theorem proving using term–rewriting systems. *Artificial Intelligence 25,* 255–300.

[10] Huet, G. (1980). Confluent reductions: abstract properties and applications to term rewriting systems. *J. ACM 27,* 797–821.

[11] Huet, G. and Hullot, J.M. (1982). Proofs by induction in equational theories with constructors. *J. of Comp. and System Sciences 25,* 239–266.

[12] Jouannaud, J.–P., and Munoz, M. (1984). Termination of a set of rules modulo a set of equations, *Proc. 7th Int. Conf. on Automated Deduction,* R. Shostak, ed., Lect. Notes in Comp. Science 170, Berlin, Springer, 175–193.

[13] Kamin, S., and Levy, J.J. (1980). Two generalizations of the recursive path ordering. Unpublished manuscript, Univ. of Illinois at Urbana–Champaign.

[14] Kapur, D., and Sivakumar, G. (1984). Architecture of and experiments with RRL, a rewrite rule laboratory. *Proc. NSF Workshop on the Rewrite Rule Laboratory,* Rensellaerville, New York, 33–56.

[15] Lankford, D.S. (1979). On proving term rewriting systems are noetherian. Memo MTP–3, Mathematics Department, Louisiana Tech. Univ., Ruston, Louisiana.

[16] Manna, Z., and Ness, S. (1970). On the termination of Markov algorithms. *Proc. Third Hawaii Int. Conf. on System Science,* 789–792.

[17] Musser, D.R. (1980). On proving inductive properties of abstract data types. *Proc. 7th ACM Symp. on Principles of Programming Languages,* Las Vegas, 154–162.

[18] O'Donnell, M.J. (1985). *Equational logic as a programming language.* MIT Press, Cambridge, Massachusetts.

[19] Plaisted, D.A. (1984). Associative path orderings, *Proc. NSF Workshop on the Rewrite Rule Laboratory,* Rensellaerville, New York, 123–126.

[20] Raoult, J.C., and Vuillemin, J. (1980). Operational and semantic equivalence between recursive programs, *J. ACM 27,* 772–796.

[21] Rosen, B. (1973). Tree–manipulating systems and Church–Rosser theorems, *J. ACM 20,* 160–187.

FULL-COMMUTATION AND FAIR-TERMINATION IN EQUATIONAL (AND COMBINED) TERM-REWRITING SYSTEMS

by

Sara Porat, Nissim Francez
Computer Science Department
Technion - Israel Institute of Technology
Haifa 32000, Israel

ABSTRACT

In [PF-85] the concepts of *fair derivations* and *fair-termination* in term-rewriting systems were introduced and studied. In this paper, we define the notion of fairness in *equational term-rewriting systems*, where a derivation step is a composition of the equality generated by a (finite) set of equations with one step rewriting using a set of rules. A natural generalization of E-termination (termination of equational term-rewriting systems), namely *E-fair-termination*, is presented. We show that fair-termination and E-fair-termination are the same whenever the underlying rewriting relation is *E-fully-commuting*, a property inspired by Jouannaud and Muñoz' E-commutation property. We obtain analogous results for *combined* term-rewriting systems.

1. INTRODUCTION

One of the basic motivations behind the definition of a rewrite-rule in term-rewriting-systems is the *simplification* obtained by rewriting a given term. However, there are rewritings that are intrinsically not simplifying. For example, the rule $x+y \to y+x$ defines a rewriting that is not simplifying a term; however it defines an important relation among terms in a given system - commutativity. So, the notion of derivation in term-rewriting systems was extended to E-derivation, wherein a derivation step is obtained by composition of an equality generated by a set of equations with a rewriting relation using rules. Consider, for example, a system where + is a binary infix operator with the following properties: commutativity, expressed by the equation $x+y = y+x$, and absorption of the left identity expressed by the simplifying rule $e+x \to x$. The term $x+e$ is in normal form with respect to the rewriting relation generated by the rule. However, if we allow commutative rewritings, then $x+e$ rewrites to x. In other words, there is no derivation that starts from $x+e$, but there is an E-derivation that starts from $x+e$. Thus, one can easily see that the concept of reducing terms modulo a set of equations affects the termination property. In this work we prove that it affects also the fair-termination property as defined in [PF-85].

Jouannaud and Munoz introduce in [JM-84] the notion of E-commuting and show that termination of a set of rules R and termination of this set R modulo a set E of equations, namely E-termination, are the same wherever R is E-commuting. Bachmair and Dershowitz use in [BD-85] the notion of commutation in order to prove termination of a combined rewrite system $R_1 \cup R_2$ by proving termination of R_1 and R_2 separately. We introduce the natural generalization of E-termination, namely E-fair-termination, and show that E-commuting of the rewriting relation together with fair termination are not sufficient for E-fair-termination. We define another property of a rewriting relation, called full-commutation. This property together with fair termination provide a sufficient condition for E-fair-termination, and analogously connects fair termination of R_1 and R_1-fair termination of $R_1 \cup R_2$.

2. EQUATIONAL TERM-REWRITING SYSTEMS: TERMINATION VERSUS E-TERMINATION

Let F be a set of function symbols, and X a set of variables. A ***rewrite rule*** is an ordered pair (r, s), denoted by $r \to s$, where r and s are terms in the free algebra over (F,X). Let R be a set of rewrite rules.

The following definitions are similar to the corresponding definitions in other works like [JM-84], [JK-84], and [BP-85].

Definition 1: *An Equational Term-Rewriting System* (ETRS) is a pair (S,E) where

- S is a term-rewriting system (F,R) and
- E is a finite set of equations (axioms) of the form $\alpha=\beta$ where α and β are terms in the free algebra over (F,X).

[]

A "computation" in an ETRS is called an E-derivation, and it combines application of rules together with "application" of equations.

Definition 2:

1) $t\,|{-}|^E t'$ iff there is a substitution σ and an equation $(\alpha=\beta) \in E$, s.t. $t=\alpha\sigma$ and $t'=\beta\sigma$.
2) The reflexive-transitive closure of the relation $|{-}|^E$ is the ***equivalence-relation*** denoted by $=^E$.

[]

The E-rewriting relation, denoted by $\xrightarrow[R/E]{}$, is defined in the following way:

Definition 3. $t \xrightarrow[R/E]{} t'$ iff there is a term t'', s.t. $t =^E t'' \xrightarrow[R]{} t'$.

[]

Definition 4: An *E-derivation* is a finite or infinite sequence of the form

$$t_1 \xrightarrow[R/E]{} t_2 \xrightarrow[R/E]{} t_3 \xrightarrow[R/E]{} \cdots \; t_n \xrightarrow[R/E]{} \cdots \quad n \geq 1$$

The reflexive-transitive closure of $\xrightarrow[R/E]{}$ corresponds to the finite E-derivations, and is denoted by $\xrightarrow[R/E]{*}$.

[]

Definition 5:

1) An ETRS (S,E) is *terminating* iff the TRS S is terminating, that is to say, every derivation (in S) is finite.

2) An ETRS (S,E) is *E-terminating* iff every E-derivation is finite.

[]

We start with some observations about the connection between termination and E-termination. First we present some notations for describing terms.

Given a term t in the free algebra over (F,X):

1) $v(t)$ is the set of variables of t $(v(t) \subseteq X)$.

2) The *positions* within t are finite dotted lists of natural numbers, i.e. expressions of the form $n_1 \cdot n_2 \cdot \cdots \cdot n_k$ for some $k \geq 0$. In case $k=0$, we use the notation λ for the empty sequence. The position u defines the subterm t/u in the following way:

   i) $t/\lambda = t$

   ii) If $t/u = f(t_1,...,t_n)$, then for every j, $1 \leq j \leq n$, $t/u \cdot j = t_j$.

3) $t[u \leftarrow t']$ is the term obtained by replacing t/u by t' in t. Thus, $t[u \leftarrow t']/u = t'$.

Since every derivation in S corresponds to an E-derivation in (S,E) (as $\xrightarrow[R]{} \subseteq \xrightarrow[R/E]{}$), if a given system is E-terminating, then it is terminating. The converse is not necessarily true.

The following example presents a terminating ETRS that is not E-terminating.

Example 1: Let

$F = \{+, s, 0\}$

$R = \{+(s(x), y) \rightarrow +(x, s(y))\}$

$E = \{+(x, y) = +(y, x)\}$

Using Dershowitz' second termination theorem [D-82], one can easily prove that this system is terminating. There is an infinite E-derivation:

$$+(0,s(s(0))) =^E +(s(s(0)),0) \xrightarrow[R]{} +(s(0),s(0)) \xrightarrow[R]{} +(0,s(s(0))) =^E \cdots$$

Thus, the system is not E-terminating

[]

Jouannaud and Munoz provide in [JM-84] some simple restrictions on the set E of equations, as necessary conditions for the E-termination of a system (S,E):

1) If there is an equation $(\alpha=\beta) \in E$, s.t. $\nu(\alpha) \neq \nu(\beta)$ then (S,E) is not E-terminating.

2) If there is an equation $(x=t) \in E$, and two different subterm positions u, v, s.t. $t/u = t/v = x$, then (S,E) is not E-terminating.

One of the important contributions in [JM-84] is a sufficient condition for the E-termination of an ETRS (S,E), given that S is a terminating TRS.

Definition 6: R *commutes* with a set of equations E (or R is *E-commuting)* iff for every s, s' and t, there is t', s.t.

if $s' =^E s \xrightarrow[R]{+} t$

then $s' \xrightarrow[R]{+} t' =^E t$.

[]

This definition assures that if s' is reducible using the E-rewriting relation, then it is reducible using the rewriting relation. Moreover, if there is a derivation, from some term that is E-equivalent to s', that ends with t, then there is a derivation from s' that ends with some term that is E-equivalent to t. The E-commutation of R is described in Figure 1.

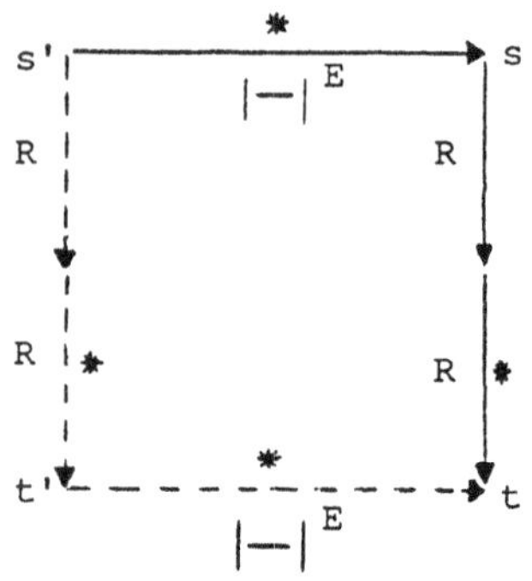

Figure 1:
A commutative diagram for E-Commutation

Theorem: (Sufficiency of E-commutation for E-termination) [JM-84]

Let $S = (F,R)$ be a terminating TRS. If R commutes with a set of equations E, then the system (S,E) is E-terminating.

The main idea in the proof is that in case there is an infinite E-derivation, then by the E-commutation of R, each rewriting using rules can be "pushed back" through the preceding equality. Thus, we can get an infinite derivation, contradicting the termination assumption.

[]

In the sequel, we shift the discussion to a new notion of termination, called E-fair-termination. We introduce necessary conditions and a sufficient one for E-fair-termination. These conditions are compared to those stated above for E-termination.

3. FAIRNESS IN ETRS

The notion of fair derivation in a term-rewriting system was introduced in [PF-85], whereby every rewrite rule enabled infinitely often along a derivation is infinitely-often applied along that derivation.

We now introduce the new definition of fairness in ETRS.

Definition 7: An E-derivation

$$d : t_1 =^E t'_1 \underset{R}{\dashrightarrow} t_2 =^E t'_2 \underset{R}{\dashrightarrow} t_3 \ldots$$

is a *fair E-derivation* iff it is finite, or it is infinite and for every rule $r \in R$, if r is applied only finitely often along d, then there is an $i \geq 1$, s.t. for every t'_j, $j \geq i$, r is not enabled in t'_j.

[]

Note that the checking of enabledness is not done before the "applications" of equations (on $t_1, t_2, t_3 \cdots$) but before the "applications" of rules (on $t'_1, t'_2, t'_3 \cdots$).

Following are examples of an infinite fair E-derivation and an infinite unfair E-derivation.

Example 2: Let

R:: 1) $g(0,y,z) \to g(0,f(y),z)$

2) $g(0,y,z) \to g(f(0),y,z)$

3) $g(f(x),f(y),z) \to g(f(x),y,z)$

4) $g(f(x),f(y),z) \to g(f(x),f(y),f(z))$

and $E = \{g(x,y,z) = g(y,x,z)\}$.

The following infinite E-derivation:

$$\begin{array}{l} g(f(0),0,0) =^E g(0,f(0),0) \underset{\{1\}}{\dashrightarrow} g(0,f(f(0)),0) \to \\ \underset{\{2\}}{\dashrightarrow} g(f(0),f(f(0)),0) \underset{\{4\}}{\dashrightarrow} g(f(0),f(f(0)),f(0)) \to \\ \underset{\{3\}}{\dashrightarrow} g(f(0),f(0),f(0)) \underset{\{3\}}{\dashrightarrow} g(f(0),0,f(0)) =^E \\ =^E g(0,f(0),f(0)) \underset{\{1\}}{\dashrightarrow} \cdots \end{array}$$

is a fair E-derivation, as every rule is infinitely often applied.

Consider the following finite E-derivation:

$$d : g(f(0),0,0) =^E g(0,f(0),0) \underset{\{2\}}{\dashrightarrow} g(f(0),f(0),0) \underset{\{3\}}{\dashrightarrow} g(f(0),0,0).$$

By repeating d infinitely, we get an infinite unfair E-derivation, as the rules (1) and (4) are infinitely often enabled and never applied.

[]

The set of equations affects also the notion of fair termination.

Definition 8:

1) An ETRS (S,E) is ***fairly-terminating*** iff the underlying TRS S is fairly terminating, that is to say, every fair derivation is finite.

2) An ETRS (S,E) is *E-fairly-terminating* iff every fair E-derivation is finite.

[]

In [P-86] sound and semantically complete proof-rules are introduced, for proving fair termination of a TRS, and E-fair-termination of an ETRS. These proof rules are based on well-foundedness arguments. Such methods for proving various notions of termination are beyond the scope of this paper, and we omit their presentations.

Claim 1: Figure 2 shows the relations between the set of terminating, E-terminating, fairly-terminating and E-fairly-terminating ETRS's.

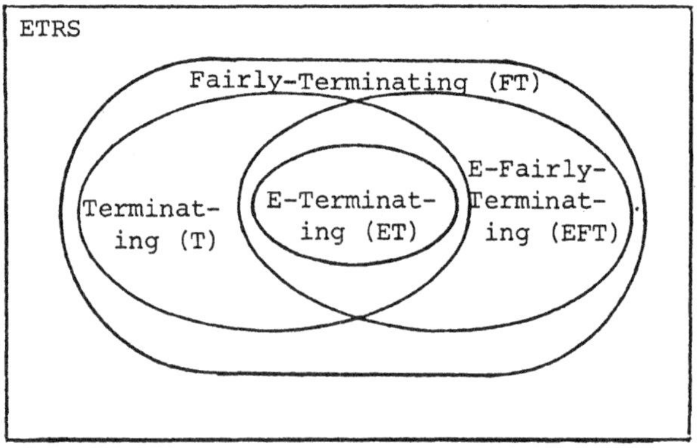

Figure 2:
Termination, E-termination, fair-termination and E-fair-termination.

Proof: By the definitions, every ETRS that is E-terminating is E-fairly-terminating and terminating, every terminating system is fairly-terminating, and every E-fairly-terminating system is also fairly-terminating.

If the set of rules R_0 consists of the single rule $\{a \to f(a)\}$ then there is an infinite fair derivation $a \to f(a) \to f(f(a)) \to \ldots$, thus for every $F \supseteq \{f,a\}$ and E, $((F,R_0),E) \notin FT$.

If the set of rules is $R_1 = \{a \to f(a), a \to b\}$, then the system is obviously not terminating, but there is no infinite fair derivation [PF-85]. For every $F \supseteq \{f,a,b\}$, $((F,R_1),\varphi) \in EFT-T$.

The system S_2 in Example 1 is terminating but not E-terminating. Since this system has a single rule, every infinite E-derivation is a fair E-derivation, thus $S_2 \in T-EFT$.

The following example presents an ETRS S_3, s.t. $S_3 \in (T \cap EFT)-ET$.

Example 3: Let

$F = \{+,s,0\}$

$R = \{+(s(x),y) \to +(x,s(y)),\ +(s(x),y) \to s(+(x,y)),\ +(0,x) \to 0\}$

$E = \{+(x,y) = +(y,x)\}$

This system is terminating (the proof is simple and uses Dershowitz' termination theorems [D-82].) Since the set of rules in this example is a subset of the one in example 1, and the set of equations in both systems is the same, this system is obviously not E-terminating. Using an appropriate proof rule [P-86] it can be proved that this system is E-fairly-terminating.

[]

The following example presents an ETRS S_4, s.t. $S_4 \in FT-(T \cup EFT)$.

Example 4: Let $F = \{a,b,c,d,e\}$

$R = \{a \to b,\ b \to a,\ a \to c,\ d \to e,\ e \to d,\ d \to c\}$

$E = \{a=e,\ d=b\}$.

There is an infinite derivation:

$$a \xrightarrow[R]{} b \xrightarrow[R]{} a \xrightarrow[R]{} b \xrightarrow[R]{} a \longrightarrow \cdots$$

Consider the following infinite E-derivation:

$$a =^E e \xrightarrow[R]{} d =^e b \xrightarrow[R]{} a =^E \cdots$$

Since there is only one rule enabled in the term **e**, and one in term **b**, this E-derivation is fair.

Using an appropriate decision algorithm [PF-85], it can be easily proved that this system is fairly terminating.

[]

4. FAIR TERMINATION VERSUS E-FAIR TERMINATION

We first prove that the simple restrictions on the set E of equations, that were introduced above as necessary conditions for the E-termination of an ETRS, are also necessary conditions for the E-fair termination of an ETRS.

Claim 2: If there is an equation $(\alpha=\beta) \in E$, s.t. $\nu(\alpha) \neq \nu(\beta)$, then (S,E) is not E-fairly-terminating.

Proof: Let $R = \{l_i \rightarrow r_i \mid 1 \le i \le n\}$. Assume $x \in \nu(\alpha)$, $x \notin \nu(\beta)$, and $\alpha/u=x$.

As we may instantiate the extra variable in α to <u>*any*</u> term, in particular we may choose a l.h.s. l_i, and obtain the following infinite E-derivation that starts from β:

$$\beta =^E \alpha[u \leftarrow l_1] \xrightarrow[R]{} \alpha[u \leftarrow r_1]| - |^E \beta | - |^E \alpha[u \leftarrow l_2] \xrightarrow[R]{}$$
$$\xrightarrow[R]{} \alpha[u \leftarrow r_2]| - |^E \beta | - |^E \alpha[u \leftarrow l_3] \xrightarrow[R]{} \cdots \ \alpha[u \leftarrow l_n] \xrightarrow[R]{}$$
$$\xrightarrow[R]{} \alpha[u \leftarrow r_n]| - |^E \beta | - |^E \alpha[u \leftarrow l_1] \xrightarrow[R]{} \cdots$$

This is a description of an E-derivation, in which every rule is applied infinitely-often, hence this E-derivation is an infinite fair E-derivation.

[]

Claim 3: If there is an equation $(x=t) \in E$, with two different subterm positions u, v s.t. $t/u=t/v=x$, then (S,E) is not E-fairly-terminating.

Proof: Let $R = \{l_i \rightarrow r_i \mid 1 \le i \le n\}$.

We construct an infinite E-derivation, starting from some reducible term t_1^0. In other words, there is some rule that is enabled in t_1^0.

For every term t_i^j $(i \geq 1, j \geq 0)$, we denote by t_i^{j+1} the term $t[u \leftarrow t_i^j][v \leftarrow t_i^j]$.

As $(x=t) \in E$, $t_i^j| - |^E t_i^{j+1}$ is always true. Since t_i^j is a subterm of t_i^{j+1}, the set of enabled rules in t_i^j is a subset of those enabled in t_i^{j+1}. Since the set of rules is finite, for every $i \geq 1$, there is k_i, s.t. the set of enabled rules in $t_i^{k_i}$ is equal to the set of enabled rules in $t_i^{k_i+1}$.

For a given term s, a derivation step from $t[u \leftarrow s][v \leftarrow s]$, s.t. the rule is applied on the subterm $t[u \leftarrow s][v \leftarrow s]/u$, is called *a derivation step that assures infinity*. If the enabled rules in s are exactly those in $t[u \leftarrow s][v \leftarrow s]$, then every rule which is enabled in $t[u \leftarrow s][v \leftarrow s]$ can be applied on the subterm $t[u \leftarrow s][v \leftarrow s]/u$.

The constructed E-derivation consists of derivation steps that assure infinity, applied on $t_i^{k_i+1}$.

We denote by t_{i+1}^0 some term that is derived from $t_i^{k_i+1}$, by applying a rule, that is chosen according to a certain method defined below, in a derivation step that assures infinity.

The E-derivation is in the following form:

$$t_1^0 =^E t_1^{k_1+1} \underset{R}{\dashrightarrow} t_2^0 =^E t_2^{k_2+1} \underset{R}{\dashrightarrow} t_3^0 =^E t_3^{k_3+1} \underset{R}{\dashrightarrow} \cdots$$

Since the enabled rules in $t_i^{k_i}$ are exactly those in $t_i^{k_i+1}$, and $t_i^{k_i}$ is a subterm of t_{i+1}^0, we get that the set of enabled rules in $t_i^{k_i+1}$ is a subset of those enabled in t_{i+1}^0.

As we already mentioned, the set of enabled rules in t_{i+1}^0 is a subset of those enabled in $t_{i+1}^{k_i+1}$, which are exactly those enabled in $t_{i+1}^{k_{i+1}+1}$.

Thus, joining these two remarks, we get that the set of enabled rules in $t_i^{k_i+1}$ is a subset of those enabled in $t_{i+1}^{k_{i+1}+1}$.

The rules that are applied along the E-derivation are chosen in the following way:

For $t_i^{k_i+1}$, the chosen rule to be applied on the derivation step from $t_i^{k_i+1}$, is the one that was applied the minimum number of times (possibly not at all) along the E-derivation starting from t_1^0, from amongst the applications on $t_1^{k_1+1}$, $t_2^{k_2+1}$ and so on, up to $t_{i-1}^{k_{i-1}+1}$. In case there are several rules enabled in $t_i^{k_i+1}$, that were applied the same minimum number of times, the rule with the smallest index is chosen.

If the rule $l_i \to r_i$ is applied along the E-derivation, then there is $l \geq 1$, s.t. this rule is enabled in $t_j^{k_j+1}$, for every $j \geq l$. According to the method of chosing the rules to be applied, and since the set of rules is a finite set, this rule is chosen for application infinitely many times. So, by definition of fairness, this E-derivation is an infinite fair E-derivation.

In order to clarify this proof, let us consider the following example:

Example 5: Let

R:: 1) $\neg\neg\alpha \to \alpha$

2) $\neg(\alpha \vee \beta) \to (\neg\alpha \wedge \neg\beta)$

3) $\neg(\alpha \wedge \beta) \to (\neg\alpha \vee \neg\beta)$

E:: $\alpha \wedge \alpha = \alpha$

Let $t_1^0 = \neg(\neg a \vee b)$.

Only rule (2) is enabled in t_1^0. Since the set of enabled rules in t_1^0 is equal to this in $t_1^1 = \neg(\neg a \vee b) \wedge \neg(\neg a \vee b)$, $k_1 = 0$. By applying a derivation step (that assures infinity), we obtain the term $t_2^0 = (\neg\neg a \wedge \neg b) \wedge \neg(\neg a \vee b)$, where both rules (1) and (2) are enabled. Again, since the enabled rules in t_2^1 ($t_2^1 = t_2^0 \wedge t_2^0$) are the same as in t_2^0, $k_2 = 0$. The chosen rule to be applied on the next derivation step, from t_2^1, is (1), and we get the term t_3^0. The constructed derivation is continued by applying rules (2) and (1) indefinitely.

[]

Next, we prove that the commutation of R with a set of equations E, which is a sufficient condition for the E-termination of a terminating ETRS, is not a sufficient condition for the E-fair-termination of a fairly terminating ETRS. This is shown using the system in example 4 that is fairly terminating and not E-fairly terminating.

Proving the commutation of R with E is very simple in this case, as we have to check only a finite set of possibilities. For example, if the E-derivation is $a =^E e \xrightarrow[R]{+} d$, then the term **b** satisfies the condition as $a \xrightarrow[R]{+} b =^E d$.

The set of equations provides the possibility to "jump" on pathological states like **a** or **d**, where a fair choice among the enabled rules to be applied imposes termination.

In order to introduce a sufficient condition for E-fair-termination of a fairly terminating ETRS, we define a stronger version of commutation and prove some lemmas related to it.

Definition 9: The derivation $t_1 \underset{R}{\dashrightarrow} t_2 \underset{R}{\dashrightarrow} \cdots \underset{R}{\dashrightarrow} t_n$, for $n \geq 2$, is a *full-derivation* iff every rule enabled in some t_i, $1 \leq i \leq n-1$, is applied along this derivation. We denote such full-derivation (that starts from t_1 and ends with t_n) by $t_1 \overset{f}{\underset{R}{\dashrightarrow}} t_n$.

[]

Definition 10: If $t_i \overset{f}{\underset{R}{\dashrightarrow}} t_{i+1}$, for every $i \geq 1$, then the derivation obtained by concatenating all these full-derivations is a *chain of full-derivations*. Note that every finite chain is itself a full-derivation.

[]

Lemma 1: If a TRS S = (F,R) is fairly-terminating, then for every t there is no infinite chain of full-derivations that starts from t.

Proof: Assume, by way of contradiction, that there is some t_1 and an infinite chain of full derivations that starts from t_1: $t_1 \overset{f}{\underset{R}{\dashrightarrow}} t_2 \overset{f}{\underset{R}{\dashrightarrow}} t_3 \overset{f}{\underset{R}{\dashrightarrow}} \cdots$. This derivation is a fair derivation, as every rule enabled infinitely often along it is enabled in infinitely many full-derivations along this chain. By definition of full-derivation, this rule is applied infinitely many times along the chain. This derivation contradicts the assumption that the given system is fairly-terminating. (Actually, since $\underset{R}{\dashrightarrow}$ is locally finite, for every t, there is a natural number n_t, s.t. every chain of full-derivations that starts from t is a concatenation of no more than n_t full-derivations.)

[]

Definition 11: The E-derivation $t_1 =^E t'_1 \underset{R}{\dashrightarrow} t_2 =^E t'_2 \underset{R}{\dashrightarrow} \cdots \underset{R}{\dashrightarrow} t_n$, for $n \geq 2$, is a *full-E-derivation* iff every rule enabled in some t'_i, $1 \le i \le n-1$, is applied along this E-derivation. We denote such a full-E-derivation (that starts from t_1 and ends with t_n) by $t_1 \overset{f}{\underset{R/E}{-\!\!-\!\!-\!\!\to}} t_n$.

[]

Definition 12: If $t_i \overset{f}{\underset{R/E}{-\!\!-\!\!-\!\!\to}} t_{i+1}$, for every $i \ge 1$, then the E-derivation obtained by concatenating all these full-E-derivations is a *chain of full-E-derivations*.

[]

Lemma 2: For an infinite fair E-derivation

$$t_1 =^E t'_1 \underset{R}{-\!\!-\!\!\to} t_2 =^E t'_2 \underset{R}{-\!\!-\!\!\to} t_3 =^E \cdots$$

there is $i \ge 1$, s.t. the E-derivation that starts from t_i (a tail of the given one) is an infinite chain of full-E-derivations.

Proof: Let $R = \{l_i \to r_i \mid 1 \le i \le n\}$.

By the definition of an infinite fair E-derivation, for every rule $l_j \to r_j$, if it is applied only finitely often along the given E-derivation, then there is $i_j \ge 1$, s.t. for every t'_k, $k \ge i_j$, this rule is not enabled in t'_k. For the sake of the formal definition of the desired i, for every rule $l_j \to r_j$, applied infinitely often along the given E-derivation, the corresponding i_j is zero. Let $i = \max_{1 \le j \le n} i_j$.

We denote by d_{t_i} the infinite fair E-derivation that starts from t_i. Every rule enabled along d_{t_i}, is infinitely often enabled (due to the way i was chosen). Thus, by the fairness assumption, this rule is infinitely often applied.

Claim: For every $k_1 \ge i$, there is k_2, $k_2 > k_1$, s.t. the E-derivation that starts from t_{k_1} and ends with t_{k_2} is a full E-derivation.

Proof of Claim: For every rule $l_j \to r_j$, if the rule is enabled along the E-derivation d_{t_i}, then there is $m_j > k_1$, s.t. this rule is applied along the E-derivation that starts from t_{k_1} and ends with t_{m_j} (since such rule is applied infinitely many times along d_{t_i}). Let m_j be zero for every rule $l_j \to r_j$ that is not enabled along d_{t_i}. So, let $k_2 = \max_{1 \le j \le n} m_j$. Every rule, enabled along d_{t_i}, is applied along the E-derivation that starts from t_{k_1} and ends with

t_{k_2}. Thus, this E-derivation is a full-E-derivation.

The proof of the claim completes the proof of the lemma.

[]

Definition 13: R is *fully commuting* with a set of equations E (or R is *E−fully−commuting*) iff for every s and t, there is t', s.t.

if $s \xrightarrow[R/E]{f} t$

then $s \xrightarrow[R]{f} t' =^E t$.

[]

This property of R is expressed by the diagram in Figure 3.

Theorem : (Sufficiency of E-full-commutation for E-fair termination)

Let S = (F,R) be a fairly terminating TRS. If R is fully commuting with a set of equations E, then the system (S,E) is E-fairly terminating.

Proof: By Lemma 1, the relation $\xrightarrow[R]{f}$ is Noetherian.

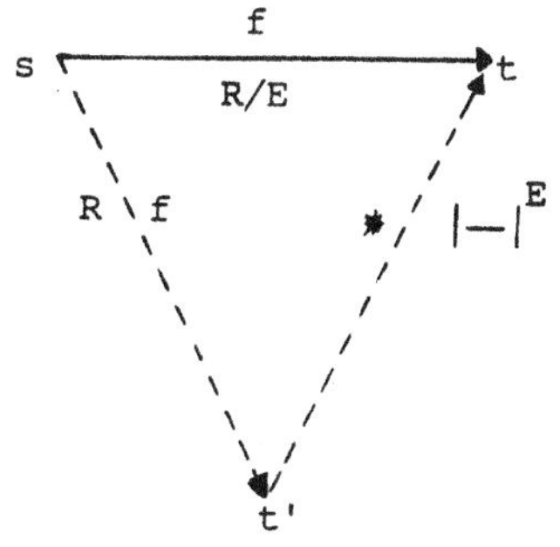

Figure 3:
A diagram for E-full-commutation

We first prove, by Noetherian induction on $\overset{f}{\underset{R}{\dashrightarrow}}$, that under the assumptions, for every term t, there is no infinite chain of full-E-derivations starting from t.

Let $t_0 \overset{f}{\underset{R/E}{\dashrightarrow}} t_1 \overset{f}{\underset{R/E}{\dashrightarrow}} t_2 \overset{f}{\underset{R/E}{\dashrightarrow}} \cdots$, be a chain of full-E-derivations, starting from t_0. By E-full-commutation there exists a term t'_0, s.t. $t'_0 =^E t_1$ and $t_0 \overset{f}{\underset{R}{\dashrightarrow}} t'_0$. As $t'_0 =^E t_1$, the rest of the chain of full-E-derivations starting from t_1 is also a chain of full-E-derivations starting from t'_0. By the induction hypothesis, this chain is finite, because $t_0 \overset{f}{\underset{R}{\dashrightarrow}} t'_0$.

By Lemma 2, existence of an infinite fair E-derivation implies the existence of an infinite chain of full-E-derivations. Hence, the system ((F,R),E) is E-fairly-terminating.

[]

The following example proves that the E-full-commutation property is not a necessary condition for E-fair-termination of a fairly-terminating ETRS.

Example 6: Let

$R = \{f(x,x) \to f(a,b),\ b \to a,\ b \to c\}$

and

$E = \{f(a,b) = f(b,a)\}$

The system is fairly terminating and E-fairly-terminating [P-86].

Consider the following E-derivation:

$$d = f(a,b) \underset{R}{\dashrightarrow} f(a,a) \underset{R}{\dashrightarrow} f(a,b) =^E f(b,a) \underset{R}{\dashrightarrow} f(c,a)$$

d is a full-E-derivation, as every rule is applied along d. But, there is no full derivation from $f(a,b)$ to $f(c,a)$. (Note that $t =^E f(c,a) \Rightarrow t = f(c,a)$.) Hence, R is not E-fully-commuting.

[]

5. FAIR TERMINATION OF COMBINED TRS

Bachmair and Dershowitz have proved [BD-85] that commutation between rewrite relations allows proving termination of a combined rewrite system $(F, R_1 \cup R_2)$ by proving termination of (F,R_1) and (F,R_2) separately. In this section we consider the notion of full commutation between rewrite relations, and get analogous results to those in the previous section.

Let (F,R_1) and $F,R_2)$ be two TRSs, and assume $R_1 \cap R_2 = \varphi$.

Definition 14: R_1 *commutes* with R_2 iff for every s, s' and t, there is t', s.t.

if $\quad s' \xrightarrow[R_2]{+} s \xrightarrow[R_1]{+} t$

then $\quad s' \xrightarrow[R_1]{+} t' \xrightarrow[R_2]{*} t.$

[]

The commutation of R_1 with R_2 is described in Figure 4.

The definition is analogous to the definition of commutation between a set of rules and a set of equations. This definition of commutation between two sets of rules is not exactly the same as in [BD-85] (which is actually analogous to local-commutation between R and E as defined in [JM-84]). But, one can easily prove that our definition of commutation implies that of [BD-85].

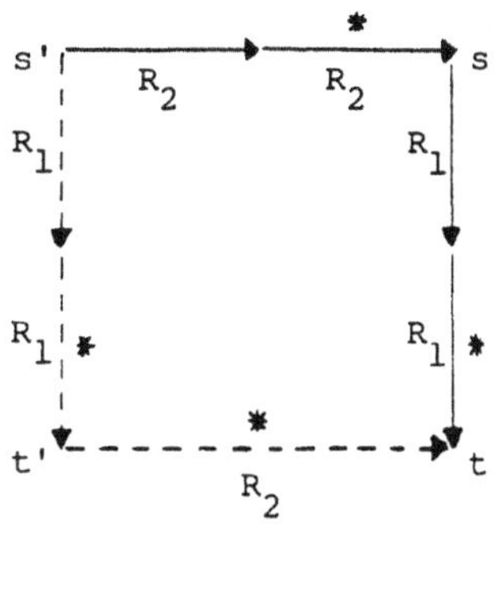

Figure 4:
A commutative diagram for commutation of R_1 with R_2

Theorem: (Sufficiency of commutation for termination in combined system) [BD-85]
Let $S_1 = (F,R_1)$ and $S_2 = (F,R_2)$ be two TRS's. Assume R_1 commutes with R_2. Then, the combined system $(F,R_1 \cup R_2)$ is terminating iff S_1 and S_2 both are.

Trivially, if $(F,R_1 \cup R_2)$ is terminating, so are (F,R_1) and (F,R_2). For the other direction, the main idea in the proof is that in case there is an infinite derivation

$$t_1 \xrightarrow[R_2]{*} t_2 \xrightarrow[R_1]{+} t_3 \xrightarrow[R_2]{+} t_4 \xrightarrow[R_1]{+} t_5 \xrightarrow[R_2]{+} \cdots$$

then, by the fact that R_1 commutes with R_2, each derivation $\xrightarrow[R_1]{+}$ can once again be "pushed back" through the preceding $\xrightarrow[R_2]{+}$. Thus, there must be an infinite derivation in (F,R_1).

[]

We would like to consider the results of the previous section about ETRS's as special cases of the following results dealing with combined TRS's. As we require in the definition of an E-derivation only finite sequences of consecutive equational replacements, one of the rewriting relations in the combined system, for example R_2, is required to be applied repeatedly only a finite number of times. Thus, an R_1-derivation in $(F,R_1 \cup R_2)$ is defined in the following way:

Definition 15: An R_1*-derivation* is a finite or infinite sequence of the form

$$t_1 \xrightarrow[R_2]{*} t'_1 \xrightarrow[R_1]{} t_2 \xrightarrow[R_2]{*} t'_2 \xrightarrow[R_1]{} t_3 \xrightarrow[R_2]{*} \cdots$$

[]

The following new definitions are similar to those defining fairness in E-derivation and E-fair-termination.

Definition 16: an R_1-derivation in the system $(F,R_1 \cup R_2)$

$$d : t_1 \xrightarrow[R_2]{*} t'_1 \xrightarrow[R_1]{} t_2 \xrightarrow[R_2]{*} t'_2 \xrightarrow[R_1]{} t_3 \xrightarrow[R_2]{*} t'_3 \xrightarrow[R_1]{} \cdots$$

is an R_1*-fair derivation* iff it is finite, or it is infinite and for every rule $r \in R_1$, if r is applied only finitely often along d, then there is an $i \geq 1$, s.t. for every t'_j, $j \geq i$, r is not enabled in t'_j.

[]

Definition 17: The system $(F, R_1 \cup R_2)$ is *R_1-fairly-terminating* iff every R_1-fair derivation is finite.

[]

Thus, one can easily describe a proof rule (similar to the rule introduced in [P-86], for proving E-fair-termination of TRS) in order to prove R_1-fair termination of a system (F,R), where $R_1 \subseteq R$.

As in the case of rewriting modulo equations, the commutation of R_1 with R_2 is not sufficient for the R_1-fair-termination of $(F, R_1 \cup R_2)$, even if the system (F, R_1) is fairly terminating.

Definition 18: An R_1-derivation in $(F, R_1 \cup R_2)$

$$d : t_1 \xrightarrow[R_2]{*} t'_1 \xrightarrow[R_1]{} t_2 \xrightarrow[R_2]{*} t'_2 \xrightarrow[R_1]{} t_3 \ldots\ldots \longrightarrow t_n, \quad n \geq 2,$$

is an *R_1-full derivation* iff every rule from R_1 enabled in some t'_i, $1 \leq i \leq n-1$, is applied along d. We denote such R_1-full derivation (that starts from t_1 and ends with t_n) by $t_1 \xrightarrow[R_1 \cup R_2]{R_1 - f} t_n$.

[]

Definition 19: R_1 is *fully commuting* with R_2 iff for every s and t, there is t', s.t.

if $\quad s \xrightarrow[R_1 \cup R_2]{R_1 - f} t$

then $\quad s \xrightarrow[R_1]{f} t' \xrightarrow[R_2]{*} t$

[]

Using similar lemmas, the proof of the following theorem is just the same as that of the sufficiency of E-full-commutation for E-fair termination theorem .

Theorem: (Sufficiency of full-commutation for fair termination in combined system) If R_1 is fully commuting with R_2, then the system (F,R_1) is R_1-fairly terminating iff (F,R_1) is fairly terminating.

[]

6. CONCLUSIONS

In this paper, we extended the notion of fair termination of term rewriting systems, as introduced in [PF-85], to equational term rewriting systems. It is rare that a computational problem is formulated in a completely free algebra. Most systems of practical interest have some properties and connections between their operators, and in many cases these can be formulated in an equational theory, e.g. associativity, commutativity and distributivity. Thus, such an extension is important for real applications of fairness to TRS.

The particular contribution of the paper is in providing a sufficient condition for E-fair termination, replacing the commuting condition of [JM-84] by a stronger notion of full-commutation. However, we consider the importance of the approach beyond this specific contribution. We believe that full-commutation, in some sense, embodies the essence of fairness. An analogous notion plays a crucial role in completeness proofs of proof rules for fair termination of nondeterministic programs [GFMdR-81, LPS-81], and also in characterizations of fairness in CCS [CS-85].

The asymmetry of the extension to combined TRSs leaves room for some further research, arriving at a more symmetric notion of relativized fairness in such systems, but still preserving the equational case as a special case.

ACKNOWLEDGEMENT

The work of the second author was partially supported by the fund for the promotion of research in the Technion.

REFERENCES

[BD-85] L. Bachmair, N. Dershowitz: "Commutation, transformation and termination", 1985, (submitted).

[BP-85] L. Bachmair, D.A. Plaisted: "Associative path ordering", *Proceedings of the First International Conference on Rewriting Techniques and Applications*, Dijon, France, May 1985, (LNCS 202, J.P. Jouannaud - ed., Springer, Berlin).

[CS-85] G. Costa, C. Stirling: "Weak and strong fairness in CCS", Internal Report, CSR-167-85, University of Edinburgh, Jan. 1985.

[D-82] N. Dershowitz: "Orderings for term rewriting systems", *J. Theoretical Computer Science*, Vol. 17, No. 3, March 1982, pp. 279-301 (previous version appeared in *Proceedings of the Symposium on Foundations of Computer Science*, San Juan, PR, pp. 123-131, Oct. 1979).

[GFMdR-81]
O. Grumberg, N. Francez, J.A. Makowsky, W.P. de Roever: "A proof rule for fair termination of guarded commands", *Proceedings of the Int. Symp. on Algorithmic Languages*, Amsterdam, Oct. 1981, North-Holland.

[JK-84] J.P. Jouannaud, H. Kirchner: 'Completion of a set of rules modulo a set of equations", *11th Ann. ACM Symp. on Principles of Programming Languages*, Salt Lake City, Utah, January 1984, pp. 83-92.

[JM-84] J.P. Jouannaud, M. Munoz: "Termination of a set of rules modulo a set of equations", *Proceedings of the 7th Intern. Conf. on Automated Deduction*, Napa, CA. May 1984, pp. 175-193 (LNCS 170, R.E. Shostak - ed., Springer, Berlin).

[LPS-81] D. Lehmann, A. Pnueli, J. Stavi: "Impartiality, justice and fairness: the ethics of concurrent termination", *Proceedings 8th ICALP*, Acre, Israel, LNCS 115 (O. Kariv, S. Even- eds.), Springer Verlag, 1981.

[P-86] S. Porat: "Fairness in models for nondeterministic computations", Ph.D. Thesis, Computer Science Dept., Technion, Haifa, Israel, Jan. 1986.

[PF-85] S. Porat, N. Francez: "Fairness in term rewriting systems", *Proceedings of the 1st Inter. Conf. on Rewriting Techniques and Applications*, Dijon, France, May 1985 (LNCS 202, J.P. Jouannaud - ed., Springer, Berlin).

An actual implementation of a procedure that mechanically proves termination of rewriting systems based on inequalities between polynomial interpretations

Ahlem Ben Cherifa   Pierre Lescanne

Centre de Recherche en Informatique de Nancy
Campus Scientifique BP 239
54506 Vandoeuvre, FRANCE

1. The origin of the problem

Termination is central in programming and in particular in term rewriting systems the later being both a theoretical and a practical basis for functional and logic languages.

Indeed the problem is not only a key for ensuring that a program and its procedures eventually produce the expected result, it is also important in concurrent programming where liveness results rely on termination of the components. Term rewriting systems are also used for proving equational theorems and are a basic tool for checking specifications of abstract data types. Again, the termination problem is crucial in the implementation of the Knuth-Bendix algorithm, which tests the local confluence and needs the termination to be able to infer the total confluence. Termination is also necessary to direct equations properly. Until now, methods based on recursive path ordering were satisfactory [Dersh82],[Jouan.*etal*.82], but when we recently ran experiments on transformation of **FP** programs [Belle84], we were faced with a problem that the recursive path ordering could not handle. The problem, a simple example of code optimization, is just Associativity + Endomorphism.

$$f(x_1) * f(x_2) \rightarrow f(x_1 * x_2)$$
$$(x_1 * x_2) * x_3 \rightarrow x_1 * (x_2 * x_3)$$
$$f(x_1) * (f(x_2) * x_3) \rightarrow f(x_1 * x_2) * x_3$$

where the third rule also decreases the number of occurrences of **f**. Fortunately, thanks to Lankford [Lankf79] this system can be easily proved terminating by using the polynomial interpretation $[f](X_1)= 2X_1$ and $[*](X_1,X_2)= X_1X_2 + X_1$, and this motivates our attempt for implementing a nice and efficient method for mechanically proving termination based on polynomial interpretations. Our purpose is not to extend his method in any way, but strictly implement what is presented by Huet and Oppen in their survey [Hue&Opp80]. They say (p.367): "the proof of "(inequality of

the form $[i(x*y)](X,Y) > [i(y)*i(x)](X,Y)$" is not straightforward, since it involves showing for instance ($\forall$ x, y $\in$ **N**) $x^2(1+2y)^2 > y^2(1+2x^2)$".
We also characterize the polynomial interpretations for associative-commutative operators, restricting drastically the space of interpretations for such operators. Finally we extend the method to Cartesian product providing a technique for proving termination of associative commutative systems, like that using Natural numbers, for which no implemented method exists. This paper contains few theory and no new theoretical idea. It rather emphasizes on the actual implementation in REVE, and the examples that were mechanically proved by REVE. It describes procedures for checking polynomial inequalities, which is both efficient and general. The reader is invited to consider essentially this aspect, since we feel it is important to propose algorithms, and people who have used our software REVE are usually grateful to notice the system performs all tedious computations required by the polynomial interpretation method at their place.

2. Interpretation by Functions over the Naturals and Termination

Let $T(F, \{x_1,...,x_m\})$ be the set of terms on $\{x_1,...,x_m\}$ and $\mathbf{N}^m \to \mathbf{N}$ the set of m-ary functions on natural numbers. Suppose that for each k-ary functions $\mathbf{f} \in F_k \subseteq F$, we define an interpretation of $\mathbf{f}$ as a polynomial with k variables. The interpretation of f will be written $\lambda X_1...X_k[f](X_1,...,X_k)$ and often we simply write $[f](X_1,...,X_k)$. This interpretation allows us to define on $\mathbf{N}^m \to \mathbf{N}$ an F-algebra $\mathbf{N}_m$, in the following way: if $p_1,...,p_k \in \mathbf{N}^m \to \mathbf{N}$, then
$f_{\mathbf{N}_m}(p_1,...,p_k)(X_1,...,X_m) = [f](p_1(X_1,...,X_m),...,p_k(X_1,...,X_m))$
If we define $[x_i](X_1,...,X_m) = X_i$, then there exists a unique extension to a morphism from $T(F,\{x_1,...,x_m\})$ to the F-algebra $\mathbf{N}_m$ imposed on $\mathbf{N}^m \to \mathbf{N}$ that we will also write $[.]$.
There exists on $\mathbf{N}^m \to \mathbf{N}$ a natural partial ordering which is defined as $h < k$ if and only if $(\forall a_1 \in \mathbf{N})...(\forall a_m \in \mathbf{N})\ h(a_1,...,a_m) < k(a_1,...,a_m)$. This ordering is obviously well-founded, otherwise a sequence $h_1 > ... > h_n > ...$ would exist which by instantiation would produce an infinite sequence

$$h_1(a_1,...,a_m) > ... > h_n(a_1,...,a_m) >...$$

On $T(F, \{x_1,...,x_m\})$, we define an ordering $<_{[.]}$ by $s <_{[.]} t$ if and only if $[s] < [t]$. This ordering is well-founded by definition. It is also stable by instantiation, which means that for all substitution σ, $\sigma(s) <_{[.]} \sigma(t)$. Indeed since $[.]$ is a morphism, $[\sigma(s)](X_1,...,X_m) < [\sigma(t)](X_1,...,X_m)$ is equivalent to $[s]([\sigma(x_1)],...,[\sigma(x_m)]) < [t]([\sigma(x_1)],...,[\sigma(x_m)])$ and this inequality holds if the inequality $[s] < [t]$ holds. It will be said to be compatible if $s <_{[.]} t$ implies $f(...,s,...) <_{[.]} f(...,t,...)$ Let us recall now a main termination criterion:

Proposition 1:[Man&Nes70, Dersh85]

A term rewriting system R terminates on a set $T(F,\{x_1,...,x_m\})$ of terms if there exists a well-founded and compatible quasi-ordering such that, for all rules $g \rightarrow d$ in R and for all substitution σ , $\sigma(g) > \sigma(d)$

So, any interpretation associated with a compatible ordering could be used to prove termination.

3. Polynomial Interpretation for proving Termination

Given a term rewriting system $\{g_1 \rightarrow d_1,...,g_n \rightarrow d_n\}$, the problem consists of first guessing a compatible interpretation **[.]** and then proving that $g_1 >_{[.]} d_1,...,g_n >_{[.]} d_n$. We will here not address the first problem since currently we do not have good heuristics. We expect that computer experiments will lead to a progress in this direction. Instead we will focus essentially on proving the inequalities between the interpretations of terms. As many authors do and as many classical examples demonstrate, we will restrict ourselves to polynomial interpretations, although it is known that there are term rewriting systems that cannot be proven to terminate using this method. However, we feel that our method could be extended to other classical recursive functions like the exponentials.

3.1. An overview of the problem

Notice first there is no algorithm to decide inequalities between n polynomials over **N**. Otherwise this algorithm could be used to solve the tenth Hilbert Problem. We do not want to decide inequalities and leave open the problem of guessing an adequate **[.]**. Our aim is a procedure that just checks properties that insure the wanted inequalities between polynomials and thereby directs equations into rules. We want to base these computations on really elementary and basic principles such that a simple and efficient implementation can be easily devised.

Our first idea will be to restrict the domain of polynomials to **N-{0,1}**, in other words to consider the set $\mathbf{N}\text{-}\{0,1\}^m \rightarrow \mathbf{N}$. Now the idea behind this becomes easy, and generalizes the fact that $XY > Y$ if $X > 1$. To guarantee the stability by instantiation, we have also to be sure that the values $[\sigma(x_i)](X_1,...,X_n)$ are in **N-{0,1}**. Thus we have to check that the interpretation of each term is a function in $\mathbf{N}\text{-}\{0,1\}^m \rightarrow \mathbf{N}\text{-}\{0,1\}$. This will be easily satisfied, if for all i in [1..m] $X_i > 1$ then $[f](X_1,...,X_m) > 1$, and this will be true if $\forall a_1 \in \mathbf{N}\text{-}\{0,1\} ... \forall a_m \in \mathbf{N}\text{-}\{0,1\}$ $(\forall i \in [1..m])$ $[f](a_1,...,a_m) \geq a_i$. This last monotonicity condition is quite similar to the subterm property of the simplification orderings and both conditions can obviously be checked with our algorithm. Usually it is enough to ensure that the coefficients of the interpretations are natural numbers. Together with the degree of each variable being greater than 1, this also implies the compatibility.

3.2. The Principles of the procedure used for proving positiveness

This is the key of our method. We prove inequalities one at a time and starting from a polynomial P_0 we build a sequence of inequalities such that "$P_0 \geq \ldots P_n -1 \geq P_n \geq 0$". The positiveness of P_n is supposed to be checked by a basic principle like "all coefficients are positive". At each step we transform some coefficients of P_i (actually two) such that $P_i = P_{i+1} + Q_i$ where Q_i is a positive polynomial. More precisely, we propose the algorithm shown in Figure-1. We suppose that

$$P_i = \sum_{p_1,\ldots,p_m \in \mathbb{N}^m} a^{[i]}_{p_1\ldots p_m} X^{p_1}\ldots X^{p_m}$$

The procedure will now rely on how we choose the function change. We propose two solutions for the body of change $(a_{p_1\ldots p_m}, a_{q_1\ldots q_m})$

```
Positive = proc(P: polynomial) returns(string)
        while there exists a negative coefficient do
              if there exist a_{p1,...,pm} > 0 and a_{q1,...,qm} < 0,
                          with p_i ≥ q_i for all i ∈ [1..m]
              then change(a_{p1,...,pm}, a_{q1,...,qm})
              else return("no-answer")
              end
        end
  return("positive")
end
```

Figure-1: A procedure for checking positiveness of a polynomial

---

Solution 1:

if $a^{[i]}_{p_1\ldots p_m} > |a^{[i]}_{q_1\ldots q_m}|$

then

$$a^{[i+1]}_{p_1\ldots p_m} := a^{[i]}_{p_1\ldots p_m} + a^{[i]}_{q_1\ldots q_m} \;;\; a^{[i+1]}_{q_1\ldots q_m} := 0$$

else

$$a^{[i+1]}_{p_1 \dots p_m} := 0 \;;\; a^{[i+1]}_{q_1 \dots q_m} := a^{[i]}_{q_1 \dots q_m} + a^{[i]}_{p_1 \dots p_m}$$

Solution 2:

if $a^{[i]}_{p_1 \dots p_m} > |a^{[i]}_{q_1 \dots q_m} 2^{q_1-p_1} \dots 2^{q_m-p_m}|$

<u>then</u>

$$a^{[i+1]}_{p_1 \dots p_m} := a^{[i]}_{p_1 \dots p_m} + a^{[i]}_{q_1 \dots q_m} 2^{q_1-p_1} \dots 2^{q_m-p_m} \;;$$

$$a^{[i+1]}_{q_1 \dots q_m} := 0$$

<u>else</u>

$$a^{[i+1]}_{p_1 \dots p_m} := 0 \;;$$

$$a^{[i+1]}_{q_1 \dots q_m} := a^{[i]}_{q_1 \dots q_m} + a^{[i]}_{p_1 \dots p_m} 2^{p_1-q_1} \dots 2^{p_m-q_m}$$

We may now prove the two following propositions

Proposition 2:

When one takes the first change function,

if $a^{[i]}_{p_1 \dots p_m} > |a^{[i]}_{q_1 \dots q_m}|$

<u>then</u>

$$P_i = P_{i+1} - a^{[i]}_{q_1 \dots q_m} X_1^{q_1} \dots X_m^{q_m} (X_1^{p_1-q_1} \dots X_m^{p_m-q_m} - 1)$$

<u>else</u>

$$P_i = P_{i+1} + a^{[i]}_{p_1 \dots p_m} X_1^{q_1} \dots X_m^{q_m} (X_1^{p_1-q_1} \dots X_m^{p_m-q_m} - 1)$$

Proposition 3:

When one takes the second change function,

if $a^{[i]}_{p_1 \dots p_m} > |a^{[i]}_{q_1 \dots q_m} 2^{q_1-p_1} \dots 2^{q_m-p_m}|$

<u>then</u>

$$P_i = P_{i+1} - a^{[i]}_{q_1 \dots q_m} X_1^{q_1} \dots X_m^{q_m} [(X_1/2)^{p_1-q_1} \dots (X_m/2)^{p_m-q_m} - 1]$$

<u>else</u>

$$P_i = P_{i+1} + a^{[i]}_{p_1 \dots p_m} X_1^{q_1} \dots X_m^{q_m} [(X_1/2)^{p_1-q_1} \dots (X_m/2)^{p_m-q_m} - 1]$$

Since $a^{[i]}_{q_1 \dots q_m}$ is negative and $a^{[i]}_{p_1 \dots p_m}$ is positive, the quantities are positive and satisfy the requirements. Currently these two methods have been implemented in

REVE.

4. Polynomial Interpretations of Associative-Commutative Operators

Associative-commutative operators often occur in rewriting system and it is really important to have methods to prove termination of associative-commutative rewriting systems. In the presence of associative-commutative equations (written AC), one interprets the quotient algebra $T(F, \{x_1,\ldots,x_m\})/AC$. Therefore any interpretation has to be consistent with the laws. The purpose of this section is to give a characterization of when polynomial interpretations are consistent in this sense. It turns out that this criterion is very simple and can be tested simply. Surprisingly enough we are not aware of any mention of it in the literature. Thus if a polynomial Q interprets an associative-commutative operator, it satisfies the two conditions:

$$Q(X, Y) = Q(Y, X) \quad \text{and} \quad Q(Q(X, Y), Z) = Q(X, Q(Y, Z)).$$

The first equation says that Q is symmetric, the second one gives a bound on the degree. Indeed if the highest degree of X in Q is m, then m has to satisfy the identity $m^2 = m$, since m^2 is the highest degree of X in $Q(Q(X,Y),Z)$. Therefore $m=0,1$ and the general form of Q is $Q(X,Y) = aXY + b(X+Y) + c$

then $Q(Q(X,Y),Z) - Q(X,Q(Y,Z)) = (ac + b - b^2)(Z - X)$

and we have this criterion:

Proposition 4:

The polynomials that satisfy associative-commutative equations, i.e., the polynomials that interpret associative-commutative operators, are the polynomials of the form: $aXY + b(X+Y) + c$ with $ac + b - b^2 = 0$

5. Interpretations by a Cartesian product of polynomials, or why Lankford's example 3 works?

When dealing with a really classical example, namely the Naturals with addition and product defined in terms of the function "successor", we arrived at rather difficult problem that neither the classical polynomial interpretations nor the other approaches could handle [Bach&Plai85]. Indeed, such a specification uses a rewriting system with "+" and "." associative and commutative.

$$0 + x \rightarrow x \qquad s(x) \,.\, y \rightarrow (x \,.\, y) + y$$

$$s(x) + y \rightarrow s(x + y) \qquad x \,.\, (y + z) \rightarrow (x \,.\, y) + (x \,.\, z)$$

$$0 \,.\, x \rightarrow 0$$

Because of the associativity and commutativity and the restrictions on the polynomial interpretations of "+" and ".", we have to choose their interpretations of degree one. A simple computation made on the degree of X in the interpretation of the distributivity shows that the interpretation of "+" has to be of the form "(X + Y) + c", but this cannot work with any interpretation proving the termination of the rule $s(x) + y \rightarrow s(x + y)$.

Thus the classical interpretations do not work in this case and we propose to use a p-tuple of polynomials for interpreting the operators instead of a unique one. We will use the notation

$$[t](X_1,\ldots,X_n) = ([t]_1(X_1,\ldots,X_n),\ldots,[t]_p(X_1,\ldots,X_n))$$

where the $[t]_i(X_1,\ldots,X_n)$ are the same kind of polynomials as defined in the previous sections. A lexicographical comparison will allow us to handle scale of ordering that polynomials cannot.

Let us first define this on the previous example. The interpretation of an operator of arity n is a pair of polynomial of same arity n. The interpretation of a term is made component-wise. The comparison of two terms is made lexicographically by first comparing the first components and if they are equal their second components. Since the lexicographical product of well-founded ordering is well-founded, we obtained that way a well-founded ordering. For instance, let us take:

$[0] = (2, 2)$, $[s](X) = (X + 2, X + 1)$

$[+](X, Y) = (X + Y + 1, XY)$, $[.](X, Y) = (XY, XY)$.

Since the components of the interpretations of "+" and "." satisfy the conditions for the polynomial interpretations of associative and commutative operators, the whole interpretations are constant on each equivalence class modulo associativity and commutativity and can be used for proving the termination of the previous associative commutative rewriting system. After computing polynomial interpretations we get

for the first rule $(X + 3, 2X) > (X, X)$

for the second rule $(X + Y + 3, XY + Y) > (X + Y + 3, XY + 1)$

for the third rule $(2X, 2X) > (2, 2)$

for the fourth rule $(XY + 2Y, XY + Y) > (XY + Y + 1, XY^2)$

for the fifth rule $(XY + XZ + X, XYZ) > (XY + XZ + 1, X^2YZ)$.

Therefore the rewriting system is terminating.

The extension to Cartesian products with more than two components are straightforward and will not be presented here. However, we feel this provides us with the only implemented method for proving such associative commutative systems. As we mention in the title this method is already present in [Lankf79] in rule (18) and rule (19). However we feel our presentation gives the conceptual framework behind rules that are only given through an example, and allows extension to more than two levels in the cartesian product. In addition, the method is actually implemented in REVE.

6. Conclusion

A method based on polynomial interpretations is currently implemented in REVE. Though this termination check procedure is absolutely necessary in order to run some of the examples we know, like Associativity+Endomorphism, we do not think it will replace the current methods based on recursive path ordering [Dersh82] or recursive decomposition ordering [Jouan.*etal*.82], since they have shown to have a large scope and to be really easy to use in many practical cases [Forg&Detl85], and Dershowitz has exhibited examples where polynomial interpretation fail [Dersh83]. So, we think we will keep both methods in REVE and let the user choose the method he or she wants to use. However in the case of associative-commutative operators, the methods based on extensions of the recursive path ordering either fail or are not ready for being incorporated in a rewrite rule laboratory like REVE. So, it is the only method currently available and we are incorporating it into REVE-3 (the general equational rewriting laboratory) as the mechanism for proving termination in the associative-commutative case.

In conclusion, we would like to mention a limit of our criterion. It cannot prove the positiveness of the polynomial $X_1^2 + X_2^2 - 2X_1X_2 + 1 = (X_1-X_2)^2 + 1$. Since we never encountered such a polynomial in termination proofs we feel that would not be an obstacle for using it.

Acknowledgement

We are pleased to thank all the people who give us helpful suggestions, especially Francoise Bellegarde, Dave Detlefs, Harald Ganzinger, Jean-Pierre Jouannaud, Dallas Lankford and Laurence Puel. We are grateful to Bachmair and Plaisted to have mention our contribution in [Bach&Plai85b]. We also thank the Greco de Programmation for its support.

Review of Systems whose termination was proved by our algorithm and actually run on REVE.

The first example we proved was obviously the Associativity+Endomorphism (see Section 1). Then we study the equations for the groups using an interpretation proposed by Huet [Huet80]. to complete these equations into the ten classical Knuth-Bendix rules. On groups we also try the same set of equations with the interpretation:

$$[e] = 2 \ , \ [i](X) = X^2 \ , \ [*](X, Y) = 1 + X + Y^4 \ , \ [/](X, Y) = X + Y^2$$

to complete them in the set of 10 rules we already discovered [Lesc83].

$(e / x) \rightarrow i(x)$ $\qquad\qquad$ $i((y / x)) \rightarrow x / y$

$i(e) \rightarrow e$ $\qquad\qquad$ $(x / y) / i(y) \rightarrow x$

$(x / e) \rightarrow x$ $\qquad$ $(x / i(y)) / y \rightarrow x$

$(x / x) \rightarrow e$ $\qquad$ $(x / (y / z)) \rightarrow ((x / i(z)) / y)$

$i(i(x)) \rightarrow x$ $\qquad$ $(x * y) \rightarrow (x / i(y))$

We also studied the examples left unfinished by Knuth and Bendix in their paper, namely axiomatization of groups due to Taussky.

$x * (y * z) = (x * y) * z$ $\qquad$ $g((x * y), y) = f((x * y, x)$

$e * e = e$ $\qquad$ $f(e, x) = x$

$x * i(x) = e$

using the interpretation:

$[e] = 2$, $[i](X) = X^2$, $[*](X, Y) = 2XY + Y$

$[f](X, Y) = X + Y$, $[g](X, Y) = X + 2XY^2 + Y^2 + 1.$

We are currently making other experiments.

References

[Bach&Plai85].

L. Bachmair and D. Plaisted, "Associative Path Orderings," in Proc. 1st Conference on Rewriting Techniques and Applications, Lecture Notes in Computer Science, vol. 202, pp. 241-254, Springer Verlag, Dijon (France), 1985.

[Bach&Plai85b].

L. Bachmair and D.A. Plaisted, "Termination Orderings For Associative-Commutative Rewriting Systems," Journal of Symbolic Computation, 1985.

[Belle84].

F. Bellegarde, "Rewriting Systems on FP Expressions that Reduce the Number of Sequences they yield," in Symposium on LISP and Functional Programming, ACM, Austin, USA, 1984.

[Dersh82].

Nachum Dershowitz, "Orderings for Term-Rewriting Systems," Theoretical Computer Science, vol. 17, pp. 279-301, 1982.

[Dersh83].

N. Dershowitz, "Well-Founded Orderings," ATR-83(8478)-3, Information Science Research Office, The Aerospace Corporation, El Segundo, California (USA), May 1983.

[Dersh85].

N. Dershowitz, "Termination," in Proc. 1rst Conf. Rewriting Techniques and Applications, Lecture Notes in Computer Science, vol. 202, pp. 180-224,

Springer Verlag, Dijon (France), May 1985.

[Forg&Detl85].

R. Forgaard and D. Detlefs, "An incremental algorithm for proving termination of term rewriting sytems," in Proc. lrst International Conference on Rewriting Techniques and Applications., Lecture Notes in Computer Science, vol. 202, Springer Verlag, Dijon (France), 1985.

[Huet80].

G. Huet, "Confluent reductions: abstract properties and applications to term rewriting systems," J. of ACM, vol. 27, no. 4, pp. 797-821, Oct. 1980.

[Hue&Opp80].

G. Huet and D. Oppen, "Equations and Rewrite Rules: A Survey," in Formal Languages: Perspectives And Open Problems, ed. Book R., Academic Press, 1980.

[Jouan.*etal*.82].

J.P. Jouannaud, P. Lescanne, and F. Reinig, "Recursive Decomposition Ordering," in Formal Description of Programming Concepts 2, ed. Bjorner D., pp. 331-346, North Holland, Garmish Partenkirschen, RFA, 1982.

[Lankf79].

D.S. Lankford, "On Proving Term Rewriting Systems Are Noetherian," Report Mtp-3, Math. Dept., Louisiana Tech University, May 1979.

[Lesc83].

P. Lescanne, "Computer Experiments with the REVE Term Rewriting System Generator," in 10th ACM Conf. on Principles of Programming Languages, pp. 99-108, Austin Texas, January 1983.

[Man&Nes70].

Z. Manna and S. Ness, "On the Termination of Markov Algorithms," Third Hawaii International Conference on System Sciences, pp. 789-792, 1970.

PROVING TERMINATION OF ASSOCIATIVE COMMUTATIVE REWRITING SYSTEMS BY REWRITING

Isabelle GNAEDIG Pierre LESCANNE

CRIN
Campus scientifique BP 239
54506 Vandoeuvre CEDEX France

ABSTRACT

We propose in this paper a special reduction ordering for proving termination of Associative Commutative (AC in short) rewriting systems. This ordering is based on a transformation of the terms by a rewriting system with rules similar to distributivity. We show this is a reduction ordering which works in the AC case since it is AC-commuting, and which provides an automatizable termination tool, since it is stable by instantiation. Thereafter, we show cases where this ordering fails, and propose an extension of this method to other transformation rules such as endomorphism.

1. INTRODUCTION

Rewriting systems are an adequate framework for automatizing proofs in equational theories. They give indeed a decision procedure for the word problem, provided the rewriting system is confluent and terminating [9].

Here, we address the problem of proving termination, in the case of equational rewriting systems, especially when the set E of equations contains only associativity and commutativity axioms.

A classical method for proving the termination consists in using a well founded ordering on terms, which includes the rewriting relation. A survey of the ordering based methods is given in [5].

The starting point of our study is a set of theorems on orderings proposed by Jouannaud and Muñoz. They give restrictions on termination orderings (i.e. $E = \emptyset$) to be applicable to equational rewriting systems (i.e. $E \neq \emptyset$)[8]. We propose here to construct an ordering that fulfills these restrictions, for the associative-commutative equational theory (AC in short). This approach gives a new light to the work of Plaisted [12]. The ordering is based on an interpretation of terms of the free term algebra T(F,X) by terms of the "flattened term algebra" FL(F,X), an approach somewhat more natural than Lankford's one, which interprets the terms with polynomials [10] , [7] , [3].

Bachmair and Plaisted address the problem in about the same way. They base their simplification ordering on term transformations that "flatten" terms and distribute operators in them. In addition, they impose a restriction on the precedence: the pair condition [1]. But our approach is different in many respects. We try to be more precise and rigorous in the definition of the manipulations involved in the method. For this reason, we work with only one distributivity axiom. We feel, this is not a real restriction when dealing with practical situations where only one distributivity occurs. In addition, we make a strong difference between the flattening and the distributing processes. Therefore we don't intercalate flattening and distributing steps. In a first time, distributivity rules map the terms in the same class modulo distributivity, in order to insure the F-compatibility. In a second time, flattening allows us to interpret "pure" terms by "flattened" terms, in order to satisfy the AC-commutation.

Furthermore, this ordering, which first was designed for dealing with rewriting systems that contain distributivity, can easily be extended to rewriting systems that contain similar rules like endomorphism. In this case, we propose a different ordering based on the same principle, replacing the distributivity transformation by an endomorphism one. In the same way, we assume that this ordering is also a reduction ordering and generalize the principle with other transformation rules. Last, we relax restrictions imposed by Bachmair and Plaisted when comparing non ground terms.

Recently, Bachmair and Dershowitz defined conditions on rewriting in order to prove termination of rewriting systems [2]. Their approach has much similarity with ours, since it is also based on transformations.

In Section 2, the construction of the ordering is explained. Section 3 establishes it is a reduction ordering, and in Section 4, the problem of non-ground terms is expounded and a solution is proposed. Finally, Section 5 presents an extension of the method by replacing distributivity by other rewriting rules.

2. CONSTRUCTION OF AN AC-ORDERING

Let T(F,X) be the free algebra of terms with variables in X, and operators in F. We suppose the reader knows the usual notions of rewriting and basic notations.

Let us now state a general theorem due to Muñoz [11].

Definition 1 : A relation > is E-commuting if for each s, t, s' in T(F,X) such that s' $=_E$ s > t, there exists a t' in T(F,X) such that s' > t' $=_E$ t.

Theorem 1 : (Muñoz) Let (R,E) be an equational rewriting system defined in T(F,X). (R,E) is noetherian if there exists an E-commuting reduction ordering in T(F,X) which orients all instances of R.

The ordering presented here satisfies the requirements of Muñoz theorem, when E is AC. We start with a common reduction ordering, namely the Recursive Path Ordering of Dershowitz [4]. This ordering is founded on a precedence ordering on the set F of operators.

However, the RPO does not work with AC because it is not AC-commuting. In fact, if s > t and s $=_{AC}$ s', it could be the case that no t' with s' > t $=_{AC}$ t' exists. For example, let s be (x+y)+z, t be f(x+y), and s' be x+(y+z), where + ≃ f for the precedence.

We will then force the AC-commutation of the ordering by projecting the terms to be compared on a particular representative of their AC congruence class, and then comparing the representatives by the RPO. A representative of a term can be obtained by "flattening" this term along its AC operators (to represent the associativity axioms), and by assigning a multiset status to the AC operators (to represent the commutativity axioms). Then, if s $=_{AC}$ t, the representatives of the two terms have the same semantic (they are equal by permutation ≃ under their AC operators). Therefore, the used RPO will be the original multiset RPO defined on the "varyadic" term algebra. Let be F = $F_{AC} \cup F_{NAC}$, where F_{AC} is the set of all AC operators. The varyadic term algebra TV(F,X) is the algebra of terms, where the AC operators have a variable arity.

Definition 2 : The flattening operation is a mapping from TV(F,X) into TV(F,X), such that for each term s of TV(F,X), the flattened form of s noted [s] is the normal form of s using an infinite system described by the scheme of rules:

$f(y_1,..y_{i-1},f(x_1,..x_m),y_{i+1},..y_n) \to f(y_1,..y_{i-1},x_1,..x_m,y_{i+1},..y_n)$

for every f in F_{AC}, for every i>0, for every m,n >1.

Remark : The normal form of s exists and is unique. Indeed we claim that the previous system is confluent and terminating.

Sometimes, we write $s_1+s_2...s_{m-1}+s_m$ for $+(s_1,...s_m)$. The set of flattened terms is the subset FL(F,X) of the algebra TV(F,X). This set has also a structure of F-algebra.

Let us now consider the AC termination of a rewriting system which contains the distributivity rule

$$x * (y+z) \rightarrow (x*y) + (x*z).$$

With the RPO, we have: x*(y+z) > (x*y)+(x*z), extending the precedence to $* >_F +$. This precedence condition produces an ordering which is not F-compatible. Indeed, x*(y+z) > (x*y)+(x*z), but u*(x*(y+z)) < u*((x*y)+(x*z)) because [u*(x*(y+z))] = u*x*(y+z), [u*((x*y)+(x*z))] = u*((x*y)+(x*z)), and u*((x*y)+(x*z)) $>_{RPO}$ u*x*(y+z).

Consequently, we may force the F-compatibility of >, using the intuitive idea that the smallest term has also to be reduced when the biggest one is reduced by flattening. Therefore, following Plaisted, we introduce an additional transformation of the terms based on the distributivity rule itself. Comparing the two terms z*(x*y) and z*(x+y), the first one is flattened in z*x*y, the second one becomes (z*x)+(z*y) by rewriting with the distributivity rule. In this case we obtain z*x*y $>_{RPO}$ (z*x)+(z*y), which preserves the compatibility.

Remark that in order to obtain symmetrically (x*y)*z > (x+y)*z, we have to introduce the symmetric distributivity rule (x+y)*z→(x*z)+(y*z) in the transformation.

Let us now define formally the distributing transformation.

<u>Definition 3</u> : The distributivity operation is a transformation from T(F,X) into T(F,X), such that for each term s of T(F,X), a distributed form of s, written s↓, is obtained by rewriting s to a irreducible form with the system (D) described by the two rules:

$$x*(y+z) \rightarrow (x*y)+(x*z) \quad (1)$$

$$(x+y)*z \rightarrow (x*z)+(y*z). \quad (2)$$

Indeed, for each s of T(F,X), there exists an irreducible form since this system terminates. But this irreducible form is not unique because the system (D) is not confluent. For example, the term (x+y)*(u+v) rewrites into ((x*u)+(x*v))+((y*u)+(y*v)) using (2) then (1) twice, and ((x*u)+(y*u))+((x*v)+(y*v)) using (1) then (2) twice. However, we can state the following property.

Proposition 1 : If $s \overset{*}{\rightarrow}_D s'$ and $s \overset{*}{\rightarrow}_D s''$ and s' and s'' are irreducible, then $[s'] \simeq [s'']$.

We can now define the complete transformation of a term s of T(F,X) by distributing and flattening.

Definition 4 : The flattening distributing operation is a transformation from the subalgebra T(F,X) of TV(F,X) onto the algebra FL(F,X) such that for each term s of T(F,X), a flattened distributed irreducible form of s is the form $[s\downarrow]$ where $s\downarrow$ is a irreducible D-form of s.

Remark : All the forms $[s\downarrow]$ are equivalent under the permutation $\simeq$ in FL(F,X).

Remark : A form $[s\downarrow]$ can be viewed as the interpretation on TV(F,X) of a distributed form $s\downarrow$.

Definition 5 : Let s and t be two terms of T(F,X). We say that s>t iff

$[s\downarrow] >_{RPO} [t\downarrow]$ or

$[s\downarrow] \simeq [t\downarrow]$ and $s \overset{+}{\rightarrow}_{D/AC} t$ by the distributivity rule $x*(y+z) \rightarrow (x*y)+(x*z)$,

where $\rightarrow_{D/AC}$ is the relation $=_{AC} . \rightarrow_D . =_{AC}$.

Remark : In the second case of the preceding definition, one of the two distributivity rules suffices, by definition of the relation $\rightarrow_{D/AC}$.

From now on, + and * are generic notations for operators in F_{AC}, that satisfy $* >_F +$.

3. A REDUCTION ORDERING

Let us first prove that our relation > called NFLO is AC-commuting, and in addition, it satisfies a stronger property, namely the AC-completeness.

Theorem 2 : The NFLO is AC-complete, i.e. for each s, t, s', t' in T(F,X) such that $s' =_{AC} s > t =_{AC} t'$, then $s' > t'$.

Let us now deal with the reduction ordering notion.

Definition 6 : An ordering on terms is F-compatible iff s>t => h(...s...) > h(...t...) for each h in F.

Definition 7 : A reduction ordering on terms is a F-compatible well-founded ordering on terms.

The irreflexivity, transitivity and well-foundedness of our relation > derive directly from the irreflexivity, transitivity and well-foundedness of both relations RPO and $\text{-+}\!\!\rightarrow_{D/AC}$. The relation > is therefore a well-founded ordering.

In order to establish the compatibility, lemmas are necessary according to the operator h of Definition 6.

Lemma 1 : In the case $[s\downarrow] \simeq [t\downarrow]$ and $s \text{-+}\!\!\rightarrow_{D/AC} t$, we have
$s>t \Rightarrow h(\ldots s\ldots) > h(\ldots t\ldots)$ for each h in F.

From now on, we consider only the case where s>t means that $[s\downarrow] >_{RPO} [t\downarrow]$.

Lemma 2 : If h is in F_{NAC}, then $s>t \Rightarrow h(\ldots s\ldots) > h(\ldots t\ldots)$.

The notion of flattening by a single AC operator is now introduced to make the proof of compatibility easier.

Definition 8 : The k-flattened form $[s]_k$ of s is obtained by rewriting s to the normal form with an infinite system described by the scheme of rules:
$k(y_1,..y_{i-1},k(x_1,..x_m),y_{i+1},..y_n) \rightarrow k(y_1,..y_{i-1},x_1,..x_m,y_{i+1},..y_n)$
for every i>0, for every m,n >1, for k in F_{AC}.

The next lemma gives an important property on flattening used in the proofs of both lemmas 4 and 5.

Lemma 3 : If k is in F_{AC} and is minimal for the precedence,
then $s>_{RPO}t \Rightarrow [s]_k >_{RPO} [t]_k$.

Lemma 4 : If $k \neq *$ is in F_{AC} and is minimal for the precedence,
then $s>t \Rightarrow k(s,u) > k(t,u)$.

Lemma 5 : If + is minimal, and there exists no $f \neq +$ such that $* >_F f$,
then $s>t \Rightarrow s*u > t*u$.

Theorem 3 : With the following precedence:
$*>+$
if k and k' are in F_{AC}, $k >_F k' \Rightarrow k{=}*$ and $k'{=}+$
if k is in F_{AC} and $k \neq *$, then k is minimal
if $* >_F f$ then f=+

the ordering > is F-compatible.

Proof: The proof immediately comes from the lemmas 1, 2, 4, and 5. □

4. THE VARIABLE CASE

Let us recall the conditions of the termination proof according to Theorem 1. A rewriting system is AC-terminating if for every rule $l \rightarrow r$ of the system, and every substitution σ, then $\sigma l > \sigma r$. The previous condition is believed to be undecidable. In order to automatize proofs of AC-termination, we attempt to modify it, in proving a weaker property that expresses the behaviour of the ordering > with respect to the substitution operation.

Definition 9 : An ordering > on terms is stable by instantiation iff $s > t \Rightarrow \sigma s > \sigma t$ for every substitution σ.

In order to prove the stability by instantiation of the ordering >, let us state obvious properties.

Lemma 6 : $[\sigma s\downarrow] = [(\sigma\ [s\downarrow])\downarrow]$ for every term s of T(F,X) and every substitution σ.

Definition 10 : The flattened D-irreducible form of any substitution σ is the substitution denoted by $[\sigma\downarrow]$ and defined by $[\sigma\downarrow]\ x = [\sigma x\downarrow]$.

Lemma 7 : $[\sigma s\downarrow] = [([\sigma\downarrow]s)\downarrow]$ for every term s in T(F,X) and every substitution σ.

Lemma 8 : If $s > t$ when $[s\downarrow] \simeq [t\downarrow]$ and $s \overset{+}{\rightarrow}_{D/AC} t$, then $\sigma s > \sigma t$.

Definition 11 : A substitution is said to be elementary iff its domain is reduced to a single variable.

Let us recall that a substitution is closed if the instantiation of every variable of this substitution is a term without variable.

Lemma 9 : Let σ be a closed substitution whose domain is $DOM=\{x_1, \ldots x_n\}$. Then $\sigma = \sigma_1 \ldots \sigma_n$ where σ_i is the closed elementary substitution whose domain is {xi}.

The following lemmas express several forms of stability by instantiation according to the form of σ. Let σ be an elementary closed substitution with domain {x}. Suppose the precedence satisfies conditions of Theorem 3.

Lemma 10 : $s>t \Rightarrow \sigma s>\sigma t$ if σx is of the form $k(u_1,\ldots u_p)$, where $k \neq +,*$ and $u_1,\ldots u_p$ are closed terms in $T(F,X)$.

Lemma 11 : $s>t \Rightarrow \sigma s>\sigma t$ if σx is of the form u_1*u_2, where u_1 and u_2 are closed terms in $T(F,X)$.

Lemma 12 : $s>t \Rightarrow \sigma s>\sigma t$ if σx is of the form u_1+u_2, where u_1 and u_2 are closed terms in $T(F,X)$.

Theorem 4 : $s>t \Rightarrow \sigma s > \sigma t$ for every closed substitution σ.

Proof: The proof is immediate with the lemmas 8, 9, 10, 11, and 12. □

Full proofs of lemmas can be found in the thesis of the first author [6].

5. AN EXTENSION OF THE METHOD TO OTHER TRANSFORMATION RULES

The ordering based on the distributivity transformation does not work with every AC rewriting system. For instance, let us consider a rewriting system including the endomorphism rule $f(x)+f(y) \to f(x+y)$, where f is in F_{NAC} and + is in F_{AC}. We cannot orient this rule with the previous ordering >, since + is minimal.

Let us use the same approach as in Section 2 by attempting to construct an AC-commuting reduction ordering orienting the endomorphism rule. We obtain the same ordering as in Section 2, where the distributivity rule system is replaced by the convergent system (PLUS), described by the following rules:

$$f(x)+y \to f(x+y)$$
$$x+f(y) \to f(x+y)$$

The normal form of a term s for (PLUS) is written $s\downarrow$. We can here define the flattened normal form of each term s in T(F,X) by $[s\downarrow]$. The definition of the new ordering can be expressed as follows:

Definition 12 : Let s and t be two terms of T(F,X). We say that s > t iff
$[s\downarrow] >_{RPO} [t\downarrow]$ or
$[s\downarrow] \simeq [t\downarrow]$ and $s \xrightarrow{+}_{PLUS/AC} t$

Under the same kind of precedence constraints as in Section 3 (such as +>f), we can prove that > is a reduction ordering stable by instantiation. By this way, a family of AC-commuting reduction orderings can be constructed, to prove AC-termination on different kinds of rewriting systems. Indeed, we can suggest to treat other rules in the same way like the definition of the operation + in terms of successors: s(x)+y → s(x+y); or a property of * w.r.t -(x)*y → -(x*y).

6. CONCLUSION

We have improved the RPO in an AC-commuting reduction ordering based on transformation by rewriting rules (distributing and flattening). Separating strongly distributivity and flattening allows us to use simply two distributivity rules on binary operators, instead of an infinite system (see [1]). In addition, we have given a proof for the stability by instantiation based on simple observations on the substitution mechanism. This last property allows us to provide an automatizable AC-termination tool. Finally, some ideas have been given in order to generalize this method by replacing the distributivity transformation by rules like endomorphism.

ACKNOWLEDGMENTS

We are indebted to Dave Plaisted for giving us a good starting point in this study. Jean-Pierre Jouannaud is surely the person who gave us the largest help during the construction of this theory and we gratefully acknowledge his contribution. This research benefited also from discussions with too many people in the community in rewrite rules especially in our research group EURECA at Nancy, to be able to name them all, and we are pleased to thank all of them.

References

1. L. Bachmair and D. Plaisted, "Associative Path Orderings," in Proc. 1st Conference on Rewriting Techniques and Applications, Lecture Notes in Computer Science, vol. 202, pp. 241-254, Springer Verlag, Dijon (France), 1985.

2. L. Bachmair and N. Dershowitz, "Commutation, Transformation, and Termination," Inter. Report U. of Illinois, 1985.

3. A. Ben Cherifa and P. Lescanne, "An actual implementation of a procedure that mechanically proves termination of rewriting systems based on inequalities between polynomial interpretations," in CADE, 1986.

4. Nachum Dershowitz, "Orderings for Term-Rewriting Systems," Theoretical Computer Science, vol. 17, pp. 279-301, 1982.

5. N. Dershowitz, "Termination," in Proc. 1rst Conf. Rewriting Techniques and Applications, Lecture Notes in Computer Science, vol. 202, pp. 180-224, Springer Verlag, Dijon (France), May 1985.

6. I. Gnaedig, "Preuves de Terminaison des Systemes de Reecriture Associatifs-commutatifs: une Methode Fondee sur la Reecriture Elle-meme," These de troisieme cycle, Nancy, 1986.

7. G. Huet and D. Oppen, "Equations and Rewrite Rules: A Survey," in Formal Languages: Perspectives And Open Problems, ed. Book R., Academic Press, 1980.

8. J.P. Jouannaud and M. Munõz, "Termination of a set of rules modulo a set of equations," Proceedings 7th Conference on Automated Deduction, vol. 170, Napa Valley (California, USA), 1984.

9. D. Knuth and P. Bendix, "Simple Word Problems in Universal Algebras," Computational Problems in Abstract Algebra Ed. Leech J., Pergamon Press, pp. 263-297, 1970.

10. D.S. Lankford, "On Proving Term Rewriting Systems Are Noetherian," Report Mtp-3, Math. Dept., Louisiana Tech University, May 1979.

11. M. Munõz, "Probleme de terminaison finie des systemes de reecriture equationnels," These 3eme Cycle, Universite de Nancy 1, 1984.

12. D.A. Plaisted, "An associative path ordering," in Proc. NSF Workshop on the Rewrite Rule Laboratory, pp. 123-136, General Electric, Schenectady, New-York, April 1984.

Relating Resolution and Algebraic Completion for Horn Logic*

Roland Dietrich
GMD Research Laboratory at the University of Karlsruhe
Haid-und-Neu-Straße 7
D 7500 Karlsruhe 1
Germany (West)
CSNET: dietrich@germany

Abstract

Since first order logic and, especially, Horn logic is used as a programming language, the most common interpretation method for such programs is resolution. Algebraic completion and term rewriting techniques were recently proposed as an alternative to resolution oriented theorem provers. In this paper, the relation between resolution and algebraic completion (restricted to Horn logic) is closely analysed. It is shown, that both methods are equivalent in terms of inference steps and unifications, and, that using the completion method for interpreting (Horn) logic programs, no efficiency can be gained as compared with resolution.

1. Introduction

Since first order logic is used as a programming language, the most common interpretation method for that language is ***resolution*** [Robi65]. Especially, *Horn logic* together with linear resolution led to PROLOG, the most well-known logic programming language [ClMe81].

Originally, resolution arose from the field of automated theorem proving. J. Hsiang [Hsia83] presented an alternative to resolution for theorem proving which is based on ***term rewriting systems*** and the ***Knuth-Bendix completion procedure***: together with a canonical term rewriting system for Boolean algebra, an extension of the completion procedure working on a rewrite rule representation of logic formulae leads to a refutation complete proof procedure for first order logic. Other authors also have shown that algebraic completion can be successfully used in automated theorem proving [DeJo84, HsDe83, Paul84, KaNa85]. In some example cases of [Hsia83] his method seems to be much more efficient than resolution, but there is no general statement about the relations between resolution and completion for theorem proving taking effiency into account[1].

* This work has been done within a joint project of the GMD and the SFB 314 (artificial intelligence) at the University of Karlsruhe

1 Kapur and Narendran [KaNa85] stated that resolution is subsumed by their approach which is similar to Hsiang's approach, but efficiency is not compared.

Thus, some questions remain open: Where and why is completion "better" than resolution or vice versa? What is common to both methods? The answers to these questions are, on the one side, of theoretical interest and, on the other side, they may be of considerable practical importance, e.g. for the construction of efficient PROLOG interpreters.

In this paper we try to answer the above questions for the case of Horn logic, the subset of first order logic PROLOG is based on. For this purpose we describe resolution and the completion method of [Hsia83] as inference systems. We have three criteria for talking about efficiency: (1) the number of inference steps, (2) the number of unifications needed for a successful inference (because unification is the central operation for generating resolvents and critical pairs as well) and, finally, (3) the size of the search space spanned by an inference system.

The main results can be summarized as follows: for every proof of a theorem in one system there exists a proof in the other one and vice versa. Moreover, in either system the amount of inference steps and unifications to be performed is the same. Further we can show that the search space spanned by the completion method is bigger and even contains more blind alleys than the search space spanned by the resolution method.

In the next section, we will describe the resolution and the completion method for Horn logic as inference systems and, section 3 contains a formal description of relations between the two systems. Section 4 gives a discussion of the results, and section 5 contains a summary and points out some directions for further research.

2. Resolution and Algebraic Completion for Horn Logic — Definition

2.1 *Basic Notions*

We assume that the reader is familiar with the basic features of first order logic, automated theorem proving (see, e.g. [Love78]) and term rewriting systems [HuOp80,Huet80]. We only agree upon the following notations for convenience:

- If S is a set of terms or a set of literals, $mgu(S)$ denotes the set of most general unifiers for S, which is either empty or has one element up to renamings of variables [Robi65].
- A clause logically is a ***disjunction*** of literals and may be denoted by the ***set*** of its literals, too. The empty clause is denoted by $\square$.
- If R is a term rewriting system and t a term, $nf_R(t)$ denotes the set of R-normalforms for t which can be empty, finite or infinite. If R is canonical, $nf_R(t)$ has exactly one element for any t.

In this section, we are describing two methods for interpreting Horn logic programs in terms of ***inference systems***. An inference system I is a pair (D_I,R_I), where D_I is a decidable set of (data) objects (its ***data domain***) and R_I is a finite set of inference rules $R_I = \{\rho_i \mid 1 \leq i \leq n\}$ which ***infer*** data objects from other data objects. These inference rules can be considered as relations $\longrightarrow_{\rho_i}$ on the set of data objects: $d_1 \longrightarrow_{\rho_i} d_2$ iff d_2 may be infered from d_1 by ρ_i. We say that $d_1 \longrightarrow_I d_2$ iff there exists $\rho \in R_I$ such that $d_1 \longrightarrow_\rho d_2$ (the ***inference relation*** of I). As usual, $\stackrel{*}{\longrightarrow}$ denotes the reflexive transitive closure of a relation $\longrightarrow$.[1]

If we consider all possible inferences which can be done starting at a data object, a ***search space*** is spanned by an inference system with respect to this object. Usually, in this search space some special inferences (***solution inferences***) must be found, if they exist at all (e.g. inferences of the empty clause from a set of clauses). The complexity of a problem essentially depends on the size of the corresponding search space.

In the following two subsections we use inference systems to first describe the resolution approach for interpreting Horn logic programs which is standard, and then the completion approach based on Hsiang's work. In the former case, the data objects to be transformed are sets of Horn clauses, and in the latter case they are sets of special rewrite rules.

2.2 Resolution for Horn Logic

A Horn logic program or *Horn specification* is a set of *facts* (clauses which are single atomic formulae) and *implications* (clauses of the form $l_0 \vee \neg l_1 \vee \cdots \vee \neg l_n$, where each l_i is an atomic formula). Interpreting a Horn specification S with respect to a *goal* G (a clause of the form $\neg l_1 \vee \cdots \vee \neg l_m$, where each l_i is an atomic formula) consists in proving $S \cup \{G\}$ to be inconsistent.

Most PROLOG interpreters[2] are based on the *resolution principle* [Robi65], especially linear resolution (cf. [Lloyd84]). We now define an inference system which describes the application of linear resolution to Horn specifications.

Definition 1 (*RHL, Resolution for Horn Logic*).
Let *RHL* be the inference system the domain set of which is the set of all sets of Horn clauses and, the inference rules of which are $R_{RHL} = \{impl, fact\}$, where

- $M \longrightarrow_{impl} M \cup \{k\}$ iff there is an implication $\{l_0, \neg l_1, \ldots, \neg l_n\} \in M$, a goal $\{\neg l'_1, \ldots, \neg l'_m\} \in M$ and a substitution θ such that $\theta \in mgu(\{l_0, l'_1\})$ and $k = \theta\{\neg l'_2, \ldots, \neg l'_m, \neg l_1, \ldots, \neg l_n\}$.
- $M \longrightarrow_{fact} M \cup \{k\}$ iff there is a fact $\{l_0\} \in M$, a goal $\{\neg l'_1, \ldots, \neg l'_m\} \in M$ and a substitution θ such that $\theta \in mgu(\{l_0, l'_1\})$ and $k = \theta\{\neg l'_2, \ldots, \neg l'_m\}$.

∎

Notation. Occasionally we write $M_1 \longrightarrow_{impl,\sigma} M_2$ ($M_1 \longrightarrow_{fact,\sigma} M_2$) if σ is the most general unifier computed for the application of *impl* (*fact*).

The following proposition immediately follows from the above definition and the refutation completeness of the resolution calculus [Robi65]:

Proposition 1. Let G be a goal and S a Horn specification. $S \cup \{G\}$ is inconsistent iff there exists a set of Horn clauses M such that $S \cup \{G\} \stackrel{*}{\longrightarrow}_{RHL} M$ and $\square \in M$.

∎

1 Later on we will use the symbol $\longrightarrow$ also for the relation induced by a term rewriting system, but confusion will be avoided by context.

2 Pure PROLOG programs are Horn specifications.

If S is a Horn specification and G a goal, we call any inference $S \cup \{G\} \xrightarrow{*}_{RHL} M$ where $\Box \in M$ a *solution inference* for $S \cup \{G\}$ via *RHL*.

2.3 Completion for Horn Logic

In this subsection we define an inference system describing a method for interpreting Horn specifications based on a variant of the Knuth-Bendix completion procedure [KnBe70]. This variant is due to Hsiang [Hsia83, Hsia85] and called *N-strategy*[1].

The N-strategy works on a rewrite rule representation of clauses: If we consider the Boolean connectives and the predicate symbols as function symbols, the first order formulae without quan-

BA	b_1: $x \vee y \rightarrow x \wedge y + x + y$	b_6: $x + x \rightarrow 0$
	b_2: $x \Rightarrow y \rightarrow x \wedge y + x + 1$	b_7: $x \wedge 1 \rightarrow x$
	b_3: $x \Leftrightarrow y \rightarrow x + y + 1$	b_8: $x \wedge x \rightarrow x$
	b_4: $\neg x \rightarrow x + 1$	b_9: $x \wedge 0 \rightarrow 0$
	b_5: $x + 0 \rightarrow x$	b_{10}: $x \wedge (y + z) \rightarrow x \wedge y + x \wedge z$

Figure 1: Canonical term rewriting system for Boolean algebra. The '+'-symbol denotes the "exclusive or" connection, 1 and 0 are constants denoting the values "true" and "false" respectively; '+' and '$\wedge$' are considered to be *AC*-operators.

tifiers are a special kind of terms, hereafter called *Boolean terms*. The term rewriting system *BA* shown in Figure 1 reduces every Boolean term into unique[2] normalform

$$(*) \quad l_{11} \wedge l_{12} \wedge \cdots \wedge l_{1n_1} + \cdots + l_{m1} \wedge l_{m2} \wedge \cdots \wedge l_{mn_m},$$

which is a "sum" of conjunctions of atomic formulae[3]. We call these normalforms *S-terms* and every conjunction of atomic formulae *N-term*.

Now clauses can be represented as rewrite rules as follows: a clause k is represented as the rule $nf_{BA}(k) \rightarrow 1$ or, equivalently[4], $nf_{BA}(k) + 1 \rightarrow 0$. The N-strategy takes a set of clauses, transforms it into a set of rewrite rules and runs the Knuth-Bendix completion procedure on this rules, computing a special kind of critical pairs called *N-critical pairs*. If a rule $1 \rightarrow 0$ is generated during completion, the initial set of clauses is inconsistent.

Henceforth, we call a rewrite rule $l \rightarrow \delta$, where l is a Boolean term and irreducible with respect to *BA* (i.e. l is a S-term) and $\delta \in \{0, 1\}$, a *Boolean rule*. The rule is called N-rule iff l is a N-term and δ=0. We restrict ourselves to a specialization of the N-strategy in Horn logic. A full

1 The term "N-strategy" is somewhat misleading: it is not really a *strategy* but rather a proof *procedure* which may be performed under several different (inference-) strategies. We leave it at "N-strategy" to be consistent with [Hsia83].

2 Up to associativity and commutativity of $+$ and $\wedge$

3 Boolean ring form

4 Note that $\neg x$ is logically equivalent to $x+1$, cf. the BA-System of Figure 1.

definition of the N-strategy for general clausal logic is given in [Hsia83, HsDe83] and, more briefly, in [Hsia85]. We first define the transformation of Horn clauses into their rewrite rule representation.

Definition 2. Let T be the function from Horn clauses into Boolean rewrite rules defined as follows:
(1) $T(l_0) = l_0 \rightarrow 1$,
(2) $T(l_0 \vee \neg l_1 \vee \cdots \vee \neg l_n) = l_0 \wedge l_1 \wedge \cdots \wedge l_n + l_1 \wedge \cdots \wedge l_n \rightarrow 0$,
(3) $T(\neg l_1 \vee \cdots \vee \neg l_m) = l_1 \wedge \cdots \wedge l_m \rightarrow 0$,
(4) $T(\Box) = 1 \rightarrow 0$,
where each l_i is an atomic formula.

∎

Due to the fact that rewrite rules of the forms described above originate from Horn clauses, we occasionally call them *Horn rules* and, especially, *fact rules* (1), *implication rules* (2) and *goal rules* (3).

T may easily be extended to Horn specifications, and we are using T to denote the transformation of a Horn clause and a Horn specification as well.

In the following definition which is taken from [Hsia85], we make extensive use of the properties of the logical *and*-operator: Two N-terms are considered to be equal, if they are identical up to permutations of literals, identical literals occurring only once and every '1' removed; a subterm of a N-term is any and-combination of literals of the N-term. Furthermore, we use the notation l_1l_2 instead of $l_1 \wedge l_2$. Recall that an N-term is a conjunction of positive literals with '$\wedge$' beeing associative, commutative and idempotent.

Definition 3 (*P-unification, overlap, N-critical pairs*).
Let t_1 and t_2 be N-terms
(a) t_1 and t_2 are *P-unifiable under* σ iff σ is a most general unifier for t_1 and t_2 up to permutations of literals.

(b) Let u_1 and u_2 be subterms of t_1 and t_2, respectively, so that $t_1 = u_1s_1$ and $t_2 = u_2s_2$. Then $\sigma(s_1t_2)$ is an *overlap* of t_1 and t_2 iff u_1 and u_2 are P-unifiable under σ. (Note that $\sigma(s_1t_2) = \sigma(s_1u_1s_2) = \sigma(s_1u_2s_2) = \sigma(t_1s_2) = \sigma(t_1t_2)$ by idempotence.)

(c) Let $s \rightarrow 0$ be a N-rule, $s = s_1s_2$, and $t_1 + t_2 + \cdots + t_n \rightarrow \delta$ a Boolean rule. $<\sigma(s_2t_2 + \cdots + s_2t_n + s_2\delta), 0>$ is a *N-critical pair* of the two rules iff s has an overlap $\sigma(s_2t_1)$ with t_1.

∎

P-unification is a certain kind of AC-unification restricted to N-terms. An algorithm for computing P-unifiers is given in [Hsia83]. It is called BN-unification. An overlap is a certain kind of superposition.

Note that we may, without loss of generality, choose the first conjunction of $t_1 + \cdots + t_n$ for overlapping with s, because of the commutativity of '+'. In the case of Horn logic, n is always less than 3.

Example. (a) $p(x,z)q(x,y)p(y,z)$ and $q(\bar{x},\bar{y})p(\bar{y},b)$ have the overlaps $q(\bar{x},\bar{y})p(\bar{y},b)q(\bar{y},y)p(y,b)$ with $\{x \leftarrow \bar{y}, z \leftarrow b\}$ and $p(\bar{x},b)q(\bar{x},\bar{y})p(\bar{y},b)$ with $\{x \leftarrow \bar{x}, y \leftarrow \bar{y}, z \leftarrow b\}$.

(b) Consider the Boolean rewrite rule $r = p(x,z)q(x,y)p(y,z)+q(x,y)p(y,z) \rightarrow 0$ and the N-rule $r' = p(\bar{x},b) \rightarrow 0$. $p(\bar{x},b)$ has an overlap with the left side of r ($s_1=p(\bar{x},b)$, $s_2=1$, $\sigma=\{\bar{x} \leftarrow x, z \leftarrow b\}$), and so $<q(x,y)p(y,b)+0, 0>$ is an N-critical pair which further reduces to $<q(x,y)p(y,b), 0>$ with respect to BA.

■

We are now ready to give the definition of an inference system which describes the interpretation of Horn specifications by the N-strategy:

Definition 4 (*CHL, Completion for Horn Logic*).
Let *CHL* be the inference system whose domain set D_{CHL} is the set of all sets of Horn rules and whose inference rules are $R_{CHL} = \{cp,\ reduce\}$ where

- $R \longrightarrow_{cp} R \cup \{t \rightarrow 0\}$ iff there are two rules in R which determine an N-critical pair $<t', 0>$ and $t = nf_{BA}(t') \neq 0$.
- $R \longrightarrow_{reduce} R \cup \{t \rightarrow \delta\}$ iff there exists $l \rightarrow \delta \in R$ such that $l \longrightarrow_R l'$ and $t = nf_{BA}(l') \neq \delta$.[1]

■

Proposition 2. Let G be a goal and S a Horn specification. $S \cup \{G\}$ is inconsistent iff there exists a set of Horn rules R such that $T(S) \cup \{T(G)\} \xrightarrow{*}_{CHL} R$ and $1 \rightarrow 0 \in R$.

■

Proof. *CHL* is a generalization of Hsiangs N-strategy which requires everything to be kept in normalform with respect to *BA* ***and*** the current rewrite system. *CHL* ***may*** apply reduction steps, but not necessarily. Clearly, if the N-strategy produces $1 \rightarrow 0$ there is also an inference in *CHL* which produces $1 \rightarrow 0$. Then, the proposition follows from the refutation completeness of Hsiangs N-strategy, cf. [Hsia83].

■

If S is a Horn specification, G a goal and $R = T(S) \cup \{T(G)\}$ we call any inference $R \xrightarrow{*}_{CHL} R'$ where $1 \rightarrow 0 \in R'$ a ***solution inference*** via *CHL* for R.

2.4 Example

The following example is taken from [Hsia83] (example 2.4). We want to prove the following proposition about group theory, henceforth referred to by **P**:

P If P is a subset of a Group $(G, *, e)$, such that $x \in P$ and $y \in P$ implies $x*y^{-1} \in P$, then $b \in P$ implies $b^{-1} \in P$ for any b.

Figure 2 gives an encoding of the negation of **P** as a Horn specification $S=\{c_1, \ldots, c_6\}$ together with a goal c_7 as well as a set of Horn rules $R=\{r_1, \ldots, r_7\}$ such that $T(r_i)=c_i$ for $i=1, \ldots, 7$. The '$*$'-operator is represented as a three placed predicate M, $i(x)$ denotes x^{-1} (i is a unary function symbol) and $P(x)$ means $x \in P$ (P is an unary predicate symbol). The first four clauses (rewrite rules) are group axioms, c_5 and c_6 (r_5 and r_6) describe the hypothesis of **P** and c_7 (r_7) represents the negation of the consequence of the hypothesis. For convenience, variables are indexed with the number of the rules in which they occur.

---

1 Note that $\longrightarrow_R$ here means the (one step) reduction relation with respect to R.

c_1: $M(e,x_1,x_1)$.	r_1: $M(e,x_1,x_1) \rightarrow 1$.
c_2: $M(x_2,e,x_2)$.	r_2: $M(x_2,e,x_2) \rightarrow 1$.
c_3: $M(x_3,i(x_3),e)$.	r_3: $M(x_3,i(x_3),e) \rightarrow 1$.
c_4: $M(i(x_4),x_4,e)$.	r_4: $M(i(x_4),x_4,e) \rightarrow 1$.
c_5: $P(z_5) \vee \neg P(x_5) \vee \neg P(y_5) \vee \neg M(x_5,i(y_5),z_5)$.	r_5: $P(z_5)P(x_5)P(y_5)M(x_5,i(y_5),z_5) + P(x_5)P(y_5)M(x_5,i(y_5),z_5) \rightarrow 0$.
c_6: $P(b)$.	r_6: $P(b) \rightarrow 1$.
c_7: $\neg P(i(b))$.	r_7: $P(i(b)) \rightarrow 0$.

Figure 2: $\neg$**P** as a set of Horn clauses and a set of Horn rules.

Figure 3 shows a refutation of $\neg$**P** both via *RHL* and via *CHL*. For each inference step the generated rewrite rule or clause, the inference rule which has been applied, and the substitution which has been computed, are shown. As above, variables are indexed with the number of the rules or clauses in which they occur. Substitution components which only affect the appropriate renamings

	RHL	*CHL*	Substitution
	$S \cup c_7$	$R = T(S) \cup T(c_7)$	
(1)	$\downarrow$ $impl(c_5,c_7)$	$\downarrow$ $cp(r_5,r_7)$	$\{z_5 \leftarrow i(b)\}$
	c_8: $\neg P(x_8) \vee \neg P(y_8) \vee \neg M(x_8,i(y_8),i(b))$	r_8: $P(x_8)P(y_8)M(x_8,i(y_8),i(b)) \rightarrow 0$	
(2)	$\downarrow$ $fact(c_1,c_8)$	$\downarrow$ $cp(r_1,r_8)$	$\{x_8 \leftarrow e, y_8 \leftarrow b, x_1 \leftarrow i(b)\}$
	c_9: $\neg P(e) \vee \neg P(b)$	r_9: $P(e)P(b) \rightarrow 0$	
(3)	$\downarrow$ $fact(c_6,c_9)$	$\downarrow$ $reduce(r_6,r_9)$	ϵ
	c_{10}: $\neg P(e)$	r_{10}: $P(e) \rightarrow 0$	
(4)	$\downarrow$ $impl(c_5,c_{10})$	$\downarrow$ $cp(r_5,r_{10})$	$\{z_5 \leftarrow e\}$
	c_{11}: $\neg P(x_{11}) \vee \neg P(y_{11}) \vee \neg M(x_{11},i(y_{11}),e)$	r_{11}: $P(x_{11})P(y_{11})M(x_{11},i(y_{11}),e) \rightarrow 0$	
(5)	$\downarrow$ $fact(c_3,c_{11})$	$\downarrow$ $cp(r_3,r_{11})$, *BA*	$\{x_{11} \leftarrow x_3, y_{11} \leftarrow x_3\}$
	c_{12}: $\neg P(x_{12})$	r_{12}: $P(x_{12}) \rightarrow 0$	
(6)	$\downarrow$ $fact(c_6,c_{12})$	$\downarrow$ $cp(r_6,r_{12})$	$\{x_{12} \leftarrow b\}$
	$\square$	$1 \rightarrow 0$	

Figure 3: Refutation of $\neg$**P** via *RHL* and *CHL*.

of variables are not shown explicitly.

Note that for the two solution inferences in Figure 3 there is a certain kind of analogy: The inference via *RHL* use as many inference steps as the inference via *CHL* and, moreover, the same unifiers are computed. If a *CHL*-step uses the *reduce*-rule, the unifier used in the corresponding *RHL*-step is also a match.

3. Resolution and Completion for Horn Logic — Relations

The main purpose of this section is to formally show a certain kind of equivalence between *CHL* and *RHL*, as just observed for the example of section 2.4. This is done in three steps: First, we give a closer analysis of the inference rules of *CHL*. Secondly, we restrict *CHL* to a simpler inference system CHL_0 and show that for every inference step in CHL_0 a corresponding one exists in *RHL* and vice versa. Finally, we show that we do not lose efficiency against *CHL* if we restrict ourselves to CHL_0. Details of proofs which are ommitted or only sketched due to lack of space can be found in [Diet85].

Notation. Throughout this section, the letters p and q always denote atomic formulae, r and s denote N-terms and R denotes a set of Horn rules. Greek letters (ϕ, θ, σ etc.) denote substitutions. All letters can be subscripted.

3.1 A Closer Look at CHL

By means of Definition 2 we are able to partition the cp-rule of *CHL* into four cases, according to the forms of the rules which are involved in the inference step.

Definition 5. The inference rules cp_i, $1 \leq i \leq 4$, are defined by

- $R \longrightarrow_{cp_1} R \cup \{t \rightarrow 0\}$ iff there exist $pr+r \rightarrow 0$, $qs \rightarrow 0 \in R$, and θ such that $p,q,r \notin \{0,1\}$, $\theta \in mgu(\{p,q\})$ and $t = nf_{BA}(\theta sr) \neq 0$.
- $R \longrightarrow_{cp_2} R \cup \{t \rightarrow 0\}$ iff there exist $pr_1r_2+r_1r_2 \rightarrow 0$, $qs_1s_2 \rightarrow 0 \in R$ and θ such that $p,q,r_1,s_1 \notin \{0,1\}$, $\theta \in mgu(\{pr_1,qs_1\})$ and $t = nf_{BA}(\theta s_2r_1r_2) \neq 0$.
- $R \longrightarrow_{cp_3} R \cup \{t \rightarrow 0\}$ iff there exist $pr_1r_2+r_1r_2 \rightarrow 0$, $s_1s_2 \rightarrow 0 \in R$ and θ such that $p,r_1,s_1 \notin \{0,1\}$, $\theta \in mgu(\{r_1,s_1\})$ and $t = nf_{BA}(\theta s_2r_1r_2) \neq 0$ or $t = nf_{BA}(\theta s_2pr_1r_2) \neq 0$.
- $R \longrightarrow_{cp_4} R \cup \{t \rightarrow 0\}$ iff there exist $p \rightarrow 1$, $qs \rightarrow 0 \in R$ and θ such that $p,q \notin \{0,1\}$, $\theta \in mgu(\{p,q\})$ and $t = nf_{BA}(\theta s) \neq 0$.

■

cp_1, cp_2 and cp_3 describe superposition between implication rules and goal rules. For an implication rule $pr+r \rightarrow 0$ there are three cases for getting an overlap with the left hand side of a N-rule: either only the literal p (which does not occur in r) is involved in the overlap (cp_1) or the literal p together with some other literals of r (cp_2) or p is not involved in the overlap (cp_3). Note that p originates from the "head" of an implication, whereas r originates from the "body" of an implication. cp_4 describes superposition between fact rules and goal rules. Superposition between two goal rules determine only trival critical pairs $<0,0>$. Therefore, we have:

Proposition 3. $R_1 \longrightarrow_{cp} R_2$ iff $R_1 \longrightarrow_{\{cp_1, \ldots, cp_4\}} R_2$.

■

In the following, we consider cp to be $\{cp_1, \ldots, cp_4\}$.

Next we restrict the *reduce*-rule to a simpler rule $reduce_0$, which is the only case where a *reduce*-application is really needed. All other cases are special cases of the cp-rules where the most general unifier involved is a match.

Definition 6. The inference rule *reduce*$_0$ is defined by

- $R \longrightarrow_{reduce_0} R \cup \{t \rightarrow 0\}$ iff there exist $p_1 p_2 r + p_2 r \rightarrow 0$, $q \rightarrow 1 \in R$ and σ such that $p_1, p_2, q \notin \{0,1\}$, $\sigma q = p_2$ and $t = nf_{BA}(p_1 r + r) \neq 0$.

■

Proposition 4. If $R \longrightarrow_{reduce} R \cup \{t \rightarrow \delta\}$ then $R \longrightarrow_{\{reduce_0, cp\}} R \cup \{t \rightarrow \delta\}$

Proof. If $pr + r \rightarrow 0 \in R$ is reduced by $s \rightarrow 0 \in R$ to $t \rightarrow 0$, then $t \rightarrow 0$ may also be inferred by cp_1, cp_2 or cp_3. If $pr + r \rightarrow 0 \in R$ is reduced by $q \rightarrow 1 \in R$ where $q \rightarrow 1$ is applied to a subterm of r, this is the case of *reduce*$_0$. If $p \rightarrow 1 \in R$ is reduced by $s \rightarrow 0 \in R$ to $0 \rightarrow 1$, then s is an Atom and cp_4 can also be applied yielding $1 \rightarrow 0$. If $s \rightarrow 0 \in R$ is reduced by $p \rightarrow 1 \in R$ this again is a special case of cp_4. All other cases only produce trivial rules $\delta \rightarrow \delta$. In all cases most general unifiers involved really are matches.

■

Notation. Occasionally we write $R_1 \longrightarrow_{cp_i, \sigma} R_2$ ($R_1 \longrightarrow_{reduce_0, \sigma} R_2$) if σ is the unifier (match) computed for the application of cp_i (*reduce*$_0$).

Finally, we mention that using the *BA*-System, especially b_5 - b_{10} which are Boolean ring axioms, for simplification of Boolean terms, has the same effect as using set notation for clauses[1]. Especially we have by idempotence of '$\wedge$':

Proposition 5. Let s be an N-term where neither 1 nor 0 occurs. For any N-term t let Lt denote the set of (different) atomic formulae occurring in t. Then $nf_{BA}(s) = r$ iff $Ls = Lr$.

■

3.2 *CHL*$_0$ *vs RHL*

We are now ready to define the restriction of *CHL* mentioned at the beginning of this section and to show its equivalence to *RHL* by means of a function on the respective data domains.

Definition 7 (*CHL*$_0$).
Let *CHL*$_0$ be the inference system (D_{CHL_0}, R_{CHL_0}), where $D_{CHL_0} = D_{CHL}$ and $R_{CHL_0} = \{cp_1, cp_4\}$.

■

The following Theorem 1 precisely describes the relations between *CHL*$_0$ and *RHL*: For any inference step in one system there is a corresponding one in the other system and vice versa. Moreover, looking at the unifiers computed by corresponding inference steps we observe that they are identical. The data objects involved in corresponding inference steps are related by the function T, which maps sets of Horn clauses (D_{RHL}) into sets of rewrite rules (D_{CHL_0}) (cf. Definition 2).

Theorem 1.
(a) $M_1 \longrightarrow_{impl, \sigma} M_2$ iff $T(M_1) \longrightarrow_{cp_1, \sigma} T(M_2)$.
(b) $M_1 \longrightarrow_{fact, \sigma} M_2$ iff $T(M_1) \longrightarrow_{cp_4, \sigma} T(M_2)$.

---

1 This has also been conjectured by [Deus85]; [GoMe84] gives an axiomatization of sets corresponding to b_5 - b_{10} of *BA* where '$\wedge$' corresponds to intersection and '$+$' corresponds to symmetric difference.

Proof.

(a) $M_1 \longrightarrow_{impl,\sigma} M \cup \{k\}$

$\Leftrightarrow\ \exists\ \{l_0, \neg l_1, \ldots, \neg l_n\},\ \{\neg l'_1, \ldots, \neg l'_m\} \in M$:

$\sigma \in mgu(\{l_0, l'_1\}),\ k = \sigma\{\neg l'_2, \ldots, \neg l'_m, \neg l_1, \ldots, \neg l_n\}$

$\Leftrightarrow\ \exists l_0 l_1 \cdots l_n + l_1 \cdots l_n \to 0,\ l'_1 \cdots l'_m \to 0 \in T(M)$:

$\sigma \in mgu(\{l_0, l'_1\}),\ T(k) = nf_{BA}(\sigma l'_2 \cdots l'_m l_1 \cdots l_n)$ (Proposition 5)

$\Leftrightarrow\ T(M) \longrightarrow_{cp_1,\sigma} T(M \cup \{k\})$.

(b) $M_1 \longrightarrow_{fact,\sigma} M_1 \cup \{k\}$

$\Leftrightarrow\ \exists\{l_0\},\ \{\neg l'_1, \ldots, \neg l'_m\} \in M$: $\sigma \in mgu(\{l_0, l'_1\}),\ k = \sigma\{\neg l'_2, \ldots, \neg l'_m\}$

$\Leftrightarrow\ \exists l_0 \to 1,\ l'_1 \cdots l'_m \to 0 \in T(M)$: $\sigma \in mgu(\{l_0, l'_1\}),\ T(k) = nf_{BA}(\sigma l'_2 \cdots l'_m)$ (Prop. 5)

$\Leftrightarrow\ T(M) \longrightarrow_{cp_4,\sigma} T(M \cup \{k\})$.

■

By this theorem, the refutation completenes of CHL_0 directly follows from the refutation completeness of *RHL* (Proposition 1):

Corollary 1. CHL_0 is refutation complete, that is, if S is a Horn specification and G a goal then $S \cup \{G\}$ is inconsistent iff there exists a set of Horn rules R such that $T(S) \cup \{T(G)\} \xrightarrow{*}_{CHL_0} R$ and $1 \to 0 \in R$.

■

Remark. Knowing the equivalence of CHL_0 and *RHL* even on the inference rule level and knowing that equivalent inference rules compute the same unifiers if applied to equivalent data objects, it is not difficult to extend CHL_0 in such a way that a solution inference outputs a "correct answer substitution". That is, as it is the case for PROLOG interpreters, the goal is not only refuted but also a substitution is deliverd, which makes the negation of the goal, together with the program, satisfiable. This answer substitution is the composition of all unifiers computed during a solution inference, restricted to the variables occurring in the initial goal. It is neither a problem to realize the standard PROLOG search strategy for CHL_0 (cf. [Lloyd84]).

3.3 CHL vs CHL_0

Theorem 2. Let S be a Horn specification and G a goal. Then there exists R such that $T(S) \cup \{T(G)\} \xrightarrow{*}_{CHL} R$ and $1 \to 0 \in R$ iff there exists a R' such that $T(S) \cup \{T(G)\} \xrightarrow{*}_{CHL_0} R'$ and $1 \to 0 \in R'$.

Proof. Both, *CHL* and CHL_0 are refutation complete (Proposition 2 and Corollary 1).

■

By Theorem 2 we know that CHL_0 is as powerfull as *CHL* in terms of solution inferences. If we restrict ourselves to CHL_0, the question arises whether we are losing other kinds of "power" such as efficiency. By the following lemmata we want to describe more exactly how CHL_0 and *CHL* are related. We are showing how inference steps in *CHL* using rules which are not part of CHL_0 can be replaced by inference steps using only rules of CHL_0.

Lemma 1. If $R \longrightarrow_{\rho} R \cup \{1 \to 0\}$, $\rho \in R_{CHL}$, then $\rho \neq reduce_0$.

Proof. The structure of rules generated by the $reduce_0$ rule is $pr + r \to 0$ where $p \notin \{0, 1\}$ (cf. Definition 6).

■

Lemma 2. Let S be a Horn specification, G a goal, l a $N\text{-}term$ and $n \in \mathbb{N}$.
If $T(S \cup \{G\}) \xrightarrow{*}_{CHL} R_1 \xrightarrow{n}_{reduce_0} R_2 \longrightarrow_{cp} R_2 \cup \{l \rightarrow 0\}$
then there exist R_3 and $k \in \mathbb{N}$ such that $R_1 \xrightarrow{k+1}_{cp} R_3$ where $l \rightarrow 0 \in R_3$ and $k \leq n$.

Proof. (Sketch.) Induction on n. Consider, for instance, the case $n=1$ and $cp=cp_1$:

$$R_1 \longrightarrow_{reduce_0, \sigma} R_1 \cup \{t \rightarrow 0\} \longrightarrow_{cp_1, \phi} R_1 \cup \{t \rightarrow 0, l \rightarrow 0\}$$

implies that $\exists p_1 p_2 r + p_2 r \rightarrow 0,\ q \rightarrow 1 \in R_1$ such that $\sigma q = p_2$ and $t = p_1 r + r$. If $t \rightarrow 0$ is not involved in the cp_1 step, we trivially have $R_1 \longrightarrow_{cp_1} R_1 \cup \{l \rightarrow 0\}$ and $k=0$. Otherwise $\exists q's \longrightarrow 0 \in R_1$ and ϕ such that $\phi \in mgu(\{p_1, q'\})$ and $l = nf_{BA}(\phi s r)$. Therefore we also have

$$R_1 \longrightarrow_{cp_1, \phi} R_1 \cup \{t' \rightarrow 0\}$$

where $t' = nf_{BA}(\phi s p_2 r)$. Since $\sigma q = p_2$ we have $\phi\sigma q = \phi p_2$, that is, $\phi\sigma$ is a match for ϕp_2 via q. Considering the ranges and domains of ϕ and σ with respect to the variables occurring in $p_1 p_2 r$, $q's$ and q we can infer that $\phi\sigma\phi p_2 = \phi p_2$ and $\phi\sigma \in mgu(\{q, \sigma p_2\})$. Then we can show that applying cp_4 to $t' \rightarrow 0$ and $q \rightarrow 1$ with mgu $\phi\sigma$ yields

$$R_1 \cup \{t' \rightarrow 0\} \longrightarrow_{cp_4, \phi\sigma} R_1 \cup \{t' \rightarrow 0, l \rightarrow 0\}$$

which concludes the proof for this case. For all other cases and more details see [Diet85].

■

Lemma 2 tells us that, in a *CHL*-inference, we can replace any sequence of n $reduce_0$ steps which are followed by a cp step by no more than n cp steps. Since the last step of a solution inference is always a cp step (Lemma 1), we can replace any $reduce_0$ step of a solution inference via *CHL* by cp steps, getting an other solution inference (Corollary 2).

Corollary 2. Let S be a Horn specification, G a goal and $k \in \mathbb{N}$. If $T(S \cup \{G\}) \xrightarrow{k}_{CHL} R$ and $1 \rightarrow 0 \in R$ then there exists R_0 and $k' \in \mathbb{N}$ such that $T(S \cup \{G\}) \xrightarrow{k'}_{cp} R_0$, $1 \rightarrow 0 \in R_0$ and $k' \leq k$.

Proof. Proposition 4, Lemma 1, Lemma 2.

■

As sketched in the proof of Lemma 2, the same unifiers must be computed as before, if all $reduce_0$-steps are replaced by cp-steps. So, no efficiency concerning unifications is lost by this process.

Lemma 3 states that cp_3 is superfluous:

Lemma 3. Let S be a Horn specification and G a goal. If $T(S \cup \{G\}) \xrightarrow{*}_{cp} R$ and $1 \rightarrow 0 \in R$ then there exists R_0 such that $T(S \cup \{G\}) \xrightarrow{*}_{cp \setminus \{cp_3\}} R_0$ and $1 \rightarrow 0 \in R_0$.

Proof. If cp_3 generates the rule $t \rightarrow 0$ by overlapping some rule $pr + r \rightarrow 0$ and some N-rule $s \rightarrow 0$, t can be rewritten to 0 by $s \rightarrow 0$, so $t \rightarrow 0$ is redundant[1].

■

Finally, we want to show that we can also omit cp_2 from *CHL*. The result of the following lemmata is that we can replace an cp_2 step by an cp_1 step, generating a data object which is in a certain sense "more general" as the data object generated by the cp_2 step (Lemma 4). Then we show that any rule which can be applied to a certain data object can also be applied to every "more gen-

1 Hsiangs N-Strategy requires N-critical pairs to be added as new rewrite rules only if they are not reducible with respect to *BA and* the current rewrite system. The reason why we do not require to reduce N-critical pairs *at once* with respect to the current rewrite system is, that we want to single out these reductions as seperate inference steps because they correspond to resolution steps in *RHL*.

eral" data object again generating a "more general" data object (Lemma 5). But first, we make precise what "more general" means.

Definition 8. Let s,t be N-terms, Ls,Lt the set of (different) atomic formulae appearing in s and t, respectively, and ϕ a substitution. Then the relation $\leq_\phi$ on N-terms is defined by $s \leq_\phi t$ iff $L\phi s = Lt$.

■

In other words, $s \leq_\phi t$ means that every literal of t is an instance by ϕ of some literal of s and s can have some more literals than t which are unified by ϕ. Then a data object R_1 of *CHL* can be considered to be "more general" than a data object R_2 if ϕ exists such that for any rule $l_2 \rightarrow \delta \in R_2$ a rule $l_1 \rightarrow \delta \in R_1$ exists where $l_1 \leq_\phi l_2$.

Lemma 4. If $R \longrightarrow_{cp_2} R \cup \{t \rightarrow 0\}$ by $pr_1r_2 + r_1r_2 \rightarrow 0 \in R$ and the N-rule $qs_1s_2 \rightarrow 0 \in R$ then there exist ϕ and an N-term t' such that $R \longrightarrow_{cp_1} R \cup \{t' \rightarrow 0\}$ and $t' \leq_\phi t$.

Proof. If $R \longrightarrow_{cp_2,\theta} R \cup \{t \rightarrow 0\}$ we have $\theta \in mgu(\{pr_1, qs_1\})$ and $t = nf_{BA}(\theta s_2 r_1 r_2)$. Then θ is also a (not necessarily most general) unifier for p and q. So σ and ϕ exist such that $\sigma \in mgu(\{p,q\})$ and $\phi\sigma = \theta$, which implies $R \longrightarrow_{cp_1,\sigma} R \cup \{t' \rightarrow 0\}$, $t' = nf_{BA}(\sigma s_1 s_2 r_1 r_2)$. Furthermore, since $\theta r_1 = \theta s_1$ and by Proposition 5 we have $\phi Lt' = L\phi nf_{BA}(\sigma s_1 s_2 r_1 r_2) = L\phi\sigma s_1 s_2 r_1 r_2 = L\theta s_1 s_2 r_1 r_2 = L\theta s_2 r_1 r_2 = Lt$ which proves $t' \leq_\phi t$.

■

Lemma 5. Let $R \longrightarrow_{\{cp_1,cp_4\}} R \cup \{t \rightarrow 0\}$ and $qs \rightarrow 0 \in R$ the N-rule involved. If the N-term l and the substitution ϕ are such that $l \leq_\phi qs$ then there exist R_0, ψ, $i \in \mathbb{N}$ and a N-term t' such that $R \setminus \{qs \rightarrow 0\} \cup \{l \rightarrow 0\} \xrightarrow{i}_{\{cp_1,cp_4\}} R_0 \cup \{l \rightarrow 0, t' \rightarrow 0\}$ where $R \setminus \{qs \rightarrow 0\} \subset R_0$ and $t' \leq_\psi t$.

■

Proof. We may assume that the domain of ϕ is restricted to variables occurring in l. Let

$$R \longrightarrow_{cp_1,\theta} R \cup \{t \rightarrow 0\}$$

by $pr + r \rightarrow 0$ and $qs \longrightarrow 0$, that is, $\theta \in mgu(\{p,q\})$ and $t = nf_{BA}(\theta sr)$. Because $L\phi l = Lqs$ an atom q' and a N-term s' exist such that $l = q's'$ and $\phi q' = q$. Because $\phi p = p$ we have $\theta\phi q' = \theta q = \theta p = \theta\phi p$, i.e. q' and p are unifiable. Thus ψ and σ exist such that $\sigma \in mgu(\{q',p\})$ and $\psi\sigma = \theta\phi$. This implies whith $R_0 := R \setminus \{qs \rightarrow 0\}$ and $t' = nf_{BA}(\sigma s'r)$

$$R \setminus \{qs \rightarrow 0\} \cup \{l \rightarrow 0\} \longrightarrow_{cp_1,\sigma} R_0 \cup \{l \rightarrow 0, t' \rightarrow 0\}.$$

Now suppose that $\phi q'$ does *not* occur in $\phi s'$. Then $L\phi s' = Ls$ and because $\phi r = r$ and with Proposition 5 we have

(*) $L\psi t' = L\psi nf_{BA}(\sigma s'r) = L\psi\sigma s'r = L\theta\phi s'r = Lt$

which proves $t' \leq_\psi t$.

If $\phi q'$ occurs once in $\phi s'$ then $\theta\phi q' = \psi\sigma q'$ occurs in $\psi t'$ and thus $\sigma q'$ in t'. Since $\psi\sigma q' = \psi\sigma p$ we can again apply cp_1 to $t' \rightarrow 0$ and $pr + r \rightarrow 0$ generating a new rule $t'_1 \rightarrow 0$ where the occurrence of $\sigma q'$ is deleted from t'. Then (*) holds for t'_1 with some ψ_1. If $\phi q'$ occurs i times in $\phi s'$, we may achieve (*) for some t'_i and some ψ_i by i times applying cp_1. Then we get

$$R \setminus \{qs \rightarrow 0\} \cup \{l \rightarrow 0\} \longrightarrow_{cp_1} R_0 \cup \{l \rightarrow 0, t' \rightarrow 0\} \xrightarrow{i}_{cp_1} R_i \cup \{l \rightarrow 0, t' \longrightarrow 0, t'_i \rightarrow 0\}$$

where R_1 contains R_0 and all $i-1$ intermediately generated rules and $t'_i \leq_{\psi_i} t$. (Note that i is bound by the number of literals of l which are unified by ϕ).

If $R \longrightarrow_{cp_4} R \cup \{t \rightarrow 0\}$ this is a special case of cp_1 where $r = 1$.

■

Lemma 1 - 5 can be used to prove Theorem 2 without knowing the completeness of CHL_0 (which then, of course, is a consequence).

Replacing a cp_2 step by a cp_1 step (Lemma 4) can have the consequence that later on, during an inference, some more inference steps will be needed to get a solution (Lemma 5). But these "some more" steps must only compute unifiers, whereas a cp_2 step must compute an overlap which is much more expensive than computing unifiers. (Recall that overlaps correspond to AC-unification for Boolean terms). The number of additional steps depends on the overlap the cp_2 step would have made: the "greater" (and thus more expensive) the overlap would have been, the more additional steps are needed when replacing the cp_2 step. By these arguments, we can say again that we do not lose efficiency in simplifying *CHL*. Moreover, we may argue that, in general, in logic programming we are interested in most general solution inferences, which are not guaranteed to be found when using the rule cp_2.

4. Discussion

The example given in section 2.4 clearly illustrates the equivalence of the resolution and the completion method for Horn logic as formally described by Theorem 1: The inference via *CHL* of Figure 3 is really a CHL_0-inference (we may replace the shown sequence of applied inference rules by $(cp_1, cp_4, cp_4, cp_1, cp_4, cp_4)$ where the rewrite rules and the involved unifiers remain the same ones). We have the same number of inference steps and the same most general unifiers are computed in both inferences. Further we can make the following observations concerning:

(1) *Data representation*: The representation of clauses as Boolean rewrite rules obviously contains a lot of redundancy. Every negative literal of an implication appears twice in the corresponding rewrite rule. Further we can observe that the canonical rewrite rule system for Boolean algebra *BA* (Figure 1), especially the Boolean ring axioms b_5 - b_{10}, accomplishes the same task for the N-strategy as the set notation of clauses for the resolution calculus (cf. Proposition 5). In step (5) of Figure 3 the N-critical pair beeing computed is $<P(x_{12})P(x_{12}), 0>$ which then further reduces by *BA* to $<P(x_{12}), 0>$. Restricting the representation to the information necessary to perform inferences results in sets of signed literals, which is exactly the representation of clauses in resolution calculus, and computing N-critical pairs is nothing more than computing resolvents.

(2) *Reduction power*: Applying reductions as often as possible during an inference seems to be very efficient, because it only needs to compute matches instead of unifiers and, in general, reduces complexity of terms. In [Hsia83] this is claimed to be the most important advantage of the N-strategy against resolution. We have replaced every reduction step by corresponding *cp* steps (Proposition 4 and Lemma 2). But in these cases, the *cp* step uses a unifier which is really a match, and the *cp* step also reduces complexity of terms. So, if matching can be performed much more efficiently than unification, we can change our strategy (even in *RHL*) and always look for matches (which are also most general unifiers) before looking for unifiers in order to apply an inference rule. Then we can achieve the same "reduction power" as by integrating special preferenced reduction rules. However, note that in the case of the N-stategy we need AC-matching, which is known to be *NP*-complete in general [BKNa85].

(3) *Search space:* Comparing *CHL* and CHL_0 we can state that the search space spanned by *CHL* is bigger than the search space spanned by CHL_0 because the inference rules of CHL_0 are also part of *CHL*. Figure 4 shows two more solution inferences for refutations of ¬**P** which cannot be performed by CHL_0. As shown in section 3.3, these inferences are not more efficient than those of CHL_0: Both solution inferences of Figure 4 are one step shorter than the solution inferences of Figure 3, but instead of this in step (3) of Figure 4 two *pairs* of literals must be unified. The generated rewrite rule corresponds to the rewrite rule generated by step (4) of Figure 3 which is still "more general" (more precisely: $P(x)P(y)M(x,i(y),e) \leq_\phi P(b)P(y)M(b,i(y),e)$ where $\phi = \{x \leftarrow b\}$, cf. Definition 8 and Lemma 4 - 5).

After step (2) in Figure 4 there are two more possibilities of applying cp_2 to $P(e)P(b) \rightarrow 0$ and r_5: the rule $P(e)P(y)M(e,i(y),b) \rightarrow 0$ may be generated with the substitution $\{z_5 \leftarrow b, x_5 \leftarrow e\}$ as well as $P(x)P(e)M(x,i(e),b) \rightarrow 0$ with $\{z_5 \leftarrow b, y_5 \leftarrow e\}$. Both rules will never lead to $1 \rightarrow 0$, because $M(x,i(e),b)$ and $M(e,i(y),b)$ will never disappear. So there are more blind alleys in the search space spanned by *CHL* than in the search space spanned by CHL_0. This is a real drawback, because the often unmanageable size of the search trees is one of the severest problems automated

$P(i(b)) \rightarrow 0$

(1) $\downarrow$ $cp_1,\ r_5,\ \{z_5 \leftarrow i(b)\}$

$P(x)P(y)M(x,i(y),i(b)) \rightarrow 0$

(2) $\downarrow$ $cp_4,\ r_1,\ \{x \leftarrow e, y \leftarrow b, x_1 \leftarrow i(b)\}$

$P(e)P(b) \rightarrow 0$

(3) $\downarrow$ $cp_2,\ r_5,\ \{z_5 \leftarrow e, x_5 \leftarrow b\}$	$\downarrow$ $cp_2,\ r_5,\ \{z_5 \leftarrow e, y_5 \leftarrow b\}$
$P(b)P(y)M(b,i(y),e) \rightarrow 0$	$P(x)P(b)M(x,i(b),e) \rightarrow 0$
(4) $\downarrow$ $cp_4,\ r_3,\ \{y \leftarrow b, x_3 \leftarrow b\}$	$\downarrow$ $cp_4,\ r_3,\ \{x \leftarrow b, x_3 \leftarrow b\}$
$P(b) \rightarrow 0$	$P(b) \rightarrow 0$
(5) $\downarrow$ $cp_4,\ r_6,\ \epsilon$	$\downarrow$ $cp_4,\ r_6,\ \epsilon$
$1 \rightarrow 0$	$1 \rightarrow 0$

Figure 4: Two refutations of ¬**P** via *CHL*.

theorem provers are faced to.

(4) *Strategy*: It is our opinion, that the efficiency of theorem proving is not a question of the method (resolution or completion) but rather a question of *strategy*[1]. The results of this paper confirm this opinion. In [Hsia83] comparisons with resolution proofs are made in some examples, where resolution seems to be very inefficient compared with completion. We conjecture that we can find resolution proofs for most or even all of these examples which are as efficient as the completion proofs if we change the strategy. Thus, the question arises, whether or not the equivalence of the two methods can be generalized to full first order clausal logic.

1 *Strategy* here means the way in which clauses or rewrite rules are chosen for building resolvents or critical pairs, respectively.

In [Hsia83] the above example is also solved by a variant of the N-strategy called RN-strategy. The characteristics of this method are that a canonical term rewriting system is "built in", which specifys a special domain set (in this case a canonical system for group theory). Then, not only N-critical pairs are computed but also critical pairs arising from superposition of rules which represent clauses and rules of the canonical built in system. We conjecture that similar relations between this approach and resolution with paramodulation (cf. [Love78]) and/or resolution with narrowing (cf. [Hull80, Slag74]) can be established as done in this paper for completion and resolution for Horn logic.

5. Summary and Future Prospects

Our main motivation for closely analysing the relations between resolution and algebraic completion for Horn logic came from the search for alternative concepts for interpreting logic programs. Hsiang [Hsia83] showed that algebraic completion is an alternative to resolution oriented theorem provers which works effectively. We have specialized his method for pure Horn logic (the inference system CHL) in order to see whether algebraic completion works not only effectively but also efficiently in this field compared with resolution (the inference system RHL).

First we have seen that we can restrict CHL to a simpler inference system CHL_0 which is still refutation complete. By Theorem 1 we know, that in CHL_0 and RHL we need exactly the same number of inference steps for solution inferences and, looking at the corresponding inference rules, we can state that exactly the same unifiers are computed. That is, no efficiency is gained by CHL_0 in terms of inference steps and unifications.

Then we analysed the relations between CHL and CHL_0 and we have seen that (1) we do not lose (unification) efficiency in simplifying CHL, and (2) the search space spanned by CHL is really bigger than the search space spanned by CHL_0.

In total we can state that we can at most achieve the same efficiency in terms of unification and inference steps when using algebraic completion instead of resolution for interpreting logic programs. In its general form, algebraic completion for Horn logic even may cause or increase search space problems.

But surely we are losing efficiency in terms of data representation: Recall that at first every negative literal of an implication appears twice in its rewrite rule representation. Secondly, we can observe that the Boolean algebra system BA accomplishes the same task as the set notation of clauses for the resolution calculus. However, implementing the BA system needs to realize some kind of AC-matching which is known to be expensive [BKNa85] .

Some remaining questions may be subject of further study: Until now we do not know whether the equivalence of resolution and completion can be generalized to full first order clausal logic. The relations between the completion method with built-in theories (RN-strategy in [Hsia83]) and resolution with paramodulation or narrowing can also be analysed in a similar way. Finally, the applied inference strategy is of considerable impact for both the resolution and the completion approach. Further research in this area may lead to improvements on both sides.

Acknowledgements. I would like to thank my colleagues P. Kursawe and I. Varsek for many fruitful discussions and their extensive comments on this paper as well as G. Goos and J. Siekmann for carefully reading and commenting earlier drafts of this paper.

References

BKNa85 D. Benanav, D. Kapur, P. Narendran: *Complexity of Matching Problems*, 1st Int. Conf. on Term Rewriting Techniques and Applications, Dijon, France, Springer LNCS 202, 1985.

ClMe81 W. F. Clocksin, C. S. Mellish: *Programming in Prolog*, Springer, 1981.

DeJo84 N. Dershowitz, N. A. Josephson: *Logic Programming by Completion*, Proc. 2nd Int. Logic Programming Conf., Uppsala, 1985.

Deus83 P. Deussen: *Control of and Algorithms for Reduction systems*, Unpublished report, University of Karlsruhe, Institut für Informatik I, 1983.

Deus85 P. Deussen: Private Communication, 1985.

Diet85 R. Dietrich: *Relating Resolution and Algebraic Completion for Horn Logic*, Arbeitspapiere der GMD Nr. 177, Gesellschaft für Mathematik und Datenverarbeitung mbH, Bonn, 1985.

GoMe84 J. A. Goguen, J. Meseguer: *Equality, Types, Modules and (Why not?) Generics for Logic Programming*, J. Logic Programming 1984, 2: 179-210.

Hero83 A. Herold: *Some Basic Notions of First-Order Unification Theory*, Universität Karlsruhe, Fakultät für Informatik, Interner Bericht Nr. 15/83, 1983.

HsDe83 J. Hsiang, N. Dershowitz: *Rewrite Methods for Clausal and Non-Clausal Theorem Proving*, Proc. 10th Int. Conf. on Automata, Languages and Programming, Springer LNCS 154, 1983.

Hsia83 J. Hsiang: *Topics in Automated Theorem Proving and Program Generation*, Ph. D. Thesis, University of Illinois at Urbana-Champaign, 1983.

Hsia85 J. Hsiang: *Two Results in Term Rewriting Theorem Proving*, 1st Int. Conf. on Term Rewriting Techniques and Applications, Dijon, France, Springer LNCS 202, 1985.

Huet80 G. Huet: *Confluent Reductions: Abstract Properties and Application to Term Rewriting Systems*, JACM, Vol. 27, No. 4, October 1980.

HuOp80 G. Huet, D. C. Oppen: *Equations and Rewrite Rules: A survey*, R. Book (Ed.): Formal Language Theory. Perspectives and Open Problems. Academic Press, 1980.

Hull80 J.-M. Hullot: *Canonical Forms and Unification*, Proc. 5th Int. Conf. on Automated Deduction, Springer LNCS 87, 1980.

KaNa85 D. Kapur, P. Narendran: *An Equational Approach to Theorem Proving in First-Order Predicate Calculus*, Proc. IJCAI 85, Los Angeles, 1985.

KnBe70 D. E. Knuth, P. B. Bendix: *Simple Word Problems in Universal Algebra*, J. Leech (Ed.): Computational Problems in Universal Algebra, Pergamon Press, 1970.

Lloyd84 J. W. Lloyd: *Foundations of Logic Programming*, Springer, 1984.

Love78 D. W. Loveland: *Automated Theorem Proving: A Logical Basis*, North-Holland, 1978.

Nils82 N. J. Nilsson: *Principles of Artificial Intelligence*, Springer, 1982.

Patt78 D. A. Waterman, F. Hayes-Roth (Eds.): *Pattern Directed Inference Systems*, Academic Press, 1978.

Paul84 E. Paul: *A New Interpretation of the Resolution Principle*, Proc. 7th Int. Conf. on Automated Deduction, Springer LNCS 170, 1984.

Robi65 J. A. Robinson: *A Machine Oriented Logic Based on the Resolution Principle*, JACM, Vol 12, No. 1, January 1965, pp. 23-41.

Slag74 J. R. Slagle: *Automated Theorem-Proving for Theories with Simplifiers, Commutativity and Associativity*, JACM, Vol 21, No. 4, October 1974, pp. 622-642.

A SIMPLE NON-TERMINATION TEST FOR THE KNUTH-BENDIX METHOD

David A. Plaisted
Department of Computer Science
University of North Carolina at Chapel Hill
Chapel Hill, North Carolina 27514

Abstract

We propose a simple test for nontermination in the Knuth-Bendix completion algorithm. This test has the property that if there exists a simplification ordering that can generate a completion S of a set R of rules, then S may be generated from R using this test. Also, this test is "user friendly" in that it does not require any detailed knowledge of termination orderings. However, this technique may generate completions S that are not terminating; therefore, traditional methods for proving termination need to be used on the locally confluent sets S of rewrite rules that are obtained. We show that this test may be implemented in reasonable time and space bounds.

1. Introduction

The Knuth-Bendix method for completing term rewriting systems (Knuth and Bendix[70], Huet[81]) has a number of practical applications. For a discussion of term rewriting, see Huet and Oppen[80] and Bundy[83]; for a discussion of the Knuth-Bendix method and its applications see Dershowitz[83, 85b] and Hsiang and Dershowitz[83]. Given an input set R of rewrite rules, the Knuth-Bendix method generates a locally confluent set S of rewrite rules; it may also halt without generating S, or compute forever. The set S is generated in a way that ensures termination, so S is confluent by Newman's theorem (see Huet[80]) and therefore canonical. Thus S can be used as a decision procedure for the equational theory of R. That is, we can decide if s = t is a logical consequence of R, by reducing s and t to normal form using S, and testing if these normal forms are equal. This is usually much more efficient than directly using a resolution theorem prover to decide if (R $\supset$ s=t) is a theorem of first-order logic with equality.

The general outline of the Knuth-Bendix algorithm, adapted from Winkler[84], is as follows, where R is the given set of rewrite rules.

R' ← R
C ← critical pairs of R'
while C not empty do
    (r, s) ← an element of C, chosen fairly;
    C ← C - {(r, s)};
    (r', s') ← R' normal forms of r and s, respectively;

if r' ≠ s' then [[orient the equation r' = s']] choose one of

R' ← R' U {r' → s'};

R' ← R' U {s' → r'};

if R' is nonterminating then back up to a previous orientation choice, fail if no more to back up to;

C ← C together with new critical pairs from the rule just added to R';

S ← R';

This procedure is said to fail if it backs up all the way to the beginning and no more orientation choices are possible. It may also halt without failure, in which case S is a canonical term rewriting system equivalent to R. Or, the procedure may run forever.

In order to ensure that R' is a terminating term rewriting system, some termination ordering > is typically used. Then an equation r' = s' is oriented r' → s' if r' > s' and is oriented s' → r' if s' > r'. If neither r' > s' nor s' > r' then the Knuth-Bendix method fails for the ordering >. For a sample of some such orderings, see Plaisted[78], Dershowitz[82], Jouannaud et al[82], and Rusinowitch[85]. For a survey, see Dershowitz[85a]. Unfortunately, these orderings are not fully satisfactory. This is true for several reasons. First, because termination is undecidable (Huet and Lankford[78]), no ordering can work in all cases, and so there are many different orderings that are needed for various special cases. Second, some of the more general orderings such as the recursive path ordering of Dershowitz[82] and the recursive decomposition ordering of Jouannaud et al[82] and their various extensions and combinations (Rusinowitch[85]), are difficult to understand, even for experts in the area of termination. This becomes even more true when status is considered. Polynomial orderings, in contrast, are fairly easy to understand, and have been successfully used. However, it requires considerable expertise to choose the polynomials. Sometimes (Dershowitz[82]) this choice can be done automatically, but the procedure to do it has a very high worst case asymptotic running time. Also, there are term rewriting systems that terminate but for which this cannot be shown using any polynomial ordering. The need to use these orderings makes term rewriting systems such as REVE (Lescanne[83]) difficult to understand for the common user, in some cases, and also makes these systems more complicated to implement. Also, there may be special cases in which no implemented ordering will work, but in which a canonical set S can still be generated using an ordering that has not been implemented.

We propose a simple non-termination test that will work whenever there exists a simplification ordering that will work. That is, if there exists a simplification ordering which permits the completion S of R to be generated, then our method also permits S to be generated from R. Therefore, in a sense, our method is as powerful as all simplification orderings put together. Note, however, that there are some sets S that are terminating but for which no simplification ordering can be used to show this. For example, consider the one rule system ff(x) → fgf(x). Our method cannot be used to show such systems are terminating. The major orderings used for proving termination are all simplification orderings, and hence are similarly incapable of showing termination of this one rule system. Our method also permits S to be generated that are not terminating, in some cases. However, if S or an intermediate set R' is nonterminating, this will not cause problems with nonterminating reductions of critical pairs. Our method will detect this nontermination after a finite number of reductions have been done, in a manner to be made precise below. Therefore, after a locally confluent set S is generated, traditional methods can be used to show S is terminating. There are advantages to this approach, despite the need for proving S is terminating. First, the termination test can be done "off line"; the final set S can be circulated to colleagues, for example. Second, S is often fairly simple in structure, and it may be easier to prove termination of S than to find an ordering for the equations obtained during the completion process. Third, since we are trying to prove termination of a fixed set S, we do not need the "incremental property" of a termination ordering. This permits simpler orderings. For example, for the recursive path ordering and the recursive decomposition ordering, we can choose a total ordering on function symbols; for such orderings, several of the termination orderings considered in the literature are identical.

2. The non-termination test

Definition. Suppose s and t are terms. Then we say s is *homeomorphically embedded in* t, written $s \leq_{he} t$, if

1) s and t are identical, possibly identical variables, or
2) $t = g(t_1, \cdots, t_n)$ and $s \leq_{he} g_i$ for some i, or
3) $s = f(s_1, \cdots, s_m)$ and $t = f(t_1, \cdots, t_n)$ and $(s_1, \cdots, s_m) \leq_{he} (t_1, \cdots, t_n)$.

Suppose $\overline{s}$ and $\overline{t}$ are (possibly empty) lists of terms. Suppose $\overline{s} = (s_1, \cdots, s_m)$ and $\overline{t} = (t_1, \cdots, t_n)$. Then $\overline{s} \leq_{he} \overline{t}$ iff (length($\overline{s}$) ≤ length($\overline{t}$) and)

4) $\overline{s} = ()$, the empty list, or
5) $s_1 \leq_{he} t_1$ and $(s_2, \cdots, s_m) \leq_{he} (t_2, \cdots, t_n)$ or
6) $\overline{s} \leq_{he} (t_2, \cdots, t_n)$

Thus we define the homeomorphic embedding relation for terms as well as for lists of terms. Note that the empty list is homeomorphically embedded in every list of terms.

For example, $f(a, b) \leq_{he} g(f(h(a), g(a, b)), c)$. To see this, note that $a \leq_{he} a$ by 1), so $a \leq_{he} h(a)$ by 2). Similarly, $b \leq_{he} g(a, b)$. Therefore, by 3), $f(a, b) \leq_{he} f(h(a), g(a, b))$. Therefore, by 2) again, $f(a, b) \leq_{he} g(f(h(a), g(a, b)), c)$. Note that if s is a subterm of t then $s \leq_{he} t$.

Definition. A partial ordering > on terms is called a *simplification ordering* if it has the following properties:

1. (Subterm property) f(... s ...) > s
2. (Monotonicity) s > t implies f(... s ...) > f(... t ...)
3. f(... s ...) > f(... ...)

The last property is only relevant when f has variable arity.

Theorem (Kruskal[60]). Suppose $r_1, r_2, r_3, \cdots,$ is an infinite sequence of ground terms over a finite set of function symbols. Then there exist i and j, $i < j$, such that $r_i \leq_{he} r_j$.

Theorem (Dershowitz[82]). Suppose > is a simplification ordering, and s and t are terms and $s \leq_{he} t$. Then $t \geq s$.

Theorem (Dershowitz[82]). If > is a simplification ordering and T is a set of ground terms over a finite set F of function symbols then > is well-founded on T.

Definition. Suppose R is a term rewriting system and r is a term. Then an *R-reduction sequence* for r is a sequence $r_1, r_2, r_3, \cdots,$ of terms such that $r = r_1$ and for all $i > 0$, $r_i \Rightarrow_R r_{i+1}$. We abbreviate R-reduction sequence by reduction sequence when R is understood.

The first theorem above is not enough for our purposes, since we are not restricted to ground terms. For example, the infinite sequence $x_1, x_2, x_3, \cdots,$ does not have any i and j such that $i < j$ and $x_i \leq_{he} x_j$. Therefore we have the following result.

Theorem . Suppose $r_1, r_2, r_3, \cdots,$ is an infinite R-reduction sequence for term rewriting system R. Then there exists i and j with $i < j$ such that $r_i \leq_{he} r_j$.

Proof. Let V(r) be the set of variables that occur in a term r. Since R is a term rewriting system, if $s \rightarrow t$ is a rule in R then $V(t) \subset V(s)$. Therefore $V(r_i) \subset V(r_1)$ for all i. Since the number of variables is bounded, there exist i and j with $i < j$ such that $r_i \leq_{he} r_j$.

Our nontermination test is the following: When reducing a term r, if the reduction sequence $r_1, r_2, \cdots,$ is generated, call this sequence *self-embedding* if there exists i and j, $i < j$, such that $r_i \leq_{he} r_j$. If a self-embedding R'-reduction sequence is generated, then we consider R' to be nonterminating. By the last theorem above, this self-embedding test will succeed on some finite prefix of any infinite reduction sequence; by the second theorem, if the self-embedding test succeeds, then any simplification ordering would have failed to orient the current set R' of rewrite rules. However, it is possible that a nonterminating set of rules can be generated, even using our homeomorphic embedding test, because the only reduction sequences we will see will be those generated when reducing critical pairs to their normal forms. Also, only one r-reduction sequence will be generated for a term r, corresponding to the choice of reduction strategy. There may be other nonterminating (hence self-embedding) reduction sequences that we do not find. For some results about the undecidability of the existence of self-embedding R-reduction sequences, see Plaisted[85].

If self-embedding is detected, then the user would be given r_i and r_j such that $i < j$ and $r_i \leq_{he} r_j$; also, the user would be given all rewrite rules used to derive r_j from r_i. We call these rules used to derive r_j from r_i a *self-embedding subset* of R. Then the user would be given a choice which of these rules to re-orient. Depending on his or her choice, the completion would be backed up to the appropriate point, and resumed from there.

To sum up, our modified Knuth-Bendix procedure is as follows:

R' ← R
C ← critical pairs of R'
while C not empty do
    (r, s) ← an element of C, chosen fairly;
    C ← C - {(r, s)};
    (r', s') ← R' normal forms of r and s, respectively;
    if one of the reduction sequences for r' and s' is self-embedding
        then let R'' be a self-embedding subset of R and back up
        to a previous choice where an element of R'' was oriented;
    if r' ≠ s' then [[orient the equation r' = s']] choose one of
        R' ← R' U {r' → s'};
        R' ← R' U {s' → r'};
    C ← C together with new critical pairs from the rule just
        added to R';
S ← R';

Proposition . Suppose a canonical term rewriting set S can be generated from R using the Knuth-Bendix method, in which the same simplification ordering is used throughout to orient all new equations generated. Then S can also be generated from R using the modified Knuth-Bendix method, assuming the correct orientation of all new equations is chosen.

2.1 Example

Suppose R is the following term rewriting system:

f(c) → gh(c)
hg(x) → ghf(x)
k(x, h(x), c) → h(x)
k(f(x), y, x) → f(x)

There is a critical pair (hf(c), f(c)) arising from superposition of the last two rules. For f(c) we have the reduction sequence f(c), gh(c). For hf(c) we have the reduction sequence hf(c), hgh(c), ghfh(c). This latter sequence is self-embedding since hf(c) $\leq_{he}$ ghfh(c). The self-embedding subset for this sequence is the set of rules

f(c) → gh(c)
hg(x) → ghf(x)

Therefore no simplification ordering can orient both of these rules as shown, and it is necessary to back up and orient one of them in the other direction.

3. Extensions

In a simplification ordering, if s > t then sΘ > tΘ for all substitutions Θ. Therefore, if $s\Theta \leq_{he} t\Theta$ then the equation s = t cannot be oriented s → t in any simplification ordering. This is not automatically detected by our method, since we only test whether $s \leq_{he} t$, not whether $s\Theta \leq_{he} t\Theta$. For example, consider the equation x * e = x. If Θ replaces x by e, then (x * e = x)Θ is e * e = e. Since $e \leq_{he} e*e$, this equation must be oriented as x * e → x. Therefore it is possible to call a reduction sequence r_i self-embedding if there exists substitution Θ and positive integers i and j with i < j such that $r_i\Theta \leq_{he} r_j\Theta$. However, this extended self-embedding test seems to be more complex than the regular self-embedding test, and we have not analyzed its complexity.

An intermediate solution, which does not increase the time complexity of the self-embedding test, is to replace all variables in a reduction sequence r_i by a fixed, user specified constant symbol. Then if $r_i \leq_{he} r_j$, $r_i\Theta \leq_{he} r_j\Theta$ where Θ replaces all variables by the constant symbol. Therefore this replacement does not decrease the power of the self-embedding test. Another possibility is for the user to give a set of constant symbols to be used in this manner, or some other finite set of ground terms.

Using these techniques, all equations generated in the standard completion of the three group axioms may be ordered automatically, except for (x*y)*z = x*(y*z) and g(x*y) = g(y)*g(x), where * is the group operation and g is the inverse operation. See for example Hullot[80] for the completion of the three group axioms using the Knuth-Bendix method. For example, e * x = x must be oriented e * x → x since $x \leq_{he} e * x$. Also, g(x) * x = e must be oriented g(x) * x → e since $e \leq_{he} g(e) * e$. In general if $r \leq_{he} s$ then the equation r = s must be oriented as s → r. Also, if for some Θ, $r\Theta \leq_{he} s\Theta$ then the equation r = s must be oriented s → r. In general, it seems that a large majority of ordering decisions can be made automatically in this way, based on the completions given in Hullot[80].

Although the Knuth-Bendix method is typically used interactively, the modified Knuth-Bendix method may be used non-interactively. This may be useful for theorem proving applications. The usual reason for human interaction is to choose a termination ordering. With the modified method, this is not necessary. Therefore, heuristics can be used to decide how to orient new equations and how to back up when self-embedding is detected. For example, the user can orient rules in R as he or she chooses, and can orient new rules r → s if r is more complex than s in some crude ordering. This orientation can be reversed if it is necessary to back up. This approach may help to make the Knuth-Bendix procedure more automatic.

Even when the modified method is used, the sequences r_i may become very long before the self-embedding test succeeds. For example, it is possible to compute Ackermann's function using a term rewriting sytem R whose termination can be shown using a simplification ordering of Kamin and Levy[80]. The computations of R, though very long, will not be self-embedding. It may be useful for the user to set a limit, so that any reduction sequences that contain more than say 1000 terms are considered as nonterminating.

3.1 Equational term rewriting systems

Suppose (R, E) is an equational term rewriting system. For example, E may be the associative and commutative equations for certain operators. Methods for proving termination for such systems are quite complicated and not fully understood; for some approaches to associative and commutative operators see Dershowitz et al[83] and Bachmair and Plaisted[85]. For another approach see Jouannaud and Munoz[84]. However, these approaches are complicated and hard to extend to terms with variables in the general case. One approach to equational term rewriting is to consider E-equivalence classes of terms rather than terms themselves. If r is a term, let [r] be its E-equivalence class, that is, the set of terms s such that r = s is a logical consequence of E. Then we can extend the homeomorphic embedding relation to equivalence classes by $[r] \leq_{he} [s]$ if there exist u in [r] and v in [s] such that $u \leq_{he} v$. Now, let $r_1, r_2, r_3, \cdots,$ be a reduction sequence of (R, E) using equational rewriting. We call this sequence self-embedding if there exist i and j with i < j such that $[r_i] \leq_{he} [r_j]$. As before, we can show that any infinite (R, E) reduction sequence must be self-embedding. In this way, the nontermination test can be extended to equational rewriting in full generality.

This approach has some disadvantages. For example, it is necessary to compute the E-equivalence classes of r and s to decide if $[r] \leq_{he} [s]$. Also, this may require many homeomorphic embedding tests. We give a more efficient method for associative and commutative operators. Suppose s is a term with AC (associative-commutative) operators. Let flat(s) be s with these operators flattened in the usual way. Thus flat(f(x, f(a, y))) = f(x, a, y) if f is AC. If s is a flattened term, let [s]' be the set of terms obtained by permuting arguments of AC operators. Thus [f(x, a, y)]' contains f(x, a, y), f(a, x, y), f(y, a, x) et cetera.

Definition. $[s]' \leq_{he} [t]'$ iff there exist terms u in [s]' and v in [t]' such that $u \leq_{he} v$.

Proposition . Suppose E consists of associative and commutative axioms for some operators. Suppose s and t are terms containing AC operators. Then if $[flat(s)]' \leq_{he} [flat(t)]'$, $[s] \leq_{he} [t]$.

Note that [flat(s)]' is a set of flattened terms with arguments permuted in all possible ways, while [s] is an E-equivalence class of s. It is possible to have $[s] \leq_{he} [t]$ but not $[flat(s)]' \leq_{he} [flat(t)]'$. For example, if f is AC then $f(a, f(b, c)) \leq_{he} f(a, g(f(b, c)))$ but not $[f(a, b, c)]' \leq_{he} [f(a, g(f(b, c)))]'$. Despite this, we have the following easy result.

Proposition . Suppose $r_1, r_2, r_3, \cdots,$ is an infinite reduction sequence and E consists of AC equations for some operators. Then there exists i and j, i < j, such that $[flat(r_i)]' \leq_{he} [flat(r_j)]'$.

Therefore the test $[flat(s)]' \leq_{he} [flat(t)]'$ is sufficient to prevent infinite reduction sequences. Also, we have the following result:

Theorem . Given terms s and t possibly containing AC operators, and given E consisting of the AC axioms, it is possible to decide whether $[flat(s)]' \leq_{he} [flat(t)]'$ in polynomial time.

Proof. The only hard case is when the top-level operators of s and t are identical and AC. Suppose flat(s) is $f(s_1, \cdots, s_m)$ and flat(t) is $f(t_1, \cdots, t_n)$ and f is AC. If $[flat(s)]' \leq_{he} [t_i]'$ for some i then $[flat(s)]' \leq_{he} [flat(t)]'$. Otherwise, if $[flat(s)]' \leq_{he} [flat(t)]'$ then there is a one-to-one mapping ϕ from integers to integers such that $[s_i]' \leq_{he} [t_{\phi(i)}]'$ for all i, $1 \leq i \leq m$. To test if such a mapping exists, we recursively compute all self embedding relations between

$[s_i]'$ and $[t_j]'$. This reduces the problem to a matching problem on a bipartite graph, which can be solved in polynomial time using known algorithms (Gabow[76]). Therefore it is not necessary to explicity enumerate [flat(s)]' or [flat(t)]' to decide whether [flat(s)]' $\leq_{he}$ [flat(t)]'. We do not know the complexity of the full test $[s] \leq_{he} [t]$ for AC operators but it may be NP-complete.

4. Time and space complexity

We now describe methods by which the nontermination test may be implemented fairly efficiently for non-equational term rewriting. First, we assume terms are represented as integers in a unique way; that is, the same term is represented by the same integer wherever it occurs. This representation may be obtained by a kind of "hash-cons" mechanism in expected constant time per term. We can represent terms as list structures in a straightforward way. Thus, f(a, b) is represented as the list (f a b), which is the list structure CONS(f, CONS(a, CONS(b, NIL))). Then these list structures can be represented as integers in a unique way, so that the same list structure is represented as the same integer wherever it occurs. To obtain these integers, the list structures are processed recursively. Atoms A are given unique integer representations I(A). Also, a list structure CONS(x, y) is processed by first processing x and y, then giving an integer representation I(CONS(x, y)) to CONS(x, y). If CONS(x, y) has already been seen, nothing is done. Otherwise, CONS(x, y) is given a new integer representation, randomly chosen from some range of integers. This information is all stored in a hash table, so that I(x) and I(y) can be computed in expected constant time from I(CONS(x, y)), and I(CONS(x, y)) can be computed in expected constant time from I(x) and I(y).

Whenever a new term is generated during reduction, its integer representation is computed as described above, together with the integer representation of all new subterms of the term. Suppose we are reducing a term r, obtaining the sequence r_1, r_2, $\cdots$, . Whenever r_j is generated, we test for all $i < j$ whether $r_i \leq_{he} r_j$. If some such test succeeds, then this sequence is self-embedding. We compute whether $r_i \leq_{he} r_j$ in a straightforward way, using the definition of homeomorphic embedding given above. However, this computation is done on the numeric representations I(r) of terms r, rather than on the terms r themselves. Also, the computation is done using a "memo function". That is, whenever a computation $r_i \leq_{he} r_j$ is done, the result is remembered, so that if the call $r_i \leq_{he} r_j$ is again encountered, the previous value is returned without any recomputation. In addition, these values are remembered across reduction sequences. Thus, the longer the completion method runs, the more values $r \leq_{he} s$ for various r and s will be stored in a hash table, and so repeated recomputation will be avoided. We may obtain the following complexity bounds on this method for implementing the self-embedding test.

Theorem . Suppose U is the set of all subterms of terms in reduction sequences in the Knuth-Bendix method. Suppose n is the maximum arity of any function symbol. Suppose C is the total number of calls made to the self-embedding test $r \leq_{he} s$; C will be quadratic in the length of the reduction sequences. Then the total expected time for these self-embedding tests is $O(n \cdot |U|^2 + C)$ and the space required is $O(n \cdot |U|^2)$.

5. A further extension

It may make sense to look at reduction sequences other than those that happen to arise during the reduction of critical pairs. This may be done as follows: Suppose a set V of terms is stored at the beginning of the completion method. All homeomorphic embedding relations among elements of V are computed at the beginning. Also, the reducibility relations among elements of V are continually updated as the completion method progresses. That is, suppose R' is the set of rules derived by the completion algorithm so far. Then we keep a table of the values of the relation $s \Rightarrow^{+}_{R'} t$ for s and t in R'. (We write $s \Rightarrow^{+} t$ if t may be obtained from s by one or more reductions.) The maintaining of this table may be done by a kind of "incremental transitive closure" algorithm. If we at any time find s and t such that $s \Rightarrow^{+} t$ and $s \leq_{he} t$ then we detect nontermination. The advantage of this method is that the homeomorphic embedding relation is computed once at the beginning and only the reducibility relation is updated. This may be simpler. Of course, we still need a self-embedding test for reduction sequences that arise in the reduction of critical pairs. Note that if R' is nonterminating then there exist terms s and t such that $s \Rightarrow^{+} t$ and $s \leq_{he} t$. Therefore this method is "complete" in a sense, subject to choice of V.

6. References

Bachmair. L., and Plaisted, D., Termination orderings for associative-commutative rewriting systems, J. Symbolic Computation 1 (1985), 329-349.

Bundy, A., The Computer Modelling of Mathematical Reasoning (Academic Press, New York, 1983).

Dershowitz, N., Hsiang, J., Josephson, A., and Plaisted, D., Associative-commutative rewriting, International Joint Conference on Artificial Intelligence, 1983, pp. 940-944.

Dershowitz, N., Orderings for term-rewriting systems, Theoretical Computer Science 17(1982)279-301.

Dershowitz, N., Applications of the Knuth-Bendix completion procedure, technical report, The Aerospace Corporation, 1983.

Dershowitz, N., Termination of rewriting, Report No. UIUCDCS-R-85-1220, University of Illinois at Urbana-Champaign, August, 1985.

Dershowitz, N., Computing with rewrite systems, Information and Control (1985).

Gabow, H., An efficient implementation of Edmond's algorithm for maximum matching on graphs, J. ACM 23 (1976) 221 - 234.

Hsiang, J. and Dershowitz, N., Rewrite methods for clausal and non-clausal theorem proving, Proc. 10th EATCS Intl. Colloq. on Automata, Languages, and Programming, Barcelona, Spain, 1983.

Huet, G.,and Lankford, D.S., On the uniform halting problem for term rewriting systems, Rapport Laboria 283, IRIA, March, 1978.

Huet, G. and Oppen, D., Equations and rewrite rules: a survey, in Formal Languages: Perspectives and Open Problems (R. Book, ed.), Academic Press, New York, 1980.

Huet, G., Confluent reductions: abstract properties and applications to term rewriting systems, J. ACM 27(1980) 797-821.

Huet, G., A complete proof of correctness of the Knuth-Bendix completion algorithm, J. Computer and System Sciences 23 (1981) 11-21.

Hullot, J.-M., A catalogue of canonical term rewriting systems, SRI Technical Report CSL-113, SRI International, April, 1980.

Jouannaud, J., Lescanne, P., and Reinig, F., Recursive decomposition ordering, Proceedings 2nd IFIP Workshop on Formal Description of Programming Concepts, Garmish Partenkirchin, W. Germany, 1982.

Jouannaud, J.-P., and Munoz, M., Termination of a set of rules modulo a set of equations, Proc. 7th Conf. on Automated Deduction (R. Shostak, Ed.), Lecture Notes in Computer Science 170 (Springer-Verlag, New York, 1984), pp. 175-193.

Kamin, S. and Levy, J.J., Two generalizations of the recursive path ordering, unpublished note, Department of Computer Science, University of Illinois, Urbana, Illinois, February, 1980.

Knuth, D. and Bendix, P., Simple word problems in universal algebras, Computational Problems in Abstract Algebra (J. Leech, ed.), Pergamon Press, Oxford, 1970, pp. 263-297.

Kruskal, J.B., Well-quasi-ordering, the Tree Theorem, and Vazsonyi's conjecture, Transactions of the American Mathematical Society 95 (1960) 210-225.

Lescanne, P., Computer experiments with the REVE term rewriting system generator, 10th POPL Conference, 1983.

Plaisted, D., A recursively defined ordering for proving termination of term-rewriting systems, Dept. of Computer Science Report No. 943, University of Illinois at Urbana-Champaign, 1978.

Plaisted, D., The undecidability of self-embedding for term rewriting systems, Information Processing Letters, February, 1985.

Rusinowitch, M., Plaisted ordering and reursive decomposition ordering revisited, Proceedings of the First International Conference on Rewriting Techniques and Applications, Dijon, France, May, 1985.

Winkler, F., The Church-Rosser Property in Computer Algebra and Special Theorem Proving: An Investigation of Critical Pair, Completion Algorithms, Dissertation der Johannes Kepler-Universitat, Linz (VWGO, 1984).

A New Formula for the Execution of Categorical Combinators

R.D.Lins
Computing Laboratory - Unikent - Canterbury - CT2 7NF - England
Chesf - Comp. Hidro Elétrica do São Francisco - Brazil

Abstract:

Categorical Combinators form a formal system similar to Curry's Combinatory Logic. It was developed by Curien [2] inspired by the equivalence of the theories of typed λ-calculus and Cartesian Closed Categories as shown by Lambek [3] and Scott [8]. In this paper we show how to "execute" Categorical Combinators in an efficient way using a rewriting system. This efficiency is achieved by cutting down the number of laws, by introducing constants in a different way and choosing a more compact notation for the code.
Keywords: Categorical Combinators, lambda calculus, functional programming, complexity.

1.Introduction

Turner [9] showed that a rewriting system based on Curry's combinators gives an efficient implementation basis for functional languages. In a similar way we will analyse the operational behaviour of categorical combinators.

The aim of this paper is to analyse the reductions of Categorical Combinators in order to get a minimal set of rewriting rules necessary to execute the Categorical Combinators compiled code using DeBruijn numbers. In doing so our aim is to reduce the processing time taken to reduce an expression to its normal form, and, as it was shown in [4], get linear size of code with an infinite number of combinators. In [4] the reader can find details of the compilation algorithm, and the analysis of the size complexity of a categorical combinator expression compared with its λ-calculus equivalent.

2.Categorical Combinators as a Rewriting System

In [2] Curien presents several different rewriting systems and their properties. The following system, which he calls CCL_β, simulates λ-calculus β-reductions:

$$
\begin{array}{ll}
(\mathrm{r.1}) & (x \circ y) \circ z \Rightarrow x \circ (y \circ z) \\
(\mathrm{r.2}) & \mathbf{Id} \circ x \Rightarrow x \\
(\mathrm{r.3}) & x \circ \mathbf{Id} \Rightarrow x \\
(\mathrm{r.4}) & \mathit{Fst} \circ \langle x, y\rangle \Rightarrow x \\
(\mathrm{r.5}) & \mathit{Snd} \circ \langle x, y\rangle \Rightarrow y \\
(\mathrm{r.6}) & \langle x, y\rangle \circ z \Rightarrow \langle x \circ z, y \circ z\rangle \\
(\mathrm{r.7}) & \mathit{App} \circ \langle \Lambda(x), y\rangle \Rightarrow x \circ \langle \mathbf{Id}, y\rangle \\
(\mathrm{r.8}) & \Lambda(x) \circ y \Rightarrow \Lambda(x \circ \langle y \circ \mathit{Fst}, \mathit{Snd}\rangle)
\end{array}
$$

We will introduce constants to this system according to the approach presented in [5,7], where constants and functions over constants form a new class of polymorphically typed arrows. The interactions between functions over constants and constants themselves is ruled by application. Using this approach to introduce constants to Categorical Combinators we will need to add just one extra law to the system:

$$(\mathrm{r.9}) \quad c \circ x \Rightarrow c, \quad \textit{where } c \textit{ is a constant}$$

Each function over constants will have particular laws associated with it. For instance, the **addition** of a constant **x** to a constant **y** will be defined by the law:

$$App \circ \langle App \circ \langle add, x\rangle, y\rangle \Rightarrow x + y$$

3.Introducing a New Law

If we have in a categorical combinator expression a sub-expression type $App \circ \langle \Lambda(x) \circ y, z\rangle$ it can be rewritten thus:

$$\begin{aligned}
App \circ \langle \Lambda(x) \circ y, z\rangle &\overset{(r.8)}{\Rightarrow} App \circ \langle \Lambda(x \circ \langle y \circ Fst, Snd\rangle), z\rangle \\
&\overset{(r.7)}{\Rightarrow} (x \circ \langle y \circ Fst, Snd\rangle) \circ \langle \mathbf{Id}, z\rangle \\
&\overset{(r.1)}{\Rightarrow} x \circ (\langle y \circ Fst, Snd\rangle \circ \langle \mathbf{Id}, z\rangle) \\
&\overset{(r.6)}{\Rightarrow} x \circ \langle (y \circ Fst) \circ \langle \mathbf{Id}, z\rangle, Snd \circ \langle \mathbf{Id}, z\rangle\rangle \\
&\overset{(r.1)}{\Rightarrow} x \circ \langle y \circ (Fst \circ \langle \mathbf{Id}, z\rangle), Snd \circ \langle \mathbf{Id}, z\rangle\rangle \\
&\overset{(r.4)}{\Rightarrow} x \circ \langle y \circ \mathbf{Id}, Snd \circ \langle \mathbf{Id}, z\rangle\rangle \\
&\overset{(r.3)}{\Rightarrow} x \circ \langle y, Snd \circ \langle \mathbf{Id}, z\rangle\rangle \\
&\overset{(r.5)}{\Rightarrow} x \circ \langle y, z\rangle
\end{aligned}$$

We will introduce in our rewriting system the following law, which will simulate the sequence of rewritings above:

$$App \circ \langle \Lambda(x) \circ y, z\rangle \Rightarrow x \circ \langle y, z\rangle$$

4.Analysis of the Rewriting Rules

The rewriting strategy will be leftmost-outermost, i.e. we will look from left to right in an expression for the syntactically outermost pattern matching the left hand side of any of the rewriting rules. When we find it this pattern will be rewritten and the rewriting will resume from the outermost level of the new expression. We examine the relationship between leftmost-outermost reduction of λ-terms and of categorical combinators expressions in section 5 below. The possible ambiguities of application of rules will be analysed later.

We will call an expression which is the Categorical combinator translation of a λ-expression a **λ-equivalent** expression. An **intermediate expression** is a non λ-equivalent categorical combinator expression which we get by rewriting a λ-equivalent expression or another intermediate expression.

In order to reduce the processing time necessary for pattern matching in the execution of categorical combinators we aim to remove rules which are not necessary for the execution of the code. Working with variables in DeBruijn notation brings us the advantage of getting linear size expressions in relation to their λ-equivalent.

4.1 Removing (r.8)

$$\Lambda(x) \circ y \Rightarrow \Lambda(x \circ \langle y \circ Fst, Snd\rangle)$$

If we examine our set of rewriting laws we see the only way of removing Λ's from our code is rewriting via

$$App \circ \langle \Lambda(x), y\rangle \Rightarrow x \circ \langle \mathbf{Id}, y\rangle$$
$$App \circ \langle \Lambda(x) \circ y, z\rangle \Rightarrow x \circ \langle y, z\rangle$$

In practical implementations of functional programming languages the result of the execution of a correct program is of ground type, i.e. not of function type or containing an embedded function type. If we assume the result of the evaluation of an expression is of ground type, in the result of a rewritten expression there are no Λ's. For each Λ combinator in an expression there will be an *App* combinator in a context type the lefthandside of one of the rules above. We can conclude that every time a subexpression $\Lambda(x) \circ y$ appears it will be syntatically enclosed in a context rewritable in a leftmost-outermost strategy via

$$App \circ \langle \Lambda(x) \circ y, z\rangle \Rightarrow x \circ \langle y, z\rangle$$

This allows us to delete rule (r.8) from our set of rewriting laws.

4.2 Replacing Variables by DeBruijn Numbers

In a λ-equivalent expression the *Fst* combinator just appears in the construction of variables. A variable $n > 0$ is represented thus:

$$n = (\ldots(Snd \circ \underbrace{Fst) \circ Fst) \circ \cdots \circ Fst)}_{n}$$

A DeBruijn number is a complex expression built up from subexpressions as shown above. A non-trivial interaction of such an expression with a rule is a case where variables (on the lefthand side of the rule) are matched with sub-expressions of the expression, rather than the expression itself. The possible non-trivial interaction between a variable and the set of rules we have are the cases in which execution will depend on decomposing a variable via associativity, thus we have: (We are interested in the case where $n \neq 0$.)

$$\text{(r.1)} \quad ((n-1) \circ Fst) \circ z \Rightarrow (n-1) \circ (Fst \circ z)$$

$$\text{(r.6)} \begin{cases} \langle n, y\rangle \circ z \Rightarrow \langle n \circ z, y \circ z\rangle \overset{(r.1)}{\Rightarrow} \langle (n-1) \circ (Fst \circ z), y \circ z\rangle \\ \langle x, n\rangle \circ z \Rightarrow \langle x \circ z, n \circ z\rangle \overset{(r.1)}{\Rightarrow} \langle x \circ z, (n-1) \circ (Fst \circ z)\rangle \end{cases}$$

$$\text{(r.7)} \begin{cases} App \circ \langle \Lambda(n), y\rangle \Rightarrow n \circ \langle \mathbf{Id}, y\rangle = ((n-1) \circ Fst) \circ \langle \mathbf{Id}, y\rangle \\ \qquad \overset{(r.1)}{\Rightarrow} (n-1) \circ (Fst \circ \langle \mathbf{Id}, y\rangle) \\ \qquad \overset{(r.4)}{\Rightarrow} (n-1) \circ \mathbf{Id} \\ \qquad \overset{(r.3)}{\Rightarrow} (n-1) \\ App \circ \langle \Lambda(x), n\rangle \Rightarrow x \circ \langle \mathbf{Id}, n\rangle \end{cases}$$

$$\text{(r.10)} \begin{cases} App \circ \langle \Lambda(n) \circ y, z\rangle \Rightarrow n \circ \langle y, z\rangle = ((n-1) \circ Fst)\langle y, z\rangle \\ \qquad \overset{(r.1)}{\Rightarrow} (n-1) \circ (Fst \circ \langle y, z\rangle) \\ \qquad \overset{(r.4)}{\Rightarrow} (n-1) \circ y \end{cases}$$

By observing the behaviour of a variable in the expressions above we see that when a variable appears the expression is either not reduced or will always follow the path, if $n > 0$

$$\begin{aligned} n \circ \langle x, y \rangle &= ((n-1) \circ Fst) \circ \langle x, y \rangle \\ &\overset{(r.1)}{\Rightarrow} (n-1) \circ (Fst \circ \langle x, y \rangle) \\ &\overset{(r.4)}{\Rightarrow} (n-1) \circ x \end{aligned}$$

Since this is the only possible reduction of a *Fst* combinator or a variable in an expression we will replace

$$Fst \circ \langle x, y \rangle \Rightarrow x \qquad , by$$

$$n \circ \langle x, y \rangle \Rightarrow (n-1) \circ x \qquad , if\ n > 0$$

and use the DeBruijn number 0 to represent the *Snd* combinator, so (r.5) will be,

$$0 \circ \langle x, y \rangle \Rightarrow y$$

4.3 New Set of Rewriting Laws

With the alterations of previous subsections our set of rewriting laws now is :

(r'.1) $(x \circ y) \circ z \Rightarrow x \circ (y \circ z)$

(r'.2) $\mathbf{Id} \circ x \Rightarrow x$

(r'.3) $x \circ \mathbf{Id} \Rightarrow x$

(r'.4) $n \circ \langle x, y \rangle \Rightarrow (n-1) \circ x,\ if\ n > 0$

(r'.5) $0 \circ \langle x, y \rangle \Rightarrow y$

(r'.6) $\langle x, y \rangle \circ z \Rightarrow \langle x \circ z, y \circ z \rangle$

(r'.7) $App \circ \langle \Lambda(x), y \rangle \Rightarrow x \circ \langle \mathbf{Id}, y \rangle$

(r'.8) $App \circ \langle \Lambda(x) \circ y, z \rangle \Rightarrow x \circ \langle y, z \rangle$

(r'.9) $c \circ x \Rightarrow c$, *where* c *is a constant*

4.4 Application as a Pair Constructor

In a λ-equivalent expression the *App* combinator always appears composed with pairs. Let us analyse the interactions of expressions involving the *App* combinator, with the set of rewriting rules we have got. We can see that the only non trivial match (a match in which the expression $App \circ \langle\ ,\ \rangle$ is not matched with a variable but with a non-trivial sub-expression) is with associativity.

$$\begin{aligned} (App \circ \langle x, y \rangle) \circ z &\overset{(r'.1)}{\Rightarrow} App \circ (\langle x, y \rangle \circ z) \\ &\overset{(r'.6)}{\Rightarrow} App \circ \langle x \circ z, y \circ z \rangle \end{aligned}$$

As we can see the *App* combinator in an expression, either disappears via (r'.7) or (r'.8), or is not reduced, or follows the path above. The sequence of reductions above is the only case of use of (r'.6), if the expression is not of product type. This will allow us to introduce a new pair $\triangleleft\ ,\ \triangleright$ to represent $App \circ \langle\ ,\ \rangle$, and replace (r'.6) by

$$\triangleleft x, y \triangleright \circ z \Rightarrow \triangleleft x \circ z, y \circ z \triangleright$$

Now our set of rules is:

$$
\begin{array}{ll}
(\mathrm{r}".1) & (x \circ y) \circ z \Rightarrow x \circ (y \circ z) \\
(\mathrm{r}".2) & \mathbf{Id} \circ x \Rightarrow x \\
(\mathrm{r}".3) & x \circ \mathbf{Id} \Rightarrow x \\
(\mathrm{r}".4) & n \circ \langle x, y\rangle \Rightarrow (n-1) \circ x, \ if\ n > 0 \\
(\mathrm{r}".5) & 0 \circ \langle x, y\rangle \Rightarrow y \\
(\mathrm{r}".6) & \lhd x, y \rhd \circ z \Rightarrow \lhd x \circ z, y \circ z \rhd \\
(\mathrm{r}".7) & \lhd \Lambda(x), y \rhd \Rightarrow x \circ \langle \mathbf{Id}, y\rangle \\
(\mathrm{r}".8) & \lhd \Lambda(x) \circ y, z \rhd \Rightarrow x \circ \langle y, z\rangle \\
(\mathrm{r}".9) & c \circ x \Rightarrow c, \quad where\ \mathbf{c}\ is\ a\ constant
\end{array}
$$

4.5 A New Compilation Algorithm

The modifications presented in last sections, and the way we work with constants [5,7] will allow us to use a simplified compilation algorithm from DeBruijn λ-calculus to Categorical Combinators.

$$
\begin{array}{rl}
[\![\lambda.a]\!] & \rightarrow \Lambda([\![a]\!]) \\
[\![ab]\!] & \rightarrow \lhd [\![a]\!], [\![b]\!] \rhd \\
[\![n]\!] & \rightarrow n, \quad where\ n\ is\ a\ variable \\
[\![c]\!] & \rightarrow c, \quad where\ c\ is\ a\ constant
\end{array}
$$

4.6 Removing (r".1)

$$(x \circ y) \circ z \Rightarrow x \circ (y \circ z)$$

Let us analyse the interactions between the set of rules we have got to see how a λ-equivalent or intermediate expression can match the lefthandside of (r".1). In a λ-equivalent expression there are no compositions. We can see that (r".2), (r".3), (r".4), (r".5), and (r".9) do not introduce any composition. Rule (r".4) works with the DeBruijn number itself as an atomic combinator, therefore no reduction is possible at this level unless the expression **x**, in (r".4) is a pair. The rules (r".7) and (r".8) themselves do not generate an expression type $(x \circ y) \circ z$, but it is necessary to analyse if this pattern can arise by interactions between (r".6) and (r".7) or (r".8). So we have

$$
\begin{array}{ll}
\text{(i)} & \lhd \Lambda(x), y \rhd \circ z \overset{(\mathrm{r}''.6)}{\Rightarrow} \lhd \Lambda(x) \circ z, y \circ z \rhd \\
\text{(ii)} & \lhd \Lambda(x) \circ y, z \rhd \circ w \overset{(\mathrm{r}''.6)}{\Rightarrow} \lhd (\Lambda(x) \circ y) \circ w, z \circ w \rhd
\end{array}
$$

The first expression is of type (r".8), where **x** and **y** are λ-equivalent. The latter expression introduces a pattern $(x \circ y) \circ z$. Let us analyse how this pattern can be generated. Compositions are introduced in the code only by (r".7)

$$\lhd \Lambda(x), y \rhd \Rightarrow x \circ \langle \mathbf{Id}, y\rangle$$

We can see that the only way of generating an expression type the lefthandside of (ii), using the leftmost-outermost strategy, is by

$$
\begin{array}{rl}
\lhd \Lambda(\lhd \Lambda(x) \circ y, z \rhd), w \rhd & \overset{(\mathrm{r}''.7)}{\Rightarrow} \lhd \Lambda(x) \circ y, z \rhd \circ \langle \mathbf{Id}, w\rangle \\
& \overset{(\mathrm{r}''.6)}{\Rightarrow} \lhd (\Lambda(x) \circ y) \circ \langle \mathbf{Id}, w\rangle, z' \rhd
\end{array}
$$

But the sequence of reductions above is not valid since it is not obtainable by a leftmost-outermost strategy. The first composition was generated by an innermost reduction instead of the outermost one. We can conclude that the pattern $(x \circ y) \circ z$ will never be generated, therefore we can delete (r".1) from our set of rules.

4.7 Deleting (r".2)

$$\mathbf{Id} \circ x \Rightarrow x$$

The **Id** combinator is just introduced by (r".7),

$$\lhd \Lambda(x), y \rhd \Rightarrow x \circ \langle \mathbf{Id}, y \rangle$$

where **x** is a λ-equivalent expression. Analysing the possible reductions of expressions with **Id** we can observe that the **Id** combinator will always appear in the righthandside when interacts with compositions. Since we perform no associative rewrites this relative position will remain unchanged. For this reason we can remove (r".2) from our set of rules.

4.8 Removing (r".3)

$$x \circ \mathbf{Id} \Rightarrow x$$

The number a variable assumes in DeBruijn's λ-calculus [1] is equal to the number of binders (λ's) which are between the variable itself and the λ by which it is bound in the parse tree of a λ-expression. This is equivalent to saying that in a λ-equivalent expression in categorical combinators for each variable **n** there will be at least **(n+1)** nested Λ combinators syntactically enclosing it. By observing the set of rules we have, one can see that the only way of removing Λ's from an expression is by (r".7) and (r".8), where applicative pairs are replaced by ordinary pairs. This means that for a variable **n** to reach the leftmost-outermost position not less than **(n+1)** nesting levels of ordinary pairs were created.

As was said before the **Id** combinator is just introduced by (r".7)

$$\lhd \Lambda(x), y \rhd \Rightarrow x \circ \langle \mathbf{Id}, y \rangle$$

Since this is the only rule which matches with a λ-equivalent expression or subexpression the pair $\langle \mathbf{Id}, y \rangle$ will always be the first ordinary pair to appear during the execution, and it will take the innermost position in the nesting structure of ordinary pairs created, as we can see in the example below. Therefore we can say that when a variable **n** reaches the leftmost-outermost position we will have at least **(n+1)** nested levels of ordinary pairs where the **Id** combinator is in the lefthandside of the innermost one. This implies that we will have, in the worst case,

$$n \circ \underbrace{\langle \cdots \langle \mathbf{Id}, a \rangle, b \rangle \cdots \rangle}_{(n+1)\ levels}$$

reducing to

$$0 \circ \langle \mathbf{Id}, a \rangle \Rightarrow a$$

Let us make an example. The λ-expression $(((\lambda x.\lambda y.x(yx))m)n)$ will be translated into categorical combinators thus:

$$\Rightarrow \lhd \lhd \Lambda(\Lambda(\lhd 1, \lhd 0, 1 \rhd \rhd)), m \rhd, n \rhd$$
$$\stackrel{(r''.7)}{\Rightarrow} \lhd \Lambda(\lhd 1, \lhd 0, 1 \rhd \rhd) \circ \langle \mathbf{Id}, m \rangle, n \rhd$$
$$\stackrel{(r''.8)}{\Rightarrow} \lhd 1, \lhd 0, 1 \rhd \rhd \circ \langle \langle \mathbf{Id}, m \rangle, n \rangle$$
$$\stackrel{(r''.6)}{\Rightarrow} \lhd 1 \circ \langle \langle \mathbf{Id}, m \rangle, n \rangle, \lhd 0, 1 \rhd \circ \langle \langle \mathbf{Id}, m \rangle, n \rangle \rhd$$
$$\stackrel{(r''.6)}{\Rightarrow} \lhd 1 \circ \langle \langle \mathbf{Id}, m \rangle, n \rangle, \lhd 0 \circ \langle \langle \mathbf{Id}, m \rangle, n \rangle, 1 \circ \langle \langle \mathbf{Id}, m \rangle, n \rangle \rhd \rhd$$

We can interpret this behaviour in the following way: During execution there is a formation of an "environment" and to each variable is associated a local "environment". The rôle of the **Id** combinator is just a place holder in the formation of an environment without any active meaning. For this reason we will delete (r".2) from our set of rules.

4.9 The Minimal Set of Rules

A set of rewriting rules is minimal, in relation to a compilation algorithm, if all the rewriting laws are necessary in the execution of the code. Due to the simplifications introduced in the previous sections our new set of rules is :

$$\begin{array}{ll}
(\mathrm{R.1}) & n \circ \langle x, y\rangle \Rightarrow (n-1) \circ x \\
(\mathrm{R.2}) & 0 \circ \langle x, y\rangle \Rightarrow y \\
(\mathrm{R.3}) & \lhd x, y \rhd \circ z \Rightarrow \lhd x \circ z, y \circ z \rhd \\
(\mathrm{R.4}) & \lhd \Lambda(x), y \rhd \Rightarrow x \circ \langle \mathbf{Id}, y\rangle \\
(\mathrm{R.5}) & \lhd \Lambda(x) \circ y, z \rhd \Rightarrow x \circ \langle y, z\rangle \\
(\mathrm{R.6}) & c \circ x \Rightarrow c, \quad \textit{where } \mathbf{c} \textit{ is a constant}
\end{array}$$

Let us show an example to illustrate that we need all the rules of this set. If we have the λ-expression $(((\lambda x.\lambda y.(xy))(\lambda z.c))n)$, where **c** is a constant, this expression will be translated into categorical combinators according to the algorithm of section (4.5) as

$$\lhd \lhd \Lambda(\Lambda(\lhd 1, 0 \rhd)), \Lambda(c) \rhd, n \rhd$$

which reduces to,

$$\begin{array}{l}
\overset{(\mathrm{R.4})}{\Rightarrow} \lhd \Lambda(\lhd 1, 0 \rhd) \circ \langle \mathbf{Id}, \Lambda(c)\rangle, n \rhd \\
\overset{(\mathrm{R.5})}{\Rightarrow} \lhd 1, 0 \rhd \circ \langle\langle \mathbf{Id}, \Lambda(c)\rangle, n\rangle \rhd \\
\overset{(\mathrm{R.3})}{\Rightarrow} \lhd 1 \circ \langle\langle \mathbf{Id}, \Lambda(c)\rangle, 0\langle\langle \mathbf{Id}, \Lambda(c)\rangle, n\rangle \rhd \\
\overset{(\mathrm{R.1})}{\Rightarrow} \lhd 0 \circ \langle \mathbf{Id}, \Lambda(c)\rangle, 0 \circ \langle\langle \mathbf{Id}, \Lambda(c)\rangle, n\rangle \rhd \\
\overset{(\mathrm{R.2})}{\Rightarrow} \lhd \Lambda(c), n \rhd \\
\overset{(\mathrm{R.4})}{\Rightarrow} c \circ \langle \mathbf{Id}, n\rangle \\
\overset{(\mathrm{R.6})}{\Rightarrow} c
\end{array}$$

As we can see all the rules are needed in the execution of the code. This allows us to say that, using the compilation algorithm in section 4.5, this set of rules is minimal.

5. Reduction Order

Rewriting categorical combinators in a leftmost-outermost strategy we simulate leftmost outermost β-reduction in the λ-calculus. If we have a general λ-expression type

$$(\lambda x_0 \lambda x_1 \ldots \lambda x_n.(y_0 y_1 \ldots y_m))\ z_0\ z_1 \ldots z_n\ z_{n+1} \ldots z_k$$

leftmost-outermost β-reduction will lead us to:

$$([z_0/x_0](\lambda x_1 \ldots \lambda x_n.(y_0 y_1 \ldots y_m)))\ z_1 \ldots z_n\ z_{n+1} \ldots z_k$$

where variable substitution $[z_i/x_i]$ is defined by:

$$\begin{aligned}[z_i/x_i](\lambda x_j.y) &= \lambda x_j.([z_i/x_i]y)\ if\ j \neq i\\ &= \lambda x_j.y\\ [z_i/x_i](y_0\ y_1) &= ([z_i/x_i]y_0)\ ([z_i/x_i]y_1)\\ [z_i/x_i]y &= z_i\ if\ y = x_i\\ &= y\end{aligned}$$

Instead of performing the full substitution as described by the equations, we can see the equations themselves as rewriting rules for a system which explicitly involves a substitution operator

$$[z_i/x_i]\ y$$

for example

$$[z_i/x_i](y_0\ y_1) \Rightarrow ([z_i/x_i]y_0)\ ([z_i/x_i]y_1)$$

In such a system the substitution is only performed on demand. It is such a mechanism which is implemented by the rewriting rules for categorical combinators. The λ-expression above will be successively leftmost-outermost rewritten to:

$$([z_n/x_n](\ldots([z_1/x_1]([z_0/x_0](y_0 y_1 \ldots y_m)))\ldots))\ z_{n+1}\ldots z_k$$

which leftmost-outermost rewrites to

$$([z_n/x_n](\ldots([z_0/x_0]y_0)))\ ([z_n/x_n](\ldots([z_0/x_0]y_1)))\ldots([z_n/x_n](\ldots([z_0/x_0]y_m)))\ z_{n+1}\ldots z_k$$

Translating the original λ-expression above into categorical combinators we get

$$\lhd\ldots\lhd\ \lhd\mathbf{\Lambda}(\mathbf{\Lambda}(\ldots(\mathbf{\Lambda}(\lhd\ \lhd\ldots\lhd\ y_0, y_1\ \rhd\ldots\ \rhd\ \rhd))\ldots)), z_0\ \rhd, z_1\ \rhd\ldots\ \rhd\ z_n\ \rhd\ldots\ \rhd, z_k\ \rhd$$

Rewriting this expression in a leftmost-outermost strategy we will generate an "environment", as analysed in section 4.8, leading us to an expression type

$$\lhd\ldots\lhd\ (\lhd\ldots\lhd\ y_0, y_1\ \rhd,\ldots y_m\ \rhd\ \circ\langle\ldots\langle\mathbf{Id}, z_0\rangle, z_1\rangle,\ldots z_n\rangle), z_{n+1}\ \rhd,\ldots\ \rhd\ z_k\ \rhd$$

the "environment" will be distributed via (R.3) successively until reaching an expression of type

$$\lhd\ldots\lhd\ (y_0\circ\langle\ldots\langle\mathbf{Id}, z_0\rangle,\ldots z_n\rangle), (y_1\circ\langle\ldots\langle\mathbf{Id}, z_0\rangle,\ldots z_n\rangle)\ \rhd,\ldots(y_n\circ\langle\ldots\langle\mathbf{Id}, z_0\rangle,\ldots z_n\rangle)\ \rhd,\ldots\ \rhd$$

The sequence of nested substitutions in λ-calculus are equivalent to the notion of the "environment" in categorical combinators. The effective substitution performed in the λ-calculus using the rewriting rules

$$\begin{aligned}[z_i/x_i]y &\Rightarrow z_i\quad if\ \ y = x_i\\ &\Rightarrow y\end{aligned}$$

are simulated in categorical combinators by composing a variable with its "environment". For example, rewriting the λ-expression:

$$\begin{aligned}(\lambda x\lambda y.yx)\ a\ b &\Rightarrow ([a/x](\lambda y.yx))\ b\\ &\Rightarrow (\lambda.([a/x](yx))\ b\\ &\Rightarrow [b/y]([a/x](yx))\\ &\Rightarrow [b/y](([a/x]y)([a/x]x))\\ &\Rightarrow ([b/y]([a/x]y))([b/y]([a/x]x))\\ &\Rightarrow ([b/y]y)([b/y]([a/x]x))\\ &\Rightarrow b([b/y]([a/x]x))\\ &\Rightarrow b([b/y]a)\\ &\Rightarrow ba\end{aligned}$$

The categorical equivalent of the expression above will lead us to the following leftmost-outermost sequence of rewritings:

$$\begin{aligned}
\lhd \lhd \Lambda(\Lambda(\lhd 0,1 \rhd)), a \rhd, b \rhd &\Rightarrow \lhd(L(\lhd 0,1 \rhd)) \circ \langle \mathbf{Id}, a\rangle, b \rhd \\
&\Rightarrow \lhd 0,1 \rhd \circ \langle\langle \mathbf{Id}, a\rangle, b\rangle \\
&\Rightarrow \lhd 0 \circ \langle\langle \mathbf{Id}, a\rangle, b\rangle, \mathit{1} \circ \langle\langle \mathbf{Id}, a\rangle, b\rangle \rhd \\
&\Rightarrow \lhd b, 1 \circ \langle\langle \mathbf{Id}, a\rangle, b\rangle \rhd \\
&\Rightarrow \lhd b, 0 \circ \langle \mathbf{Id}, a\rangle \rhd \\
&\Rightarrow \lhd b, a \rhd
\end{aligned}$$

As one can observe in the two sequences of reductions above $[b/y]([a/x]y)$ and $0 \circ \langle\langle \mathbf{Id}, a\rangle, b\rangle$ play the same rôle in their own systems.

There are two cases in which substitutions are not completely performed. As we will see with examples, in these cases there are symmetries between the behaviour of rewritings in the λ-calculus and its categorical counterpart. The first case is the one in which the expression is a selector. For instance (where a is a constant) :

$$\begin{aligned}
(\lambda x.xx)\ (\lambda y\lambda z.z)\ a &\Rightarrow ([(\lambda y\lambda z.z)/x](xx))\ a \\
&\Rightarrow ([(\lambda y\lambda z.z)/x]x)([(\lambda y\lambda z.z)/x]x)\ a \\
&\Rightarrow (\lambda y\lambda z.z)([(\lambda y\lambda z.z)/x]x)\ a \\
&\Rightarrow ([([(\lambda y\lambda z.z)/x]x)/y](\lambda z.z))\ a \\
&\Rightarrow (\lambda z.([([(\lambda y\lambda z.z)/x]x)/y]z))\ a \\
&\Rightarrow [([([(\lambda y\lambda z.z)/x]x)/y]z)/z]a \\
&\Rightarrow a
\end{aligned}$$

The innermost substitution dissapears, therefore will never be performed. Now let us see how this happens in categorical combinators.

$$\begin{aligned}
\lhd \lhd \Lambda(\lhd 0,0 \rhd), \Lambda(\Lambda(0)) \rhd, a \rhd &\Rightarrow \lhd \lhd 0,0 \rhd \circ \langle \mathbf{Id}, \mathit{\Lambda}(\mathit{\Lambda}(\mathit{0}))\rangle, a \rhd \\
&\Rightarrow \lhd \lhd 0 \circ \langle \mathbf{Id}, \mathit{\Lambda}(\mathit{\Lambda}(\mathit{0}))\rangle, \mathit{0} \rhd \circ \langle \mathbf{Id}, \mathit{\Lambda}(\mathit{\Lambda}(\mathit{0}))\rangle, a \rhd \\
&\Rightarrow \lhd \lhd \Lambda(\Lambda(0)), 0 \rhd \circ \langle \mathbf{Id}, \mathit{\Lambda}(\mathit{\Lambda}(\mathit{0}))\rangle, a \rhd \\
&\Rightarrow \lhd \Lambda(0) \circ \langle \mathbf{Id}, \mathit{0} \rhd \circ \langle \mathbf{Id}, \mathit{\Lambda}(\mathit{\Lambda}(\mathit{0}))\rangle\rangle, a \rhd \\
&\Rightarrow 0 \circ \langle\langle \mathbf{Id}, \mathit{0} \rhd \circ \langle \mathbf{Id}, \mathit{\Lambda}(\mathit{\Lambda}(\mathit{0}))\rangle\rangle, a\rangle \\
&\Rightarrow a
\end{aligned}$$

Expressions without a normal form are the second case, in which some substitutions may never take place, rewriting the code in a leftmost-outermost sequence. For example :

$$\begin{aligned}
(\lambda x.xxx)(\lambda x.xxx) &\Rightarrow [(\lambda x.xxx)/x]xxx \\
&\Rightarrow ([(\lambda x.xxx)/x]xx)([(\lambda x.xxx)/x]x) \\
&\Rightarrow ([(\lambda x.xxx)/x]x)([(\lambda x.xxx)/x]x)([(\lambda x.xxx)/x]x) \\
&\Rightarrow (\lambda x.xxx)([(\lambda x.xxx)/x]x)([(\lambda x.xxx)/x]x) \\
&\Rightarrow ([([(\lambda x.xxx)/x]x)/x]xxx)([(\lambda x.xxx)/x]x)
\end{aligned}$$

As we can observe the left sub-expression above will grow and under leftmost-outermost rewriting the rightmost substitution will never be performed. The categorical combinator equivalent of the expression above will have a similar behaviour, as follows

$$\lhd \Lambda(\lhd \lhd 0,0 \rhd,0 \rhd), \Lambda(\lhd \lhd 0,0 \rhd,0 \rhd) \rhd$$
$$\Rightarrow \lhd \lhd 0,0 \rhd,0 \rhd \circ \langle \mathbf{Id}, \Lambda(\lhd \lhd 0,0 \rhd,0 \rhd)\rangle$$
$$\Rightarrow \lhd \lhd 0,0 \rhd \circ \langle \mathbf{Id}, \Lambda(\lhd \lhd 0,0 \rhd,0 \rhd)\rangle, 0 \circ \langle \mathbf{Id}, \Lambda(\lhd \lhd 0,0 \rhd,0 \rhd)\rangle \rhd$$
$$\Rightarrow \lhd \lhd 0 \circ \langle \mathbf{Id}, \Lambda(\lhd \lhd 0,0 \rhd,0 \rhd)\rangle, 0 \circ \langle \mathbf{Id}, \Lambda(\lhd \lhd 0,0 \rhd,0 \rhd)\rangle \rhd, \quad 0 \circ \langle \mathbf{Id}, \Lambda(\lhd \lhd 0,0 \rhd,0 \rhd)\rangle \rhd$$
$$\Rightarrow \lhd \lhd \Lambda(\lhd \lhd 0,0 \rhd,0 \rhd), 0 \circ \langle \mathbf{Id}, \Lambda(\lhd \lhd 0,0 \rhd,0 \rhd)\rangle \rhd, 0 \circ \langle \mathbf{Id}, \Lambda(\lhd \lhd 0,0 \rhd,0 \rhd)\rangle \rhd$$
$$\Rightarrow \lhd \lhd \lhd 0,0 \rhd,0 \rhd) \circ \langle \mathbf{Id}, 0 \circ \langle \mathbf{Id}, \Lambda(\lhd \lhd 0,0 \rhd,0 \rhd)\rangle\rangle, 0 \circ \langle \mathbf{Id}, \Lambda(\lhd \lhd 0,0 \rhd,0 \rhd)\rangle \rhd$$

The rewriting of the rightmost expression in the applicative pair will never occur, because the leftmost expression will grow indefinitely.

6.Conclusions

Reducing the set of rules and using DeBruijn numbers to represent variables should mean that we need less time and space to execute categorical combinators as a rewriting system, because less pattern matching is necessary, and we have a linear relation between source and translated codes. This is borne out by experimental results in [6].

Acknowledgements

The author owes very much to his supervisor, Dr. Simon Thompson, for several discussions, suggestions, and comments.

The financial support for this work was provided by CNPq, Brazil, Grant 20.2744/84.

References

1. N.G.DeBruijn, Lambda Calculus Notation with Nameless Dummies, a Tool for Automatic Formula Manipulation, Indag.Math. 34, 381-392 (1972).
2. P-L.Curien, Categorical Combinators, Sequential Algorithms and Functional Programming, Draft version of a monograph submitted to publication, Université Paris VII, LITP (1985).
3. J.Lambek, From Lambda-calculus to Cartesian Closed Categories, in To H.B.Curry: Essays on Combinatory Logic, Lambda-Calculus and Formalism, ed J.P.Seldin and J.R.Hindley, Academic Press (1980).
4. R.D.Lins, The Complexity of a translation of λ-calculus to Categorical Combinators, Computing Lab. Report N. 27 - The University of Kent at Canterbury (1985).
5. R.D.Lins, A New Way of Introducing Constants in Categorical Combinators, privately circulated, The University of Kent at Canterbury (1985).
6. R.D.Lins, On The Efficiency of Categorical Combinators as a Rewriting System, Computing Lab.Report N.34 - The University of Kent at Canterbury (1985).
7. R.D.Lins, Constants and Higher-Order Data Types in Categorical Combinators, in preparation.
8. D.Scott, Relating Theories of the Lambda-Calculus, in To H.B.Curry: Essays on Combinatory Logic, Lambda-Calculus and Formalism, ed. J.P.Seldin and J.R.Hindley, Academic Press (1980).
9. D.A.Turner, A New Implementation Technique for Applicative Languages, Software Practice and Experience, Vol 9, 31-49 (1979).

PROOF BY INDUCTION USING TEST SETS

Deepak Kapur, Paliath Narendran, and
Corporate Research & Development
General Electric Co.
Schenectady, NY, USA

Hantao Zhang
Dept. of Computer Science
Rensselaer Polytechnic Institute
Troy, NY, USA

Abstract

A new method for proving an equational formula by induction is presented. This method is based on the use of the Knuth-Bendix completion procedure for equational theories, and it does not suffer from limitations imposed by the inductionless induction methods proposed by Musser and Huet and Hullot. The method has been implemented in RRL, a Rewrite Rule Laboratory. Based on extensive experiments, the method appears to be more practical and efficient than a recently proposed method by Jouannaud and Kounalis. Using ideas developed for this method, it is also possible to check for sufficient completeness of equational axiomatizations.

Key Words: Inductionless Induction, Proof by Induction, Equational Theory, Knuth-Bendix Completion Procedure, Consistency, Sufficient-Completeness, Induction.

---

† Partially supported by the National Science Foundation Grant no. DCR-8408461.

PROOF BY INDUCTION USING TEST SETS

1. Introduction

We present a new method for proving inductive properties using the Knuth-Bendix completion procedure [12]. For an equational theory having a complete (canonical) term rewriting system, proving an equation by induction reduces to comparing its *test set*, which is a finite description of irreducible ground terms of a rewriting system, with the test set of the complete rewriting system generated after adding the equation being proved to the original complete rewriting system. This method has an advantage over the methods proposed by Musser [14], Goguen [2], and Huet and Hullot [5] as it does not require an equality predicate and also that constructor symbols can be related using equations in an axiomatization. Further, there is no need to distinguish between constructors and non-constructors, although making that distinction can improve the efficiency of the method to prove or disprove an equation. The method is as powerful as a method based on the concept of quasi-reducibility recently proposed by Jouannaud and Kounalis [7]. We have implemented both Jouannaud and Kounalis's method as well as our method in RRL, a Rewrite Rule Laboratory [10, 11]. Our experience with a number of examples so far indicates that our method is more practical and efficient than Jouannaud and Kounalis's method.

We first briefly review the approach for proving inductive properties using the Knuth-Bendix completion procedure, which is also known as the *inductionless induction* approach. Section 2 introduces preliminary concepts needed in the paper. In Section 3, we discuss how the set of irreducible ground terms of a complete term rewriting system can be used for proving properties by induction using the Knuth-Bendix completion procedure. In Section 4, we introduce the concept of a test set of a rewriting system and discuss its properties in relation to the set of irreducible ground terms of a rewriting system. The next section is a discussion of how to compute the test set for a rewriting system. In Section 6, we discuss in considerable detail a practical method for computing test sets of left-linear complete rewriting systems. This test set can also be used to check quasi-reducibility and is more efficient than a method in [1,7,13]. Section 7 is a comparison of the test set method and Jouannaud and Kounalis's method as implemented in RRL; we discuss some examples showing why our method appears to be more practical than Jouannaud and Kounalis's method.

1.1 Overview of the Inductionless Induction Method

Musser [14] showed that for an equational axiomatization that satisfies a certain completeness property, which he called the *full specification property*, proving an equation by induction is equivalent to checking the consistency of the extended theory obtained by adding the equation to be proved to the equational axiomatization itself. He also observed that the consistency of an equational theory can be decided if we are able to generate a complete system for it, say using the Knuth-Bendix completion procedure [14, 15]. He required the equality predicate in an equational axiomatization so that the inconsistency of a theory is equivalent to generating the rule *true* $\rightarrow$ *false*.

This method was further refined by Goguen [2] and Huet and Hullot [5]. Huet and Hullot in particular classified function symbols into constructors and non-constructors, and for theories with free constructors, defined inconsistency of a theory to be the one in which two distinct ground terms built from constructors were made equivalent.

Dershowitz [1] related the inductionless induction method to the set of irreducible ground terms of a rewriting system. He showed that given an equational axiomatization that has a complete set of rewrite rules **R**, an equation $t \equiv u$ is a theorem by induction in **R** if and only if the set of irreducible ground terms of a complete system obtained from $\mathbf{R} \cup \{t \rightarrow u\}$ remains the same as that of **R**.

Kapur and Musser [8] generalized these results by introducing the notion of a system with three components - L, a language; C, a subset of ground terms such that making any two of them equivalent causes inconsistency; **E**, a set of equations. While using the rewriting approach for checking for consistency, C can be viewed as a subset of the set of irreducible ground terms with respect to a rewriting system **R** for **E**. The work presented in the present paper is closely related to [1, 8].

The recent work of Jouannaud and Kounalis [7] is discussed later in the paper.

2. Preliminaries

For simplicity and without any loss of generality, we will consider untyped (single-sorted) equational axiomatizations and term rewriting systems. The results reported in the paper also apply with minor modifications to multi-sorted equational axiomatizations and term rewriting systems, as well as to equational term rewriting systems.

Let F be a finite set of function symbols in an axiomatization. Let X be a set of variables. Let $GT(F)$ be the set of all ground terms constructed using F and let $T(F, X)$ be the set of terms constructed using F and variables in X. Obviously $GT(F)$ is non-empty only if F includes a nullary function symbol. Let C be a subset of F which is designated to be a set of constructors; the set $F' = F - C$ is the set of nonconstructor function symbols. Whenever C is not explicitly identified, we assume that $C = F$.

Let **E** be a finite set of pairs of terms in $T(F, X)$; we will call **E** an *equational specification* or *equational axiomatization.* Let **R** be a term rewriting system associated with **E**. Let $\rightarrow$ be the reduction relation on $T(F, X)$ induced by **R** (see [3] for definitions). The congruence relation induced by **E** on $T(F, X)$ is the reflexive, symmetric and transitive closure of $\rightarrow$, denoted by $\leftrightarrow^*$. Let $\rightarrow^+$ and $\rightarrow^*$ respectively denote the transitive closure and reflexive and transitive closure of $\rightarrow$. The reader can consult [3] for definitions of substitution, position, subterm, etc, as well as the definitions of a Noetherian, confluent, locally-confluent, canonical relation. A set **R** of rules is Noetherian (confluent, locally-confluent, canonical, complete) if the reduction relation $\rightarrow$ induced by **R** is Noetherian (confluent, locally-confluent, canonical, complete).

A term t is *reducible* iff there is a term t' such that $t \rightarrow t'$. A term t is *irreducible* or in *normal form* iff there is no term t' such that $t \rightarrow t'$. Let **IRG**(**R**) be the set of all ground terms in $GT(F)$ in normal form with respect to **R**.

A term t is *linear* iff no variable in $Vars(t)$ appears more than once in t. A rule $L \rightarrow R$ is called *left-linear* if L is linear, *right-linear* if R is linear, and *linear* if both L and R are linear. A set **R** of rules is *left-linear* iff every rule in **R** is left-linear.

A substitution σ is a *ground substitution* if for every variable x such that $\sigma(x) \neq x$, $\sigma(x)$ is a ground term. A substitution σ is a *constructor substitution* if for every variable x such that $\sigma(x) \neq x$, $\sigma(x)$ is a constructor term. A substitution σ is an *irreducible* or *normalized substitution* if for every variable x such that $\sigma(x) \neq x$, $\sigma(x)$ is irreducible.

For a given set **R** of rules, we now define the notion of quasi-reducibility. A term t is *quasi-reducible* with respect to a set S of terms iff for every substitution σ: $Vars(t) \rightarrow S$, $\sigma(t)$ is reducible. We simply say that t is quasi-reducible if t is quasi-reducible with respect to $GT(F)$. A term t is *quasi-reducible* with respect to constructors iff for every constructor ground substitution σ: $Vars(t) \rightarrow GT(C)$, $\sigma(t)$ is reducible.

An equation $s \equiv t$ is in the equational theory (or is an equational consequence) of **E** iff $s \leftrightarrow^* t$. An equation $s \equiv t$ is an inductive theorem in **E** iff for every ground substitution σ, $\sigma(s) \leftrightarrow^* \sigma(t)$. If a set of constructors C in **E** is explicitly identified, then an equation $s \equiv t$ is an inductive theorem in **E** wrt C iff for every ground constructor substitution σ, $\sigma(s) \leftrightarrow^* \sigma(t)$.

An equational axiomatization **E** is said to be *sufficiently complete* (with respect to a set of constructors C) iff for every ground term g in $GT(F)$, there is a ground term c in $GT(C)$ such that $g \leftrightarrow^* c$. A set **R** of rules is *sufficiently complete* if its associated equational theory **E** is sufficiently complete.

3. Induction Proof Using Irreducible Ground Terms

The following theorem serves as the basis of the proposed method.

Theorem 3.1 [1]: Let **R1**, **R2** be canonical rewriting systems such that **R2** is generated from **R1** $\cup \{ l \equiv r \}$. Then, $l \equiv r$ is an inductive theorem in **R1** iff **IRG(R1)** = **IRG(R2)**.

Proof: If $l \equiv r$ is an inductive theorem of **R1**, then for any ground substitution σ, $\sigma(l)$ and $\sigma(r)$ must have the same normal form wrt **R1**. If **IRG(R1)** and **IRG(R2)** are not the same, there exist $g1$ and $g2$ in **IRG(R1)** such that $g1$ and $g2$ are equivalent in the equational theory of **R2**; a ground instance of the equation $l \equiv r$ must have been used to show this equivalence. However, every ground instance of $l \equiv r$ belongs to the equational theory of **R1**, which gives us a contradiction.

If $l \equiv r$ is not an inductive theorem, then there exists a ground substitution σ such that $\sigma(l)$ and $\sigma(r)$ do not have the same normal form wrt **R1**. So, at least one of these two normal forms is reducible in **R2**, thus **IRG(R1)** $\neq$ **IRG(R2)**. □

The above result is used to design a new version of the Knuth-Bendix completion procedure to prove inductive theorems; this procedure is elaborated later when we replace line 5 by an algorithm for comparing **IRG**'s.

Method 3.2 (*Inductive Completion Procedure*):

Input: **R1**: a canonical system; $>$: a reduction ordering; $l \equiv r$: the equation to be proven.

1. **E** := $\{ l \equiv r \}$; **R2** := **R1**;
2. Repeat steps 3 to 9 below as long as **E** is non-empty;
   if **E** becomes empty, stop with *"proved"*.
3. Remove an equation $s' \equiv t'$ (or $t' \equiv s'$) from **E**;
   compute their normal forms wrt **R2**, say s and t, respectively.
   If they are the same, then discard the equation and repeat this step. Otherwise,
   make a rule $s \rightarrow t$ (or $t \rightarrow s$); if that is not possible, stop with *"method failed"*.
4. Add the rule $s \rightarrow t$ to **R2**.
5. If **IRG(R1)** $\neq$ **IRG(R2)**, stop with *"not true"*.
6. Remove all the old rules from **R2** whose left-hand side contains an instance of s and
   put them into **E** as equations.
7. Use $s \rightarrow t$ (followed by any rules in **R2**) to reduce the right-hand-sides of
   existing rules to their normal forms.
8. Add to **E**, all critical pairs formed from **R2** using $s \rightarrow t$.

The only difference between the Knuth-Bendix completion procedure and ours is in line 5 which is added to check whether the sets of irreducible ground terms are equivalent. The correctness of the procedure is guaranteed by the following theorem and the correctness of the Knuth-Bendix completion procedure [4].

Theorem 3.3: Assume **R1** is canonical and the completion procedure on **R1** $\cup \{l \equiv r\}$ neither stops with "*method failed*" nor runs forever. Then $l \equiv r$ is an inductive theorem of **R1** iff the procedure stops with *"proved"*.

Proof: Method 3.2 can stop with (1) *"method failed"*, (2) *"proved"* or (3) "*not true*". We need to consider only the last two cases. If the method stops with "*proved*", by the correctness of the completion procedure [4], **R2** is canonical. Moreover, because of the check at Line 5, **IRG(R1)** = **IRG(R2)**; so the result holds by Theorem 3.1. If the method stops with "*not true*", then **IRG(R1)** $\neq$ **IRG(R2)**. More precisely, **IRG(R1)** strictly contains **IRG(R2)**. For any canonical system **R2'** containing **R2**, **IRG(R1)** also strictly contains **IRG(R2')**. Again by Theorem 3.1, we get the result. □

If the procedure runs forever, we do not know whether there exists a finite canonical system for **R1** $\cup \{ l \rightarrow r \}$. Without **R2** being canonical, we could say nothing about whether $l \equiv r$ is an inductive theorem. If the procedure employs a complete strategy for generating critical pairs, which means that the completion procedure will eventually consider all critical pairs among all the rules generated at any point, then if $l \equiv r$ is not an inductive theorem, then the procedure is guaranteed to stop with *"not true"*; if it does not terminate, then $l \equiv r$ is an inductive theorem. Thus, the above procedure is a co-semi-decision procedure for inductive theorems or equivalently, a semi-decision procedure for proving whether an equation is *not* an inductive theorem.

The above result can be carried over to equational rewriting systems consisting of a set **R** of rules and a set **E** of equations. We borrow the concept of *E-canonicity* from [6] and give results similar to [7]. Analogous to Theorem **3.1**, we have the following theorem.

Theorem 3.4: Let **R1** and **R2** be **E**-canonical rewriting systems such that **R2** is generated from **R1** $\cup \{l \equiv r\}$. Then $l \equiv r$ is an inductive theorem iff **IRG** (**R** 1)/**E** = **IRG** (**R** 2)/**E** .

The proof is similar to that of Theorem 3.1 and is thus omitted. In addition, we also have:

Theorem 3.5: If **R** is both **E**-canonical and (**E** $\cup \{l \equiv r\}$)-canonical, then $l \equiv r$ is an inductive theorem of (**R**, **E**) iff **IRG** (**R**)/**E** = **IRG** (**R**)/(**E** $\cup \{l \equiv r\}$).

Proof: Let $[t]_{\mathbf{E}}$ denote the equivalence class of t modulo **E**, where t is a ground term. If $l \equiv r$ is an inductive theorem over **R**/**E**, then for any ground substitution σ, $[\sigma(l)]_{\mathbf{E}} = [\sigma(r)]_{\mathbf{E}}$. In that case, it is obvious that for any distinct $t\,1$, $t\,2$ in **IRG(R)/E**, $[t\,1]_{\mathbf{E}} \neq [t\,2]_{\mathbf{E}}$ implies $[t\,1]_{\mathbf{E} \cup \{l \equiv r\}} \neq [t\,2]_{\mathbf{E} \cup \{l \equiv r\}}$.

If $l \equiv r$ is not an inductive theorem over **R**/**E**, then there exits a ground substitution σ such that $[\sigma(l)]_{\mathbf{E}} \neq [\sigma(r)]_{\mathbf{E}}$. However, these two equivalence classes are merged into one in **IRG** (**R**)/(**E** $\cup \{l \equiv r\}$). □

The above theorem allows us to prove non-orientable equations as inductive properties by incorporating these equations into unification and matching algorithms.

4. Test Set

In this section, we introduce the concept of a test set which is in essence a finite description of **IRG**. Given two rewriting systems **R1** and **R2**, we would like to reduce the problem of comparing **IRG(R1)** and **IRG(R2)** to comparing their suitably constructed test sets. If **IRG(R)** is finite, then **IRG(R)** itself could be a test set of **R**. However, if **IRG(R)** is infinite, then there can be many ways to finitely describe it. Further, we would like a test set to be complete and minimal as defined below.

Definition 4.1: (i) A set T of terms is *complete* with respect to **R** if for all s in **IRG(R)**, there exists t in T such that $s = \sigma(t)$ for some ground substitution σ.

(ii) T is *minimal* if two distinct terms of T are not unifiable and for each t' in T, there exists a ground substitution σ such that $\sigma(t')$ is in **IRG(R)**.

When a test set of **R** is both complete and minimal, we will call it *standard.*

Lemma 4.2: No term in a minimal test set of a system **R** is quasi-reducible.

Proof: Since every term in a minimal test set has a ground instance that is irreducible, it cannot be quasi-reducible. □

Lemma 4.3T: For every s in **IRG(R)**, there exists at most one term t in a minimal test set T such that $s = \sigma(t)$ for some ground substitution σ.

Proof: If there exist two distinct terms $t1$ and $t2$ in T such that $s = \sigma_1(t1) = \sigma_2(t2)$ for ground substitutions σ_1 and σ_2, then $t1$ and $t2$ are unifiable, which cannot be true for a minimal T. □

A test set of a rewriting system **R** characterizes some properties of **R**, which distinguish it from others. For example, we have a result on sufficient completeness for systems whose rules are constructor-preserving (a rule $L \rightarrow R$ is called *constructor-preserving* iff whenever L is a constructor term, i.e., a term made from constructors, R is also a constructor term).

Theorem 4.4: A canonical constructor-preserving set **R** of rules is sufficiently complete iff no $f \in F - C$ appears in a standard test set of **R**.

A similar result can be found in [13].

5. Computing Test Set

We now discuss how to compute a test set of a rewriting system **R**. Let us first introduce a result given in [9]. Intuitively, the result says that for a term rewriting system, there is a finite bound on the substitutions of ground terms that need to be considered in order to check for any term, whether the result of applying any ground substitution on the term is irreducible. Before giving the result, some definitions are needed.

Let *depth* be a function on terms which gives the maximum depth of the tree representation of terms. The *depth* of a substitution is the depth of a term with maximum depth in the substitution. The *depth* of a rewriting system **R** is defined as the depth of a left-hand side with maximum depth in **R**. Further, define $MaxD(x,t)$ as the maximum depth of a variable x in a term t, i.e.,

$$MaxD(x,t) = \max \{ \ |p| \ \mid t/p = x \ \},$$

where t/p is the subterm of t at position p, which is a sequence of natural numbers; $|p|$ is the length of position p.

Now we state the theorem:

Theorem 5.1: [9] Let t be a term and x be a variable in t. There exists a number $b(x, t, \mathbf{R})$, which depends on x, t and the rewriting system **R**, satisfying the following property: if θ is a ground substitution such that $\theta(t) \in IRG(\mathbf{R})$ and $depth(\theta(x)) > b(x, t, \mathbf{R})$, then there exists a (smaller) substitution θ', where

$$depth(\theta'(x)) < depth(\theta(x)) \text{ and}$$
$$\theta'(y) = \theta(y) \text{ if } y \neq x,$$

such that $\theta'(t) \in IRG(\mathbf{R})$.

The function b is explicitly given in [9]; it depends on the depth of the deepest left-hand side in **R**, the number of times x appears in a term t and the positions of its occurrences in t.

Let **TS1(R,** k**)** denote the set of the irreducible ground terms of depth at most k. i.e.,

$$\mathbf{TS1(R}, k) = \{ t \mid t \in \mathbf{IRG}(\mathbf{R}), depth(t) \leq k \}.$$

We show below that **TS1(R,** k**)** can serve as a test set for **R** if k is suitably chosen.

Theorem 5.2: Let **R1**, **R2** be the same as in the inductive completion procedure and $K = depth(\mathbf{R2}) + b(x, l_i, \mathbf{R1})$, where l_i is the left-hand side of a rule in **R2** - **R1** such that $(n * MaxD(x, l_i))$, n being the number of occurrences of x, is maximal.
Then **TS1(R1,** K**)** = **TS1(R2,** K**)** iff **IRG(R1)** = **IRG(R2)**.

Proof: The if-part is trivial. For the only-if part, since **IRG(R2)** $\subseteq$ **IRG(R1)**, it needs to be shown that **TS1(R1,** K**)** = **TS1(R2,** K**)** implies **IRG(R1)** $\subseteq$ **IRG(R2)**. The proof is by contradiction.

Suppose **TS1(R1,** K**)** = **TS1(R2,** K**)** and **IRG(R1)** $\neq$ **IRG(R2)**. Let t be the smallest term in **IRG(R1)** - **IRG(R2)**. Obviously, t is reducible by **R2** at the root and $depth(t) > K = depth(\mathbf{R2}) + b(x, l_i, \mathbf{R1})$. Suppose $l \to r$ in **R2** reduces t with the substitution σ, i.e., $\sigma(l) = t$, then σ is a "big" substitution in the sense of Theorem 5.1. By Theorem 5.1, there exists a small substitution σ' such that $\sigma'(l)$ is not reducible by **R1** and $depth(\sigma'(l)) \leq b(x, l_i, \mathbf{R1}) + depth(\mathbf{R}2)$. That is, $\sigma'(l)$ is in **TS1(R1,** K**)**, hence in **TS2(R2,** K**)**. However, $\sigma'(l)$ is reducible by **R2** at the root, which is a contradiction.
□

Corollary: Let **R1**, **R2**, and K be as in the above theorem. **IRG(R1)** = **IRG(R2)** iff no term in **TS1(R1,** K**)** is reducible by **R2** at the root.

The above results allow us to implement the inductive completion procedure, since **TS1(R1,** k**)** is computable for any k. In line 5 of Method 3.2, **TS1(R1,** $b(x, l_i, \mathbf{R1})$**)** can be computed; every time a new rule is added, we add more terms to **TS1** if the depth of the new rule is greater than the depth of previous rules added. Then it is checked whether the new rule reduces some terms of **TS1(R1,** K**)** at the root.

6. Left-Linear Systems

We now turn our attention to left-linear rewriting systems since their test sets can be computed (relatively) efficiently. From now on in this section, every rewriting system is assumed to be left-linear.

6.1 Superterm

Definition 6.1: A superterm is a mapping from N^* to subsets of F, where N is the set of natural numbers. The domain of this mapping is called the set of *positions* of the superterm just as for terms.

A superterm corresponding to a term is obtained by replacing its variable occurrences by the empty set and replacing its operator occurrences by the set containing that single operator.

Remark: The difference between a superterm and a term is that a position in a superterm corresponds to a set of operators, while a position in a term corresponds to exactly one operator or a variable; further there are no variables in a superterm.

For terms and superterms, let $t(p)$ stand for the label of the node at position p in the tree representation of t, where t is a term or a superterm; $t(p)$ is defined only if p is a legal position of t. If t is a term, then $t(p)$ is a function symbol or a variable; if t is a superterm, then $t(p)$ is a subset of function symbols.

Definition 6.2: A superterm $s1$ *covers* another superterm $s2$ if for each position p in $s2$, $s2(p) \subseteq s1(p)$.

Given two superterms $s1$, $s2$, the *merge* of $s1$ and $s2$ is the minimal superterm that covers both $s1$ and $s2$.

Examples of term, superterm and the merge of superterms:

term	superterm	merge of superterms
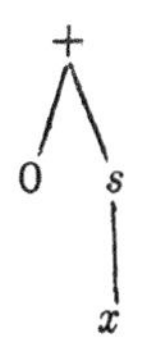	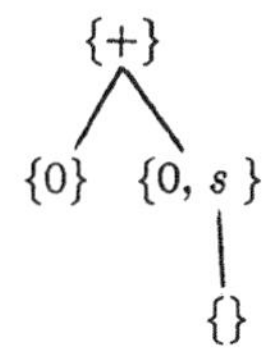	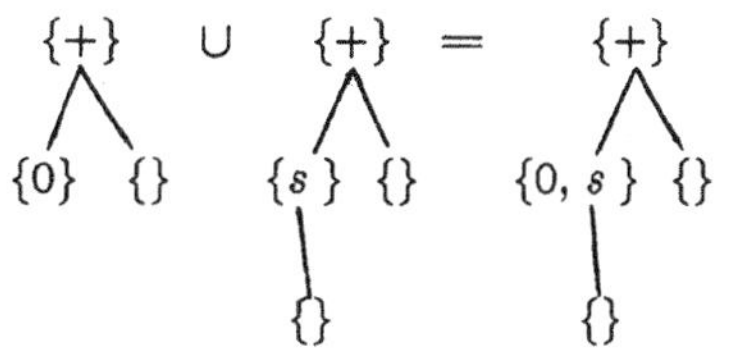

6.2 Defining Domain, Skeleton and TS2

Definition 6.3: Given a rewriting system **R**, the defining domain of an operator f, denoted by *Def-domain*(f, **R**), is the merge of all superterms corresponding to subterms of left-hand sides of **R** such that the root symbol of each subterm is f.

Definition 6.4: Let t be a ground term with the root symbol f and S_f be a (finite) superterm covering *Def-domain(f,* **R**). The *skeleton* of t with respect to S_f is a term t', maximal in size, such that

$t'(\lambda) = f$, where λ is the empty position, and

for all positions $p.i$ in t',

$$t'(p.i) = \begin{cases} t(p.i) & \text{if } t(p) \in S_f(p) \\ x & \text{otherwise, where } x \text{ is a distinct variable.} \end{cases}$$

We denote t' by *Ske(t,* S_f) and S_f is called an *extending domain* of f.

Note that *Ske(t, S)* is unique; it is a linear term; further, there exists a substitution σ such that $\sigma(Ske(t, S_f)) = t$.

Example: Let $\mathbf{R} = \{ s(s(0)) \rightarrow 0,\ x + 0 \rightarrow x,\ x + s(0) \rightarrow s(0) \}$.

Def-domain(+, **R**)	*Def-domain*(s, **R**)	skeletons
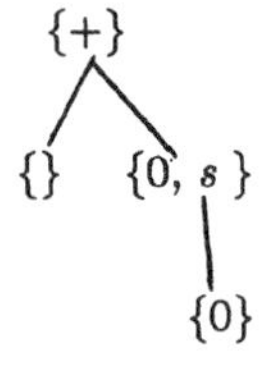	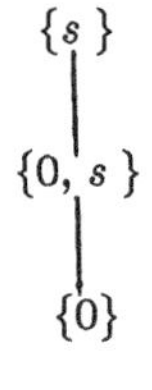	$Ske(x + s(0), S_+) = x + s(0)$ $Ske(s(x) + s(y), S_+) = s(x1) + s(y)$ $Ske(0 + s(s(0)), S_+) = 0 + s(s(x1))$ where $S_+ =$ *Def-domain*(+, **R**)

A set S of superterms is an *extending domain* of **R** if (a) for each function symbol f appearing in a left-hand side in **R**, there exists a unique $S_f \in S$ such that S_f covers *Def-domain(f,* **R**), and (b) for each $s \in S$ there exists a unique function symbol f such that s covers *Def-domain(f,* **R**), i.e., $S = \{ S_f \mid S_f$ is an extending domain of f in **R** $\}$.

Let S be an extending domain of **R**. We define **TS2(R,** S) as:

$$\mathbf{TS2}(\mathbf{R}, S) = \{ Ske(t, S_f) \mid t \in \mathbf{IRG}(\mathbf{R}),\ S_f \in S \text{ and the root symbol of } t \text{ is } f \}.$$

For the previous example, let $S = \{$ *Def-domain*(+, **R**), *Def-domain*(s, **R**) $\}$, then

$$\mathbf{TS2}(\mathbf{R}, S) = \mathbf{IRG}(\mathbf{R}) = \{ 0, s(0) \}.$$

Lemma 6.5: Let $t \in$ **TS2(R,** S) and let f be its root symbol. Then,
(i) if p and q are positions in t such that $p = q.i$ for some integer i and $t(p)$ is a variable, then $t(q)$ is not in $S_f(q)$;
(ii) *Ske* $(\sigma(t), S_f) = t$ for all substitutions σ.

Proof: (i) follows from the definitions of *Ske* and **TS2** and (ii) follows from (i). □

Using the above lemma, we get:

Theorem 6.6: Let S be an extending domain of **R**.
(i) **TS2(R,** S) is a standard test set (with respect to S) of **R**.
(ii) Let s be a subterm of a left-hand side of **R** and t be in **TS2(R,** S). Then, if s matches any instance of t, then s matches t.
(iii) For any $t \in$ **TS2(R,** S), for any σ, $\sigma(t)$ cannot be reduced at its root.

Proof: (i) **TS2(R,** S) is complete since for any t in **IRG(R)**, there exists *Ske* (t, S_f) in **TS2(R,** S), where S_f in S, and $\sigma($*Ske* $(t, S_f)) = t$ for some substitution σ. **TS2(R,** S) is minimal since each term has an irreducible ground instance and for any $t1$, $t2$ in **TS2(R,** S), if $t1$ and $t2$ are unifiable, then $t1 = t2$ by Lemma 6.5.
(ii) Since s and t are linear terms, if s is unifiable with t and s does not match t, then there exists a position p such that $t(p)$ is a variable and $s(p)$ is not a variable. Since t is a non-variable term, there must be a q such that $p = q.i$ for an integer i. Thus $t(q) = s(q)$, which is a contradiction, since $t(q)$ is not in $S_f(q)$ by Lemma 6.5.
(iii) Let t be in **TS2(R,** S) and σ be any substitution. If $\sigma(t)$ is reducible by a rule $l \to r$ in **R** at the root, then by (ii), t is also reducible by $l \to r$, which is a contradiction to t being not quasi-reducible. □

Theorem 6.7: If S covers the defining domains of **R1** and **R2**, then **TS2(R1,** S) = **TS2(R2,** S) iff **IRG(R1)** = **IRG(R2)**.

Proof: The if-part is obvious since the skeleton of a term t, *Ske*(t, S_f), is unique (after renaming the variables) with respect to S_f, where $S_f \in S$. The only-if part is proven by contradiction as follows.

Suppose **TS2(R1,** S) = **TS1(R2,** S) and **IRG(R1)** $\neq$ **IRG(R2)**. Without any loss of generality, suppose **IRG(R1)** - **IRG(R2)** is not empty. Let t be the smallest term (in terms of depth) in **IRG(R1)** - **IRG(R2)**. Then t is reducible by **R2** at the root. Since t is not reducible by **R1**, there exists t' in **TS1(R1,** S), and hence in **TS2(R2,** S), such that $t = \sigma(t')$ for some σ. That is, $\sigma(t')$ is reducible by **R2** at its root, which is a contradiction.
□

6.3 Computing TS2

(i) We start with the set $T0$ of terms, called *structural schema.*

$T0 = \{ f(x1, ..., xn) \mid f$ in F, arity$(f) = n$ and $x1, ..., xn$ are distinct variables$\}$

For example, $T0 = \{0, s(x1), (x2 + x3)\}$ for $F = \{0, s, +\}$. Obviously, $T0$ is complete with respect to **IRG(R)**.
(ii) We gradually expand each term in $T0$ ensuring the completeness property.

Definition 6.8: Given **R** and a term t such that the root of t is f, let S_f be an extending domain of f and $T0$ defined as above. The term t is said to be *extensible at position p* with respect to S_f if $t(p)$ is a variable, $S_f(p)$ is not empty and for each ancestor position p' of p, $t(p')$ is in $S_f(p')$.

We define the *patterns* of t at position p as the set

$$Pattern(t, p) = \{ t[p \leftarrow t0] \mid t0 \in T0 \}.$$

Definition 6.9: The set *Cand(***R***, S)* of terms obtained using the following procedure is called *candidates of TS2*, where S is an extending domain of **R**:

Step 1. $Cand := T0$

Step 2. While there is a term t in *Cand* and a position p in t such that
t is extendable at p, do
$Cand := Cand - \{t\} + Pattern(t, p)$

Step 3. Remove all terms from *Cand* that are reducible wrt **R**.

The termination of the above method is guaranteed since S is finite.

Lemma 6.10: (i) **TS2(R,** S**)** $\subseteq$ *Cand(***R***, S)*.
(ii) Let s be a subterm of a left-hand side of **R** and $t \in Cand$ (**R**, S). Then if s matches any instance of t, then s matches t.

Proof: (i) is obvious from the definitions of **TS2** and *Cand*. (ii) the proof is the same as that of Theorem 6.6 (ii). □

For each term t in *Cand(***R***, S)*, if t is a ground term, then t is in **TS2(R,** S**)**. If t contains variables, we are not sure whether t is in **TS2(R,** S**)** since t may be quasi-reducible.

The last step in our construction is to remove all quasi-reducible terms from *Cand(***R***, S)*.

Lemma 6.11: Let $T1$ be a subset of *Cand(***R***, S)*, including every ground term of *Cand(***R***, S)*. Further, suppose that no term t in $T1$ is quasi-reducible, i.e., each one has a ground instance in **IRG(R)**, and let $T2$ = *Cand(***R***, S) - T1*.

(a) For any t in $T2$, if t is not quasi-reducible with respect to $T1$, then t is not quasi-reducible (with respect to $GT(F)$).

(b) If every term in $T2$ is quasi-reducible with respect to $T1$, then every term in $T2$ is quasi-reducible.

Proof: (a) Suppose $t \in T2$ is not quasi-reducible with respect to $T1$. That is, if $p1, p2, ..., pk$ are the variable positions of t, then there exist $t1, t2, ..., tk$ in $T1$ such that $t' = t[p1 \leftarrow t1][p2 \leftarrow t2] \cdots [pk \leftarrow tk]$ is not reducible. By the hypothesis, there exists a ground instance ti' of ti, such that $ti' \in IRG(\mathbf{R})$, for $1 \leq i \leq k$. Suppose that $t'' = t[p1 \leftarrow t1'][p2 \leftarrow t2'] \cdots [pk \leftarrow tk']$ is reducible by $l \rightarrow r$ in **R**. If a subterm s of l matches ti' at the position pi, by Lemma 6.10 (ii), s matches ti too. So l rewrites $t''[pi \leftarrow ti]$ also. Repeat the above the process, until each ti' is replaced by ti. Eventually, we get that l rewrites t', which is a contradiction.

(b) Suppose every term in $T2$ is quasi-reducible with respect to $T1$. If $t \in T2$ is not quasi-reducible, then let σ be such a ground substitution that $\sigma(t)$ is minimal in size and not reducible. Then the range of σ is the set of the instances of a subset of $T1$. Since $T2$ is quasi-reducible with respect to $T1$, $\sigma(t)$ should be reducible too, which is a contradiction.
□

This lemma allows us to design an algorithm to remove quasi-reducible terms from *Cand*(**R**, S).

Algorithm 6.12: *Quasi-Check*($T1$, $T2$, **R**);

1. $ADL := \emptyset$;
2. Repeat
3. $T1 := T1 \cup ADL$;
4. $T2 := T2 - ADL$;
5.   For t in $T2$ do
6.   Suppose $Vars(t) = \{x1, x2, ..., xk\}$
7.   For $< t1, ..., tk >$ in $T1^k$ do
8.   Construct $\sigma = \{ x1 \leftarrow t1, x2 \leftarrow t2, ..., xk \leftarrow tk \}$
9.   If $\sigma(t)$ is not reducible by **R** then
10. $ADL := ADL \cup \{ t \}$;
11.   go to 14.
12.   end if
13.   end for;
14.   Continue;
15.   end for
16. until $ADL = \emptyset$;
17. return($T1$);

In the call to Quasi-Check, $T1$ is initialized to be the ground terms in *Cand*(**R**, *S*) and $T2$ by *Cand*(**R**, *S*) - *T1*; the returned result is **TS2(R,** ***S*****)**.

The loop 7-13 is always finite. The loop 5-15 is executed at the most $| T2 |^2$ times and the main loop is executed at the most $| T2 |$ times. Thus the algorithm always terminates. The correctness of the algorithm is guaranteed by Lemma 6.11.

Example: $\mathbf{R} = \{\ s\ (s\ (0)) \rightarrow 0,\ p\ (0) \rightarrow s\ (0)\ \}$.

By Definition 6.9, the candidates of test set $Cand\,(\mathbf{R},S\,) = \{\ 0,\ s\ (0),\ p\ (s\ (x\ )),\ p\ (p\ (x\ )),\ s\ (p\ (x\ )),\ s\ (s\ (s\ (x\ ))),\ s\ (s\ (p\ (x\ )))\ \}$

When *Quasi-Check* is invoked, our inputs are

$T\,1 = \{\ 0,\ s\ (0)\ \}$

$T\,2 = \{\ p\ (s\ (x\ )),\ p\ (p\ (x\ )),\ s\ (s\ (p\ (x\ ))),\ s\ (s\ (s\ (x\ )))\ \}$

Consider $p\ (s\ (x\ ))$ in $T\,2$. Since $p\ (s\ (0))$ is not reducible, $p\ (s\ (x\ ))$ is transferred from $T\,2$ into $T\,1$. Replacing the variables in the other terms of $T\,2$ by $p\ (s\ (x\ ))$, we find none of them reducible, so they are all put into $T\,1$.

Now *Cand(***R***, S)* = *T1* = **TS2(R**, *S)*.

Similar to the method introduced in [13], there is yet another method for computing the test set.

Theorem 6.13: Let $B_i = \{\ Ske(t,\ S)\ |\ depth\,(t\,)\ \leq\ i$ and t in **IRG(R)** $\}$. Then **TS2(R, S)** $= B_k$, where k is the least number such that $B_k = B_{k+1}$.

The proof consists of showing by induction that once $B_j = B_{j+1}$ then $B_j = B_{j+k}$ for all $k > 0$.

We have implemented both the methods discussed above. Our experience shows that the first method for computing a test set is more efficient in most cases than the second method.

6.4 Using TS2 in the Inductive Completion Procedure

The method introduced above to compute **TS2(R**, *S*) is used to implement line 5 of the inductive completion procedure (Method 3.2).

For efficiency, we could start up by computing **TS2(R1**, *S*), where S is the set of the defining domains of the operators in **R1**. When each new rule $s \rightarrow t$ is generated, we check at first whether any term in **TS2(R1**, *S*) is reducible by $s \rightarrow t$. If some term is reducible, then stop with "*not true*". For every non-variable subterm $s\,'$ of s, if a superterm in S having the same root as $s\,'$, say S_f, does not cover the superterm corresponding to $s\,'$, then let $S\,' = S - \{S_f\} \cup \{$the merge of S_f and the superterm corresponding to $s\,'\}$. Next, extend terms in **TS2(R1**, *S*) as much as possible, and call *Quasi-Check* (Algorithm 6.12) to remove quasi-reducible terms, and get **TS2(R1**, $S\,'$). If a term in **TS2(R1**, $S\,'$) is reducible by $s \rightarrow t$, **IRG**(**R**1) cannot be equal to **IRG**(**R**1 $\cup \{\ s \rightarrow t\ \}$) and hence the equation being proved is not true. If the new terms in **TS2(R1**, $S\,'$) are not reducible by $s \rightarrow t$, then the algorithm continues.

That is, line 5 in the inductive completion procedure

5. If **IRG(R1)** $\neq$ **IRG(R2)**, stop with *"not true"*.

is replaced by the following lines:

Suppose S is initialized to { *Def-domain(f,* **R**) | f in F } and T is initialized to **TS2(R1,** *S*).

3. ... ...
4. Add the rule $s \rightarrow t$ to **R2**.
5.1. If $s \rightarrow t$ reduces some terms of T, then stop with "*not true*";
5.2. For every non-variable subterm s' of s, repeat 5.2.1 and 5.2.2.
  5.2.1. Let $S_f \in S$ have the same root as s'. If S_f covers s', then go to 6.
  5.2.2. $S := S - \{S_f\} + \{$Merge(S_f, superterm corresponding to s')$\}$;
    $T1 := \{u \mid u \in T$ and u is extensible with respect to $S\}$;
    $T2 := T - T1$;
5.3. While there exists an extensible term u at position p in $T1$ do
    $T1 := T1 - \{u\} + Pattern(u, p)$
5.4. $T := $ *Quasi-Check* ($T1$, $T2$, **R1**);
5.5. If $s \rightarrow t$ reduces some terms of T, then stop with "*not true*";
6. ... ...

The correctness of the above algorithm is guaranteed by the following result:

Lemma 6.14:
(a) T = **TS2(R1,** *S*) after the execution of 5.1 - 5.5;
(b) If $s \rightarrow t$ reduces some term of T, then **IRG(R1)** $\neq$ **IRG(R2)**;
(c) If no terms in T is reduced by $s \rightarrow t$ after line 5.5, then **IRG(R1)** = **IRG(R2)**.

The proofs of (a) and (b) are easy. The proof of (c) is similar to the last paragraph of the proof of Theorem 6.7.

Example: Let **R** = $\{ s(s(0)) \rightarrow 0,\ x + 0 \rightarrow x,\ x + s(0) \rightarrow s(x) \}$.
S = { *Def-domain*(s, **R**), *Def-domain*(+, **R**) }, **TS2(R,** S) = $\{0, s(0)\}$.
$x + x = 0$ is an inductive theorem of **R**, since **TS2(R** $\cup$ $\{x + x \rightarrow 0\}$) = **TS2(R,** S).
$x + x = x$ is not an inductive theorem since the rule $s(0) \rightarrow 0$ is generated when the completion procedure is run with **R** $\cup$ $\{x + x = x\}$. This reduces $s(0)$ in **TS2(R,** S). See also the next section.

7. Implementation and Comparison

We have implemented the test set method for proving inductive properties in RRL, a Rewrite Rule Laboratory [10,11]. Below, we give partial transcripts of some examples run on RRL using this method.

****

```
... ...
Your system is canonical:

  [1] s(s(0)) -> 0  [user, 1]
  [2] (x + 0) -> x  [user, 2]
  [3] (x + s(0)) -> s(x)  [user, 3]

Type Add, Akb, Auto, Break, Clean, Delete, Dump, Grammar, Init, Kb, List,
     Log, Norm, Order, Option, Operator, Prove, Quit, Read, Statis, Suffic
     Undo, Unlog, Write or Help.
RRL-> prove
Type equation to prove in the format:  L == R (if C)
Enter a ']' to exit when no equation is given.
s(s(x)) == x

No, it is not equational theorem.
Normal form of the left hand side is: s(s(x))
Normal form of the right hand side is: x
Do you want to see it is an inductive theorem ? y

        Current Constructor Set  =  { }
To prove the equation with the constructors ? (y/n) n

Adding to testset: s(0)
Adding to testset: 0

Proving equation: s(s(x)) == x  [user, 4]
Adding rule:   [4] s(s(x)) -> x  [user, 4]
Deleting rule:  [1] s(s(0)) -> 0  [user, 1]

Your system is canonical:

  [2] (x + 0) -> x  [user, 2]
  [3] (x + s(0)) -> s(x)  [user, 3]
  [4] s(s(x)) -> x  [user, 4]

Following equation
    s(s(x)) == x  [user, 4]
    is an inductive theorem in the current system.
Processor time used      =  0.25 sec
      ...   ...
RRL-> prove  x + x == x
No, it is not equational theorem.

Normal form of the left hand side is: (x + x)
Normal form of the right hand side is: x
Do you want to see it is an inductive theorem ? y
```

```
    Current Constructor Set  =  { }
To prove the equation with the constructors ? (y/n) y
Note: Constructor set is empty.
Type operators you wish to be constructors: 0 s
    Constructor Set =  { s, 0 }
Specification of '+' is complete relative to { s, 0 }

Proving equation: (x + x) == x  [user, 5]
Adding rule:   [4] (x + x) -> x  [user, 5]
Equation:  s(0) == 0  [3, 4] violates the consistency condition.
Following equation
    (x + x) == x  [user, 5]
    is not an inductive theorem in the system.
Processor time used      =  0.13 sec
```

****

From the above simple example, we see that an inductive theorem could be proven (or disproved) with or without constructors. Constructors are in general helpful to compute test sets since we know the operators in $F - C$ do not appear in test sets if the system is sufficiently complete.

The method proposed by Huet and Hullot [5] requires that no relations exist between constructors, that is, every constructor is free, and the system is sufficiently complete. Under these conditions, The structural schema for the constructors are then a test set in our method. Thus, our method is a strict extension of that of Huet and Hullot.

Our method is as powerful as the method based on the concept of quasi-reducibility proposed in [7]. When a new rule is generated in the inductive completion procedure, instead of checking whether the test set is changed by this rule, Jouannaud and Kounalis' method checks whether the left-hand side of the rule is quasi-reducible. They gave a method to decide the quasi-reducibility of a term in left-linear systems. We implemented their method in RRL and compared it with ours on several examples. We discuss below two examples.

Examples:

R1 = { [1] (b + a) ---> (a + b)
[2] ((x + y) + z) ---> (x + (y + z))
[3] (b + (a + z)) ---> (a + (b + z))
[4] f(a, x) ---> a
[5] f(b, x) ---> b
[6] f((x + y), z) ---> (f(x, z) + f(y, z)) }

To prove that $f(x, y) = x$ is an inductive theorem of **R1** using constructors a, b, and +, our method needs 0.25 seconds while the quasi-reducibility method needs 4.7 seconds.

R2 = { [1] s(s(y)) ---> y
[2] p(0) ---> s(0)
[3] g(0, x) ---> 0
[4] g(s(x), y) ---> f(g(x, y), 0)
[5] f(0, y) ---> y
[6] f(s(x), y) ---> s(f(x, y))
[7] f(f(g(y, y1), 0), 0) ---> g(y, y1) }

To disprove that $f(p(x),0) = 0$ is an inductive theorem of **R2** using constructors 0, s, p, and g, our method takes 4.8 seconds. The quasi-reducibility method fails on this example due to lack of space.

Our method is better because the left hand sides of the new rules are usually covered by the defining domains of the initial system **R**, and therefore $TS\,2(\mathbf{R}, S)$ has to be computed only once. The only thing to do then is to check whether the new rules reduce some terms of $TS\,2(\mathbf{R}, S)$ at the root. For the quasi-reducibility method, we must check whether the left-hand side of each new rule is quasi-reducible.

8. References

[1] Dershowitz, N., "Applications of the Knuth-Bendix Completion Procedure," Laboratory Operation, Aerosapce Corporation, Aerospace Report No. ATR-83(8478)-2, May 15, 1983.

[2] Goguen, J., "How to Prove Algebraic Inductive Hypotheses without Induction," *Proc. of the Fifth Conference on Automated Deduction*, Les Arces, France, LNCS 87, Springer Verlag, pp. 356-372, July 1980.

[3] Huet, G., "Confluent Reductions: Abstract Properties and Applications to Term Rewriting Systems," *JACM* 27, 4, pp. 797-821, October 1980.

[4] Huet, G., "A Complete Proof of Correctness of the Knuth-Bendix Completion Procedure," *JCSS* 23, 1, 1981.

[5] Huet, G., and Hullot, J.M., "Proof by Induction in Equational Theories with Constructors," *JCSS* 25, 2, 1982.

[6] Jouannaud, J.-P., and Kirchner, H., "Completion of a Set of Rules modulo a Set of Equations," Proc. of the *11th Symp. on Principles of Programming Languages*, 1984.

[7] Jouannaud, J.-P., and Kounalis, E., "Proofs by Induction in Equational Theories Without Constructors," CRIN, University of Nancy, France, May 1985. To appear in the *Proc. of IEEE Conf. on Logic in Computer Science*, Cambridge, MA, June 1986.

[8] Kapur, D., and Musser, D.R., "Proof by Consistency," *Proc. of an NSF Workshop on the Rewrite Rule Laboratory,* Sept. 4-6, 1983, Schenectady, G.E. R&D Center Report GEN84008, April 1984. To appear in the *AI Journal.*

[9] Kapur, D., Narendran, P., and Zhang, H., "On Sufficient Completeness and Related Properties of Term Rewriting Systems," Unpublished Manuscript, General Electric R&D Center, Schenectady, NY, Oct. 1985. Submitted to *Acta Informatica.*

[10] Kapur, D., and Sivakumar, G., "Experiments with and Architecture of RRL, a Rewrite Rule Laboratory," Proc. of *An NSF Workshop on the Rewrite Rule Lab., Sept. 1983,* General Electric R&D Center Report 84GEN008, pp. 33-56, April 1984.

[11] Kapur, D., Sivakumar, G., and Zhang, H., "RRL: A Rewrite Rule Laboratory," to appear in the *Proc. of the 8th International Conf. on Automated Deduction (CADE-8)*, Oxford, England, July 1986.

[12] Knuth, D., and Bendix, P., "Simple Word Problems in Universal Algebras," in *Computational Problems in Abstract Algebra* (ed. Leech), Pergamon Press, pp. 263-297, 1970.

[13] Kounalis, E., and Zhang, H., "A General Completeness Test for Equational Specifications," CRIN [85-R-05], Nancy, France, November 1985.

[14] Musser, D.R., "On Proving Inductive Properties of Abstract Data Types," Proc. of the *7th Symp. on Principles of Programming Languages,* Las Vegas, Jan. 1980.

[15] Musser, D.R., and Kapur, D., "Rewrite Rule Theory and Abstract Data Type Analysis," *Computer Algebra, EUROSAM 1982,* LNCS 144 (ed. Calmet), Springer Verlag, pp. 77-90, April 1982.

How to prove Equivalence of Term Rewriting Systems without Induction

Yoshihito TOYAMA

NTT Electrical Communications Laboratories
3-9-11 Midori-cho, Musashino-shi, Tokyo 180 Japan

Abstract

A simple method is proposed for testing equivalence in a restricted domain of two given term rewriting systems. By using the Church-Rosser property and the reachability of term rewriting systems, the method allows us to prove equivalence of these systems without the explicit use of induction; this proof usually requires some kind of induction. The method proposed is a general extension of *inductionless induction* methods developed by Musser, Goguen, Huet and Hullot, and allows us to extend *inductionless induction* concepts to not only term rewriting systems with the termination property, but also various reduction systems: term rewriting systems without the termination property, string rewriting systems, graph rewriting systems, combinatory reduction systems, and resolution systems. This method is applied to test equivalence of term rewriting systems, to prove the inductive theorems, and to derive a new term rewriting system from a given system by using equivalence transformation rules.

1. Introduction

We consider how to prove equivalence in a restricted domain of two term rewriting systems [5][6] without induction. Equivalence in a restricted domain means that the equational relation (or the transitive reflexive closure) generated by the reduction relation of one system is equal in the restricted domain to that of another system.

We first explain the concept of equivalence in a restricted domain through simple examples. Consider the term rewriting system R_1 computing the addition on the set N of natural numbers represented by $0, s(0), s(s(0)), \ldots$;

$$\begin{aligned} R_1 : \ & x + 0 \rhd x, \\ & x + s(y) \rhd s(x + y). \end{aligned}$$

By adding the associative law to R_1, we can obtain another system R_2 computing the same function;

$$\begin{aligned} R_2 : \ & x + 0 \rhd x, \\ & x + s(y) \rhd s(x + y), \\ & x + (y + z) \rhd (x + y) + z. \end{aligned}$$

Then, R_2 can reduce $(M + N) + P$ and $M + (N + P)$ to the same normal form for any terms M, N, P, but R_1 cannot reduce them unless M, N, P can be reduced to natural numbers. Thus, equivalence of R_1 and R_2 must be regarded as equivalence in the domain in which terms can be reduced into natural numbers.

We show another example concerning equivalence of recursive programs. Assuming rules for primitive functions, we define the factorial function $f(n) = n!$ by using the term rewriting systems R_1 and R_2;

$R_1 : f(x) \triangleright \text{if} \;\; \text{equal}(x,0) \text{ then } s(0) \text{ else } x * f(x - s(0)),$

$R_2 : f(0) \triangleright s(0),$

$\quad\;\; f(s(x)) \triangleright s(x) * f(x).$

Since the rewriting rule of R_1 can be infinitely applied to the function symbol f, R_1 has an infinite reduction sequence starting with the term $f(M)$ for any term M. On the other hand, R_2 cannot reduce $f(M)$ unless M can be reduced to a natural number. Thus, R_1 and R_2 generally produce different reduction sequences, although they can reduce the term $f(M)$ to the same result if M can be reduced to a natural number. Therefore, equivalence for the recursive programs may be regarded as equivalence in the restricted domain of term rewriting systems.

Thus, the concept of equivalence in a restricted domain of term rewriting systems frequently appears in computer science: automated theorem proving, semantics of functional programs, program transformation, verification of programs, and specification of abstract data types. However, this equivalence cannot, in general, be proved by mere equational reasoning: some kind of induction on the domain is necessary.

In this paper, we present a new very simple method for proving equivalence in a restricted domain of two term rewriting systems without explicit induction. Our approach to this problem was inspired by *inductionless induction* methods developed by Musser [13], Goguen [4], Huet and Hullot [7]. However, our method is more general than their *inductionless induction* methods, and allows us to extend *inductionless induction* concepts to not only term rewriting systems with the termination property [4][7][13], but also various reduction systems: term rewriting systems without the termination property [5][6], string rewriting systems (Thue systems) [2], graph rewriting systems [17], combinatory reduction systems [10], and resolution systems [8][16], etc.

The key idea of our method is that equivalence in a restricted domain can be easily proved by using the Church-Rosser property and the reachability of reduction systems. We first explain this idea in an abstract framework. Simple sufficient conditions for equivalence in a restricted domain of two given abstract reduction systems are shown. Our results are carefully partitioned between abstract properties depending solely on the reduction relation and properties depending on the term structure. We show how one can formally validate equivalence for term rewriting systems by using these abstract results, and how *inductionless induction* methods by [4][7][13] can be naturally extended to our method. Finally, we propose an equivalence transformation technique for term rewriting systems.

2. Reduction Systems

We explain notions of reduction systems and give definitions for the following sections. These reduction systems have only an abstract structure, thus they are called abstract reduction systems [5][10].

A reduction system is a structure $R = \langle A, \rightarrow \rangle$ consisting of some object set A and some binary relation $\rightarrow$ on A (i.e., $\rightarrow \subset A \times A$), called a reduction relation. A reduction (starting with x_0) in R is a finite or infinite sequence $x_0 \rightarrow x_1 \rightarrow x_2 \rightarrow \ldots$. The identity of elements of A (or syntactical equality) is denoted by $\equiv$. $\xrightarrow{*}$ is the transitive reflexive closure of $\rightarrow$ and $=$ is the equivalence relation generated by $\rightarrow$ (i.e., the transitive reflexive symmetric closure of $\rightarrow$). If $x \in A$ is minimal with respect to $\rightarrow$, i.e., $\neg \exists y \in A[x \rightarrow y]$, then we say that x is a normal form; let NF be the set of normal forms. If $x \xrightarrow{*} y$ and $y \in NF$ then we say x has a normal form y and y is a normal form of x.

Definition. $R = \langle A, \rightarrow \rangle$ is strongly normalizing (denoted by $SN(R)$), or R has the termination property, iff every reduction in R terminates, i.e., there is no infinite sequence $x_0 \rightarrow x_1 \rightarrow x_2 \rightarrow \ldots$. R is weakly normalizing (denoted by $WN(R)$) iff any $x \in A$ has a normal form.

Definition. $R = \langle A, \rightarrow \rangle$ has the Church-Rosser property (denoted by $CR(R)$) iff $\forall x, y, z \in A[x \xrightarrow{*} y \wedge x \xrightarrow{*} z \Rightarrow \exists w \in A, y \xrightarrow{*} w \wedge z \xrightarrow{*} w]$.

The following proposition is well known [1][5][10].

Proposition 2.1. Let R have the Church-Rosser property, then,
(1) $\forall x, y \in A[x = y \Rightarrow \exists w \in A, x \overset{*}{\rightarrow} w \wedge y \overset{*}{\rightarrow} w]$,
(2) $\forall x, y \in NF[x = y \Rightarrow x \equiv y]$,
(3) $\forall x \in A \forall y \in NF[x = y \Rightarrow x \overset{*}{\rightarrow} y]$.

3. Basic Results

Let $R_1 = \langle A, \underset{1}{\rightarrow} \rangle$ and $R_2 = \langle A, \underset{2}{\rightarrow} \rangle$ be two abstract reduction systems having the same object set A, and let $\underset{i}{\overset{*}{\rightarrow}}$, $\underset{i}{=}$ and NF_i (i=1,2) be the transitive reflexive closure, the equivalence relation and the set of normal forms in R_i respectively. Note that $\overset{*}{\rightarrow}$ and $=$ are subsets of $A \times A$; for example, $\underset{1}{=} \subset \underset{2}{=}$ means that $\forall x, y \in A[x \underset{1}{=} y \Rightarrow x \underset{2}{=} y]$.

Definition. Let A' and A'' be any nonempty subsets of the object set A. Let $\underset{i}{\sim}$ $(i = 1, 2)$ be any two binary relations on A. Then $\underset{1}{\sim} = \underset{2}{\sim}$ in $A' \times A''$ iff $\forall x \in A' \forall y \in A''[x \underset{1}{\sim} y \iff x \underset{2}{\sim} y]$. We write this as $\underset{1}{\sim} = \underset{2}{\sim}$ in A' if $A' = A''$. A' is reachable to A'' under $\underset{1}{\sim}$ iff $\forall x \in A' \exists y \in A''[x \underset{1}{\sim} y]$. A' is closed under $\underset{1}{\sim}$ iff $\forall x \in A' \forall y \in A[x \underset{1}{\sim} y \Rightarrow y \in A']$.

We first show sufficient conditions for $\underset{1}{=} = \underset{2}{=}$ in A'.

Lemma 3.1. Let R_1, R_2 satisfy the following conditions:
(1) $\underset{1}{=} \subset \underset{2}{=}$,
(2) $\underset{1}{=} = \underset{2}{=}$ in A'',
(3) A' is reachable to A'' under $\underset{1}{=}$.
Then $\underset{1}{=} = \underset{2}{=}$ in A'.

Proof. Prove $\forall x, y \in A'[x \underset{1}{=} y \iff x \underset{2}{=} y]$. $\Rightarrow$ is trivial from condition (1), hence we will show $\Leftarrow$. Assume $x \underset{2}{=} y$, where $x, y \in A'$. By using condition (3), there are some elements $z, w \in A''$ such that $x \underset{1}{=} z$ and $y \underset{1}{=} w$. Since $x \underset{2}{=} z$ and $y \underset{2}{=} w$ are obtained from condition (1), $z \underset{2}{=} w$ can be derived from $z \underset{2}{=} x \underset{2}{=} y \underset{2}{=} w$. From condition (2), $z \underset{1}{=} w$ holds. Therefore $x \underset{1}{=} y$ from $x \underset{1}{=} z \underset{1}{=} w \underset{1}{=} y$. □

If R_2 has the Church-Rosser property, we can modify condition (2) of Lemma 3.1 as follows.

Theorem 3.1. Assume the following conditions:
(1) $\underset{1}{=} \subset \underset{2}{=}$,
(2) $CR(R_2)$, $\underset{1}{\rightarrow} = \underset{2}{\rightarrow}$ in A'', and A'' is closed under $\underset{2}{\rightarrow}$,
(3) A' is reachable to A'' under $\underset{1}{=}$.
Then $\underset{1}{=} = \underset{2}{=}$ in A'.

Proof. Show condition (2) of Lemma 3.1, i.e., $\forall x, y \in A''[x \underset{1}{=} y \iff x \underset{2}{=} y]$. $\Rightarrow$ is trivial from condition (1), hence we will prove $\Leftarrow$. Assume $x \underset{2}{=} y$, where $x, y \in A''$. From $CR(R_2)$, the closed property of A'' under $\underset{2}{\rightarrow}$, and Proposition 2.1(1), there exists some $z \in A''$ such that $x \underset{2}{\overset{*}{\rightarrow}} z$ and $y \underset{2}{\overset{*}{\rightarrow}} z$. By using $\underset{1}{\rightarrow} = \underset{2}{\rightarrow}$ in A'' and the closed property of A'' under $\underset{2}{\rightarrow}$, $x \underset{1}{\overset{*}{\rightarrow}} z$ and $y \underset{1}{\overset{*}{\rightarrow}} z$ can be derived. Therefore $x \underset{1}{=} y$. □

Theorem 3.2. Assume the following conditions:

(1) $\underset{1}{=} \subset \underset{2}{=}$,

(2) $CR(R_2)$ and $A'' \subset NF_2$,

(3) A' is reachable to A'' under $\underset{1}{=}$.

Then $\underset{1}{=} = \underset{2}{=}$ in A'.

Proof. Show condition (2) of Lemma 3.1, i.e., $\forall x, y \in A''[x \underset{1}{=} y \iff x \underset{2}{=} y]$. $\Rightarrow$ is trivial. $\Leftarrow$: By using condition (2) of this theorem and Proposition 2.1(2), $x \underset{2}{=} y \Rightarrow x \equiv y$ for any $x, y \in A''$. Therefore $x \underset{1}{=} y$. □

Corollary 3.1. Assume the conditions:

(1) $\underset{1}{=} \subset \underset{2}{=}$,

(2) $CR(R_2)$ and $NF_1 = NF_2$,

(3) $WN(R_1)$.

Then $\underset{1}{=} = \underset{2}{=}$ is obtained.

Proof. Set $A' = A$ and $A'' = NF_1 = NF_2$ in Theorem 3.2. □

Next, we consider sufficient conditions for $\overset{*}{\underset{1}{\rightarrow}} = \overset{*}{\underset{2}{\rightarrow}}$ in $A' \times A''$.

Theorem 3.3. Assume the following conditions:

(1) $\underset{1}{=} \subset \underset{2}{=}$,

(2) $CR(R_2)$ and $A'' \subset NF_2$,

(3) A' is reachable to A'' under $\overset{*}{\underset{1}{\rightarrow}}$.

Then $\overset{*}{\underset{1}{\rightarrow}} = \overset{*}{\underset{2}{\rightarrow}}$ in $A' \times A''$.

Proof. Prove $\forall x \in A' \forall y \in A''[x \overset{*}{\underset{1}{\rightarrow}} y \iff x \overset{*}{\underset{2}{\rightarrow}} y]$. $\Rightarrow$: Let $x \overset{*}{\underset{1}{\rightarrow}} y$. Then $x \underset{2}{=} y$ from condition (1). Thus $x \overset{*}{\underset{2}{\rightarrow}} y$ is obtained from condition (2) and Proposition 2.1(3). $\Leftarrow$: Let $x \overset{*}{\underset{2}{\rightarrow}} y$. Then, from condition (3), there exists some $z \in A''$ such that $x \overset{*}{\underset{1}{\rightarrow}} z$. By condition (1), $x \underset{2}{=} z$; hence, $y \underset{2}{=} z$ can be derived from $y \underset{2}{=} x \underset{2}{=} z$. Thus, $y \equiv z$ is obtained from condition (2) and Proposition 2.1(2). Therefore $x \overset{*}{\underset{1}{\rightarrow}} y$. □

Corollary 3.2. Assume the conditions:

(1) $\underset{1}{=} \subset \underset{2}{=}$,

(2) $CR(R_2)$ and $NF_1 = NF_2$,

(3) $WN(R_1)$.

Then $\overset{*}{\underset{1}{\rightarrow}} = \overset{*}{\underset{2}{\rightarrow}}$ in $A \times NF_1$.

Proof. Set $A' = A$ and $A'' = NF_1 = NF_2$ in Theorem 3.3. □

In the following sections, we will explain how to apply the above abstract results to term rewriting systems that are reduction systems having a term set as the object set A. However, note that the above abstract results can be applied to not only term rewriting systems, but also various reduction systems.

4. Term Rewriting Systems

Assuming that the reader is familiar with the basic concepts concerning term rewriting systems, we briefly summarize the important notions below [5][6].

Let F be an enumerable set of function symbols denoted by $f, g, h, \ldots$, and let V be an enumerable set of variable symbols denoted by $x, y, z, \ldots$ where $F \cap V = \phi$. By $T(F, V)$ (abbreviated by T) we denote the set of terms constructed from F and V. If V is empty, $T(F, V)$, denoted as $T(F)$, is the set of ground terms.

A substitution θ is a mapping from a term set T to T such that for term M, $\theta(M)$ is completely determined by its values on the variable symbols occurring in M. Following common usage, we write this as $M\theta$ instead of $\theta(M)$.

Consider an extra constant $\square$ called a hole and the set $T(F \cup \{\square\}, V)$. Then $C \in T(F \cup \{\square\}, V)$ is called a context on F. We use the notation $C[\ ]$ for the context containing precisely one hole, and if $N \in T(F, V)$ then $C[N]$ denotes the result of placing N in the hole of $C[\ ]$. N is called a subterm of $M \equiv C[N]$. Let N be a subterm occurrence of M, then, write $N \subset M$.

A rewriting rule on T is a pair $\langle M_l, M_r \rangle$ of terms in T such that $M_l \notin V$ and any variable in M_r also occurs in M_l. The notation $\triangleright$ denotes a set of rewriting rules on T and we write $M_l \triangleright M_r$ for $\langle M_l, M_r \rangle \in \triangleright$. The set $\triangleright$ of rewriting rules on T defines a reduction relation $\rightarrow$ on T as follows:

$M \rightarrow N$ iff $M \equiv C[M_l\theta]$, $N \equiv C[M_r\theta]$, and $M_l \triangleright M_r$ for some $M_l, M_r, C[\ ]$, and θ.

Definition. A term rewriting system R on T is a reduction system $R = \langle T, \rightarrow \rangle$ such that the reduction relation $\rightarrow$ is defined by a set $\triangleright$ of rewriting rules on T.

$R \cup \{M \triangleright N\}$ (resp. $R - \{M \triangleright N\}$) denotes the term rewriting system obtained by adding the rule $M \triangleright N$ to R (resp. by removing the rule $M \triangleright N$ from R).

If every variable in term M occurs only once, then M is called linear. We say that R is linear iff for any $M_l \triangleright M_r \in R$, M_l is linear.

Let $M \triangleright N$ and $P \triangleright Q$ be two rules in R. We assume that we have renamed variables appropriately, so that M and P share no variables. Assume $S \notin V$ is a subterm occurrence in M, i.e., $M \equiv C[S]$, such that S and P are unifiable, i.e., $S\theta \equiv P\theta$, with a minimal unifier θ [5][11]. Since $M\theta \equiv C[S]\theta \equiv C\theta[P\theta]$, two reductions starting with $M\theta$, i.e., $M\theta \rightarrow C\theta[Q\theta] \equiv C[Q]\theta$ and $M\theta \rightarrow N\theta$, can be obtained by using $P \triangleright Q$ and $M \triangleright N$. Then we say that the pair $\langle C[Q]\theta, N\theta \rangle$ of terms is critical in R [5][6]. We may choose $M \triangleright N$ and $P \triangleright Q$ to be the same rule, but in this case we shall not consider the case $S \equiv M$, which gives trivial pair $\langle N, N \rangle$. If R has no critical pair, then we say that R is nonoverlapping [5][6][11][19].

The following sufficient conditions for the Church-Rosser property are well known [5][6][11].

Proposition 4.1. Let R be strongly normalizing, and let for any critical pair $\langle P, Q \rangle$ in R, P and Q have the same normal form. Then R has the Church-Rosser property.

Proposition 4.2. Let R be linear and nonoverlapping. Then R has the Church-Rosser property.

For more discussions concerning the Church-Rosser property of term rewriting systems having overlapping or nonlinear rules, see [5][19][21].

There are several sufficient conditions for the reachability of term rewriting systems to hold. However, we will omit all proofs of the reachability in the following examples, because they are mostly technical. For discussions of techniques for proving the reachability, see [7][12][15][18].

5. Examples

We now illustrate how to prove equivalence in a restricted domain of two term rewriting systems R_1 and R_2 by using Theorems 3.1, 3.2, and 3.3.

Example 5.1. Let $F' = \{+, s, 0\}$ and $F'' = \{s, 0\}$. Consider the term rewriting systems R_1 and R_2 computing the addition on the set N:

$$R_1 : x + 0 \triangleright x, \qquad R_2 : x + 0 \triangleright x,$$
$$x + s(y) \triangleright s(x + y), \qquad x + s(y) \triangleright s(x + y),$$

$$x + (y + z) \rhd (x + y) + z.$$

We will prove that $\underset{1}{=} = \underset{2}{=}$ in $T(F')$ by using Theorem 3.2. Let $A' = T(F'), A'' = T(F'')$ in Theorem 3.2. We must show conditions (1), (2), (3) of Theorem 3.2 for R_1 and R_2. Since $\underset{1}{\rhd} \subset \underset{2}{\rhd}$, condition (1), i.e., $\underset{1}{=} \subset \underset{2}{=}$, is obvious. By using $SN(R_2)$ and Proposition 4.1, $CR(R_2)$ is obtained. Condition (2) holds, since $T(F'') \subset NF_2$. $T(F')$ is reachable to $T(F'')$ under $\underset{1}{=}$. Therefore, $\underset{1}{=} = \underset{2}{=}$ in $T(F')$.

It is also possible to prove $\overset{*}{\underset{1}{\rightarrow}} = \overset{*}{\underset{2}{\rightarrow}}$ in $T(F') \times T(F'')$ by using Theorem 3.3. □

Example 5.2. Let $F' = \{+, s, 0\}$ and $F'' = \{s, 0\}$. Consider the term rewriting systems R_1 and R_2 computing the addition on Z_3:

$$\begin{aligned} R_1 : {} & s(s(s(x))) \rhd x, \\ & x + 0 \rhd x, \\ & x + s(y) \rhd s(x + y), \end{aligned} \qquad \begin{aligned} R_2 : {} & s(s(s(x))) \rhd x, \\ & x + 0 \rhd x, \\ & x + s(y) \rhd s(x + y), \\ & x + (y + z) \rhd (x + y) + z. \end{aligned}$$

We will prove that $\underset{1}{=} = \underset{2}{=}$ in $T(F')$ by using Theorem 3.1. Let $A' = T(F')$, $A'' = T(F'')$. We must show conditions (1), (2), (3) of Theorem 3.1 for R_1 and R_2. Condition (1), i.e., $\underset{1}{=} \subset \underset{2}{=}$, is obvious. By using $SN(R_2)$ and Proposition 4.1, $CR(R_2)$ is obtained. Condition (2) holds, since $\underset{1}{\rightarrow} = \underset{2}{\rightarrow}$ in $T(F'')$, and $T(F'')$ is closed under $\underset{2}{\rightarrow}$. $T(F')$ is reachable to $T(F'')$ under $\underset{1}{=}$. Therefore, $\underset{1}{=} = \underset{2}{=}$ in $T(F')$.

Note that it is also possible to prove $\underset{1}{=} = \underset{2}{=}$ in $T(F')$ by letting $A'' = \{0, s(0), s(s(0))\}$ and using Theorem 3.2. □

We next show an example in which R_2 does not have the strongly normalizing property.

Example 5.3. Consider the following term rewriting systems R_1 and R_2 computing the *double* function $d(n) = 2 * n$:

$$\begin{aligned} R_1 : {} & d(0) \rhd 0, \\ & d(s(x)) \rhd s(s(d(x))), \end{aligned} \qquad \begin{aligned} R_2 : {} & d(x) \rhd \mathrm{if}(x, 0, s(s(d(x - s(0))))), \\ & \mathrm{if}(0, y, z) \rhd y, \\ & \mathrm{if}(s(x), y, z) \rhd z, \\ & x - 0 \rhd x, \\ & s(x) - s(y) \rhd x - y. \end{aligned}$$

The term rewriting system R_2 does not have the strongly normalizing property, since the first rewriting rule in R_2 can be applied infinitely to the function symbol d.

Let $F' = \{d, s, 0\}$ and $F'' = \{s, 0\}$. We will show that the function d of R_1 equals that of R_2 in the restricted domain $T(F')$, that is, $\underset{1}{=} = \underset{2}{=}$ in $T(F')$. For this purpose, Theorem 3.2 is used. Let $A' = T(F')$, $A'' = T(F'')$. We must show conditions (1), (2), (3) of Theorem 3.2. Since $d(0) \underset{2}{=} 0$ and $d(s(x)) \underset{2}{=} s(s(d(x)))$, condition (1), i.e., $\underset{1}{=} \subset \underset{2}{=}$, is obtained. It is obvious that R_2 is linear and nonoverlapping. Hence, by using Proposition 4.2, R_2 has the Church-Rosser property. Since some function symbol not in F'' appears in the left hand side of any rewriting rule in R_2, we can obtain that $T(F'') \subset NF_2$. Thus, condition (2) holds. $T(F')$ is reachable to $T(F'')$ under $\underset{1}{=}$. Therefore, $\underset{1}{=} = \underset{2}{=}$ in $T(F')$ holds.

Note that $T(F')$ is also reachable to $T(F'')$ under $\underset{1}{\rightarrow}$. Hence, by Theorem 3.3, we can prove $\overset{*}{\underset{1}{\rightarrow}} = \overset{*}{\underset{2}{\rightarrow}}$ in $T(F') \times T(F'')$ in the same way as the above proof. □

6. Inductionless Induction

By using Theorem 3.2, an equation whose proof usually requires induction on some data structure can be proved without the explicit use of induction. In this section, we will explain how to prove an equation with *inductionless induction* method [4][7][9][12][13][14][15][16].

Let R_1 be a term rewriting system with reachability from $T(F')$ to $T(F'')$ under $\underset{1}{=}$. For a term set T, let $M \underset{1}{=} N$ in T denote $\forall\theta[M\theta, N\theta \in T \Rightarrow M\theta \underset{1}{=} N\theta]$. Now, for given terms $M, N \in T(F', V)$ such that any variable in N also occurs in M, consider the validity of $M \underset{1}{=} N$ in $T(F')$. Note that this validity cannot be proved by merely equational reasoning: some kind of induction on $T(F')$ usually becomes necessary [4][7][13]. However, we can prove that $M \underset{1}{=} N$ in $T(F')$ by using the following theorem, without induction.

Theorem 6.1. Let $R_2 = R_1 \cup \{M \triangleright N\}$. If R_2 has the Church-Rosser property and $T(F'') \subset NF_2$, then $M \underset{1}{=} N$ in $T(F')$.

Proof. It is obvious that R_1 and R_2 satisfy conditions (1), (2), (3) of Theorem 3.2 by letting $A' = T(F')$, $A'' = T(F'')$. Thus $\underset{1}{=} \; = \; \underset{2}{=}$ in $T(F')$. Since $M \triangleright N \in R_2$, we can show that $M \underset{2}{=} N$ in $T(F')$. Therefore, $M \underset{1}{=} N$ in $T(F')$. □

Example 6.1. Consider R_1 defining the *half* function $h(n) = n/2$ and the *double* function $d(n) = 2 * n$:

$$\begin{aligned} R_1 : \; & h(0) \triangleright 0, \\ & h(s(0)) \triangleright 0, \\ & h(s(s(x))) \triangleright s(h(x)), \\ & d(0) \triangleright 0, \\ & d(s(x)) \triangleright s(s(d(x))). \end{aligned}$$

Let $F' = \{h, d, s, 0\}$ and $F'' = \{s, 0\}$. $T(F')$ is reachable to $T(F'')$ under $\underset{1}{=}$. Now, let us prove $h(d(x)) \underset{1}{=} x$ in $T(F')$ by using Theorem 6.1. Take $R_2 = R_1 \cup \{h(d(x)) \triangleright x\}$. Then, $CR(R_2)$ by Proposition 4.1 and $T(F'') \subset NF_2$. Therefore $h(d(x)) \underset{1}{=} x$ in $T(F')$. □

When $R_2 = R_1 \cup \{M \triangleright N\}$ does not satisfy the conditions in Theorem 6.1, we may find R_3 instead of R_2 such that $CR(R_3)$, $T(F'') \subset NF_3$, and $\underset{2}{=} \; = \; \underset{3}{=}$ in $T(F')$. If term rewriting systems are strongly normalizing, then the effective search for R_3 can be done by using the Knuth-Bendix completion algorithm [11]. Thus, this method allows us to automatically prove inductive theorems.

The original idea of this method was proposed by Musser [13], and has been extended by Goguen [4], Huet and Hullot [7], and others [9][12][14][15]. However, their *inductionless induction* methods have many limitations. In particular, the requirement for the strongly normalizing property [4][7][13][14] (or the strongly normalizing property on equivalence classes of terms if there are associative/commutative laws [9][12][15]) restricts its application, since most term rewriting systems in which functions are denoted by recursive definitions, such as recursive programs, do not satisfy this property. On the other hand, since Theorem 6.1 holds under very weak assumptions, our method allows us to overcome these limitations.

We next show an example in which R_1 does not have the strongly normalizing property.

Example 6.2.

$$\begin{aligned} R_1 : \; & d(x) \triangleright \mathrm{if}(x, 0, s(s(d(x - s(0))))), \\ & h(x) \triangleright \mathrm{if}(x, 0, \mathrm{if}(x - s(0), 0, s(h(x - s(s(0)))))), \\ & \mathrm{if}(0, y, z) \triangleright y, \\ & \mathrm{if}(s(x), y, z) \triangleright z, \\ & x - 0 \triangleright x, \\ & s(x) - s(y) \triangleright x - y. \end{aligned}$$

Note that the term rewriting system R_1 does not have the strongly normalizing property, since the first and the second rules in R_1 can be infinitely applied to the function symbols d and h respectively.

Let $F' = \{d, h, \text{if}, -, s, 0\}$ and $F'' = \{s, 0\}$. $T(F')$ is reachable to $T(F'')$ under $\underset{1}{=}$. Now, we show that $h(d(x)) \underset{1}{=} x$ in $T(F')$. Take $R_2 = R_1 \cup \{h(d(x)) \rhd x\}$. To easily show the Church-Rosser property of the term rewriting system obtained by adding the rule $h(d(x)) \rhd x$, we consider R_3 instead of R_2:

$$\begin{aligned} R_3 : \ & d(0) \rhd 0, \\ & d(s(x)) \rhd s(s(d(x))), \\ & h(0) \rhd 0, \\ & h(s(0)) \rhd 0, \\ & h(s(s(x))) \rhd s(h(x)), \\ & \text{if}(0, y, z) \rhd y, \\ & \text{if}(s(x), y, z) \rhd z, \\ & x - 0 \rhd x, \\ & s(x) - s(y) \rhd x - y, \\ & h(d(x)) \rhd x. \end{aligned}$$

Then, $\underset{2}{=} = \underset{3}{=}$ in $T(F')$ can be proved in the same way as for Example 5.3. It is shown from Proposition 4.1 that R_3 has the Church-Rosser property. Clearly, $T(F'') \subset NF_3$. Hence, R_3 satisfies the conditions in Theorem 6.1. Therefore, $h(d(x)) \underset{1}{=} x$ in $T(F')$. □

7. Equivalence Transformation Technique

In this section, we propose the equivalence transformation rules for term rewriting systems. We show that equivalence of term rewriting systems, to which it is difficult to apply Theorem 3.2 and 3.3 directly, can be easily proved by an equivalence transformation technique.

Let $R_0 = \langle T(F, V), \underset{0}{\rightarrow} \rangle$ with $\underset{0}{\rhd}$, and let F' be a subset of F. Now, we give the equivalence transformation rules in $T(F')$ for R_0. Let F_0 be the union of the set F' and the set of all function symbols appearing in the rewriting rules of R_0. Then, we transform $R_n = \langle T(F, V), \underset{n}{\rightarrow} \rangle$ with $\underset{n}{\rhd}$ $(n \geq 0)$ to $R_{n+1} = \langle T(F, V), \underset{n+1}{\rightarrow} \rangle$ with $\underset{n+1}{\rhd}$ by using the following rules:

(*D*) **Definition**: Add a new rewriting rule $g(x_1, \ldots, x_k) \rhd Q$ to R_n, where $g \in F - F_n$, $g(x_1, \ldots, x_k)$ is linear, and $Q \in T(F_n, V)$. Thus, $R_{n+1} = R_n \cup \{g(x_1, \ldots, x_k) \rhd Q\}$. Set $F_{n+1} = F_n \cup \{g\}$.
(*A*) **Addition**: Add a new rule $P \rhd Q$ to R_n, where $P \underset{n}{=} Q$ and $P, Q \in T(F_n, V)$. Thus, $R_{n+1} = R_n \cup \{P \rhd Q\}$. Set $F_{n+1} = F_n$.
(*E*) **Elimination**: Remove a rule $P \rhd Q$ from R_n. Thus, $R_{n+1} = R_n - \{P \rhd Q\}$. Set $F_{n+1} = F_n$.

Remarks. The above three rules are a natural extension of the program transformation rules suggested by Burstall and Darlington [3]. Note that their program transformations can be seen as special cases of equivalence transformations for term rewriting systems in restricted domains. Thus, we can give formal proofs to correctness of program transformations by the technique developed in this section [20].

$R_n \Rightarrow R_{n+1}$ shows that R_n is transformed to R_{n+1} by rule(*D*), (*A*), or (*E*). $\overset{*}{\Rightarrow}$ denote the transitive reflexive closure of $\Rightarrow$.

Theorem 7.1. Let $R_0 \overset{*}{\Rightarrow} R_n$, where R_0 is a linear system and $CR(R_0)$. Let $F'' \subset F'$ and $T(F'') \subset NF_0$. Assume that $T(F')$ is reachable to $T(F'')$ under $\underset{0}{=}$ and under $\underset{n}{=}$ (resp. under $\underset{0}{\rightarrow}$ and under $\underset{n}{\rightarrow}$). Then $\underset{0}{=} = \underset{n}{=}$ in $T(F')$ (resp. $\underset{0}{\overset{*}{\rightarrow}} = \underset{n}{\overset{*}{\rightarrow}}$ in $T(F') \times T(F'')$).

Proof. See [20]. □

Example 7.1 (Summation). Consider the following term rewriting systems R_1 and R_2 computing the summation function $f(n) = n + \ldots + 1 + 0$:

$$\begin{aligned} R_1 : & f(0) \rhd 0, \\ & f(s(x)) \rhd s(x) + f(x), \\ & x + 0 \rhd x, \\ & x + s(y) \rhd s(x + y), \end{aligned} \qquad \begin{aligned} R_2 : & f(0) \rhd 0, \\ & f(s(x)) \rhd g(x, s(x)), \\ & g(0, y) \rhd y, \\ & g(s(x), y) \rhd g(x, y + s(x)), \\ & x + 0 \rhd x, \\ & x + s(y) \rhd s(x + y). \end{aligned}$$

Let $F' = \{f, +, s, 0\}$ and $F'' = \{s, 0\}$. By using the equivalence transformation rules, we will show that $\underset{1}{=} = \underset{2}{=}$ in $T(F')$. To transform R_1 to R_2, we first add the associative law for $+$ to R_1: take $R_3 = R_1 \cup \{x + (y + z) \rhd (x + y) + z\}$. Then $R_3 \Rightarrow R_1$ by rule(E). From Proposition 4.1, $CR(R_3)$. Clearly $T(F'') \subset NF_3$. $T(F')$ is reachable to $T(F'')$ under $\underset{1}{=}$ (and also under $\underset{3}{=}$). By Theorem 7.1, $\underset{1}{=} = \underset{3}{=}$ in $T(F')$ is obtained.

Now, let us transform R_3 to R_2 by using the transformation rules. By using rule(D), we introduce a new function g,

(1) $g(x, y) \rhd y + f(x)$.

Let $R_4 = R_3 \cup \{(1)\}$, then we can prove $f(s(x)) \underset{4}{=} g(x, s(x))$, $g(0, y) \underset{4}{=} y$, and
$g(s(x), y) \underset{4}{=} y + f(s(x)) \underset{4}{=} y + (s(x) + f(x)) \underset{4}{=} (y + s(x)) + f(x) \underset{4}{=} g(x, y + s(x))$.

By using rule(A), we can obtain $R_5 = R_4 \cup \{(2), (3), (4)\}$:

(2) $f(s(x)) \rhd g(x, s(x))$,
(3) $g(0, y) \rhd y$,
(4) $g(s(x), y) \rhd g(x, y + s(x))$.

Finally, by using rule(E), remove unnecessary rules $x+(y+z) \rhd (x+y)+z$, $f(s(x)) \rhd s(x)+f(x)$, and $g(x, y) \rhd y + f(x)$ from R_5. Thus, we can obtain R_2. $T(F')$ is reachable to $T(F'')$ under $\underset{2}{=}$. Hence, $\underset{3}{=} = \underset{2}{=}$ in $T(F')$ is obtained by Theorem 7.1.

Now, we obtain an equivalence transformation sequence from R_1 to R_2: $R_1 \Leftarrow R_3 \Rightarrow R_4 \overset{*}{\Rightarrow} R_5 \overset{*}{\Rightarrow} R_2$. Therefore, $\underset{1}{=} = \underset{2}{=}$ in $T(F')$.

Note that it is also possible to prove $\underset{1}{\overset{*}{\rightarrow}} = \underset{2}{\overset{*}{\rightarrow}}$ in $T(F') \times T(F'')$ by this transformation technique. □

8. Conclusion

In this paper, we have proposed a new simple method to prove equivalence in a restricted domain for reduction systems without the explicit use of induction. The key idea is that equivalence in the restricted domain can be easily tested by using the Church-Rosser property and the reachability of reduction systems. We have shown that this technique can be effectively applied to test the equality of term rewriting systems and to prove the inductive theorems without induction. We believe firmly that our method provides us with systematic means of proving equivalence which arises in various formal systems: automated theorem proving, program transformation, program verification, and semantics of abstract data types.

Acknowledgments

The author is grateful to Hirofumi Katsuno, Shigeki Goto, and other members of the First Research Section for their suggestions.

References

[1] Barendregt,H.P." The lambda calculus, its syntax and semantics", North-Holland (1981).

[2] Book,R." Confluent and other types of Thue systems", J.ACM, Vol.29 (1982), pp.171-182.

[3] Burstall,R.M. and Darlington,J." A transformation system for developing recursive programs", J.ACM, Vol.24 (1977), pp.44-67.

[4] Goguen,J.A." How to prove algebraic inductive hypotheses without induction, with applications to the correctness of data type implementation", Lecture Notes in Comput. Sci., Vol.87, Springer-Verlag (1980), pp.356-373.

[5] Huet,G." Confluent reductions: abstract properties and applications to term rewriting systems", J.ACM, Vol.27 (1980), pp.797-821.

[6] Huet,G. and Oppen,D.C." Equations and rewrite rules: a survey", Formal languages: perspectives and open problems, Ed.Book,R., Academic Press (1980), pp.349-393.

[7] Huet,G. and Hullot,J.M." Proofs by induction in equational theories with constructors", J. Comput. and Syst.Sci., Vol.25 (1982), pp.239-266.

[8] Kapur,D. and Narendran,P." An equational approach to theorem proving in first-order predicate calculus", General Electric Corporate Resarch Development Report, No.84CRD322, (1985).

[9] Kirchner,H."A general inductive completion algorithm and application to abstract data types", Lecture Notes in Comput. Sci., Vol.170, Springer-Verlag (1985), pp.282-302.

[10] Klop,J.W." Combinatory reduction systems", Dissertation, Univ. of Utrecht (1980).

[11] Knuth,D.E. and Bendix,P.G." Simple word problems in universal algebras", Computational problems in abstract algebra, Ed.Leech,J., Pergamon Press (1970), pp.263-297.

[12] Kounalis,E."Completeness in data type specifications", Lecture Notes in Comput. Sci., Vol.204, Springer-Verlag (1985), pp.348-362.

[13] Musser,D.R." On proving inductive properties of abstract data types", Proc. 7th ACM Sympo. Principles of programming languages (1980), pp.154-162.

[14] Nipkow,T. and Weikum,G."A decidability results about sufficient-completeness of axiomatically specified abstract data type", Lecture Notes in Comput. Sci., Vol.145, Springer-Verlag (1983), pp.257-267.

[15] Paul,E." Proof by induction in equational theories with relations between constructors", 9th Colloquium on trees in algebra and programming, Ed. Courcelle,B., Cambridge University Press (1984), pp.211-225.

[16] Paul,E."On solving the equality problem in theories defined by Horn clauses", Lecture Notes in Comput. Sci., Vol.204, Springer-Verlag (1985), pp.363-377.

[17] Raoult,J.C."On graph rewriting", Theoretical Comput. Sci. Vol.32 (1984), pp.1-24.

[18] Thiel,J.J." Stop losing sleep over incomplete data type specifications", Proc. 11th ACM Sympo. Principles of programming languages (1984), pp.76-82.

[19] Toyama,Y." On commutativity of term rewriting systems", Trans. IECE Japan, J66-D, 12, pp.1370-1375 (1983), in Japanese.

[20] Toyama,Y." On equivalence transformations for term rewriting systems", RIMS Symposia on Software Science and Engineering, Kyoto (1984), Lecture Notes in Comput. Sci., Vol.220, Springer-Verlag (1986), pp.44-61.

[21] Toyama,Y." On the Church-Rosser property for the direct sum of term rewriting systems", to appear in J.ACM.

Sufficient Completeness, Term Rewriting Systems and "Anti-Unification"

Hubert COMON
LIFIA, Grenoble, BP 68
38402 Saint Martin d'Hères, France

Abstract
We propose an "anti-unification" algorithm to solve inequations in an algebra of terms. It enables us to decide the "convertibility" property without assuming any linear hypothesis on the left hand sides of the rules. Since this property is connected with the sufficient completeness of algebraic specifications, we may decide of the latter in the same way.

1 Introduction

Algebraic specifications play an important role in the design of software systems and it has been recognized that the problem of their correctness is a crucial point [6,8,15]. More precisely, when we deal with stepwise (or hierarchical) specifications [5,9], we want to be sure that a new step does not modify the previous ones. This condition is insured by two properties which, depending on the framework , are known as either "no confusion" or "consistency" for the first one and either "no junk" or "sufficient completeness" for the second one [5,6,9,17,18]. Roughly speaking, consistency means that one cannot derive true = false from the new specification [17,18], or, more generaly, that the "new" equations do not "add equalities" in the already defined algebras [5]. Sufficient completeness means that the new specification does not introduce "new terms" in the already defined algebras [5,9].
Another point of view for the correctness of algebraic specifications is their "operational" correctness. Indeed, specification languages like OBJ [7] or LPG [2] use Term Rewriting Systems (TRS) as their operational semantics. The "correctness" of the specification as a program is then equivalent to the canonicity [11] of the associated TRS.
Unfortunately, both sufficient completeness and canonicity are undecidable [9,11,13]. It is then important to study the links between these properties in order to avoid redundant conditions when one wants to satisfy both properties simultaneously.
Also, recent papers on term rewriting techniques and inductive proofs illustrate the importance of the concept of sufficient completeness in this domain [12,21]. Indeed, if a set of constructors is not declared in the specification, then one would like to know whether any ground term involving a given operator may be reduced to a term which does not include it (this is the "quasi-reducibility" property in [12,13,14] and the "convertibility" property in our own work). More generaly, it would be interesting to deduce from the specification a usable recursive definition of all ground terms which are in normal form. Also, convertibility is equivalent to a completeness property for "constructor preserving" canonical TRS [13].

The sufficient completeness was first studied by Guttag & Horning [9]. In [3,9] sufficient conditions on the form of equations within the specifications are given.

Other contributions [12,13,14,19,21] give, under various hypothesis, algorithms allowing to decide of the convertibility. The key idea of these algorithms is that it is sufficient to check for the reducibility on a finite set of ground terms instead of on all ground terms. But in [12,13,19,21]

this set is very large and the theoretical results seem to be of little practicable use. Only [14] exhibits an usable algorithm but its hypothesis are restrictive.
This paper presents a new method which permits to decide sufficient completeness whithout assuming left-linearity (which is needed in [12,14,19]) and which is no longer based on testing methods. Intuitively, the key idea of this method is, given a rule, to find the ground terms which are "not covered" by this rule. This is achieved by solving "inequations" (this is why we may refer to this algorithm as "anti-unification"). Then, an operator f is convertible iff there exists no solution in normal form to an inequation system $t_i \# f(x_1, ..., x_n)$. The advantage of this technique is that to solve an inequation may be done in polynomial time and thus our algorithm seems to be more easely usable than those given in [12,13,21]. Moreover, the algorithm gives all the "reasons" for incompleteness (when it occurs). More precisely, it may be used for computing a set which describes all the irreducible ground terms.

Proofs are not given in this paper. They can be found, together with a much more detailed presentation, in [4].

In section 2 we introduce our framework and we show in section 3 how the problem of sufficient completeness may be turned into the resolution of a so-called "inequation system". We do not reproduce all the basic definitions which are the usual ones. In section 4 we introduce a preorder on the substitutions which are the solutions of an inequation (this is very similar to the unification case). Then we show that we may compute only linear solutions of an inequation. Finaly, the results are extended to the inequation system case. In section 5, we return to the problem of sufficient completeness and the required hypothesis are derived.

2 Framework

We assume that the reader is familiar with Term Rewriting Systems and algebraic specifications. The missing definitions can be found in [11]. Here, we precise only our notations and own definitions.

2.1 Algebraic Specifications and Sufficient Completeness

An algebraic specification of an abstract data type is classicaly a triplet (S,Σ,E). S is a (finite) set of sorts. Σ is a signature (operators together with a typing function τ). In the following, Σ is also used to identify the underlying set of operators. E is a (finite) set of equations $E = \{l_i = r_i\}$ where both l_i and r_i are terms in the algebra $T_\Sigma(X)$ of "well-formed" terms constructed on the signature Σ and a set of variables $X = \bigcup_{s \in S} X_s$ where X_s is a denumerable set of variables of sort s. We shall use T(X) instead of $T_\Sigma(X)$ and T instead of $T_\Sigma(\emptyset)$ when there is no ambiguity.

If A is a set of operators of Σ, we denote by Σ_{-A} the signature obtained from Σ by removing the operators of A. T_{-A} (resp. $T_{-A}(X)$) denotes the algebra $T_{\Sigma_{-A}}$ (resp. $T_{\Sigma_{-A}}(X)$). Finaly, $T_{\Sigma,E}$ denotes an initial algebra in the category of Σ-algebras satisfying E.

Example 1:
We consider a specification of the integers with equality, where the equation $s(p(x)) == x$ is removed. In this example which will be used again in the following , S, Σ, E are given by:
S = {bool,intl}
$\Sigma = \{f, t : \longrightarrow bool;\ \ not : bool \longrightarrow bool;\ \ and, or : bool \times bool \longrightarrow bool;$
$\quad 0 : \longrightarrow intl;\ \ p, s : intl \longrightarrow intl;\ \ eq : intl \times intl \longrightarrow bool\ \}$
E is the union of the classical set defining the booleans with the set:
$\{\ p(s(x)) == x;\qquad eq(x,x) == t;\qquad eq(s(x),x) == f;$
$eq(x,s(x)) == f;\ \ eq(s(x),s(y)) == eq(x,y);\ \ eq(x,p(x)) == f;$
$eq(p(x),x) == f;\ \ eq(p(x),p(y)) == eq(x,y)\quad \}$

We are interested in the following problem: D being a subset of Σ, is any term in T congruent (modulo the congruence $=_E$ generated on T by E) to a term in T_{-D}?
This problem has other formulations:

- May we use structural induction over $\Sigma - D$ to prove equational theorems in $T_{\Sigma,E}$?
- (Σ, E) being viewed as an "enrichment" of $(\Sigma - D, E')$, is the unique $(\Sigma - D)$-homomorphism from $T_{\Sigma-D,E'}$ to $T_{\Sigma,E}$ surjective ? (See [1] or [17] for example).

This question is also connected with the "sufficient completeness" [9] and "no junk" properties [17]. We shall use the first one since it looks to be more often used although it is more awkward than the second one and it is mainly suitable for a first order logic framework.

Definition 2.1 *Assuming that Σ is split in two sets C and D, we shall say that the specification (S,Σ,E) is* **sufficiently complete** *with respect to C iff:*

$$\forall f \in D,\ \forall f(t_1,...,t_n) \in T,\ \exists t' \in T_{-D},\ f(t_1,...,t_n) =_E t'$$

($=_E$ is the congruence on T generated by E).

Example 1:
We want to check whether the specification is sufficiently complete taking D = {eq}. Here, this is not the case since $\forall t' \in T_{-D}$, eq(s(s(0)),0) and t' are not in the same congruence class (see below for more details).

2.2 Term Rewriting Systems

As mentioned in the introduction, we shall assume some hypothesis on the associated TRS.
We assume that R_E, the TRS associated with E which is obtained by orienting the equations in E from the left to the right, is canonical (see [11] for complete definitions). The corresponding reduction relation is denoted by "$\rightarrow$". And, if $e \in E$," $\rightarrow_{-e}$ " is the reduction relation generated by the TRS associated with E-{e}.

We assume moreover that for each equation $e \in E$, both left hand side (lhs) and right hand side (rhs) of e are in normal form w.r.t $\rightarrow_{-e}$ (which is noetherian since $\rightarrow$ is so).

We reformulate the sufficient completeness problem, using the canonicity property of the associated TRS. In that respect we introduce a definition of the concept of convertibility and then an hypothesis which is crucial for our approach.
If $t \in T(X)$, let $type(t)$ denote the sort of t, $V(t)$ the set of variables occuring in t and $root(t)$ the operator which is at the root of t.

Definition 2.2 *An operator f of Σ is* **convertible** *iff*
$\forall f(t_1,...,t_n) \in T,\ f(t_1,...,t_n)$ is not in normal form (w.r.t $\rightarrow$).

As shown in [13] the convertibility of a given operator remains undecidable. This is the reason why we introduce the new hypothesis which expresses that an operator f does not "suddenly appear on the right".

From now on, we assume that: the subset D of Σ has the property:

$$\textbf{(P1)}\ \ \forall f \in D,\ \forall (l_i \rightarrow r_i) \in R_E,\ f = root(r_i)\ \Rightarrow\ f = root(l_i).$$

C will denote the complement of D in Σ.

We may now state how the problem is reformulated:

Proposition 2.3 *The specification is sufficiently complete w.r.t C iff each operator of D is convertible.*

This proposition is very similar to the theorem in section 4 of [12] and theorem 4.1 in [13]. But the so-called "constructor-preserving" condition is not really needed. It is replaced by the weaker condition (P1). The proof is given in [4].

Example 1: Orienting the equations of E from left to right, we get a canonical TRS (assuming that this is also true for the missing equations). Moreover, D={eq} and Σ have the property (P1). Thus, from proposition 2.3, we must only check for the convertibility of eq. Here, eq is not convertible since eq(s(s(0)),0) is in normal form. Hence, the above specification is not sufficiently complete (w.r.t C).

3 Convertibility and Inequations

The main problem remains to decide the convertibility of each operator of D. In this section we express this problem in terms of "inequations". Let us first introduce some notations.
Ω is the set of substitutions from T(X) to T(X), and Ω_g is the set of substitutions from T(X) to T. (so-called "ground substitutions").

For each $\sigma \in \Omega$ and each $t \in T(X)$ let:

$$Dom(\sigma) = \{x \in X, \sigma(x) \neq x\},$$
$$Im(\sigma) = \{x \in X,\ \exists y \in Dom(\sigma) \ / \ x \in V(\sigma(y))\}.$$

If $t \in T$, let $type(t)$ be the sort of t. If $x_1, ..., x_n$ are variables and $t_1, ..., t_n \in T_\Sigma(X)$ with $type(t_i) = type(x_i)$ for every i, let $(x_1 \leftarrow t_1; ...; x_n \leftarrow t_n)$ denote the substitution σ such that : $Dom(\sigma) = \{x_1, ..., x_n\}$ and, for every x_i, $\sigma(x_i) = t_i$.

Definition 3.1 *A* **linear term** *is a term where each $x \in V(t)$ occurs only once. $\sigma \in \Omega$ is a* **linear substitution** *if, for every linear term t, $\sigma(t)$ is linear.*

Let LHS(E) denote the set of left hand sides of the rules of R_E. LHS(E,f) denotes the subset of terms in LHS(E) having the operator f as their root. In order to check the convertibility of f, we must look only at LHS(E,f) and answer to the question: are all the ground terms having f as their root instances of a term in LHS(E,f)?

The next proposition reduces the problem to a subset of LHS(E,f); LHS1(E,f) denotes the subset of LHS(E,f) of terms where f occurs only once. Unification,unifiers ..., must be understood in the following as unification,unifiers,... in the empty theory.

Proposition 3.2 *$f \in D$ is convertible iff, $\forall \sigma \in \Omega_g$, either $\sigma(f(x_1, ..., x_n))$ is unifiable with a term in LHS1(E,f) or $\sigma(x_i)$ is not in normal form (w.r.t $\rightarrow$) for some index i. ($x_1, ..., x_n \in X$ are distinct variables).*

Thus the problem is now:

- firstly to find terms $\sigma(f(x_1, ..., x_n))$ which are not unifiable with a term of LHS1(E,f). More precisely: find every $\sigma \in \Omega$ such that, $\forall \theta \in \Omega$, $\forall u \in LHS1(E, f)$, $\theta(u) \neq \sigma(f(x_1, ..., x_n))$ where $\neq$ is the syntactic inequality. This is our concept of inequalities (see the next definition).
- then to check for the reducibility of the solutions. This point will be studied in section 5.

Definition 3.3 *Let* $(\alpha_1, ..., \alpha_m), (\beta_1, ..., \beta_m) \in T(X)^m$ *such that:*

$$(\bigcup_{i=1}^{m} V(\alpha_i)) \cap (\bigcup_{i=1}^{m} V(\beta_i)) = \emptyset \text{ and } B \subseteq \Sigma. \quad (1)$$

We shall say that:

- $\theta \in \Omega_g$ *such that* $Dom(\theta) \subseteq \bigcup_{i=1}^m V(\beta_i)$ *is a* **solution in T_{-B} of the inequation system** $\alpha_i \# \beta_i$ $(i \in \{1, ..., m\})$ *if, for any substitution* $\sigma \in \Omega_g$ *such that* $Dom(\sigma) \subseteq \bigcup_{i=1}^m V(\alpha_i)$,

$$\sigma(\alpha_i) \neq \theta(\beta_i) \text{ and } \theta(X) \subseteq T_{-B}. \quad (2)$$

- *More generaly, a substitution* $\theta \in \Omega$ *such that:* $Dom(\theta) \subseteq \bigcup_{i=1}^m V(\beta_i)$ *and* $Im(\theta) \cap \bigcup_{i=1}^m V(\alpha_i) = \emptyset$ *is a solution in* $T_{-B}(X)$ *to the system* $\alpha_i \# \beta_i$ *if, for any* $\sigma \in \Omega_g$,

$$\sigma \circ \theta \text{ is a solution in } T_{-B} \text{ of } \alpha_i \# \beta_i \text{ (in the above sense)}. \quad (3)$$

We may identify a set A $\subseteq \Omega$ of solutions to an inequation system $\alpha_i \# \beta_i$ with the set $\{\sigma(\beta_i), \sigma \in A\}$.

Remark: The condition (1) allows another formulation of (3): $\theta \in \Omega$ is a solution in $T_{-B}(X)$ of the inequation system $t \# t'$ iff

- t and $\theta(t')$ are not unifiable
- $V(t) \cap Im(\theta) = \emptyset$

Proposition 3.4 *Let f∈D and LHS1(E,f)* $= \{t_1, ..., t_n\}$. *f is convertible iff the inequation system* $t_i \# f(x_1, ..., x_n)$ *, where* $x_1, ..., x_n \in X$ *are distinct variables, has no solution in normal form in* $T_{-\{f\}}$.

Hence, the problem of sufficient completeness is nothing else but a particular case of solving inequation systems such as those we propose to study here.

Example 2:
In the example 1, the inequation system:

$$\left\{ \begin{array}{rcl} eq(x, x) & \# & eq(x_1, x_2) \\ eq(x, s(x)) & \# & eq(x_1, x_2) \\ eq(s(x), x) & \# & eq(x_1, x_2) \\ eq(s(x), s(y)) & \# & eq(x_1, x_2) \\ & \cdots & \end{array} \right.$$

has a solution in T(X) : $(x_1 \leftarrow s(s(y)); x_2 \leftarrow y)$ since no left hand side can be unified with $eq(s(s(y)), y)$. The case $eq(s(s(y)), y)$ is not "covered" by the left hand sides of the rules.

4 Resolution of Inequations

From now on, for simplification sake, we assume that S is made of only one sort. The results may be extended to the multi-sorted case [4].

4.1 Resolution of a Single Inequation

The problem is to find the solutions of: $t \# t'$ where $t, t' \in T(X)$ and $V(t) \cap V(t') = \emptyset$.

We must first remark that, if Σ contains at least a non constant operator, say g, the solutions of the inequation $f(x,x) \# f(y,z)$ with $x, y, z \in$ X can not be described by a finite set A of substitutions. Indeed, to solve the inequation would be equivalent to find σ such that: $\sigma(y) \neq \sigma(z)$, and there exists an infinite set of "minimal" solutions: $(y \leftarrow c;\ z \leftarrow g(x_2)), (y \leftarrow g(c); z \leftarrow g(g(x_2)))$,..., where c is a constant operator.

We shall thus give an algorithm permitting to describe A with the help of:

- a finite set of substitutions
- a finite set of constraints $x_i \not\approx y_i$ where $x_i, y_i \in X$.

We shall assume moreover that t' is linear. Note that this does not constraint the inequations problem of section 3 since $f(x_1, ..., x_n)$ (in proposition 3.4) is always linear.
Note finaly that one may call this algorithm "anti-unification" since we must find substitutions such that t and $\sigma(t')$ cannot be unified.

As in the unification case, it is possible to define a preordering relation on the set of substitutions "obstructing" unification of t and t' . We shall say that $\sigma \succ_I \theta$, where (I) is an inequation, if:

(i) σ and θ are solutions of (I).

(ii) there exists a substitution σ_2 such that $\sigma = \sigma_2 \circ \theta$

As in the unifiers case, this preodering relation has minimal elements but, here, they are not unique (up to renaming). Solving (I) amounts to find minimal elements for $\succ_I$, but this task can be very difficult, in particular in cases like:

Example 3:
$t = f(x, g(x))$; $t' = f(y, z)$. $\sigma = (z \leftarrow y)$ is minimal with respect to $\succ_{t \# t'}$. If we want to find σ, we must recognize that x and g(x) cannot be unified.

The first result is that we may only find linear minimal solutions:

Theorem 4.1 *Let t,t'$\in T(X)$ and t' be linear. If σ is a linear solution of t # t', then there exists a minimal substitution σ', with respect to $\succ_{t\#t'}$, such that $\sigma'(t')$ is linear and $\sigma \succ_{t\#t'} \sigma'$*

Corollary 4.2 *Let t' be a linear term. If t # t' has a non-linear solution, then it has at least a minimal solution (w.r.t $\succ_{t\#t'}$) which is linear.*

On the other hand, we are able to compute the linear minimal solutions of an inequation:

Theorem 4.3 *If t,t'$\in T(X)$ where t' is linear and $V(t) \cap V(t') = \emptyset$, then there exists an algorithm computing a set A of pairs (σ,c) where $\sigma \in \Omega$ and c is a set of constraints $x_i \not\approx y_i$ such that :*

- *if $\theta \in \Omega$ is a solution in T(X) of the inequation and $\theta(t')$ is linear, then there exists (σ,c)$\in$A such that $\theta \succ_I \sigma$.*

- $\forall\,\theta \in \Omega,\ \forall\,(\sigma,c) \in A,\ (\forall i,\ \theta(x_i) \neq \theta(y_i)) \Rightarrow \theta \circ \sigma$ *is a solution in T of the inequation.*

Moreover, the complexity of this algorithm is O(nk) where n is the cardinal of Σ *and k the maximal number of symbols occuring in t and t' (each variable and each operator is regarded as one symbol).*

The algorithm is given in the appendix. Note that t is still allowed to be not linear.

Example 4
The minimal linear solutions in T(X) of eq(s(x),x) # eq(x_1, x_2) are given by the set :
A = $\{\ (\ (x_1 \leftarrow 0),\ \emptyset),\ ((x_1 \leftarrow p(x_3)),\ \emptyset),\ ((x_1 \leftarrow s(x_3)),\ \{x_3 \not\approx x_2\})\ \}$

4.2 Inequation Systems

This algorithm may be applied recursively to an inequation system, extending the definition of $\succ_I$ to the case of inequation systems:

Lemma 4.4 *If C is an inequation system, each minimal (resp. minimal and linear) substitution w.r.t.* $\succ_{C\cup\{t\#u\}}$ *is the composite of a minimal (resp. minimal and linear) substitution* σ *w.r.t.* $\succ_C$ *and a minimal (resp. minimal and linear) substitution w.r.t.* $\succ_{t\#\sigma(u)}$.

Therefore, we may apply recursively the algorithm of theorem 3.3. to find all linear solutions of an inequation system.

But we cannot describe such a set of solutions in the same way as we do for a single inequation since we must take into account all sets of constraints which arise at each solution of each inequation. This is shown by the example:

$$\begin{cases} a(x,x) & \#\ \ a(x_1,x_2) \\ a(f(x,y),x) & \#\ \ a(x_1,x_2) \end{cases}$$

The set of substitutions solving the first inequation is described by a constraint: $x_1 \not\approx x_2$. A solution of the second inequation is $((x_1 \leftarrow f(x_3,x_4)),\ \{x_3 \not\approx x_2\})$.
A solution of the system is then: $(\ (x_1 \leftarrow f(x_3,x_4)),\ \{f(x_3,x_4) \not\approx x_2\ ;\ x_3 \not\approx x_2\})$.
The above constraints cannot be turned into another form. We must therefore allow a more general form for the constraints:

Definition 4.5 *A set of constraints c is a set of pairs* $\{\ (x_i \not\approx u_i),\ 1 \leq i \leq n,\ x_i \in T(X),$ $u_i \in T(X),\ u_i$ *and* x_i *are linear and* $V(x_i) \cap V(u_i) = \emptyset\ \}$. *c is* **normalized** *if c may be split into m sets of constraints* $c_1,\ldots,c_m$ *such that:*

(i) $\forall i,\ \exists x_i \in X,\ \forall (x \not\approx u) \in c_i,\ x = x_i$

(ii) $\forall i,\ \forall j > i,\ \forall (x \not\approx u) \in c_j,\ x_i \neq x$ *and* $x_i \notin V(u)$

The "normalization" of a set of constraints is important for our problem since, if the set of ground terms in normal form is infinite and if c is a normalized set of constraints, then it exists at least one substitution σ on the variables occuring in c such that:

$$\forall\,(x \not\approx u) \in c,\ \sigma(x) \neq \sigma(u).$$

Theorem 4.6 *Let $C = \{t_i \# t'_i,\ 1 \le i \le k\}$ be an inequation system such that:*

- $(\bigcup_{i=1}^{k} V(t_i)) \cap (\bigcup_{i=1}^{k} V(t'_i)) = \emptyset$
- $\forall i$, t'_i *is linear*

Then, there exists an algorithm which computes a set A of pairs (σ, c) where $\sigma \in \Omega$ and c is a normalized set of constraints, $c = \{(x_i \not\approx u_i),\ 1 \le i \le n\}$ such that:

(i) $\forall i$, $x_i \in Im(\sigma)$, $u_i \in T(X)$ *and* $V(u_i) \subseteq Im(\sigma)$

(ii) *if θ is a solution in T(X) to the system and θ is linear, then $\exists(\sigma, c) \in A$, $\theta \succ_C \sigma$.*

(iii) $\forall \theta \in \Omega$, $(\forall i,\ \theta(x_i) \neq \theta(u_i)) \Rightarrow$ *$\theta \circ \sigma$ is a solution in T(X) to the system.*

Informally, (i) says that the constraints lay on $Im(\sigma)$ only, (ii) says that every solution is obtained from an element of A, (iii) precises in which sense an element of A is a solution of the inequation system.
Thus, the set of all linear solutions to the system is described by the set A.
Note moreover, that if σ' is a non-linear solution to the system, then $\forall \theta \in \Omega_g$, $\theta \circ \sigma' \in \Omega_g$ and is hence linear. Thus, from point (ii), $\exists \omega \in \Omega_g, \exists(\sigma, c) \in A$, $\theta \circ \sigma' = \omega \circ \sigma$. This means that every "ground instance" of a solution is also an instance of a solution in A.

This algorithm, refered to as algorithm 2, is based upon the algorithm given in appendix 1 and may be sketched as follows:

solve($\{ t_i \# t'_i,\ 1 \le i \le k\}$) :=

1. if k=0 then {id, $\emptyset$}
2. if k=1 then use algorithm 1
3. otherwise
   (a) solve($\{ t_i \# t'_i,\ 1 \le i \le [k/2] \}$) and get a set $A_1 = \{ (\sigma_{i,1}, c_{i,1}) \}$ of solutions.
   (b) for each $(\sigma_{i,1}, c_{i,1}) \in A_1$, solve($\{ t_j \# \sigma_{i,1}(t'_i),\ 1 + [k/2] \le j \le k\}$) and get a set $A_{i,2} = \{ (\sigma_{i,j}, c_{i,j}) \}$ of solutions
   (c) Return $\bigcup_{i,j} Normalize(\sigma_{i,j}, c_{i,j} \cup \sigma_{i,j}(c_i))$

where *Normalize* is the normalization fonction given in appendix 2 and [k/2] is the integer part of k/2.

Example 5:
Let us come back to the example 2. We wanted to check for the convertibility of eq and got from proposition 3.4 an inequation system. Then, algorithm 2 applied to this system gives a set of 13 solutions. We give some of them:

$$s1 = ((x_1 \leftarrow p(p(x_4));\ x_2 \leftarrow 0),\ \emptyset)$$

$$s2 = ((x_1 \leftarrow p(s(x_4));\ x_2 \leftarrow 0),\ \emptyset)$$

$$s10 = ((x_1 \leftarrow 0;\ x_2 \leftarrow p(x_3)),\ \{x_3 \not\approx 0\})$$

The solution $(x_1 \leftarrow s(s(y));\ x_2 \leftarrow y)$ is not given since it is not a linear one. But, as noticed above, each ground instance of it is also an instance of at least one of the 13 solutions.

5 Decision of Convertibility

Let us first come back to the specification. Σ was split in two sets C and D and we wanted to check the convertibility of all $f \in D$. We may again split C into CF and CL where CF is the set of operators g such that LHS(E,g) $= \emptyset$ ("free-constructors") and CL is the complement of CF.

Proposition 5.1 *We may use the algorithm 2 to decide whether all the operators which do not belong to CF are convertible.*

But, in the general case, after solving the inequation system of proposition 3.4, one must decide, for each solution (σ, c) whether it exists a substitution θ such that $\theta \circ \sigma(x_i)$ is in normal form and $\theta(x) \neq \theta(u)$ for each constraint $x \not\approx u \in$ c. Informally, the question is: which of the solutions are in normal form and "satisfy" the constraints?
To do that, the idea is to use recursively our algorithm on the solutions in order to decide which solutions are not "covered" by the other equations. Note that we are not wrapping around since $\sigma(x_i)$ belongs to $T_{-\{f\}}(X)$ and thus the set of operators is "decreasing". But, in the general case, we may not use recursively our algorithm since the propositions 3.2 and 3.4 are no longer true if we replace $f(x_1, ..., x_n)$ by $\sigma(x_i)$.

This is the reason why we need to introduce some new hypothesis. We assume now that the property (P) holds:

$$(\mathbf{P}) \begin{cases} 1) & \textit{It exists an order } g_1 < ... < g_k \textit{ on the set} \\ & CL = \{g_1, ..., g_k\} \textit{ such that :} \\ & \forall i, j \in \{1, ..., k\},\ j > i \Rightarrow g_j \textit{ does not occur in any } t \in LHS1(E, g_i) \\ 2) & \forall i \in \{1, ..., k\},\ LHS(E, g_i) = LHS1(E, g_i) \end{cases}$$

Let NF(f) denote the set of ground terms in normal form and having the operator f as their root. Then, the property (P) implies, in particular, that $\forall g_i \in CL$, $NF(g_i)$ is either infinite or empty. In the latter case, this means that g_i is convertible.
Moreover, from property (P) again, if g_i is convertible for some i, then every g_j, $j \leq i$, is also convertible.

Examples 6:

(i) In our example 1, the property (P) holds since CL = {p}.

(ii) In The example :

$$\begin{aligned} \Sigma &= \{tt, ff : \to s;\ not : s \to s;\ f : s \times s \times s \to s\} \\ E &= \{\, not(tt) == ff\,;\ \ not(ff) == tt\,;\ \ f(x,x,y) == tt\,; \\ &\quad\ f(x,y,x) == tt\,;\ \ f(y,x,x) == tt\ \ \} \end{aligned}$$

CF = {tt,ff} ; CL = {not} ; D = {f} . The property (P) thus holds.

Theorem 5.2 *Assuming that the property (P) holds, we may use the algorithm 2 to decide the convertibility of all $f \in D$.*

In the same way we may find a "description" of all the ground terms in normal form.
To do that, we first look for the convertibility of the operators of CL (by solving some inequation systems). Then, we solve the inequation system of proposition 3.4. If a solution (σ, c) is found, we look for the reducibility of $\sigma(f(x_1, ..., x_n))$. More precisely, if $root(\sigma(x_i)) = g_j \in CL$ and $depth(\sigma(x_i)) > 1$, we solve the inequation system $\{\ t \# \sigma(x_i),\ t \in LHS1(E, g_i)\ \}$ (note that this system satisfies the conditions of theorem 4.6). Then the solution (σ, c) is replaced by the

solutions of the above system and we apply again the same transformation until for each $\sigma(x_i)$, either $root(\sigma(x_i)) \in CF$ or $depth(\sigma(x_i)) = 1$.
The demonstration of theorem 5.2 is given in [4]. It is quite long and involved. We are looking for a simpler proof.
Practicaly, this method seems to be efficient when card(CL) = 1, 2 or 3. But the complexity seems to be at least exponential in card(CL).

Example 7:
When looking for the convertibility of eq (example 1),we had first to solve an inequation system (example 5). We must then check for the reducibility of the solutions. First, we look for the convertibility of p. p is found to be not convertible since p(0) is in normal form. Then NF(p) is described by: p(0)∈ NF(p) and, if t∈ NF(p), then p(t)∈ NF(p).
The solutions s1,s2,s10 of example 5 become, when looking for their reducibility:

- **s1** is unchanged since $p(s(x)) \# p(p(x_4))$ has the identity as a solution .
- **s2** is deleted since $p(s(x)) \# p(s(x_4))$ has no solution.
- **s10** is turned into $((x_1 \leftarrow 0; x_2 \leftarrow p(p(x_4))), \emptyset)$ since $p(s(x)) \# p(x_3)$ has two solutions: $(x_3 \leftarrow 0)$ and $(x_3 \leftarrow p(x_4))$. Normalizing the first set of constraints, a contradiction is found and only the second one is kept.

6 Conclusion

We have presented a new method to decide the sufficient completeness and related properties. This method does not require any longer the use of tests. Moreover, it may be used for finding all normal forms of ground terms. This may be usefull for a theorem prover based on rewriting techniques. Finaly, a complete step by step test of the algorithm on some examples has shown that it is fairly efficient. Hence the algorithm is well suited for a specification analyzer.

However, some interesting questions have yet to be answered and are under investigation:

(i) Is the property (P) really needed ?

(ii) May we drop some of the hypothesis of Theorem 4.6 to get a more general procedure for solving inequations ?

(iii) May we use a similar approach to prove inductive theorems without using completion methods ?

Finaly, it must be pointed out that the work could be completed without using Theorem 4.1. But this theorem is usefull to set the framework we rely on (and has a potential usefull applications in algebraic specifications).

Aknowledgments:
The author thanks Deepak Kapur for a very stimulating and usefull discussion and Jacques Calmet for supervising this work.

References:

[1] Bert D. *La Programmation Generique.* Thèse d'Etat (1979).

[2] Bert D. *Refinement of Generic Specifications with Algebraic Tools.* IFIP 1983 (ed. R.EA.Mason)

[3] Bidoit M. *Une méthode de présentation des types abstraits : applications.* Thèse de troisieme cycle. Orsay (1981)

[4] Comon H. *An anti-unification approach to decide the sufficient completeness of algebraic specification.* To appear as a LIFIA report.

[5] Ehrig H., Kreowsky H. & Padawitz P. *Stepwise specifications and implementations of Abstract Data Types.* LNCS 62 (1978) pp 205-226.

[6] Goguen J.A., Thatcher J.W. & Wagner E.G. *An Initial Algebra Approach to the Specification, Correctness and Implementation of Abstract Data Types.* Current trends in programming methodology vol. 4, pp 80-149, Ed. Prentice Hall (1978).

[7] Goguen J.A., Futatsugi K., Jouannaud JP. & Messeguer J. *Principles of OBJ2.* Proc. POPL 1985.

[8] Guttag J.V., Horowitz E. & Musser D.R. *The design of data type specification.* Current trends in programming methodology, vol. 1, ch.4. Prentice Hall (1977).

[9] Guttag J.V. and Horning J.J. *The Algebraic Specification of Abstract Data Types.* Acta Informatica 10 (1978).

[10] Huet G. & Hullot JM. *Proofs by induction in equational theories with constructors.* JCSS 25-2 (1982).

[11] Huet G. & Oppen D.C. *Equations and rewrite Rules : a survey.* Technical Report, SRI International, 1980.

[12] Jouannaud JP. & Kounalis E. *Proofs by Induction in Equational Theories Without Constructors.* CRIN, Nancy, France, May 1985.

[13] Kapur D., Narendran P. & Zhang H. *On sufficient completeness and related properties of term rewriting systems.* Preprint October 1985.

[14] Kounalis E. & Zhang H. *A general Completeness Test for Equational Specifications* CRIN Nancy, France, Nov. 1984.

[15] Liskov B. & Zilles S. *An introduction to formal specifications of data abstractions.* Current trends in programming methodology vol. 1. Prentice Hall (1977).

[16] Martelli A. & Montanari U. *An efficient Unification algorithm* ACM TOPLAS, vol.4, pp 258-282 (1982).

[17] Meseguer J. & Goguen J.A. *Initiality, Induction and Computability.* To appear in "Application of Agebra to Language Definition and Compilation", M.Nivat & J.Reynolds (eds), Cambridge U.P.

[18] Musser D. *Convergent Sets of Rewrite Rules for Abstract Data Types.* Report of USC Information Sciences Institute (1979).

[19] Nipkov T. & Weikum G. *A decidability result about sufficient completeness of axiomaticaly specified abstract data types.* LNCS 145 (1982).

[20] Paterson M.S. & Wegman M.N. *Linear Unification.* Journal of Computer and Systems Sciences. vol.16. pp 158-167 (1978).

[21] Plaisted D. *Semantic Confluence and Completion Methods.* Information and Control 65 (1985) pp 182-215.

[22] Remy J.L. *Etude de systèmes de réecriture conditionnels & application aux types abstraits algébriques.* Thèse, Nancy, 1982.

APPENDIX 1 : The algorithm of theorem 4.3.

Let us first introduce some notations:

(i) $\Gamma = <\gamma_1, ..., \gamma_n>$ and $E = <\epsilon_1, ..., \epsilon_n>$ denote sequences in T(X).

(ii) *id* denotes the identity of T(X).

(iii) S being a sequence of terms in T(X), $[S;\ s_1 \rightarrow S_1, ..., s_k \rightarrow S_k]$ is the sequence of terms obtained by replacing each s_i by S_i

(iv) If $\Lambda = \{(\sigma_1, c_1), .., (\sigma_k, c_k)\}$ where $\sigma_i \in \Omega$ and, if $\theta \in \Omega$, then let $\Lambda \otimes \theta$ denote $\{(\theta \circ \sigma_1, c_1), ..., (\theta \circ \sigma_k, c_k)\}$.

The set A of theorem 4.3 is obtained by performing $\eta(<t>, <t'>)$, if t # t' is the inequation to be solved. t and t' are supposed to satisfy the conditions of theorem 4.3.

$\eta(\Gamma, E)$:= case of:

(0) $\Gamma = \emptyset$,
Then $\emptyset$

(1) $\exists j$, $(root(\gamma_j \neq root(\epsilon_j))$ and $(root(\gamma_j) \notin X)$ and $(root(\epsilon_j) \notin X)$,
Then $\{(\text{id}, \emptyset)\}$.

(2) $\exists j$, $(root(\gamma_j) \notin X)$ and $(root(\epsilon_j) \notin X)$ and $(root(\gamma_j) = root(\epsilon_j))$,
$\gamma_j = f(\varsigma_1, ..., \varsigma_r)$ and $\epsilon_j = f(\xi_1, ..., \xi_r)$,
Then $\eta([\Gamma;\ \gamma_j \rightarrow <\varsigma_1, .., \varsigma_r>], [E;\ \epsilon_j \rightarrow <\xi_1, ..., \xi_r>])$.

(3) $\exists j, (\gamma_j \notin \bigcup_{i \neq j} V(\gamma_i))$ and $(\gamma_j \in X)$,
Then $\eta([\Gamma;\ \gamma_j \rightarrow \emptyset], [E;\ \epsilon_j \rightarrow \emptyset])$

(4) $\exists j$, $(\gamma_j \notin X)$,
let $\Sigma = \{f_1, ..., f_k\}$ and, for $l \leq k$, let X_l be a set of variables (not yet used),
Then $\bigcup_{l=1}^{k}(\eta(\Gamma, [E;\ \epsilon_j \rightarrow f_l(X_l)]) \otimes [\epsilon_j \rightarrow f_l(X_l)])$

(5) $\exists j$, $(\gamma_j = \gamma_1)$
Then $\eta([\Gamma;\ \gamma_j \rightarrow \emptyset], [E;\ \epsilon_j \rightarrow \emptyset]) \cup \eta([\Gamma;\ \gamma_1 \rightarrow \emptyset], [E;\ \epsilon_1 \rightarrow \emptyset]) \cup \mu(\epsilon_1, \epsilon_j)$

where μ is recursively defined by: $\mu(\epsilon_1, \epsilon_j)$:= case of:

(6) $(\epsilon_1 \in X)$ and $(\epsilon_j \in X)$
Then $\{(id, \{\epsilon_1 \not\approx \epsilon_j\})\}$

(7) $(\epsilon_1 \notin X)$ and $(\epsilon_j \in X)$
Then $\eta(<\epsilon_1>, <\epsilon_j>)$

(8) $(\epsilon_1 \in X)$ and $(\epsilon_j \notin X)$
Then $\eta(<\epsilon_j>, <\epsilon_1>)$

(9) ϵ_1 and ϵ_j are both non-variables
Then

(9-1) if $root(\epsilon_1) \neq root(\epsilon_j)$
Then $\{(id, \emptyset)\}$

(9-2) if $\epsilon_1 = f(\varsigma_1, ..., \varsigma_r)$ and $\epsilon_j = f(\varsigma_1, ..., \varsigma_r)$ and $\epsilon_j = f(\xi_1, ..., \xi_r)$
Then $\bigcup_{i=1}^{r} \mu(\varsigma_i, \xi_i)$

The proof of termination and corectness of this algorithm can be found in [4].

APPENDIX 2: the Normalize function of theorem 4.6

We shall use the notations:

- If c $= \{ u_1 \not\approx v_1, ..., u_m \not\approx v_m\}$ is a set of constraints and $\sigma \in \Omega$, then $\sigma(c)$ denotes $\{ \sigma(u_1) \not\approx \sigma(v_1), ..., \sigma(u_m) \not\approx \sigma(v_m)\}$.
- If $\sigma_1, ..., \sigma_k, \sigma \in \Omega$ and $c_1, ..., c_k, c$ are sets of constraints, let $\{ (\sigma_1, c_1), ..., (\sigma_k, c_k) \} \circ (\sigma, c)$ denote the set $\{ (\sigma_1 \circ \sigma, c_1 \cup \sigma_1(c)), ..., (\sigma_k \circ \sigma, c_k \cup \sigma_k(c)) \}$
- If c $= \{ u_1 \not\approx v_1, ..., u_m \not\approx v_m \}$ is a set of constraints, let V(c) be $\bigcup_{i=1}^{m}(V(u_i) \cup V(v_i))$

Normalize(σ, c) :=

(1) if c $= \emptyset$ then $\{(\sigma, \emptyset)\}$

(2) otherwise, let $x \in V(c)$,
$c_1(x) = \{ (u \not\approx v) \in c,\ x \in V(u) \cup V(v)\}$, $c_2(x) = c - c_1(x)$,
$c_3(x) = \{ (u \not\approx v) \in c_1(x),\ x = u \text{ or } x = v \}$, $c_4(x) = c_1(x) - c_3(x)$

(2-1) If $c_4(x) = \emptyset$ then (Normalize$(id, c - c_3(x))) \circ (\sigma, c_3(x))$.

(2-2) If $\exists (u \not\approx v) \in c_4(x),\ x = v$, then Normalize$(\sigma, (c - \{u \not\approx v\}) \cup \{v \not\approx u\})$

(2-3) If $\exists (u \not\approx v) \in c_4(x), i\, x \in V(v)$
then Normalize$(\sigma, (c - \{u \not\approx v\}) \cup \{v \not\approx u\})$

(2-4) If $\exists (u \not\approx v) \in c_4(x),\ v \notin X,\ root(u) \neq root(v)$, then Normalize$(\sigma, c - \{u \not\approx v\})$

(2-5) If $\exists (u \not\approx v) \in c_4(x),\ v \notin X$,
let $u = f(\varsigma_1, ..., \varsigma_r)$ and $v = f(\xi_1, ..., \xi_r)$,
then $\bigcup_{i=1}^{r}$Normalize$(\sigma, (c - \{u \not\approx v\}) \cup \{\varsigma_i \not\approx \xi_i\})$

(2-6) If $(u \not\approx v) \in c_4(x),\ x \in V(u),\ v \in X$,
let $\Sigma = \{f_1, ..., f_k\}$ and $\forall i \leq k,\ y_{i,1}, ..., y_{i,r_i}$ be still unused variables (r_i is the arity of f_i) ,
then $\bigcup_{i=1}^{r}$Normalize($(v \leftarrow f_i(y_{i,1}, ..., y_{i,r_i})) \circ \sigma$, $(v \leftarrow f_i(y_{i,1}, ..., y_{i,r_i}))$ (c))

A New Method for Establishing Refutational Completeness in Theorem Proving

Jieh Hsiang[1]
Department of Computer Science
SUNY at Stony Brook
Stony Brook, NY 11794
U.S.A.

Michael Rusinowitch
CRIN
B.P. 239
54506 Vandoeuvre-les-Nancy
France

Abstract

We present here a new technique for establishing completeness of refutational theorem proving strategies. This method employs semantic trees and, in contrast to most of the semantic tree methods, is based on proof by refutation instead of proof by induction. Thus, it works well on transfinite semantic trees as well as on finite ones. This method is particularly useful for proving the completeness of strategies with the presence of the equality predicate. We have used the method to prove the completeness of the following strategies (without the need of the functional reflexive axioms), where the precise definition of *oriented paramodulation* will be given later.

- Resolution + oriented paramodulation
- P1-resolution + oriented paramodulation
- Resolution with ordered predicates + oriented paramodulation using clauses only containing the equality predicate
- unfailing Knuth-Bendix-Huet algorithm
- The EN-Strategy ([Hsi85])

1. Introduction

The question of completeness of resolution and paramodulation without the functional reflexive axioms has been a long standing problem in theorem proving. In [Bra75] an indirect proof (as a corollary of the completeness of the modification method) was given. A direct proof was given in ([Pet83]). However, Peterson's proof requires the use of a simplification ordering which is also order isomorphic to ω on ground terms. This puts a serious restriction on the type of orderings which can be used for ordering equalities. Peterson conjectured that lifting this restriction requires a proof method that can successfully deal with transfinite semantic trees. In this paper we present such a method. Our method is different from the usually semantic tree methods (e.g. [ChL73, KoH69, Pet83]) in that it "shrinks" the tree *refutationally* as opposed to inductively. This eliminates the difficulty with transfinite induction.

2. First Order Logic with Equality

An *inference rule* is a rule for deducing a consequence from a set of formulas. Unless otherwise specified, most of the inference rules in this paper build a new clause from a set of clauses.

Let *Inf* be a set of inference rules and S a set of clauses, $Inf(S)$ denotes the set of clauses obtained by adding to S all clauses generated by applying some rule in *Inf* to S. Let $Inf^{n+1}(S)=Inf(Inf^n(S))$ and $Inf*(S)$ be the limit of $Inf^n(S)$ when n approaches infinity. When there is no ambiguity about *Inf*, we simply use $S*$ for $Inf*(S)$.

We call a set of inference rules *Inf* **(refutationally) complete** if, given any unsatisfiable set of clauses S, *Inf* can deduce *NIL*, the empty clause. The following trivial result will serve as the basis of our development.

Proposition: *Inf is (refutationally) complete iff for every unsatisfiable set of clauses S, NIL (the empty clause) belongs to S*.*

As mentioned before, we are particularly interested in theorem proving methods with the presence of the equality predicate. The equality predicate cannot be treated as any other predicate since the following axioms about equality are assumed:

$(x=x)$
$(x=y)\supset(y=x)$
$(x=y)\wedge(y=z)\supset(x=z)$
Given any P, $(x=y)\wedge P(\cdots,x,\cdots)\supset P(\cdots,y,\cdots)$
Given any f, $(x=y)\supset f(\cdots,x,\cdots)\supset f(\cdots,y,\cdots)$,

where all free variables are universally quantified. These axioms can be satisfied in a special class of interpretations called the *E-interpretations*. Let *TERM* (resp. *GT*) be the set of terms (resp. ground terms), *ATOM* (resp. *GA*) the set of atomic formulas (resp. ground atomic formulas). An *E-interpretation* is a function I with domain *GA* and range $\{true, false\}$ satisfying :

$I(a=a)=true$,
If $I(a=b)=true$ then $I(B[a])=I(B[b])$,

where a is a subterm of B and $B[b]$ replaces an instance of a in $B[a]$ by b. Note that we consider $a=b$ and $b=a$ as the same atom.

We have the following well-known result (see, e.g., [ChL73]):

Theorem: *Call the set of equational axioms K, and let S be a set of clauses. Then $S\cup K$ is unsatisfiable (we say that S is E-unsatisfiable) iff S is not valid in any E-interpretation on GA.*

This motivates us to build semantic trees which capture the E-interpretations. For that purpose, we need *GA* to be well-ordered in a way such that every equality atom occurs before any atom it may reduce.

3. Strong Simplification Orderings

Let $<$ be an ordering on $TERM \cup ATOM$ such that:

S1. $<$ is well-founded.

S2. $<$ is total on $GT \cup GA$.

S3. for every $w, v \in ATOM \cup TERM$ and every substitution θ:
$w < v$ implies $w\theta < v\theta$.

S4. for every t, s in $TERM$ and w in $ATOM \cup TERM$,
$t < s$ implies $w[t] < w[s]$

S5. for every t, s, a, b in $TERM$, where $t \leq s$, and w in $ATOM$,
1 if s is a subterm of w and w is not an equality atom,
then $(s = t) < w$
2 if s is a strict subterm of a or b then
$(s = t) < (a = b)$

We call an ordering satisfying these properties a **strong simplification ordering**. There are some minor differences between the above ordering and the one given in [Pet83], such as we treat $s = t$ and $t = s$ as the same. The *major* difference, however, is that we only require $<$ to be total on ground terms ($GT \cup GA$), *not* order isomorphic to ω. As a consequence, our ordering allows limit ordinals. With this improvement, we can now use almost all the *well-behaved* orderings in the term rewriting literature (e.g., the original Knuth-Bendix ordering [KnB70], recursive path ordering [Der82], recursive decomposition ordering [JLR82], the path of subterm ordering [Pla78]) since all of them can be modified to satisfy this property.

Note the following property which we shall use when we construct the semantic trees corresponding to the E-interpretations.

An ordering satisfying S1 through S5 is order-isomorphic to some ordinal number α on GA.

4. Transfinite E-Semantic Trees

The set of E-interpretations can be represented using an *E-semantic tree*, which is *unique* with respect to the ordering $<$ on the set GA.

Given an ordering $<$ which satisfies the above requirements, and an E-interpretation I, suppose B is a ground atom in GA, then a **partial E-interpretation of I at B** is a partial interpretation of I which is defined on all the members of GA which are smaller than B (not including B). Given two partial E-interpretations I_{B1} and J_{B2}, we say J_{B2} is an **extension** of I_{B1} if $B1 < B2$ and I_{B1} is a partial interpretation of J at $B1$ (i.e. I_{B1} and J_{B2} are identical when they are both defined).

The **E-semantic tree** (denoted ET) is a (downward) tree whose nodes at level B (where B is an element in GA) are all the partial E-interpretations at B. The E-semantic tree with respect to a specific ordering $<$ can be defined inductively as follows:

- the root is the empty interpretation
- according to the definition of E-interpretations, the successors of a partial interpretation I at level B are the extensions of I by one of the following cases:

Case 1: If $B = (a = a)$ then I has only one successor J and J satisfies $J(a = a) = true$.

Case 2: If $B = B[s]$, $s = t < B$, $B[t] < B$, and $I(s = t) = true$, then I has one successor J which satisfies $J(B) = I(B[t])$.

Case 3: Otherwise, I has two successors L and R with: $L(B) = true$ and $R(B) = false$.

Case 2, in which the major difference between this definition of a semantic tree and the other definitions occurs, is explained as follows: If $s > t$, then by property S4 of the ordering which ensures that $B[t] < B[s]$, the atom $B[t]$ must have appeared before $B[s]$. By the way the E-semantic tree is defined, $I(B[t])$ must have already been assigned a value. If $I(s = t) = true$, then by the definition of E-interpretations, $I(B[s])$ must have the same value as $I(B[t])$. Therefore there is only one consistent extension of I to the atom B, not two. Case 2 also indicates how paramodulations are done in the form of reduction: If $I(s = t) = true$ and $s > t$, then by definition of the ordering, $B[s] > (s = t)$. Thus, $B[s]$ can be *reduced* (using I) to $B[t]$ using $s \rightarrow t$, and the two atoms ($B[s]$ and $B[t]$) must have the same truth value. We say $B[s]$ is **I-reduced** to $B[t]$ when this happens. A atom B which can be reduced using some equality which is true in I is called **I-reducible**. Otherwise, it is *I-irreducible*. The well-definedness of *I-reducibility* is given in [Pet83] (Theorem 2).

Our definition of E-semantic tree is essentially the same as the one in [Pet83]. However, by the way we define the orderings, the semantic trees so constructed can be transfinite.

The **closed E-semantic tree** of an unsatisfiable set of clauses S, denoted by $ET(S)$, is the maximal subtree of ET such that for every node I in $ET(S)$, every clause C in S, and every ground substitution θ such that the atoms of $C\theta$ are in the domain of I, $I(C\theta) \neq false$. In other words, if I is the last node of a **maximal path** in a closed semantic tree, then any extension of I will refute some ground instance of some clause C in S. If I refutes some $C\theta$, that is, $I(C\theta) = false$, then we call I a **failure node**. The crucial property of the closed semantic trees is that they are, indeed, topologically closed.

Closure Lemma: *The limit of an increasing sequence of nodes of ET(S) belongs to ET(S).*

The significance of the Closure Lemma is that no maximal path can stop just before a limit ordinal. Thus, no failure node can occur at a limit ordinal. Note that a closed E-semantic tree for an unsatisfiable set of clauses may *not* be finite. Thus the usual Herbrand's theorem, which states that S is unsatisfiable if and only if there is a *finite* closed semantic tree, is not true under our definition of semantic tree. As a simple example, consider the set of clauses $S = \{\neg P(g(x)), P(y)\}$. Suppose the language has two unary function symbols f and g, a constant a, and we use the recursive path ordering with $a < f < g$. Then the set GA will be ordered as $\{P(a), P(fa), P(ffa), \cdots, P(ga), P(gfa), P(gffa), \cdots, P(gga), \cdots\}$. The closed semantic tree corresponds to this ordering and S will refute the set of ground instances $\{P(a), P(fa), \cdots, P(ga), \neg P(ga)\}$, which is an infinite set (see Figure 1).

The concept of semantic tree makes it easy to understand the notion of completeness. First we define the ***closed E-semantic tree** of S to be $ET(S*)$. (Recall that $S*$ with respect to a set of inference rules *Inf* includes all the consequences which can be deduced from S using the inference rules.)

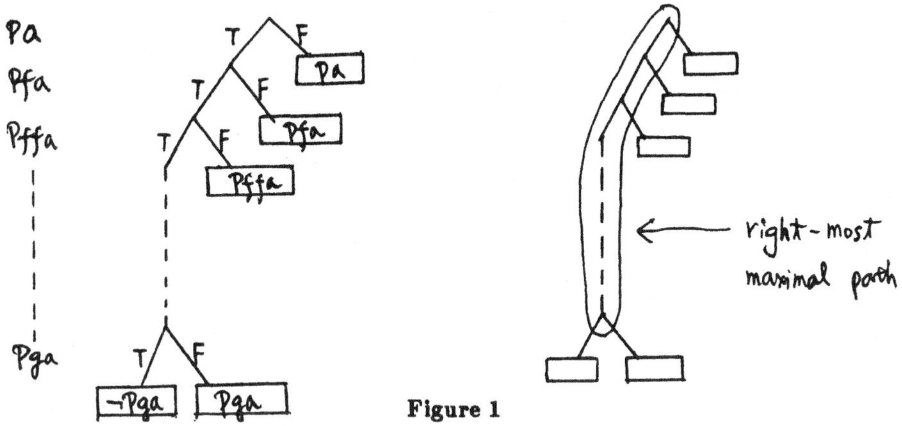

Figure 1

Fundamental Theorem: *Inf is complete iff $ET(S*)$ is empty whenever S is unsatisfiable.*

5. Completeness of Resolution and Oriented Paramodulation

With the above preliminary notions, we now present our proof method. Recall that the purpose of our proof method is to show that a set of inference rules *Inf* is complete. In other words, given any unsatisfiable set of clauses S, we shall show that $ET(S*)$ is empty.

The intuitive idea behind the method is as follows: Given an arbitrary unsatisfiable set of clauses S, if $ET(S*)$ is empty, then by the Fundamental Theorem, *Inf* is complete. If not, find a maximal path in the closed E-semantic tree. If we can show that in all cases, there is a rule in *Inf* which can be applied to the failure node to obtain a "smaller" closed semantic tree, then we have a contradiction. This is because $S*$, by definition, includes *all* the clauses that can be obtained using the inference rules in *Inf*. Thus, *Inf* is complete.

We illustrate this method by establishing the completeness of resolution (with factoring) and oriented paramodulation without the functional reflexive axioms. First we give the definition of these inference rules:

Let S be a set of clauses, and $>$ a strong simplification ordering on the atom set.

Factoring

Let $L_1, L_2 \cdots L_k$ be literals of a clause C which are unifiable with an mgu θ, and let $D = C\theta - (L_2\theta \vee \cdots \vee L_k\theta)$, then D is a *factor of C*.

BinaryResolution

If $C_1 = L_1 \vee D_1$ and $C_2 = L_2 \vee D_2$ are clauses such that L_1 and $\neg L_2$ are unifiable with mgu θ, then $D_1\theta \vee D_2\theta$ is a ***resolvent*** of C_1 and C_2.

Oriented Paramodulation

If $C_1 = (s = t) \vee D$ and $C_2 = C_2[r]$ (where r is a nonvariable subterm of C_2), and

(1) $s\theta = r\theta$ for some mgu θ, and

(2) $s\theta \not\leq t\theta$,

then $C = (C_2[t] \vee D)\theta$ is an ***(oriented) paramodulant of*** C_1 ***into*** C_2 ***at*** r.

Let *Inf* be the set of the above three inference rules. We have

Theorem (Completeness of Resolution+Oriented Paramodulation): *Given a set of clauses S', if S' is E-unsatisfiable then* $NIL \in Inf*(S' \cup \{x=x\})$.

Equivalently, if S' is E-unsatisfiable then $ET((S' \cup \{x=x\})*)$ is empty.

Proof of Completeness (Ground Case):

We first give a proof for the ground case. Then we shall show some Lifting Lemmas which we use to extend the proof to the non-ground case. Let S' be an E-unsatisfiable set of (ground) clauses. Let $S=S' \cup \{x=x\}$. Recall that $S*$ is the closure of S from applying the inference rules on S. If $ET(S*)$ is empty, then by the Fundamental Theorem, *Inf* is complete. Assume not, let us consider the ***rightmost maximal path*** of $ET(S*)$. Let N be the last node of this maximal path. For convenience, from now on we use I_N for the partial interpretation defined at node N. Recall that by definition of a maximal path, I_N does not falsify any $C \in S*$ while any extension of N falsifies some $C \in S*$. Several situations may occur at the end of the path, according to the type of the nodes that follow N in ET. (Recall that ET denotes the E-semantic tree.)

Case 1: N has one child L and the edge is labeled by $a=a$ for some term a.

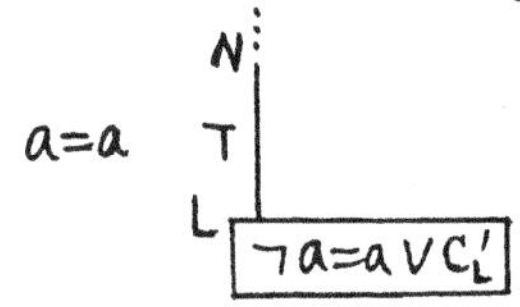

Since L is not in $ET(S*)$, there is a clause C_L in $S*$ for which $I_L(C_L)$=*false*. Since I_N does not falsify C_L, $C_L=\neg(a=a) \vee C_L'$ for some clause C_L'. Moreover, $I_N(C_L')$=*false* since $a=a$ does not appear in C_L', and I_N and I_L differ only on the value for $a=a$.[1] Now, C_L' is a resolvent of C_L and $x=x$, therefore by the definition of $S*$, $C_L' \in S*$. However, this implies that N is a failure node, contradicting our assumption that it is in $ET(S*)$, whose nodes, by construction, do not falsify any $C \in S*$. Therefore case 1 cannot happen.

Case 2: The edge follows N is B and it is I_N-irreducible. Thus, N has two children, L and R.

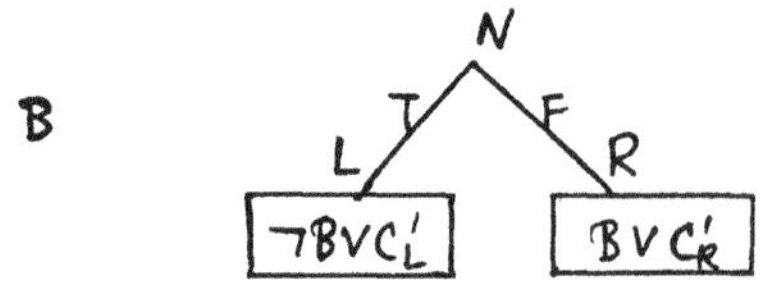

Since L and R are not in $ET(S*)$, there are C_L and C_R in $S*$ for which $I_L(C_L)$=*false* and $I_R(C_R)$=*false*. However, I_N does not falsify C_L or C_R. Hence

[1]Recall that we are working on the ground case, therefore we do not need to consider multiple occurrences of the same literal. In non-ground case, this will be taken care of by factoring.

$C_L = \neg B \vee C_L{}'$
$C_R = B \vee C_R{}'$, for some clauses $C_L{}'$ and $C_R{}'$.

and $I_N(C_L{}') = I_N(C_R{}') = false$. Now once again, C_L and C_R form a resolvent $C_L{}' \vee C_R{}'$, which is in $S*$ and is falsified by I_N. Thus, I_N must be a failure node, which contradicts our assumption.

Case 3: The edge following N is labeled by an atom B which is I_N-reducible.

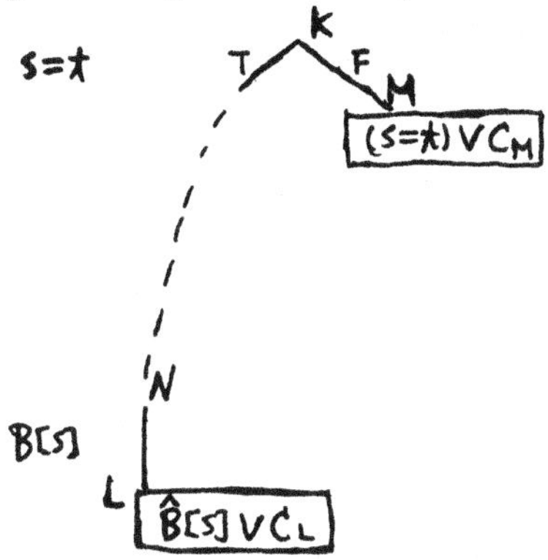

Let $s = t$ be the *smallest* equation such that $s > t$, s is a subterm of B, and $I_N(s = t) = true$. Such a smallest element must exist since B is reducible and the ordering on atoms is well-founded and total on ground terms. Note that the atom $s = t$ has to be irreducible since it is the smallest equation which reduces B. In the following, we represent B as $B[s]$. Let M, L, and K be nodes as indicated in the figure. We have the following facts:

(1) M is a failure node, which falsifies some clause $(s = t) \vee C_M \epsilon S*$.

(2) L is a failure node, which falsifies some clause $\hat{B}[s] \vee C_L \epsilon S*$, where $\hat{B}[s]$ is either $B[s]$ or $\neg B[s]$.

(3) $I_K(C_M) = I_N(C_M) = I_N(C_L) = false$.

(4) $I_L(\hat{B}[s]) = I_L(\hat{B}[t]) = I_N(\hat{B}[t]) = false$.

(1) is true since N is the last node of the *right-most* maximal path. If M were not a failure node, there would be another maximal path to the right of N. (2) is obvious, for the same reason as in the previous two cases. (3) is true by the definition of paths and failure nodes (as in the previous case). (4) is true by the definition of the semantic tree. To be more precise, since $I_N(s = t) = true$, the value of $B[s]$ in I_N has to be the same as the value of $B[t]$ (in I_N). Therefore (4) is true.

By the oriented-paramodulation inference rule, since $s > t$, there is an oriented paramodulant, $\hat{B}[t] \vee C_M \vee C_L \epsilon S*$, between $s = t \vee C_M$ and $\hat{B}[s] \vee C_L$. By (3) and (4) above, $I_N(\hat{B}[t] \vee C_M \vee C_L) = false$, which contradicts the assumption that I_N is not a failure node.

Thus, we have shown that in each of the three cases, there is a contradiction. Therefore, $ET(S*)$ has to be empty, or, equivalently, *Inf* (in this case, oriented paramodulation, resolution, and factoring) is complete.

Note that this proof is considerably simpler than the one given in [Pet83]. Also, since the proof is not based on induction, we do not have to worry about the problems which may come with transfinite E-semantic trees. Consequently, our method will work nicely on transfinite E-semantic trees, as well as on the finite ones.

6. The Lifting Arguments

To lift our proof from the ground case to non-ground case, we can simply use the Lifting Lemmas given in [Pet83]. Note that in the proofs of the Lifting Lemmas in [Pet83], the assumption that the ordering is order isomorphic to ω on ground terms is never used. To avoid repetition, we only state the key lemmas ([Rob65, WoR70]) needed for the lifting.

Resolution Lifting Lemma If C_1' and C_2' are instances of C_1 and C_2, respectively, and if C' is a resolvent of C_1' and C_2', then there is a resolvent C of C_1 and C_2 such that C' is an instance of C.

A proof can be found in [ChL73]. This lemma is used for lifting in Cases 1 and 2 given above.

The major difference between the lifting lemma for paramodulation and for resolution is that in paramodulation the lifting can be done only when the term which is being paramodulated into already exists in the original clause.

Paramodulation Lifting Lemma Let C_1 and C_2 be two clauses and θ be a ground substitution. Also let r be a proper (i.e. nonvariable) subterm of C_2 and C' be an oriented paramodulant from paramodulating $C_1\theta$ into $C[r]_2\theta$ at $r\theta$. Then there is a paramodulant of C_1 into $C_2[r]$ at r.

In order to use the paramodulation lifting lemma, we need to ensure that the clause $\hat{B}[s] \vee C_L$ used in the above proof has the desired property. This can be ensured by the following lemma:

Lemma Suppose θ is a ground substitution and $C\theta$ is a clause such that $I_L(C\theta)=\mathit{false}$. Then there exists an I_L-irreducible ground substitution σ such that $I_L(C\sigma)=\mathit{false}$.

This is basically Theorem 4 in [Pet83]. Note that with this lemma, we can choose $\hat{B}[s] \vee C_L$ so that the substitutions corresponding to its non-ground clause (call it C_L') are I_L-irreducible. Thus, the oriented paramodulation which is performed on $\hat{B}[s] \vee C_L$ has to be on a subterm corresponding to a nonvariable subterm of C_L'.

Thus, we have completed the lifting arguments.

7. Completeness of Other Strategies

We now use the method to prove the completeness of some other strategies.

7.1. Ordered Predicate Strategy

This strategy puts restrictions on resolution and on paramodulation. First we assume that predicates are totally ordered, with the equality predicate as the smallest. That is, there is an ordering $>_P$ on the predicate symbols such that

$$P_n >_p P_{n-1} >_p \cdots >_p P_1 >_p =$$

The inference rules are the following:

Factoring

As usual.

Ordered Predicate Resolution

Given clauses $C_1 = L_1 \vee D_1$ and $C_2 = L_2 \vee D_2$. If L_1 and $\neg L_2$ are unifiable with mgu σ, *and* the predicate symbol in L_1 is the largest predicate symbol in *both* C_1 and C_2, then $D_1\sigma \vee D_2\sigma$ is an **ordered˜ predicate˜ resolvent** of C_1 and C_2.

E-P Paramodulation

Given $C_1 = (s = t) \vee D$ and $C_2 = C_2[r]$, where $s\sigma = r\sigma$, then $C = (C_2[t] \vee D)\sigma$ is an **E-P paramodulant** of C_1 into C_2 if

(1) C is an oriented paramodulant of C_1 into C_2, and

(2) the only predicate symbol in C_1 is $=$.

Then we have

Theorem The ordered predicate strategy is complete for first order predicate calculus with equality.

The *only* difference between the proof for this completeness theorem and the previous one is that we further restrict the strong simplification ordering on the atom set.

Suppose $>_F$ is a simplification ordering on the set of terms which is also total on ground terms. Define $>$ on the atom set as follows:

$s = P(s_1, \cdots, s_n) > t = Q(t_1, \cdots, t_m)$ if

(i) $P >_p Q$, or

(ii) $P = Q$, P is not the equality predicate, and $(s_1, \cdots, s_n) >_F (t_1, \cdots, t_m)$ compared lexicographically, or

(iii) $P = Q$, P is the equality predicate, and $\{s_1, s_2\} >>_F \{t_1, t_2\}$, where $>>_F$ is the multiset ordering of $>_F$.

It is not hard to verify that $>$ is indeed a strong simplification ordering. A similar ordering is also used in [Hsi85]. Another interesting property of this ordering is that it orders the ground atoms first by their predicate symbols, then by their arguments. Therefore in the E-semantic tree, no atom of the form $P_1[r]$ will be enumerated before all atoms of the form $s = t$ are exhausted; similarly no atom of the form $P_2[u]$ will appear before all $P_1[v]$ are exhausted.

The proof of completeness is very similar to the previous one, and for convenience we will use the same names for nodes and clauses as before. If $ET(S*)$ is empty, then the strategy is complete. Assume not, we choose the right-most maximal path in $ET(S*)$, and we have three cases to consider.

In Case 2 (Case 1 is similar), the literals to be resolved (B and $\neg B$) are the last literals in the respective partial interpretations (I_R and I_L). By the definition of the ordering, B

must have the largest predicate symbol among literals in both C_L and C_R. Therefore the resolution done is indeed an ordered resolution. Case 3 is similar. Since $(s=t)$ is the largest literal (with respect to $>$) in $(s=t)\vee C_M$, and since no predicate symbol can appear in the ordering before all equality atoms are exhausted, C_M can contain only equality literals. Thus, we have a contradiction in all three cases.

Note that the completeness of this strategy ***cannot*** be proved using Peterson's method if there are predicate symbols other than $=$ in the language. It is because the ordering given above cannot be order-isomorphic to ω.

7.2. Ordered Resolution Strategy

Noting that in the above proof the atom B is the *largest* literal in both C_L and C_R, we may further improve the aforementioned strategy to restrict resolution to the following (the other inferences remain the same):

Ordered Resolution

Given clauses $C_1=L_1\vee D_1$ and $C_2=L_2\vee D_2$. If L_1 and $\neg L_2$ are unifiable with mgu σ, no literal in D_1 is larger than L_1, ***and*** no literal in D_2 is larger than L_2, then $D_1\sigma\vee D_2\sigma$ is an **ordered resolvent** of C_1 and C_2.

The usefulness of this strategy is demonstrated by the following example: Given two clauses $P(0)$ and $P(x)\supset P(s(x))$. Using unrestricted resolution (or ordered predicate resolution), infinitely many resolvents (P(s(0)), P(s(s(0))), $\cdots$) will be generated. On the other hand, ***no*** resolvent will be generated using ordered resolution, since $P(s(x))>P(x)$ and there is no resolvent between $P(s(x))$ and $P(0)$. (This also implies that the two clauses are satisfiable.)

7.3. P1-Resolution with Ordered Paramodulation

A clause is a **P1-clause** if it has only positive literals. The strategy we describe here has the following inference rules:

Factoring

As usual.

Oriented Paramodulation

As before.

P1-Resolution

Two clauses are **P1-resolvable** if they are resolvable and if one of them is a P1-clause. A resolvent from P1-resolution is called a **P1-resolvent.**

The proofs for Case 1 and Case 3 remain the same. Case 2 goes as follows: If C_R is a P1-clause, then P1-resolution can be performed and we are done. If not, let $\neg B_1$ be a negative literal in C_R. Then by the definition of the right-most maximal path, there is a node M such that $I_M(B_1)$=*false* and M is a failure node (see figure below).

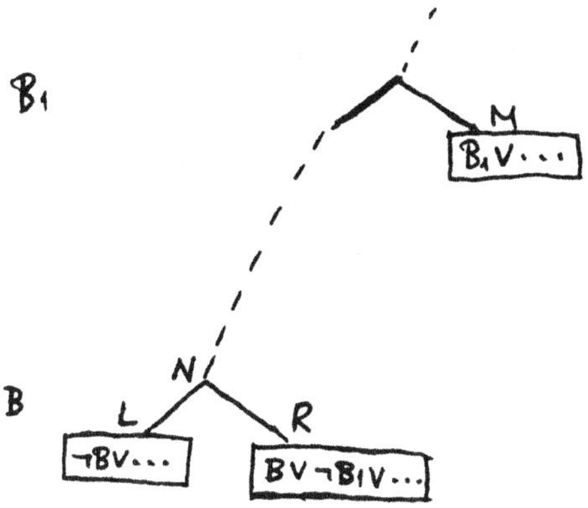

If C_M is a P1-clause, then C_M and C_R can P1-resolve and obtain another clause $C_R{}'$ which also refutes I_R but does not contain $\neg B_1$. Otherwise, we apply the same method on C_M and try to P1-resolve away its negative literals. Because the ordering is well-founded, such an upward search is finite. Then since each clause has only finitely many (negative) literals, by the Koenig Lemma, we will find, in finite time, a clause C (using P1-resolution) which does not contain any negative literal and which also refutes I_R. Then C and C_L can perform P1-resolution on B and derive the desired contradiction. Note that the literal B must still be in C, since all the P1-resolutions performed above can only eliminate literals (from C_R) which are smaller than B.

7.3.1. P1-resolution with Ordered Predicates

An immediate consequence of the above proof is the completeness of a mixture of the two strategies. If the ordering on predicates as given in the ordered predicate strategy is used in the above proof, then we have the following strategy:

Factoring

As usual.

PO–Resolution

Given two clauses, one is a P1-clause. Then they are **PO–resolvable** if they are resolvable and the predicate being resolved in the P1-clause contains the largest predicate symbol in that clause.

E–P Paramodulation

Same as before.

The completeness of this strategy follows immediately from the proof given above.

7.4. More Strategies

We have used this proof method to prove the completeness of some other strategies. One of them is the EN-strategy as described in [Hsi85]. The proof is very similar to the proof for the P1-strategy given above. Another method that we have proved to be complete is an extension of the Knuth-Bendix procedure for equational theories, which lifts the noetherian requirement. The method is the subject of an exposition by itself and it will not be described here. We shall simply mention that the proof requires the selection of a different maximal path. Also note that all of the above strategies can be modified to incorporate some demodulation technique ([LaB79, Sla74, WoR70]), namely, a newly generated clause can always be fully simplified using the existing orientable equalities before being added to the clause set.

Acknowledgement

This research is supported in part by the NSF grant DCR-8401624 of U.S, and the Greco de Programmation of CNRS, France. The authors would like to thank Jean-Pierre Jouannaud and Pierre Lescanne for the discussions and material support.

1. References

[Bra75] D. Brand, "Proving Theorems with the Modification Method", *SIAM J. of Computing*, **4**, (1975), 412-430.

[ChL73] C. L. Chang and C. T. Lee, *Symbolic Logic and Mechanical Theorem Proving*, Academic Press, 1973.

[Der82] N. Dershowitz, "Orderings for Term Rewriting Systems", *J.TCS*, **17**, 3 (1982), 279-301.

[Hsi85] J. Hsiang, "Two Results in Term Rewriting Theorem Proving", *Proc. of 1st International Conference in Rewrite Techniques and Applications*, May, 1985.

[JLR82] J. P. Jouannaud, P. Lescanne and F. Reinig, "Recursive Decomposition Ordering", *Conf. on Formal Description of Programming Concepts II*, 1982, 331-346.

[KnB70] D. E. Knuth and P. B. Bendix, "Simple Word Problems in Universal Algebras", in *Computational Algebra*, J. Leach, (ed.), Pergamon Press, 1970, 263-297.

[KoH69] R. A. Kowalski and P. Hayes, "Semantic Trees in Automatic Theorem Proving", in *Machine Intelligence*, vol. 5, B. Meltzer and D. Michie, (eds.), American Elsevier, 1969, 181-201.

[LaB79] D. S. Lankford and A. M. Ballantyne, "The Refutation Completeness of Blocked Permutative Narrowing and Resolution", *4th Conf. on Automated Deduction*, Austin, TX, 1979.

[Pet83] G. E. Peterson, "A Technique for Establishing Completeness Results in Theorem Proving with Equality", *SIAM J. of Computing*, **12**, 1 (1983), 82-100.

[Pla78] D. A. Plaisted, "A Recursively Defined Ordering for Proving Termination of Term Rewriting Systems", UIUCDCS-R-78-943, Univ. of Illinois, Urbana, IL, 1978.

[Rob65] J. A. Robinson, "A Machine Oriented Logic based on the Resolution Principle", *J. ACM*, **12**, 1 (January 1965), 23-41.

[Sla74] J. Slagle, "Automated Theorem Proving with Simplifiers, Commutativity, Associativity", *J. ACM*, **21**, (1974), 622-642.

[WoR70] L. Wos and G. A. Robinson, "Paramodulation and Set of Support", Lecture Notes in Math. No 125, Springer-Verlag, 1970.

A Theory of Diagnosis from first Principles

R.Reiter
Dept. of Computer Science
University of Toronto
Toronto M5S 1 A7

CANADA

SOME CONTRIBUTIONS TO THE LOGICAL

ANALYSIS OF CIRCUMSCRIPTION

Gerhard Jaeger

ETH-Zuerich

Mathematik

ABSTRACT

After sketching the first and second order version of McCarthy's predicate circumscription, we introduce the notion of positive disjunctive circumscription as an approach to (this form of) non-monotonic reasoning which guarantees consistency. We define the positive disjunctive extension $PD(T)$ of a first order theory T and show that it is conservative over T. Then we turn to sets defined by positive disjunctive circumscription and state a boundedness theorem concerning their stages. The last considerations refer to generalizations of positive disjunctive circumscription. We discuss the inclusion of (intersective) priority relations and extensions by iteration.

0. Introduction

The concept of circumscription goes back to McCarthy and is an approach to the formalization of non-monotonic reasoning. The main references are [11] and [12] where he introduces various forms of circumscription, gives some motivation and provides examples of its application. In contrast to the work of McDermott-Doyle [13], Moore [14], Parikh [16] et al., his approach is not based on modal logic but represents non-monotonicity by augmenting first order predicate logic in a suitable way. The circumscription of the predicate P in the formula $A(P)$ is the first order theory $Circum(A(P))$ which consists of

(1) $A(P)$;

(2) $A(B) \mathbin{\&} \forall x(B(x) \rightarrow P(x)) \quad \rightarrow \quad \forall x(P(x) \rightarrow B(x))$

for all formulas $B(x)$. Roughly speaking, these axioms can be regarded as an implicit definition of P, stating that it is a minimal solution of A.

In spite of its expressive power and elegance, the general concept of circumscription is problematic from a logical point of view. $Circum(A(P))$ may lead to inconsistencies even if $A(P)$ is consistent. And, yet more serious for applications, it is often very complicated to decide whether this is the case or not. In order to overcome this problem, we will develop a formalism, hereafter called positive disjunctive circumscription, which is guaranteed to be consistent and has the same range of applicability as the original versions of circumscription.

Sections 1 and 2 of this paper collect the basic notions and contain a theorem concerning the second order version of predicate circumscription. The new concept of the positive disjunctive extension $PD(T)$ of a first order theory T is introduced in section 3 where we also prove that it is consistent, provided that T is consistent. In section 4 we stratify the predicates defined by positive disjunctive circumscription and study their mathematical complexity. Thereby we also obtain some information about the proof-theoretic strength of our approach. Section 5 finally is dedicated to generalizations of positive disjunctive circumscription.

Questions concerning the consistency of circumscription are also considered by Etherington, Mercer and Reiter [5], who are interested in universal theories. A theorem closely related to Corollary 1 below has been found by Lifschitz independently and is proved in his very recent [10]. Reiter's closed world assumption [17] and Clark's completion of theories [3] are other forms of non-monotonic reasoning which have some connections to circumscription (cf. Lifschitz [9] and Reiter [18] for partial results). The investigations of this article will be used in a subsequent paper to establish the exact relationship between these concepts.

1. The formal framework

We begin with setting up the formal framework. Let L be a language of first order predicate calculus with equality $=$ whose non-logical symbols consist of countably many individual variables $x, y, z, x_1, y_1, z_1, \ldots$, finitely many function symbols and finitely many relation symbols; 0 ary function symbols are called constants. To keep notation simple, we claim that all relation symbols are unary and that we have a binary pairing function $<.,.>$ together with a distinguished constant 0. The terms $s, t, s_1, t_1, \ldots$ and formulas $A, B, C, D, A_1, B_1, C_1, D_1, \ldots$ of L are defined as usual; a sentence is a formula without free variables. We write $\underline{t}$ for a finite string $t_1, \ldots, t_n$ of L terms and use the notation $A[\underline{x}]$ to indicate that all free variables of A come from the list $\underline{x}$; $A(\underline{x})$ may contain other free variables besides $\underline{x}$.

If V and W are relation symbols, we sometimes write $t \in V$ for $V(t)$, $s \in (V)_t$ for $<s,t> \in V$, $V \subset W$ for $\forall \underline{x}(V(\underline{x}) \rightarrow W(\underline{x}))$ and $V = W$ for $V \subset W \;\&\; W \subset V$. This notation is also extended to sequences $\underline{V} = V_1, \ldots, V_n$ and $\underline{W} = W_1, \ldots, W_n$ of relation symbols: $\underline{V} \subset \underline{W}$ is an abbreviation for $V_1 \subset W_1 \;\&\; \cdots \;\&\; V_n \subset W_n$ and $\underline{V} = \underline{W}$ for $\underline{V} \subset \underline{W} \;\&\; \underline{W} \subset \underline{V}$. Given a relation symbol V, a formula $A(V)$

and a formula $B(x)$, perhaps with other free variables, $A(B)$ denotes the result of substituting $B(t)$ for each occurrence of $V(t)$ in $A(V)$.

A theory in L is a (possibly infinite) collection of L formulas. We write $Th \vdash A$, if the formula A can be deduced from the theory Th by the usual axioms and rules of predicate logic with equality. The theory *Pair* of elementary pairing consists of the axioms

$(P1)$ $0 \neq \langle x,y \rangle$;

$(P2)$ $\langle x_1,y_1 \rangle = \langle x_2,y_2 \rangle \rightarrow x_1 = x_2 \;\&\; y_1 = y_2$.

An L structure $\mathbf{M}$ is a tuple of the form $\langle M,... \rangle$, $M \neq \emptyset$, where the set M is the range of the individual variables and ... is a list of functions and relations (on M) of the appropriate arities which serve as interpretations of the function and relation symbols of L; the interpretation of the function or relation symbol J is denoted by $\mathbf{M}[J]$. The truth of a sentence A in the structure $\mathbf{M}$ is defined as usual and denoted by $\mathbf{M} \models A$. $\mathbf{M}$ is called a model of the theory T, if every sentence which is provable in T is true in $\mathbf{M}$.

In L we introduce as abbreviations: $S(t) := \langle t,0 \rangle$, $1 := S(0)$, $2 := S(1)$, $3 := S(2)$, ... In *Pair* one can easily prove that the terms 0, 1, 2, ... are all distinct. Consequently, in every model $\mathbf{M}$ of *Pair*, the set $\{ \mathbf{M}[0], \mathbf{M}[1], \mathbf{M}[2], \ldots \}$ will be identified with the usual set N of natural numbers and S with the successor function on N.

In the following T will be a consistent theory which is formulated in L and contains *Pair*. We also assume that P, Q, P_1, ..., P_n are new unary relation symbols which do not belong to L. The language $L(P)$ is like L except that the new relation symbol P has been added which allows to obtain new atomic formulas of the form $P(t)$. $L(P_1, \ldots ,P_n)$ is the analogous extension of L by new relation symbols $P_1, \ldots ,P_n$.

The terminology introduced so far is standard and presented in most textbooks on mathematical logic and theoretical computer science (cf. e.g. [2], [19]).

2. The notion of circumscription

In this section we introduce the first and second order version of McCarty's predicate circumscription. By circumscribing the predicate P in the sentence $A(P)$ one attempts to capture an idea which is inherent in Occam's razor: Only those elements should belong to P which are minimally required by the context (cf. [4]). It is our intention to circumscribe the predicates $\underline{P} = P_1, \ldots ,P_n$ for suitable formulas $A(\underline{P})$ of $L(\underline{P})$. However, to motivate the axioms, we first present the model-theoretic counterpart of circumscription (cf. [11]).

Definition. Let $A(\underline{P})$ be a formula of $L(\underline{P})$ where $\underline{P} = P_1, \ldots, P_n$.

1. The $L(\underline{P})$ structure $\mathbf{M} = <M,...>$ is called a proper $\underline{P}$ substructure of the $L(\underline{P})$ structure $\mathbf{M}' = <M',...>$ if

(i) $M = M'$;

(ii) $\mathbf{M}[J] = \mathbf{M}'[J]$ for every function and relation symbol J of L;

(iii) $\mathbf{M}[P_i] \subset \mathbf{M}'[P_i]$ for all $i = 1,\ldots,n$;

(iv) $\mathbf{M}[P_i] \neq \mathbf{M}'[P_i]$ for some $i = 1,\ldots,n$;

2. $\mathbf{M}$ is a $\underline{P}$ minimal model of the $L(\underline{P})$ theory Th if $\mathbf{M}$ is a model of Th and there are no proper $\underline{P}$ substructures of $\mathbf{M}$ which are models of Th.

3. The L theory Th minimally entails the $L(\underline{P})$ sentence B with respect to the $L(\underline{P})$ formula $A(\underline{P})$, in symbols $Th \models_{A(\underline{P})} B$, if B is true in all $\underline{P}$ minimal models of $Th \cup \{A(\underline{P})\}$.

Example 1. We define in $L(P)$:

$A_1(P)$:<=> $\forall x(P(x) \rightarrow P(S(x)))$ & $\exists y(P(y)$ & $\forall x(P(x) \rightarrow y \neq S(x)))$;

$A_2(P)$:<=> $P(0)$ & $\forall x(P(x) \rightarrow P(S(x)))$;

$A_3(P)$:<=> $P(0) \vee (P(1)$ & $\forall x(P(x) \rightarrow P(S(x))))$.

Then we have: $A_1(P)$ is consistent but there is no P minimal model of $Pair \cup \{A_1(P)\}$. In every P minimal model of $Pair \cup \{A_2(P)\}$, P is interpreted as the set of natural numbers. Every model of $Pair$ can be extended to a P minimal model of $Pair \cup \{A_3(P)\}$ if we interpret P as $\{0\}$ or $\{1,2,3, \cdots\}$.

Circumscription of a single predicate P in an $L(P)$ formula $A(P)$ is given by the axioms

(1) $A(P)$;

(2) $A(B)$ & $B \subset P \rightarrow P \subset B$

for all $L(P)$ formulas $B(x)$. By the axiom scheme (2) we formalize that there are no definable proper subset X of P such that $A(X)$. The following definition extends this idea to the parallel circumscription of several predicates.

Definition. Let T be an L theory and $A(\underline{P})$ an $L(\underline{P})$ sentence, $\underline{P} = P_1, \ldots, P_n$. Then $Circum(T,A(\underline{P}))$ is the $L(\underline{P})$ theory which has as axioms

$(CS.1)$ *all axioms of* T;

$(CS.2)$ $A(\underline{P})$;

$(CS.3)$ $A(\underline{B})$ & $\underline{B} \subset \underline{P} \rightarrow \underline{P} \subset \underline{B}$

for all $L(\underline{P})$ formulas $\underline{B}(x) = B_1(x),\ldots,B_n(x)$.

Interpretation. The following interpretation is possible: T is a collection of general facts (possibly represented as a logical database) which do not involve the predicate P. $A(P)$ formalizes some (partial) information concerning P. Then the circumscription of P in $A(P)$ has the effect of implicitly defining P as a minimal solution of A. If we gain some additional information about P, we have to update our knowledge and replace $A(P)$ by a new sentence $A'(P)$. Now circumscription has to be started again with A replaced by A'. This process of updating our knowledge yields non-monotonicity. Let $A(P)$ be the trivial sentence $P(0)$. Then $Circum(T,A(P))$ proves $\forall x(P(x) \longleftrightarrow x = 0)$. If we later learn that 1 belongs to P as well, we obtain the new formula $A'(P)$ defined as $P(0) \vee P(1)$. Circumscribing P in $A'(P)$ then implies $\forall x(P(x) \longleftrightarrow (x = 0 \vee x = 1))$.

Theorem 1. We have for all $L(\underline{P})$ sentences $A(\underline{P})$ and B: $Circum(T,A(\underline{P})) \vdash B \Rightarrow T \models_{A(\underline{P})} B$.

This theorem is obvious from the previous definitions. Together with the usual completeness theorem of first order logic it implies that the theory $Circum(Pair,A_2(P))$, with A_2 defined as in Example 1, is inconsistent.

We end this section with an observation concerning circumscription in a second order context which is propagated by McCarthy in order to obtain a more uniform approach. If set variables are allowed, the circumscription of the predicates $\underline{P} = P_1, \ldots ,P_n$ in $A(\underline{P})$ can be expressed by the single formula $A_{CS}(\underline{P})$

$$A_{CS}(\underline{P}) \quad :\Longleftrightarrow \quad A(\underline{P}) \ \& \ \forall \underline{X}(A(\underline{X}) \ \& \ \underline{X} \subset \underline{P} \ \rightarrow \ \underline{X} = \underline{P}).$$

The second order language L_2 is the extension of L by adding set variables X, Y, Z, ... and the membership relation $\in$. The L_2 formulas are generated from the L formulas and the new atomic formulas ($t \in X$) by $\neg$, $\vee$, $\&$, $\rightarrow$, $\forall$ and $\exists$. An L_2 structure is a pair $(\mathbf{M},H)$ where $\mathbf{M} = \langle M,\ldots\rangle$ is an L structure and H a non-empty collection of subsets of M, the range of the set variables.

An $L_2(\underline{P})$ formula A is called elementary, if it contains no bound set variables; A may have free set variables. The elementary comprehension axioms are all formulas of the form

$(E\text{–}CA)$ $\exists X \forall \underline{x}(x \in X \longleftrightarrow B(x))$,

where $B(x)$ is an elementary formula of $L_2(P)$.

Definition. Let T be an L theory and $A(\underline{P})$ an $L(\underline{P})$ formula. The theory $Circum^2(T,A(\underline{P}))$ is formulated in $L_2(\underline{P})$ and has as axioms:

(1) *axioms of* T;

(2) *elementary comprehension* $(E\text{–}CA)$;

(3) $A_{CS}(\underline{P})$

Evidently, $Circum(T,A(\underline{P}))$ is a subtheory of $Circum^2(T,A(\underline{P}))$. Using a standard trick of proof theory (cf. [6] for a similar application) we can show that $Circum^2(T,A(\underline{P}))$ is a conservative extension of $Circum(T,A(\underline{P}))$.

Theorem 2. $Circum^2(T,A(\underline{P}))$ is a conservative extension of $Circum(T,A(\underline{P}))$; i.e. we have for every $L(\underline{P})$ sentence B:

$$Circum^2(T,A(\underline{P})) \vdash B \quad \Longleftrightarrow \quad Circum(T,A(\underline{P})) \vdash B.$$

Proof. The direction "<=" is obvious. So assume that $Circum(T,A(\underline{P}))$ does not prove B. Hence, by completeness, there exists a model $\mathbf{M} = \langle M,\dots \rangle$ of $Circum(T,A(\underline{P})) \cup \{\neg B\}$. Now take $Def(\mathbf{M})$ to be the collection of all $L(\underline{P})$ definable subsets of M. It is easy to see that $(\mathbf{M},Def(\mathbf{M}))$ is a model of $Circum^2(T,A(\underline{P})) \cup \{\neg B\}$ and so $Circum^2(T,A(\underline{P}))$ cannot prove B.

3. Positive disjunctive circumscription

Now we turn to the central notion of this paper: positive disjunctive circumscription. Motivated by the fact that $Circum(T,A(P))$ may be inconsistent, even if $T \cup \{A(P)\}$ is consistent, we suggest the following definition.

Definition. A class K of $L(\underline{P})$ sentences is called suitable (for circumscription with respect to T) if $Circum(T,A(\underline{P}))$ is consistent for all $A(\underline{P})$ in K.

From the experience in proof theory one knows that it is nearly impossible to characterize the largest suitable class. Hence the problem consists in singling out natural suitable classes which are general enough to include many interesting applications of circumscription. We claim that this aim is achieved by positive disjunctive circumscription.

First we need some terminology. Given a sequence $\underline{P} = P_1, \dots, P_n$ of relation symbols, we define the classes $Pos(\underline{P})$, $Neg(\underline{P})$, $Pb(\underline{P})$ and $Nb(\underline{P})$ of $L(\underline{P})$ formulas by the following inductive clauses:

(i) all formulas of L are in $Pos(\underline{P})$ and $Neg(\underline{P})$;

(ii) the formulas $P_i(t)$ are in $Pos(\underline{P})$ and $Pb(\underline{P})$;

(iii) if A is in $Pos(\underline{P})$ [$Neg(\underline{P})$, $Pb(\underline{P})$, $Nb(\underline{P})$], then $\neg A$ is in $Neg(\underline{P})$ [$Pos(\underline{P})$, $Nb(\underline{P})$, $Pb(\underline{P})$];

(iv) if A and B are in $Pos(\underline{P})$ [$Neg(\underline{P})$, $Pb(\underline{P})$, $Nb(\underline{P})$], then $A \& B$ and $A \vee B$ are in $Pos(\underline{P})$ [$Neg(\underline{P})$, $Pb(\underline{P})$, $Nb(\underline{P})$];

(v) if A is in $Neg(\underline{P})$ [$Pos(\underline{P})$, $Nb(\underline{P})$, $Pb(\underline{P})$] and B in $Pos(\underline{P})$ [$Neg(\underline{P})$, $Pb(\underline{P})$, $Nb(\underline{P})$], then $A \rightarrow B$ is in $Pos(\underline{P})$ [$Neg(\underline{P})$, $Pb(\underline{P})$, $Nb(\underline{P})$];

(vi) if $A(x)$ is in $Pos(\underline{P})$ [$Neg(\underline{P})$], then $\forall x A(x)$ and $\exists x A(x)$ are in $Pos(\underline{P})$ [$Neg(\underline{P})$].

A formula A is called $\underline{P}$ positive or positive in $\underline{P}$, if it belongs to the class $Pos(\underline{P})$. The formulas in $Pb(\underline{P})$ are the positive boolean connections of the atomic formulas $P_i(t)$. The following lemma is obvious.

Lemma 1. If $\mathbf{M} = \langle M,...\rangle$ is an L structure and $A[P,\underline{x}]$ a P positive $L(P)$ formula, then we have for all $\underline{a} \in M$ and $X \subset Y \subset M$: $\quad \mathbf{M} \models A[X,\underline{a}] \rightarrow A[Y,\underline{a}]$.

Definition.

1. For every sequence $\underline{t} = t_1, \ldots, t_m$ of L terms we introduce the following abbreviation:

$$\underline{t} \in_d Q \quad :\Longleftrightarrow \quad Q(t_1) \vee \cdots \vee Q(t_m)$$

2. An $L(Q)$ sentence $D(Q)$ is called positive disjunctive (in Q) if it is of the form

$$\forall \underline{x}(A[Q,\underline{x}] \rightarrow \underline{x} \in_d Q)$$

where $A[Q,\underline{x}]$ is a Q positive $L(Q)$ formula.

3. An $L(\underline{P})$ sentence $B(\underline{P})$ is called positive boolean (in $\underline{P}$) if it is of the form

$$\forall \underline{x}(A[\underline{P},\underline{x}] \rightarrow E[\underline{P},\underline{x}])$$

where $A[\underline{P},\underline{x}]$ belongs to $Pos(\underline{P})$ and $E[\underline{P},\underline{x}]$ to $Pb(\underline{P})$.

Now we use the pairing facilities of T in order to reduce the circumscription of the predicates $\underline{P} = P_1, \ldots, P_n$ in a positive boolean sentence $B(\underline{P})$ to the circumscription of a unary predicate Q in a suitably chosen positive disjunctive sentence $D(Q)$. The proof of the following lemma is a straigthforward but tedious coding argument and left to the reader.

Lemma 2. For every positive boolean $L(P_1, \ldots, P_n)$ sentence $B(P_1, \ldots, P_n)$ there exists a positive disjunctive $L(Q)$ sentence $D(Q)$ such that $Circum(T,D(Q))$ proves for all $L(Q)$ formulas $C_1(x), \ldots, C_n(x)$:

(1) $B((Q)_1, \ldots, (Q)_n)$;

(2) $B(C_1, \ldots, C_n)\ \&\ C_1 \subset (Q)_1\ \&\ \cdots\ \&\ C_n \subset (Q)_n \ \rightarrow\ (Q)_1 \subset C_1\ \&\ \cdots\ \&\ (Q)_n \subset C_n$.

Definition. Let $\mathbf{M} = \langle M,...\rangle$ be an L structure and $D(P)$ a positive disjunctive $L(P)$ sentence. We say that a subset $I \subset M$ is D closed (over $\mathbf{M}$) if $\mathbf{M} \models D(I)$; I is called a minimal D closed set (over $\mathbf{M}$) if I is D closed and $\mathbf{M} \models \neg D(I')$ for all proper subsets I' of I.

Remark. M is D closed for every L structure $\mathbf{M} = <M,...>$ and positive disjunctive $L(P)$ sentence $D(P)$.

Lemma 3. For every L structure $\mathbf{M} = <M,...>$ and positive disjunctive $L(P)$ sentence $D(P)$ there exists at least one minimal D closed set $I \subset M$.

Proof. Let Φ be the collection of all D closed subsets of M. First we choose an ordinal σ and a function $f : \sigma \to \Phi$ which is one-to-one and onto. By recursion on α we then define a decreasing sequence of D closed sets $M_\alpha \subset M$.

(i) $M_0 := M$;

(ii) $M_\lambda := \bigcap\{M_\xi : \xi < \lambda\}$, if λ is a limit ordinal;

(iii) Now let $\alpha = \beta+1$. If there is no proper subset X of M_β such that $\mathbf{M} \models D(X)$, then $M_\alpha := M_\beta$. Otherwise we set

$m_\beta := \min\{\xi : f(\xi) \subset M_\beta \ \& \ f(\xi) \neq M_\beta\}$.

$M_\alpha := f(m_\beta)$.

Then it is $M_\beta \subset M_\alpha$ for $\alpha < \beta$; by an easy cardinality argument one can show that there exists a smallest ordinal γ such that

(1) $M_\gamma = M_{\gamma+1}$.

By induction on α we now prove that each set M_α is D closed. This is clear for M_0 and $M_{\alpha+1}$, provided that M_α is D closed. So assume that λ is a limit ordinal and

(2) $\mathbf{M} \models D(M_\xi)$

for all $\xi < \lambda$. Since $D(P)$ is positive disjunctive it can be written as

(3) $\forall x_1 \cdots \forall x_n(A[P,x_1, \ldots ,x_n] \to x_1 \in P \vee \cdots \vee x_n \in P)$

for some P positive formula $A[P,x_1, \ldots ,x_n]$. Hence we have to prove that

(4) $\mathbf{M} \models A[M_\lambda,x_1, \ldots ,x_n] \to x_1 \in M_\lambda \vee \cdots \vee x_n \in M_\lambda$

for all $x_1, \ldots ,x_n \in M$. To do this, we take $\underline{a} = a_1, \ldots ,a_n \in M$ and suppose

(5) $\mathbf{M} \models A[M_\lambda,\underline{a}]$.

It is our aim to show that $a_i \in M_\lambda$ for some $i = 1,...,n$. The P positivity of $A[P,\underline{x}]$, Lemma 1 and (5) imply

(6) $\mathbf{M} \models A[M_\xi,\underline{a}]$

for all $\xi < \lambda$. By the pigeonhole principle and (2) there exists an $m \in \{1,...,n\}$ and a subset d of λ such that

(7) $sup(d) = \lambda$,

(8) $\mathbf{M} \models A[M_\xi,\underline{a}] \to a_m \in M_\xi$

for all $\xi \in d$. In the view of (6), (7) and (8) we conclude that

(9) $a_m \in \bigcap\{M_\xi : \xi \in d\} = M_\lambda$.

This finishes the proof that each M_α is D closed. Now choose $I = M_\gamma$ to obtain a minimal D closed set.

Remark. The construction of I in this proof depends on the enumeration $f : \sigma \rightarrow \Phi$ of M. Different enumerations may give different minimal D closed sets. Examples will follow later.

Theorem 3. If $D(P)$ is a positive disjunctive $L(P)$ sentence, then *Circum*(T,$D(P)$) is a conservative extension of T. This means that we have for all L sentences B:

$$Circum(T,D(P)) \vdash B \iff T \vdash B.$$

Proof. Only the direction "=>" requires any special consideration. So let us assume that B is not provable in T. Then there exists a model $\mathbf{M} = \langle M,...\rangle$ of $T \cup \{\neg B\}$ and, according to Lemma 3, a minimal D closed subset I of M. We extend $\mathbf{M}$ by I as interpretation for P and obtain a model of *Circum*(T,$D(P)$) $\cup \{\neg B\}$. Hence B is not provable in *Circum*(T,$D(P)$).

This theorem is important for applications: Assume again that T is a collection of general facts and $D(P)$ represents our present knowledge about P. Now we circumscribe P in $D(P)$ and draw conclusions from *Circum*(T,$D(P)$). If we later discover some new facts concerning P, not all of our previous work will be worthless. According to Theorem 3, all consequences of *Circum*(T,$D(P)$) which do not refer to P will follow from the new situation as well.

Corollary 1. The class of all positive disjunctive sentences is suitable.

Example 2. Let a and b be to different constants of L and suppose that

$$D(P) \quad :\Longleftrightarrow \quad a \in P \vee b \in P.$$

Then $\{a\}$ and $\{b\}$ are the only minimal D closed sets.

Example 3. There are particular interesting instances of the positive disjunctive sentences which have been studied in proof theory and recursion theory the so called positive inductive definitions (cf. [1] and [15]). They are given by sentences

$$D(P) \quad :\Longleftrightarrow \quad \forall x(A[P,x] \rightarrow x \in P)$$

where $A[P,x]$ is a P positive $L(P)$ formula with only one free variable x. Over every L structure $\mathbf{M} = \langle M,...\rangle$ there exists exactly one minimal D closed set I which can be defined by the following transfinite recursion:

$$I^{<\alpha} := \bigcup \{I^{\xi} : \xi < \alpha\};$$
$$I^{\alpha} := \{x \in M : \mathbf{M} \models A[I^{<\alpha},x]\};$$
$$I := \bigcup \{I^{\xi} : \xi \ \textit{ordinal}\}.$$

This set I is called the minimal fixed point of the inductive definition given by A, since I is the least subset of M such that

$$x \in I \iff \mathbf{M} \models A[I,x]$$

for all $x \in M$. The closure ordinal of this inductive definition is the smallest α such that $I^{<\alpha} = I^{\alpha} = I$.

Finally we introduce the positive disjunctive extension $PD(T)$ of the L theory T. We write L^* for the first order language which extends L by a unary relation symbol Q_D for every positive disjunctive sentence $D(P)$. The axioms of $PD(T)$ are the axioms of T plus the following

(Q_D.1) $D(Q_D)$,

(Q_D.2) $D(B)\ \&\ B \subset Q_D \ \rightarrow\ Q_D \subset B$

for all positive disjunctive $D(P)$ and all L^* formulas B. Obviously $PD(T)$ is an extension of $Circum(T,D(P))$ for every positive disjunctive sentence $D(P)$, if P is translated into Q_D. From Lemma 3 and Theorem 3 we conclude that $PD(T)$ is a conservative extension of T.

Theorem 4. $PD(T)$ is a conservative extension of T and hence consistent.

In view of Lemma 2 we are allowed to say that the circumscription in positive boolean sentences is covered by positive disjunctive circumscription. This shows the great generality of the theory $PD(T)$ and means that our approach is well suited to treat the (most) examples which have been studied in the literature on this subject.

4. The norm of minimal disjunctively closed sets

In this section we introduce and study the norm of minimal disjunctively closed sets. Our definition is motivated by the theory of positive inductive definitions and tries to follow this approach as closely as possible (cf. Example 3).

We assume that $\mathbf{M} = \langle M,...\rangle$ is an L structure and $D(P)$ the positive disjunctive $L(P)$ formula $\forall x_1 \cdots \forall x_n(A[P,x_1, \ldots ,x_n] \rightarrow x_1 \in P \vee \cdots \vee x_n \in P)$. For every minimal D closed set $I \subset M$ we define its stages I_α by the following recursion on the ordinals:

$$I_{<\alpha} := \bigcup \{I_\xi : \xi < \alpha\};$$
$$I_\alpha := \{y \in I : \exists x_1 \cdots \exists x_n(A[I_{<\alpha},x_1, \ldots ,x_n]\ \&\ (y = x_1 \vee \cdots \vee y = x_n))\}.$$

Since $A[P,x_1, \ldots ,x_n]$ is P positive, we have $I_\alpha \subset I_\beta$ for all $\alpha < \beta$, and the existence of a minimal α with the property $I_\alpha = I_{<\alpha}$ follows by cardinality considerations. This set I_α is an D closed subset of I; the minimality of I therefore implies $I_\alpha = I$. Hence the following definitions make sense for all $\underline{a} \in_d I$:

$$\| \underline{a} \|_I := \min\{\alpha : \underline{a} \in_d I_\alpha\}$$
$$\| I \| := \min\{\alpha : I_\alpha = I\}$$

We call $\| I \|$ the norm of I and $\| \underline{a} \|_I$ the norm of $\underline{a}$ (with respect to I). It is easy to see that $\| I \| = sup\{\| \underline{a} \|_I + 1 : \underline{a} \in_d I\}$.

Remark. In the definition of the stages I_α we explicitly refer to the minimal D closed I. The set I must be given first and is split into its stages afterwards. This is a significant difference to the case of positive inductive definitions (cf. Example 3) where the stages are defined from the scratch and used in order to define the least fixed point. However, if $A[P,x_1, \ldots ,x_n]$ is of the form $B[P,y]$, then both definitions coincide.

Example 4. This example shows that different minimal D closed sets may have different norms. Consider an arbitrary model $\mathbf{M} = \langle M,\ldots\rangle$ of *Pair* and the following $L(P)$ formulas

$$A[P,x,y] \;:\Longleftrightarrow\; (x = 0) \;\&\; (y = 1 \vee \exists z(z \in P \;\&\; y = S(z)));$$
$$D(P) \;:\Longleftrightarrow\; \forall x \forall y(A[P,x,y] \rightarrow x \in P \vee y \in P).$$

$D(P)$ is a a positive disjunctive sentence, and both sets $I := \{0\}$ and $I' := \{1,2,3, \cdots\}$ are minimal D closed sets; but we also have: $\| I \| = 1 < \omega = \| I' \|$.

Open question. For a moment we consider the standard structure $\mathbf{N} = \langle N,\ldots\rangle$ of the natural numbers and the corresponding language L_N. In view of Lemma 3, every positive disjunctive $D(P)$ has one or more minimal D closed sets I over $\mathbf{N}$. A standard theorem in recursion theory says that the norm of I is less than or equal to the least non-recursive ordinal ω_1^{ck} provided that $D(P)$ has the form $\forall x(A[P,x] \rightarrow x \in P)$ (cf. [15]). It would be interesting to know whether a similar result holds for the general case as well. Is the norm of an arbitrary minimal disjunctively closed set $\leq \omega_1^{ck}$ if we work over the ground structure $\mathbf{N}$?

Although the norms of minimal disjunctively closed sets may be very large in general, we will see that the stages $\geq \omega$ do not matter as soon as provability is taken into account. Before going into more details, we consider an example.

Example 5. We define in $L(P)$:

$Suc(x) \;:\Longleftrightarrow\; \exists y(x = S(y));$

$x <_0 y \;:\Longleftrightarrow\; Suc(x) \;\&\; (y = S(x) \vee \neg Suc(y));$

$A[P,x] \quad :\Leftrightarrow \quad \forall y(y <_0 x \rightarrow y \in P);$

$D(P) \quad :\Leftrightarrow \quad \forall x(A[P,x] \rightarrow x \in P).$

If $\mathbf{M} = <M,...>$ is a Herbrand model of *Pair*, then M is the only minimal D closed set (over this model). In this case it is $\| M \| = \omega+1$ and $\| 0 \|_I = \omega$. However, $0 \in Q_D$ is not provable in $PD(Pair)$. There exist (non-standard) models $\mathbf{M}' = <M',...>$ of *Pair* such that the set $GT := \{S(t) : t \text{ ground term of } L\}$ is minimal D closed. In this case we have $\| GT \| = \omega$ and $0 \notin GT$.

The next results concern the provable parts of minimal disjunctively closed sets. They are obtained by proof-theoretic methods and require some terminology. As usual we take ε_o to denote the least ordinal ξ such that $\omega^\xi = \xi$. For every limit ordinal λ less than ε_o we then define its fundamental sequence $(\lambda[n] : n < \omega)$ by the following recursion:

(i) $\omega[n] := n;$

(ii) if λ is the ordinal $\omega^\alpha+\beta$, where $\beta < \omega^{\alpha+1}$, then $\lambda[n] := \omega^\alpha+\beta[n];$

(iii) $\omega^{\alpha+1}[n] := \omega^\alpha + \cdots + \omega^\alpha$ (n summands);

(iv) if λ is of the form ω^α and α is a limit ordinal, then $\lambda[n] := \omega^{\alpha[n]}$.

This covers all possible cases. As a consequence we obtain that $\lambda = sup\{\lambda[n] : n < \omega\}$. Using these fundamental sequences we now introduce a hierarchy $(f_\alpha : \alpha < \varepsilon_o)$ of number-theoretic functions from ω to ω:

$$\begin{aligned} f_0(n) &:= n+1; \\ f_{\alpha+1}(n) &:= f_\alpha(f_\alpha(\cdots f_\alpha(n) \cdots)) \quad \text{(n applications of } f_\alpha); \\ f_\lambda(n) &:= f_{\lambda[n]}(n), \end{aligned}$$

if λ is a limit ordinal. The hierarchy $(f_\alpha : \alpha < \varepsilon_o)$ is known as the fast growing hierarchy and is important for the classification of number-theoretic functions; f_ω corresponds to the Ackermann function.

Theorem 5 (Boundedness Theorem). Let $A[\underline{x},\underline{y}]$ be an arbitrary L formula and $D(P)$ a positive disjunctive $L(P)$ formula. Besides this we assume that $\mathbf{M} = <M,...>$ is a model of T and I a minimal D closed subset of M. If

$$PD(T) \vdash (\forall \underline{x} \in_d Q_D)(\exists \underline{y} \in_d Q_D) A[\underline{x},\underline{y}],$$

then there exists an ordinal $\alpha < \varepsilon_o$ such that for all $n < \omega$:

$$\mathbf{M} \models (\forall \underline{x} \in_d I_{<n})(\exists \underline{y} \in_d I_{<f_\alpha(n)}) A[\underline{x},\underline{y}].$$

This theorem is proved in [7]. It is also shown there that it is best possible: in general ε_o cannot be replaced by an ordinal smaller than ε_o.

Corollary 2. If $PD(T)$ proves $\underline{a} \in_d {}^{\backprime}Q_D$ for some sequence of constants $\underline{a}$, then $\| \underline{a} \|_I < \omega$ for every set I which is minimal D closed over some model of T.

Theorem 5 and Corollary 2 have an important consequence: For $PD(T)$ only the finite stages of minimal disjunctively closed sets are relevant. Hence $PD(T)$ provides a framework for reasoning about information which is coded in the finite stages of minimal disjunctively closed sets; even a very powerful ground theory T cannot exploit more. On the other hand, $PD(T)$ contains Peano arithmetic PA in all interesting cases. For the notion of proof-theoretic strength mentioned below cf. [1] or [7].

Theorem 6. Peano arithmetic PA can be interpreted in the theory $PD(Pair)$. Moreover, $PD(Pair)$ and PA have the same proof-theoretic strength.

Proof. We choose a formulation of PA with the constant 0, the successor function $'$ and 3 ary relation symbols Pl and Ti defining the functions of addition and multiplication. In order to find an interpretation of PA in $PD(Pair)$, we take $B[P,a]$ to be the disjunction of the following formulas (1) - (6):

(1) $a = \langle 0,0\rangle$,

(2) $\exists x(\langle x,0\rangle \in P \;\&\; a = \langle S(x),0\rangle)$,

(3) $\exists x(a = \langle\langle x,0,x\rangle,1\rangle)$,

(4) $\exists x\exists y\exists z(\langle\langle x,y,z\rangle,1\rangle \in P \;\&\; a = \langle\langle x,S(y),S(z)\rangle,1\rangle)$,

(5) $\exists x(a = \langle\langle x,0,0\rangle,2\rangle)$,

(6) $\exists w\exists x\exists y\exists z(\langle\langle x,y,z\rangle,2\rangle \in P \;\&\; \langle\langle z,x,w\rangle,1\rangle \in P \;\&\; a = \langle\langle x,S(y),w\rangle,2\rangle)$.

By $D(P)$ we denote the positive disjunctive $L(P)$ formula $\forall x(B[P,x] \rightarrow x \in P)$. The translation A^* of a PA sentence A is obtained by the following manipulations: Replace each term t' by $S(t)$; replace all subformulas $Pl(a,b,c)$ and $Ti(a,b,c)$ by $\langle\langle a,b,c\rangle,1\rangle \in Q_D$ and $\langle\langle a,b,c\rangle,2\rangle \in Q_D$, respectively; replace all quantifiers $\forall x(...)$ and $\exists x(...)$ by $\forall x(\langle x,0\rangle \in Q_D \rightarrow ...)$ and $\exists x(\langle x,0\rangle \in Q_D \;\&\; ...)$, respectively. Then it is not hard to show that $PA \vdash A \;\Rightarrow\; PD(Pair) \vdash A^*$.

The second assertion is proved in [7]. It follows from the proof-theoretic analysis of $PD(Pair)$ and is given in nuce by Theorem 5.

5. Generalizations of positive disjunctive circumscription

In this section we study two forms of generalizing positive disjunctive circumscription. One is by introducing priority relations $\leq_R$ and minimizing the predicates with respect to $\leq_R$. The other is by iterating positive disjunctive circumscription.

Prioritized circumscription has been introduced by McCarthy [12] and Lifschitz [8]. In [12] it is used to establish priorities between different kinds of abnormality on a particular common sense data base. Lifschitz [8] is interested in the special cases where prioritized circumscription is equivalent to a first order formula. We recommend these two papers for further details, motivation and a variety of examples.

Example 6. Consider the $L(P_1,P_2)$ formula

$$A(P_1,P_2) \quad :\Leftrightarrow \quad \forall x(x \in P_1 \vee x \in P_2).$$

Then the theory $Circum(T,A(P_1,P_2))$ proves $\forall x(x \notin P_1 \vee x \notin P_2)$, and P_1, P_2 are minimized with respect to the relation $\leq$ defined by

$$(P_1,P_2) \leq (P_1',P_2') \quad :\Leftrightarrow \quad P_1 \subset P_1' \;\&\; P_2 \subset P_2'.$$

If P_1 has to be minimized with higher priority than P_2, we can choose the relation $\leq'$,

$$(P_1,P_2) \leq' (P_1',P_2') \quad :\Leftrightarrow \quad P_1 \subset P_1' \;\&\; (P_1 = P_1' \rightarrow P_2 \subset P_2')$$

and work with the axioms

(1) $A(P_1,P_2)$,

(2) $A(B_1,B_2) \;\&\; (B_1,B_2) \leq' (P_1,P_2) \quad \rightarrow \quad (P_1,P_2) \leq' (B_1,B_2)$

for all $L(P_1,P_2)$ formulas B_1 and B_2. Then it is provable that $\forall x(x \notin P_1 \;\&\; x \in P_2)$. $\mathbf{M} = \langle M,\mathbf{M}[P_1],\mathbf{M}[P_2],\ldots\rangle$ is a model of $Circum(T,A(P_1,P2))$ provided that $\langle M,\ldots\rangle$ is a model of T, $\mathbf{M}[P_1] \cap \mathbf{M}[P_2] = \emptyset$ and $\mathbf{M}[P_1] \cup \mathbf{M}[P_2] = M$. If $\mathbf{M}$ has to be a model of $T + (1) + (2)$, then we need in addition that $\mathbf{M}[P_1] = \emptyset$ and $\mathbf{M}[P_2] = M$.

Every $L(P_1,P_2)$ sentence $R(P_1,P_2)$ defines a binary relation on the powerset $Pow(M)$ of every L structure $\mathbf{M} = \langle M,\ldots\rangle$; it consists of all pairs X, $Y \subset M$ such that $\mathbf{M} \models R(X,Y)$. We will write $P_1 \leq_R P_2$ for $R(P_1,P_2)$, even if the induced binary relation is not an ordering relation.

Definition. Take an $L(P_1,P_2)$ sentence $R(P_1,P_2)$. The R prioritized positive disjunctive extension of T is the first order theory $PD_R(T)$ which is formulated in the language L^* and given by the following axioms:

(1) *all axioms of* T;

(2) $D(Q_D)$;

(3) $D(A) \;\&\; A \leq_R Q_D \quad \rightarrow \quad Q_D \leq_R A$

for all positive disjunctive $L(P)$ formulas $D(P)$ and all L^* formulas $A(x)$.

Remark. $PD(T)$ is the theory $PD_R(T)$ for $R(P_1,P_2)$ defined by $\forall x(x \in P_1 \;\rightarrow\; x \in P_2)$.

Example 7. One has to be careful with the choice of the priority relation $\leq_R$. If $A(P)$ is the positive disjunctive formula $\forall x(x \in P \rightarrow (x+1) \in P)$ of first order number theory and $R(P_1,P_2)$ is defined by

$$R(P_1,P_2) \quad :<=> \quad P_2 = \emptyset \ \vee \ (P_1 \subset P_2 \ \& \ P_1 \neq \emptyset),$$

then the R prioritized circumscription of P in $A(P)$ is inconsistent with the axioms of Peano arithmetic. Hence $PD_R(T)$ may be inconsistent for badly chosen relations $\leq_R$.

The next definition introduces priority relations $\leq_R$ such that the R prioritized positive disjunctive extensions of T will be consistent.

Definition. An $L(P_1,P_2)$ sentence $R(P_1,P_2)$ is called an intersective priority relation for T if there exists a model $\mathbf{M} = <M,...>$ of T which satisfies for all X, X_1, X_2, $X_3 \subset M$, $Y \subset Pow(M)$ and $Z \in Y$:

$(IPR.1)$ $X_1 \leq_R X_2 \ \& \ X_2 \leq_R X_3 \ \rightarrow \ X_1 \leq_R X_3$;

$(IPR.2)$ $X \leq_R X$;

$(IPR.3)$ $\bigcap Y \leq_R Z$

In the most cases the axioms $(IPR.1)$ – $(IPR.3)$ will be logically valid or consequences of T (see example below). Our formulation is slightly more general in claiming only that they are consistent with T. Then there exists a model of T and $(IPR.1)$ – $(IPR.3)$ which can be extended to a model of $PD_R(T)$ by using similar techniques as in Lemma 3 and Theorem 3. We omit a detailed proof of th theorem below and turn to an interesting intersective priority relation.

Theorem 7. If R is an intersective priority relation for T, then $PD_R(T)$ is a conservative extension of T. Hence $PD_R(T)$ is consistent for such R.

Example 8. Prioritized circumscription of abnormalities a la Lifschitz [8] and McCarthy [12] can be studied in the following abstract framework. We use our standard pairing facilities and define for unary relation symbols P_1 and P_2 and every natural number n:

$(P_1)_n \subset (P_2)_n \quad :<=> \quad \forall x(<x,n> \in P_1 \ \rightarrow \ <x,n> \in P2)$;

$P_1 \subset_n P_2 \quad :<=> \quad (P_1)_0 \subset (P_2)_0 \ \& \ \cdots \ \& \ (P_1)_n \subset (P_2)_n$;

$P_1 \equiv_n P_2 \quad :<=> \quad P_1 \subset_n P_2 \ \& \ P_2 \subset_n P_1$;

$P_1 \leq_0 P_2 \quad :<=> \quad P_1 \subset_0 P_2$;

$P_1 \leq_{n+1} P_2 \quad :<=> \quad P_1 \leq_n P_2 \ \& \ (P_1 \equiv_n P_2 \rightarrow P_1 \subset_{n+1} P_2)$.

Every relation $\leq_n$ is an intersective priority relation for T. In the $\leq_n$ prioritized circumscription of P in a formula $A(P)$, the projection $(P)_i$ is minimized at higher priority than $(P)_j$ provided that $0 \leq i < j \leq n$.

Open question. The theories $PD_R(T)$ are also interesting from a purely proof-theoretic standpoint. In view of Theorem 6 we know that $PD_R(Pair)$ has the proof-theoretic strength of Peano arithmetic PA, provided that $R(P_1,P_2)$ is the trivial priority relation $\forall x(x \in P_1 \rightarrow x \in P_2)$. It is unknown what happens in more complex situations. Later we will find an upper bound for the proof-theoretic strength of the theories $PD_{\leq_n}(Pair)$. However, we do not think that this bound is sharp.

The theories ID_n of n-times iterated inductive definitions show another way how to generalize positive disjunctive circumscription. These systems have been introduced in mathematical logic in order to study the foundations of constructive mathematics and to analyze the proof-theoretic strength of various formalizations of classical mathematics. For more information on ID_n and related topics we refer to [1].

In this paper the more general notion of iterated positive disjunctive circumscription will be developed. Now we circumscribe predicated P in positive disjunctive formulas $D(P)$ which may contain constants for previously defined disjunctively closed sets as additional parameters; and it is important that these constants may occur positively and negatively in D. More precisely:

(i) $L^{(0)}$ is the language L.

(ii) For $n > 0$, $L^{(n)}$ is the extension of $L^{(n-1)}$ by unary relation symbols U_D for each positive disjunctive formula $D(P)$ of $L^{(n-1)}(P)$.

Given the first order theory T, we formulate $PD_n(T)$ in the language $L^{(n)}$; the axioms of $PD_n(T)$ are the axioms of T plus

(1) $D(U_D)$;

(2) $D(A) \;\&\; A \subset U_D \;\rightarrow\; U_D \subset A$

for all formulas $A(x)$ of $L^{(n)}$. These axioms formalize that U_D is a D closed set which is minimal with respect to all $L^{(n)}$ definable subsets of the universe. If U_D is a relation symbol introduced at level m of the definition of $L^{(n)}$, then all relation symbols of its definition clause D belong to the language $L^{(m-1)}$. Hence the consistency of $PD_n(T)$ follows by an iterated application of Lemma 3.

Theorem 8. $PD_n(T)$ is consistent.

The stratified definition of the disjunctively closed sets in $PD_n(T)$ also captures some aspects of prioritized circumscription. We consider a particular case and prove the following theorem.

Theorem 9. $PD_{\leq_n}(T)$ is contained in $PD_n(T)$.

Proof. Let $D(P)$ be the positive disjunctive $L(P)$ sentence $\forall \underline{x}(A[P,\underline{x}] \rightarrow \underline{x} \in_d P)$. Working in $L^{(n)}$, we have to find an interpretation for the constant Q_D of $PD_{\leq_n}(T)$. For every m, $0 \leq m \leq n$, we put:

$G_m(x) \;:\!\!<\!\!=\!\!>\; \exists y(x = \langle y,m\rangle)$;

$H_m(x) \;:\!\!<\!\!=\!\!>\; G_0(x) \vee \cdots \vee G_m(x)$;

$H_m(x_1, \ldots, x_k) \;:\!\!<\!\!=\!\!>\; H_m(x_1) \;\&\; \cdots \;\&\; H_m(x_k)$.

By induction on m, $0 \leq m \leq n$, we then introduce the $L^{(m)}(P)$ formulas $B_m(P,\underline{x})$ and define

$C_m(P,\underline{x}) \;:\!\!<\!\!=\!\!>\; \forall \underline{x}(B_m(P,\underline{x}) \;\&\; \underline{x} \notin_d U_0 \;\&\; \cdots \underline{x} \notin_d U_{m-1} \;\&\; H_m(\underline{x}) \;\rightarrow\; \underline{x} \in_d P)$

where we write U_m instead of U_{C_m}.

1. The formula $B_0(P,\underline{x})$ is the result of replacing each subformula $t \in P$ of $A(P,\underline{x})$ by $(G_0(t) \;\&\; t \in P) \vee \neg G_0(t))$.
2. The formula $B_{m+1}(P,\underline{x})$ is the result of replacing each subformula $t \in P$ of $A(P,\underline{x})$ by the disjunction of the following formulas:

   $G_i(t) \;\&\; t \in U_i$ for $i = 1,\ldots,m$,

   $G_{m+1}(t) \;\&\; t \in P$,

   $\neg G_{m+1}(t)$.

Obviously $B_m(P,\underline{x})$, $C_m(P)$ and $t \in U_m$ are $L^{(n)}$ formulas which depend on $D(P)$. Finally we interpret Q_D by the following $L^{(n)}$ formula E_D:

$E_D(x) \;:\!\!<\!\!=\!\!>\; (G_0(x) \;\&\; x \in U_0) \vee \cdots \vee (G_n(x) \;\&\; x \in U_n) \vee \neg H_n(x)$.

Then $PD_n(T)$ proves:

(1) $D(E_D)$;

(2) $D(F) \;\&\; F \leq_n E_D \;\rightarrow\; E_D \leq_n F$

for every $L^{(n)}$ formula $F(x)$. Therefore E_D satisfies the relevant axioms of $PD_{\leq_n}(T)$, and the rest of the proof is obvious.

For the special case $T = Pair$, a complete proof-theoretic analysis of $PD_n(T)$ is possible. To finish this paper, we state the following result without further comment and refer to [1,7] for unexplained notions and proofs.

Theorem 10. The proof-theoretic ordinal of $PD_1(Pair)$ is ε_o; if $n > 1$, then $PD_{n+1}(Pair)$ has the proof-theoretic ordinal $\theta\varepsilon_{\Omega_n+1}0$.

In particular we can show that the theories $PD_{n+1}(Pair)$ are proof-theoretically equivalent to the theories ID_n for all $n > 0$. $PD_1(Pair)$ is the same as $PD(Pair)$ and therefore of the same proof-theoretic strength as Peano arithmetic PA.

REFERENCES.

[1] Buchholz, W., Feferman, S., Pohlers, W. and Sieg, W. Iterated Inductive Definitions and Subsystems of Analysis: Recent Proof-Theoretical Studies. Lecture Notes in Mathematics 897. Springer-Verlag, Berlin, Heidelberg, New York (1981).

[2] Chang C.L. and Lee R.C.T. Symbolic Logic and Mechanical Theorem Proving. Academic Press, New York (1973).

[3] Clark, K. Negation as failure. In: Logic and Databases. Gallaire, H. and Minker, J. (eds.). Plenum Press, New York (1978).

[4] Davis, M. The mathematics of non-monotonic reasoning. Artificial Intelligence 13 (1980).

[5] Etherington, D.W., Mercer, R.E. and Reiter, R. On the adequacy of predicate circumscription for closed-world reasoning. Computational Intelligence 1 (1985).

[6] Feferman, S. Monotone inductive definitions. In: The L.E.J. Brouwer Centenary Symposion. Troelstra, A.S. and van Dalen, D. (eds.). North Holland, Amsterdam (1982).

[7] Jaeger, G. The proof-theoretic analysis of positive disjunctive circumscription. Preprint, Zuerich (1985).

[8] Lifschitz, V. Computing circumscription. Proceedings IJCAI-85.

[9] Lifschitz, V. Closed world data bases and circumscription. Preprint 1985.

[10] Lifschitz, V. On the satisfiability of circumscription. Preprint 1986.

[11] McCarthy, C. Circumscription - a form of non-monotonic reasoning. Artificial Intelligence 13 (1980).

[12] McCarthy, C. Applications of circumscription to formalizing common sense knowledge. AAAI Workshop on Non-Monotonic Reasoning (1984).

[13] McDermott, D. and Doyle, J. Non-monotonic logic I. Artificial Intelligence 13 (1980).

[14] Moore , R.C. Semantical considerations on nonmonotonic logic. Artificial Intelligence 25 (1985).

[15] Moschovakis, Y.N. Elementary Induction on Abstract Structures. North Holland, Amsterdam (1974).

[16] Parikh, R. Logic of knowledge, games and dynamic logic. Lecture Notes in Computer Science 181. Springer-Verlag, Berlin, Heidelberg, New York, Tokyo (1984).

[17] Reiter, R. On closed world data bases. In: Logic and Databases. Gallaire, H. and Minker J. (eds.). Plenum Press, New York (1978).

[18] Reiter, R. Circumscription implies predicate completion (sometimes). Proceedings of the National Conference on Artificial Intelligence, AAAI-82.

[19] Shoenfield, J.R. Mathematical Logic. Addison-Wesley, Reading, Mass. (1967).

MODAL THEOREM PROVING

Martín Abadi & Zohar Manna
Computer Science Department
Stanford University

We describe resolution proof systems for several modal logics. First we present the propositional versions of the systems and prove their completeness. The first-order resolution rule for classical logic is then modified to handle quantifiers directly. This new resolution rule enables us to extend our propositional systems to complete first-order systems. The systems for the different modal logics are closely related.

1. INTRODUCTION

Modal logics ([HC]) have found a variety of uses in Artificial Intelligence (e.g., [Mc]), in Logics of Programs (e.g., [P]), and in the analysis of distributed systems (e.g., [HM]). For such applications, natural and efficient automated proof systems are very desirable. A variety of decision procedures have been proposed for propositional modal logics (e.g., [W]). The traditional proof systems for first-order modal logics are simple; this makes them appropriate for metamathematical studies ([Fi1]). However, they often require much creative help from a user or give rise to long proofs. Thus, they are not suitable for automatic implementation.

Classical clausal resolution proofs ([R]) are usually short and their discovery requires little or no human guidance. Classical nonclausal resolution ([MW1], [Mu]) has the virtue of added clarity, since formulas do not need to be rephrased in unnatural and sometimes long clausal forms.

Fariñas del Cerro ([Fa1], [Fa2], [Fa3]) proposed imitating classical clausal resolution in some modal logics. The proposed methods are rather attractive, but fail to treat the full modal logics under consideration – quantifiers are not allowed in the scope of modalities. Geissler and Konolige ([Ko], [GK]) attempted to solve this problem with the addition of a new operator, $\bullet$, and the introduction of "semantic attachment" procedures.

In this paper we extend nonclausal resolution to eight modal logics with the operators $\Box$ ("necessarily") and $\Diamond$ ("possibly"). Our approach is quite uniform and generalizes to a wide class of modal logics in different languages. For instance, this class includes logics of knowledge with a knowledge operator K_i for each knower. In fact, all "analytic logics" as well as some "non-analytic" ones (in the terminology of Fitting ([Fi1])) are tractable by these techniques. Also, similar methods can be used for more complicated logics, such as Temporal Logic ([AM1], [AM2]).

In the next section we introduce some basic definitions. In section 3 we present the propositional proof systems for K, T, K4, S4, S5, D, D4, and G; their completeness is proved in section 4. These propositional modal systems are lifted to first-order modal systems by adding some

This research was supported in part by the National Science Foundation under grant DCR-84-13230 and by the Defense Advanced Research Projects Agency under Contract N00039-84-C-0211.

quantifier rules (section 5), special auxiliary rules (section 6), and an extended resolution rule (section 7). Skolemization rules (mentioned in section 5) are optional. Section 8 contains a simple example. The completeness of the first-order systems is proved in section 9.

2. PRELIMINARIES

a. Informal syntax and semantics

The propositional modal language includes propositions, modal operators, and connectives. All propositions are *flexible*, i.e., they may change value from "world" to "world." The modal operators we consider are the usual ones: $\Box$ ("necessarily") and $\Diamond$ ("possibly"). The primitive connectives are just $\neg$, $\wedge$, $\vee$, *true*, and *false*. It is practical to regard all other connectives as abbreviations. Formulas are not restricted to any special form such as clausal form.

For the first-order versions, the quantifiers $\forall$ and $\exists$, variables, and flexible predicate symbols are added. It is convenient and natural to include flexible function symbols and world-independent, *rigid* predicate and function symbols as well. Informally, we may say that variables are also rigid. For example, the formula $\exists x.[q(x) \vee \Box p(x)]$ expresses that the same object has property q or necessarily has property p.

Models and and the satisfaction relation can be described in terms of possible worlds ([HC]). A *model* is a tuple $\langle D, W, w_0, R, I \rangle$, where

- the *domain* D is a non-empty set (note that we require that there be just one domain rather than one for each element of W);
- W is a set with a distinguished element w_0; intuitively W is the set of possible worlds and w_0 the real world;
- R is a binary *accessibility relation* on W;
- the *interpretation* I gives a meaning over D to each predicate symbol and each function symbol at each world in W; the meaning of rigid symbols is required to be the same at all worlds.

An *assignment* α is a function from the set of variables to D. The *satisfaction* relation, $\models$, is then defined inductively over formulas. In particular, the semantics of $\Diamond$ and $\exists$ are given by:

$$(\langle D, W, w_0, R, I \rangle, \alpha) \models \Diamond u \quad \text{if for some } w_1 \in W,\ w_0 R w_1 \text{ and } (\langle D, W, w_1, R, I \rangle, \alpha) \models u,$$

$$(\langle D, W, w_0, R, I \rangle, \alpha) \models \exists x.u \quad \text{if for some } d \in D,\ (\langle D, W, w_0, R, I \rangle, \alpha \cdot \langle x \leftarrow d \rangle) \models u.$$

As usual, the semantics of $\Box$ and $\forall$ are dual to those of $\Diamond$ and $\exists$, respectively, and *validity* is defined as the dual of satisfiability. Free variables are implicitly universally quantified: u is valid exactly when $\forall x.u$ is valid.

The different logics are characterized by properties of the accessibility relation R:

K: R does not need to satisfy any special conditions.

T: R is reflexive.

K4: R is transitive.

S4: R is reflexive and transitive.

S5: R is reflexive, symmetric, and transitive.

D: R is serial (i.e., there is some accessible world from every world).

D4: R is serial and transitive.

G: R^{-1} is transitive and well-founded.

b. Proofs and rules

$\vdash w$ denotes that the formula w can be proved by resolution, that is, that there is a sequence of formulas $S_0, \ldots, S_n$ such that $S_0 = \neg w$, $S_n = \mathit{false}$, and S_{i+1} is obtained from S_i by an application of a rule. We refer to $S_0, \ldots, S_n$ as a *proof* of w or a *refutation* of $\neg w$.

Our proof systems include two kinds of rules: simplification rules and deduction rules.

- The *simplification rules* have the form

$$u_1, \ldots, u_m \Rightarrow v.$$

Suppose the formulas $u_1, \ldots, u_m$ occur in some conjunction in S_i, in any order. Then we delete an occurrence of each of them and add the derived formula v to the conjuction.

Example:

The rule $u, \neg u \Rightarrow \mathit{false}$ applied to

$$S_i : (q \vee \Diamond(\neg p \wedge q \wedge p))$$

yields

$$S_{i+1} : (q \vee \Diamond(q \wedge \mathit{false})).$$ ∎

- The *deduction rules* have the form

$$u_1, \ldots, u_m \mapsto v.$$

Suppose the formulas $u_1, \ldots, u_m$ occur in some conjunction in S_i, in any order. Then the derived formula v is added to that conjunction.

Unlike simplification rules, deduction rules do not discard the premises $u_1, \ldots, u_m$. Sometimes, however, we may use the weakening rule (defined in section 3) to discard $u_1, \ldots, u_m$ immediately after applying a deduction rule.

Example:

The rule $\Box u, \Diamond v \mapsto \Diamond(u \wedge v)$ applied to

$$S_i : q \vee [\Diamond q \wedge r \wedge \Box p]$$

yields

$$S_{i+1} : q \vee [\Diamond q \wedge r \wedge \Box p \wedge \Diamond(p \wedge q)].$$ ∎

An occurrence of a subformula has *positive polarity* in a formula if it is in the scope of an even number of explicit or implicit $\neg$'s. It has *negative polarity* if it is in the scope of an odd number of $\neg$'s. For instance, $\Box p$ occurs with positive polarity and *false* occurs with negative polarity in $\Diamond \neg(\mathit{false} \vee \neg \Box p)$.

We use the following *polarity restriction* to reduce the proof search space:
Rules are applied only to positive occurrences of $u_1, \ldots, u_m$.

c. Soundness

For our proof notion to be meaningful, we require that rules be sound, i.e., that they maintain satisfiability: if S_i is satisfiable then S_{i+1} is satisfiable as well.

We say that u *entails* v (and denote it $u \hookrightarrow v$) if $(u \supset v)$ is valid. The following observation is often helpful in soundness arguments: a formula gets "truer" as its positive subformulas get "truer" and as its negative subformulas get "falser." More precisely, we can prove:

Lemma (Monotonicity of entailment):

For all u and v, if $u \hookrightarrow v$ and
w' is the result of replacing one positive occurrence of u by v in w, or
w' is the result of replacing one negative occurrence of v by u in w,
then $w \hookrightarrow w'$.

Proof sketch: The lemma is proved by complete induction on pairs of formulas, with the order $\prec$ defined by: $(w, w') \prec (z, z')$ if w and w' are proper subformulas of z and z', respectively, or of z' and z, respectively. ∎

Suppose that for any S_i and any S_{i+1} obtained from S_i by applying a given rule we have $S_i \hookrightarrow S_{i+1}$. Then soundness is clearly guaranteed for the rule under consideration. Consequently, we can use the lemma to conclude that simplification rules are sound if $v \hookrightarrow (u_1 \wedge \ldots \wedge u_m)$ for negative occurrences of $u_1, \ldots, u_m$. For positive occurrences, it suffices that $(u_1 \wedge \ldots \wedge u_m) \hookrightarrow v$. The entailment $(u_1 \wedge \ldots \wedge u_m) \hookrightarrow v$ holds for all the simplification rules we will present, except for the skolemization rules. With the polarity restriction, this guarantees the soundness of all the simplification rules except the skolemization rules. We prove the soundness of the skolemization rules with a different method.

Similarly, deduction rules are always sound for negative occurrences of $u_1, \ldots, u_m$ (since the given formulas $u_1, \ldots, u_m$ are kept); for positive occurrences, $(u_1 \wedge \ldots \wedge u_m) \hookrightarrow v$ suffices. The entailment $(u_1 \wedge \ldots \wedge u_m) \hookrightarrow v$ holds for all the deduction rules we will present. This guarantees the soundness of deduction rules, with no need for polarity arguments.

3. PROPOSITIONAL SYSTEMS

a. Simplification rules

- *true-false simplification* rules:

These are the regular *true-false* simplification rules, such as

$$\mathit{false} \vee u \;\Rightarrow\; \mathit{false} \quad \text{and} \quad \mathit{false},\, u \;\Rightarrow\; \mathit{false},$$

and the rule

$$\Diamond \mathit{false} \;\Rightarrow\; \mathit{false}.$$

- *Negation* rules:

$$\neg \Box u \;\Rightarrow\; \Diamond \neg u, \qquad \neg \Diamond u \;\Rightarrow\; \Box \neg u,$$

$$\neg(u \wedge v) \;\Rightarrow\; (\neg u \vee \neg v), \qquad \neg(u \vee v) \;\Rightarrow\; (\neg u \wedge \neg v), \qquad \neg\neg u \;\Rightarrow\; u.$$

- *Weakening* rule:

$$u,\, v \;\Rightarrow\; u.$$

The weakening rule lets us discard any conjunct v that we regard as no longer useful.

- *Distribution* rule:

$$u,\, v_1 \vee \ldots \vee v_k \Rightarrow (u \wedge v_1) \vee \ldots \vee (u \wedge v_k).$$

b. The resolution rule

We write $u\langle v\rangle$ to indicate that v occurs in u, and then $u\langle w\rangle$ denotes the result of replacing exactly one occurrence of v by w in u. Similarly, $u[v]$ indicates that if v occurs in u then $u[w]$ denotes the result of replacing all occurrences of v by w in u.

The nonclausal resolution rule for classical propositional logic is:

$$A\langle u,\ldots,u\rangle,\; B\langle u,\ldots,u\rangle \;\mapsto\; A\langle \mathit{true}\rangle \vee B\langle \mathit{false}\rangle.$$

That is, if the formulas $A\langle u,\ldots,u\rangle$ and $B\langle u,\ldots,u\rangle$ have a common subformula u, then we can derive the resolvent $A\langle \mathit{true}\rangle \vee B\langle \mathit{false}\rangle$. This is obtained by substituting *true* for certain (one or more) occurrences of u in $A\langle u,\ldots,u\rangle$, and *false* for certain occurrences of u in $B\langle u,\ldots,u\rangle$, and taking the disjunction of the results.

In propositional modal logics, this rule is not sound. For instance, consider the formula $(u \wedge \Diamond \neg u)$; it is satisfied by any model where u holds in the real world and fails in some possible world. We cannot soundly deduce $\big(u \wedge \Diamond \neg u \wedge (\Diamond \neg \mathit{true} \vee \mathit{false})\big)$, as the rule would suggest, since this formula is unsatisfiable. The problem is that while u occurs in both $\Diamond \neg u$ and u, it does not need to have the same truth value in all contexts. Intuitively, different occurrences of u may refer to u at different worlds.

The resolution rule is sound in propositional modal logics under the following *same-world* restriction:

> The occurrences of u in A or B that are replaced by *true* or *false*, respectively, are not in the scope of any $\Box$ or $\Diamond$ in A or B.

Informally, this imposes that all the occurrences of u under consideration are evaluated in the same world.

c. Modality rules

These rules deal with formulas in the scope of modal operators. For each modal logic there is a set of modality rules:

- K:

$$\Box u,\ \Diamond v\ \mapsto\ \Diamond(u \wedge v).$$

- T:

$$\Box u,\ \Diamond v\ \mapsto\ \Diamond(u \wedge v), \qquad \Box u\ \mapsto\ u.$$

- K4:

$$\Box u,\ \Diamond v\ \mapsto\ \Diamond(u \wedge v), \qquad \Box u,\ \Diamond v\ \mapsto\ \Diamond(\Box u \wedge v).$$

- S4:

$$\Box u,\ \Diamond v\ \mapsto\ \Diamond(\Box u \wedge v), \qquad \Box u\ \mapsto\ u.$$

- S5:

$$\Box u,\ \Diamond v\ \mapsto\ \Diamond(\Box u \wedge v), \qquad \Box u\ \mapsto\ u,$$

$$\Diamond u,\ \Diamond v\ \mapsto\ \Diamond(\Diamond u \wedge v), \qquad u\ \mapsto\ \Diamond u.$$

- D:

$$\Box u,\ \Diamond v\ \mapsto\ \Diamond(u \wedge v), \qquad \mapsto\ \Diamond\, \mathit{true}.$$

- D4:

$$\Box u,\ \Diamond v\ \mapsto\ \Diamond(u \wedge v), \qquad \Box u,\ \Diamond v\ \mapsto\ \Diamond(\Box u \wedge v), \qquad \mapsto\ \Diamond\, \mathit{true}.$$

- G:

$$\Box u,\ \Diamond v\ \mapsto\ \Diamond(u \wedge \Box u \wedge v \wedge \neg \Diamond v).$$

4. COMPLETENESS FOR PROPOSITIONAL SYSTEMS

Theorem: The resolution systems for propositional K, T, K4, S4, S5, D, D4, and G are complete for the corresponding classes of models.

Proof sketch: We exploit some known abstract characterizations of completeness for these logics. Specifically, *model existence lemmas* (stated in terms of *consistency properties*) ([Fil]) turn out to provide simple and uniform proofs for all the systems. A consistency property is a syntactic property of sets of sentences that satisfies certain conditions depending on the logic. Typically, consistency properties have the form "is not refutable (in a given proof system)." Model existence lemmas guarantee that if a set of sentences satisfies a consistency property then all the sentences in the set are satisfiable (in fact, all the sentences are simultaneously satisfiable in some logics).

We give a proof sketch for K and point out where it should be modified to apply to the other systems. Consider restricting the proof system for K so that negation rules are applied as early as possible. It suffices to show that the restricted system is complete.

We say that a set S of sentences is *admissible* (for K) if no finite conjunction of members of S can be refuted (in the resolution system for K). More precisely, S is admissible if for all distinct $w_1, \ldots, w_k \in S$ there is a permutation $\pi : \{1, \ldots, k\} \to \{1, \ldots, k\}$ such that $w_{\pi(1)} \wedge \ldots \wedge w_{\pi(k)}$ cannot be refuted (or, as we often say for simplicity, "$w_1, \ldots, w_k \in S$ cannot be refuted"). We show that admissible is a consistency property for K. To this end we check that admissible satisfies the conditions in the definition of consistency property for K:

if S is admissible and $S^{\#} = \{u \mid \Box u \in S\} \cup \{\neg u \mid \neg \Diamond u \in S\}$ then

1) S contains no proposition and its negation; $false \notin S$, $\neg true \notin S$;
2) if $(u \wedge v) \in S$ then $S \cup \{u, v\}$ is admissible;
3) if $\neg(u \vee v) \in S$ then $S \cup \{\neg u, \neg v\}$ is admissible;
4) if $(u \vee v) \in S$ then $S \cup \{u\}$ is admissible or $S \cup \{v\}$ is admissible;
5) if $\neg(u \wedge v) \in S$ then $S \cup \{\neg u\}$ is admissible or $S \cup \{\neg v\}$ is admissible;
6) if $\Diamond u \in S$ then $S^{\#} \cup \{u\}$ is admissible;
7) if $\neg \Box u \in S$ then $S^{\#} \cup \{\neg u\}$ is admissible.

Thus, admissible is a consistency property for K. Hence, by the model-existence lemma for K, if S is admissible then each member of S is satisfiable. It follows (taking $S = \{u\}$) that if u cannot be refuted then u is satisfiable. Therefore, the propositional proof system for K is complete.

The completeness arguments for the other logics only differ from the one for K in the definition of consistency property that admissible needs to satisfy. ∎

5. QUANTIFIER RULES

Starting in this section, we consider the extension of the resolution systems to first-order modal logics. The propositional language is extended with quantifiers, variables, predicate symbols, and function symbols. The definition of models imposes that the Barcan formula $(\forall x. \Box u(x)) \supset (\Box \forall x. u(x))$ and its converse $(\Box \forall x. u(x)) \supset (\forall x. \Box u(x))$ are theorems of the first-order systems.

We first give four definitions:

- An occurrence of a quantifier $Q^{\forall}$ is of *universal force* if it is either a universal quantifier $\forall$ and has positive polarity or an existential quantifier $\exists$ and has negative polarity. An occurrence of a quantifier $Q^{\exists}$ is of *existential force* if it is either a universal quantifier $\forall$ and has negative polarity or an existential quantifier $\exists$ and has positive polarity.

- An occurrence of a modal operator $M^{\Box}$ is of *necessary force* if it is either $\Box$ and has positive polarity or $\Diamond$ and has negative polarity. An occurrence of a modal operator $M^{\Diamond}$ is of *possible force* if it is either $\Box$ and has negative polarity or $\Diamond$ and has positive polarity.

This section discusses skolemization and gives some skolemization rules. Completeness of the systems does not depend on the inclusion of the skolemization rules, but the rules may sometimes give rise to short-cuts in proofs. In general, we do not rely on skolemization to eliminate quantifiers. Instead, we describe some rules to move quantifiers; we manipulate formulas with quantifiers, and, therefore, the resolution rule presented in the next section takes quantifiers into account.

a. Skolemization

In classical logic, all quantifiers can be eliminated by applications of skolemization rules. This is elegant for quantifiers of both universal and existential force, and very practical for quantifiers of existential force. The classical skolemization rule for eliminating quantifiers of existential force is:

$$\exists x.u[x] \Rightarrow u[f(x_1, \ldots, x_n)],$$

where f is a new rigid function symbol and $x, x_1, \ldots, x_n$ are all the free variables in u.

In modal logics, this rule is sound as long as u is not in the scope of any $\Box$ or $\Diamond$. Unfortunately, this rule is not sound in general. For instance, consider the formula

$$(\forall x. \Diamond p(x)) \wedge (\Box \exists y. \neg p(y)),$$

where p is a flexible predicate symbol. The formula is satisfied by the model $\mathcal{M}$ with $D = \{0,1\}$, $W = \{0,1\}$, $w_0 = 0$, $R = W^2$, where p holds for 0 only in the real world and p fails for 1 only in the real world. The rule replaces y by a new rigid constant symbol a, yielding the formula

$$(\forall x. \Diamond p(x)) \wedge (\Box \neg p(a)),$$

which is unsatisfiable. Notice that the new formula states that there is an element in the domain that has the property $\neg p$ in all possible worlds. The original sentence, on the other hand, only claimed that in each possible world there was some element with property $\neg p$. Therefore, the classical rule does not capture implicit dependencies on worlds.

A variant of the rule with flexible skolem symbols does capture implicit dependencies on worlds and soundly eliminates some quantifiers of existential force in the scope of modal operators. Consider, for instance, the formula $\Box \exists x.p(x)$. If $\Box \exists x.p(x)$ holds then in each world there must be some element with property p. In each world, denote this element by a. Thus, we may derive that for a new flexible constant symbol a, $\Box p(a)$ holds. More generally, flexible function symbols are introduced when free variables appear. For instance, assume $\Box \exists x.p(x,y)$ holds. Then, for a new flexible function symbol f, $\Box p(f(y), y)$ holds.

We obtain a *flexible skolemization* rule of the same form as the classical skolemization rule:

$$\exists x.u[x] \Rightarrow u[f(x_1, \ldots, x_n)],$$

where f is a new flexible function symbol, x, $x_1, \ldots, x_n$ are all the free variables in u, and x does not occur in the scope of any modal operator in u.

Proposition (Soundness of flexible skolemization):

If $v\langle \exists x.u[x] \rangle$ is satisfiable, f is a new flexible function symbol, $x, x_1, \ldots, x_n$ are all the free variables in u, x does not occur in the scope of any modal operator in u, and $\exists x.u[x]$ occurs positively in v,
then $v\langle u[f(x_1, \ldots, x_n)] \rangle$ is also satisfiable.

The rule is not always satisfactory when x occurs in the scope of modal operators in u. For instance, the formula

$$\Box \exists x.(p(x) \wedge \Diamond p(x))$$

yields

$$\Box (p(a) \wedge \Diamond p(a))$$

for a flexible constant symbol a. The original formula is stronger than the one we deduce: the original formula asserts that for each world the same x satisfies $p(x)$ in the real world and in some possible world. On the other hand, since a is world-dependent, $\Box(p(a) \wedge \Diamond p(a))$ does not guarantee that the same element of the domain has property p in the real world and in some possible world.

Instead, we could deduce the formula

$$\Box \forall x.[x = a \supset (p(x) \wedge \Diamond q(x))].$$

This formula is as strong as the original one. Note that it involves a $\forall$ instead of a $\exists$.

This suggests how to eliminate all quantifiers of existential force. The price paid is that the deduced formulas involve some new equations and some new quantifiers of universal force. The general rule is

$$\exists x.u \;\Rightarrow\; \forall x.[x = f(x_1, \ldots, x_n) \supset u],$$

where f is a new flexible function symbol and x, $x_1, \ldots, x_n$ are all the free variables in u.

Proposition (Soundness of generalized flexible skolemization):

If $v\langle \exists x.u \rangle$ is satisfiable, f is a new flexible function symbol, $x, x_1, \ldots, x_n$ are all the free variables in u, and $\exists x.u$ occurs positively in v,
then $v\langle \forall x.(x = f(x_1, \ldots, x_n) \supset u)\rangle$ is also satisfiable.

b. Quantifier extraction rules

The quantifier extraction rules move quantifiers to the outside of formulas. We can always extract quantifiers of universal force:

$$u\langle Q^{\forall} x.v[x]\rangle \;\Rightarrow\; \forall x'.u\langle v[x']\rangle,$$

where x' is a new variable. ($Q^{\forall}$ is $\forall$ or $\exists$, whichever is of universal force in the context under consideration.)

Proposition (Soundness of $Q^{\forall}$ rule):

$$u\langle Q^{\forall} x.v[x]\rangle \hookrightarrow \forall x'.u\langle v[x']\rangle.$$

Sometimes we can extract quantifiers of existential force in a similar way:

$$u\langle Q^{\exists} x.v[x]\rangle \;\Rightarrow\; \exists x'.u\langle v[x']\rangle,$$

where x' is a new variable. The rule is restricted so that dependencies on other variables and implicit dependencies on worlds are not overlooked: the replaced occurrence of $Q^{\exists}x.v[x]$ should not occur in the scope of any quantifier of universal force or modal operator of necessary force in u.

Proposition (Soundness of $Q^{\exists}$ rule):

If the replaced occurrence of $Q^{\exists}x.v[x]$ is not in the scope of any quantifier of universal force or modal operator of necessary force in u,
then $u\langle Q^{\exists}x.v[x]\rangle \hookrightarrow \exists x'.u\langle v[x']\rangle$.

6. AUXILIARY RULES

a. Rigid symbols and the frame rules

It is convenient to include rigid symbols for world-independent functions and predicates in the first-order modal language. The frame rules reflect the fact that the meanings of these symbols do not depend on the world where they are evaluated:

if u is a formula with no occurrences of flexible symbols, then

$$\Diamond u \mapsto u \quad \text{and} \quad u \mapsto \Box u.$$

For instance, if p is a rigid proposition symbol, then $\Diamond p$ can yield p, and then $\Box p$.

b. Equality

As in classical logic, we can add axioms for the equality symbol. Alternatively, we can include an extension of paramodulation or E-resolution (see [MW2]).

c. The cut rule

The cut rule is

$$\mapsto u \vee \neg u.$$

Note that the cut rule requires heuristics to choose u. This may be impractical in fully automatic systems. On the other hand, the cut rule is quite convenient in interactive settings, where a user may suggest appropriate u's to obtain shorter proofs.

This rule is not essential for completeness for the propositional modal systems, but it is essential in the first-order systems. Other first-order modal systems include similar devices. In fact, there exists proof-theoretic evidence that some rule like the cut rule is necessary for the logics in question ([Fi1]).

7. THE RESOLUTION RULE

In subsections a, b, and c we describe a unification algorithm and a resolution rule for first-order modal logics. For the sake of simplicity, the language is temporarily restricted not to contain flexible function symbols. In subsection d this restriction is abandoned.

a. Unification

We extend the classical unification algorithm to handle formulas with modal operators and quantifiers. Suppose we have one of the usual recursive definitions of the function *unifier* to compute most-general unifiers of classical quantifier-free expressions. Two clauses are added to the recursive definition, one for modal operators and one for quantifiers.

- *Modality extension:* Let M be a modal operator.

$$\mathit{unifier}(Mu_1, \ldots, Mu_m) \quad \text{is} \quad \begin{cases} \mathit{unifier}(u_1, \ldots, u_m) & \text{if it exists} \\ \text{fail} & \text{otherwise} \end{cases}$$

  In other words, $\Box$ and $\Diamond$ are treated just like unary connectives as far as unification is concerned.

- *Quantifier extension:* Let Q be a quantifier and x' a new variable.

$$\mathit{unifier}(Qx_1.u_1[x_1], \ldots, Qx_m.u_m[x_m])$$
$$\text{is} \begin{cases} \mathit{unifier}(u_1[x'], \ldots, u_m[x']) & \text{if it exists and does not bind } x' \\ \text{fail} & \text{otherwise} \end{cases}$$

  For instance, $\forall x.p(x)$ and $\forall y.p(y)$ unify because $p(x')$ unifies with itself and the unifier (the empty substitution) does not bind x'. On the other hand, $\forall x.p(a)$ and $\forall y.p(y)$ do not unify, since the most-general unifier of $p(a)$ and $p(x')$ binds x' to a. The formulas $\forall x.p(x)$ and $p(y)$ do not unify: the main operator of the latter formula is not a $\forall$.

These additions to the recursive definition of *unifier* are simple enough that most-general unifiers can still be computed when unifiers exist at all.

b. The resolution rule

The classical nonclausal resolution rule can be written

$$A\langle v_1, \ldots, v_n\rangle,\ B\langle v_{n+1}, \ldots, v_m\rangle \ \mapsto\ A\theta\langle \mathit{true}\rangle \vee B\theta\langle \mathit{false}\rangle$$

where θ is a most-general unifier of $v_1, \ldots, v_m$ and replaces only variables that are (implicitly) universally quantified ([MW1]). As might be expected, the classical rule is not sound for formulas with quantifiers, modal operators, and flexible symbols.

Since we do not rely on skolemization and the quantifier extraction rules only shift quantifiers outwards, the modal nonclausal resolution rule should handle quantifiers in front of A and B. Also, the conclusion of the resolution rule, $A\theta\langle \mathit{true}\rangle \vee B\theta\langle \mathit{false}\rangle$, may be preceded by some quantifiers (obtained by mixing those in front of A and B). Moreover, the formulas A, B, and $A\theta\langle \mathit{true}\rangle \vee B\theta\langle \mathit{false}\rangle$ may contain quantifiers. Some restrictions guarantee that the presence of quantifiers does not make the rule unsound. Other restrictions deal with flexible symbols and modal operators.

The rule is:

$$\begin{aligned} &Q_1x_1 \ldots Q_hx_h.A\langle v_1, \ldots, v_n\rangle, \quad R_1y_1 \ldots R_ky_k.B\langle v_{n+1}, \ldots, v_m\rangle \\ &\quad \mapsto\ S_1z_1 \ldots S_{h+k}z_{h+k}.\left[A\theta\langle \mathit{true}\rangle \vee B\theta\langle \mathit{false}\rangle\right] \end{aligned}$$

where θ is a most-general unifier of $v_1, \ldots, v_m$ and $Q_1, \ldots, Q_h, R_1, \ldots, R_k, S_1, \ldots, S_{h+k}$ are quantifiers, under the restrictions:

(i) The variables $x_1, \ldots, x_h, y_1, \ldots, y_k$ are all different.

(ii) The sequence $S_1 z_1 \ldots S_{h+k} z_{h+k}$ is a merge of $Q_1 x_1 \ldots Q_h x_h$ and $R_1 y_1 \ldots R_k y_k$, that is, $Q_1 x_1 \ldots Q_h x_h$ and $R_1 y_1 \ldots R_k y_k$ are subsequences of $S_1 z_1 \ldots S_{h+k} z_{h+k}$.

(iii) The same-world restriction: If the replaced occurrences of $v_1\theta, \ldots, v_m\theta$ are in the scope of any modal operator in $A\theta$ or $B\theta$ then $v_1\theta, \ldots, v_m\theta$ contain only rigid symbols.

(iv) The replaced occurrences of $v_1\theta, \ldots, v_m\theta$ are not in the scope of any quantifier in $A\theta$ or $B\theta$.

(v) If $(x \leftarrow t) \in \theta$ then for some i, $1 \le i \le h+k$, $S_i = \forall$, $z_i = x$, and no variable in t occurs bound in $\forall x S_{i+1} z_{i+1} \ldots S_{h+k} z_{h+k}.(A \vee B)$.

Once all the restricticion are checked, redundant quantifiers in $S_1 z_1 \ldots S_{h+k} z_{h+k}$ may be discarded.

Restriction (iii) is necessary for modal logics, even at the propositional level. On the other hand, restrictions (i), (ii), (iv), and (v) are intended to solve classical logic problems; some of them are actually related to restrictions described by Manna and Waldinger ([MW3]) for resolution with quantifiers in classical logic. Restriction (v) is intended to enforce that the application of θ does not cause any capture of free variable, that θ only instantiates universally quantified variables, and that if $(x \leftarrow t) \in \theta$ then t does not depend on x implicitly.

Example: When we apply the resolution rule to

$$\begin{array}{l} \exists x_1 \forall x_2 \exists x_3.(\Diamond\, p(x_1, x_2) \ \vee \ q(x_2, x_3)) \\ \qquad \wedge \\ \exists y_1 \forall y_2. \neg q(y_1, y_2). \end{array}$$

with

$$\begin{array}{l} A = \neg q(y_1, y_2) \quad \text{and} \quad B = (\Diamond\, p(x_1, x_2) \ \vee \ q(x_2, x_3)), \\ v_1 = q(y_1, y_2) \quad \text{and} \quad v_2 = q(x_2, x_3), \\ \theta = \{x_2 \leftarrow y_1, y_2 \leftarrow x_3\}, \end{array}$$

restrictions (i), (iii), and (iv) are satisfied.

To satisfy the remaining restrictions, we choose

$$\exists x_1 \exists y_1 \forall x_2 \exists x_3 \forall y_2.[\neg \mathit{true} \ \vee \ (\Diamond\, p(x_1, y_1) \ \vee \ \mathit{false})]$$

as the derived formula. We delete redudant quantifiers to obtain

$$\exists x_1 \exists y_1.[\neg \mathit{true} \ \vee \ (\Diamond\, p(x_1, y_1) \ \vee \ \mathit{false})].$$

Simplification yields

$$\exists x_1 \exists y_1. \Diamond\, p(x_1, y_1). \quad \blacksquare$$

Example: Whether the resolution rule is applicable or not can be extremely sensitive to the order of the quantifiers in the premises. For instance, suppose we change the formula in the previous example to

$$\begin{array}{l} \exists x_1 \forall x_2 \exists x_3 . (\Diamond\, p(x_1, x_2) \;\vee\; q(x_2, x_3)) \\ \qquad \wedge \\ \forall y_2 \exists y_1 . \neg q(y_1, y_2) \end{array}$$

and take

$$\begin{array}{l} A = \neg q(y_1, y_2) \quad \text{and} \quad B = (\Diamond\, p(x_1, x_2) \;\vee\; q(x_2, x_3)), \\ v_1 = q(y_1, y_2) \quad \text{and} \quad v_2 = q(x_2, x_3), \\ \theta = \{x_2 \leftarrow y_1, y_2 \leftarrow x_3\}. \end{array}$$

Restrictions (i), (iii), and (iv) are still satisfied, but it is not possible to satisfy restrictions (ii) and (v) simultaneously. For instance, if we derive the formula

$$\exists x_1 \forall y_2 \exists y_1 \forall x_2 \exists x_3 . [\neg \mathit{true} \;\vee\; (\Diamond\, p(x_1, y_1) \;\vee\; \mathit{false})]$$

restriction (v) is not satisfied: $(y_2 \leftarrow x_3) \in \theta$ and x_3 is bound in

$$\forall y_2 \exists y_1 \forall x_2 \exists x_3 . [\neg q(y_1, y_2) \;\vee\; (\Diamond\, p(x_1, x_2) \;\vee\; q(x_2, x_3))].$$

Other formulas we may want to derive give rise to similar restriction violations. ▮

c. Merging the quantifiers

The resolution rule does not explicitly specify the order of the quantifiers $S_1 z_1, \ldots, S_{h+k} z_{h+k}$. A method for obtaining the sequence $S_1 z_1 \ldots S_{h+k} z_{h+k}$ is based on systematically merging the sequences $Q_1 x_1 \ldots Q_h x_h$ and $R_1 y_1 \ldots R_k y_k$ in different ways, until one of the results satisfies all the restrictions at once. Fortunately, there are less expensive implementations.

For instance, the one sketched here is based on choosing a partial order for the quantifiers and then running a topological sort. As a preliminary step, we check that conditions (i), (iii), and (iv) are satisfied. Then we build a directed graph with nodes labelled by the quantifiers from the premises of the rule, that is, $S_1 z_1, \ldots, S_{h+k} z_{h+k}$. There is an edge from $S_i z_i$ to $S_j z_j$ if $(z_j \leftarrow t\langle z_i \rangle) \in \theta$ for some term t or if $S_j z_j$ is in the scope of $S_i z_i$ in either of the premises' quantifier sequences, $Q_1 x_1 \ldots Q_h x_h$ and $R_1 y_1 \ldots R_k y_k$. An edge from $S_i z_i$ to $S_j z_j$ can be interpreted as expressing that z_j depends on z_i and implies that $S_i z_i$ should occur to the left of $S_j z_j$ in the formula derived by the rule.

If the graph is cyclic, the rule is not applicable. Otherwise, the graph can be mapped into a string by a topological sort. The output string is just $S_1 z_1 \ldots S_{h+k} z_{h+k}$. When arbitrary choices are possible, it is convenient to place $\exists$'s close to the source (that is, to the left in $S_1 z_1 \ldots S_{h+k} z_{h+k}$) in order to get a stronger conclusion. This construction respects the original order of the quantifiers and dependencies; therefore, restrictions (ii) and (v) are satisfied. Finally, redundant quantifiers may be discarded in the derived formula.

Example: The graph for the first example above is

$$\begin{array}{ccccc} \exists x_1 & \longrightarrow & \forall x_2 & \longrightarrow & \exists x_3 \\ & & \uparrow & & \downarrow \\ & & \exists y_1 & \longrightarrow & \forall y_2 \end{array}$$

It can be flattened into the string

$$\exists x_1 \;\longrightarrow\; \exists y_1 \;\longrightarrow\; \forall x_2 \;\longrightarrow\; \exists x_3 \;\longrightarrow\; \forall y_2 .$$ ▮

Example: The graph for the second example is

$$\begin{array}{ccccc} \exists x_1 & \longrightarrow & \forall x_2 & \longrightarrow & \exists x_3 \\ & & \uparrow & & \downarrow \\ & & \exists y_1 & \longleftarrow & \forall y_2 \end{array}$$

The resolution rule is not applicable because the graph is cyclic. ▮

d. Resolution with flexible function symbols

In the presence of flexible function symbols, a new restriction on the resolution rule is necessary. The following examples show that the current rule is not sound for formulas with flexible function symbols.

Example: Consider the formula

$$u: \ (\forall x. \neg \Box p(x)) \ \land \ \Box p(a),$$

where a and p are flexible. The formula u is satisfied by the model $\mathcal{M}$ with $D = \{0, 1\}$, $W = \{0, 1\}$, $w_0 = 0$, $R = W^2$, where a has value 0 in the real world and 1 elsewhere, p holds for 0 only in the real world, and p fails for 1 only in the real world. Take $A = \neg \Box p(x)$, $B = \Box p(a)$, $v_1 = \Box p(x)$, $v_2 = \Box p(a)$. The most-general unifier of v_1 and v_2 is $\theta = \{x \leftarrow a\}$. With the restrictions we have presented so far, the resolution rule allows us to deduce

$$(\forall x. \neg \Box p(x)) \ \land \ \Box p(a) \ \land \ (\neg \mathit{true} \lor \mathit{false}).$$

Simplification yields *false*. According to this proof, u is unsatisfiable. ▮

Example: Consider the formula

$$u: \ p(a) \ \land \ \Box p(a) \ \land \ \forall x. \big(\neg p(x) \lor \Diamond \neg p(x)\big),$$

where a and p are flexible. The model $\mathcal{M}$ described in the previous example satisfies u. Take $A = \big(\neg p(x) \lor \Diamond \neg p(x)\big)$, $B = p(a)$, $v_1 = p(x)$, $v_2 = p(a)$. The most-general classical unifier of v_1 and v_2 is $\theta = \{x \leftarrow a\}$. With the restrictions we have presented so far, the resolution rule allows us to deduce

$$p(a) \ \land \ \Box p(a) \ \land \ \forall x. \big(\neg p(x) \lor \Diamond \neg p(x)\big) \ \land \ \big[(\neg \mathit{true} \lor \Diamond \neg p(a)) \lor \mathit{false}\big].$$

Simplification yields

$$\Box p(a) \land \Diamond \neg p(a),$$

a clearly provably unsatisfiable formula. According to this proof, then, u is unsatisfiable. ▮

Unification in the scope of modal operators and substitution into the scope of modal operators give rise to incorrect derivations in these examples. The basic problem is simply that equals cannot be substituted for equals in modal logics. The resolution rule is restricted further in order to avoid this problem:

(vi) If $(x \leftarrow t) \in \theta$ and a flexible symbol occurs in t then x does not occur in the scope of any modal operator in either A or B.

e. Soundness of resolution

The restrictions presented in the last two subsections are actually sufficient to guarantee the soundness of the resolution rule. We first show:

Lemma (Soundness of instantiation):

Given the substitution θ, the quantifiers $T_1, \ldots, T_\ell$, and the formulas $v = T_1 w_1 \ldots T_\ell w_\ell . u$, and $v' = T_1 w_1 \ldots T_\ell w_\ell . u\theta$, such that

if $(x \leftarrow t) \in \theta$ then for some i, $1 \leq i \leq \ell$, $T_i = \forall$, $w_i = x$, and no variable in t occurs bound in $\forall x T_{i+1} w_{i+1} \ldots T_\ell w_\ell . u$,

if $(x \leftarrow t) \in \theta$ and t contains flexible symbols then x does not occur in the scope of any modal operator in u,

then $v \hookrightarrow v'$.

Theorem: The resolution rule, with restrictions (i), (ii), (iii), (iv), (v), and (vi), is sound.

Proof sketch: It suffices to show that the premises entail the conclusion, that is,

$$Q_1 x_1 \ldots Q_h x_h . A\langle v_1, \ldots, v_n\rangle \wedge R_1 y_1 \ldots R_k y_k . B\langle v_{n+1}, \ldots, v_m\rangle \\ \hookrightarrow S_1 z_1 \ldots S_{h+k} z_{h+k} . \left[A\theta\langle \mathit{true}\rangle \vee B\theta\langle \mathit{false}\rangle\right].$$

Assume the premises $Q_1 x_1 \ldots Q_h x_h . A$ and $R_1 y_1 \ldots R_k y_k . B$ hold. Conditions (i) and (ii) guarantee that the (sound) quantifier rules allow us to derive $S_1 z_1 \ldots S_{h+k} z_{h+k} . (A \wedge B)$. This formula and θ fulfill the hypotheses of the lemma by conditions (v) and (vi). Therefore, we can derive $S_1 z_1 \ldots S_{h+k} z_{h+k} . (A \wedge B)\theta$, that is, $S_1 z_1 \ldots S_{h+k} z_{h+k} . (A\theta \wedge B\theta)$. (At this point redundant quantifiers can be deleted from the conclusion without harm.)

We have shown that

$$Q_1 x_1 \ldots Q_h x_h . A \wedge R_1 y_1 \ldots R_k y_k . B \\ \hookrightarrow S_1 z_1 \ldots S_{h+k} z_{h+k} . \left(A\theta\langle v_1, \ldots, v_n\rangle \wedge B\theta\langle v_{n+1}, \ldots, v_m\rangle\right).$$

It suffices to show that

$$S_1 z_1 \ldots S_{h+k} z_{h+k} . \left(A\theta\langle v_1, \ldots, v_n\rangle \wedge B\theta\langle v_{n+1}, \ldots, v_m\rangle\right) \\ \hookrightarrow S_1 z_1 \ldots S_{h+k} z_{h+k} . \left[A\theta\langle \mathit{true}\rangle \vee B\theta\langle \mathit{false}\rangle\right].$$

This can be proved by purely propositional modal reasoning: by the monotonicity of entailment lemma, it suffices to show that

$$\left(A\theta\langle v_1, \ldots, v_n\rangle \wedge B\theta\langle v_{n+1}, \ldots, v_m\rangle\right) \hookrightarrow \left[A\theta\langle \mathit{true}\rangle \vee B\theta\langle \mathit{false}\rangle\right].$$

The formulas $A\theta$ and $B\theta$ have some subformulas in common, since $v_1\theta = \ldots = v_m\theta$. Let $v\theta$ denote $v_1\theta, \ldots, v_m\theta$. Consider occurrences of $v\theta$ not in the scope of any quantifier and, if $v\theta$ contains any flexible symbols, not in the scope of any modal operator. Assume that $A\theta\langle v_1, \ldots, v_n\rangle$ and $B\theta\langle v_{n+1}, \ldots, v_m\rangle$ hold. If $v\theta$ is true then $A\theta\langle true\rangle$ holds; otherwise, $B\theta\langle false\rangle$ holds. In either case, $A\theta\langle true\rangle \vee B\theta\langle false\rangle$ holds, as we wanted to show. ▮

8. AN EXAMPLE

We prove that

$$\Box(\forall x.p(x)) \supset (\forall x. \Box p(x))$$

in the resolution system for K. We will derive *false* from

$$S_0 : \neg[\neg \Box(\forall x.p(x)) \vee (\forall x. \Box p(x))].$$

By the negation rules, we first get

$$\Box(\forall x.p(x)) \wedge (\exists x. \Diamond \neg p(x)).$$

The rule for moving quantifiers of existential force yields

$$\exists x'.[\Box(\forall x.p(x)) \wedge \Diamond \neg p(x')].$$

The modality rule in the system for K yields

$$\exists x'.[\Box(\forall x.p(x)) \wedge \Diamond \neg p(x') \wedge \Diamond((\forall x.p(x)) \wedge \neg p(x'))].$$

Weakening reduces this sentence to

$$\exists x'. \Diamond [(\forall x.p(x)) \wedge \neg p(x')].$$

Take $A = \neg p(x')$, $B = p(x)$, $v_1 = p(x')$, $v_2 = p(x)$. Resolution yields

$$\exists x'. \Diamond [(\forall x.p(x)) \wedge \neg p(x') \wedge (\neg true \vee false)].$$

true-false simplifications yield *false*.

9. COMPLETENESS FOR FIRST-ORDER SYSTEMS

Our propositional modal resolution systems together with the quantifier rules, the auxiliary rules, and the resolution rule for the first-order language with flexible function symbols, constitute first-order resolution systems. Skolemization rules may be added, but are not essential.

Theorem: The first-order resolution systems for K, T, K4, S4, S5, D, and D4 are complete for the corresponding classes of models.

Proof sketch: Some Hilbert systems are known to be complete for these logics, at least for the language with no rigid symbols and no function symbols (e.g., [HC], [Fil]). We can extend these completeness results to the language with rigid symbols and function symbols. Then we show that each of the resolution systems is at least as powerful as one such complete Hilbert system. Specifically, we show that any Hilbert proof can be transformed into a resolution proof, by induction on the structure of Hilbert proofs. ▮

Remark: We will not discuss completeness issues for first-order G. Several notions of completeness have been proposed for this logic and none of those based on Kripke models seems fully satisfactory ([Fi2]).

Acknowledgements:

We are grateful to Bengt Jonsson and John Lamping for critical reading of the manuscript.

REFERENCES

[AM1] M. Abadi and Z. Manna, "Nonclausal temporal deduction," in *Logics of Programs* (R. Parikh, ed.), Springer-Verlag LNCS 193, 1985, pp. 1–15.

[AM2] M. Abadi and Z. Manna, "A timely resolution," Proceedings of the Symposium on Logic in Computer Science (LICS), 1986.

[Fa1] L. Fariñas del Cerro, "Temporal reasoning and termination of programs," Eighth International Joint Conference on Artificial Intelligence, 1983, pp. 926–929.

[Fa2] L. Fariñas del Cerro, "Un principe de résolution en logique modale," *RAIRO Informatique Théorique*, Vol. 18, No. 2, 1984, pp. 161–170.

[Fa3] L. Fariñas del Cerro, "Resolution modal logics," in *Logics and Models of Concurrent Systems* (K.R. Apt, ed.), Springer-Verlag, Heidelberg, 1985, pp. 27–55.

[Fi1] M. Fitting, *Proof Methods for Modal and Intuitionistic Logics*, D. Reidel Publishing Co., Dordrecht, 1983.

[Fi2] M. Fitting, private communication.

[GK] C. Geissler and K. Konolige, "A resolution method for quantified modal logics of knowledge and belief," in *Theoretical Aspects of Reasoning about Knowledge* (J. Halpern, ed.), Morgan Kaufmann Publishers, Palo Alto, 1986, pp. 309–324.

[HC] G.E. Hughes and M.J. Cresswell, *An Introduction to Modal Logic*, Methuen & Co., London, 1968.

[HM] J.Y. Halpern and Y. Moses, "Knowledge and Common Knowledge in a Distributed Environment," Third ACM Conference on the Principles of Distributed Computing, 1984, pp. 50–61. A revised version appears as IBM RJ 4421, 1984.

[Ko] K. Konolige, *A Deduction Model of Belief and its Logics*, Ph.D. Thesis, Computer Science Department, Stanford University, 1984.

[Mc] D. McDermott, "Nonmonotonic Logic II: Nonmonotonic Modal Theories," *Journal of the ACM*, Vol. 29, No. 1, Jan. 1982, pp. 33–57.

[Mu] N.V. Murray, "Completely nonclausal theorem proving," *Artificial Intelligence*, Vol. 18, No. 1, January 1982, pp. 67–85.

[MW1] Z. Manna and R. Waldinger, "A deductive approach to program synthesis," *ACM Transactions on Programming, Languages, and Systems*, Vol. 2, No. 1, Jan. 1980, pp. 90–121.

[MW2] Z. Manna and R. Waldinger, "Special relations in automated deduction," *Journal of the ACM*, Vol. 33, No. 1, Jan. 1986, pp. 1–59.

[MW3] Z. Manna and R. Waldinger, "Special relations in program-synthetic deduction," Report No. STAN-CS-82-902, Computer Science Department, Stanford University, March 1982.

[P] A. Pnueli, "The temporal logic of programs," 18th Annual Symposium on Foundations of Computer Science, 1977, pp. 46–57.

[R] J.A. Robinson, "A machine-oriented logic based on the resolution principle," *Journal of the ACM*, Vol. 12, No. 1, January 1965, pp. 23–41.

[W] P. Wolper, "Temporal Logic can be more expressive," 22nd Annual Symposium on Foundations of Computer Science, 1981, pp. 340–348.

Computational aspects of three-valued logic

P. H. Schmitt
IBM Deutschland GmbH, TK LILOG
D-7000 Stuttgart 1

Abstract

This paper investigates a three-valued logic L_3, that has been introduced in the study of natural language semantics. A complete proof system based on a three-valued analogon of negative resolution is presented. A subclass of L_3 corresponding to Horn clauses in two-valued logic is defined. Its model theoretic properties are studied and it is shown to admit a PROLOG-style proof procedure.

Introduction

Many-valued logics, a topic that has been studied extensively in the past in Mathematical and Philosophical Logic, has recently obtained increased interest in Computer Science and in particular in Artificial Intelligence. As a good survey we still recommend (Rescher 1969), which also contains an extensive bibliography. For a more recent account we refer to (Urquhart 1986) and (Dunn and Epstein 1977). Our interest into three-valued logics arose from occupation with "situation semantics", an approach to natural language semantics proposed in (Barwise and Perry 1983), that has lately received much attention. The basic objects in this theory are called situations. The basic facts in describing a situation come together with a polarity, 1 (if the fact holds true in the given situation) and 0 (if the fact is false). Assigning the truth value "undefined" to facts that do not appear in the description of a situation we arrive at a three-valued structure. We adopt in this paper the particular system L_3 for three-valued logic, proposed in (Fenstad et al. 1985), also treated with an eye on situation semantics in (Blamey 1986). The truth tables for conjunction and disjunction in L_3 coincide with those of Lukasiewicz and of Kleene. A particular feature is the presence of two negation symbols, that may be combined to yield the four basic types of literals: "A is true", "A is false", "A is not true" and "A is not false". Similar systems have been investigated e.g. in (Slupecki 1936) for the propositional calculus and in (Blau 1978) p. 222 ff.
We are mainly interested in the proof theory of L_3. Various complete axiomatizations of three-valued logics have been given; in addition to the literature already quoted see e.g. (Rosser and Turquette 1952) for an early example or (Ebbinghaus 1969) for a Gentzen style system. In particular (Fenstad et al. 1985) contains an axiomatization of L_3. We will in this paper present a complete resolution calculus for L_3 as a basis for an automated proof system. This calculus corresponds to negative resolution in two-valued logic. A resolution calculus for three-valued logic, as a special case of a calculus for many-valued logic, using a set of literals which is different from ours, is described in (Morgan 1973).
Considering the great importance of Horn clauses in the computational theory of two-valued logic, it would be of interest to find a subclass of L_3 with similar properties. The definition we pursue in this paper arises from the observation, that the literals of the form "A is true" and "A is not false" behave like positive literals, the remaining two types like negative literals. L_3-Horn clauses are now taken to be those containing at most one positive literal. We show, that there is a PROLOG-style proof procedure for this class of clauses and its model theoretic properties are analogous to two-valued Horn clauses, e.g. with respect to direct products and the existence of initial term structures.

Basic Definitions and Notation.

The syntax of the system L_3 of three-valued logic is the same as for the ordinary two-valued first order predicate calculus with the only exception, that it contains two negation symbols: $\neg A$ for the strong negation of proposition A, $\sim A$ for weak negation. We use $\vee, \wedge$ to denote disjunction and conjunction respectively and $\exists x\, A$, $\forall x\, A$ for the existential resp. universal quantification of formula A with respect to the variable x. A structure *M* consists of a universe M with interpretations of the function and constant symbols as usual and a valuation function v, if necessary denoted more precisely v_M, which associates with every n-ary predicate symbol P and every n-tupel $m_1,..,m_n$ of elements from M a truth value $v(P(m_1,...,m_n))$. The possible truth values are 0 (false), 1 (true), u (undefined). If need arises we write more precisely v(*M*) instead of v. The valuation function v can be extended to all formulas of L_3 without free variables using the following rules and truth tables.

Rules for quantifiers:

- $v(\forall x\ A) = 1$ iff for all elements m in M $v(A(m/x)) = 1$.
- $v(\forall x\ A) = 0$ iff there is an element m in M satisfying $v(A(m/x)) = 0$.
- $v(\forall x\ A) = u$ in all other cases.

- $v(\exists x\ A) = 1$ iff there is an element m in M satisfying $v(A(m/x)) = 1$.
- $v(\exists x\ A) = 0$ iff for all elements m in M $v(A(m/x)) = 0$.
- $v(\exists x\ A) = u$ in all other cases

For the sake of this definition we have assumed that constant symbols for all elements of M are available. A(m/x) is the formula that arises from A by substituting the free occurences of the variable x by a constant symbol for m.

Rules for $\wedge$ and $\vee$

- $v(A \wedge B)$ = the minimum of v(A) and v(B).
- $v(A \vee B)$ = the maximum of v(A) and v(B).

Here we think of the truth values being ordered as $0 < u < 1$.

Forms of Negation

A	¬A	~A	~¬A	~~A	¬¬A	¬~A
1	0	0	1	1	1	1
u	u	1	1	0	u	0
0	1	1	0	0	0	0

The content of this table will be more intuitive when rephrased in natural language:

v(A) = 1 iff A is true

v(¬A) = 1 iff A is false

v(~A) = 1 iff A is not true

v(~¬A) = 1 iff A is not false

We will use $A \supset B$ (A implies B) and $A \equiv B$ (A is equivalent to B) as abbreviations for $\sim A \vee B$ and $(A \supset B) \wedge (B \supset A)$.

The corresponding notions of strong implication and strong equivalence which are defined by replacing weak negation by strong negation will not be used in this paper. Then next lemma lists some basic equivalences of L_3.

Lemma 1:

1. $\neg\neg A \equiv A$
2. $\sim\sim A \equiv A$
3. $\neg\sim A \equiv A$
4. $(A \wedge B) \vee C \equiv (A \vee C) \wedge (B \vee C)$

5. $(A \vee B) \wedge C \equiv (A \wedge C) \vee (B \wedge C)$
6. $\neg(A \vee B) \equiv \neg(A) \wedge \neg(B)$
7. $\neg(A \wedge B) \equiv \neg(A) \vee \neg(B)$
8. $\sim(A \vee B) \equiv \sim(A) \wedge \sim(B)$
9. $\sim(A \wedge B) \equiv \sim(A) \vee \sim(B)$
10. $\sim(\exists x\, A) \equiv \forall x \sim(A)$
11. $\neg(\exists x\, A) \equiv \forall x\, \neg(A)$
12. $\sim(\forall x\, A) \equiv \exists x \sim(A)$
13. $\neg(\forall x\, A) \equiv \exists x\, \neg(A)$ •

Terminology:

- A *literal* is a formula of the form A, $\neg A$, $\sim A$ or $\sim\neg A$, where A is an atomic formula. In each case the literal is said to contain the atomic formula A.
- Literals of the form A or $\sim\neg A$ are called *positive*, the others *negative*.
- A *clause* is a (possibly empty) disjunction of literals.
- A clause is called *negative* (*positive*) if it contains only negative (resp. positive) literals.
- A set S of clauses is called *satisfiable* if there is a L_3-structure $\mathcal{M} = <M,v>$, such that $v(K) = 1$ for all clauses K in S. In this case $\mathcal{M}$ is called a *model* of S.
- A clause K is a consequence of a set of clauses S ($S \models K$), if for all models $\mathcal{M} = <M,v>$ of S also $v(K) = 1$.

The next lemma collects some of the basic results for manipulating formulas in predicate calculus, that are still true in L_3.

Lemma 2:

1. Every atomic formula preceeded by a (possibly empty) string of negation symbols is equivalent to a literal.
2. Every quantifier free formula is equivalent to a disjunctive normal form and also to a conjunctive normal form.
3. Every formula is equivalent to a prenex normal form.
4. For a set of formulas S and a formula K

   $S \models K$ iff $S \cup \{ \sim K \}$ is not satisfiable.•

A Resolution calculus

The calculus considered here works exclusively with clauses, implicitely thought of as beeing universally quantified and contains as its only proof rule the (non-deterministic) resolution rule. For presentational purposes we explain first the *basic resolution* rule.

Two sets S_1, S_2 of literals are called *complementary* if one of the following is true:

1. all literals in S_1 are of the form A and all literals in S_2 are of the form $\neg A$ or $\sim A$; or the same holds true with S_1 and S_2 interchanged.
2. all literals in S_1 are of the form A or $\sim\neg A$ and all literals in S_2 are of the form $\neg A$; or the same holds true with S_1 and S_2 interchanged.

Let $K_1 = L_{1,1} \vee \ldots \vee L_{1,n}$ $K_2 = L_{2,1} \vee \ldots \vee L_{2,r}$ be clauses. A clause K is obtained from K_1 and K_2 by a basic resolution step if there are complementary subsets S_1, S_2 of the literals in K_1, resp. K_2 and a unifying substitution σ for the set of atomic formulas contained in the literals in the union of S_1 and S_2, such that $K = \sigma(K_0)$, where K_0 is the disjunction of the literals appearing in K_1 or K_2 but neither in S_1 nor S_2. K is then called a *basic resolvent* of K_1 and K_2.

K is obtained from $K_1$and K_2 by a (general) *resolution* step if there are renaming substitutions μ_1, μ_2, such that K can be produced from $\mu_1(K_1)$ and $\mu_2(K_2)$ by a basic resolution step.

Here are three typical examples of the resolution rule for propositional formulas:

$$\frac{L \vee A \qquad K \vee \neg A}{L \vee K} \qquad \frac{L \vee A \qquad K \vee \sim A}{L \vee K}$$

$$\frac{L \vee \sim\neg A \qquad K \vee \neg A}{L \vee K}$$

For a set S of clauses Res(S) denotes the closure of S under resolution, Subst(S) the set of all substitution instances of clauses in S. The following easily proved compatibility property of these two operators will be used to reduce problems in the resolution calculus to the propositional (i.e.variable free) case.

Lemma 3:

1. Subst(Res(S)) = Res(Subst(S))
2. Furthermore it is possible to find for every variable free instance K of a clause in Res(S) variable free clauses K_1, K_2 in Subst(S), such that K is a resolvent of K_1 and K_2.•

Lemma 4:

A set of clauses S is satisfiable iff Res(S) is satisfiable.

Proof: Since S is a subset of Res(S) only the implication from left to right is non-trivial. Let $\mathcal{M} = \langle M,v \rangle$ be a model for S. We claim that $\mathcal{M}$ is also a model of Res(S). By the previous lemma we may, without loss of generality, assume, that S contains only variable free clauses. Let K be a resolvent, that is obtained from the clauses K_1 and K_2 using the complementary sets of literals S_1 and S_2. Since K_1 and K_2 are variable free there is exactly one atomic formula A contained in the literals of S_1 and S_2. By assumption we have $v(K_1) = v(K_2) = 1$. Since the following three combinations are impossible

$v(A) = 1$ and $v(\neg A) = 1$.
$v(A) = 1$ and $v(\sim A) = 1$.
$v(\sim\neg A) = 1$ and $v(\neg A) = 1$.

it is a consequence of the definition of complementarity, that for some literal L in K_1 and not in S_1 or L in K_2 and not in S_2 we must have $v(L) = 1$. But this also yields $v(K) = 1$.•

Remark: It is quite possible that $v(\sim\neg A)$ and $v(\sim A)$ both have the value 1 . This is the reason why it is not allowed to resolve $\sim\neg A$ against $\sim A$.

Theorem 5: Let S be a set of clauses.

S is satisfiable iff the empty clause is not an element of Res(S).

Proof: Only the implication from right to left is non-trivial. Using Lemma 3.1 S may be assumed, without loss of generality, to be variable free. Let $\{ A_n : n > 0 \}$ be an enumeration of all variable free atomic formulas. We will use Cl_n to denote the set of clauses whose literals contain only the variable free atomic formulas A_i with $i < n$.

The first part of the proof consists in the construction of a three-valued analogon of a Herbrand structure; this amounts to the specification of a truth value $v(A_n)$ for every n. We proceed inductively and assume that $v(A_i)$ has already been defined for $i < n$. The truth value of A_n is determined by the following instructions executed in the given order.

1. Set $v(A_n) = 1$, unless there is a negative clause K, $K = K_1 \vee K_2$ with K_1 in Cl_n, $v(K_1)$ different from 1 and all literals in K_2 contain A_n.
2. Set $v(A_n) = u$, unless there is a negative clause K, $K = K_1 \vee K_2$ with K_1 in Cl_n, $v(K_1)$ different from 1 and K_2 is a disjunction of occurences of $\neg(A_n)$.
3. Set $v(A_n) = 0$.

In the second part of the proof we verify $v(K) = 1$ for every clause K in Res(S). We proceed by induction on the number p of positive literals in K.

Initial case $p = 0$.

Since K cannot be the empty clause we may choose n to be the maximal index such that A_n appears in a literal of K. By instruction 1 we know already that $v(A_n)$ must be different from 1. If $\sim A_n$ occurs in K, this suffices already to conclude $v(\sim A_n) = 1$ and a fortiori $v(K) = 1$.If $\neg A_n$ is the only literal in K containing A_n, then instructions 2 and 3 require $v(A_n) = 0$ and again $v(K) = 1$ follows.

Induction step from p to $p+1$.

Choose n, such that A_n appears in apositive literal of K. If $v(A_n)$ happens to be 1, then we get immediately $v(K) = 1$. If $v(A_n)$ is different from 1, then there must be by the first instruction a negative clause $N = N_1 \vee N_2$ with $v(N_1)$ different from 1 and all literals in N_2 contain A_n.

If $\sim\neg A_n$ is not among the literals in K, we may form the resolvent R of K and N, that does not have a literal containing A_n any more. By induction hypothesis $v(R) = 1$. The restrictions on N guarantee, that already $v(K) = 1$.

Now let us consider the case, when $\sim\neg A_n$ is among the literals in K. If $v(A_n)$ happens to be u we easily arrive at $v(K) = 1$. But if $v(A_n)$ is also different from u, instruction 2 tells us that there has to be a negative clause $N = N_1 \vee N_2$ in Res(S) with $v(N_1)$ different from 1 and N_2 a disjunction of occurences of $\neg A_n$. Again a resolvent R without literals containing A_n can be derived from K and N, and we get $v(K) = 1$ as above.•

An inspection of the above proof shows, that a resolution step is only performed if one of the participating clauses is negative; this is what we call a *negative resolution*. If we denote by NRes(S) the closure of S under negative resolution, we can formulate the following corollary:

Corollary 6:

S is satisfiable iff NRes(S) does not contain the empty clause.

Horn formulas

Considering the important role that Horn formulas play in classical two-valued logic it seems worthwhile to look for analoga in the three-valued case. Among the several natural candidates we will investigate in this paper the following class:

Definition:

A *Horn clause* is a clause containing at most one positive literal.
A *Horn formula* is built up from Horn clauses by conjunction and quantification.

Let us consider, what this definition has to recommend it. It is a the well known result from two-valued logic, that the universal formulas preserved under direct products are exactly the universal Horn formulas. We will prove, that this still holds true in L_3.

Direct products for L_3-structures are given as follows. Let $M_i = <M_i,v_i>$ for i in I be a family of L_3-structures. The universe and the interpretation of the constants and function symbols of the direct product $M = <M,v>$ of the family M_i are as usual. For every n-ary predicate symbol P and every n-tupel $m_1,..,m_n$ of elements from M we set:

- $v(P(m_1,...,m_n)) = 1$ if for all i in I $v_i(P(m_{1i},...,m_{ni})) = 1$.
- $v(P(m_1,...,m_n)) = 0$ if for some i in I $v_i(P(m_{1i},...,m_{ni})) = 0$.
- $v(P(m_1,...,m_n)) = u$ in all other cases.

We only consider direct products of non-empty families of L_3-structures.

Theorem 7:

Let S be a set of formulas, that is preserved under direct products. A universal formula K is preserved under direct products of models of S iff there is a universal Horn formula H, such that $S \mid\!- (K \equiv H)$.

Proof: Let $M = \langle M,v \rangle$ be the direct product of the family $M_i = \langle M_i,v_i \rangle$ with i ranging over I.

The crucial step in proving the implication from right to left is to show that for a variable free Horn clause H with $v_i(H)=1$ for all i in I, also $v(H)$ has to equal 1.

If for some i in I and some negative literal L of H $v_i(L) = 1$, then:

- either L is of the form $\neg A$ and $v_i(A) = 0$
- or L is of the form $\sim A$ and $v_i(A)$ may be 0 or u.

By the definition of direct product we have:

- If for some i in I $v_i(A) = 0$, then $v(A) = 0$
- If for some i in I $v_i(A)$ equals 0 or u, then $v(A)$ equals 0 or u.

This shows that $v(L) = 1$ in both cases.

If for all i in I we have $v_i(L) = 1$, where L is the unique positive literal in H, then:

- either L is of the form A and for all i in I $v_i(A) = 1$
- or L is of the form $\sim\neg A$ and for all i in I $v_i(A)$ equals either 1 or u.

Using the following property of direct products

- If for all i in I $v_i(A) = 1$, then $v(A) = 1$.
- If for all i in I $v_i(A)$ equals 1 or u, then also $v(A)$ equals 1 or u.

we again arrive at $v(L) = 1$ and therefore at $v(K) = 1$.

To prove the implication from left to right we may assume, without loss of generality, that K is in conjunctive normal form:

$$
\begin{aligned}
\mathbf{K} = \ & \textbf{universal quantification of} \\
& (K_{1,0} \vee L_{1,1} \vee ... \vee L_{1,r1}) \\
& \wedge \\
& \cdot \\
& \cdot \\
& \cdot \\
& \wedge \\
& (K_{n,0} \vee L_{n,1} \vee ... \vee L_{n,rn})
\end{aligned}
$$

where $K_{i,0}$ are disjunctions of negative literals and $L_{i,j}$ are positive literals.

Let r be the product of all those r_i, that are different from 0. If $r = 1$, then K is a Horn formula. In general the i-th conjunctive component of K may be reduced to a Horn clause in r_i possible ways:

$K_{i,0} \vee L_{i,1}$
$K_{i,0} \vee L_{i,2}$

.

.

.

$K_{i,0} \vee L_{i,ri}$

We form the collection of all conjunctive combinations of these Horn clauses. There are exactly r many, which we denote by N_1 through $N(r)$. For the special case $n = 2$ and $r_1 = r_2 = 2$ this would give :

$$N_1 = (K_{1,0} \vee L_{1,1}) \wedge (K_{2,0} \vee L_{2,1})$$
$$N_2 = (K_{1,0} \vee L_{1,1}) \wedge (K_{2,0} \vee L_{2,2})$$
$$N_3 = (K_{1,0} \vee L_{1,2}) \wedge (K_{2,0} \vee L_{2,1})$$
$$N_4 = (K_{1,0} \vee L_{1,2}) \wedge (K_{2,0} \vee L_{2,2})$$

We claim, that for some j $S \vdash (K \equiv N_j)$ must be true. Obviously $(K \supset N_j)$ is a tautology for every j. If none of the reverse implications, restricted to models of S, holds, then we obtain for every j a model $M_j = \langle M_j, v_j \rangle$ of S, such that $v_j(N''_j)$ is different from 1, where N''_j is a variable free instantiation of N_j by constants for elements from M_j and $v_j(K) = 1$. If $v_j(N''_j)$ is different from 1, then at least one of its conjuctive components must have a v_j-value different from 1. A simple combinatorial argument shows, that for some i , there are $j_1,...,j_{ri}$, such that

$v(j_1)(K_{i,0} \vee L_{i,1})$ is different from 1
and
$v(j_2)(K_{i,0} \vee L_{i,2})$ is different from 1
and

.

.

.

and
$v(j_{ri})(K_{i,0} \vee L_{i,ri})$ is different from 1.

For the direct product $M = \langle M,v \rangle$ of $M(j_1),...M(j_{ri})$ we find therefore $v(K'')$ different from 1 for the appropriate instantiation of K , since its i-th conjuctive component has v-value different from 1. A fortiori v(K) is different from 1. This contradicts the assumption that K is preserved under direct product thus establishing the claim made above. •

Another characterization of universal Horn clauses in the two-valued case states, that

• A theory T can be axiomatized by a set of universal Horn clauses
• iff
• T uniformly admits initial term models.

For the unexplained notions and proofs see (Makowski 1984) and (Volger 1985). We remark without proof, that this characterization can also be lifted to the three-valued case. To help the reader familiar with the involved concepts to understand this result, we have to disclose how we define substructures in L_3.

An L_3-structure $M = \langle M,v \rangle$ is a *substructure* of another L_3-structure $N = \langle N,v'' \rangle$, if M is a subset of N, the usual substructure conditions apply for the constants and functions and for every basic predicate P containing parameters from M $v(P) = v''(P)$. Thus N does not contain new information about the objects already contained in M. This may in many contexts not be the proper notion, since $v(P) = u$ suggests, that in some extension of M we gain information about P, e.g. $v''(P) = 1$. This is more closely modelled by the following weaker notion: M is a *weak substructure* of N in case:

• $v''(P) = 1$, if $v(P) = 1$
• $v''(P) = 0$, if $v(P) = 0$
• and the remaining requirements as above.

Preservation results with respect to weak substructures have been obtained in (Fenstad et al. 1985) and (Langholm forthcoming).

Let us now turn to the computational aspect of Horn clauses. As in the two-valued case we restrict attention to *strict Horn clauses*, i.e. those clauses with exactly one positive literal. These are of the form

$A \quad \vee L_1 \vee \ldots \vee L_n$
or
$\sim\neg A \vee L_1 \vee \ldots \vee L_n$
where the L_i are of the form $\neg(A)$ or $\sim(A)$ and $n = 0$ is permitted.

or rewritten in more familiar looking implications:

$L_1 \wedge \ldots \wedge L_n \supset A$
or
$L_1 \wedge \ldots \wedge L_n \supset \sim\neg A$
where the L_i are of the form $\sim\neg A$ or A and $n = 0$ is permitted.

As in PROLOG we use strict Horn clauses thought of as being universally quantified to built up a knowledge base. It is in keeping with the two-valued case, that we can only represent "positive" knowledge in this way.

Querries to this knowledge base are posed as conjunctions of positive literals implicitely thought of as being existentially quantified.

Given a set D of strict Horn clauses and a query Q we define the *proof tree* of D and Q, denoted by $PT(D,Q)$. Because of the similarity of this concept with its two-valued counterpart the following explanations should suffice. Every node in $PT(D,Q)$ will be labeled by a (possibly empty) conjunction of positive literals. The root is labeled by Q. For a node labeled by $L_1 \wedge \ldots \wedge L(r)$ there will be as many successor nodes as there are clauses in D, whose head unifies with one of the literals L_i. In particular, a node labeled by the empty conjunction has to be a leaf. Let e.g. $N_1 \wedge \ldots \wedge N_s \supset N$ be a clause in D and μ the most general unifier of N and, say, L_1, then the corresponding successor node will be labeled by

$\mu(N_1 \wedge \ldots \wedge N_s \wedge L_2 \wedge \ldots \wedge L_r)$.

One would of course like to prove that "$D \vdash Q$ iff there is a leaf in the proof tree $PT(D,Q)$ labeled with the empty conjunction". Unfortunately there is a small problem with this, which in its simplest form reads as follows. Let D contain as its only element the atomic formula A and $Q = \sim\neg A$. Evidently $D \vdash Q$, but $PT(D,Q)$ consists only of one node and this is labeled by Q. When we form $D_1 = D \cup \{ \sim Q \}$ (or equivalently $D_1 = D \cup \{ \neg A \}$, then the empty clause is in $Res(D_1)$. The problem only arises, when we want to shortcut this procedure and show the effects of resolution in the conjunctive format of the goal statement. Here is the solution:

Theorem 8:

Let D be a set of universal strict Horn clauses and Q an existential conjunction of positive literals.Then:

$D \vdash Q$ iff the prooftree $PT(D^*,Q)$ contains a leaf labeled by the empty conjunction,

where D^* is the union of D with the set of all clauses $A \supset \sim\neg A$ for all atomic formulas contained in Q or in the head of a clause in D.

The proof now follows along the usual lines using Theorem 5.•

References

Barwise and Perry 1983 J.Barwise and J.Perry. Situations and Attitudes. Bradford Books, Cambridge, Mass.

Blamey 1986 S.Blamey. Partial logic. In: (Gabbay and Guenthner 1986), p.1-70.

Blau 1978 U.Blau. Die dreiwertige Logik der Sprache. Walter de Gruyter.

Dunn and Epstein 1977 J.M.Dunn and G.Epstein (eds.). Modern Uses of Multiple-Valued Logic. D.Reidel Publish. Co., Dordrecht.

Ebbinghaus 1969	H.-D.Ebbinghaus. Über eine Prädikatenlogik mit partiell definierten Prädikaten und Funktionen. Archiv für mathematische Logik 12, p.39-53.
Fenstad et al. 1985	J.E.Fenstad, P.-K.Halvorsen, T.Langholm and J.van Benthem. Equations, Schemata and Situations: A framework for linguistic semantics. CSLI Report No. 29. Center For The Study Of Language And Information, Stanford, CA 94305.
Gabbay and Guenthner 1986	D.Gabbay and F.Guenthner (eds.). Handbook of Philosophical Logic, Vol.III: Alternatives in Classical Logic. D. Reidel Publishing Co., Dordrecht.
Langholm forthcoming	T.Langholm. Characterizations of persistent formulas in a three-valued logic. University of Oslo, Sweden.
Makowski 1984	J.A.Makowski. Why Horn formulas matter in computer science: initial structures and generic examples. Techn.Report No.329, Technion, Haifa.
McCall 1967	S.McCall (ed.). Polish Logic 1920-1939. Oxford.
Morgan 1973	C.G.Morgan. A Resolution Procedure For A Class Of Many-valued Logics. Journal of Symbolic Logic vol. 39, 199-200 (abstract).
Rescher 1969	N.Rescher. Many-Valued Logic. McGraw Hill Book Co., New York.
Rosser and Turquette 1952	J.B.Rosser and A.R.Turquette. Many-Valued Logics. North-Holland Publishing Co., Amsterdam.
Slupecki 1936	J.Slupecki. Der volle dreiwertige Aussagenkalkül. Comptes rendues des s&eaccr.ance de la Soci&eaccr.t&eaccr. des Science de Lettres de Varsovie, Classe III, vol. 29, 9-11. English translation in (McCall) 335-337.
Urquhart 1986	A.Urquhart. Many-valued logic. In: (Gabbay and Guenthner 1986), p.71-116.
Volger 1985	H.Volger. On Theories which Admit Initial Structures. FNS-Bericht 85-1. Universität Tübingen, Forschungsstelle für natürlich-sprachige Systeme.

RESOLUTION AND QUANTIFIED EPISTEMIC LOGICS

Kurt Konolige*
Artificial Intelligence Center and CSLI
SRI International
Menlo Park, California 94025/USA

Abstract

Quantified modal logics have emerged as useful tools in computer science for reasoning about knowledge and belief of agents and systems. An important class of these logics have a possible-world semantics from Kripke. Surprisingly, there has been relatively little work on proof theoretic methods that could be used in automatic deduction systems, although decision procedures for the propositional case have been explored. In this paper we report some general results in this area, including completeness, a Herbrand theorem analog, and resolution methods. Although they are developed for epistemic logics, we speculate that these methods may prove useful in quantified temporal logic also.

1 Introduction

Quantified modal logics (QML) have emerged as an important tool for reasoning about knowledge and belief in Artificial Intelligence (AI) systems. The idea of formalizing the basic properties of knowledge and belief in QML originated with Hintikka [7], who was interested in the analysis of several epistemic paradoxes. Subsequently he reformulated the semantics of his work using Kripke's notion of relative accessibility between possible worlds [8]. In the computer science community, McCarthy [13], Sato [16], Moore [14], Levesque [11], Halpern and Moses [5] and others have used variations of his approach to formalize and reason about knowledge and belief.

Whether quantified modal logics of this sort are appropriate as *epistemic* logics is controversial, both in philosophy and AI. The major objection is that they assume agents are perfect reasoners, so that they know all the logical consequences of their knowledge. Several attempts have been made to modify the possible-world semantics to avoid this assumption [12,2], and there are also other formal approaches which take into account the limited reasoning power of agents (for example, [9]). It is not the purpose of this paper to comment on the relative merits of these approaches; quantified modal logics with Kripke semantics are an important research tool for epistemic reasoning in computer science at present, and will probably remain so. Here we will be concerned with proof methods for these logics that could be used in automatic deduction systems.

*This research was made possible in part by a gift from the System Development Foundation, and was also supported by Contract N00014O-85-C-0251 from the Office of Naval Research.

Surprisingly, there has been relatively little work in this area, although decision procedures for the propositional case have been explored (see Halpern and Moses [6]).

In this paper we lay the theoretical groundwork leading to the derivation of a resolution procedure for certain quantified modal logics. The procedure has been implemented and successful solves a version of a standard benchmark in epistemic reasoning, the Wise Man Puzzle [4]. In outline, the derivation is as follows.

Kripke [10] has given a completeness proof for quantified $S5$, based on the analytic tableaux method. We build on his results, extending them in two main areas. First, Kripke's methods work only with constants which have a fixed interpretation relative to a single domain (*rigid designators*). For expressivity it is important to have terms whose interpretation varies over possible worlds; it is also technically necessary for the development of resolution methods, since skolem functions, by their very nature, cannot be rigid designators. However, there are well-known problems with the substitution of nonrigid terms into modal contexts. We solve these by the technical device a *bullet constructor*, essentially a method for turning nonrigid terms into rigid ones in the proper context.

Second, we isolate a concept, that of reducing the unsatisfiability of a set of modal atoms to the unsatisfiability of some set of their arguments, that is the key step in proving completeness. We will show how any system which possesses a reduction theorem of the proper form is complete.

Third, using a method from Smullyan [17], we put analytic tableaux for prenex sentences into a form in which all operations on quantifiers precede those on truth-functional connectives. Then, as a corollary to the completeness proof, we have a version of the Skolem-Herbrand-Gödel theorem for modal systems: a sentence is unsatisfiable if and only if a finite conjunction of its instances is. This theorem is the foundation of all automatic deduction procedures for first-order logic, including Robinson's resolution method [15].

With the help of the bullet constructor, it is possible to eliminate existential operators from a sentence in prenex normal form, a process referred to as *skolemization* in first-order systems. Finally, using our Skolem-Herbrand-Gödel theorem and drawing on the technique of theory resolution (from Stickel [18]), we show how the reduction theorem for a modal system leads to a sound and complete resolution system.

Because of space limitations, it is impossible to give proofs of the theorems in this paper.

2 Logical preliminaries

2.1 Epistemic logics

We consider six quantified modal logics that are typically used in reasoning about knowledge and belief (see Halpern and Moses [6]); we call these collectively *epistemic* logics. All of the logics have the following properties. Their language is first-order with the addition of a modal operator of the form $B\phi$, where ϕ is a formula of the language. Informally, $B\phi$ means that the agent believes or knows the proposition expressed by ϕ. All of the results of this paper are easily extended to the case where there is a sequence of modal operators B_i indexed by agent, but for simplicity we present the single-agent case.

Both arbitrary nesting of operators and "quantifying in" (*i.e.*, statements of the form $\exists x.B\phi(x)$ or $\forall x.B\phi(x)$) are allowed in the language. In addition, there is a bullet construction $\bullet t$, where t is a term not containing any bullet operators. A ***sentence*** is a formula which has no free variables, and whose bullet constructions are all under the scope of a modal operator. A ***modal atom*** is a formula $B\phi$; if ϕ contains no variables, it is a ***ground*** modal atom. A ***modal literal*** is either a modal atom or its negation.

We will use uppercase Greek letters (Γ, Δ, *etc.*) to stand for denumerable sets of formulas; if $\Gamma = \gamma_1, \gamma_2, \ldots$, then $B\Gamma$ abbreviates $B\gamma_1$, $B\gamma_2$, ..., and $\neg B\Gamma$ abbreviates $\neg B\gamma_1$, $\neg B\gamma_2$, ...

2.2 Semantics

The semantics of these logics is the standard Kripke possible-worlds model. A ***frame*** is a structure $\langle W, R\rangle$, where W is a set of possible worlds, and R is a binary relation on W. A particular logic will often place restrictions on the type of relation allowed in frames, *e.g.*, in some epistemic logics (see below) R is transitive. In this paper we will restrict ourselves to a single relation for simplicity; the generalization to families of operators is straightforward.

A ***model*** consists of a frame, a special world $w_0 \in W$ (the ***actual world***), a domain D_i for each world $w_i \in W$, and a valuation function V. At each possible world, V assigns a value to each term and sentence of the language. V obeys first-order truth-recursion rules; it also obeys particular rules for the modal operators, depending on the logic.

If $V(w, \phi) = \mathbf{true}$, then we write $\models_m^w \phi$. $\models_m \phi$ is an abbreviation for $\models_m^{w_0} \phi$. If ϕ is true in all models of a logic A, we write $\models_A \phi$ or simply $\models \phi$ if the logic is understood.

The bullet construction has a special semantics. No matter where it occurs in a formula, $\bullet t$ always refers to the actual individual denoted by t, so that for all $w \in W$, $V(w, \bullet t) = V(w_0, t)$.

Different constraints on R yield different versions of epistemic logic. We consider the following variations:

Logic	Restriction on R
K	none
$K4$	transitive
$K45$	transitive, euclidean
T	reflexive
$S4$	reflexive, transitive
$S5$	equivalence

The first three logics (K, $K4$, $K45$) have belief as their intended interpretation. K is the simplest of these, placing the fewest restrictions on beliefs. $K4$ and $K45$ represent various types of introspective properties. In $K4$, if one believes something, one believes one believes it ($B\phi \supset BB\phi$). $K45$ has this and its converse: if one doesn't believe something, one believes one doesn't believe it ($\neg B\phi \supset B\neg B\phi$).

The three logics which have reflexive R are logics of knowledge. The distinguishing characteristic here is that knowledge must be true ($B\phi \supset \phi$). T, $S4$ and $S5$ are the epistemic logics corresponding to K, $K4$ and $K45$.

It must be stressed that the purpose of the paper is not to argue for the appropriateness of these logics for modeling epistemic concepts. Indeed, it is easy to find problems here; for example, there are good reasons for denying that knowledge is only *true* belief, since it also seems to involve some complex notion of justification; and this is not formalized in T, $S4$, or $S5$.

Truth-recursion equations for these logics are the same. Along with rules for the boolean operators and quantifiers, we add the following rule for the modal operators:

$$V(w, B\phi) = \mathbf{true} \quad \text{iff} \quad \forall w'.\ wRw' \rightarrow V(w', \phi) = \mathbf{true} \tag{1}$$

2.3 Substitution

Substitution of terms for quantified-in variables is problematic, since it does not preserve validity. Consider the following example of an agent's beliefs.

$$\begin{array}{l} P(m(c)) \\ \neg BP(m(c)) \\ \forall x.Px \supset BPx \end{array} \tag{2}$$

We can construct a model as follows. Let P be the property of being non-Italian, let $m(x)$ denote the mayor of the city x, and c denote New York. Suppose the agent believes the mayor of New York is Fiorello LaGuardia (and not Ed Koch, the actual mayor); it is easy to confirm that all the sentences are satisfied.

Now if we substitute $m(c)$ for x in the third sentence, the resulting set is unsatisfiable. The reason is that, although x must refer to the same individual in all possible worlds, the substituted expression $m(c)$ need not. So even if a universal sentence is true in a model, some of its instances can be false.

Our solution to this problem is to redefine the meaning of "instance" by introducing a bullet construction ($\bullet$) whenever there is a substitution for variables inside the context of modal operators. In the above example, substituting $m(c)$ for x yields

$$P(m(c)) \supset BP(\bullet m(c))\ , \tag{3}$$

which is still satisfied by the original model, since $\bullet m(c)$ refers to Ed Koch even in the context of the belief operator.

We revise the substitution rule in the following way. Let ϕ^a_x stand for the substitution of a for the free variable x in ϕ.

$$(B\phi)^t_x = \begin{cases} B\phi^{\bullet t}_x & \text{if } t \text{ is not a bullet construction} \\ B\phi^t_x & \text{otherwise.} \end{cases} \tag{4}$$

3 Reduction theorems

A key notion for our development is that of a *reduction theorem* for a modal logic A. Basically, such a theorem shows how to reduce the unsatisfiability of a set of modal literals Z to the unsatisfiability of a set of sentences W whose modal depth is strictly less than that of Z. For example, consider the simplest case, the propositional belief logic K for a single agent. It is easy to prove that the set of modal atoms $Z = \{B\Gamma, \neg B\Delta\}$ is K-unsatisfiable if and only if for some $\delta \in \Delta$ the set $W = \{\Gamma, \neg\delta\}$ is K-unsatisfiable. Hence the unsatisfiability of Z is reducible to the unsatisfiability of W, and the modal depth of W is at least one less than than of W.

For some logics, such as $S4$ and $S5$, it is not easy to find a reduction in terms of the modal depth of formulas, which is a syntactic property. Instead, we define a more semantic characterization of reduction in the next subsection.

3.1 Unsatisfiability depth for Kripke models

Consider a logic A and a sentence S. Suppose a model m satisfies S, so that $A \models_m S$. Now we may only have to search a certain part of m's possible-world structure to establish the truth of S; for example, in the epistemic logic $S4$, the truth of $S = Bp \supset BB\phi$ can be established for any m by traversing paths on the accessibility relation only to a depth of two. Paths longer than this have no role in determining the truth value of S.

To make this more precise, we introduce the concept of *agreement trees*. Let m be a model $\langle W, R, w_0, D, V\rangle$. A model m' is an agreement tree for m to depth n if the following conditions hold:

1. The structure of R' is a tree.

2. There is a one-one correspondence between paths of length less than or equal to n in the two models.

3. If $w_0 \ldots w_{j-1}w_j$ is a path in m (with $j \leq n$), and $w'_0 \ldots w'_{j-1}w'_j$ is its corresponding path in m', then the domain and valuation of w_j and w'_j are the same, and $w_0 \ldots w_{j-1}$ and $w'_0 \ldots w'_{j-1}$ are also corresponding paths.

The agreement tree "unwinds" any cyclic structure of R to a depth of n. Note that the rest of the agreement tree can be arbitrary, *i.e.*, it need not correspond to m.

Definition 1 *A set of sentences Γ is A-unsatisfiable at depth n if for every A-model m, every agreement tree of depth n falsifies some element of Γ.*

We will write *unsat$_n$* to indicate unsatisfiability at depth n. Note that if a set is unsat$_n$, it is also unsat$_k$ for all $k > n$, and also (simply) unsatisfiable.

3.2 Reduction theorems for epistemic logics

We now give reduction theorems for the six epistemic logics.

Definition 2 *The* bullet transform *of a set of formulas W is a set $W^\bullet$ derived from W by replacing all occurences $\bullet t$ of the bullet construction with either $\bullet n(t)$ (if $\bullet t$ is under the scope of a modal operator) or $n(t)$ (if it is not), where n is function not occurring in W; e.g., $\phi(\bullet a) \wedge B\phi(\bullet a) \rightarrow \phi(n(a)) \wedge B\phi(\bullet n(a))$. The* identity transform *of W is a set W^I formed by deleting all bullet constructors not under the scope of a modal operator, e.g., $\phi(\bullet a) \wedge B\phi(\bullet a) \rightarrow \phi(a) \wedge B\phi(\bullet a)$.*

Theorem 1 *Let Z be a first-order satisfiable set of literals* $\{\Sigma, B\Gamma, \neg B\Delta\}$ *of an epistemic logic A, where* Σ *are nonmodal, and all bullet terms occurring in* Δ *also occur in* Γ. *Z is* $unsat_n$ *if and only if for some* $\delta \in \Delta$,

$$\left.\begin{array}{ll} (K) & \{\Gamma, \neg\delta\}^\bullet \\ (K4) & \{\Gamma, \neg\delta, B\Gamma\}^\bullet \\ (K45) & \{\Gamma, \neg\delta, B\Gamma, \neg B\Delta\}^\bullet \\ (T) & \{\Gamma, \neg\delta\}^\bullet \text{ or } \{\Gamma', \Sigma\} \\ (S4) & \{\Gamma, \neg\delta, B\Gamma\}^\bullet \text{ or } \{\Gamma', \Sigma\} \\ (S5) & \{\Gamma, \neg\delta, B\Gamma, \neg B\Delta, \neg B\neg\Sigma\}^\bullet \text{ or } \{\Gamma', \Sigma\} \end{array}\right\} \textit{ is } unsat_{n-1}.$$

An example:

$\{\neg B\neg B(p \wedge q), \neg Bp\}$	is $S5$-unsat$_2$
$\{\neg B\neg B(p \wedge q), B(p \wedge q), \neg Bp\}$	is $S5$-unsat$_1$
$\{\neg B\neg B(p \wedge q), B(p \wedge q), p \wedge q, \neg Bp, \neg p\}$	is $S5$-unsat$_0$

4 Analytic tableaux and completeness

We now give a brief overview of prenex analytic tableaux, which are defined in Smullyan [17]. Let S be a finite set of sentences in prenex form (all quantifiers precede other operators). A prenex tableau for S is a sequence of sentences starting with S, and containing instances derivable by the rules:

$$\frac{\forall x.\phi}{\phi^t_x} \qquad \frac{\exists x.\phi}{\phi^t_x}, \text{ with proviso.}$$

In the existential rule, the proviso is that the term t has not yet been introduced in the tableau.

A prenex tableau is *closed* if some finite subset of its ground sentences is truth-functionally unsatisfiable. It is provable that the (perhaps infinite) set of sentences of an open prenex tableau are first-order satisfiable. This yields a version of the Skolem-Herbrand-Gödel theorem for first-order logic: a set of sentences in prenex form is unsatisfiable if and only if a finite set of its instances is.

For a modal logic A, prenex form is the same as in first-order logic, taking modal formulae as unanalyzed predications. Thus $\forall x B \exists y Pxy$ is in prenex form; note that quantifiers which are under the scope of modal operators are *not* affected. We modify the definition of *closed prenex tableau* to be: some finite subset of its ground sentences is A-unsatisfiable. The key theorem for modal prenex tableaux is the following.

Theorem 2 *If a finite set S of prenex sentences is A-unsatisfiable, then there exists a closed prenex tableau for S.*

We give a brief proof sketch of Theorem 2. The proof is by induction on the unsatisfiability level of the prenex tableau. If S is unsat$_0$, then by the results of Smullyan [17], its prenex tableau closes. Now assume that all sets that are unsat$_{n-1}$ have closed prenex tableau. Let S be unsat$_n$. Suppose S has an open prenex tableau; consider the set W of all ground sentences on this branch.

This set is first-order satisfiable. By elementary rules of propositional logic, and the reduction theorem for A, some set W' must be A-unsat$_{n-1}$. Now we convert this set to prenex form, and again set up a prenex tableau for W'; this must close by the induction hypothesis for some finite $W'' \subseteq W'$; hence S cannot have an open tableau. Thus every set S which is A-unsat$_n$ for finite n has a closed prenex tableaux. Finally, if S is A-unsatisfiable, it is A-unsat$_n$ for some n less than the maximum depth of embedding of modal operators; hence it must have a closed prenex tableau.

As an obvious corollary, we have the Skolem-Herbrand-Gödel theorem for A.

5 B-resolution

Using the results of the previous section, we can now give a resolution method for the epistemic logics, which we call B-resolution.

5.1 Clause form

Converting to clause form is the same as for first-order logic, with modal atoms having different argument structures treated as if they were different predicate symbols. Thus $B\forall x.P(x)$, BPa, and $B\exists x.P(x)$ are all considered to be different nilary predicates. Modal atoms with n free variables are n-ary predicates, *e.g.*, $B(P(x) \wedge \exists y.P(y))$ and $B(\exists y.P(y) \wedge P(x))$ are different unary predicates with the free variable x. Variables quantified under the scope of the modal operator remain unanalyzed or inert in B-resolution, and do not interact with variables quantified outside the operators. An example:

$$\forall x \exists y.P(x,y) \supset B\exists z.Q(x,y,z) \;\Rightarrow\; \neg P(x, f(x)) \vee B\exists z.Q(\bullet x, \bullet f(x), z)$$

Note that substitution of $f(x)$ for y in the modal context is done with $\bullet f(x)$. Also, in clause form we automatically insert a bullet operator before quantified-in variables (like x), to distinguish them from variables whose quantifiers are inside the scope of modal operators (like z).

We have proven the following theorem:

Theorem 3 *A sentence is A-unsatisfiable if and only if its clause form is.*

5.2 B-resolution

Our resolution method is based on Stickel's *total narrow theory resolution* rule [18], which has the following form. Let L be a language that embeds a theory T, that is, the axioms of T contain a set of predicates P of L (but not necessarily all predicates of L). Suppose there is a decision procedure for determining a set of ground literals W in P to be unsatisfiable (according to T). Then

$$\frac{\begin{array}{l} L_1 \vee A_1 \\ L_2 \vee A_2 \\ \quad\vdots \\ L_n \vee A_n \end{array}}{A_1 \vee A_2 \vee \ldots \vee A_n}, \quad \text{when } \{L_1, L_2, \ldots L_n\} \text{ is } T\text{-unsatisfiable} \tag{5}$$

is a resolution rule that is sound and complete for the theory T. This rule includes binary resolution as a special case, where L_1 and L_2 are complementary literals.

For epistemic logic A, the reduction theorem tells us when a set of literals will be A-unsatisfiable. Hence we can rephrase this rule is rephrased as follows. Let $\Gamma = \{\gamma_1, \gamma_2, \ldots\}$ and $\Delta = \{\delta_1, \delta_2, \ldots\}$ be finite sets of sentences, and $\Sigma = \{\sigma_1, \sigma_2, \ldots\}$ a finite set of literals. In the case of ground clauses, we have the following two resolution rules:

$$\frac{\begin{array}{c} B\gamma_1 \lor A_1 \\ B\gamma_2 \lor A_2 \\ \vdots \\ \neg B\delta_1 \lor A'_1 \\ \neg B\delta_2 \lor A'_2 \\ \vdots \\ \sigma_1 \lor A''_1 \\ \sigma_2 \lor A''_2 \\ \vdots \end{array}}{A_1 \lor A_2 \lor \cdots \lor A'_1 \lor A'_2 \lor \cdots \lor A''_1 \lor A''_2 \lor \cdots,} \quad \text{where} \tag{6}$$

$$\left.\begin{array}{ll} (K,T) & \{\Gamma, \neg\delta_1\}^\bullet \\ (K4,S4) & \{\Gamma, B\Gamma, \neg\delta_1\}^\bullet \\ (K45) & \{\Gamma, B\Gamma, \neg\delta_1, \neg B\Delta\}^\bullet \\ (S5) & \{\Gamma, B\Gamma, \neg\delta_1, \neg B\Delta, \neg B\neg\Sigma\}^\bullet \end{array}\right\} \text{is unsat}$$

$$\frac{B\phi \lor A}{\phi^I \lor A} \tag{7}$$

In the first rule, we have listed all of the possibilities for the different epistemic logics. For the simplest case, K, only the clauses with Γ and δ_1 are used. The second rule is applied only for the knowledge logics T, $S4$, and $S5$.

5.3 Lifting

The resolution rules have been given only for the ground case. Because of Theorem 2, these rules will be complete if we are allowed to derive instances of any clause. Of course, this is a very inefficient way to do resolution, which is why unification is such an important concept. In this respect, B-resolution is more complicated than ordinary binary resolution, because there may be no "most general" unifier covering all possible ground resolutions. For example, consider the following two clauses:

$$\begin{array}{l} B(p(\bullet a) \land p(\bullet b)) \\ \neg Bp(\bullet x) \end{array} \tag{8}$$

There are two substitutions for x which yield a resolvent (a/x and b/x), but no most general unifier.

A second problem is that (6) and (7) are not true deduction rules, in the sense that they are not effective. The solution to this and the instantiation problems lies in how we check the unsatisfiability conditions. Suppose, each time we wish to do a B-resolution, we start another refutation procedure using the indicated sets of sentences. Then we intermix the execution of deductions in the main refutation proof with execution in the subsidiary ones being used to check unsatisfiability. If at some point a subsidiary refutation succeeds, we can construct a resolvent in the main refutation. If in addition we use a subsidiary refutation procedure that allows free variables in the input (essentially doing schematic refutations), then it is possible to subsume many instances of the application of the resolution rules in one unsatisfiability check. The details of this approach are discussed in Geissler and Konolige [4].

6 Discussion

We are interested in general methods for finding resolution proof procedures for quantified modal logics. As this paper shows, one such method is to prove a reduction theorem for the logic. The nature of the reduction is apparent in the resolution rules, where unsatisfiability of a set of modal literals is reexpressed in terms of unsatisfiability of their arguments. We believe that such resolution methods are a natural and conceptually transparent means of finding refutations. A large part of the advantage comes from being able to strip off the modal operator and perform deductions on its arguments.

For the epistemic logics, reduction theorems are available. It is not clear that reduction theorems will always be provable for a modal logic. For example, if we add a common knowledge operator to an epistemic logic (see Halpern and Moses [5]), the resulting system is much more complicated, and it is an open question as to whether a reduction theorem exists.

Temporal logics are another important class of modal systems. Abadi and Manna [1] and Fariñas-del-Cerro [3] have both defined resolution systems for propositional temporal logics. However, their methods are not readily extendable to the quantified case. It would be interesting to try to use the techniques of this paper to formulate an alternative resolution system, and compare them.

References

[1] Abadi, M. and Manna, Z. (1985). Nonclausal temporal deduction. Report No. STAN–CS–85–1056, Computer Science Department, Stanford University, Stanford, California.

[2] Fagin, R. and Halpern, J. Y. (1985). Belief, awareness, and limited reasoning. In Proceedings of the Ninth International Joint Conference on AI, Los Angeles, California, pp. 491–501.

[3] Fariñas-del-Cerro, L. (1983). Temporal reasoning and termination of programs. In Proceedings of the Eighth International Joint Conference on Artificial Intelligence, Karlsruhe, West Germany, pp. 926-929.

[4] Geissler, C. and Konolige, K. (1986). A resolution method for quantified modal logics of knowledge and belief. In Proceedings of the Conference on Theoretical Aspects of Reasoning about Knowledge, Monterey, California.

[5] Halpern, J. Y. and Moses, Y. (1984). Knowledge and common knowledge in a distributed environment. In Proceedings of the 3rd ACM Conference on Principles of Distributed Computing, pp. 50–61.

[6] Halpern, J. Y. and Moses, Y. (1985). A guide to the modal logics of knowledge and belief: preliminary draft. In Proceedings of the Ninth International Joint Conference on AI, Los Angeles, California, pp. 479–490.

[7] Hintikka, J. (1962). *Knowledge and Belief.* Cornell University Press, Ithaca, New York.

[8] Hintikka, J. (1969). Semantics for propositional attitudes. In L. Linsky (ed.), *Reference and Modality,* Oxford University Press, London (1971), pp. 145–167.

[9] Konolige, K. (1984). *A Deduction Model of Belief and its Logics.* Doctoral thesis, Stanford University Computer Science Department Stanford, California.

[10] Kripke, S. A. (1959). A Completeness Theorem in Modal Logic. *Journal of Symbolic Logic* **24**, pp. 1–14.

[11] Levesque, H. J. (1982). A Formal Treatment of Incomplete Knowledge Bases. FLAIR Technical Report No. 614, Fairchild Laboratories, Palo Alto, California.

[12] Levesque, H. J. (1984). A logic of implicit and explicit belief. In Proceedings of the National Conference on Artificial Intelligence, Houston, Texas, pp. 198–202.

[13] McCarthy, J. *et. al.* (1978). On the Model Theory of Knowledge. Memo AIM-312, Stanford University, Stanford.

[14] Moore, R. C. (1980). Reasoning About Knowledge and Action. Artificial Intelligence Center Technical Note 191, SRI International, Menlo Park, California.

[15] Robinson, J. A. (1965). A machine-oriented logic based on the resolution principle. *J. Assoc. Comput. Mach. 12*, pp. 23–41.

[16] Sato, M. (1976). *A study of Kripke-type models for some modal logics by Gentzen's sequential method.* Research Institute for Mathematical Sciences, Kyoto University, Kyoto, Japan.

[17] Smullyan, R. M. (1971). *First-order logic.* Springer-Verlag, New York.

[18] Stickel, M. E. (1985). Automated deduction by theory resolution. *Proceedings of the Ninth International Joint Conference on Artificial Intelligence,* Los Angeles, California.

A COMMONSENSE THEORY OF NONMONOTONIC REASONING

Dr. Frank M. Brown
Department of Computer Science
University of Kansas
Lawrence, KS 66045/USA

Abstract

A commonsense theory of nonmonotonic reasoning is presented which models our intuitive ability to reason about defaults. The concepts of this theory do not involve mathematical fixed points, but instead are explicitly defined in a monotonic modal quantificational logic which captures the modal notion of logical truth. The axioms and inference rules of this modal logic are described herein along with some basic theorems about nonmonotonic reasoning. An application to solving the frame problem in robot plan formation is presented.

1. Introduction

The basic idea of out theory of nonmonotonicity is that nonmonotonicity is already encompassed in the normal intensional logic of everyday commonsense reasoning and can be explained precisely in that terminology.

For example, a knowledgebase consisting of a simple default axiom expressing that a particular bird flies whenever that bird flies is possible with respect to what is assumed is stated as:

(that which is assumed is
(if (A is possible with respect to what is assumed) then A))

where A stands for the proposition that that particular bird flies.

Reflection of the meaning of this knowledgebase leads immediately to the conclusion that either A is logically possible and the knowledgebase is synonymous to A, or A is not logically possible and the knowledgebase is synonymous to logical truth. This conclusion is obtained by simple case analysis: for if A is possible with respect to what is assumed then, since truth implies A is A, that which is assumed is indeed A. Since that which is assumed is A, A is possible with respect to what is assumed only if A is logically possible. On the other hand, if A is not possible with respect to what is assumed then since falsity implies A is just truth, that which is assumed is truth. Since that which is assumed is truth, A is not possible with respect to what is assumed only if A is not logically possible.

Thus if it is further assumed that A is logically possible, then it follows that the knowledgebase is synonymous to A itself.

The nonmonotonic nature of these expressions becomes apparent if an additional proposition that that particular bird does not fly is added to the knowledgebase:

```
(that which is assumed is
    (and (not A)
         (if (A is possible with respect to what is assumed) then A)))
```

Reflection on this new knowledgebase leads immediately to the conclusion that it is synonymous to not A. This conclusion is again obtained by simple case analysis: for if A is possible with respect to what is assumed then, since truth implies A is just A, that which is assumed is indeed not A and A which is falsity. Since that which is assumed is falsity A is possible with respect to what is assumed only if A and falsity is logically possible which it is not. Thus A is not possible with respect to what is assumed. On the other hand, if A is not possible with respect to what is assumed then, since falsity implies A is just truth, that which is assumed is just not A. Since that which is assumed is (not A), A is not possible with respect to what is assumed only if A and (not A) is not logically possible which is the case. Thus it follows that the knowledgebase is synonymous to (NOT A).

Therefore whereas the original knowledgebase was synonymous to A the new knowledgebase, obtained by adding (not A), is synonymous, not to falsity, but to (not A) itself.

These simple intuitive nonmonotonic arguments involve logical concepts such as not, inplies, truth, falsity, logical possibility, possibility with respect to some assumed knowledgebase, and synonymity to a knowledgebase. The concepts: not, implies, truth(i.e.T), and falsity (i.e. NIL) are all concepts of (extensional)quantificational logic and are well known. The remaining concepts: logical possibility, possibility with respect to something, and synonymity of two things can be defined in a very simple modal logic extension of quantificational logic, which we call Z[Brown76,78a,79,78b]. The axiomatization of the modal logic Z is described in detail in section 2. But briefly, it consists of (extensional) quantificational logic plus the intensional concept of something being logically true written as the unary predicate: (LT P). The concept of a proposition P being logically possible and the concept of two propositions being synonymous are then defined as:

```
(POS P) = (NOT(LT(NOT P)))        ;P is logically possible
(SYN P Q) = (LT(IFF P Q))         ;P is synonymous to Q
```

The above knowledgebases and arguments can be formalized in the modal logic Z quite simply by letting some letter such as K stand for the knowledgebase under discussion. The idiom "that which is assumed is X" can be rendered to say that K is synonymous to X, and the idiom "X is possible with respect to what is assumed" can be rendered to say that K and X is possible:

```
(that which is assumed is X)                          = (SYN K X)
(X is possible with respect to what is assumed)       = (POS(AND K X))
```

These two idioms are indexial symbols referring implicitly to some particular knowledgebase K under discussion. This knowledgebase referenced by the (X is possible with respect to what is assumed) idiom is always the meaning of the symbol generated by the enclosing (that which is assumed is X) idiom. Each occurrence of the (that which is assumed is X) idiom always generates a symbol (unique to the theory being discussed) to stand for the database under discussion.

The first knowledgebase is then expressed as:

```
(SYN K(IMPLY(POS(AND K A))A))
```

Its commonsense argument could be carried out in the following steps where (IF p l r) means if p then l else r:

```
(IF(POS(AND K A))
   (SYN K(IMPLY T A))
   (SYN K(IMPLY NIL A)) )
(IF(POS(AND K A))
   (SYN K A)
   (SYN K T))
(OR (AND      (POS(AND K A))   (SYN K A))
    (AND (NOT(POS(AND K A)))  (SYN K T)) )
(OR (AND      (POS(AND A A))   (SYN K A))
    (AND (NOT(POS(AND T A)))  (SYN K T)) )
(OR (AND      (POS A)  (SYN K A))
    (AND (NOT(POS A)) (SYN K T)) )
```

by equality substitution using the following derived rules of inference of the modal quantificational logic: Z.

```
(g(POS P))=(IF(POS P)(g T)(g NIL))
(IMPLY T A)=A
(IMPLY NIL A)=T
(IF P L R)=(OR(AND P L)(AND(NOT P)R))
(AND(P X Y)(SYN X Y))=(AND(P Y Y)(SYN X Y))
(AND A A)=A
(AND T A)=A
```

Furthermore if A is logically possible: (POS A) then further simplification using laws about AND and OR yields the fact that the knowledgebase K is synonymous to A:

```
(OR (AND  T   (SYN K A))
    (AND  NIL (SYN K T)) )
(SYN K A)
```

The second knowledgebase is then expressed as:

```
(SYN K(AND(NOT A)(IMPLY(POS(AND K A))A)))
```

Its commonsense argument could be carried out in the following steps:

```
(IF(POS(AND K A))
   (SYN K(AND(NOT A)(IMPLY T A)))
   (SYN K(AND(NOT A)(IMPLY NIL A))) )
(IF(POS(AND K A))
   (SYN K NIL)
   (SYN K(NOT A)) )
(OR (AND     (POS(AND K A))    (SYN K NIL))
    (AND (NOT(POS(AND K A)))   (SYN K(NOT A))) )
(OR (AND     (POS(AND NIL A))       (SYN K NIL))
    (AND (NOT(POS(AND(NOT A)A)))   (SYN K(NOT A))) )
(OR (AND    NIL    (SYN K NIL))
    (AND (NOT NIL) (SYN K(NOT A))) )
(SYN K(NOT A))
```

These knowledgebases have been expressed solely in terms of the modal quantificational logic Z. In particular, the nonmonotonic concepts were explicitly defined in this logic. The intuitive arguments about the meaning of these nonmonotonic knowledgebases have been carried out solely in the modal quantificational logic Z. Most importantly, our commonsense understanding and reasoning about nonmonotonicity is directly represented by the inference steps of this formal theory. Therefore, it is clear that nonmonotonic reasoning needs no special axioms or rules of inference because it is already inherent in the normal intentional logic of everyday commonsense reasoning as modeled by the modal quantificational logic Z.

The modal quantificational logic Z is described in section 2. This is followed in section 3 by the presentation of some of the basic theorems of our nonmonotonic theory. Due to space restrictions most proofs will be omitted and no comparison will be made to the fixed point theories of nonmonotonicity [McDermott&Doyle,McDermott,Moore,and Reiter]. An application of this theory of nonmonotonic reasoning to the frame problem in robot plan formation is given in section 4.

2. The Modal Quantificational Logic: Z

Our theory of commonsense intensional reasoning is a simple modal logic that captures the notion of logical truth. The symbols of this modal logic consist of the symbols of (extensional) quantificational logic plus the primitive modal symbolism: (LT p) which is truth whenever the proposition p is logically true. Propositions are intuitively the meanings of sentences. For example, the sentence: '(IMPLY p q) and '(OR (NOT p)q) both mean that p implies q. Thus, although these two sentences are different, the two propositions: (IMPLY p q) and (OR(NOT p)q) are the same. Propositions may be true or false in a given world, but with the exception of the true proposition (i.e. the meaning of '(IMPLY p p)) and the false proposition (i.e. the meaning of '(AND p(NOT p))), propositions are not inherently true or false. Thus mathematically, propositions may be thought of as being the elements of a complete atomic Boolean algebra with an arbitrary (possibly infinite) number of generators.

The axioms and inference rules of this modal logic include the axioms and inference rules of (extensional) quantificational logic similar to that used by Frege in Begriffsschrift[Frege], plus the following inference rule and axioms about the concept of logical truth.

The Modal Logic Z

R0: from p infer (LT p)
A1: (IMPLY(LT P) P)
A2: (IMPLY(LT P Q) (IMPLY(LT P)(LT Q)))
A3: (OR(LT P) (LT(NOT(LT P))))
A4: (IMPLY(ALL Q(IMPLY(WORLD Q)(LT(IMPLY Q P)))) (LT P))
A5: (ALL S(POS(meaning of the generator subset S)))

The inference rule R0 means that P is logically true may be inferred from the assertion of P to implicitly be logically true. The consequence of this rule is that a proposition P may be asserted to be logically true by writing just:

P

and that a proposition P is asserted to be true in a prticular world or state of affairs W by writing:

(LT(IMPLY W P))

The axiom A1 means that if P is logically true then P. Axiom A2 means that if it is logically true that P implies Q then if P is logically true then Q is logically true. Axiom A3 means that P is logically true or it is logically true that P is not logically true. The inference rule R0 and the axioms A1, A2 and A3 constitute an S5 modal logic. A good introduction to modal logic in general and in particular to the properties of S5 modal logic is given in [Hughes and Cresswell]. Minor variations of the axioms A1, A2, and A3 were shown in [Carnap] to hold for the modal concept of logical truth. We believe that the additional axioms, namely A4 and A5, are needed in order to precisely capture the notion of logical truth.

The axiom A4 states that a proposition is logically true if it is true in all worlds. We say that a proposition P is a world iff P is possible and P is complete, that P is complete iff for all Q, P determines Q, that P determines Q iff P entails Q or P entails not Q, that P entails Q iff it is logically true that P implies Q, and that P is possible iff it is not the case that not P is logically true. These definitions are given below:

```
(WORLD P)      =df (AND(POS P)(COMPLETE P))          ;P is a world
(COMPLETE P)   =df (ALL Q(DET P Q))                  ;P is complete
(DET P Q)      =df (OR(ENTAIL P Q)(ENTAIL P(NOT Q))) ;P determines Q
(ENTAIL P Q)   =df (LT(IMPLY P Q))                   ;p entails Q
(POS P)        =df (NOT(LT(NOT P)))                  ;P is possible
```

Thus a world is a possible proposition which for every proposition entails it or its negation. Axiom A4 therefore eliminates from the interpretations of the modal logic Z those complete Boolean algebras which are not atomic.

The axiom A5 states that the meaning of every conjunction of the generated contingent propositions or their negations is possible. We call this axiom "The Axiom of the Possibility of Contingent facts" or simply the "Possibility Axiom". The need for this axiom follows from the fact that the other axioms of the modal logic do not imply certain elementary facts about the possibility of conjunctions of distinct possibly negated atomic expressions consisting of nonlogical symbols. For example, if we have a theory formulated in our modal logic which contains the nonlogical expression (ON A B) then since (ON A B) is not logically true, it follows that (NOT(ON A B)) must be possible. Yet (POS(NOT(ON A B))) does not follow from these other axioms. Likewise, since (NOT (ON A B)))is not logically true (ON A B) must be possible. Yet (POS(ON A B)) does not follow from the other axioms. Thus these contingent propositions (ON A B) and (NOT(ON A B)) need to be asserted to be possible. There are a number of ways in which this may be done and these ways essentially correspond to different ways the idiom: (P is a meaning combination of generators) may be rendered. In this paper we have chosen a general method which is applicable to just about any contingent theory one wishes. This rendering is given below:

```
(meaning of the generator subset S) =df
        (ALL G(IMPLY(GENERATORS G)
                    (IFF(S G)(GMEANING G)) ))

(GMEANING  (p ,X1...,XN)) =df (p(GMEANING X1)...(GMEANING XN))
    for every contingent symbol p of arity n.

(GENERATORS) =df (LAMBDA(A)(A is a contingent variable-free simple
                                sentence))
```

We say that the meaning of the generator subset S is the conjunction of the GMEANINGs of every generator in S and the negation of the GMEANINGS of all the generators not in S. The generator meaning of any expression beginning with a contingent symbol 'p is p of the GMEANING of its arguments. The generators are simply any contingent variable-free atomic sentences we wish to use. The GMEANINGS of the generators may be interpreted essentially as being the generators of a complete atomic Boolean algebra.

For example, a contingent language with a single contingent propositional function 'P and names 'A and 'B gives rise to two contingent generators: '(P A) and '(P B). The GENERATORS and GMEANING functions for this language are defined as:

```
(GENERATORS)=df {'(P A) '(P B)}
{P1...Pn}   =df (LAMBDA(X)(OR(EQUAL X P1)...(EQUAL X Pn)))

(GMEANING `(P ,X)) = (P (GMEANING X))
(GMEANING 'A) = A
(GMEANING 'B) = B
```

and the Possibility Axiom simplifies to :

```
(AND(POS(AND(P A)(P B)))
    (POS(AND(P A)(NOT(P B))))
    (POS(AND(NOT(P A))(P B)))
    (POS(AND(NOT(P A))(NOT(P B)))) )
```

If the set of GENERATORS is finite then the possibility axiom reduces, in a manner similar to the above derivation, to a conjunction of sentences stating that any conjunction of simple sentences or their negations is possible, and this resulting sentence is entirely expressed within the modal logic Z based on an underlying (extensional) first order quantificational logic. However, it is important to note that finiteness of the generator set is not required by our modal logic and that the possibility axiom A5 will provide the necessary possibilities as theorems for any contingent language. For example, the fact that the conjunction of the P of all natural numbers is possible can be derived as follows from the infinite generator set consisting of all simple sentences of the form `(P ,N) where N is a numeral.

```
(GENERATORS) =df (LAMBDA(X)(EX N(AND(NUMERAL N)(EQUAL X(P N)))))
(GMEANING `(P ,N)) = (P (GMEANING N))
```

For this contingent language the possibility axiom simplifies to:

```
(ALL S(POS(ALL N(IMPLY(NUMERAL N)
                      (IFF(S `(P ,N))(P(GMEANING N))) ))))
```

Thus if S is the universe it follows that:

```
(POS(ALL N(IMPLY(NUMERAL N) (P(GMEANING N))) ))
```

which intuitively is: (POS(AND(P 1)...))

3. The Reflexive Theory

One of the most striking features of nonmonotonic knowledgebases is that they are sometimes described in terms of themselves. Such knowledgebases are said to be reflexive[Hayes79]. For example, the knowledgebase K purportedly defined by the axiom:

```
(SYN K (IMPLY(POS(AND K A))A) )
```

is defined as being synonymous to the default: (IMPLY(POS(AND K A))A) which in turn is defined in terms of K. Thus this purported definition of K is not actually a definition at all but is merely an axiom describing the properties possessed by any knowledgebase K satisfying this axiom. In general, a purported definition of a knowledgebase:

```
(SYN K(f K))
```

will be implied by zero or more explicit definitions of the form:

```
(SYN K g)
```

where K does not occur in g. The explicit definitions which imply a purported definition of a knowledgebase are called the solutions of that purported definition. In general a purported definition may have zero or more solutions. For example, (SYN K(NOT K)) is (LT(IFF K(NOT K))) which is (LT NIL) which is NIL and therefore has no solutions, and (SYN K K) is (LT(IFF K K)) which is (LT T) which is T and therefore has all solutions. Finally, (SYN K G) where K does not accur in G is an explicit definition of K and therefore has only one solution namely itself.

Because K is the knowledgebase under discussion. it is not itself a contingent proposition of that knowledgebase. Thus 'K is not a GENERATOR and the possibility axiom A5 of section 2 will not apply to it. This is verified by the above example (SYN K NIL) where K consists of the false expression, and thus is not possible.

In order to make use of a knowledgebase it is helpful to know what is actually in that knowledgebase. For a non-reflexive knowledgebase (i.e. a knowledgebase defined by an explicit definition) this is no problem because there is obviously only one solution, namely that explicit definition itself. However, in the more general case of a purported definition there may be any number of solutions. Thus the basic goal of a theory of nonmonotonicity of reflexive knowledgebases must be to describe the solutions for various kinds of purported definitions.

The first kind of purported definition we consider is a knowledgebase K consisting of (a conjunction of)axioms G not containing K plus one additional standard default axiom. A standard default axiom is an axiom of the form:

```
(IMPLY(POS(AND K A))(IMPLY B A))
```

This structure contains as instances default axioms such as:

```
(IMPLY(POS(AND K(CAN-FLY ENTERPRISE)))
      (IMPLY(IS-SPACE-SHUTTLE ENTERPRISE)(CAN-FLY ENTERPRISE)))
```

T1: A knowledgebase containing exactly one variable-free standard default has precisely one solution.

```
(IFF(SYN K(AND G(IMPLY(POS(AND K A))(IMPLY B A))))
    (SYN K(AND G(IMPLY(POS(AND G A))(IMPLY B A)))))
```

proof

```
(SYN K(AND G(IMPLY(POS(AND K A))(IMPLY B A))) )
(IF(POS(AND K A))
   (SYN K(AND G(IMPLY(AND B T)A)))
```

```
      (SYN K(AND G(IMPLY(AND B NIL)A))) )
   (IF(POS(AND K A))
      (SYN K(AND G(IMPLY B A)))
      (SYN K G))
   (OR(AND(POS(AND K A))(SYN K(AND G(IMPLY B A))))
      (AND(NOT(POS(AND K A)))(SYN K G)) )
   (OR(AND(POS(AND G(IMPLY B A)A))(SYN K(AND G(IMPLY B A))))
      (AND(NOT(POS(AND G A)))(SYN K G)))
   (OR(AND(POS(AND G A))(SYN K(AND G(IMPLY B A))))
      (AND(NOT(POS(AND G A)))(SYN K G)) )
   (IF(POS(AND G A))
      (SYN K(AND G(IMPLY B A)))
      (SYN K G))
   (SYN K(IF(POS(AND G A))(AND G(IMPLY B A))G))
   (SYN K(AND G(IF(POS(AND G A))(IMPLY B A)T)))
   (SYN K(AND G(IMPLY(POS(AND G A))(IMPLY B A))))
```

The solutions to the two purported definitions descussed in section 1 are obtained from theorem T1 as corollaries for if G is T, B is T, and A is possible it follows that:

```
   (IFF(SYN K(IMPLY(POS(AND K A))A))
       (SYN K A))
```

and if G is (NOT A) and B is T it follows that:

```
   (IFF(SYN K(AND(NOT A)(IMPLY(POS(AND K A))A)))
       (SYN K(NOT A)))
```

T1 shows that a knowledgebase with only one variable-free standard default, has the same essential status as an explicit definition. The next theorem: T2 shows that this is not the case for a knowledgebase with 2 variable-free standard defaults.

T2: A knowledgebase consisting of two variable-free standard defaults has precisely one or two solutions.

```
   (IFF(SYN K(AND G(IMPLY(POS(AND K A1)) (IMPLY B1 A1))
                   (IMPLY(POS(AND K A2))(IMPLY B2 A2)) ))
       (IF(POS(AND G(IMPLY B2 A2)A1))
          (IF(POS(AND G(IMPLY B1 A1)A2))
             (SYN K(AND G(IMPLY B1 A1)(IMPLY B2 A2)))
             (SYN K(AND G(IMPLY B1 A1))))
          (IF(POS(AND G(IMPLY B1 A1)A2))
             (SYN K(AND G(IMPLY B2 A2)))
             (IF(POS(AND G A1))
```

```
(IF(POS(AND G A2))
   (OR(SYN K(AND G(IMPLY B1 A1)))
      (SYN K(AND G(IMPLY B2 A2))))
   (SYN K(AND G(IMPLY B1 A1))))
(IF(POS(AND G A2))
   (SYN K(AND G(IMPLY B2 A2)))
   (SYN K G)) ))))
```

The flipped Coin Corollary to T2:

If A1 is A, A2 is (NOT A), B1 is T, B2 is T, (AND G A) is possible, and (AND G(NOT A)) is possible then this proposition reduces to:

```
(IFF(SYN K(AND G(IMPLY(POS(AND K A))A)(IMPLY(POS(AND K(NOT A)))(NOT
    A))))(OR(SYN K(AND G A))(SYN K(AND G(NOT A))))) )
```

which states that K has precisely two solutions. Furthermore, if G is T, then these two solutions are direct opposites in that one says the knowledgebase is A and the other says the knowledgebase is (NOT A):

```
(IFF(SYN K(AND(IMPLY(POS(AND K A))A)(IMPLY(POS(AND K(NOT A)))(NOT
    A))))(OR(SYN K A)(SYN K(NOT A))))
```

There is nothing at all bizarre about having multiple solutions or even oppositie multiple solutions for one can easily imagine a Robot executing actions is a given state resulting in a new state K which in its planning cannot determine which solution for K is actually the case. For example let A be the proposition that a flipped coin will land with heads. The default: (IMPLY(POS(AND K A))A) then means that if it is possible for the coin to land on heads then assume it does so. Likewise, the default (IMPLY(POS(AND K(NOT A)))(NOT A)) means that if it is possible for a coin to land tails (i.e. not heads) then assume it does so. The result of the action is then one of two states K:

```
(OR(SYN K A)(SYN K(NOT A)))
```

where the coin landed heads and where the coin landed tails (i.e. not heads). It should be noted that a disjunction of solutions is altogether different from a solution which is a disjunction of alternatives such as: (SYN K(OR A(NOT A))) which in this case is equivalent to:(SYN K T) and which would be an incorrect rendering of what is intuitively meant by multiple defaults.

In planning further actions to the resulting state K in order to achieve some overall goal the robot must take into account all the different solutions for K and make its plans accordingly. For example, if the robots overall goal is to flip the coin until it lands heads then the robot should plan to do nothing for the solution:(SYN K A), but should plan to continue flipping the coin for the solution(SYN K(NOT A)).

The purpose of using these default axioms is to allow for the case of where additional information in G contradicts the defaults. For example if the coin has tails on both sides then the flipped coin will always land tails. Thus letting G be (NOT A) and assuming that (NOT A) is logically possible, the first corollary expression of T2 above reduces to the single solution:

```
(IFF(SYN K(AND(NOT A)(IMPLY(POS(AND K A))A)(IMPLY(POS(AND K(NOT A)))
     (NOT A))))(SYN K(NOT A)) )
```

Likewise if the coin has heads on both sides then the flipped coin will always land heads. Thus letting G be A and assuming that A is logically possible, the first corollary expression of T2 above reduces to the single solution:

```
(IFF(SYN K(AND(NOT A)(IMPLY(POS(AND K A))A)(IMPLY(POS(AND K(NOT A)))
      A)))(SYN K A) )
```

The flipped coin example can be generalized to a knowledgebase containing N mutually exclusive defaults. Essentially, this is done by adding N standard defaults to the theory and letting G state that the conclusions of all these defaults are mutually exclusive. Such a knowledgebase will have precisely N solutions whenever no other information is available. This fact is given below in theorem T3. The proof of T3 illustrates the smooth interaction of reasoning with complex mixtures of contingent and necessary expressions involving quantifiers in the modal quantificational logic Z.

T3: For all N there exist a knowledgebase with N standard defaults and with N solutions.

```
(IMPLY(AND(SYN G(AND G2(ALL I(ALL J(OR(EQUAL I J)(IMPLY(P I)(NOT
           (P J))))))))(ALL N(POS(P N))) )
      (ALL N(IFF(SYN K(AND G(ALL M(IMPLY(<= M N)
                                (IMPLY(POS(AND K(P M)))(P M)))) ))
                 (EX M(AND(<= M N)(SYN K(AND G(P M))))) )))
```

The following particular knowledgebase is taken from[Reiter] where it is claimed that the analogous formulation in the nonmonotonic logic of [McDermott&Doyle] has no fixed point.

T4: There exists a knowledgebase consisting of three standard defaults which has no solutions:

if 'A1, 'A2, 'A3, 'B1, 'B2, 'B3 are all generators then:

```
(IFF(SYN K(AND
     (IMPLY(POS(AND K A1))(IMPLY B1 A1))(IMPLY A1 B2)(NOT(AND A1 A2))
```

```
    (IMPLY(POS(AND K A2))(IMPLY B2 A2))(IMPLY A2 B3)(NOT(AND A2 A3))
    (IMPLY(POS(AND K A3))(IMPLY B3 A3))(IMPLY A3 B1)(NOT AND A3 A1))))
NIL)
```

Some examples dealing with more esoteric cases of nonmonotonic reasoning are now given.

T5: A knowledgebase with one default (not necessarily standard):

```
(IFF(SYN K(AND G(IMPLY(POS(AND K A))X)))
    (OR(AND(POS(AND G X A))(SYN K(AND G X)))
       (AND(NOT(POS(AND G A)))(SYN K G))) )
```

corollary if G is T and if (AND X A) is possible it follows that:

```
(IFF(SYN K(IMPLY(POS(AND K A))X))
    (SYN K X))
```

T6: A knowledgebase with two defaults (not necessarily standard):

```
(IFF(SYN K(AND G(IMPLY(POS(AND K B))X)(IMPLY(POS(AND K D))Y)))
    (OR(AND(POS(AND G X Y B))(POS(AND G X Y D))(SYN K(AND G X Y)))
       (AND(POS(AND G X B))(NOT(POS(AND G X D)))(SYN K(AND G X)))
       (AND(NOT(POS(AND G Y B)))(POS(AND G Y D))(SYN K(AND G Y)))
       (AND(NOT(POS(AND G B)))(NOT(POS(AND G D)))(SYN K G)) ) )
```

T7: A knowledgebase having a unique solution may have a subknowledgebase which has no solutions.

```
(IFF(SYN K (AND G(IMPLY(POS(AND K A))(NOT A))))
    (AND(NOT(POS(AND G A)))(SYN K G)))
```

Thus if G is T and (POS A) we find that there are no solutions:

```
(IFF(SYN K(IMPLY(POS(AND K A))(NOT A))) NIL)
```

If however, we allow G to be (NOT A), we get:

```
(AND (NOT(POS (AND A (NOT A))))(SYN K (NOT A)))
```

giving(SYN K (NOT A)).

T8: The knowledgebase: G,(NOT B)unless A, (NOT A)unless B

```
(IFF(SYN K(AND G(IMPLY(POS(AND K A))(NOT B))(IMPLY(POS(AND K B))
                                                    (NOT A))))
    (OR(AND(POS(AND G(NOT B)A))(SYN K(AND G(NOT B))))
       (AND(POS(AND G(NOT A)B))(SYN K(AND G(NOT A))))
       (AND(NOT(POS(AND G A)))(NOT(POS(AND G B)))(SYN K G)) ))
```

Letting G be T, we have three solutions:

```
(OR(AND(POS(AND(NOT B)A))(SYN K(NOT B)))
   (AND(POS(AND(NOT A)B))(SYN K(NOT A)))
   (AND(SYN A NIL)(SYN B NIL)(SYN K T)) )
```

If we now assume (POS(AND(NOT B)A)) and (POS(AND(NOT A)B)), it follows that:

```
    (OR(SYN K(NOT B))(SYN K(NOT A)))
```

4. Action and the Frame Problem

One fundamental problem in Robot plan formation is how properties which are true in a state remain true in the succeeding state obtained by applying an action unless specifically stated otherwise[Hayes72]. Besides, the need for specific defaults within a knowledgebase representing a state this indicates a need for a general default mechanism. Our law of action states that the properties which are true in a succeeding state (DO A K) obtained by applying the action A to the state K are the physical laws which are true of all (real)states, the explicitly named results of the action A, and those restricted propositions which are true in K and which are logically possible with the new state(DO A K):

```
;;;the law of action -- including automatic frame defaults:
    (IMPLY(ENTAIL K(PRECONDITIONS A))
          (SYN(DO A K)
              (AND(PHYSICAL-LAWS)
                  (RESULTS A)
                  (ALL X(IMPLY(AND(RESTRICTION X)
                                  (ENTAIL K X)
                                  (POS(AND(DO A K)X)))
                              X)) )))
```

This action law involves reflexive reasoning[Hayes79] because the new state (DO A K) is specified by a purported definition which may have 0 or more solutions. The POS symbol of our modal logic Z was first used in a formal language as a hypothesis of an action law in[Schwind78]. Our action law however differs from [Schwind78,80] in that the states are generally incomplete propositions instead of worlds, in that the resulting state is (DO A K) rather than being merely some existentially quantified state in the future, and in that our law involves reflexive reasoning. The details of this law are given below along with an example deduction illustrating how this law of action automatically handles the frame problem[Hayes73,Hayes79] by allowing properties which are true in an initial state to be carried over into the new state, even though such properties are never mentioned as being part of(or implied by) the results of the action that is applied.

```
;;;a restriction on the law of action -- other are possible
    (EQUAL(RESTRICTION X) (EX G(AND(GENERATORS G)(SYN X(GMEANING G)))))
```

```
    (EQUAL GENERATORS
        {'(AT ROBOT HOME)"(AT ROBOT OFFICE)'(AT JOHN HOME)'(AT JOHN
                                                    OFFICE)})
;;;definition of commonsense physics
    (SYN(PHYSICAL-LAWS)
        (AND(ALL X(ALL P1(ALL P2(IMPLY(AND(AT X P1)(AT X P2))(EQUAL P1
             P2))))))(NOT(EQUAL HOME OFFICE)) ))
;;;definitions of the preconditions and effects of the moving action:
    (SYN(PRECONDITIONS(MOVE ROBOT P1 P2))(AT ROBOT P1))
    (SYN(RESULTS(MOVE ROBOT P1 P2))(AT ROBOT P2))

;;;an initial state
    (SYN KSTART(AND(PHYSICAL-LAWS)(AT JOHN HOME)(AT ROBOT HOME)))
```

From the above axioms it follows that John stays at home in the state of the world where the robot performs the action of going to the office even though this fact is not mentioned as being a result of the moving action:

```
    (SYN(DO(MOVE ROBOT HOME OFFICE)KSTART)
        (AND(PHYSICAL-LAWS)(AT ROBOT OFFICE)(AT JOHN HOME)))
                        proof
;;;instantiating the law of action and then simplifying:
    (IMPLY(ENTAIL KSTART(PRECONDITIONS(MOVE ROBOT HOME OFFICE)))
          (SYN(DO(MOVE ROBOT HOME OFFICE)KSTART)
              (AND(PHYSICAL-LAWS)
                  (RESULTS(MOVE ROBOT HOME OFFICE))
                  (ALL X(IMPLY(AND(RESTRICTION X)
                                  (ENTAIL KSTART X)
                                  (POS(AND(DO(MOVE ROBOT HOME OFFICE)
                                              KSTART)X)))
                              X)) )))
    (IMPLY(ENTAIL(AND(PHYSICAL-LAWS)(AT JOHN HOME)(AT ROBOT HOME))(AT
                                      ROBOT HOME))
          (SYN(DO(MOVE ROBOT HOME OFFICE)KSTART)
              (AND(PHYSICAL-LAWS)
                  (AT ROBOT OFFICE)
                  (ALL X(IMPLY(AND(EX G(AND(GENERATORS G)(SYN X(GMEANING
                                                               G))))
                                  (ENTAIL(AND(PHYSICAL-LAWS)
                                             (AT JOHN HOME)
                                             (AT ROBOT HOME))
                                      X)
```

```
                                (POS(AND(DO(MOVE ROBOT HOME OFFICE)
                                           KSTART)X)))
                            X) ))))
(SYN(DO(MOVE ROBOT HOME OFFICE)KSTART)
    (AND(PHYSICAL-LAWS)
        (AT ROBOT OFFICE)
        (ALL X(IMPLY(AND(OR(SYN X(AT ROBOT HOME))
                            (SYN X(AT ROBOT OFFICE))
                            (SYN X(AT JOHN HOME))
                            (SYN X(AT JOHN OFFICE)))
                         (ENTAIL(AND(PHYSICAL-LAWS)
                                    (AT JOHN HOME)
                                    (AT ROBOT HOME)) X)
                         (POS(AND(DO(MOVE ROBOTHOME OFFICE)KSTART)X)))
                     X)) ))
(SYN(DO(MOVE ROBOT HOME OFFICE)KSTART)
    (AND(PHYSICAL-LAWS)
        (AT ROBOT OFFICE)
        (IMPLY(AND(ENTAIL(AND(PHYSICAL-LAWS)(AT JOHN HOME)(AT ROBOT
                                                    HOME))
                          (AT ROBOT HOME))
                   (POS(AND(DO(MOVE ROBOT HOME OFFICE)KSTART)(AT
                                                    ROBOT HOME))))
              (AT ROBOT HOME))
        (IMPLY(AND(ENTAIL(AND(PHYSICAL-LAWS)(AT JOHN HOME)(AT ROBOT
                                                    HOME))
                          (AT ROBOT OFFICE))
                   (POS(AND(DO(MOVE ROBOT HOME OFFICE)START)(AT ROBOT
                                                    OFFICE))))
              (AT ROBOT OFFICE))
        (IMPLY(AND(ENTAIL(AND(PHYSICAL-LAWS)(AT JOHN HOME)(AT ROBOT
                                                    HOME))
                          (AT JOHN HOME))
                   (POS(AND(DO(MOVE ROBOT HOME OFFICE)KSTART)(AT JOHN
                                                    HOME))))
              (AT JOHN HOME))
        (IMPLY(AND(ENTAIL(AND(PHYSICAL-LAWS)(AT JOHN HOME)(AT ROBOT
                                                    HOME))
                          (AT JOHN OFFICE))
                   (POS(AND(DO(MOVE ROBOT HOME OFFICE)KSTART)(AT JOHN
                                                    OFFICE))))
              (AT JOHN OFFICE)) ))
```

```
    (SYN(DO(MOVE ROBOT HOME OFFICE)KSTART)
        (AND(PHYSICAL-LAWS)
            (AT ROBOT OFFICE)
            (IMPLY(POS(AND(DO(MOVE ROBOT HOME OFFICE)KSTART)(AT ROBOT
                                                                  HOME)))
                  (AT ROBOT HOME))
            T
            (IMPLY(POS(AND(DO(MOVE(ROBOT HOME OFFICE)KSTART)(AT JOHN
                                                                  HOME)))
                  (AT JOHN HOME))
            T))
    (SYN(DO(MOVE ROBOT HOME OFFICE)KSTART)
        (AND(PHYSICAL-LAWS)
            (AT ROBOT OFFICE)
            (IMPLY(POS(AND(DO(MOVE ROBOT HOME OFFICE)KSTART)(AT ROBOT
                                                                  HOME)))
                  NIL)
            (IMPLY(POS(AND(DO(MOVE ROBOT HOME OFFICE)KSTART)(AT JOHN
                                                                  HOME)))
                  (AT JOHN HOME))))
    (SYN(DO(MOVE ROBOT HOME OFFICE)KSTART)
        (AND(PHYSICAL-LAWS)
            (AT ROBOT OFFICE)
            (NOT(POS(AND(DO(MOVE ROBOT HOME OFFICE)KSTART)(AT ROBOT HOME))))
            (IMPLY(POS(AND(DO(MOVE ROBOT HOME OFFICE)KSTART)(AT JOHN
                                                                  HOME)))
                  (AT JOHN HOME))))
;;;case analysis
    (IF(POS(AND(DO(MOVE ROBOT HOME OFFICE)KSTART)(AT ROBOT HOME)))
       (SYN(DO(MOVE ROBOT HOME OFFICE)KSTART)
           (AND(PHYSICAL-LAWS)
               (AT ROBOT OFFICE)
               (NOT T)
               (IMPLY(POS(AND(DO(MOVE ROBOT HOME OFFICE)KSTART)(AT JOHN
                                                                  HOME)))
                     (AT JOHN HOME)) ))
       (IF(POS(AND(DO(MOVE ROBOT HOME OFFICE)KSTART)(AT JOHN HOME)))
          (SYN(DO(MOVE ROBOT HOME OFFICE)KSTART)
              (AND(PHYSICAL-LAWS)
                  (AT ROBOT OFFICE)
                  (NOT NIL)
                  (IMPLY T(AT JOHN HOME)) ))
```

```
      (SYN(DO(MOVE ROBOT HOME OFFICE)KSTART)
          (AND(PHYSICAL-LAWS)
              (AT ROBOT OFFICE)
              (NOT NIL)
              (IMPLY NIL(AT JOHN HOME)) ))))
 (IF(POS(AND(DO(MOVE ROBOT HOME OFFICE)KSTART)(AT ROBOT HOME)))
    (SYN(DO(MOVE ROBOT HOME OFFICE)KSTART)NIL)
    (IF(POS(AND(DO(MOVE ROBOT HOME OFFICE)KSTART)(AT JOHN HOME)))
       (SYN(DO(MOVE ROBOT HOME OFFICE)KSTART)
           (AND(PHYSICAL-LAWS)(AT ROBOT OFFICE)(AT JOHN HOME)))
       (SYN(DO(MOVE ROBOT HOME OFFICE)KSTART)
           (AND(PHYSICAL-LAWS)(AT ROBOT OFFICE))) ))
 (OR(AND(POS(AND(DO(MOVE ROBOT HOME OFFICE)KSTART)(AT ROBOT HOME)))
        (SYN(DO(MOVE ROBOT HOME OFFICE)KSTART)NIL))
    (AND(NOT(POS(AND(DO(MOVE ROBOT HOME OFFICE)KSTART)(AT ROBOT
                                                          HOME))))
        (POS(AND(DO(MOVE ROBOT HOME OFFICE)KSTART)(AT JOHN HOME)))
        (SYN(DO(MOVE ROBOT HOME OFFICE)KSTART)
            (AND(PHYSICAL-LAWS)(AT ROBOT OFFICE)(AT JOHN HOME)) ))
    (AND(NOT(POS(AND(DO(MOVE ROBOT HOME OFFICE)KSTART)(AT ROBOT
                                                          HOME))))
        (NOT(POS(AND(DO(MOVE ROBOT HOME OFFICE)KSTART)(AT JOHN
                                                          HOME))))
        (SYN(DO(MOVE ROBOT HOME OFFICE)KSTART)
            (AND(PHYSICAL-LAWS)(AT ROBOT OFFICE)))) )
 (OR(AND(POS NIL)
        (SYN(DO(MOVE ROBOT HOME OFFICE)KSTART)NIL))
    (AND(NOT(POS NIL))
        (POS(AND(PHYSICAL-LAWS)(AT ROBOT OFFICE)(AT JOHN HOME)))
        (SYN(DO(MOVE ROBOT HOME OFFICE)KSTART)
            (AND(PHYSICAL-LAWS)(AT ROBOT OFFICE)(AT JOHN HOME)) )))
    (AND(NOT(POS NIL))
        (NOT(POS(AND(PHYSICAL-LAWS)(AT ROBOT OFFICE)(AT JOHN HOME))))
        (SYN(DO(MOVE ROBOT HOME OFFICE)KSTART)
            (AND(PHYSICAL-LAWS)(AT ROBOT OFFICE)))) )
 (OR(AND(POS(AND(PHYSICAL-LAWS)(AT ROBOT OFFICE)(AT JOHN HOME)))
        (SYN(DO(MOVE ROBOT HOME OFFICE)KSTART)
            (AND(PHYSICAL-LAWS)(AT ROBOT OFFICE)(AT JOHN HOME)) ))
    (AND(NOT(POS(AND(PHYSICAL-LAWS)(AT ROBOT OFFICE)(AT JOHN HOME))))
        (SYN(DO(MOVE ROBOT HOME OFFICE)KSTART)
            (AND(PHYSICAL-LAWS)(AT ROBOT OFFICE)))) )
```

```
(OR(AND T(SYN(DO(MOVE ROBOT HOME OFFICE)KSTART)
            (AND(PHYSICAL-LAWS)(AT ROBOT OFFICE)(AT JOHN HOME)) ))
   (AND NIL(SYN(DO(MOVE ROBOT HOME OFFICE)KSTART)
             (AND(PHYSICAL-LAWS)(AT ROBOT OFFICE)))) )
(SYN(DO(MOVE ROBOT HOME OFFICE)KSTART)
    (AND(PHYSICAL-LAWS)(AT ROBOT OFFICE)(AT JOHN HOME)))
```

Acknowledgements

This research was supported by the Mathematics Division of the US Army Research Office with contract: DAAG29-85-C-0022 and by the National Science Foundation grant: DCR-8402412, to AIRIT Inc., and by a grant from the University of Kansas. I wish to thank the members of the Computer Science Department and college administration at the University of Kansas for providing the research environment to carry out this research, Seung Park for his corrections, and also Glenn Veach who has collaborated with me on some of the research herein described.

References

Brown, F.M. (1976) "A Theory of Meaning," Department of Artificial Intelligence Working Paper 16, University of Edinburgh.

Brown, F.M. (1978a) "An Automatic Proof of the Completeness of Quantificational Logic", Department of Artificial Intelligence Research Report 52.

Brown, F.M. (1979) "A Theorem Prover for Metatheory," 4th CONFERENCE ON AUTOMATIC THEOREM PROVING, Austin Texas.

Brown, F.M. (1978b), "A Sequent Calculus for Modal Quantificational Logic," 3rd AISB/GI CONFERENCE PROCEEDINGS, Hamburg.

Carnap, Rudolf (1956) MEANING AND NECESSITY: A STUDY IN THE SEMANTICS OF MODAL LOGIC, The University of Chicago Press.

Frege, G. (1879) "Begriffsschrift, a formula language, modeled upon that of arithmetic, for pure thought", in FROM FREGE TO GODEL, 1967.

Hayes, P.J. (1973) "The Frame Problem and Related Problems in Artificial Intelligence" ARTIFICIAL AND HUMAN THINKING, eds. Eilithorn and Jones.

Hughes, G.E. and Creswell, M.J. (1968) AN INTRODUCTION TO MODAL LOGIC, METHUEN and Co, Ltd., London.

McDermott, D., (1982) "Nonmonotonic Logic II: Nonmonotonic Modal Theories" JACM, Vol. 29, No. 1.

McDermott, D., Doyle, J. (1980) "Nonmonotonic Logic I" ARTIFICIAL INTELLIGENCE 13.

Moore, R.C. (1985) "Semantical Considerations on Nonmonotonic Logic" ARTIFICIAL INTELLIGENCE 25.

Reiter, R. (1980) "A Logic for Default Reasoning" ARTIFICIAL INTELLIGENCE 13.
Schwind, C.B. (1978) "The theory of Actions" report TUM-INFO 7807, Technische Universitat Munchen.
Schwind, C.B. (1980) "A Completeness Proof for a Logic Action" Laboratoire D'Informatique pour les sciences de l'homme.

Negative Paramodulation*

L. Wos

and

W. McCune

Mathematics and Computer Science Division
Argonne National Laboratory
Argonne, IL 60439-4844

Abstract

In this paper, we introduce the inference rule *negative paramodulation*. This rule reasons from inequalities, in contrast to paramodulation which reasons from equalities. Negative paramodulation is recommended for use when certain conditions are satisfied; here we give those conditions. We present experimental evidence that suggests the potential value of employing the closely related inference rule *negative hyperparamodulation*.

1. Introduction

The effectiveness of a computer program that reasons logically can be sharply increased by employing inference rules that "build in" semantics. An example of such a rule is paramodulation [2,5], a rule that generalizes the usual notion of equality substitution. Paramodulation applied to the equations $x + -x = 0$ and $y + (-y + z) = z$ yields in a single step $y + 0 = -(-y)$. In clause form, from

EQUAL(sum(x,minus(x)),0)

into

EQUAL(sum(y,sum(minus(y),z)),z)

the clause

EQUAL(sum(y,0),minus(minus(y)))

is obtained by paramodulation. As the example shows, equality substitution itself is not sufficient to make the given deduction. Rather, an instantiation of variables is found that unifies the argument *sum(x,minus(x))* with the term *sum(minus(y),z)* to then permit an obvious substitution of one term for another. Combining in a single step the unification process with the equality substitution process markedly improves the efficiency of an automated reasoning program.

For various problem domains, paramodulation is the best suited rule of inference. For example, many problems from group theory and ring theory are easily solved with this inference rule. The rule does not, however, enable a reasoning program to make deductions of one obvious class. Specifically, although the use of paramodulation enables the program to take appropriate action when the term s is known to equal the term t, its use does not address the case in which it is known that s does not equal t.

*This work was supported by the Applied Mathematical Sciences subprogram of the office of Energy Research, U.S. Department of Energy, under contract W-31-109-Eng-38.

For example, what deduction is valid given the associative law, $(xy)z = x(yz)$, and the fact that the product of ab is not equal to the product ba?

EQUAL(prod(prod(x,y),z),prod(x,prod(y,z)))
¬EQUAL(prod(a,b),prod(b,a))

The group theorist would correctly accept as a valid conclusion

¬EQUAL(prod(prod(x,a),b),prod(x,prod(b,a)))

stating that $(xa)b$ is not equal to $x(ba)$ for all x. Such an algebraist would make this deduction in a single step and, if asked, give the justification that cancellation holds. With the appropriate clauses and the inference rules currently in use, two steps would be required to make this deduction in contrast to the one-step deduction produced with negative paramodulation. If in addition it were known that, in the groups under consideration, the square of every element is the identity e,

EQUAL(prod(u,u),e)

our algebraist would have immediately concluded in a single but larger step that

¬EQUAL(b,prod(a,prod(b,a)))

without stopping to note that $(xa)b$ is not equal to $x(ba)$.

To enable an automated reasoning program to make one-step deductions exemplified by the clause

¬EQUAL(prod(prod(x,a),b),prod(x,prod(b,a)))

negative paramodulation was formulated. So that one-step deductions exemplified by the clause

¬EQUAL(b,prod(a,prod(b,a)))

are possible, negative hyperparamodulation was formulated. (For those who wish history accurately recorded, negative hyperparamodulation was formulated first.) Since current evidence supports the position that negative hyperparamodulation will prove more useful, we concentrate on experiments with that inference rule. In Section 2, we give the precise conditions under which the two inference rules are sound; a glance at the first paragraph of that section summarizes the differences between positive and negative paramodulation.

Although the two inference rules presented here give a reasoning program the capacity to draw conclusions that a person might draw, they were not formulated to imitate a person's reasoning. Rather, they are intended to give an automated reasoning program the same kind of power as paramodulation does, but the power to reason from inequalities. The example of paramodulation cited earlier illustrates one of the differences between reasoning that imitates that of some person and reasoning that is tailored to a machine. A person would not ordinarily use (explicitly or implicitly) such an inference rule. Rules of this type are common in automated reasoning. They are not so common in a classical artificial intelligence approach. In the latter, the objective is to have a machine employ problem-solving techniques that closely imitate those a person might use.

A simpler illustration of the difference between automated reasoning and classical artificial intelligence is provided by the possible use of instantiation. In various mathematical proofs, instantiation is used. For example, consider the theorem that states that, in a group, if the square of every element is the identity, the group is commutative. A typical first step in proving this theorem is that of deducing $(yz)(yz)=e$ with the justification that $uu=e$. The variable u is simply instantiated to the product yz. Were one designing a program to find proofs in algebra, and were one taking a classical artificial intelligence approach, instantiation would get strong consideration as an inference rule. In the typical automated reasoning program, simply instantiating some clause to produce another is not available. Currently, no strategy is known for picking effective instantiations. The generality offered by the available inference rules seems far preferable when compared to deducing instances of clauses.

The preceding remarks explain in part why we seek rules such as negative paramodulation. The nature of such rules, at least in their more complex use, does not reflect the type of reasoning that one would expect from a person. But they are well suited to a machine. An example similar to that used earlier for paramodulation again illustrates this point.

Consider an array A of rational numbers in which numbers may occur more than once. Let its elements be denoted by $c_1, c_2, \cdots$. Assume that c_1 is such that no c_i has the property that $c_i+c_i=c_1$. Negative paramodulation applied to the clauses

$\neg ELEMENT(x,arraya)$ $\neg EQUAL(sum(x,x),c1)$
$EQUAL(sum(y,sum(minus(y),z)),z)$

yields the clause

$\neg ELEMENT(minus(y),arraya)$ $\neg EQUAL(sum(y,c1),minus(y))$

in one step. The soundness of this inference follows from the fact that the left cancellation law holds for the function *sum*.

2. Negative Paramodulation Defined

Negative paramodulation is similar to (positive) paramodulation with three exceptions. First, the literal justifying the term replacement (the *from* literal) is a negative equality rather than a positive equality literal. Second, the literal containing the term to be replaced (the *into* literal) must be a positive literal. Finally, the transformed *into* literal in the negative paramodulant is negated.

Negative paramodulation is sound provided that the *into* literal satisfies certain properties. Those properties are defined before the formal definition of negative paramodulation is presented.

2.1. Admissibility of "into" Terms

The ith argument position of an n-ary function f is *admissible for negative paramodulation* (hereafter abbreviated *admissible*) if the clause

$$\neg EQUAL(f(x_1, \cdots, x_{i-1}, x_i, x_{i+1}, \cdots, x_n), f(x_1, \cdots, x_{i-1}, y, x_{i+1}, \cdots, x_n)) \quad EQUAL(x_i, y)$$

is valid in the theory under consideration. Similarly, the ith argument position of an n-ary predicate P is *admissible* if the clause

$$\neg P(x_1, \cdots, x_{i-1}, x_i, x_{i+1}, \cdots, x_n) \quad \neg P(x_1, \cdots, x_{i-1}, y, x_{i+1}, \cdots, x_n) \quad EQUAL(x_i, y)$$

is valid in the theory under consideration. The ith argument of a term whose major function symbol is f (respectively, positive literal whose predicate is P) is *admissible* if the ith argument position of f (respectively, P) is admissible.

Admissibility of the ith argument position of a function f can be viewed as one-to-oneness of all unary functions in the family of functions obtained by fixing all but the ith argument of f. Admissibility in a function can also be viewed as a generalized cancellation law. For example, the left argument of a binary function is admissible provided that the right cancellation law holds for that function. Admissibility in a predicate can be viewed as a functional dependency. For example, the third argument of a ternary predicate is admissible provided that the third argument is a function of the first two arguments. For the predicate *FATHER*, where *FATHER*(x,y) means that x is the father of y, the first argument is a function of the second, but the second is not a function of the first.

The position of an occurrence of a subterm in a literal can be identified by a *position vector*, which is a sequence of positive integers that gives the path from the root of the literal to the occurrence of the subterm. For example, the position of the term c in the literal $P(a,x,g(f(b,c)))$ is given by the position vector (3,1,2), for c occurs in the third argument of P, the first argument of g, and the second argument of f.

The position of an occurrence of a subterm in a literal is *admissible* if the literal is positive and the literal and all terms that contain the occurrence contain it in an admissible argument position of the corresponding function or predicate symbol. In the preceding example, the position of c is admissible provided that the following three clauses hold.

$\neg EQUAL(f(x,y),f(x,z))\quad EQUAL(y,z)$
$\neg EQUAL(g(y),g(z))\quad EQUAL(y,z)$
$\neg P(x,w,y)\quad \neg P(x,w,z)\quad EQUAL(y,z)$

Notice that both argument positions of the equality predicate are admissible. In particular, the clauses

$\neg EQUAL(x,y)\quad \neg EQUAL(z,y)\quad EQUAL(x,z)$
$\neg EQUAL(y,x)\quad \neg EQUAL(y,z)\quad EQUAL(x,z)$

are what is needed. They are obviously deducible from transitivity and symmetry of *EQUAL*.

In Section 1, the example of deducing

$\neg EQUAL(prod(prod(x,a),b),prod(x,prod(b,a)))$

from the clauses

$EQUAL(prod(prod(x,y),z),prod(x,prod(y,z)))$
$\neg EQUAL(prod(a,b),prod(b,a))$

is sound because the second argument position of *prod* is admissible. The conclusion is obtained by unifying $prod(a,b)$ with $prod(y,z)$.

To see that the admissibility property is required for soundness to be present, consider the following case. Let j be a function that is not one-to-one. Therefore, the pair of clauses $EQUAL(j(a),j(b))$ and $\neg EQUAL(a,b)$ is consistent. If negative paramodulation is applied to these two clauses, an unsound inference is yielded, namely, $\neg EQUAL(j(b),j(b))$.

2.2. The Formal Definition of Negative Paramodulation

We can now give the formal definition of negative paramodulation. We simply quote the definition of (positive) paramodulation, making the appropriate modifications. See Figure 1.

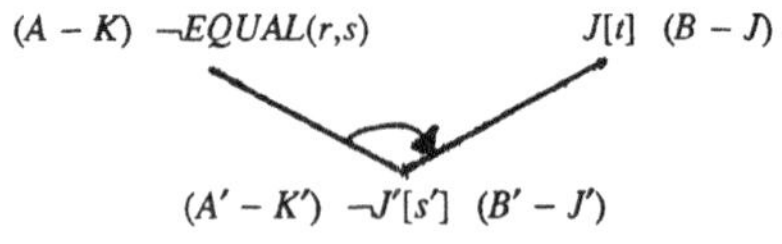

Figure 1. Negative Paramodulation

Definition. An *equality literal* is a literal whose predicate is to be interpreted as "equal". The inference rule *negative paramodulation* yields the clause C from the clauses A and B (that are assumed to have no variables in common) when A contains a negative equality literal K and B contains a positive literal J that in turn contains a term t admissible in J that unifies with one of the arguments of K. Assume without loss of generality that the chosen negative equality literal K has the form $\neg EQUAL(r,s)$ for terms r and s, and that the first argument is that which is being unified with the chosen term t in B. The clause C is obtained from A and B with the following procedure. First, find an MGU for the argument r and the term t. Second, apply the MGU to A and B, yielding A', B', K', J', and t' as the respective correspondents of A, B, K, J, and t. Third, generate J'' from J' by first replacing t' by s', where K' is of the form $\neg EQUAL(r',s')$, and then negating the literal. Finally, form the disjunction of J'' and $B'-J'$ (B' minus a single occurrence of J') and $A'-K'$ (A' minus a single occurrence of K'). Clause A is called the *from clause*, clause B the *into clause*, term t the *into term*, and clause C a *negative paramodulant*.

2.3. The Soundness of Negative Paramodulation

This subsection contains an informal justification of the soundness of negative paramodulation. Given a negative paramodulation inference that produces a clause C from clause A into clause B, one can construct a binary resolution deduction of C from A, B, and the clauses that justify the admissibility of the *into* position in B. Consider the earlier example $P(a,x,g(f(b,c)))$ in which the term c is in an admissible position. With the additional clause $\neg EQUAL(c,d)$, one can deduce $\neg P(a,x,g(f(b,d)))$ by negative paramodulation. To construct a binary resolution deduction of this negative paramodulant, first consider the following clauses which justify the admissibility of the position in which c occurs.

(1) $\neg EQUAL(f(x,y),f(x,z)) \quad EQUAL(y,z)$
(2) $\neg EQUAL(g(y),g(z)) \quad EQUAL(y,z)$
(3) $\neg P(x,w,y) \quad \neg P(x,w,z) \quad EQUAL(y,z)$

Start with $\neg EQUAL(c,d)$, resolve with clause (1), then resolve the result with clause (2), then resolve that result with clause (3) to obtain

$$\neg P(x,y,g(f(u,c))) \quad \neg P(x,y,g(f(u,d))).$$

The desired clause follows by resolving with $P(a,x,g(f(b,c)))$.

We emphasize the point that it is the predicate and functions that contain the term being replaced that must have certain properties, and not the terms that are unified. We illustrate with an example from ring theory, in which both argument positions of *sum* are admissible, but neither argument position of *prod* is admissible. (Note that groups differ from rings in that the arguments of product are admissible.) Consider the clauses

(1) $EQUAL(prod(x,sum(y,z)),sum(prod(x,y),prod(x,z)))$
(2) $\neg EQUAL(prod(a,0),0)$
(3) $EQUAL(sum(x,0),x)$

where clause (2) is the denial of an elementary theorem. The negative paramodulation inference that produces $\neg EQUAL(prod(a,sum(y,0)),sum(prod(a,y),0))$ by unifying $prod(a,0)$ in clause (2) with $prod(x,z)$ in clause (1) is sound even though *prod* does not have the admissibility property. (As an aside, notice that if clause (3) is present as a demodulator, then a contradiction is obtained in one step in the presence of reflexivity.)

3. Negative Hyperparamodulation

In order to easily introduce the inference rule "negative hyperparamodulation", we first discuss (positive) hyperparamodulation. We begin with an example taken from elementary group theory. The theorem of interest, so often quoted, says that if the square of every element is the identity e, then the group is commutative.

In one presentation of this theorem to an automated reasoning program, the set of input clauses includes the following.

(1) *EQUAL(prod(prod(x,y),z),prod(x,prod(y,z)))*
(2) *EQUAL(prod(x,e),x)*
(3) *EQUAL(prod(x,x),e)*
(4) *EQUAL(prod(a,b),c)*

If the version of (positive) hyperparamodulation under discussion in this paper is applied to clauses (1) through (4) considered simultaneously, the clause

(5) *EQUAL(prod(c,b),a)*

is deduced. Clause (5) is obtained by letting clause (1) be the nucleus and clauses (2), (3), and (4) be satellites, and applying (simultaneously) paramodulation three times. The term *prod(x,y)* of clause (1) is unified with the argument *prod(a,b)* of clause (4); the term *prod(y,z)* of clause (1) is unified with the argument *prod(x,x)* of clause (3); and the term *prod(a,e)*, which is the transform of the term *prod(x,prod(y,z))* of (1), is unified with the argument *prod(x,e)* of (2). If the set of chosen terms from the nucleus contains some term s and a subterm t of s, then any unification involving t must be considered before any involving s. Further, rather than unifying s with some argument r, hyperparamodulation requires attempting to unify s' with r, where s' is the term resulting from the paramodulations involving the subterms of s that are among the terms chosen from the nucleus. Finally, each of the chosen terms must participate exactly once.

The version of hyperparamodulation discussed here is somewhat different from that introduced in an earlier paper on the subject [4]. In the version here, new nuclei can be added during a run, any input clause can be a nucleus, various input clauses can be designated as inadmissible as satellites, and the user does not mark the terms in the nuclei to be considered for unification. From one viewpoint, once a nucleus has been chosen, all of its terms must participate. With this view, an effective strategy requires all terms that are variables to be paramodulated with the clause for reflexivity, *EQUAL(x,x)*. From another viewpoint, this version of hyperparamodulation permits some set of terms of the nucleus to be involved, but omits all terms that are variables and allows other terms to be omitted also.

Intuitively, hyperparamodulation is to paramodulation as hyperresolution is to P1-resolution. Any hyperresolvent can be obtained by applying the appropriate set of P1-resolution steps. With such an implementation, only the final positive clause, if one is found, is considered for retention. A more efficient implementation is achieved if the appropriate set of unifications is sought, processed to produce the required single unifier that unifies certain pairs of literals, and no intermediate clauses are generated. The object is, of course, to remove all negative literals from the nucleus without introducing any new negative literals. Similarly, hyperparamodulation can be implemented by a succession of paramodulation steps that produces a sequence of clauses such that the final clause is the only one considered for retention. It can also be implemented by a simultaneous application of paramodulation steps that produce no intermediate clauses. Where hyperresolution attempts to remove some designated set of literals (the negative literals of the nucleus), hyperparamodulation attempts to substitute (with equalities) into some designated set of terms of the into clause.

Negative hyperparamodulation requires that exactly one of the satellites be an inequality. The other important requirement is that the paramodulation involving the inequality must obey the constraints imposed on negative paramodulation. The remaining requirements are those of positive hyperparamodulation.

The following example is typical of negative hyperparamodulation. We again consider the same theorem from elementary group theory. We replace clause (2) with clause (6), and clause (4) with clause (7).

(1) *EQUAL(prod(prod(x,y),z),prod(x,prod(y,z)))*
(6) *EQUAL(prod(e,x),x)*
(3) *EQUAL(prod(x,x),e)*
(7) *¬EQUAL(prod(a,b),prod(b,a))*

In one negative hyperparamodulation step, an automated reasoning program can deduce

(8) *¬EQUAL(b,prod(a,prod(b,a)))*

by simultaneously applying negative paramodulation to the argument *prod(a,b)* of (7) and the term *prod(y,z)* of (1), paramodulation to the argument *prod(x,x)* of (3) and the term *prod(x,y)* of (1), and paramodulation to the argument *prod(e,x)* of (6) and the term *prod(e,b)* which is the transform of the term *prod(prod(x,y),z)* of (1).

Before turning to the algorithm employed to implement negative paramodulation and negative hyperparamodulation, we give the following rather elegant proof—a proof that was new to us and was first found by the reasoning program. Again, the focus is that same theorem from group theory. When the set of support consists of clause (7) alone, clauses (7) and (3) are used as satellites, clause (1) is used as a nucleus, and clauses (1) and (2) and (3) are used as demodulators, a single application of negative hyperparamodulation coupled with demodulation yields a proof of the theorem. We are, of course, assuming that the input set of clauses consists of the axioms for a group (expressed as unit clauses) together with the axiom of reflexivity, the hypothesis of the theorem, and clause (7) as the denial. Among the negative hyperparamodulants, the clause

(9) *¬EQUAL(e,e)*

is deduced. The set of terms from the nucleus, clause (1), consists of *prod(y,z)* and *prod(x,prod(y,z))*. The first of these is unified with the argument *prod(a,b)* of clause (7), and the resulting transform of the second is unified with the argument *prod(x,x)* of clause (3). The mathematical translation of a detailed account of this one-step negative hyperparamodulation proof focuses on an instance of associativity, namely, $((ba)a)b=(ba)(ab)$. Since ab is assumed not to equal ba, the right side of the equality becomes $(ba)(ba)$, and the equality becomes an inequality. But $(ba)(ba)=e$, from the hypothesis that the square of every element equals e. The left side of the original equality becomes e, by first reassociating, then using the hypothesis of the theorem, then using right identity, and finally using the hypothesis of the theorem again. The conclusion is that e is not equal to e, and the proof is complete.

4. Implementation and Experiments

Negative hyperparamodulation has been incorporated into *tp0*, an automated theorem-proving program based on LMA [1]. The user declares the admissible argument positions of the appropriate functions and predicates. The basic algorithm for generating negative hyperparamodulants is the following. Assume that the *given clause* is a negative unit equality clause, and that *nuclei* and *satellites* are sets of clauses.

(1) Initialize the set of negative hyperparamodulants to the empty set.
(2) For each nonvariable admissible *into* position of each member of *nuclei*, where the term in the *into* position unifies with the left argument of the given clause, perform steps (a) through (d).
    (a) Mark all nonvariable terms in the nucleus.
    (b) Generate a negative paramodulant such that the negative paramodulant inherits all marks from the nucleus except the marks on the *into* term and its subterms.

(c) Add the negative paramodulant to the set of negative hyperparamodulants, then do step (d) with the negative paramodulant.

(d) Generate a set of positive binary paramodulants by paramodulating members of *satellites* into the marked terms such that the paramodulants inherit the following marks from the into clause: marks on terms to the right of the into term, and marks on terms that properly contain the into term. Add each paramodulant to the set of negative hyperparamodulants. Do step (d) for each paramodulant.

(3) Demodulate and test for retention each clause in the set of negative hyperparamodulants.

Problem 1 is to prove that groups of exponent 2, groups in which $xx=e$ holds, are commutative. The following set of clauses is used, where *prod* is abbreviated to f, and *inv* is abbreviated to g.

(1) $EQUAL(x,x)$
(2) $EQUAL(f(e,x),x)$
(3) $EQUAL(f(x,e),x)$
(4) $EQUAL(f(g(x),x),e)$
(5) $EQUAL(f(x,g(x)),e)$
(6) $EQUAL(f(f(x,y),z),f(x,f(y,z)))$
(7) $EQUAL(f(x,x),e)$
(8) $\neg EQUAL(f(a,b),f(b,a))$

Clause (8) is the only clause in the initial set of support, and clauses (2) through (7) are used as demodulators as well as axioms.

Problem 2 is to prove that in groups of exponent 3, groups in which $xxx=e$ holds, $h(h(x,y),y)=e$, where h is the commutator function $h(x,y)=xyx^{-1}y^{-1}$. The following set of clauses is used, where clause (13) is the demodulated form of the denial of the theorem.

<clauses (1) through (6) from problem 1>
(7) $EQUAL(g(e),e)$
(8) $EQUAL(g(g(x)),x)$
(9) $EQUAL(f(x,f(g(x),y)),y)$
(10) $EQUAL(f(g(x),f(x,y)),y)$
(11) $EQUAL(g(f(x,y)),f(g(y),g(x)))$
(12) $EQUAL(f(x,f(x,x)),e)$
(13) $\neg EQUAL(f(a,f(b,f(g(a),f(b,f(a,f(g(b),f(g(a),g(b)))))))),e)$

Clause (13) is the only clause in the initial set of support, and clauses (2) through (12) are used as demodulators as well as axioms.

Problem 3 is to show that $(-x)(-y)=xy$ in rings, without using the lemmas $x0=0$ and $0x=0$. The following set of clauses is used.

(1) $EQUAL(x,x)$
(2) $EQUAL(sum(0,x),x)$
(3) $EQUAL(sum(x,0),x)$
(4) $EQUAL(sum(minus(x),x),0)$
(5) $EQUAL(sum(x,minus(x)),0)$
(6) $EQUAL(sum(sum(x,y),z),sum(x,sum(y,z)))$
(7) $EQUAL(prod(prod(x,y),z),prod(x,prod(y,z)))$
(8) $EQUAL(prod(x,sum(y,z)),sum(prod(x,y),prod(x,z)))$
(9) $EQUAL(prod(sum(y,z),x),sum(prod(y,x),prod(z,x)))$
(10) $EQUAL(sum(x,y),sum(y,x))$
(11) $\neg EQUAL(prod(minus(a),minus(b)),prod(a,b))$

Clause (11) is the only clause in the initial set of support, and clauses (2) through (9) are used as demodulators as well as axioms.

Table 1 contains the results of some initial experiments on these three problems. Negative hyperparamodulation is the sole inference rule used, and demodulation is applied to all deduced clauses. The unification counts do not include failed unification attempts or matches for demodulation or subsumption checks. "Length" is the number of steps in the proof, where one step is one negative hyperparamodulation inference, including demodulation of the result. The problems were run on a SUN 3/75 workstation.

Table 1. Experimental Results

Problem	Generated	Kept	Seconds	Unifications	Length
1. $xx=e$ groups	8	4	2	81	1
2. $xxx=e$ groups	405	115	193	5831	11
3. $-x-y=xy$ in rings	154	59	51	2088	4

To give some perspective to the results presented in Table 1, we close this section with some data gathered from related experiments. In particular, we discuss the results of attempting to prove the given three theorems when negative hyperparamodulation is replaced with either paramodulation or hyperparamodulation. Although the comparisons are in general very favorable to negative hyperparamodulation, we advise against concluding that this new inference rule is far superior to its positive relatives. The experiments are simply too few and too incomplete; this paper merely presents a preliminary study. In fact, the very nature of the field makes interpreting experiments at best somewhat suspect. The results can even be very misleading. The difficulty rests with the interconnection of representation, inference rule, and strategy. The comparisons provide only a hint of what may be true, but do suggest certain soft conjectures.

A glance at the data leads to the conjecture that the use of negative hyperparamodulation may remove some of the burden on the user of picking which demodulators to employ and deciding whether or not to use back demodulation. Of course, it is well known that back demodulation can be extremely useful, but it can also be too expensive for certain given problems. For all of the experiments cited in this paper, we chose to suppress the use of back demodulation. A second conjecture says that employment of negative hyperparamodulation avoids the need for such careful tuning of the various parameters that govern the program's reasoning; for example, it may not be crucial to choose the weights to reflect the user's insight. A third conjecture concerns the inclusion of auxiliary lemmas. For example, when negative hyperparamodulation is the sole rule of inference, the ring theory theorem was proved without the usual lemmas $x0=0$ and $0x=0$. But, as already commented, these conjectures are soft conjectures; far more data is required before any of them can be strongly considered to be true. What is obvious, however, is that the use of negative hyperparamodulation sharply increases the role of negative equality unit clauses, clauses that might otherwise play no active part in the search for a proof. Other than this obvious fact, as the following discussion shows, the situation is still too complicated for drawing many well-founded conclusions.

For all of the paramodulation and hyperparamodulation experiments, the set of demodulators was the same as in the corresponding negative hyperparamodulation experiment, but the set of support was modified.

In the $xx=e$ group problem, paramodulation produced a proof in 5 seconds, with 71 generated clauses, 11 kept clauses, 235 successful unifications, and a proof length of 4. Hyperparamodulation produced a proof in

137 seconds, with 1513 generated clauses, 49 kept clauses, 10,519 successful unifications, and a proof length of 3. For both of these experiments, the special hypothesis $xx=e$ was moved to the set of support.

For the $xxx=e$ group problem, no proofs were obtained using the default parameter settings with either paramodulation or hyperparamodulation. During each attempt, 2500 clauses were generated. Some tuning of the parameters enabled a proof to be found in 81 seconds, with 532 generated clauses, 69 kept clauses, 2002 successful unifications, and a proof length of 6. The set of support consisted of the special hypothesis $xxx=e$.

Finally, no proofs were found for the $-x-y=xy$ ring problem with either paramodulation or hyperparamodulation, regardless of whether or not the lemmas $x0=0$ and $0x=0$ were included. During each attempt, 2500 clauses were generated. The left distributivity axiom was moved to the set of support.

5. Summary

In this paper, we have introduced two new inference rules, *negative paramodulation* and *negative hyperparamodulation*. The actions of negative paramodulation complement those of (positive) paramodulation. Where (positive) paramodulation (which generalizes the usual notion of equality substitution) focuses on positive equality literals, negative paramodulation focuses on negative equality literals. Both rules permit an automated reasoning program to act directly on terms deep within some expression.

Negative hyperparamodulation was formulated to enable a reasoning program to (simultaneously) consider sets of clauses n at a time such that one of the clauses contains a negative equality literal, $n-2$ each contain a positive equality literal, and the nth (the *into clause*) contains a positive literal. We give the additional conditions that the into clause must satisfy to guarantee the *logical soundness* of the conclusions yielded by applying negative paramodulation and negative hyperparamodulation. The simultaneous consideration of clause sets containing more than two clauses permits an automated reasoning program to take larger and more significant deduction steps.

Since the most relevant measure of the value of an inference rule is its performance, we have included the results of some preliminary experiments. Each of the experiments relies solely on reasoning backward from the denial of the theorem to be proved, and employs negative hyperparamodulation as the only inference rule. Our intention is to experiment with bidirectional proof searches that simultaneously employ (positive) hyperparamodulation to reason forward and negative hyperparamodulation to reason backward. This experimentation will be conducted with both a uniprocessing and a parallel-processing reasoning program. One objective is to show that such bidirectional searches circumvent an obstacle that commonly occurs with an approach solely employing the positive variant of an inference rule for "building in" equality. With such a forward-reasoning approach, unfortunately, the denial of the theorem to be proved often plays a very passive role in the search. In fact, negative equality clauses often are considered only at the last step of the proof, that which completes the establishment of contradiction. With rules such as negative hyperparamodulation, on the other hand, negative equality clauses can play a very active role in the search for a proof.

Three properties of both negative paramodulation and negative hyperparamodulation deserve mention. First, by permitting the reasoning program to focus directly on terms deep within an expression, the potential power of the program is sharply increased. Being able to bypass so-called intermediate clauses and explore much smaller clause spaces is one of the goals of the field. Second, both rules "build in" cancellation. The importance of such a built-in treatment of cancellation is amply demonstrated by Stickel's excellent study of rings in which $x^3=x$ [3]. He gives his program access to cancellation by relying on appropriate demodulators, which is distinctly different from that which occurs with the two inference rules discussed in this paper. The third property concerns the attempt to measure the overall advancement of the fields of automated reasoning and automated theorem proving. Briefly, trivial theorems should be proved trivially. Little time or control should be required to find the proof of a simple theorem from mathematics. Our early experiments suggest that the use of negative hyperparamodulation permits the program to easily prove simple theorems. Both the extent to which this property holds and the value of the two introduced inference rules for attacking hard problems remains to be ascertained through future experimentation.

References

1. Lusk, E., McCune, W., and Overbeek, R., ''Logic Machine Architecture: Kernel Functions'', *Proceedings of the 6th Conference on Automated Deduction*, Springer-Verlag Lecture Notes in Computer Science (ed. D. Loveland), vol. 138, pp. 70-84 (June 1982).

2. Robinson, G., and Wos, L., ''Paramodulation and theorem proving in first-order theories with equality'', *Machine Intelligence* (ed. Meltzer and Michie), vol. 4, American Elsevier, New York, pp. 135-150 (1969).

3. Stickel, M., ''A case study of theorem proving by the Knuth-Bendix Method: discovering that $x^3=x$ implies ring commutativity'', *Proceedings of the 7th Conference on Automated Deduction*, Springer-Verlag Lecture Notes in Computer Science (ed. R. Shostak), vol. 170, pp. 248-258 (May 1984).

4. Wos, L., Overbeek R., and Henschen, L., ''Hyperparamodulation: A refinement of paramodulation'', *Proceedings of the 5th Conference on Automated Deduction*, Springer-Verlag Lecture Notes in Computer Science (eds. R. Kowalski and W. Bibel), vol. 87, pp. 208-219 (July 1980).

5. Wos, L., Overbeek, R., Lusk, E., and Boyle, J., *Automated Reasoning: Introduction and Applications*, Prentice-Hall, Englewood Cliffs, N.J. (Feb. 1984).

The Heuristics and Experimental Results of a New Hyperparamodulation: *HL*-resolution*

Younghwan Lim

Computer Technology Department
Electronics and Telecommunications Research Institute
P. O. Box 8
Daedog Danji, Chungnam
The Republic of KOREA

ABSTRACT

Equality is an important relation and many theorems can be easily symbolized through it's use. But it presents special strategic problems, both theoretical and practical, for theorem proving programs. A proposed inference rule in [6] called *HL*-resolution is intended to have the benefits of hyper steps while controlling the application of paramodulation. The rule is complete for E-unsatisfiable Horn sets. Here we try to find an efficient procedure for finding all *k-pd* links between two terms and try to find useful heuristics by making some experiments on ring theory. The linking process makes use of an equality graph which is constructed once at the beginning of the run. Once a pair of candidate terms for *HL*-resolution is chosen in the search, potential linkages can be found and tested for compatibility efficiently by looking at the paths in the graph. Further we try to find heuristics to prevent redundant *k-pd* links from being used to generate *HL*-resolvents. The method has been implemented on an existing theorem-proving system. A number of experiments were conducted on problems in abstract algebra and a comparison with set-of-support paramodulation was made.

1. Introduction

Equality is a very important relation, and many theorems can be easily symbolized through its use. Important research with respect to the equality relation has been carried out in several directions [2,3,4,5,7,10,11,13].

This work was supported, in part, by the research project 5EU230 of Electronics Telecommunications Research Institute.

In [5,6], we proposed a new hyperparamodulation called *HL*-resolution which generates a resolvent by building a paramodulation/demodulation link between two terms using a preprocessed plan as guide and we proved its completeness for E-unsatisfiable Horn sets. Here we try to find an efficient algorithm for finding all *k-pd* links between two terms. The linking process makes use of an equality graph which is constructed once at the beginning of the run. Once a pair of candidate terms for *HL*-resolution is chosen in the search, potential linkages can be found and tested for compatibility efficiently by looking at the paths in the graph. Further we try to find heuristics to prevent redundant *k-pd* links from being used to generate *HL*-resolvents and try to determine its practical value by experimenting with a variety of problems with the different choices of axiom sets, paramodulator sets and demodulator sets.

2. A New Hyperparamodulation: *HL*-resolution

A proposed rule is intended to have the benefits of hyper steps while controlling the use of paramodulation. The detailed description of the rule is given in [6].

Definition Let P be a set of paramodulators and D be a set of demodulators. A clause C' is called a *k-para/demod link (k-pd link)* of a clause C relative to $P \cup D$ if and only if there exists a sequence of clauses $A_0, A_1, \ldots, A_k$ such that

1) $A_0 = C$ and $A_k = C'$
2) A_i, for $0<i<k$, is a paramodulant or demodulant of A_{i-1} and a clause in $P \cup D$ under the restriction that the into-terms of A_{i-1} and A_i are from the same literal.
3) For each paramodulation/demodulation, the into term of A_i is not properly contained in the replacement of the into term of A_{i-1}.
4) If there is j such that $0<j<k$ and A_j is a demodulant of A_{j-1}, then each A_i, for all $j<i<k$, is a demodulant of A_{i-1}.

The k is called the *length* of the link. The definition implies that a clause is a 0-pd link of itself. Sets of equalities P and D need not be disjoint.

Definition A *partial unifier* of terms/literals t_1 and t_2 having the same function/predicate symbol is a substitution which unifies t_1 and t_2 from left to right, skipping over any pair of ununifiable arguments.

Definition A *function substitution link* of a clause of the form: $f(t_1,\ldots,t_n) \text{<>} f(s_1,\ldots,s_n) \vee A$ is a clause $D \vee A{*}E$, where A is a set of literals, E is a given substitution to be applied to $f(t_1,\ldots,t_n)$ and $f(s_1,\ldots,s_n)$, and D is a disjunction of inequalities formed by the pairs of arguments not unified in $f(t_1,\ldots,t_n){*}E$ and $f(s_1,\ldots,s_n){*}E$.

Definition A *predicate substitution link* of a pair of clauses of the form $P(t_1,\ldots,t_n) \vee A$ and $-P(s_1,\ldots,s_n) \vee B$, where A and B are sets of literals, is $D \vee A{*}E \vee B{*}E$, where E is a substitution to be applied to $P(t_1,\ldots,t_n)$ and $P(s_1,\ldots,s_n)$, and D is a disjunction of inequalities formed by the pairs of arguments not unified in $P(t_1,\ldots,t_n){*}E$ and $P(s_1,\ldots,s_n){*}E$.

The role of the substitution link is to simplify a clause by stripping off the outer function symbol of one of its literals. The soundness of rules of inferences which generate fuction/predicate substitution links can be derived directly by the use of the function/predicate substitution axioms. In our equality-reasoning system, the above two rules of inference will replace the use of the function and predicate substitution axioms. This, in effect, restricts the use of those axioms by not allowing the generation of clauses corresponding to arbitrary resolutions from substitution axioms. However, unlike previous attempts in this direction (e.g., [3,4,8]), a system will be proposed in which the rules themselves will be used in a very restricted way, further cutting down on the number of clauses they are allowed to generate.

Now here is the definition of a new inference rule called HL(Henschen-Lim)-resolution.

Definition Let S be a set of clauses, P be a set of paramodulators and D be a set of demodulators. Let N be the transitivity clause $\{x \lt\gt y \quad y \lt\gt z \quad x=z\}$ in the equality axioms, A_1 be a positive unit equality clause in S, and $A_2 v B$ be a clause in S, where A_2 is a negative equality literal and B is the set of the remaining literals. Let the variables in these clauses all be separated. Suppose that the set of clauses $\{A_1, A_2 v B, N\}$ satisfies one of the following conditions:

1) (forward) There exists a most general unifier(MGU) E_1 of A_1 and the literal $x \lt\gt y$ and a MGU E_2 of A_2 and L_1', where L_1' is a *k-pd* link of $\{x=z\}*E_1$ relative to $P \cup D$ with a k-linked unifier $k\text{-}E$.

2) (backward) There exists a MGU E_1 of A_2 and the literal $x=z$ and a MGU E_2 of A_1 and L_2', where L_2' is a *k-pd* link of $x \lt\gt y*E_1$ relative to $P \cup D$ with a k-linked unifier $k\text{-}E$.

Then the clause $(y \lt\gt z \text{ v } B)*E_1*E_2*k\text{-}E$ or its function substitution link on the descendent of $y \lt\gt z$ is called an *HL*-resolvent of the set.

The terms of an (in)equality in A_1, A_2, P or D are allowed to be flipped if necessary to match. In particular, paramodulation proceeds from either side of any equality in P. Further this definition can be extended in such a way that A_1 is an arbitrary clause in S containing a positive equality literal. We can also allow the equality literal in A_2 to be positive and link to $y \lt\gt z$ in N, or A_1 to link to $y \lt\gt z$, etc.

Example Consider the E-unsatisfiable set of clauses:

```
1. k<>g(a)          2. f(h(b),c)=a
3. d=h(b)           4. e=c
5. k=l              6. i=f(d,e)
7. l=g(i)
```

```
1)      x<>y          y<>z               x=z
                                          | res.
                                      [1.k<>g(a)]
        k<>y          y<>g(a)
        |____ 5.k=l
        l<>y          y<>g(a)
          res.
      [7.l=g(i)]
                  CL1: g(i)<>g(a)
                            |  function substitution
                     CL2: i<>a
```

```
2)      x<>y         y<>z         x=z
         |  res.
      [6.f(d,e)=i]
                     i<>z       f(d,e)=z
                                  |____ 2.d=h(b)
                     i<>z     f(h(b),e)=z
                                        |____ 4.e=c
                     i<>z     f(h(b),c)=z
                                      |____ 2.f(h(b),c)=a
                     i<>z           a=z
                                     | res.
                                [CL2: a<>i]
                     i<>i
                      |  unit conflict
                     proof
```

If S is an E-unsatisfiable Horn set of clauses including $x=x$ and the functional reflexive axioms FR, then S has an HL–refutation with the choice that the paramodulator set consists of the positive, unit equalities in S and positive, unit equality HL–resolvents. Its proof is given in [5].

While completeness is important to know about, it is more important in our view to develop effective proof procedures. Clearly a major part of HL–resolution is to determine if there is one or more *k–pd* links between the chosen target terms. In order to develop a mechanism to help find linkages efficiently, the previous property is important. Further, the generation of new positive equality clauses is very restrictive. Whether or not this is an advantage is controversial in view of the use of new demodulators suggested in [9]. All these properties suggest the use of an equality graph to aid in finding links, very much like regular connection graphs are used in finding resolutions.

3. Planning for the *k-pd* linking process

The basic idea is to form a graph at the beginning of the run in which terms that could potentially paramodulate at the outer level are connected and the corresponding unifiers are formed. Then two candidate terms for *HL*-resolution can be attached to the graph. Paths of length less than the bound for k which connect both the outer terms and inner terms can then be easily found and the corresponding set of unifiers tested for compatibility. Whenever a new positive equality unit is generated, it is included into the equality graph to be used as paramodulator. Now here are some definitions leading up to such a graph mechanism and the formation of *HL*-resolution plans for a pair of terms.

Definition Let P be a set of positive unit equality clauses. An equality graph (*EG*) is a graph such that

1) To every left or right term of equality, there corresponds a node whose label is the term.
2) Two nodes are connected if their terms are unifiable after renaming variables so that different clauses contain different variables. The most general unifier is the label of the edge.
3) Nodes corresponding to terms which belong to the same equality clause are grouped together in the graph. The clause number is the lable of the group.

Definition Let the linking path between two terms s_1 and t_{n+1} be

```
       E1         E2          E3        En
s1 -- t2=s2 -- t3=s3 -- ...  -- tn+1
```

where the variables in all clauses have been separated and E_i, $1 \le i \le n$, is a MGU of s_i and t_{i+1}. If $E=E_1*E_2*...*E_n$ is defined, where * is the operation of compatible composition, then the linking path is said to be *link compatible* and E is called a link compatible unifier.

Definition An *augmented equality graph* of a term t for a term s, denoted by $AEG(t,s)$, is an equality graph as above with the two extra groups of nodes $\{t\}$ and $\{s\}$, where,

1) All nodes in *EG* whose labels are unifiable with t or s are connected to t or s, respectively with a labeled link labeling the unifier.
2) All nodes in *EG* whose labels have the same outer function symbol as t but are not unifiable with t are connected to t with an unlabeled link.

The labeled link is used for checking compatibility in finding a paramodulation sequence and the unlabeled link is for finding inner level target terms, which will be described below.

Rather than trying to build *k-pd* links of a term t which are to be resolved with a term s in an ad hoc way, a systematic method like target-driven search can be devised using the restrictions on the *k-pd* link. Since a *k-pd* link has the restriction of into-term containment, any into term should not be contained in any proceeding into-term. This restriction allows an inner level linking process to occur only between the two end terms.

Suppose we try to find all k-pd links of a term t which are to be resolved with.

1) case 1: t is a constant or variable

In this case, the linking process is simple due to the into-term-containment restriction. In fact, all the position vectors of the into-terms in a sequence are the same, i.e., the outer level position vector. The set of outer level links is the set of *k–pd* links of the term t to s.

2) case 2: t is complex term

Let t be a complex term $f(t_1,t_2,\ldots,t_n)$. There will in general be many into term candidates at the beginning of the linking process. Furthermore, it seems difficult to know when to terminate an inner level linking sequence. We propose a method called target-driven search, that works backwards from s rather than forwards from t.

Step 1) Try to find outer level links of s which are either unifiable with t or have the same function symbol as t and are of length no greater than the bound on k.

Step 2) Let s' be one of the outer level links of s.

Subcase 1): s' is unifiable with $f(t_1,t_2,\ldots,t_n)$. Then, by the compatibility of unifiers on the *k–pd* link, there exists a *k–pd* link of t on the path t—s'—...—s. Therefore we have found a *k–pd* link.

Subcase 2): s' is not unifiable with t but has the same function symbol f. Let s' be of the form $f(s_1,s_2,\ldots,s_n)$. Now we can break down the linking process into sub-linking processes of finding *k–pd* links of t_1 to s_1, t_2 to s_2, ..., and t_n to s_n. The sum of the lengths of these links and the length of the link from s to s' must be bounded, which narrows the search considerably. Assume that all sub-link paths with compatible compositions E_1, E_2, ..., E_n respectively are found. If E_1, E_2, and E_n are compatible, then $f(s_1',s_2',\ldots,s_n')$, where s_i', $1 \le i \le n$, are *k–pd* links of s_i, is checked to see if it is unifiable with t. If they are unifiable, there exists a *k–pd* link of t on the path $f(t_1,t_2,\ldots,t_n)$--$f(s_1',s_2',\ldots,s_n')$--...--$f(s_1,s_2,\ldots,s_n)$--...--s.

Note that if s' is neither of those two cases, there is no link between t and s' using the path from t to s' because of the restriction of into-term containment. Detailed descriptions can be found in [5].

Here we discuss additional heuristic restrictions on *k–pd* links. We would not recommend using FR(Functional Reflexive) axioms. We have seen that for some cases where FR is required in simple paramodulation, *HL*–resolution proofs exist without FR because we are allowed to link either forward or backward. Whether *HL*–resolution is complete without FR (perhaps with some other restrictions relaxed) remains open. However, in practice we do not recommend using them. We also impose the standard restrictions of no paramodulaton from or into variables. This is given as an option in our program but in none of the experiments was paramodulation from or into variables allowed. Further, the use of the same paramodulator more than once at any term position in a *k–pd* link sequence is excluded. Again it is an open problem as to how these restrictions effect completeness.

4. Heuristics

Basically the *HL*-refutation procedure consists of generating *HL*-resolvents and processing the clauses for retention. List 1 contains the supported input clauses and list 2 contains the non-supported input clauses. Paramodulators and demodulators are kept in list 3 and list 4 respectively. List 6 is the pool of clauses waiting to be chosen as the given clause. List 5 contains weight templates for choosing a given clause from list 6. During the *HL*-refutation runs, list 1 contains the clauses that have already been used as the given clause.

Algorithm

1. Input all the clauses, set up the equality graph and read in the heuristics and options.
2. Choose a clause for the given clause from list 6 and move it to list 1.
3. Generate *HL*-resolvents of the given clause and clauses in list 2 with respect to the paramodulators and demodulators in list 3 and list 4 respectively.
4. Process each *HL*-resolvent as it is inferred and assign it to list 6 if necessary.
5. If a proof is found or no clause has been generated at step 3 or the time/space limits have been exceeded, then stop. Otherwise, go to step 2 for next given clause.

The algorithm loops until either a proof is found during the process of step 3 or some predefined limit on the number of generated clauses is reached. For the efficiency, we may utilize the following heuristics for each step.

4.1. Heuristics for the *HL*-resolution Process

Let $t_1<>t_2$ and $s_1=s_2$ be parent clauses to be resolved, and P and D be the sets of paramodulators and demodulators respectively.

Lemma Given the two clauses $s_1=s_2$ and $t_1<>t_2$, consider the following forward *HL*-resolution;

```
    u<>v          v<>w          u=w
      | res.
  [ s1=s2 ]     s2<>w          s1=w
                                 |
                                 .  a linked unifier kE
                                 .
                                 |
               s2'<>w          s1'=w
                                 | resolve with a MGU E
        (s2'<>t2)*E          [t1<>t2]
```

If s_2' can be demodulated to s_1' then $(s_2'<>t_2)*E$ is also demodulated to a clause which is subsumed by the clause $t_1<>t_2$.

(Proof) Since s_2' can be demodulated to s_1', the term $s_2'*E$ can be demodulated to the term $s_1'*E$ because $s_2'*E$ is an instance of s_2' and the same demodulators from s_2' to s_1' can be applied. Hence, $(s_2'<>t_2)*E$ can be demodulated to $(s_1'<>t_2)*E$. On the other hand, since the two terms s_1' and t_1 are unifiable with the MGU E, $s_1'*E$ is equal to t_1*E. That means that $(s_1'<>t_2)*E$ is equal to

$(t_1<>t_2)*E$ which is subsumed by $t_1<>t_2$. Q.E.D.

1) Filtering *k–pd* links

Not all generated *k–pd* links are used to generate *HL*–resolvents. Using the properties of the above Lemma, if s_2' can be simplified to s_1' in forming forward *HL*–resolvents, then do not generate the *HL*–resolvent unless it has a possibility of generating a function substitution link.

Example Given a set of clauses

```
1. f(e,x)=x
2. f(g(x),x)=e
3. f(x,x)=x
4. f(f(x,y),z)=f(x,f(y,z))
5. f(a,e)<>a
```

Assume that the clauses 1,2, and 3 are demodulators. Consider the following *HL*–resolution;

```
   u<>v                 v<>w                    u=w
     | res.
 [cl. 4]     f(x,f(y,z))<>w        f(f(x,y),z)=w
                                          <- 2. f(e,x')=x'
               f(e,f(y,z))<>w        f(y,z)=w
                                          <-2. f(e,x')=x'
               f(e,f(e,z))<>w           z=w
                                          | res.
               f(e,f(e,f(a,e)))<>a   [cl. 5]
```

Since $f(e,f(e,z))$ can be demodulated to z by 2 rewrites of the demodulator $f(e,x)=x$,$f(e,f(e,f(a,e)))<>a$ will be demodulated to $f(a,e)<>a$. So we need not use the *k–pd* link in this example to generate the *HL*–resolvent $f(e,f(e,f(a,e)))<>a$ because it will be demodulated to a clause which is subsumed by an existing clause $f(a,e)<>a$.

2) Deletion of a reversed demodulation

If the clause $s_1=s_2$ is a demodulator in *D* and the term s_2 is a given term which is linked backward to one of t_1 or t_2 with *k–pd* link such that

i) All paramodulators $r_i=r_i'$, $0 \leq i \leq k$, are also demodulators and all the from terms are r_i',

ii) All the into terms are the outer level terms only.

Then the *HL*–resolvent is, always, demodulated to a clause which is subsumed by the clause $t_1<>t_2$. But the *HL*–resolvent may have a function substitution link which is on the way to a proof. Hence it is recommended not to generate this type of *HL*–resolvent unless it has a function substitution link.

Example Given sets of clauses and demodulators the same as the sets in the previous example, respectively, consider the following HL-resolution;

```
            u<>v         v<>w               u=w
                                             | res.
            a<>v         v<>f(a,e)  [5. a<>f(a,e)]
             ⊳-1. x=f(e,x)
         f(e,a)<>v      v<>f(a,e)
             | res.
      [3. x=f(x,x)]    f(f(e,a),f(e,a))<>f(a,e)
```

Since clauses 1 and 3 are also demodulators, the resolvent $f(f(e,a),f(e,a))<>f(a,e)$ is demodulated to $a<>f(a,e)$ which is subsumed by $a<>f(a,e)$. So it seems that we need not generate the *HL*-resolvent. But it has the function substitution link $f(e,a)<>a \vee f(e,a)<>e$, so we do generate this *HL*-resolvent.

3) History checking

If the clause $s_1=s_2$ itself is one of parents of the clause $t_1<>t_2$ with the *k-pd* link such that

i) All the paramodulators $d_i=d_i'$, $0 \le i \le k$, are also demodulators whose from terms are d_i,

ii) If $r_1<>r_2$ is the other parent of $t_1<>t_2$, then s_1 is linked to r_1 in forward *HL*-resolution.

Then if t_1 and s_2 are linked with a 0-pd link, the *HL*-resolvent will be demodulated to a clause which is subsumed by the clause $r_1<>r_2$. Therefore it is desirable not to generate this kind of *HL*-resolvents unless they have a function substitution link.

Example Given set of clauses as previous example, consider the following *HL*-resolutions;

```
      u<>v              v<>w                v=w
       | res.
     [cl. 4]       f(f(x,y),z)<>w       f(x,f(y,z))=w
                                              ⊳-2. f(g(x'),x')=e
                 f(f(x,g(z)),z)<>w         f(x,e)=w
                                              | res.
         cl 6: f(f(a,g(z)),z)<>a          [f(a,e)<>a]


          u<>v            v<>w                 v=w
                                                | res.
  f(f(a,g(z)),z)<>v      v<>a         [cl.6 f(f(a,g(z)),z)<>a]
           | res.
       [cl.4]       cl.7: f(a,f(g(z),z))<>a
```

Since clause 4 is one of parent clauses of the generated clause 6 and the clause 4 is, again, linked to the clause 6 backward, the generated clause 7, $f(a,f(g(z),z))<>a$ is demodulated to $f(a,e)<>a$ using the demodulator $f(g(x),x)=e$. So we need not generate this *HL*-resolvent because it does not have a function substitution link.

4). Heuristics for Substitution Links

(1) The level of substitutions: A problem of substitution links is how many levels down into the

structure of the two terms the substitution links should be generated. It is given as an option whether substitution links generate the first level disjunction of inequalties of the deepest level disjunction. By default, it generates the deepest substitution links.

(2) The maximum number of literals : The maximum number of literals generated by a substitution link may be provided by the user as an option The default is one.

(3) Chossing a substitution unifier : The substitution unifier may be partial unifier. The default is a partial unifier.

4.2. Heuristics for *HL*-refutation

1) Choosing sets of axioms, paramodulators, demodulators and supported clauses

A general theory for this problem is not found yet. For the set of supported clauses, the denial of theorem may be used. All positive unit equalities in *S* may be used as paramodualtors. Normally, a positive unit equality t=s in *S* such that weight(t)-weight(s)>threshold should be used as a demodulator. For axiom clauses, usually relatively complex clauses are used.

2) Ordering for choosing a given clause

Because, in general, many clauses are generated between two clauses, choosing the next given clause at step 2 is very important for efficiency. Neither breadth search nor depth first search is effective for general problems. Instead, the heuristic of choosing the simplest clause using the given weighting templates in list 5 is used. If any weighting template is not provided, the complexity of a literal is the number of constant, variable, function, predicate and negation symbols it contains. The complexity of a clause is the maximum of the complexity of its literals. If there is more than one clause with the minimum complexity, then the first generated clause is used.

3) Handling positive unit equalities generated

A positive unit equality clause may be used as a paramodulator by including it in the equality graph. Further it can be used as a demodulator if the weight difference of the two arguments is greater than a threshold.

4.3. Processing generated clauses

In addition to the regular processing procedures such as tautology, demodulation, subsumption, etc, the following clause processing procedures may be used to process a generated clause.

1) Demodftsublink

If the generated clause can be rewritten to an existing clause, it is not to be kept except in the case that it has a function substitution link, in which case only the substitution link is kept.

2) Outersymbol

Delete the generated clause if either one of its outer symbols does not occur as an outer symbol in an input equality clause.

5. Experiments and Results

The problems we experiment with are taken from ring theory because they are standard problems represented in equality relation and its proof requires non-unit resolvents. The primary purpose is to compare *HL*-resolution with s et-of-support(SOS) paramodulation because it is believed that SOS paramodulation is one of the best strategies for handling the equality relation. The second purpose is to see how the different choices of axiom sets, paramodulator sets and demodulator sets effect the performance and what choices are best. The third purpose is to see what the effect of demodulation on both strategies is. And the last one is to understand the relationship between *HL*-resolution and SOS paramodulation. For *HL*-resolution, we start experimentation with the choice of axiom set, paramodulator set and demodulator set as general as we can. Then we tried to reduce the number of clauses in those sets while a proof was found within a bound on a number of generated clauses. For SOS paramodulation, we would choose one or two positive clauses as supported clauses based on McCune's experimental results[7].

5.1. Ring Theory Experiments

Legend: In the following report of experiments,the experiment **h** and **p** mean *HL*-resolution experiment and set-of-support paramodulation experiment, respectively. Further, A, P, D, S, and N represent Axiom set, Paramodulator set, Demodulator set, Supported clause set and Non-supported clause set, respectively. Here the axiom set is the set of clauses which can be used as satellite clauses.

Set of input clauses

```
1. s(x,y)=s(y,x)            2. s(s(x,y),z)=s(x,s(y,z))
3. s(x,0)=x                 4. s(0,x)=x
5. s(x,i(x))=0              6. s(i(x),x)=0
7. p(x,p(y,z))=p(p(x,y),z)
8. s(p(x,y),p(x,z))=p(x,s(y,z))
9. s(p(y,x),p(z,x))=p(s(y,z),x)
```

Problems to prove

```
10. p(a,0) <> 0             ;(Vx) p(x,0)=0
11. p(a,i(b))<>i(p(a,b))    ;(VxVy) p(x,i(y))=i(p(x,y))
```

The following problems are Boolean ring problems.

```
12. p(x,x)=x                  ;(Vx) p(x,x)=x
13.  s(a,a)<>0                ;   --> (Vy) s(y,y)=0
14. p(x,x)=x                  ;(Vx) p(x,x)=x
15. p(a,b)<>p(b,a)            ;   --> (VyVz) p(y,z)=p(z,y)
16. s(x,x)=0
```

Experiments

```
r1hd1: A:1-9,10              P:3,5              D:3-6
r1hd2: A:2,5,8,10            P:1,3,5,7,8,9      D:3,4,5,6,8,9
r1pd1: S:2,8                 N:1,3-7,9,10       D:3,4,5,6,8,9
r2hd1: A:2-9,11              P:1-9              D:2-9
r2hd2: A:2,4,5,8,11          P:1,3,5,7,8,9      D:3,4,5,6,8,9
r2pd1: S:2,8                 N:1,3-7,9 11       D:3,4,5,6,8,9
r3hd1: A:2,6,8,13            P:4,6,9,12         D:3-9,12
r3pd1: S:2,8                 N:1,3-7,9,12,13    D:3-6,8,9,12
r4hd1: A:1,2,5,9,14,15,16    P:2,3,5,8,16       D:3,4,5,6,14,16
```

rihj and ripj are similar to the experiments rihdj and ripdj, respectively, except that no demodulation is applied to inferred clauses. For the experiment r4hd1, the next-have-been clause was selected by the user

Results

	proof found	HL-res.	para-mod.	kept	gen. time	proc. time
r1hd1	yes	114	(310)	27	124	68
r1h1	yes	114	(310)	27	125	64
r1hd2	yes	19	(68)	8	27	11
r1h2	yes	19	(68)	10	27	11
r1pd1	no	na	600	112	119	206
r1p1	no	na	600	152	188	170
r2hd1	yes	84	(432)	31	289	66
r2hd2	yes	38	(143)	21	76	24
r2h2	no	240	(794)	146	1050	326
r2pd1	no	na	600	112	119	204
r2p1	no	na	600	152	191	169
r3hd1	yes	197	(605)	98	543	446
r3h1	yes	200	(598)	116	670	538
r3pd1	no	na	600	112	117	213
r3p1	no	na	600	152	187	170
r4hd1	yes	229	(520)	93	374	217

5.2. Comments

In the 16 ring theory experiments (6 *HL*-resolution experiments with demodulation, 4 *HL*-resolution experiments without demodulation, 3 SOS paramodulation experiments with demodulation and 3 SOS paramodulation experiments without demodulation), none of SOS paramodulation experiments under the same options specified succeeded in finding a proof within the bound on the number of generated clauses. In fact, we tried SOS paramodulation experiments with different choices of supported clause(s), but we failed to find a proof. The reasons may be that those problems require paramodulation from or into variables, the bound 600 on the number of generated clauses is too small or the restricted use of the symmetry clause may block a proof. At the same time we tried those experiments under the different options, but we could not find a proof because the relaxed options for SOS paramodulation experiments leads to an explosion of the search space. As a usual, we started *HL*-resolution experiments with the choices of axiom sets, paramodulator sets and demodulator sets as general as possible. With those choices, only the experiments r1hld1, r1hl1 and r2hld1 succeeded in finding a proof, but all the others failed. So we tried *HL*-resolution experiments with a reduced number of clauses in those sets. As a result, all the *HL*-resolution experiments with demodulation found a proof. In these experiments, the importance of demodulation was confirmed again. And it was demonstrated that the choices of axoim sets, paramodulator sets and demodulator sets have a significant impact on the relative performance. An interesting observation from the experiments r3 and r4 is that paramodulation from or into variables may not be a severe problem in *HL*-resolution because it is allowed only from or into an outer level term. Although, in the linking process, a paramodulation from or into variables is not allowed, *HL*-resolution may generate an *HL*-resolvent using a 0-pd link. In the case that one of from-term or to-term happens to be a variable, the *HL*-revolvent is, in fact, a paramodulant generated by paramodulation from or into a variable. In all the *HL*-resolution experiments on problem r4, non-unit *HL*-resolvents have to be generated. Further they require the generation of top most function substitution links. The experiment r4 indicates that a good heuristic strategy for choosing the next have-been-given clause from the pool of generated clauses is necessary.

6. Conclusion

We proposed a new inference rule called *HL*-resolution for equality relations that is intended to have the benefits of hyper steps and to control the uses of paramodulation. It generates a resolvent by building a paramodulation and demodulation link between two terms using a preprocessed plan as a guide. We suggested an efficient method and heuristics for implementation. The method was implemented on an existing automated theorem-proving system, and a number of experiments were conducted to compare the performance of *HL*-resolution and set-of-support paramodulation. In most cases, *HL*-resolution experiments search a larger space and keep a smaller generated clauses but take a longer time than SOS paramodulation.

But many problems remain untouched. Completeness without function reflexive axioms or with some other restrictions relaxed remains open. And we do not have a theory as to how to restrict the choice of the sets of paramodulators and demodulators and still maintain completeness. But we believe this research deservesto do further investigations.

7. Reference

1) Blasius, K. H., "Equality Reasoning in Clause Graph," IJCAI, 1983.

2) Darlington, J. L., "Automated Theorem Proving with Equality Substitutions and Mathematical Induction," in Machine Intelligence, Vol. 3 (B. Meltzer and D. Michie, eds.), American Elsevier, New York, pp. 113-127.

3) Digricoli, V. J., "Resolution by unification and Equlaity," Proceedings of 4th Workshop on Automated Deduction, 1979.

4) Harrison, M. and N. Rubin, "Another Generalization of Resolution," J. ACM, Vol. 25, No. 3, July 1978, pp.341-351.

5) Lim, Younghwan, A New Hyperparamodulation Strategy for the Equality Clauses, Ph. D. Dissertation at Northwestern University, 1985.

6) Younghwan Lim and Lawrence J. Henschen, "A New Hyperparamodulation Strategy for the Equality Relation," IJCAI 85, pp. 1138-1145.

7) McCune, W., Semantic Paramodulation for Horn Sets, Ph.D. Sissertation at Northwestern University, 1984.

8) Morris, J., "E-resolution: an Extension of Resolution to Include the Equality Relation," IJCAI, 1969.

9) Overbeek, R., J. McCharen and L. Wos, "Complexity and Related Enhancements for Atomated Theorem-proving Program," Comp. and Maths. with Appls., Vol. 2, No. 1-A, 1976, pp.1-16.

10) Robinson, G. A. and L. Wos, "Paramodulaiton and Theorem Proving in First Order Theories with Equality, " in Machine Intelligence Vol. 4 (B. Meltzer and D. Michie, eds) Amerian Elsever, New York, 1969, pp.135-150.

11) Siekmann, J. and G. Wrightson, "Paramoulated Connection Graphs," Acta Informatica, 1980

12) Wos, L., "Paramodulation and Set of Support," the IRIA Symposium on Automatic Demonstration ar Versailles, 1968.

13) Wos, L., R. Overbeek and L. Henschen, "HYPERPARAMODULATION: A Refinement of Paramodulation, "Proceedings of the 5th Conference on Automated Deduction, 1980, pp.208-219.

ECR: An Equality Conditional Resolution Proof Procedure

Tie Cheng Wang
Automatic Theorem Proving Project
The University of Texas at Austin
Austin, Texas 78712

Abstract

This paper presents an equality conditional resolution proof procedure, ECR, that incorporates a user's knowledge concerning the different roles of input equations in a proof. The input equations are separated into different classes according to the roles they will play. Each such role has rule schema into which the corresponding equations are transformed. The conditions on the application of these rules control the inference, and prevent inappropriate use of the equations. The paper will introduce the concept of potential completeness to characterize the generality of a deduction method. ECR is potentially complete for proving the set of positive Horn theorems, that is, it can finally prove any of these theorems by first proving the needed lemmas, and then using them. This procedure has been used to prove a number of theorems of group theory, ring theory, boolean algebra and field theory. The paper will give a summary of these proofs, among which, the efficient proofs of the associativity law and De.Morgan's law of boolean algebra seem interesting.

1. Introduction

The equality conditional resolution (ECR) to be presented in this paper is a goal-oriented resolution-style proof procedure. It incorporates the user's knowledge concerning the different roles of input equations in a proof to control the inference on them.

In proving a theorem by ECR, the user is responsible for helping the procedure to construct a knowledge base, by classifying the input equations and other hypotheses into a number of distinct subsets of rules. According to this classification, each input equation will be transformed into some particularly formulated reduction rules (unit clauses) and/or "if-then" rules (multi-literal clauses with exactly one positive literal in each of them). A condition on the application of an if-then rule will be set up by assigning marks to a subset of hypothesis literals of this rule. These marks are defined according to marking-assignment rules, or suggested by users according to their experience.

By means of this knowledge base, ECR will control the inference on the input equations according to their formulations and rule types. Those that are represented as if-then rules will be used for "term rewrite" or "demodulation" according to their respective types. But no consequence may be deduced from any of them unless a subset of the (instantiated) literals of this rule can be unified by some reduction rules. These literals are identified by their attached marks.

The experiments with ECR show that it usually proves a theorem quite efficiently, though may fail to obtain a proof if some important lemmas have not been supplied. But, it is capable of finally proving any positive Horn theorem*, by first proving the useful lemmas and then using them. This feature is called potential completeness.

Though completeness is important to the generality of a mechanical prover, human deduction seems to be incomplete: it usually explores a small search space before obtaining or abandoning a proof. On the other hand, the human (at least as a whole and in the long run) seems to be potentially complete theorem prover in the sense that a hard theorem will be eventually proved when sufficient knowledge relevant to a proof of the unproven theorem is discovered.

Formally speaking, we say a deduction method is potentially complete for a particular theory, if for any theorem T of this theory:

> Either it can prove T, or there is a finite sequence of lemmas, L1, ..., Ln, for $0<n$, such that it can prove L1 with the hypotheses of T, and it can prove Lj , for $1<j\leq n$, if the proceeding lemmas, L1, ..., Lj-1, are added to the hypotheses of T, and finally, it can prove T, by adding the lemmas, L1, ..., Ln.

Guessing the intermediate lemmas needed for a proof is an important ability of mathematicians. This ability is expected to be simulated by a computer in future. In this report,

---

*If every clause of a set S contains at most one positive literal, then S is called a positive Horn set. A theorem T is called a positive Horn theorem, if the set S of clauses obtained from the negation of T is a positive Horn set and S is unsatisfiable.

we suppose that the user will supply the needed lemmas, so we can concentrate on designing a potentially complete deductive component.

There exist trivial solutions. For example, linear resolution with a fixed search depth limit will be a potentially complete deduction method of first order theory. Limiting the size of a search though useful, is weak. ECR will employ a more effective control strategy: restricting the inference on if-then rules.

ECR is based on a basic conditional resolution (BCR). We first define BCR, then discuss the construction of a knowledge base for a theorem with equality relations. Following these, we present the ECR proof procedure. This procedure has been used to prove a number of theorems of group theory, ring theory, boolean algebra and field theory. A summary of these proofs will be given in a later section, among which, the efficient proofs of the associativity law and De.Morgan's law of Boolean algebra obtained by the prover seem interesting.

2. A Basic Conditional Resolution (BCR)

A control strategy often used by some natural provers [3,12] is to restrict the use of some particular if-then rules. For example, given a transitivity law, $x<y \wedge y<z \rightarrow x<z$, and a subgoal $a<b$, the consequence $a<y \wedge y < b$ is not allowed to be derived from them by back-chaining, unless there are already established some facts related to the constant a, or b, such as $a<c$, then $c<b$ is allowed to derive. This control strategy sounds quite natural, and is often useful to the efficiency of a mechanical prover. Of course, it can be applied to restrict the inference on other if-then rules. The BCR to be defined in this section is a modification of input resolution incorporating this strategy.

In this paper, unless mentioned otherwise, the symbols x, y, z, u, v, w, x1, ..., w1, represent universally quantified variables, t, t1, ..., s, s1, ..., represent terms, a, b, c, d, e, k, i, f, +, *, represent constants or function symbols. A clause is a sequence of literals. Let C_1 and C_2 be two clauses. Then $C_1 \vee C_2$ is the clause obtained by concatenating C_1 to the left of C_2, $C_1 - C_2$ is the clause obtained by deleting each element from C_1 that is an element of C_2. The empty clause is called "box".

Let S be a set of clauses. Some literals of the clauses in S may be marked with the integer 1. The marks are used to indicate some constraints applied to the unification related to these marked literals. The ordinary resolvents will be treated by BCR as intermediate resolvents

(IBC-resolvents).

IBC-resolvent. Let G and C be two clauses with no variable in common. Let G be $L_g \vee G_1$, and L_c be a unmarked literal of C. If $\sim L_g$ and L_c are unifiable, with a mgu θ, then the clause, $(C - L_c)\theta \vee G_1\theta$ is called an **IBC-resolvent** of G against C.

The BC-resolvent defined next is obtained from an IBC-resolvent by reducing away all marked literals of it by means of unification with input unit clauses.

BC-resolvent. Let R be an IBC-resolvent of G against C. Let R_0 be a clause obtained from R, $R_0 = \{L: L \in R, L \text{ is marked by } 1\}$.** Let E be a set of unit clauses of S. For each input-refutation D of the set $E \cup \{R_0\}$ with R_0 as the starting goal, such that for each goal clause R' of D, the left-most literal of R' is resolved upon, let θ be the most general substitution used by D, then $(R - R_0)\theta$ is called a BC-resolvent of G against C within S.

A basic conditional resolution (BCR). A clause G_n is said to be deduced from G_0 by BCR within a set S iff there is a deduction, $G_0, G_1,..,G_n$, such that, G_0 is a clause of S, and for each i, for $1 \leq i \leq n$, G_i is a BC-resolvent of G_{i-1} against a clause C of S. If G_n is "box", then this deduction is called a BCR-refutation of S.

Example 2.1. Use BCR to prove the following theorem: In a group, if x+x=e, then the group is commutative. The following set $S_{2.1}$ is a set of clauses obtained from this (negated) theorem:

```
 1. i(x)+x=e                            2. x+i(x)=e
 3. e+x=x                               4. x+e=x
 5. x+x=e                               6. a+b=c
 7. x+v=w  x+y≠u¹  u+z≠w¹  y+z≠v                       ; associativity law
 8. u+z=w  y+z≠v¹  x+v≠w¹  x+y≠u                       ; associativity law
 9. x=x
10. y=x  x≠y¹
11. x=z  x≠y¹  y≠z¹
12. x1+x2=y1+x2  x1≠y1¹  x2≠y2¹
13. b+a≠c                                              ; starting goal
```

Though it appears to be a simple problem, a very large search space needs to be explored by an ordinary goal-oriented deduction procedure. For example, starting from goal clause 13, there are four branches: (temporarily, let us forget the marks.)

** By this notation, we mean that R_0 is a clause consisting of the marked literals of R, and the order of these literals in R_0 is the same as the order of their corresponding literals in R.

```
14-1. b+y≠u¹ u+z≠c¹ y+z≠a          [13,7]
15-1. y+a≠v¹ x+v≠c¹ x+y≠b          [13,8]
16-1. c≠b+a¹                       [13,10]
17-1. b+a≠y¹ y≠c¹                  [13,11]
```

Among them, the resolvents 14-1 and 15-1 will finally prove to be on the optimized proof paths. From each of them, 9 different resolvents can be generated.

Now, let us use BCR to prove this theorem. By BCR, the above 4 resolvents are all intermediate resolvents. From them, only four BC-resolvents can be produced after unifying the marked literals with some input unit clauses. For example, the BC-resolvent 14 was produced from 14-1 in the following process:

```
14-1. b+y≠u¹ u+z≠c¹ y+z≠a          [13,7]
14-2. e+z≠c¹ b+z≠a                 [14-1,5]
14.   b+c≠a                        [14-2,3]
```

An optimized BCR-refutation of this example consists of the following 4 BC-resolvents.

```
14. b+c≠a      [13,7,5,3]      18. a+c≠b      [14,8,5,4]
19. a+b≠c      [18,7,5,3]      20. "box"      [19,6].
```

In comparison with the search space of a linear resolution, that consists of more than 4 to the power of 9 (262,144) resolvents in breadth-first search, the search space of a BCR deduction is fairly small: it consists of only 4 to the power of 3 (64) BC-resolvents.

Certainly, BCR is not a complete deduction method. But, for Horn sets, BCR is potentially complete. (The theorem can be simply proved in analogy with the proof of completeness theorem of semantic resolution [6].)

Consider the set $S_{2.2}$, which is the above set $S_{2.1}$ without the clauses 2 and 4. Though $S_{2.2}$ is known to be unsatisfiable, BCR fails to refute it. But, clause 2 can be proved by BCR within the set $S_{2.2}$. Then clause 4 can be proved within $S_{2.2} \cup \{2\}$. Finally, the original theorem can be proved by adding the clauses 2 and 4 to $S_{2.2}$.

3. A Knowledge Base for the ECR Procedure

It is noticed that the input equations of the theorem in example 2.1 are not treated uniformly. Some of them, such as i(x)+x=e, x+x=e, represented by unit clauses, are treated as reduction rules, some of them, represented by multi-literal clauses, such as clauses 7 and 8 of the associativity law, are treated as if-then rules. This treatment is realized by using a specific knowledge representation, "resolution-oriented representation" (ROR) [1].

With ROR, a relation may be represented by several distinct clauses, depending on how its terms are split up. Though all of these clauses are semantically equivalent in regarding to the set of equality axioms (the proof is not given here), they can play different roles in a deduction. This feature motivates us to use some particular ROR to represent the input equations for conveying our knowledge to an automated theorem prover.

In constructing knowledge bases to be used by the ECR procedure, we will mainly consider three types of ROR. One is the equation in its natural form. Others are the clauses obtained by splitting some or all of its function-terms into inequalities. For a formal discussion, we define the following:

Function-term. A term that is not a variable or a constant is called a function-term.

Simple-function-term. A function-term is called a simple-function-term if it does not contain a proper sub-term that is a function-term. Otherwise it is called a **Complex-function-term.**

Basic-ROR. Let C be a clause that contains a function-term t. A clause, C[v/t] $\vee$ t$\neq$v, is called a basic-ROR of C, if v is a variable that has no occurrence in C, and C[v/t] is the clause C with all occurrences of t replaced by the variable v.

ROR. A clause R is called a ROR of a clause C iff

1. R is identical to C; or
2. R is a basic-ROR of C; or
3. There is a clause R' that is a ROR of C, and R is a basic-ROR of R'.

Sparse-ROR. A clause R is called a sparse-ROR of a clause C iff

1. R is a ROR of C; and
2. R contains no inequality or equation whose right hand side is a function-term; and
3. R contains no complex-function-term.

Given a set S of clauses. Suppose S contains no equality axioms, and all equations of S are represented in their natural forms. Assume that each equation t1=t2 or inequality t1$\neq$t2 contained in S satisfies or has been processed satisfying $||t1|| \geq ||t2||$, where $||.||$ is a measure function. Let S_E consist of each equation t1=t2 of S, and also t2=t1 if t2 is a complex function term. We now describe the construction of a knowledge base, KB(S).

3.1. DM-rules (Demodulation rules)

The set of DM-rules is obtained by transforming a subset, S_{DM}, of equations in S_E into sparse-RORs. S_{DM} is selected by users. In our experiments, we usually transform few input equations whose both sides are complex function into DM-rules. The associativity laws and distributivity laws are examples of such equations.

A DM-rule will play the role of a conditional demodulation rule with respect to its original equation. In the example 2.1, clause 7 is a sparse-ROR of the equation a1l: x+(y+z)=(x+y)+z. The IBC-resolvent 14-1 of clause 13 against clause 7 corresponds to a consequence d-3 derived by the following process related to the equation a1l:

```
13.  b+a≠c
d-1. x+(y+z)≠b+a  (x+y)+z≠c ;Demodulating 13 against a1l
d-2  x≠b  y+z≠a  (x+y)+z≠c  ;Demodulating inequality x+(y+z)≠b+a
d-3. y+z≠a  (b+y)+z≠c       ;Erase x≠b by unifying with x=x
```

The condition on the application of a DM-rule is set up by the marks attached to a subset of its literals. We state a rule for attaching marks to DM-rules.

> Let t1=t2 be an equation, $R_1 \vee R_2$ be the sparse-ROR of t1=t2, where R_1 is a sparse-ROR of t1=w, and R_2 is a sparse-ROR of t2≠w. Attach a mark 1 to each literal of R_2. And if R_1 contains more than one inequalities, then attach a mark 2 to each inequality of R_1.

The mark 2 will be used in ECR to indicate another constraint weaker than that of the mark 1. All unit clauses and some other rules (such as TR-rules to be defined next) will be allowed to be applied to a literal marked by 2.

According to the above rule, given the following distributivity laws:

```
d1l: (x*y)+(x*z)=x*(y+z)
d2l: x*(y+z)=(x*y)+(x*z),
```

the DM-rules of them will be marked as follows:

$$\text{d1d: } u+v=w \quad x*r \neq w^1 \quad y+z \neq r^1 \quad x*y \neq u^2 \quad x*z \neq v^2$$
$$\text{d2d: } x*v=w \quad u+v \neq w^1 \quad x*y \neq u^1 \quad x*z \neq v^1 \quad y+z \neq r.$$

The marks attached to the clause 7 and 8 of example 2.1 also follow this rule. But, our experiments have shown that, using a modified version of this rule for marking the DM-rules of an associativity law is preferable. In our experiments, these rules are represented as follows:

a1d : $x+v=w \; x+y \neq u^1 \; u+z \neq w^2 \; y+z \neq v$
a2d : $u+z=w \; y+z \neq v^1 \; x+v \neq w^2 \; x+y \neq u$

3.2. TR-rules (Term-Rewrite rules)

Select a subset S_{TR} from S_E. For each equation t1=t2 in S_{TR}, if t1 is a complex-function-term and t2 is a function-term, then t1=w v R, is called a TR-rule if w is a variable that has no occurrence in t1=t2, and R is a sparse-ROR of t2≠w.

A TR-rule will play the role of a conditional term-rewrite rule with respect to its original equation. For example, from the associativity law, (the above equation a1l), we can obtain a TR-rule:

a1t: $x+(y+z)=w \; x+y \neq u \; u+z \neq w$

Then, given an inequality, t1+(t2+t3)≠s1, the left side term of it will be "rewritten" by a1t into (t1+t2)+t3. Again, the condition on this "term-rewrite" can be set up by assigning marks to some inequalities of this TR-rule.

The ECR procedure will build-in another type of constraint to the TR-rules, that requires all the inequalities **but one** inherited from a TR-rule by a resolvent to be unifiable with some reduction rules. (We noticed that, under this constraint, a term t1+(t2+t3) can always be "rewritten" by the TR-rule a1t to be (t1+t2)+t3 because of the existence of equality reflexive axiom x=x. In order to prevent this sort of unconditional "term rewriting", a mark -1 is used in place 1 to denote that the marked inequality is not allowed to be unified with x=x. In order to simplify our discussion, we will not make further comments about this problem and this new mark.)

Unlike the case of the DM-rules, we usually transform many equations with complex structure into TR-rules, because the inference on them is easier to control by using a restricted unification algorithm defined as follows:

Pattern-match. Two literals L_g and L_c are said to have a pattern-match, iff they are unifiable by such a restricted unification, that at least two function symbols of L_c are matched with some function symbols of L_g.

3.3. Others

The input equation units are used mainly as reduction rules for unifying the marked inequalities in the ECR procedure. In order to reduce the number of redundancy produced by the inference on a large number of lemmas, the pattern-match defined above is used by ECR in unifying with some particular unit clauses. These unit clauses will be called LR-rules (Lemma-reduction-rules). The other unit clauses will be called BR-rules (basic-reduction-rules).

Select a subset S_{LR} of unit clauses that have complex structure from the set of unit clauses of S. Then

LR-rules = $S_{LR} \cup \{t2=t1: \ ||t1||=||t2||, \ t1=t2 \in S_{LR}\}$.

Let S_{TR} consist of every unit clause of S that is not a member of S_{LR}. Then

BR-rules = $S_{TR} \cup \{x=x\}$
$\qquad\qquad \cup \{t2=t1: \ ||t1||=||t2||, \ t1=t2 \in S_{TR}\}$

CBR-rules. Besides unit clauses, some practical problems often provide so called conditional reduction rules, such as a typed statement, bird(x)->fly(x) or an inverse law in field theory: $x \neq 0$ -> $x/x=1$. We define the set of this sort of rule contained in S as CBR-rules (conditional BR-rules). Every hypothesis literal of a CBR-rule is marked by 1.

The ECR procedure will allow a subgoal marked by 1 to be unified by resolving against a CBR-rule if the (instantiated) hypotheses of this rule are satisfied by the set of BR-rules and LR-rules.

OI-rules (ordinary if-then rules). It consists of every multi-literal clause of S that is not a CBR-rule. Some literals of an OI-rule may be marked (with 1 or 2) by user to restrict its application. For example, according to our experience, it is beneficial to represent a cancel law as follows:

$x=y \;\; y+z \neq w^{1} \;\; x+z \neq w$.

And if the cancel law of an operator is included in the knowledge base, then we will not transform the associativity law of this operator, if it is a hypothesis, into DM-rules.

EQ-rules. It consists of following equality axioms. The equality reflexive axiom e1, $x=x$, is classified as a BR-rule. The equality symmetry axiom is implied by e1 and e2.

e2: $x=y \;\; x \neq z^{1} \;\; y \neq z$

e3: x=y $z \neq x^1$ $z \neq y$

For each function symbol f occurring in S,
A4: f(x1,...,x0,...,xn)=f(x1,...,xi,...,xn) $x0 \neq xi^1$

For each predicate P (other than =) occurring in S,
A5: P(x1,...,xi,...,xn) $\sim P(x1,...,x0,...,xn)^3$ $x0 \neq xi$

The mark 3 of A5 will be used by ECR to prevent repeated application of the predicate substitution axiom.

A Knowledge base. In a summary, the knowledge base, KB(S), constructed in the above from the input set S, consists of the following set of rules.

1. BR-rules	2. LR-rules	3. CBR-rules
4. TR-rules	5. DM-rules	6. OI-rules
7. EQ-rules		

4. The Equality Conditional Resolution Procedure (ECR)

By means of the knowledge base KB(S) constructed from an input set S of clauses, the ECR procedure will make a flexible control to the inference on the equality relations and other hypotheses of S.

The intermediate resolvents of ECR are defined similarly to the IBC-resolvents of BCR, but some of them will be produced by using the restricted unification algorithm, pattern-match. We denote these resolvents by **IEC*-resolvents.** Other intermediate resolvents are denoted by **IEC-resolvents.**

EC-resolvents. Let C1 v G1 be an IEC-resolvent, where C1 is the set of literals inherited from a rule C. Let

$C1_m$ = {L: L $\in$ C1, L is marked by 1},
$C1_f$ = {$t1 \neq t2$: $t1 \neq t2 \in (C1 - C1_m)$, t1 and t2 are function-terms of a same function symbol},
$R_0 = C1_f \text{ v } C1_m$. (Notice $C1_f$ is on the left side of R_0.)

For each input-refutation D of the set, TR-rules $\cup$ LR-rules $\cup$ CBR-rules $\cup$ $\{R_0\}$, with R_0 as the starting goal, such that each literal inherited from a CBR-rule is unified only by a TR-rule or a LR-rule, and pattern-match is applied for the unification with LR-rules, let θ be the most general substitution of D, then $\{C1 - R_0\}\theta \text{ v } G_1\theta$ is called an EC-resolvent of C1 v G1 within KB(S).

Example 4.1. Suppose the knowledge base of a theorem includes the following clauses:

```
 1. x=x                                   BR-rule
 2. x + 0 = x                             BR-rule
10. k(x)*x=0                              BR-rule
15. r*x=w u+v≠w¹ y*x≠u¹ z*x≠v¹ y+z≠r      DM-rule
29. [a+k(b)]*b≠a*b.                       Starting goal
```

There will be an IEC-resolvent 35-1 produced from 29 and 15.

```
35-1. u+v≠a*b¹ y*b≠u¹ v z*b≠v¹ y+z≠a+k(b).
```

Since the both sides of the unmarked inequality y+z≠a+k(b) are terms of a same function symbol, it is required to be unified first. Then "box" is the only possible EC-resolvent of 35-1:

```
35-2 u+v≠a*b¹ a*b≠u¹ k(b)*b≠v¹            35-1,1
35-3 a*b≠a*b¹ k(b)*b≠0¹                   35-2,2
35-4 k(b)*b≠0¹                            35-3,1
35.  "box"                                35-4,10.
```

The ECR Procedure is a modified version of BCR with best-first search. A global variable *goal-list* is used to store the candidate goal clauses. Originally, the goal-list contains only the input goal clauses of the set S. It proceeds as follows:

```
(0) Until goal-list is empty or a proof is found,
      take out the first goal G,
      let L_g be the first literal of G.

(1) CS := {IEC*-resolvent of G against C, C ∈ LR-rules};

(2) CS := CS ∪ {IEC-resolvent of G against C, C ∈ BR-rules};

(3) For each IEC-resolvent R of G against a CBR-rule do
      CS := CS ∪ {EC-resolvents of R within KB(S)};

(4) For each IEC*-resolvent R of G against a TR-rule do
      for each literal L of R[T]*** do
        Let R' be R with each literal of R[T] - L marked by 1,
        CS := CS ∪ {EC-resolvents of R' within KB(S)};

(5) If L_g has a mark 2, then go to step (8),
      else if L_g has a mark 3, then go to step (7);

(6) For each IEC-resolvent R of G against a clause in
      DM-rules ∪ EQ-rules do
        CS := CS ∪ {EC-resolvents of R within KB(S)};
```

*** R[T] is the subclause of R inherited from the TR-rule.

```
(7) For each IEC-resolvent R of G against an OI-rule do
      CS := CS ∪ {EC-resolvent of R within KB(S)};

(8) Insert each element of CS into goal-list on their
      priorities in descending order.
    Go (0).
```

The following dynamic marking-mechanism is added to the above procedure: For each EC-resolvent R just produced:

> If R is a unit clause, then strip marks off it (we allow a unit goal clause to be full developed), otherwise if L_g has a mark 2, then for each literal L inherited from a rule C, assign a mark 2 to L (because its parent is not allowed to be full developed).

The following example includes a case that a CBR-rule is used in producing EC-resolvents.

Example 4.2. Prove the theorem: In a field, $x \neq 0 \wedge x*y=x*z \rightarrow y=z$.
There are about 30 clauses contained in the knowledge base. For simplicity, we write down only a subset of the clauses that are useful for the proof obtained by an ECR prover.

5.	$x*1=x$	;BR-rule	6. $a*b=a*c$	;BR-rule
8.	$a \neq 0$	;BR-rule	10. $x=z \;\; y \neq x^1 \;\; y \neq z$	;EQ-rule
11.	$1(x)*x=1 \;\; x=0^1$	; CBR-rule		
15.	$x*r=w \;\; x*y \neq u^1 \;\; u*z \neq w^2 \;\; y*z \neq r$	;DM-rule		
27.	$x*(y*z)=w \;\; x*y \neq u^1 \;\; u*z \neq w$	;TR-rule		
30.	$b \neq c$	;Goal		

The commutative laws for * and + were built-in by using a commutative unification algorithm. The clause 11 is the inverse law for *, which is defined as a CBR-rule.

33.	$b*1 \neq c$	[30,10,5]
41.	$y1*(a*c) \neq c^2 \;\; y1*a \neq 1$	[33,15,6]
88.	$1*c \neq c^2 \;\; 1(a)*a \neq 1$	[41,27,11,8]
90.	$1(a)*a \neq 1$	[88,5]
99.	"box"	[90,11,8]

For positive Horn sets, the ECR procedure is potentially complete, if the knowledge base constructed from the given set S contains an empty set of LR-rules and no positive literals of an input clause nor any literals of an input negative clause are marked.

5. Experimentation

The ECR-prover is built on a modified Semantically Guided Hierarchical Deduction Prover (SHD-prover) [14,15], that provides a number of narrowing strategies for discarding the redundant EC-resolvents. This prover is used to prove a number of theorems of group theory, ring theory, boolean algebra and field theory. In all of these proofs, the priority of a candidate goal G is determined by a same evaluation function according to the complexity of G (||G||), and the marks of G. The measure function ||.|| is defined over all constants c, all variables v, and terms t, t1, ..., as ||c||=1, ||v||=1, ||i(t)||=4*||t||+1, ||m(t)||=4*||t||+1, ||t1+t2||=||t1||+||t2||+1, ||t1*t2||=||t1||+||t2||+1. By this measure function, given two terms i(a*b) and i(a)*i(b), we will have ||i(a)*i(b)||=(4*1+1)+(4*1+1)+1=11 < ||i(a*b)||=4*(1+1+1)+1=13. It conforms to our intuition that i(a)*i(b) is simpler than i(a*b).

In the following listing, each equation is assigned a name, each rule transformed from an equation is named by the name of this equation followed by the first character of its rule type. For example, a1d, a1t and a1l are the DM-rule, TR-rule and LR-rule, respectively, that are transformed from the equation a1 according to the rules described in section 3.

a1: x+(y+z)=(x+y)+z — a2: (x+y)+z=x+(y+z)
a3: x*(y*z)=(x*y)*z — a4: (x*y)*z=x*(y*z)
d1: (x*y)+(x*z)=x*(y+z) — d2: x*(y+z)=(x*y)+(x*z)
d3: (y*x)+(z*x)=(y+z)*x — d4: (y+z)*x=(y*x)+(y*z)
d5: (x+y)*(x+z)=x+(y*z) — d6: x+(y*z)=(x+y)*(x+z)
e1: x=x — e2: $x=y \quad x \neq z^1 \quad y \neq z$ — e3: $x=y \quad z \neq x^1 \quad z \neq y$
e4: $x1+x0=x1+x2 \quad x0 \neq x2^1$ — e5: $x0+x2=x1+x2 \quad x0 \neq x1^1$
e6: $x1*x0=x1*x2 \quad x0 \neq x2^1$ — e7: $x0*x2=x1*x2 \quad x0 \neq x1^1$

5.1. Group theory

1: i(x)+x=e — 2: x+i(x)=e — 3: e+x=x — 4: x+e=x — 5: x+x=e
6: a+b=a+c — 7: a+c=a+b — 8: x+x+x=e — 9: i(x)=x+x
G1: a+u≠e — G2: a+u≠a — G3: k(u)+u≠k(u) — G4: i(i(a))≠a
G5: i(a+b)≠i(b)+i(a)
G6: a+b≠b+a ; ∀x[x+x=e] -> ∀x,y[x+y=y+x]
G7: b≠c ; ∀x,y,z[x+y=x+z->y=z]
G8: i(a)≠a+a ; ∀x[x+x+x=e] -> ∀x[i(x)=x+x]
A_g={1b,3b,a1l,a2l,a1t,a2t,a1d,a2d,e1b,e2e,e3e,e4e,e5e)

Goal	Rules of KB	EC-accepted	EC-useful	CPU-second
G1	A_g	11	5 (7)	0.58
G2	A_gu{2b}	24	3 (4)	1.53
G3	A_gu{2b}	6	3 (4)	0.46
G4	A_gu{2b,4b}	11	4 (6)	0.63
G5	A_gu{2b,4b}	14	5 (10)	2.70
G6	A_gu{2b,4b,5b}	31	5 (9)	3.25
G7	A_gu{2b,4b,6b,7b}	23	5 (8)	1.45
G8	A_gu{2b,4b,8b}	10	3 (6)	0.81

In the above and the following tables, "EC" means EC-resolvents. Under the column "EC-useful", the entry "5 (7)" denotes that there were 5 EC-resolvents useful for an ECR-refutation, and they were produced by total 7 unifications.

5.2. Free ring

00: x+y=y+x
1: i(x)+x=0
2: 0+x=x
3: 0*x=0
4: x*0=0
5: i(x*y)=x*i(y)
6: i(x)*i(y)=x*y
7: x*x=x
8: x+x=0
9: x=y y+u≠w^1 x+u≠w
R1: 0*a≠0 ; Neg. of 3
R2: a*0≠0 ; Neg. of 4
R3: i(a*b)≠a*i(b) ; Neg. of 5
R4: i(a)*i(b)≠a*b ; Neg. of 6
R5: i(a)≠a ; ∀[x*x=x] -> ∀x[i(x)=x]

A_r={1b,2b,9o,a1l,a1t,a2l,a2t,a3l,a3t,a4l,a4t,d1l,d1t,d1d,d2t,d2d,
d3l,d3t,d3d,d4t,d4d,e1b,e2e,e3e,e4e,e6e,e7e}

A commutative unification was used for +.

Goal	Rules of KB	EC-accepted	EC-useful	CPU-second
R1	A_r	7	2 (5)	0.85
R2	A_r	6	2 (5)	0.65
R3	A_ru{3b,4b}	11	3 (9)	2.58
R4	A_ru{3b,4b}	14	4 (13)	3.85
R5	A_ru{3b,4b,7b}	76	6 (13)	19.6

5.3. Field theory

00: x+y=y+x
01: x*y=y*x
1: m(x)+x=0
2: 0+x=x
3: 1*x=x
4: 0*x=0
6: x*i(x)=1 x=0^1
7: x=y y+u≠w^1 x+u≠w
8: d≠0
9: b≠0
10: d*b=d*c
10': d*c=d*b
11: a*b=0
12: a*i(b)=c*i(d)
12': c*i(d)=a*i(b)
13: a*d=b*c
13': b*c=a*d
F1: b=c ; d≠0∧d*b=d*c->b=c
F2: a≠0 ; a*b=0 -> a=0 v b=0
F3: a*d≠b*c ; b≠0∧d≠0∧a*i(b)=c*i(d)->a*d=b*c
F4: a*i(b)≠c*i(d) ; b≠0∧d≠0∧a*d=b*c->a*i(b)=c*i(d)

A_f={1b,2b,3b,4b,6c,7o,a1l,a1t,a2l,a2t,a3l,a3t,a3d,a4l,a4t,a4d,

d1l,d1t,d1d,d2t,d2d,e1b,e2e,e3e,e4e,e6e}
A commutative unification was used for + and *.

Goal	Rules of KB	EC-accepted	EC-useful	CPU-second
F1	A_fu{8b,10b,10'b}	69	5 (10)	11.0
F2	A_fu{9b,11b}	58	4 (7)	6.20
F3	A_fu{8b,9b,12b,12'b}	46	4 (10)	14.7
F4	A_fu{8b,9b,13b,13'b}	41	4 (10)	12.2

5.4. Boolean algebra

```
00: x+y=y+x                        01: x*y=y*x
 1: 0+x=x                           2: 1*x=x
 3: k(x)+x=1                        4: k(x)*x=0
 5: k(1)=0                          6: k(0)=1
 7: 1+x=1                           8: 0*x=0
 9: x+x=x                          10: x*x=x
11: x*(x+y)=x                      12: x+(x*y)=x
13: k(k(x))*x=k(k(x))              14: k(k(x))=x
15: (x*k(y))+y=x+y                 16: (x+k(y))*y=x*y
17: (x*y)+k(y)=x+k(y)              17': x+k(y)=(x*y)+k(y)
18: (x+y)*k(y)=x*k(y)              18': x*k(y)=(x+y)*k(y)
19: [(x+y)+z]*x=x                  20: [(x*y)*z]+x=x
21: [(x+y)+z]*k(x)=(y+z)*k(x)      22: [(x*y)*z]+k(x)=(y*z)+k(x)
23: (x+y)+z=x+(y+z)                24: (x*y)*z=x*(y*z)
25: k(x+y)=k(x)*k(y)               26: k(x*y)=k(x)+k(y)
```

$A1_b$={1b,2b,3b,4b,d1l,d1t,d1d,d2t,d2d,d5l,d5t,d5d,d6t,d6d,e1b,e2e,
e3e,e4e,e6e}
$A2_b$=$A1_b$u{5b,6b,7b,8b,9b,10b,11l,12l}
$A3_b$=$A2_b$u{14l,15l,15t,16l,16t,17l,17t, 17't,18l,18t,18't}
A commutative unification was used for + and *.

For each i, 5≤i≤26, Bi is the negation of the above equation i, i.e. Bi is an inequality obtained by applying a substitution {≠/=, a/x,b/y,c/z} to the equation i.

Goal	Rules of KB	EC-accepted	EC-useful	CPU-second
B5	$A1_b$	3	2 (3)	0.13
B7	$A1_b$	14	3 (7)	2.63
B9	$A1_b$u{5b,6b,7b,8b}	10	3 (5)	1.41
B11	$A1_b$u{5b,6b,7b,8b}	3	2 (5)	1.93
B13	$A2_b$	36	3 (7)	16.7
B14	$A2_b$u{13l}	37	4 (7)	5.15
B15	$A2_b$u{14l}	6	2 (5)	3.60
B17	$A2_b$u{14l}	5	2 (5)	2.71
B19	$A3_b$	3	1 (5)	2.86
B21	$A3_b$	2	1 (5)	6.28
B23	$A3_b$u{19l,21l}	24	3 (7)	17.2
B25	$A3_b$u{a1l,a1t,a3l,a3t}	72	9 (18)	20.7

For the boolean experimentation reported in the above table, an algebraic model of boolean algebra was supplied, and was used by a semantic testing subroutine to discard the semantically unprovable goals. Without the help of a model, these theorems were also proved by the same ECR-prover, but more CPU times were used in proving some difficult theorems. For example, for proving B23, 71.2 seconds were used with 127 EC-resolvents accepted, and for proving B25, 91.3 seconds were used with 250 EC-resolvents accepted. (Our program has not been well optimized.)

According to our experience, some theorems proved by ECR-prover are quite difficult to prove automatically by the SHD-prover. But, the narrowing strategies of SHD-prover are fairly effective when used by ECR-prover for discarding the redundant EC-resolvents. This is because these strategies can usually efficiently detect and discard those redundant resolvents that contain none or few variables, and the EC-resolvents produced by ECR are usually this sort of clauses. It turns out that facilitating the detection of redundancy is another advantage of ECR. For example, in proving B25, there were total 188 BC-resolvents generated, among which 61 were discarded by semantic testing, 55 were discarded by other narrowing strategies, then only 72 BC-resolvents were accepted.

5.5. A Proof of De.Morgan's Law Found by ECR

The following listing includes the resolvents obtained from each useful unification.

```
[top ] G0  : k(a+b)≠k(a)*k(b)
[e3e ] G1-1: z≠k(a+b)^1 z≠k(a)*k(b)
[2   ] G1  : 1*k(a+b)≠k(a)*k(b)
[d5d ] G2-1: x+r≠k(a)*k(b)^1 y*z≠r^1 x+y≠k(a+b)^2 x+z≠1^2
[1b  ] G2-2: y*z≠0^1 [k(a)*k(b)]+y≠k(a+b)^2 [k(a)*k(b)]+z≠1^2
[4b  ] G2  : [)(k(a)*k(b)]+k(z)≠k(a+b)^2 [k(a)*k(b)]+z≠1^2
[17't] G3-1: k(z)≠v^1 u+v≠k(a+b)^1 [k(a)*k(b)]*z≠u^2
             [k(a)*k(b)]+z≠1^2
[e1b ] G3-2: u+k(z)≠k(a+b)^1 [k(a)*k(b)]*z≠u^2 [k(a)*k(b)]+z≠1^2
[1b  ] G3  : [k(a)*k(b)]*(a+b)≠0^2 [k(a)*k(b)]+(a+b)≠1^2
[a3t ] G4-1: k(a)*(a+b)≠u^1 u*k(b)≠0^2 [k(a)*k(b)]+(a+b)≠1^2
[181 ] G4  : [k(a)*b]*k(b)≠0^2 [k(a)*k(b)]+(a+b)≠1^2
[a3t ] G5-1: b*k(b)≠u^1 u-*k(a)≠0^2 [k(a)*k(b)]+(a+b)≠1^2
[4b  ] G5  : 0*k(a)≠0^2 [k(a)*k(b)]+(a+b)≠1^2
[8b  ] G6  : [k(a)*k(b)]+(a+b)≠1
[a1t ] G7-1: b+[k(a)*k(b)]≠v^1 a+v≠1
[151 ] G7  : a+[b+k(a)]≠1
[a1t ] G8-1: a+k(a)≠u^1 u+b≠1
[4b  ] G8  : 1+b≠1
[7b  ] G9  : "box"
```

6. Remarks

We proposed an equality conditional resolution proof procedure (ECR), that emphasizes the use of knowledge from two different aspects. One is the knowledge supplied by users concerning the different roles of input equations in a proof. Another is the known facts provided by input unit clauses and "lemmas". The implementation results are encouraging, which show that ECR-prover can prove many theorems with equality relations by exploring a fairly small search space. Of course, ECR described in this paper is only a preliminary version of the new proof procedure currently developing by us. We believe that ECR can be extended to handle the theorems of whole first order logic, and can be enhanced by incorporating many existing strategies useful in dealing with the inference on equality relation, such as term-rewriting technique, building-in associativity law, ac-unification algorithm, etc.

Our project is based to a great extent on that of others. Previous works to which this is closely related include Bledsoe's peeking and conditional rewriting techniques, Lusk and Overbeek's tagging literal strategy, Wos and McCune's linked inference, Stickel's theory resolution, Winker's work on qualification, Nevins's human oriented logic, etc. The notion of indexing literals goes back to Boyer's thesis on locking.

Acknowledgments

I wish to thank Prof. W. W. Bledsoe for his support and valuable suggestions, Dr. D. Plummer for correcting the earlier draft, and Prof. J. Siekmann for his help related to this paper. This work was supported by NSF Grant DCR-8313499.

References

[1] Bledsoe, W. W., and Henschen, L. J. What is automated theorem Proving? J. Automated Reasoning, Vol. 1, No. 1, pp. 23-28 (1985).

[2] Bledsoe, W. W. Non-resolution theorem proving. Artificial Intelligence 9, pp. 1-35 (1977).

[3] Bledsoe, W. W. The UT interactive prover. Tech. Report ATP-17B. Univ. of Texas at Austin (1983).

[4] Boyer, R. S. Locking: a restriction of resolution. Ph.D. Thesis. Univ. of Texas at Austin (1971).

[5] Boyer, R. S., and Moore, J S. A computational logic. Academic Press (1979).

[6] Chang, C., and Lee, R. C. Symbol logic and mechanical theorem proving. Academic Press (1973).

[7] Digricoli, V.J. The management of heuristic search in boolean experiments with RUE resolution. Proc. IJCAI-9, pp. 1156-1161 (1985).

[8] Lusk, E. L., and Overbeek, R. A. A portable environment for research in automated reasoning. Proc. CADE-7, Lecture notes in computer science, Vol. 170, Springer-Verlag, New York, pp. 43-52 (1984).

[9] Lenat, D. B. Automated theory formation in mathematics. Proc. IJCAI-5 (1977).

[10] Lim, Y. and Henschen, L. A New hyperparamodulation strategy for the equality relation. Proc. IJCAI-9, pp. 1139-1145 (1985).

[11] McCharen, J. D., Overbeek, R. A. and Wos, L. A. Problems and experiments for and with automated Theorem-proving Programs. IEEE trans. on compt. C-25, NO.8 (Aug. 1976).

[12] Nevins, A. J. A human oriented logic for automating theorem-proving. J. ACM, Vol. 21, No. 4 (Oct. 1974).

[13] Stickel, M. A unification algorithm for associative-commutative functions, J. ACM, Vol. 28, pp. 423-434 (1981).

[14] Wang, T.-C. Hierarchical Deduction. Tech. Report ATP-78A, Univ. of Texas at Austin (March 1984).

[15] Wang, T.-C. Designing Examples for semantically guided hierarchical deduction. Proc. IJCAI-9, 1201-1207 (1985).

[16] Wos, L., Overbeek, R., Lusk, E., and Boyle. J. Automated Reasoning: Introduction and Application, Prentice-Hall, Englewood Cliffs (1984).

[17] Wos, L., Veroff, R., Smith, B., and McCune, W. The linked inference principle, II: the user's viewpoint, CADE-7, Proc. CADE-7, Lecture notes in computer science, Vol. 170, Springer-Verlag, New York, pp. 316-332 (1984).

[18] Wos, L., Robinson, G.A., and Carson. The concept of demodulation in theorem proving. J. ACM, Vol. 14, N0. 4 (Oct. 1967).

USING NARROWING TO DO ISOLATION IN SYMBOLIC EQUATION SOLVING
- an experiment in automated reasoning.

A. J. J. Dick
Informatics Division
Rutherford Appleton Lab.
Chilton, Didcot
OXON OX11 0QX

R. J. Cunningham
Dept. of Computing
Imperial College
LONDON SW7

ABSTRACT

The PRESS symbolic equation solving system [STE82], and other algebraic manipulation packages, use a method known as *isolation* for equations containing only a single occurrence of an unknown. In effect, equations of the form $f(t) = t'$ are rewritten by $t = f^{-1}(t')$, where f^{-1} is the inverse of f and t is a term containing the unknown. Meta-level inference is used to decide when isolation is applicable as opposed to other methods such as *attraction* and *collection* in which multiple occurrences of an unknown are drawn together.

This paper demonstrates how the technique of *narrowing*[RET85] implicitly performs isolation to solve equations. Narrowing involves the unification of the left-hand side of a rule with the equation to be solved, followed by rewriting by that rule (and others if applicable). Rewrite rules for isolation are provided, along with other properties of the functions involved, in the form $f^{-1}(f(v)) \Longrightarrow v$, where v is a simple variable.

The potential advantage of this is that it may be possible to avoid the need for meta-level inference by mixing rewrite rules for isolation with rules expressing other methods to form a single term rewriting system, in which the conditions for applicability of all rules are the same; namely, matching and unification.

A very preliminary investigation is made on the basis of a single example in which some of the limitations of narrowing are highlighted. In particular, it is not always possible to find a *finite confluent* term rewriting system containing all the necessary rules.

1. INTRODUCTION

Isolation is a well established method for solving symbolic equations in which a single occurrence of the unknown is isolated by applying rules about the inverses of functions.

For instance, in the PRESS equation solving system [STE82], rewrite rules of the following form are provided to describe isolation:-

$$[f(t) = t'] \Longrightarrow [t = f^{-1}(t')]$$

where f^{-1} is the inverse of f, and t is a term containing the unknown. Other methods are used to eliminate multiple occurrences of unknowns prior to isolation, such as *collection* and *attraction*. Where there are a variety of methods available, applicable under differing circumstances, a certain amount of meta-level reasoning is required, e.g. counting the number of occurrences of unknowns.

Consider an example borrowed from [STE82] of how PRESS solves the following equation for x:-

$$\log_e(x+1) + \log_e(x-1) = 3.$$

Isolation is not an immediately applicable method, because there is more than one occurrence of the unknown. A rule for attraction, $\log_e u + \log_e v = \log_e u.v$, is first applied, drawing together the two occurrences as follows:-

$$\log_e(x+1).(x-1) = 3;$$

then a rule for collection, $(u + v).(u - v) = u^2 - v^2$, produces a single occurrence of x:-

$$\log_e(x^2 - 1) = 3$$

with 1^2 simplified to 1. Now isolation is used to eliminate the outermost function symbols from the left-hand side in three stages:-

$$x^2 - 1 = e^3$$

$$x^2 = e^3 + 1$$

$$x = \pm\sqrt{e^3+1}.$$

PRESS treats unknowns as constants, and finds a possibly *over-general* solution which must be checked against the original equation. In the example above, for instance, the negative square root is not a valid solution.

In this paper, we compare PRESS with *narrowing*[RET85], a recently proposed method of solving equations which makes use of unification and rewriting in its reasoning process. In particular, we wish to draw attention to how isolation is performed implicitly in the narrowing process, and to make a very preliminary investigation into the potential of narrowing as a heuristic to replace the need for meta-level inference.

Narrowing uses techniques closely related to the Knuth-Bendix completion algorithm [KNU70, HUE80b], and requires the properties of the functions involved to be expressed as a confluent set of rewrite rules. In this framework, isolation rules appear as pairs of rewrite rules of the form $f^{-1}(f(x)) \Longrightarrow x$ embedded in the confluent set.

The example quoted above was solved using ERIL (Equational Reasoning: an Interactive Laboratory) [DIC86a, DIC86b, CUN85], configured for narrowing. Being equipped for the treatment of partial functions such as log, ERIL is particularly suited to handle this problem.

2. NARROWING

The reader is referred to [RET85] for a clear exposition of narrowing as an algorithm for solving symbolic equations in equational theories. Its correctness and completeness are proved, and an implementation based on the REVE term rewriting system [FOR84, LES83] is described with examples. Problems associated with infinite derivations are discussed, and solutions proposed. We give here a brief overview of the narrowing process:

Narrowing is a complete method for solving equations in a theory described by a confluent and noetherian term rewriting system, R (see [HUE80b]). Given an equation

$$E = F,$$

the method attempts to find a substitution σ which satisfies

$$\sigma(E) =_R \sigma(F);$$

i.e. narrowing attempts to unify E and F within the equational theory described by the confluent set of rewrite rules, R. This is achieved by a series of derivation steps called *U-reductions* which reduce E and F until they are identical, in doing so, constructing σ.

A U-reduction, written $T \rightsquigarrow T'$, consists of three stages:-

1) The left-hand side of some rule in R is unified with a sub-term t of T. The unifier, γ, is called the narrowing substitution.

2) $\gamma(T)$ is rewritten to S by the same rule.

3) The normal form of S is found, forming T'.

Narrowing is performed on both sides of an equation in parallel by combining E and F with equality predicate symbol, $=$, and making derivations from the initial term $E = F$

$$E = F \rightsquigarrow_{\gamma_1} E_1 = F_1 \rightsquigarrow_{\gamma_2} E_2 = F_2 \rightsquigarrow \cdots \rightsquigarrow_{\gamma_n} E_n = F_n$$

until E_n and F_n are unifiable by σ; i.e. $\sigma(E_n) = \sigma(F_n)$ for some σ. Then the solution to the equation lies in the composite substitution $\sigma \cdot \gamma_1 \cdot \gamma_2 \cdot \cdots \gamma_n$.

In practice, the composite substitution is constructed as the derivation proceeds by associating with the term $E = F$ another which contains as arguments the unknowns for which we are solving, i.e. the variables occurring in $E = F$, in the form $< E = F,\ answer(x,y,z, \cdots)>$. As substitutions are made in $E = F$, they are also made in $answer(x,y,z, \cdots)$, thus constructing the solution.

There may be, of course, more than one, or an infinite number of solutions to an equation. The narrowing process will actually search the whole tree of U-reductions until all solutions are found. There may be infinite derivations in the tree, and some technique must be employed to avoid these.

3. ERIL

ERIL is an experimental interactive laboratory for equational reasoning based on Knuth-Bendix superposition. Two features that distinguish it from similar facilities are

1) The ability to construct specialist completion algorithms;
2) The ability to treat certain types of partial algebra.

A variety of types of equality are permitted, including rewrite rules, bi-directional rules and undirected equations. The whole equation base is divided into sets, whose number and function are specified by the user. The relationships between sets or individual rules can be dynamically controlled. The user may supply orderings to be associated with particular sets of directed rules, as well as algorithms for sorting equations within a set. Various forms of the Knuth-Bendix algorithm fall naturally into this framework, including, in this instance, narrowing.

Underlying ERIL is a lattice-structured typing method which allows the treatment of certain classes of partial algebra that would otherwise require the use of conditional rewrite rules. The undefined type is used to describe the result of applying a function to an argument outside its domain of definition, and derivations that yield the undefined type are considered as meaningless.

The reader is referred to [DIC86b, CUN85, DIC86a] for a more extended exposition of ERIL.

In the framework provided by ERIL, narrowing can be achieved by providing equations that express the properties of the equational theory, including the mutual invertibility of functions for the purposes of isolation. From these equations is found (if possible) a confluent set of rules to be used in the actual narrowing process. Another set is established containing the one rule $x = x \Rightarrow true$ which is used to cause an attempted unification of E and F for an immediate solution. Another set of undirected equalities is used to hold equations to be solved.

An equation $E = F$ is represented as the equivalence

$$(E = F) \Leftrightarrow ans(x_1, x_2, \cdots)$$

where $x_1, x_2, \cdots$ are the set of variables that occur in E and F. The predicate *ans* is used simply to record the instantiations given to each variable. Superposition is then performed between the left-hand sides of the confluent set of rules and the left-hand side of the equivalence. The resulting critical pairs are equivalences of the above form, and, after they have been normalised, represent partial solutions of the equation. Superposition is now performed on each new equivalence, and the process repeated until a equivalence of the form

$$true \Leftrightarrow ans(S_1, S_2, \cdots)$$

is generated, at which stage the solution is described by $x_1 = S_1, x_2 = S_2, \cdots$.

4. THE EXAMPLE

To begin with, we shall demonstrate how narrowing effects the isolation part of the solution, leaving a discussion of the problems associated with attraction and collection for later.

We are to find solutions to $\log_e(x^2 - 1) = 3$. A set of rules is provided which expresses the mutual invertibility of log and exp, square and square-root, and addition and subtraction; this set is combined with other general simplification rules, and is then completed using Knuth-Bendix completion. The completed set of rules is used to apply narrowing to the equation to be solved.

The algebra definitions used are listed in the appendix. Reflecting the use of partial functions and generic function signatures in ERIL, they declare the names of types to be used, the relationships between types (by means of the partial ordering, $<$), and function signatures. Note that ERIL requires several signatures for most functions to define behaviour on sub-types of the domain.

The example is presented in two stages:

1) the completion of the set of rules, and
2) the narrowing process.

For the first stage, two equality sets are used, A for the initial axioms supplied for the equation theory, and C for the confluent set. An initial set of 13 axioms is placed in A, and Knuth-Bendix completion is performed by moving axioms from A to C, producing a confluent set of 18 rules:-

A6	$psqrt(1)=1$	becomes	C1	$psqrt(1) \Rightarrow 1$
A10	$abs(p)=p$		C2	$abs(p) \Rightarrow p$
A7	$exp(x,1)=x$		C3	$exp(x,1) \Rightarrow x$
A8	$exp(1,x)=1$		C4	$exp(1,x) \Rightarrow 1$
A9	$log(x,x)=1$		C5	$log(x,x) \Rightarrow 1$
A13	$x-x=0$		C6	$x-x \Rightarrow 0$
A3	$exp(nsqrt(p),2)=p$		C7	$exp(nsqrt(p),2) \Rightarrow p$
A4	$exp(psqrt(p),2)=p$		C8	$exp(psqrt(p),2) \Rightarrow p$
A5	$psqrt(exp(x,2))=abs(x)$		C19	$psqrt(exp(x,2)) \Rightarrow abs(x)$
A1	$log(x,exp(x,x1))=x1$		C10	$abs(nsqrt(p)) \Rightarrow psqrt(p)$
A2	$exp(x,log(x,x1))=x1$		C11	$exp(abs(x),2) \Rightarrow exp(x,2)$
A11	$x+x1-x1=x$		C12	$log(x,exp(x,x1)) \Rightarrow x1$
A12	$x-x1+x1=x$		C13	$log(psqrt(p),p) \Rightarrow 2$
			C14	$exp(x,log(x,x1)) \Rightarrow x1$
			C15	$x+x1-x1 \Rightarrow x$
			C16	$x-x1+x1 \Rightarrow x$
			C17	$0+x \Rightarrow x$
			C18	$log(abs(x),exp(x,2)) \Rightarrow 2$

The confluent set of axioms now in C are used in the second stage, where altogether four equality sets are used:-

C	The confluent set of rules from stage 1.
W	The working set of equivalences waiting to be processed.
E	The single rule $x = x \Rightarrow true$.
P	The solution path selected.

The equation to be solved is placed in W. The narrowing process repeated selects the first equation from W, superposes its left-hand side on the left-hand sides of rules in C and P wherever possible, placing critical pairs back in W. The strategy is controlled by placing the equivalences with the smallest left-hand side at the top of W, but forcing "dead-end" solutions (where all variables have been eliminated from the equation) to the bottom.

The set P is named the "Solution path" because the chronological sequence of equivalences in P describe, in a sense, the path of instantiations that lead to the solution. There are in fact two solutions to the equation in question, one containing a negative square root, and the other a positive square root. The process of narrowing finds them both by, in a sense, back-tracking until all alternative equivalences have been considered.

Thus the initial state for the second stage is:-

C1	$psqrt(1) \Rightarrow 1$	W1	$log(e,exp(x,2)-1)=3 \Leftrightarrow ans(x)$
C2	$abs(p) \Rightarrow p$		
C3	$exp(x,1) \Rightarrow x$		
C4	$exp(1,x) \Rightarrow 1$	E1	$x=x \Rightarrow true$
C5	$log(x,x) \Rightarrow 1$		
C6	$x-x \Rightarrow 0$		
C7	$exp(nsqrt(p),2) \Rightarrow p$		
C8	$exp(psqrt(p),2) \Rightarrow p$	P empty	
C9	$psqrt(exp(x,2)) \Rightarrow abs(x)$		
C10	$abs(nsqrt(p)) \Rightarrow psqrt(p)$		
C11	$exp(abs(x),2) \Rightarrow exp(x,2)$		
C12	$log(x,exp(x,x1)) \Rightarrow x1$		
C13	$log(psqrt(p),p) \Rightarrow 2$		
C14	$exp(x,log(x,x1)) \Rightarrow x1$		
C15	$x+x1-x1 \Rightarrow x$		
C16	$x-x1+x1 \Rightarrow x$		
C17	$0+x \Rightarrow x$		
C18	$log(abs(x),exp(x,2)) \Rightarrow 2$		

W1 represents the original equation to be solved. Whilst the sets C and E remain constant, equivalences are moved one by one from W to P. Shown below are the contents of W and P after each step. Initially, three narrowings are found:-

W2	$log(e,p-1)=3 \Leftrightarrow ans(nsqrt(p))$	P1	$log(e,exp(x,2)-1)=3 \Leftrightarrow ans(x)$
W3	$log(e,p-1)=3 \Leftrightarrow ans(psqrt(p))$		
W4	$log(e,exp(x,2)-1)=3 \Leftrightarrow ans(abs(x))$		

In W2 and W3, the left-hand sides of C7 and C8 (respectively) have been unified with the subexpression $exp(x,2)$, causing x to be isolated by one step from the inside out. The default typing of variable p indicates that this unknown must be a positive real number. W4 is the critical pair resulting from the superposition of C11 on P1. The fact that the left-hand side of W4 is identical to that of W1 signals a possible infinite solution path (see later).

W5	$log(e,p)=3 \Leftrightarrow ans(nsqrt(p+1))$	P1	$log(e,exp(x,2)-1)=3 \Leftrightarrow ans(x)$
W3	$log(e,p-1)=3 \Leftrightarrow ans(psqrt(p))$	P2	$log(e,p-1)=3 \Leftrightarrow ans(nsqrt(p))$
W4	$log(e,exp(x,2)-1)=3 \Leftrightarrow ans(abs(x))$		

In W5, p has been isolated by using C16.

W6	$x=3 \Leftrightarrow ans(nsqrt(exp(e,x)+1))$	P1	$log(e,exp(x,2)-1)=3 \Leftrightarrow ans(x)$
W3	$log(e,p-1)=3 \Leftrightarrow ans(psqrt(p))$	P2	$log(e,p-1)=3 \Leftrightarrow ans(nsqrt(p))$
W4	$log(e,exp(x,2)-1)=3 \Leftrightarrow ans(abs(x))$	P3	$log(e,p)=3 \Leftrightarrow ans(nsqrt(p+1))$

The unknown is now fully isolated in W6, the final step being to unify with E1 to cause x to be instantiated to 3 in $ans(nsqrt(exp(e,x)+1))$.

W7	$ans(nsqrt(exp(e,3)+1))$	P1	$log(e,exp(x,2)-1)=3 \Leftrightarrow ans(x)$
W3	$log(e,p-1)=3 \Leftrightarrow ans(psqrt(p))$	P2	$log(e,p-1)=3 \Leftrightarrow ans(nsqrt(p))$
W4	$log(e,exp(x,2)-1)=3 \Leftrightarrow ans(abs(x))$	P3	$log(e,p)=3 \Leftrightarrow ans(nsqrt(p+1))$
		P4	$x=3 \Leftrightarrow ans(nsqrt(exp(e,x)+1))$

W3	$log(e,p-1)=3 \Leftrightarrow ans(psqrt(p))$	P1	$log(e,exp(x,2)-1)=3 \Leftrightarrow ans(x)$
W4	$log(e,exp(x,2)-1)=3 \Leftrightarrow ans(abs(x))$	P2	$log(e,p-1)=3 \Leftrightarrow ans(nsqrt(p))$
		P3	$log(e,p)=3 \Leftrightarrow ans(nsqrt(p+1))$
		P4	$x=3 \Leftrightarrow ans(nsqrt(exp(e,x)+1))$
		P5	$ans(nsqrt(exp(e,3)+1))$

At this stage, a solution has been found, namely $x = -\sqrt{e^3+1}$, represented by P5. Now the remaining equivalences in W will be processed, to find alternative solutions.

W8	$log(e,p)=3 \Leftrightarrow ans(psqrt(p+1))$	P1	$log(e,exp(x,2)-1)=3 \Leftrightarrow ans(x)$
W4	$log(e,exp(x,2)-1)=3 \Leftrightarrow ans(abs(x))$	P2	$log(e,p-1)=3 \Leftrightarrow ans(nsqrt(p))$
		P3	$log(e,p)=3 \Leftrightarrow ans(nsqrt(p+1))$
		P4	$x=3 \Leftrightarrow ans(nsqrt(exp(e,x)+1))$
		P5	$ans(nsqrt(exp(e,3)+1))$
		P6	$log(e,p-1)=3 \Leftrightarrow ans(psqrt(p))$

W9	$x=3 \Leftrightarrow ans(psqrt(exp(e,x)+1))$	P1	$log(e,exp(x,2)-1)=3 \Leftrightarrow ans(x)$
W4	$log(e,exp(x,2)-1)=3 \Leftrightarrow ans(abs(x))$	P2	$log(e,p-1)=3 \Leftrightarrow ans(nsqrt(p))$
		P3	$log(e,p)=3 \Leftrightarrow ans(nsqrt(p+1))$
		P4	$x=3 \Leftrightarrow ans(nsqrt(exp(e,x)+1))$
		P5	$ans(nsqrt(exp(e,3)+1))$
		P6	$log(e,p-1)=3 \Leftrightarrow ans(psqrt(p))$
		P7	$log(e,p)=3 \Leftrightarrow ans(psqrt(p+1))$

W10	$ans(psqrt(exp(e,3)+1))$	P1	$log(e,exp(x,2)-1)=3 \Leftrightarrow ans(x)$
W4	$log(e,exp(x,2)-1)=3 \Leftrightarrow ans(abs(x))$	P2	$log(e,p-1)=3 \Leftrightarrow ans(nsqrt(p))$
		P3	$log(e,p)=3 \Leftrightarrow ans(nsqrt(p+1))$
		P4	$x=3 \Leftrightarrow ans(nsqrt(exp(e,x)+1))$
		P5	$ans(nsqrt(exp(e,3)+1))$
		P6	$log(e,p-1)=3 \Leftrightarrow ans(psqrt(p))$
		P7	$log(e,p)=3 \Leftrightarrow ans(psqrt(p+1))$
		P8	$x=3 \Leftrightarrow ans(psqrt(exp(e,x)+1))$

W4	$log(e,exp(x,2)-1)=3 \Leftrightarrow ans(abs(x))$	P1	$log(e,exp(x,2)-1)=3 \Leftrightarrow ans(x)$
		P2	$log(e,p-1)=3 \Leftrightarrow ans(nsqrt(p))$
		P3	$log(e,p)=3 \Leftrightarrow ans(nsqrt(p+1))$
		P4	$x=3 \Leftrightarrow ans(nsqrt(exp(e,x)+1))$
		P5	$ans(nsqrt(exp(e,3)+1))$
		P6	$log(e,p-1)=3 \Leftrightarrow ans(psqrt(p))$
		P7	$log(e,p)=3 \Leftrightarrow ans(psqrt(p+1))$
		P8	$x=3 \Leftrightarrow ans(psqrt(exp(e,x)+1))$
		P9	$ans(psqrt(exp(e,3)+1))$

Now a second solution has been found, namely $x = +\sqrt{e^3+1}$, represented by P9. There still remains an untried equivalence, W4, so the final state is:-

W	empty	P1	$log(e,exp(x,2)-1)=3 \Leftrightarrow ans(x)$
		P2	$log(e,p-1)=3 \Leftrightarrow ans(nsqrt(p))$
		P3	$log(e,p)=3 \Leftrightarrow ans(nsqrt(p+1))$
		P4	$x=3 \Leftrightarrow ans(nsqrt(exp(e,x)+1))$
		P5	$ans(nsqrt(exp(e,3)+1))$
		P6	$log(e,p-1)=3 \Leftrightarrow ans(psqrt(p))$
		P7	$log(e,p)=3 \Leftrightarrow ans(psqrt(p+1))$
		P8	$x=3 \Leftrightarrow ans(psqrt(exp(e,x)+1))$
		P9	$ans(psqrt(exp(e,3)+1))$
		P10	$log(e,exp(x,2)-1)=3 \Leftrightarrow ans(abs(x))$

P10 represents an unsuccessful attempt to find more solutions. All the critical pairs formed on moving W4 into P are rejected, since they are identical to equivalences already processed. In this way, we have been able to avoid infinite loops int he search space.

An examination of the the solution path represented by members of P (e.g. P1 to P5) reveals how narrowing, in contrast to PRESS, rather elegantly isolates the unknown by removing the innermost function symbol first. Unknowns are treated as variables rather than constants, and in this framework, narrowing finds specific solutions using backtracking to find the general solution.

5. INCORPORATING OTHER METHODS

Now let us consider the problems involved in attempting to incorporate rules for attraction and collection into the confluent set. If this can be achieved, then the distinction between isolation, attraction and collection can be removed, and the need for meta-level inference is unnecessary.

Unfortunately, in this instance, it has not proved possible to incorporate either of the rules

$$log(x,x1) + log(x,x2) \Rightarrow log(x,x1*x2) \quad \text{or}$$

$$(x+x1)*(x-x1) \Rightarrow exp(x,2) - exp(x1,2)$$

into C without the confluent set becoming infinite. Our inability to achieve this is perhaps due to the limited nature of the simplification ordering available in ERIL (the ordering described in [KNU70]). It is possible that with a more powerful method of orienting rules (e.g. the Recursive Decomposition Ordering [LES82]), the limitation could be overcome.

The only way to incorporate these rules, in this instance, is to place them in a new set used simply to rewrite equivalences before narrowing takes place, not taking part in the narrowing process at all.

In this framework, the equivalence $log(e,x+1) + log(e,x-1) = 3 \Leftrightarrow ans(x)$ is rewritten in two stages and simplified by rule C4 to form $log(e,exp(x)-1) = 3 \Leftrightarrow ans(x)$ before narrowing begins.

Observing that both the positive and negative square roots are valid solutions to $\log_e(x^2-1) = 3$. The over-general solution does not arise as a result of the narrowing process, but because we have applied the attraction using matching only, thus treating the unknown as a constant.

In the algebraic definitions of the narrowing process, we have defined log as being undefined for negative arguments. This rule should perhaps be more correctly expressed by replacing the xs by ps, i.e. making the rule applicable only for positive arguments of log.

The revised attraction rule still cannot be incorporated in the confluent set C, for the same reasons as before. If it could have been, application of the rule by unification in the narrowing process would have coerced the arguments of our original equivalence to be positive, and only the valid solution would have been found. However, as it stands, we are limited to applying the correctly typed rule by matching only and, for the rule to be applicable at all, we have to assume that the arguments to log in the original equation to be solved are positive. This could be viewed as requiring some prior knowledge about the nature of the solution, i.e. that the unknown x is greater than 1.

6. CONCLUSIONS

We have compared the methods used by the PRESS system with narrowing. By using rewriting by pattern matching only, PRESS treats unknowns as constants, and finds possibly over general solutions. Narrowing, on the other hand, treats unknowns as variables by using unification in rule application, finds specific solutions, and uses backtracking to find the general solution.

In the scope of the single example chosen, an attempt to incorporate other methods into the narrowing process was unsuccessful. Thus the example has fallen short of demonstrating the ideal of completely removing the division of rules into various methods.

As soon as we begin to divide rules, applying some by use of matching rather than by unification, we either have to accept over general solutions, or assume prior knowledge about the nature of the solution.

An idea worth pursuing as future work, is to treat infinite confluent sets by generating them simultaneously with the narrowing process, creating more of the set as needed to find the next solution. One may never know with this method when all the solutions have been found.

ACKNOWLEDGEMENT

We would like to thank Peter Ruffhead for his invaluable advice, and the referees, whose comments have helped us to improve the paper considerably.

APPENDIX

Algebraic definitions.

TYPES

true < bool. false < bool.
lt-zero < lt-one < r. ge-two < ge-one < ge-zero < r.
zero < lt-one. zero < ge-zero. one < ge-one. two < ge-two. three < ge-two.

FUNCTION SIGNATURES

true: true. false: false. 0: zero. 1: one. 2: two. 3: three. e: ge-one.

log: r x r -> r.	exp: r x r -> r.
log: r x gt-zero -> ge-zero.	exp: ge-zero x r -> ge-zero.
log: r x le-zero -> !.	exp: zero x r -> !.
log: r x one -> zero.	exp: r x zero -> one.
log: one x r -> !.	exp: r x two -> ge-zero.
log: lt-one x r -> !.	exp: r x ! -> !.

(Positive square root.)
psqrt: r -> ge-zero.
psqrt: zero -> !.
psqrt: lt-zero -> !.

(Negative square root.)
nsqrt: r -> lt-zero.
nsqrt: zero -> !.
nsqrt: lt-zero -> !.

(Abs: Maps any positive type T<r onto itself; m(T,r) is the meet of T and r, etc.)
abs: T -> (if m(T,r)=T then (if m(T,ge-zero)=T then T else ge-zero) else !)

(Product)
*: r x r -> r.
*: T x one -> (if m(T,r)=T then T else !).
*: one x T -> (if m(T,r)=T then T else !).
*: zero x r -> zero.
*: r x zero -> zero.
*: ! x r -> !.
*: r x ! -> !.

(Addition)
+: r x r -> r.
+: ge-zero x ge-zero -> ge-zero.
+: lt-zero x lt-zero -> lt-zero.
+: T x zero -> (if m(T,r)=T then T else !).
+: zero x T -> (if m(T,r)=T then T else !).
+: one x one -> two.
+: two x one -> three.
+: one x two -> three.
+: r x ! -> !.
+: ! x r -> !.

(Subtraction)
-: r x r -> r.
-: one x one -> zero.
-: two x two -> zero.
-: three x three -> zero.
-: ge-one x one -> ge-zero.
-: ge-two x one -> ge-one.
-: ge-two x two -> ge-zero.
-: T x zero -> (if m(T,r)=T then T else !).
-: r x ! -> !.
-: ! x r -> !.

(Equality: Undefined if the meet of the argument types is undefined.)
=: r x r -> bool.
=: T1 x T2 -> (if m(T1,T2)=! then ! else bool).

ans: r -> bool.
ans: ! -> !.

GENERIC VARIABLE TYPES

b*: bool.   x*: r.   n*: lt-zero.   p*: ge-zero.

REFERENCES

CUN85. R. J. Cunningham and A. J. J. Dick, "Rewrite Systems on a Lattice of Types," *Acta Informatica* **22**, pp. 149-169 (1983).

DIC86a. A. J. J. Dick, "ERIL - Equational Reasoning: an Interactive Laboratory," Report No. RAL86010, Rutherford Appleton Laboratory, Chilton, OX11 0QX (1986).

DIC86b. A. J. J. Dick, "Equational Reasoning and Rewrite Systems on a Lattice of Types," PhD. Thesis, Dept. Computing, Imperial College, London (1986).

FOR84. R. Forgaard and J.V. Guttag, "REVE: A term Rewriting System Generator with Failure Resistant Knuth-Bendix ," pp. 33-56 in *Proceedings of an NSF workshop on the rewrite rule laboratory* (Apr. 1984).

HUE80b. G. Huet and D. C. Oppen, "Equations and Rewrite Rules - A Survey," STAN-CS-80-785 (1980).

LES82. J-P. Jouannoud, P. Lescanne, and F. Reinig, "Recursive Decomposition Ordering," Conf. on Formal Description of Programming Concepts, North Holland (1982).

KNU70. D. E. Knuth and P. B. Bendix, "Simple Word Problems in Universal Algrebras," pp. 263-297 in *Computational Problems in Abstract Algebra*, ed. J. Leech , Pergamon Press (1970).

LES83. P. Lescanne, "Computer Experiments with the REVE Term Rewriting System Generator," pp. 99-108 in *Proc. 10th Symp. on Principles of Programming Languages, ACM Austin, TX* (1983).

RET85. P. Rety, C. Kirchner, H. Kirchner, and P. Lescanne, "NARROWER: a new algorithm for unification and its application to Logic Programming," To appear in Procs. of the First Internat. Conf. on Rewriting Techniques and Applications, LNCS, Springer-Verlag (May 1985).

STE82. L. Sterling, A. Bundy, L. Byrd, R. O'Keefe, and B. Silver, "Solving Symbolic Equations with PRESS," Univ. of Edinburgh, Dept. of Artificial Intelligence, Research Paper No: 171 (Jan. 1982).

Formulation of Induction Formulas in Verification of Prolog Programs

Tadashi KANAMORI and Hiroshi FUJITA

Mitsubishi Electric Corporation

Central Research Laboratory

Tsukaguchi-Honmachi 8-1-1, Amagasaki, Hyogo, JAPAN 661

Abstract

How induction formulas are formulated in verification of Prolog programs is described. The same problem for well-founded induction was investigated by Boyer and Moore in their verification system for terminating LISP programs (BMTP). The least fixpoint semantics of Prolog provides us with the advantages that computational induction can be easily applied and can generate simpler induction schemes. We investigate how computational induction is applied to a class of first order formulas called S-formulas, in which specifications in our verification system are described. In addition, we show how equivalence preserving program transformation can be utilized to merge induction schemes into a simpler one when more than two induction schemes are suggested.

Keywords : Program Verification, Induction, Prolog, Program Transformation.

Contents

1. Introduction

The intimacy of Prolog to first order logic is expected to bring advantages to first order inference in verification ([11],[17]). But how about induction, which plays an important role in verification of programs? As we resort to the usual mathematical induction on the natual numbers in order to prove $1+3+\cdots+(2N-1)=N^2$, or more general induction (either explicitly or implicitly) in order to show the validity in the initial models, we need some kind of induction to prove properties of Prolog programs. The least fixpoint semantics of Prolog suggests fixpoint induction. But how can we find such induction schemes from theorems to be proved ? In addition, when more than two different induction schemes are suggested, how can we manage them?

In this paper, we answer these questions. So far the same problem was investigated by Boyer and Moore [3] in their verification system for terminating LISP programs (BMTP) where they adopted quantifier-free formulas as specifications and applied well-founded induction. The least fixpoint semantics of Prolog provides us with the advantage that computational induction is applicable while staying within usual first order formulas and moreover it generates simpler induction schemes [5]. Here we investigate how computational induction is applied to a class of first order formulas called S-formulas, in which specifications in our verification system are described. In addition, we show how Tamaki-Sato's Prolog program transformation can be utilized to merge computational induction schemes into a simpler one when more than two induction schemes are suggested. Because Tamaki-Sato's transformation preserves equivalence, the manipulation of induction schemes is justified immediately, and it is easy to grasp the meaning.

After summarizing preliminary materials in Section 2 and a framework of verification in Section 3, we describe in Section 4 how computational induction schemes are found from theorems to be proved. In Section 5, we show how induction schemes are merged into a more appropriate one when more than two induction schemes are suggested. Lastly in Section 6, we discuss relations to other works.

2. Preliminaries

In the following, we assume familiarity with the basic terminology of first order logic such as term, atom (atomic formula), formula, substitution, most general unifier (m.g.u.) and so on. We also assume knowledge of the semantics of Prolog such as completion P^*, minimum Herbrand model M_0 and transformation T of Herbrand interpretations (see [1],[5],[7],[10]). We follow the syntax of DEC-10 Prolog [15]. Variables appearing in the head of a definite clause are called *head variables*. Other variables are called *internal variables*. As syntactic variables, we use X, Y, Z for variables, s, t for terms, A, B for atoms and $\mathcal{F}, \mathcal{G}, \mathcal{H}$ for formulas, possibly with primes and subscripts. In addition, we use σ, τ, μ, ν for substitutions and $\mathcal{F}_{\mathcal{G}_1, \mathcal{G}_2, \ldots, \mathcal{G}_n}[\mathcal{H}_1, \mathcal{H}_2, \ldots, \mathcal{H}_n]$ for a replacement of an occurrence of $\mathcal{G}_i$ in $\mathcal{F}$ with $\mathcal{H}_i$ for all $1 \leq i \leq n$. An atom $p(X_1, X_2, \ldots, X_n)$ is said to be *in general form* when $X_1, X_2, \ldots, X_n$ are distinct variables.

2.1. Polarity of Subformulas

We generalize the distinction of positive and negative goals. The *positive* and *negative subformula* of a formula $\mathcal{F}$ is defined as follows (Murray [14]).

(a) $\mathcal{F}$ is a positive subformula of $\mathcal{F}$.

(b) When $\neg\mathcal{G}$ is a positive (negative) subformula of $\mathcal{F}$, then $\mathcal{G}$ is a negative (positive) subformula of $\mathcal{F}$.

(c) When $\mathcal{G}\wedge\mathcal{H}$ or $\mathcal{G}\vee\mathcal{H}$ is a positive (negative) subformula of $\mathcal{F}$, then $\mathcal{G}$ and $\mathcal{H}$ are positive (negative) subformulas of $\mathcal{F}$.

(d) When $\mathcal{G}\supset\mathcal{H}$ is a positive (negative) subformula of $\mathcal{F}$, then $\mathcal{G}$ is a negative (positive) subformula of $\mathcal{F}$, and $\mathcal{H}$ is a positive (negative) subformula of $\mathcal{F}$.

(e) When $\forall X\,\mathcal{G}$, $\exists X\,\mathcal{G}$ are positive (negative) subformulas of $\mathcal{F}$, then $\mathcal{G}_X(t)$ is a positive (negative) subformula of $\mathcal{F}$.

Example 2.1. Let $\mathcal{F}$ be

∀ A,B (reverse(A,B) ⊃reverse(B,A)).

Then *reverse(A, B)* is a negative subformula of $\mathcal{F}$.

2.2. S-formulas and Goal Formulas

Let $\mathcal{F}$ be a closed first order formula. When $\forall X\,\mathcal{G}$ is a positive subformula or $\exists X\,\mathcal{G}$ is a negative subformula of $\mathcal{F}$, X is called a *free variable* of $\mathcal{F}$. When $\forall Y\,\mathcal{H}$ is a negative subformula or $\exists Y\,\mathcal{H}$ is a positive subformula of $\mathcal{F}$, Y is called an *undecided variable* of $\mathcal{F}$. In other words, free variables are variables quantified universally and undecided variables are those quantified existentially when $\mathcal{F}$ is converted to prenex normal form.

Example 2.2.1. Let $\mathcal{F}$ be

∀ A,V,C (append(A,[V],C) ⊃∃B reverse(C,[V|B])).

Then A, V, C are all free variables, while B is an undecided variable.

A closed first order formula S is called a *specification formula* (or S-formula for short) when

(a) no free variable in S is quantified in the scope of quantification of an undecided variable in S and

(b) no undecided variable appears among the negative atoms of S.

In other words, S-formulas are formulas convertible to prenex normal form $\forall X_1, X_2, \ldots, X_n\ \exists Y_1, Y_2, \ldots, Y_m \mathcal{F}$

and none of $Y_1, Y_2, \ldots, Y_m$ appears among the negative atoms of $\mathcal{F}$. Note that S-formulas include both universal formulas $\forall X_1, X_2, \ldots, X_n \mathcal{F}$ and usual execution goals $\exists Y_1, Y_2, \ldots, Y_m\ (A_1 \wedge A_2 \wedge \cdots \wedge A_k)$.

Example 2.2.2. Let S be

$\forall$ A,V,C (append(A,[V],C) $\supset\exists$B reverse(C,[V|B])).

Then S is an S-formula, because free variables A, V, C are quantified outside $\exists B$, and B appears only in the positive atom $reverse(C, [V|B])$. A universal formula $\forall A, B(reverse(A, B) \supset reverse(B, A))$ and an execution goal $\exists C append([1, 2], [3], C)$ are also S-formulas.

A formula G obtained from an S-formula S by leaving free variable X as it is, replacing undecided variable Y with $?Y$ and deleting all quantifiers is called a ***goal formula*** of S. Note that S can be uniquely restorable from G. In the following, we use goal formulas in stead of original S-formulas. Goal formulas are denoted by F, G, H.

Example 2.2.3. An S-formula

$\forall$ A,V,C (append(A,[V],C) $\supset\exists$B reverse(C,[V|B])).

is represented by a goal formula

append(A,[V],C)$\supset$reverse(C,[V|?B]).

A universal formula $\forall A, B(reverse(A, B) \supset reverse(B, A))$ and an execution goal $\exists C append([1, 2], [3], C)$ are represented by $reverse(A, B) \supset reverse(B, A)$ and $append([1, 2], [3], ?C)$ respectively.

2.3. Manipulation of Goal Formulas

Lastly we introduce two manipulations of goal formulas. One is an application of a class of substitutions. To avoid conflict of variable names, we introduce a class of substitutions. A substitution σ is called a *substitution away from* A when σ instantiates each free variable X in A to t such that every variable in t is a fresh free variable not in A.

Example 2.3.1. $< A \Leftarrow [X|L], B \Leftarrow M >$ is a substitution away from $reverse(A, B)$ and unifies it with $reverse([X|L], M)$, where X, L and M are considered fresh free variables. $< C \Leftarrow [X|L], U \Leftarrow Y, B \Leftarrow [X|N] >$ is a substitution away from $append(C, [U], B)$ and unifies it with $append([X|L], [Y], [X|N])$ where X, L, Y and N are considered fresh free variables.

Another manipulation is a *reduction* of goal formulas with the logical constants *true* and *false*. The *reduced form* of a goal formula G, denoted by $G\downarrow$, is the normal form in the reduction system defined as follows :

$$\neg true \rightarrow false, \qquad \neg false \rightarrow true,$$

$$true \wedge G \rightarrow G, \qquad false \wedge G \rightarrow false,$$

$$G \wedge true \rightarrow G, \qquad G \wedge false \rightarrow false,$$

$$true \vee G \rightarrow true, \qquad false \vee G \rightarrow G,$$

$$G \vee true \rightarrow true, \qquad G \vee false \rightarrow G,$$

$$true \supset G \rightarrow G, \qquad false \supset G \rightarrow true,$$

$$G \supset true \rightarrow true, \qquad G \supset false \rightarrow \neg G.$$

Example 2.3.2. Let G_1 and G_2 be $false \supset reverse(B,A)$ and $true \supset reverse(B,A)$. Then $G_1 \downarrow$ is *true* and $G_2 \downarrow$ is $reverse(B,A)$.

3. Framework of Verification of Prolog Programs

3.1. Programming Language

Our programming language is an extension of Prolog. We introduce **type** construct into Prolog to separate definite clauses defining data structures from others defining procedures, e.g.,

```
type.
    list([ ]).
    list([X|L]) :- list(L).
end.
```

The body of **type** is a conjunction of definite clauses whose head is with a unary predicate defining a data structure. (**type** in our verification system corresponds to *shell* in BMTP.) Procedures are defined by following the syntax of DEC-10 Prolog [15], e.g.,

```
append([ ],K,K).
append([X|L],M,[X|N]) :- append(L,M,N).
reverse([ ],[ ]).
reverse([X|L],M) :- reverse(L,N),append(N,[X],M).
```

Throughout this paper, we study pure Prolog consisting of definite clauses "B :- $B_1, B_2, \ldots, B_m$" ($m \geq 0$) and regard a finite set of definite clauses P as their conjunction. We assume that variables in each definite clause are renamed at each use so that there occurs no conflict of variable names.

3.2. Specification Language

The main construct of our specification language is "**theorem**" to state a theorem to be proved, e.g.,

```
theorem(halting-theorem-for-append).
    ∀ A:list,B ∃ C append(A,B,C).
end.
```

The body of **theorem** must be a closed S-formula. Any variable X in quantification may be followed by a type qualifier : p (e.g., : *list* above). $\forall X{:}p\mathcal{F}$ and $\exists X{:}p\mathcal{F}$ are abbreviations of $\forall X(p(X) \supset \mathcal{F})$ and

$\exists X(\mathcal{F} \wedge p(X))$, respectively.

3.3. Formulation of Verification

Let S be a specification in an S-formula, M_0 be the minimum Herbrand model of P and P^* be the completion of P. We adopt a formulation as follows : Model-theoretically speaking, verification of S with respect to P is showing $M_0 \models S$. Proof-theoretically speaking, it is proving S from P^* using first order inference and computational induction. (Of course, the proof-theoretical formulation is weaker than the model-theoretical formulation.)

The most important difference between our verification system and BMTP is that specifications in BMTP are quantifier-free (i.e., universal) formulas while ours are S-formulas. Though we prove quantifier-free specifications of the form $\forall X_1, X_2, \ldots, X_n\ (A_1 \wedge A_2 \wedge \cdots \wedge A_m \supset A_0)$ in most cases, the consideration of existential quantifiers is inevitable because of the effects of internal variables in Prolog. For example, suppose we prove $\forall X, Y(condition(X,Y) \supset p(X,Y))$ with respect to a program $p(X,Y)$:- $q(X,Z), r(Z,Y)$. Then we must prove $\forall X, Y\ (condition(X,Y) \supset \exists Z(q(X,Z) \wedge r(Z,Y))$ substantially.

4. Generation of Computational Induction Schemes

4.1. Computational Induction

For pure Prolog, the transformation T of Herbrand interpretations is always continuous and there holds $\bigcup_{i=0}^{\infty} T^i(\emptyset) = M_0$. This suggests use of fixpoint induction to prove $M_0 \models \psi$. Here we explain it rather intuitively following Clark [5] p.75-76.

Example 4.1.1. Let *reverse* be a relation defined by the definite clauses before. The *reverse* relation is the smallest set of pairs of terms that includes a pair $([\,],[\,])$ and that, for any term s,includes $([s|t_1], t_2)$ whenever it includes (t_1, t) and $(t, [s], t_2)$ is in the *append* relation. Hence, suppose $Q(A,B)$ is a formula with free variables A, B. For any Herbrand interpretation, $Q(A,B)$ will denote some binary relation over terms. If this relation includes $([\,],[\,])$,i.e.,

Q([ ],[ ])

is true, and if it includes $([s|t_1], t_2)$ whenever it includes (t_1, t) and $(t, [s], t_2)$ is in the *append* relation, i.e.,

∀ A,B,C,U (Q(A,C)∧append(C,[U],B)⊃Q([U|A],B))

is true, then the relation $Q(A,B)$ includes all the pairs of terms in the *reverse* relation. In other words,

∀ A,B (reverse(A,B)⊃Q(A,B))

is true of the *reverse* relation and such $Q(A,B)$. Hence we get the following computational induction scheme

$$\frac{Q([\,],[\,]) \qquad \forall A,B,C,U\ (Q(A,C) \wedge append(C,[U],B) \supset Q([U|A],B))}{\forall A,B\ (reverse(A,B) \supset Q(A,B))}$$

This is a Prolog version of de Bakker and Scott's computational induction [2]. In order to apply such induction, we need to know (a) how we select the key atom ($reverse(A,B)$ above), (b) what conditions we have to observe for soundness and (c) how we can make the generated induction schemes simple.

First of all, we introduce the notion "dominant atom". An atom A in goal formula G is said to be *dominant* in G when G is tautologically equivalent to a goal formula of the form $A \supset H$. Note that A is dominant in G if and only if A is negative and $G_A[false] \downarrow$ is *true*. Then H is $G_A[true] \downarrow$.

Example 4.1.2. Let G be $reverse(A,B) \supset reverse(B,A)$. Then the left $reverse(A,B)$ is dominant in G, but the right $reverse(B,A)$ is not.

4.2. Closed Atom and Instantiated Program

Must the key atom be always in general form $p(X_1, X_2, \ldots, X_n)$? An atom A is said to be *closed with respect to* a definite clause C, when, for any ground instance of C such that the head is a ground instance of A, any recursive call in the body is also a ground instance of A. (Hence an atom A is closed w.r.t. any non-recursive definite clause; A is closed w.r.t. any definite clause whose head is not unifiable with A). When A is closed w.r.t. every definite clause in P, A is said to be *closed with respect to P*. This means that the set of all ground atoms in M_0 that are instances of A is computable by some instances of definite clauses with respect to which A is closed. Note that an atom $p(X_1, X_2, \ldots, X_n)$ in general form is always closed.

Example 4.2.1. Consider an atom $append(A, [U], C)$. This atom is closed, which means that

$\{append(t_1, [t_2], t_3) \mid t_1, t_2$ and t_3 are ground terms$\} \bigcap M_0$

is computable by some instances of definite clauses, i.e.,

```
append([ ],[Y],[Y]).
append([X|L],[Y],[X|N]) :- append(L,[Y],N).
```

The closedness w.r.t. "B_0 :- $B_1, B_2, \ldots, B_m$" can be checked as follows.

(a) Check whether the head B_0 is unifiable with A by a substitution away from A (see 2.3). If it is, decompose the m.g.u. to $\sigma \circ \tau_0$ where σ is the restriction to variables in B_0 and τ_0 is the restriction to variables in A. If it is not, A is closed w.r.t. the definite clause.

(b) Check whether each instance of the recursive call in the body $\sigma(B_i)$ is an instance of A and if it is, compute the instantiation τ_i. If it is not, A is not closed w.r.t. the definite clause.

The set of all instances of definite clauses by σ is called an *instantiated program* for A.

Example 4.2.2. Consider an atom, $append(A, [U], C)$. Then the first head $append([], L, L)$ is unifiable with $append(A, [U], C)$ by

$< L \Leftarrow [Y] > \circ < A \Leftarrow [\,], U \Leftarrow Y, C \Leftarrow [Y] >$.

The second head $append([X|L], M, [X|N])$ is unifiable with $append(A, [U], C)$ by

$< M \Leftarrow [Y] > \circ < A \Leftarrow [X|L], U \Leftarrow Y, C \Leftarrow [X|N] >$

and the instance $append(L, [Y], N)$ in the body is also an instance of $append(A, [U], C)$ by

$< A \Leftarrow L, U \Leftarrow Y, C \Leftarrow N >$.

4.3. Generation of Induction Schemes

Let G be a goal formula, $A{=}p(t_1,t_2,\ldots,t_n)$ be a dominant and closed atom in G and Q be $G_A[true]\downarrow$. Let "B_0 :- $B_1,B_2,\ldots,B_m$" be a definite clause of the instantiated program for A. By $\sigma(B_0$:- $B_1,B_2,\ldots,B_m)_p(Q)$, we denote a formula obtained by replacing $\sigma(B_i) = p(s_1,s_2,\ldots,s_n)$ in $\sigma(B_1{\wedge}B_2{\wedge}\cdots{\wedge}B_m{\supset}B_0)$ with $\tau_i(Q)$ when B_0 is unifiable with A and *true* otherwise.

We generate an induction scheme as follows. All free variables are quantified universally at the outermost, which keeps the generated subgoals within S-formulas.

$$\frac{\sigma_1(\mathrm{B}_{10} \text{ :- } \mathrm{B}_{11},\mathrm{B}_{12},\ldots,\mathrm{B}_{1m_1})_p(Q),\ \sigma_2(\mathrm{B}_{20} \text{ :- } \mathrm{B}_{21},\mathrm{B}_{22},\ldots,\mathrm{B}_{2m_2})_p(Q),\ \ldots,\ \sigma_k(\mathrm{B}_{k0} \text{ :- } \mathrm{B}_{k1},\mathrm{B}_{k2},\ldots,\mathrm{B}_{km_k})_p(Q)}{G}$$

Example 4.3.1. Suppose we prove a theorem

theorem(reverse-reverse).

∀A,B (reverse(A,B) ⊃reverse(B,A)).

end.

This S-formula is represented by

reverse(A,B)⊃reverse(B,A).

Then $reverse(A,B)$ is dominant, because $false{\supset}reverse(B,A)\downarrow$ is *true*. Let $Q(A,B)$ be ($true{\supset}reverse$ $(B,A))\downarrow = reverse(B,A)$. $reverse(A,B)$ is closed with its instantiated program

reverse([ ],[ ]).

reverse([U|A],B) :- reverse(A,C),append(C,[U],B).

By replacing $reverse([\,],[\,])$ in the first clause with $Q([\,],[\,])$, the first induction formula is

reverse([ ],[ ]).

Similarly, by replacing $reverse([U|A],B)$ and $reverse(A,C)$ in the second clause with $Q([U|A],B)$ and $Q(A,C)$ respectively, the second induction formula is

reverse(C,A)∧append(C,[U],B)⊃reverse(B,[U|A]).

The induction scheme generated is the one in example 4.1, i.e.,

$$\frac{\mathrm{reverse}([\,],[\,]) \qquad \forall A,B,C,U\ (\mathrm{reverse}(C,A)\ \wedge \mathrm{append}(C,[U],B) \supset \mathrm{reverse}(B,[U|A]))}{\forall A,B\ (\mathrm{reverse}(A,B)\ \supset \mathrm{reverse}(B,A))}$$

Example 4.3.2. Suppose we prove a theorem

theorem(appended-single-element-is-the-last).

∀A,U,B (append(A,[U],B) ⊃last(B,U)).

end.

Then $append(A,[U],B)$ is dominant and closed. Let $Q(A,U,B)$ be $(true{\supset}last(B,U))\downarrow = last(B,U)$. The induction scheme generated is

$$\frac{\mathrm{last}([U],U) \qquad \forall U,B\ (\mathrm{last}(B,U) \supset \mathrm{last}([V|B],U))}{\forall A,U,B\ (\mathrm{append}(A,[U],B)\ \supset \mathrm{last}(B,U))}$$

4.4. Use of Generalization

Must we always give up applications of computational induction, when the key atom A is not closed w.r.t. some definite clause ?

Example 4.4.1. Consider an atom, $reverse(A, [V|B])$. Then A is not closed w.r.t.

reverse([X|L],M) :- reverse(L,N),append(N,[X],M).

because the instance of the recursive call by σ in the definition above may have the form $reverse(L, N)$ which is not necessarily an instance of $reverse(A, [V|B])$.

A set of replacement $\mathcal{M}$ of occurrences of subterms in A with fresh free variables is called a *generalization mask* when no different subterm is replaced with a same fresh free variable. An atom obtained from A by replacing the occurrence of t_{i_k} with a fresh free variable X_{i_k} for all $< t_{i_k} \Leftarrow X_{i_k} > \in \mathcal{M}$ is called a *generalization* of A with respect to $\mathcal{M}$ and is denoted by $A_{\mathcal{M}}$. Then $X_{i_1} = t_{i_1} \wedge X_{i_2} = t_{i_2} \wedge \cdots \wedge X_{i_k} = t_{i_m}$ is called a *generalization equation*. (When $\mathcal{M}=\emptyset$, $A_\emptyset$ is A itself and the generalization equation is *true*.)

Example 4.4.2. $\mathcal{M}=\{ [V|B] \Leftarrow B' \}$ is a generalization mask of $reverse(A, [V|B])$ in

∀ A,V,B (reverse(A,[V|B]) ⊃member(V,A)).

The generalization of $reverse(A, [V|B])$ w.r.t. $\mathcal{M}$ is $reverse(A, B')$ where B' is a fresh free variable. The generalization equation is $B' = [V|B]$. Intuitively this corresponds to modify the theorem to

∀ A,V,B,B′ (reverse(A,B′)∧B′=[V|B] ⊃member(V,A)).

A is said to be *closed with* $\mathcal{M}$ *w.r.t.* "B_0 :- $B_1, B_2, \ldots, B_m$", when $A_{\mathcal{M}}$ is closed w.r.t. it.

Example 4.4.3. Let $\mathcal{M}$ be $\{ [V|B] \Leftarrow B' \}$. Then $reverse(A, [V|B])$ is closed with $\mathcal{M}$ w.r.t.

reverse([X|L],M) :- reverse(L,N),append(N,[X],M).

because the head $reverse([X|L], M)$ is unifiable with $reverse(A, B')$ by $<> \circ < A \Leftarrow [X|L], B' \Leftarrow M >$ and the recursive call $reverse(L, N)$ in the body is also an instance of $reverse(A, B')$.

One naturally expects that $A_{\mathcal{M}}$ be the most specific one. An atom $\overline{A}$ is called a *closure* of A with respect to P when

(a) $\overline{A}$ is closed with respect to P,

(b) A is an instance of $\overline{A}$ and

(c) There is no strict instance of $\overline{A}$ satisfying (a) and (b).

The closure is unique up to renaming, and A is closed if and only if $A = \overline{A}$ modulo renaming. (See appendix for the proof of uniqueness and the algorithm to compute the closure.) A generalization mask $\mathcal{M}$ corresponding to $\overline{A}$ is called a *computational induction mask*. Examples in 4.1–3 are cases with the empty mask $\emptyset$.

Example 4.4.4. Let A be $reverse(A, [V|B])$ and $\mathcal{M}$ be $\{[V|B] \Leftarrow B'\}$. Then the first head $reverse([\,], [\,])$ is unifiable with $reverse(A, B')$ by $<> \circ < A \Leftarrow [\,], B' \Leftarrow [\,] >$. A is closed with $\mathcal{M}$ w.r.t. the second definite clause. Moreover $reverse(A, B')$ is a closure of $reverse(A, [V|B])$ and $\mathcal{M}$ is a computational induction mask.

Then computational induction schemes are generated as they are in 4.3 by replacing A with A_M and Q with $G_A[e]\downarrow$ where e is the generalization equation.

Example 4.4.5. Let *member* be defined by the following programs.

member(X,[X|L]).

member(X,[Y|L]) :- member(X,L).

Suppose we prove

theorem(last-is-a-member).

$\forall$A,V,B (reverse(A,[V|B]) $\supset$member(V,A)).

end.

Note the dominant atom *reverse*$(A,[V|B])$. Because $\emptyset$ is not a computational induction mask, we can't apply the induction immediately without generalization. We need a nonempty mask $M=\{[V|B] \Leftarrow B'\}$ and generated formulas are

$\forall$ V,B ([ ]=[V|B] $\supset$member(V,[ ])),

$\forall$ U,V,A,B,C,D ((C=[V|B] $\supset$member(V,A)) $\wedge$append(C,[U],D) $\supset$(D=[V|B] $\supset$member(V,[U|A]))).

Note that the second subgoal remains within S-formulas. The first goal is reduced to *true* by the definition $Z=Z$. The second goal is reduced to

$\forall$ U,V,A,B,C ((C=[V|B] $\supset$member(V,A)) $\wedge$append(C,[U],[V|B]) $\supset$member(V,[U|A])).

and it is proved using the definition of *append*.

4.5. Examples of Induction Schemes

We show how computational inductions are applied by several examples.

Example 4.5.1. We need not impose the termination condition, i.e., execution of $p(t_1,t_2,\ldots,t_n)$ may run indefinitely for some ground terms $t_1,t_2,\ldots,t_n$, even if an ideal, fair non-deterministic interpreter is used. This means that we can separate M_0 from other models of P^* and show properties specific to M_0. Let p and q be defined by the following programs (Apt and van Emden [1] p.846).

p(a) :- p(X),q(X).

p(suc(X)) :- p(X).

q(b).

q(suc(X)) :- q(X).

The execution of ?- $p(suc^i(a))$ never terminates, while those of ?- $p(suc^j(b))$ and ?- $q(suc^i(a))$ fail finitely. The minimum (and the maximum) Herbrand model is $M_0=\{q(b),q(suc(b)),q(suc(suc(b))),\ldots\}$. But we can't infer $\neg p(suc^i(a))$ by first order inference from the completion P^*. This is shown by the existence of a non-Herbrand model M. Let the domain of M be a set consisting of red natural numbers, green natural numbers and blue integers and the interpretation of symbols on M be one interpreting a as red 0, b as green 0, suc be a function mapping X to $X+1$ in the usual arithmetic, $p(i)$ is *true* iff i is either a red natural number or a blue integer and $q(j)$ is *true* iff j is either a green natural number or a blue integer. Then M is a model of P^* not isomorphic to M_0 and $p(suc^i(a))$ is valid in M (cf. Jaffar et al

[10]). Suppose we prove a theorem

theorem(p-does-not-hold).

$\forall$A $\neg$p(A).

end.

which is valid in M_0 but not valid in M, hence not provable by first order inference from P^*. Note the dominant atom $p(A)$. Because it is closed with $\emptyset$ w.r.t. each definite clause, we can apply computational induction immediately. Because $Q(A)$ is *false*, we have two generated formulas

$\forall$ A (false $\wedge$q(A) $\supset$false)

false $\supset$false.

Both of them are trivially *true*.

Example 4.5.2. Computational induction in the example above has separated Herbrand models from non-Herbrand models. It is also possible to separate minimum Herbrand models from general Herbrand models. Suppose we prove

theorem(even-pair).

$\forall$ A,B (even-difference(A,B)$\supset$even(A)).

end.

where *even-difference* and *even* are defined by

even-difference(0,0).

even-difference(suc(suc(X)),Y) :- even-difference(X,Y).

even-difference(X,suc(suc(X))) :- even-difference(X,Y).

even-difference(X,Y) :- even-difference(suc(suc(X)),Y).

even-difference(X,Y) :- even-difference(X,suc(suc(Y))).

even(0).

even(suc(suc(X))) :- even(X).

Then the minimum Herbrand model of P^* is

M_{00} = { even-difference(suc^{2i}(0),suc^{2j}(0))| i,j$\in$N}

and the maximum Herbrand model is

M_{11} = { even-difference(suc^{i}(0),suc^{j}(0))| i,j$\in$N}.

There are another two Herbrand models of P^*,

M_{10} = M_{00} $\cup$ { even-difference(suc^{2i+1}(0),suc^{2j}(0))| i,j$\in$N}

M_{01} = M_{00} $\cup$ { even-difference(suc^{2i}(0),suc^{2j+1}(0))| i,j$\in$N}.

The theorem holds in M_{00}, but in neither of M_{10},M_{01} and M_{11}. Note the dominant atom *even-difference*(A,B). Because it is closed w.r.t. each definite clause, we have five generated formulas

even(0).

$\forall$ A (even(A)$\supset$even(suc(suc(A)))).

$\forall$ A (even(A)$\supset$even(A)).

$\forall$ A (even(suc(suc(A)))$\supset$even(A)).

$\forall$ A (even(A)$\supset$even(A)).

Example 4.5.3. When the theorem to be proved includes a dominant atom with a type predicate, we have the effect to perform the usual structural induction. Suppose we prove (cf.Kowalski [13] pp.221-222)

theorem(right-identity-of-append).

$\forall$ A:list append(A,[ ],A).

end.

Note the dominant atom *list(A)*. Because it is closed with $\emptyset$ w.r.t. each definite clause, we can apply computational induction immediately. Because $Q(A)$ is $append(A, [\,], A)$, we have two generated formulas

append([ ],[ ],[ ]),

$\forall$ U,A (append(A,[ ],A) $\supset$append([U|A],[ ],[U|A])).

Example 4.5.4. Theorems may include existentially quantified variables. Suppose we prove

theorem(last-can-be-first-of-reversed).

$\forall$ A,V,C (append(A,[V],C) $\supset\exists$B reverse(C,[V|B])).

end.

Note the dominant atom $append(A, [V], C)$. Because it is closed with $\emptyset$ w.r.t. each definite clause, we can apply computational induction immediately. Because $Q(A, V, C)$ is $\exists B reverse(C, [V|B])$, we have two generated formulas

$\forall$ V $\exists$ B reverse([V],[V|B]),

$\forall$ U,A,V,C ($\exists$ B reverse(C,[V|B]) $\supset\exists$ B reverse([U|C],[V|B])).

Note that the generated subgoals still remain within S-formulas.

5. Merging of Computational Induction Schemes

5.1. Mergeable Schemes

In Section 4, we cited examples with only one dominant atom intentionally. But sometimes more than two different induction schemes are suggested from theorems to be proved.

Example 5.1. Suppose we prove the second formula generated from *reverse-reverse*

theorem(first-last).

$\forall$A,B,C,U (reverse(C,A) $\wedge$append(C,[U],B) $\supset$reverse(B,[U|A])).

end.

Then both $reverse(C, A)$ and $append(C, [U], B)$ are dominant. Let Q_1 and Q_2 be

Q_1(A,B,C,U) : append(C,[U],B) $\supset$reverse(B,[U|A]),

Q_2(A,B,C,U) : reverse(C,A) $\supset$reverse(B,[U|A]).

Then $reverse(C, A)$ and $append(C, [U], B)$ suggests two different induction schemes.

$$\frac{\forall B,U\ Q_1([\,],B,[\,],U)\quad \forall A,B,C,U,A_1,V,\ (Q_1(A_1,B,C,U)\wedge append(A_1,[V],A)\supset Q_1(A,B,[V|C],U))}{\forall A,B,C,U\ (reverse(C,A)\wedge append(C,[U],B)\supset reverse(B,[U|A]))}$$

$$\frac{\forall A,U\ Q_2(A,[U],[\,],U)\quad \forall A,B,C,U,V\ (Q_2(A,B,C,U)\supset Q_2(A,[V|B],[V|C],U))}{\forall A,B,C,U\ (reverse(C,A)\wedge append(C,[U],B)\supset reverse(B,[U|A]))}$$

In the example above, the substitution to C in each scheme coincides and it suggests a possibility to apply both inductions simultaneously. In order to justify such manipulation, we describe an equivalence preserving transformation system in the next subsection.

5.2. Tamaki-Sato's Transformation

Tamaki-Sato's transformation system was developed for Prolog programs based on unfold/fold transformations. The entire process proceeds as follows [16].

P_0 :=the initial program ; D_0 := {};
mark every clause in P_0 "foldable";
for i := 1 **to** arbitrary N
    apply any of the transformation rules to obtain P_i and D_i from P_{i-1} and D_{i-1};

Figure 1. Process of Tamaki-Sato's Transformation

Example 5.2.1. Before starting the transformation, the initial program is given, e.g.,

P_0 : C_1. append([ ],L,L).
C_2. append([X|L],M,[X|N]) :- append(L,M,N).
C_3. reverse([ ],[ ]).
C_4. reverse([X|L],M) :- reverse(L,N),append(N,[X],M).

and D_0 is initialized to {}.

The basic part of Tamaki-Sato's transformation system consists of three transformation rules, i.e., definition,unfolding and folding.

Definition : Let C be a clause of the form $p(X_1, X_2, \ldots, X_n)$:- $A_1, A_2, \ldots, A_m$ where

(a) p is an arbitrary predicate appearing neither in P_{i-1} nor in D_{i-1},
(b) $X_1, X_2, \ldots, X_n$ are distinct variables,and
(c) $A_1, A_2, \ldots, A_m$ are atoms whose predicates all appear in P_0.

Then let P_i be $P_{i-1} \bigcup \{C\}$ and D_i be $D_{i-1} \bigcup \{C\}$. Do not mark C "foldable".

Example 5.2.2. Suppose we need conjunction of atoms each induction scheme is accounting for in example 5.1. Then we introduce it by the following definition.

C_5. new-p(L,M,N,X) :- reverse(N,L),append(N,[X],M).

Then $P_1 = \{\underline{C_1}, \underline{C_2}, \underline{C_3}, \underline{C_4}, C_5\}$ and $D_1 = \{C_5\}$. The underlines indicate "foldable" clauses.

Unfolding : Let C be a clause in P_{i-1},A be an atom in its body and $C_1, C_2, \ldots, C_k$ be all the clauses in P_{i-1} whose heads are unifiable with A. Let C'_i be the result of resolving C with C_i on A. Then let P_i be $(P_{i-1} - \{C\}) \bigcup \{C'_1, C'_2, \ldots, C'_k\}$ and D_i be D_{i-1}. Mark each C'_j "foldable" unless it is already in P_{i-1}.

Example 5.2.3. When C_5 is unfolded at its first atom in the body, we obtain $P_2 = \{\underline{C_1}, \underline{C_2}, \underline{C_3}, \underline{C_4}, \underline{C_6}, \underline{C_7}\}$

and $D_2 = \{C_5\}$ where

C_6. new-p([ ],M,[ ],X) :- append([ ],[X],M).

C_7. new-p(L,M,[Y|N],X) :- reverse(N,L$_1$),append(L$_1$,[Y],L),append([Y|N],[X],M).

C_6 and C_7 are still unfoldable into

C_6'. new-p([ ],[X],[ ],X).

C_7'. new-p(L,[Y|M],[Y|N],X) :- reverse(N,L$_1$),append(L$_1$,[Y],L),append(N,[X],M).

and we get $P_3 = \{\underline{C_1}, \underline{C_2}, \underline{C_3}, \underline{C_4}, \underline{C_6'}, \underline{C_7'}\}$ and $D_3 = \{C_5\}$.

Folding : Let C be a clause in P_{i-1} of the form A_0 :- $A_1, A_2, \ldots, A_n$ and C_{folder} be a clause in D_{i-1} of the form B_0 :- $B_1, B_2, \ldots, B_m$. Suppose there is a substitution σ and a subset $\{A_{i_1}, A_{i_2}, \ldots, A_{i_m}\}$ of the body of C such that the following conditions hold.

(a) $A_{i_j} = \sigma(B_j)$ for $j = 1, 2, \ldots, m$,

(b) σ substitutes distinct variables for the internal variables of C_{folder} and moreover those variables occur neither in A_0 nor in $\{A_1, A_2, \ldots, A_n\} - \{A_{i_1}, A_{i_2}, \ldots, A_{i_m}\}$,and

(c) C is marked "foldable" or $m < n$.

Then let P_i be $(P_{i-1} - \{C\}) \bigcup \{C'\}$ and D_i be D_{i-1} where C' is a clause with head A_0 and body $(\{A_1, A_2, \ldots, A_n\} - \{A_{i_1}, A_{i_2}, \ldots, A_{i_m}\}) \bigcup \{\sigma(B)\}$. Let C' inherit the mark of C.

Example 5.2.4. By folding the first and the third atom of the body of C_7' by C_5, we obtain $P_4 = \{\underline{C_1}, \underline{C_2}, \underline{C_3}, \underline{C_4}, \underline{C_6}, \underline{C_8}\}$ and $D_4 = \{C_5\}$ where

C_8. new-p(L,[Y|M],[Y|N],X) :- new-p(L$_1$,M,N,X),append(L$_1$,[Y],L).

The most important property of this transformation system is stated as follows [16].

Equivalence Preservation Theorem

P_N is equivalent to $P_0 \bigcup D_N$ in the minimum Herbrand model semantics, i.e., the minimum Herbrand models of them are identical.

Example 5.2.5. Through the previous transformation process, we reach a program

P_4 : C_1. append([ ],L,L).

C_2. append([X|L],M,[X|N]) :- append(L,M,N).

C_3. reverse([ ],[ ]).

C_4. reverse([X|L],M) :- reverse(L,N),append(N,[X],M).

C_6'. new-p([ ],[X],[ ],X).

C_8. new-p(L,[Y|M],[Y|N],X) :- new-p(L$_1$,M,N,X),append(L$_1$,[Y],L).

which is equivalent to

P_0 : C_1. append([ ],L,L).

C_2. append([X|L],M,[X|N]) :- append(L,M,N).

C_3. reverse([ ],[ ]).

C_4. reverse([X|L],M) :- reverse(L,N),append(N,[X],M).

D_4 : C_5. new-p(L,M,N,X) :- reverse(N,L),append(N,[X],M).

5.3. Derivation of Merged Schemes

We keep each induction scheme in a triple $(A, Q, \{\sigma_i(C_i)\})$, i.e., an atom A, a goal formula Q and an instantiated program $\{\sigma_1(C_1), \sigma_2(C_2), \ldots, \sigma_k(C_k)\}$.

Example 5.3.1. Suppose we are trying to prove the theorem *first-last*. Before merging, we have two induction schemes I and J whose instantiated programs are

I : reverse([ ],[ ]).
reverse([X|L],M) :- reverse(L,N),append(N,[X],M).

J : append([ ],[Y],[Y]).
append([X|L],[Y],[X|M]) :- append(L,[Y],M).

Let I and J be induction schemes accounting for $p(t_1, t_2, \ldots, t_n)$ and $q(s_1, s_2, \ldots, s_m)$,respectively. Let A_0 :- $A_1, A_2, \ldots, A_\alpha$ be a recursive definite clause of I and B_0 :- $B_1, B_2, \ldots, B_\beta$ be a recursive definite clause of J and suppose A_0 is $\tau_0(p(t_1, t_2, \ldots, t_n))$ and B_0 is $\tau'_0(q(s_1, s_2, \ldots, s_m))$. (See 4.2 for τ_0 and τ'_0.) Then these two recursive definite clauses are said to be *mergeable* when the following condition is satisfied : When τ_0 and τ'_0 substitute an identical term to each common variable of $p(t_1, t_2, \ldots, t_n)$ and $q(s_1, s_2, \ldots, s_m)$, there is a one-to-one correspondence between recursive calls in the bodies such that the corresponding τ_i and τ'_j substitute an identical term to each common variable of $p(t_1, t_2, \ldots, t_n)$ and $q(s_1, s_2, \ldots, s_m)$.

Induction scheme I and induction scheme J are said to be *mergeable* when any pair of a recursive definite clause of I and a recursive definite clause of J are mergeable. Note that the criteria for mergeability is much looser than that of BMTP (cf. [3] pp.191-194).

Example 5.3.2. The previous two induction schemes are obviously mergeable, because, for the second clause,

(a) When *reverse*(C, A) is unified with *reverse*$([X|L], M)$ by $\tau_0 = < C \Leftarrow [X|L], A \Leftarrow M >$, *append*$(C, [U], B)$ is unifiable with *append*$([X|L], [Y], [X|N])$ by $\tau'_0 = < C \Leftarrow [X|L], B \Leftarrow [X|N], U \Leftarrow Y >$

(b) There is only one recursive calls *reverse*(L, N) and *append*$(L, [Y], N)$ in each second definite clause and they are instances of *reverse*(C, A) by $\tau_1 = < C \Leftarrow L, A \Leftarrow N >$ and of *append*$(C, [U], B)$ by $\tau'_1 = < C \Leftarrow L, B \Leftarrow N, U \Leftarrow Y >$.

The mergeability guarantees that a transformation sequence like the one stated in example 5.2.1–5 is applicable. Note that the transformation is a routine and we can apply it mechanically. When two schemes are mergeable, we derive a new scheme as follows.

(a) Define *new-p*$(X_1, X_2, \ldots, X_l)$ by the conjunction of $p(t_1, t_2, \ldots, t_n)$ and $q(s_1, s_2, \ldots, s_m)$.

(b) Unfold at $p(t_1, t_2, \ldots, t_n)$ and $q(s_1, s_2, \ldots, s_m)$ once in the bodies of definite clauses.

(c) Fold if possible.

Then the obtained definite clauses represent a new scheme accounting for *new-p*$(X_1, X_2, \ldots, X_l)$ with $Q = G_{p(t_1,t_2,\ldots,t_n),q(s_1,s_2,\ldots,s_m)}[true, true] \downarrow$.

Example 5.3.3. As shown in example 5.2.1—5, we can obtain a definite clause program for *new-p* from two induction schemes in example 5.1.

new-p([ ],[X],[ ],X).

new-p(L,[Y|M],[Y|N],X) :- new-p(L_1,M,N,X),append(L_1,[Y],L).

and the theorem to be proved is now

∀ A,B,C,U (new-p(A,B,C,U)⊃reverse(B,[U|A])).

Hence the new scheme accounting for *new-p* represents

$$\frac{\forall U\ reverse([U],[U]) \quad \forall A,B,C,U,A_1,V\ (reverse(B,[U|A_1]) \wedge append(A_1,[V],A) \supset reverse([V|B],[U|A]))}{\forall A,B,C,U\ (reverse(C,A) \wedge append(C,[U],B) \supset reverse(B,[U|A]))}$$

Merging of computational induction schemes sometimes results in surprisingly simple schemes.

Example 5.3.4. Suppose we prove a theorem

theorem(equivalence-of-flatten-and-mc-flatten).

∀W,A,B,C (flatten(W,A) ∧mc-flatten(W,B,C) ⊃append(A,B,C)).

end.

where *flatten* and *mc-flatten* are procedures collecting leaves of binary trees defined as follows [3]. (*mc-flatten*(*W*, [ ], *A*) performs the same task as *flatten*(*W*, *A*) more efficiently.)

flatten('X',['X']).

flatten(tree(X,Y),L) :- flatten(X,L_1),flatten(Y,L_2),append(L_1,L_2,L).

mc-flatten('X',M,['X'|M]).

mc-flatten(tree(X,Y),M,N) :- mc-flatten(Y,M,L),mc-flatten(X,L,N).

Then by defining a new predicate by

new-p(Z,L,M,N) :- flatten(Z,L),mc-flatten(Z,M,N).

and applying the transformations,we obtain a new definition

new-p('X',['X'],M,['X'|M]).

new-p(tree(X,Y),L,M,N) :- append(L_1,L_2,L),new-p(X,L_1,L_3,N),new-p(Y,L_2,M,L_3).

The scheme accounting for *new-p*(W, A, B, C) is

$$\frac{\begin{array}{c}\forall U,B\ append(['U'],B,['U'|B]) \\ \forall A,B,C,A_1,A_2,A_3\ (append(A_1,A_2,A) \wedge append(A_1,A_3,C) \wedge append(A_2,B,A_3) \supset append(A,B,C))\end{array}}{\forall W,A,B,C\ (flatten(W,A) \wedge mc\text{-}flatten(W,B,C) \supset append(A,B,C))}$$

The second formula is the associativity of *append* for A_1,A_2 and B.

6. Discussion

Computational induction is due to de Bakker and Scott [2]. Its use in mechanical verification was investigated by Weyrauch and Milner [18] and Gordon etc [8]. The simplest form of its use in Prolog was pointed out by Clark [5]. (Integration of induction into a logic programming system as a rule for execution was investigated by Hagiya and Sakurai [9].) But our use is more general in the sense that atoms accounted for need not to be in general form $p(X_1, X_2, \ldots, X_n)$, e.g., *Example 4.4.5* and *4.5.4*. Our method to merge computational induction schemes and its justification are new as far as we know.

Compared with BMTP, use of computational induction (a) need not guarantee termination by any well-founded ordering, (b) does not require the restriction that procedures be terminating, (c) generates induction goals 1 less than BMTP-like well-founded induction, (d) generates simpler induction formulas more first order inferences are already performed to and (e) accommodates naive structural induction and well-founded induction in many cases.

7. Conclusions

We have shown how to formulate computational induction formulas in verification of Prolog programs and its advantages. This method to formulate induction formulas is an element of our verification system Argus/V for proving properties of Prolog programs, the first version of which was developed between April,1984 and March,1985. It consists of about 7000 lines in DEC-10 Prolog and takes about 9.5 seconds (CPU time of DEC2060 with 384 kw main memory) to prove *reverse-reverse* automatically. More than 50 theorems have already been proved automatically and the number is increasing.

Appendix. Closure of Atom

Theorem. Closure is unique up to renaming.

Proof. Suppose A has two closures $\overline{A}'$ and $\overline{A}''$. Then from the condition (b), they are unifiable. Let its most general instance be $\overline{A} = \mu'(\overline{A}') = \mu''(\overline{A}'')$. Suppose a head of a recursive definite clause B_0 is unifiable with $\overline{A}$ by an m.g.u. $\tau_0 \circ \sigma$. Hence $(\tau_0 \circ \mu') \circ \sigma$ is a unifier of $\overline{A}'$ and B_0 and $(\tau_0 \circ \mu'') \circ \sigma$ is a unifier of $\overline{A}''$ and B_0. Because $\overline{A}'$ and $\overline{A}''$ are closures of A, B_0 is unifiable with $\overline{A}'$ by an m.g.u. $\tau_0' \circ \sigma'$ and unifiable with $\overline{A}''$ by an m.g.u. $\tau_0'' \circ \sigma''$. This means that for some ν' and ν'', $\sigma = \nu' \circ \sigma' = \nu'' \circ \sigma''$. For all i such that B_i is a recursive call, $\sigma(B_i) = \nu' \circ \sigma'(B_i) = \nu' \circ \tau_i'(\overline{A}') = \nu'' \circ \sigma''(B_i) = \nu'' \circ \tau_i''(\overline{A}'')$. Hence $\sigma(B_i)$ is a common instance of $\overline{A}'$ and $\overline{A}''$. Because $\mu' \circ \mu''$ is an m.g.u. of $\overline{A}'$ and $\overline{A}''$, there exists a substitution τ_i such that $\sigma(B_i) = \tau_i \circ \mu'(\overline{A}') = \tau_i \circ \mu''(\overline{A}'')$. Then τ_i satisfies the condition of the closedness and $\overline{A}$ is a closure of A. Because of the condition (c), $\overline{A}'$ and $\overline{A}''$ are variants. Hence the closure is unique up to renaming.

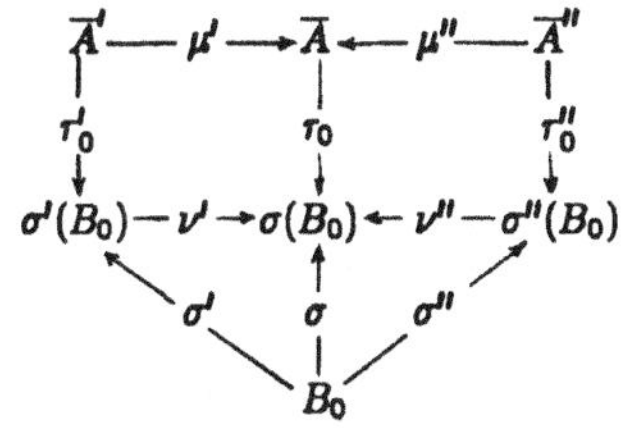

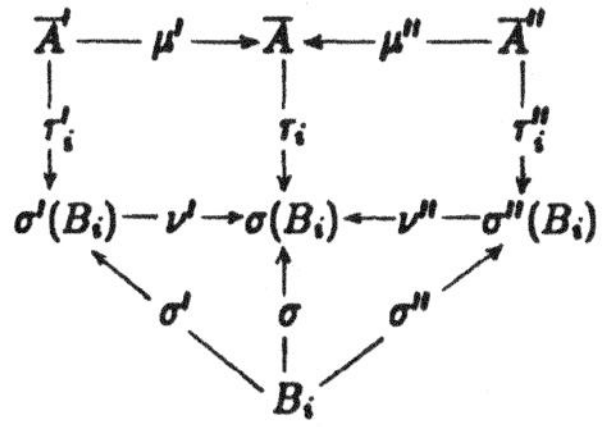

Computation of $\overline{p(t_1, t_2, \ldots, t_n)}$

$i := -1$; $A_0 := p(t_1, t_2, \ldots, t_n)$; $P_0 :=$ the set of recursive definite clauses defining p;

repeat

$i := i + 1$; select a recursive definite clause C in P_i fairly;

if the head of C is unifiable with A_i

then $A_{i+1} := closure1(A_i, C)$; $P_{i+1} := P_i - \{C\}$;

else $A_{i+1} := A_i$; $P_{i+1} := P_i$;

until all heads of definite clauses in P_i are not unifiable with A_i

return A_{i+1}

$closure1(A, \text{``}B_0 \text{ :- } B_1, B_2, \ldots, B_m\text{''})$;

$i := -1$; $A_0 := A$;

repeat forever

$i := i + 1$;

let $\tau_0 \circ \sigma$ be an m.g.u. of A_i and B_0

where τ_0 and σ are the restrictions to A_i and B_0;

let $B'_0, B'_1, B'_2, \ldots, B'_l$ be variants of atoms with p in $\sigma(B_0 \text{ :- } B_1, B_2, \ldots, B_m)$

without shared variables by an appropriate renaming;

$B :=$ most specific common generalization of $A_i, B'_0, B'_1, B'_2, \ldots, B'_l$;

if B is an instance of A_i **then** return A_i **else** $A_{i+1} := B$;

where most specific common generalization is the dual of most general common instantiation, i.e., E is a most specific common generalization of $E_1, E_2, \ldots, E_l$ when

(a) $E_1, E_2, \ldots, E_l$ are instances of E,

(b) E is an instance of any E' satisfying (a).

It is easily obtained by comparing the corresponding subexpressions.

Acknowledgements

The authors would like to express deep gratitude to Dr.T.Sato (Electrotechnical Laboratory) and Prof.H.Tamaki (Ibaraki University) for their stimulating and perspicuous works.

Our verification system Argus/V is a subproject of the Fifth Generation Computer System (FGCS) "Intelligent Programming System". The authors would like to thank Dr.K.Fuchi (Director of ICOT) for the opportunity of doing this research and Dr.K.Furukawa (Chief of ICOT 1st Laboratory) and Dr.T.Yokoi (Chief of ICOT 2nd Laboratory) for their advice and encouragement.

References

[1] Apt,K.R. and M.H.van Emden, "Contribution to the Theory of Logic Programming", J. ACM, Vol. 29, No. 3, pp. 841-862, 1982.

[2] de Bakker,J.W. and D.Scott, "A Theory of Programs", Unpublished Notes, IBM Seminar, Vienna,

1969.

[3] Boyer,R.S. and J.S.Moore, "Computational Logic", Chap.14-15, Academic Press, 1979.

[4] Burstall,R., "Proving Properties of Programs by Structural Induction", Comput.J., Vol. 12, No. 1, pp. 41-48, 1969.

[5] Clark,K.L., "Predicate Logic as a Computational Formalism", pp.75-76, Research Monograph : 79/59, TOC, Imperial College, 1979.

[6] Clark,K.L. and S-Å.Tärnlund, "A First Order Theory of Data and Programs", in Information Processing 77 (B.Gilchrist Ed), pp. 939-944, 1977.

[7] van Emden,M.H. and R.A.Kowalski, "The Semantics of Predicate Logic as Programing Language", J. ACM, Vol. 23, No. 4, pp. 733-742, 1976.

[8] Gordon,M.J.,A.J.Milner and C.P.Wadsworth, "Edinburgh LCF — A Mechanized Logic of Computation' Lecture Notes in Computer Science 78, Springer, 1979.

[9] Hagiya,M. and T.Sakurai, "Foundation of Logic Programming Based on Inductive Definition", New Generation Computing, Vol.2, pp.59-77, 1984.

[10] Jaffar,J.,J-L.Lassez and J.Lloyd, "Completeness of the Negation as Failure Rule", Proc. 8th Internation: Joint Conference on Artificial Intelligence, Vol.1, pp.500-506, 1983.

[11] Kanamori,T.and H.Seki, "Verification of Prolog Programs Using An Extension of Execution", ICOT Technical Report, TR-093, 1984. Also Proc. 3rd International Conference on Logic Programming, 1986.

[12] Kanamori,T.and K.Horiuchi, "Type Inference in Prolog and Its Applications", ICOT Technical Report, TR-095, 1984. Also Proc. 9th International Joint Conference on Artificial Intelligence, Vol. 2, pp. 704-707, 1985.

[13] Kowalski,R.A., "Logic for Problem Solving", Chap. 10-12, North Holland,1980.

[14] Murray,N.V., "Completely Non-Clausal Theorem Proving", Artificial Intelligence, Vol. 18, pp. 67-85, 1982.

[15] Pereira,L.M.,F.C.N.Pereira and D.H.D.Warren, "User's Guide to DECsystem-10 Prolog", Occasional Paper 15, Dept. of Artificial Intelligence, Edinburgh, 1979.

[16] Tamaki,H. and T.Sato, "Unfold/Fold Transformation of Logic Programs", Proc. 2nd International Logic Programming Conference, pp. 127-138, 1984.

[17] Tärnlund,S-Å., "Logic Programming Language Based on A Natural Deduction System", UPMAIL Technical Report, No. 6, 1981.

[18] Weyrauch,R.W. and R.Milner, "Program Correctness in A Mechanized Logic", Proc. 1st USA-Japan Computer Conference, 1972.

Program Verifier "Tatzelwurm": Reasoning about Systems of Linear Inequalities

Thomas Käufl

Institut für Informatik 1
Universität Karlsruhe
Kaiserstraße 12
7500 Karlsruhe 1

Abstract: An algorithm is presented deciding whether a system of linear inequalities over the rationals is unsatisfiable. If this is not the case the procedure eliminates subsumed inequalities and determines the implied equations. It is proved that the strongest conjunction of implied equations is obtained by the algorithm.

1. Introduction

"Tatzelwurm" is a program verifier working like the program verifier of King [KIN69] or the verifier of Stanford [LUC]. It accepts an asserted Pascal program and generates a set of theorems - verification conditions - which are sufficient for the (partial) correctness of the program. (The rules used for the generation of verification conditions may be found in [KÄU1]. A sketch of the simplifier is presented in [KÄU2].) The conditions obtained are submitted to the simplifier.
Systems of linear inequalities are simplified by the sup-inf-reduction which applies the sup-inf-method. The sup-inf-method was developed by Bledsoe [BLE] and refined by Shostak [SHO]. It determines whether a system of inequalities is unsatisfiable. If this is not the case, the subsumed inequalities are removed (see section 3) and then the equations implied by the system are determined. It is shown in section 4 that if a system S of linear inequalities implies equations, a set G of equations is found by the sup-inf-reduction with the property: Each equation of G is implied by S and any equation implied by S is implied by G too. Then a short description of the algorithm used in the simplifier follows. Some remarks concerning the incompleteness of the sup-inf-reduction when a system of linear inequalities over the integers is treated and an example of the reduction of a verification condition will conclude the paper.

2. Notation

A system of linear inequalities (in the following linear is omitted) is a conjunction of inequalities $t_1\ \varrho_1\ a_1 \wedge \ldots \wedge t_n\ \varrho_n\ a_n$ where the t_i are linear arithmetical terms, the a_i are reals and each of the ϱ_i is one of the relation symbols $\geq$, $\leq$, $>$, $<$, $\neq$. One can show that one can restrict to the case where only $\leq$ occurs as relation symbol. If $t < a$ is an inequality over the rationals then it is replaced by $t + \varepsilon \leq a$. ε denotes a positive "infinitesimal" number. (For further details see [BUN].)

Frequently we shall write systems of inequalities as sets and use $S \wedge t \varrho a$ as shorthand for $S \cup \{t \varrho a\}$.
$S \Vdash t \varrho a$ means that $t \varrho a$ may be inferred from S.
In section 4 we shall frequently use definitions and results of linear algebra. A system S of linear inequalities where only $\leq$ occurs as relation symbol determines a convex subset of the euclidean space R^n provided that n is the number of distinct variables occuring in S. Points of the euclidean space are denoted by bold faced typed letters. By $t(\mathbf{x})$ the scalar product of the coefficients of t and the components of $\mathbf{x}$ is denoted. If $t(\mathbf{x}) \leq a$ holds then the inequality $t \leq a$ is satisfied by $\mathbf{x}$.

3. Elimination of Subsumed Inequalities

Let S be a system of inequalities. An inequality $t \leq a$ is termed <u>subsumed</u> if $t \leq a \in S$ and $S\setminus\{t \leq a\} \Vdash t \leq a$. ($\setminus$ is the set theoretical difference.)
Because of $S\setminus\{t \leq a\} \Vdash t \leq a$ iff $(S \setminus \{t \leq a\}) \wedge \neg t \leq a$ is unsatisfiable subsumed inequalities can be discovered using the sup-inf-reduction.

4. Deduction of Implied Equalities

Let S be a satisfiable system of inequalities. An equation t=a is termed <u>implied</u> if $S \Vdash t=a$. If S implies an equation then it implies an infinite set of equations. For example the system

$$y \leq z+1 \wedge x+1 \leq y \wedge x \geq z$$

implies

$$[*]\; x = z \wedge y = z+1$$

but also $y+x = 2z + 1$ and in general $\lambda(x-z) + \mu(y-z) = \mu$, i.e. every "linear combination" of [*].
Hence we must show: If S implies equations then a set $\{g_1 = a_1, \dots, g_n = a_n\}$ is implied by S such that $S \Vdash g = a$ iff $\{g_1 = a_1, \dots, g_n = a_n\} \Vdash g = a$.
In the following we show the existence of such a set and the principles of determining it are presented. The subsequent definitions and theorems are taken from [SHO]. For the remainder of this section suppose that S is a system of inequalites where only $\leq$ occurs as relation symbol. t, possibly indexed denotes a linear term and a, b, c denote reals.

<u>Definition 1:</u>
1. <u>t can have the value b in S</u> if there is a point $\mathbf{x}$ satisfying S such that $t(\mathbf{x})=b$.

2.
$$\max_S t = \begin{cases} b & \text{if t can have the value b in S but no greater value} \\ \text{undefined} & \text{otherwise} \end{cases}$$

3. <u>S bounds t at b</u> if $\max_S t = b$.
4. <u>S bounds t</u> if for some b S bounds t at b.

A subset of a system of inequalities is termed a <u>subsystem</u>.

<u>Definition 2</u>:

1. <u>S minimally bounds t at b</u> if S bounds t at b and no proper subsystem of S bounds t at b.
2. <u>S minimally bounds t</u> if S bounds t and no proper subsystem of S bounds t.

<u>Theorem 3</u>: (Shostak)

1. Suppose S bounds t. Then there exists a subsystem T of S such that T minimally bounds t and $\max_S t = \max_T t$.
2. Let S minimally bound t at b. Then any point $\mathbf{x}$ satisfying S such that $t(\mathbf{x})=b$ satisfies S_E. (S_E is obtained from S by $\leq$ replacing by =.)

<u>Definition 4</u>:

1. $t=c$ is a <u>linear combination</u> of the equations $t_1=c_1,\ldots, t_n=c_n$ iff there are reals $\lambda_1,\ldots,\lambda_n$ with $\lambda_1 t_1 + \ldots + \lambda_n t_n = t$ and $\lambda_1 c_1 + \ldots + \lambda_n c_n = c$.
2. $t = c$ <u>intersects</u> with the solution of $t_1 = c_1, \ldots, t_n = c_n$ if any point satisfying $t_1 = c_1, \ldots, t_n = c_n$ satisfies $t = c$ too.

<u>Theorem 5</u>: Suppose the system $t_i = c_i$ $(1 \leq i \leq n)$ has a nonempty solution and $t = c$ intersects with the solution of this system. Then $t=c$ is a linear combination of the $t_i=c_i$.

Sketch of proof: As $t = c$ intersects with the solution of $t_i = c_i$ $(1 \leq i \leq n)$ every point satisfying $t_i = c_i$ satisfies $t = c$. Hence the system $t = c$, $t_i = c_i$ $(1 \leq i \leq n)$ has the same solution as the system $t_i = c_i$. By consideration of the ranks of the matrices of the systems of equations, the theorem is proved. ∎

<u>Theorem 6</u>: Let S be satisfiable and $S \Vdash t = b$. Then there exists a subsystem S' of S such that $t = b$ is a linear combination of the equations of S'_E and $S \Vdash S'_E$.

Proof: If $S \Vdash t = b$ then S bounds t at b. Hence theorem 3(1) guarantees the existence of a subsystem S' of S bounding t minimally at b. By part (2) of the same theorem we have

(1) If $\mathbf{x}$ satisfies S then it satisfies S'_E too

Now we prove

(2) If $\mathbf{x}$ satisfies S'_E then $\mathbf{x}$ satisfies $t = b$

As S is satisfiable, there exists an $\mathbf{x}_0$ such that $t(\mathbf{x}_0) = b$ and $\mathbf{x}_0$ satisfies S'_E. Suppose there is a second point $\mathbf{x}_1$ satisfying S' with $t(\mathbf{x}_1) \neq b$. As S' bounds t at b, $t(\mathbf{x}_1) < b$ must hold. We shall show that the line $\mathbf{x}_0\mathbf{x}_1$ contains points $\mathbf{x}$ with $t(\mathbf{x}) > b$. Each point of the line $\mathbf{x}_0\mathbf{x}_1$ satisfies

$$\mathbf{x} = \mathbf{x}_0 + \alpha(\mathbf{x}_1 - \mathbf{x}_0) \quad \text{where } \alpha \text{ is an arbitrary real}$$

Now

$$t(\mathbf{x}) = t(\mathbf{x}_0) + \alpha(t(\mathbf{x}_1) - t(\mathbf{x}_0)) = b + \alpha(t(\mathbf{x}_1) - b)$$

Because of $t(\mathbf{x}_1) < b$ any assignment of a negative value to α yields a point $\mathbf{x}$ satisfying S'_E such that $t(\mathbf{x}) > b$. This contradicts the fact that S' bounds t at b. Hence $t = b$ intersects with the solution of S'_E.
By theorem 5 we can infer that $t = b$ is a linear combination of the equations of S'_E. ■

Theorem 7: If S implies an equation then there exists a subsystem S' of S such that any equation implied by S is a linear combination of the equations of S'_E and S'_E is implied by S.

Proof: If $t = c$ is an equation implied by S then by theorem 6 there exists a subsystem T such that $t = c$ is a linear combination of the equations of T_E. The union of all of these subsystems T is the asserted subsystem S'. By theorem 3 we have that each of the equations of T_E is implied by S and hence their union S'_E too. ■

Theorem 7 allows us to prove the equivalence stated at the beginning of the section. S'_E is the system of equations looked for and as any equation $t = c$ implied by S is a linear combination of the equations of S'_E, $t = c$ is implied by S'_E.
Now we must show how S'_E is determined. In order to do this we need an additional result from Shostak [SHO].

Theorem 8: If S bounds a variable v, $\sup_S(v, \emptyset) = \max_S(v)$.

Suppose t is an arbitrary term and S is satisfiable. Bledsoe [BLE] has proven that

$$\sup_S(t, \emptyset) \geq \max_S(t) \quad \text{if S bounds t}$$

and $\sup_S(t, \emptyset) = \infty$ if S does not bound t. The definition of $\sup_S$ is given in [SHO] too.

Theorem 9: Suppose S is a system of inequalities $t_i \leq c_i$ $(1 \leq i \leq n)$. Let z be a fresh variable, $1 \leq j \leq n$ and T the system consisting of $t_i \leq c_i$ $(1 \leq i \leq n)$ and $-t_j + z \leq -c_j$.
Then $S \Vdash t_j = c_j$ iff $\max_T(z) = 0$.

Proof:
If: Suppose $(\mathbf{x}_0 \mid z_0)$ is a point satisfying T. Then $\mathbf{x}_0$ satisfies S and as S implies $t_j = c_j$, this equation is satisfied by $\mathbf{x}_0$ too. Hence $z_0 \leq 0$. By this $\max_T(z) = 0$ follows.
Only if: Suppose $\mathbf{x}_0$ satisfies S then $\mathbf{x}_0$ satisfies $t_j \leq c_j$. Because of $\max_T(z) = 0$, $\mathbf{x}_0$ satisfies $t_j \geq c_j$ too. Hence $t_j = c_j$ is implied by S. ■

5. A Sketch of the Algorithm

If a system S of inequalities is to be simplified by the sup-inf-reduction the following actions are performed.

1. Determine whether S is unsatisfiable. If this is the case then the sup-inf-reduction returns false. If not the algorithm continues with step 2.
2. Eliminate all subsumed inequalities. The step yields a new system S'.
3. Determine the implied equalities. Suppose S' consists of the inequalities $t_1 \leq c_1, \ldots, t_n \leq c_n$. For all i ($1 \leq i \leq n$) construct the system $T = \{t_1 \leq c_1, \ldots, t_n \leq c_n, -t_i + z \leq -c_i\}$ where z is a fresh variable. Compute $\sup_T(z, \emptyset)$. If the result is 0 then by theorems 8 and 9, S' implies $t_i = c_i$.

The behaviour of the algorithm is improved by some modifications. If S implies an equation x = c where c is a number then this equation can be determined during execution of step 1. After computation of an implied equation this equation is used to eliminate a variable before step 3 is begun resp. continued.

6. Two Examples

The sup-inf-method is incomplete for systems of inequalities over the integers.

$4x + 4y \leq 3 \wedge 2x + 2y \geq 1$

for example is not satisfiable for the integers. (This because 4 resp. 2 are not divisors of 3 resp. 1. Complete algorithms for systems of inequalities over integers are given by [KR] or [COO].) King [KIN69] suggests replacing the system given above by the equivalent

$x + y \leq 0 \wedge x + y \geq 1.$

By this the domain of systems the sup-inf-method recognizes as unsatisfiable is enlarged.

Verification conditions are theorems where subformulae belonging to diverse theories occur. Assume we are given

$\forall x \forall y\, (z \leq x \wedge x \leq z + 1 \wedge z \leq y \wedge y \leq z + 1 \wedge x \leq y - 1 \rightarrow a(x+1) = a(y))$

Formulae of this structure are encountered when one verifies programs dealing with arrays. The simplifier of Nelson and Oppen [NO] would split the formula above into one part containing the arithmetical subformulas and into a second containing formulas treating arrays. This is not done by the simplifier of "Tatzelwurm". When the if-part of the bound formula is submitted to the sup-inf-reduction, it reduces the premise to

$z \leq x \wedge y \leq z + 1 \wedge x \leq y - 1$

(Step 2 of the algorithm eliminates the subsumed inequalities $x \leq z + 1$ and $z \leq y$.) Step 3 of the procedure computes $y = z + 1$ as implied equation. The sup-inf-procedure is interrupted and the premise simplified to

$z \leq x \wedge x \leq z \wedge y = z + 1$

(The simplifier always tries to eliminate variables bound by a quantifier.) The formula obtained is reduced to

$x = z \wedge y = z + 1$

These equations allow to replace the then-part of the implication by

$a(z+1) = a(z+1)$

and thus the formula is proved.

7. Conclusion

The program verifier is implemented in Interlisp running on a Siemens S7561 computer. The system comprises one million bytes of compiled code at present.
Using special techniques for storing, for example structure sharing and special data structures the simplifier works with high performance.
By use of the technique of data-driven programming it is possible to add further reduction procedures without the necessity of changing the system. Thus the behaviour of the extended simplifier can be tested without change of the system.

Acknowledgement: I am greatly indebted to A. Bockmayr who read an earlier version of this paper.

References

[BLE] W.W. Bledsoe: A New Method for Proving Certain Presburger Formulas
4th Int. Joint Conference on Artificial Intelligence, Tiblisi, pp.15-21: 1975

[BUN] A. Bundy: The Computer Modelling of Mathematical Reasoning
London: 1983

[COO] D.C. Cooper: Programs for Mechanical Program Verification
Machine Intelligence 7, B. Meltzer and D. Michie, eds., New York: 1971

[KÄU1] Th. Käufl: Automated Construction of Verification Conditions
in Schriften zur Informatik und angewandten Mathematik Nr. 87
RWTH Aachen: 1983

[KÄU2] Th. Käufl: The Simplifier of the Program Verifier "Tatzelwurm"
Österreichische Artificial Intelligence-Tagung 1985
Informatik-Fachberichte; Berlin, Heidelberg, New York: 1985

[KIN69] J.C. King: A Program Verifier
Ph.D. Thesis, Carnegie Mellon University, Pittsburgh: 1969

[KIN72] J.C. King, R.W. Floyd: An Interpretation-Oriented Theorem Prover over Integers
Journal of Computer and System Sciences 6, pp. 305 - 323: 1972

[KR] G. Kreisel, J.-L. Krivine: Modelltheorie
Berlin, Heidelberg, New York: 1972

[LUC] D.C. Luckham, e.a.: Stanford Verifier User Manual
Report No. Stan-CS-79-731
Computer Science Department, University of Stanford: 1979

[NO] C.G. Nelson, D.C. Oppen: Simplification by Cooperating Decision Procedures
ACM TOPLAS 1, pp. 245 - 257: 1979

[SHO] R. Shostak: On the SUP-INF Method for Proving Presburger Formulas
JACM 24/4, pp. 529 - 543: 1977

AN INTERACTIVE VERIFICATION SYSTEM BASED ON DYNAMIC LOGIC

R.Hähnle, M.Heisel, W.Reif, W.Stephan
Universität Karlsruhe
Institut für Informatik I
Postfach 6980
D-7500 Karlsruhe 1

ABSTRACT

An interactive verification system based on dynamic logic is presented. This approach allows to strengthen the role of "dynamic reasoning", i.e. reasoning in terms of state transitions caused by programs.
The advantages of the approach are: (i) dynamic logic is more expressive than HOARE's logic, e.g. termination and program implications can be expressed; (ii) user-defined rules enable reasoning in a very natural way; (iii) simpler verification conditions are obtained; (iv) many proofs can be performed schematically.
The problem of rule validation is discussed.
An example demonstrates the style of reasoning supported by the system.

1 INTRODUCTION

This paper is a first report on the Karlsruhe Interactive Verifier which is currently under development at the University of Karlsruhe. The system is based on dynamic logic and is an attempt to strengthen the role of reasoning in terms of state transitions (caused by programs). We concentrate on the basic ideas and motivations and present only the most important features of the system which has been implemented in the last two years. For a more detailed description see [HHRS86].

Dynamic logic (**DL**) extends first-order logic by formulas $[\alpha]\varphi$, where α is a program and φ again is a formula. $[\alpha]\varphi$ has to be read "if α terminates, φ holds". The formal treatment of semantics as it goes back to PRATT [Prat76] uses concepts from modal logic. We note that HOARE's language of partial correctness assertions is a sublanguage of **DL**: $\varphi\{\alpha\}\psi$ becomes $\varphi \rightarrow [\alpha]\psi$. Compared to the full-fledged system of **DL** HOARE's logic is a significant restriction. For example neither $\neg[\alpha]\varphi$ nor $[\alpha]\varphi \rightarrow [\beta]\psi$ can be expressed. For a survey of dynamic logic see [Har84].

As in most applications we deal with *concrete* programs, i.e. there is a fixed domain together with an interpretation of the symbols occurring in the programs. HAREL has shown in [Har79] that for a certain class of structures (arithmetical universes) all **DL** formulas can be translated into equivalent first-order formulas using a set of simple rules and axioms. In the case of HOARE-formulas we are lead to the well-known HOARE-calculus [Ho69].

Although this translation can be carried out in a structured manner we are restricted to a style of proof which reduces "dynamic reasoning" to a minimum. For a moment we therefore forget about the datastructure(s) we deal with and concentrate only on the programming language constructs. We want to use the full power of so-called first-order uninterpreted reasoning to derive program-free formulas (verification conditions) which have to be proved elsewhere. The main advantages of this approach are:

- A more flexible way in performing proofs is achieved by the use of many new (derived) rules. Reasoning about programs can be done in a natural and direct way.
- The verification conditions generated are simpler than those obtained by weaker calculi.
- There are many proofs which can be carried out schematically in the sense that we are able to prove formulas which contain *metavariables*.

It is quite obvious that large proofs using only the basic rules are by no means feasible. Therefore one of the main features of the KIV system is the use of a large database of *user-defined rule schemes*. To preserve soundness as well as to check additional constraints the user has to supply a *validation* for each rule he wants to add to the rule base. Thus the KIV system is a *metasystem* which in particular uses similar concepts as the Edinburgh LCF system [GMW79].

After an introduction into the logical basis we give an example to demonstrate the style of reasoning which is supported by our system. The main system facilities are presented in chapter 4.

2 LOGICAL BASIS

The formula syntax is based on a first-order language over a many-sorted signature Σ. The underlying programming language considered so far is made up of assignments ($x:=\tau$), compositions ($\alpha;\beta$), conditionals **if** ε **then** α **else** β **fi**, and iterations **while** ε **do** α **od**.

The semantics is defined relative to a given Σ-algebra $\mathbf{A}$ which provides an interpretation for the symbols of Σ. States over $\mathfrak{A}$ are abstract entities to which $\mathfrak{A}$-valuations υ of variables are attached.

With each program α there is associated a binary relation $[\![\alpha]\!]$ on the set of states to model its input-output behaviour. By so-called standard-model conditions the programming language constructs are given their intended meanings.

For an $\mathfrak{A}$-based standard model $\mathfrak{M}$ the validity of a formula $[\alpha]\varphi$ in a state s is defined by :

$$\mathfrak{M} \models_s [\alpha]\varphi \text{ iff for all } t : s[\![\alpha]\!]t \text{ implies } \mathfrak{M} \models_t \varphi .$$

Our language is powerful enough to provide a formalization of DIJKSTRA's predicate transformers [Di76]. The *weakest precondition* for a program α that corresponds to a postcondition φ, $wp(\alpha,\varphi)$, can be expressed as

$$\neg[\alpha]\textbf{false} \wedge [\alpha]\varphi \ .$$

However, this identification is appropriate only for deterministic programs.

GOLDBLATT [Go82] gives an axiomatization of the valid formulas. It includes the following infinitary *Omega-rule* in order to deal with admissible statements $\Phi(..)$ about loops :

$$\text{From } \{\Phi(\varphi_n(\varepsilon,\alpha)) : n \in \mathbb{N}\} \text{ infer } \Phi([\textbf{while } \varepsilon \textbf{ do } \alpha \textbf{ od}]\varphi) \ ,$$

where $\varphi_0(\varepsilon,\alpha) \equiv (\neg\varepsilon \rightarrow \varphi)$ and $\varphi_{n+1}(\varepsilon,\alpha) \equiv (\varepsilon \rightarrow [\alpha]\varphi_n(\varepsilon,\alpha))$.

The Omega-rule cannot be replaced by one or more finitary rules without losing completeness. For an effective proof system however, we are restricted to finitary rules and therefore the main problem consists of appropriatly "approximating" the Omega-rule without losing too much of deductive strength.

Our approach is based on the observation that in concrete proofs requiring the Omega-rule, its premisses are shown by induction on the metalevel. The idea is to incorporate this into the logic itself.

In order to be able to carry out induction proofs in the object language we extend the logic by a *counter-structure* with only zero and a successor function. Adding such a counter-structure does not yield an arithmetical universe. It is too weak to enable an encoding of statements about programs into first-order statements.

We further introduce a new construct **loop** ε **times** δ , where δ is a counter-expression. Starting in a state s, α is executed $\upsilon_s(\delta)$ times. This construct is axiomatized by

$$[\textbf{loop } \alpha \textbf{ times zero}]\varphi \leftrightarrow \varphi$$

$$[\textbf{loop } \alpha \textbf{ times suc}(c)]\varphi \leftrightarrow [\alpha][\textbf{loop } \alpha \textbf{ times } c]\varphi \ .$$

With the aid of loop-programs we are now able to express the definition of $\varphi_n(\varepsilon,\alpha)$ in the object language in contrast to the one which is given above by induction on the metalevel. For all standard models $\mathfrak{M}$, states s, and $n \in \mathbb{N}$

$$\mathcal{M} \models_s \varphi_n(\varepsilon,\alpha) \quad \text{iff}$$

$$\mathcal{M} \models_s [\textbf{loop} \;\; \textbf{if}\ \varepsilon\ \textbf{then}\ \alpha\ \textbf{else}\ \textbf{abort}\ \textbf{fi} \;\; \textbf{times}\ \bar{n}](\neg\varepsilon \rightarrow \varphi)$$

where $\bar{n}$ stands for $\underbrace{\textbf{suc}(\ldots(\textbf{suc}}_{n\text{-times}}(\textbf{zero}))\ldots)$.

This leads to an axiom scheme corresponding to the Omega-rule:

$$[\textbf{while}\ \varepsilon\ \textbf{do}\ \alpha\ \textbf{od}]\varphi \leftrightarrow$$
$$\forall c.[\textbf{loop} \;\; \textbf{if}\ \varepsilon\ \textbf{then}\ \alpha\ \textbf{else}\ \textbf{abort}\ \textbf{fi} \;\; \textbf{times}\ c](\neg\varepsilon \rightarrow \varphi) \ .$$

The induction axioms now necessary to prove statements about while-programs take the form:

$$(\varphi^c_{\textbf{zero}} \wedge \forall c.(\varphi \rightarrow \varphi^c_{\textbf{suc}(c)})) \rightarrow \forall c.\varphi \ .$$

The counter-structure is part of our basic logical system. Its only purpose is to prove properties of loop-programs by induction. Loop-programs are used to "approximate" while-programs.

HAREL has given arithmetically complete axiomatizations of **DL** and **DDL** (deterministic dynamic logic) [Har79]. HAJEK has shown [Haj81] that admitting non-standard interpretations of the arithmetical notions and of iteration, HAREL's axioms and rules for **DL** together with the axioms of Peano-arithmetic are sufficient to obtain all valid formulas.

Except the rule of convergence all of HAREL's axioms and rules for **DDL** can be derived in our logic. The reader is referred to chapter 4 for some remarks on the treatment of derived rules. For the rule of convergence we need an additional (non-logical) axiom. For example

$$\forall y.\neg[x:=o;\textbf{while}\ \neg x=y\ \textbf{do}\ x:=x+1\ \textbf{od}]\textbf{false}$$

will do.

But in contrast to HAREL's logic our system is powerful enough to prove statements like

$$(\forall x.h(f(x))=x) \wedge z=a \wedge x=y \rightarrow$$
$$[\textbf{while}\ \neg z=x\ \textbf{do}\ z:=f(z);y:=h(y)\textbf{od}]y=a$$

without any appeal to arithmetical notions.

As more realistic concepts of a programming language we have axiomatized arrays and recursive procedures [St85],[RS84].

3 AN EXAMPLE

The KIV system is based on a sequent calculus [Praw65], [Ri78]. For sake of readability we do not use exactly the system syntax and concentrate on the main proof steps. In particular we omit all intermediate steps which are carried out automatically by the system as for example the elimination of propositional connectives and "simple" programming language constructs.

The example is taken from DIJKSTRA's book "A Discipline of Programming", [Di76]. If initialized properly (r:=a and r:=a;dd:=d respectively) the programs given below both compute the smallest nonnegative remainder after division of a by d. α_o is the simplest program one can think of, while β_o is somewhat more elaborate.

```
α_o: while r≥d do          β_o: while r≥dd do          ⎫
         r:=r-d                     dd:=dd*3            ⎬ β_1
     od                         od ;                    ⎭
                                while ¬dd=d do          ⎫
                                    dd:=dd/3 ;      ⎫   ⎪
                                    while r≥dd do   ⎬ β_3 ⎬ β_2
                                       r:=r-dd ;    ⎪   ⎪
                                    od              ⎭   ⎪
                                od                      ⎭
```

We want to prove:

$$\Rightarrow (dd=d \wedge [\alpha_o][dd:=o]\varphi_o) \rightarrow [\beta_o][dd:=o]\varphi_o \qquad (G) ,$$

where φ_o is left unspecified. "⇒" stands for the sequent arrow. Obviously G is true for all postconditions φ_o. Meaningful instantiations of φ_o are for example

$$\varphi_o \equiv o \leq r < d \wedge d \mid (a-r) ,$$

where "|" is to be read "is a divisor of", and

$$\varphi_o \equiv \textbf{false} .$$

In the latter case G can be spelled out as: "If dd=d and β_o terminates, then α_o terminates." We remark that using similar techniques we may as well prove

$$(dd=d \wedge [\beta_o]\textbf{false}) \rightarrow [\alpha_o]\textbf{false}$$

which states: "If dd=d and α_o terminates, then β_o terminates."

We prove G in a goal-directed manner. Starting out with G the rules are applied backwards until we have reached an axiom, a lemma or a sequent which is program-free.

Having eliminated "→" and "∧" **R1** (see Figure1) yields the new subgoal:

$$dd=d, [\alpha_o]\varphi_1 \Rightarrow [\beta_o]([\alpha_o]\varphi_1 \wedge \neg r \geq d) \qquad (G1) ,$$

where $\varphi_1 \equiv [dd:=o]\varphi_o$. To use **R2** ($\beta_o \equiv \beta_1;\beta_2$) backwards the user has to supply the intermediate assertion ψ. We set

$$\psi \equiv \psi_o \wedge [\alpha_o]\varphi_1 \wedge \neg r \geq dd \text{ , where}$$

$$\psi_o \equiv \exists i. dd=d*3^i .$$

R2 yields two subgoals of G1:

$$dd=d, [\alpha_o]\varphi_1 \Rightarrow [\beta_1](\psi_o \wedge [\alpha_o]\varphi_1 \wedge \neg r \geq dd) \quad (G11) ,$$

$$\psi_o, [\alpha_o]\varphi_1, \neg r \geq dd \Rightarrow [\beta_2]([\alpha_o]\varphi_1 \wedge \neg r \geq d) \qquad (G12) .$$

Using **R3** (a version of HOARE's rule of invariance) with $\psi \equiv \psi_o \wedge [\alpha_o]\varphi_1$

we get from G11 the nontrivial subgoals:

$$dd=d, [\alpha_o]\varphi_1 \Rightarrow \psi_o \qquad (G111)\ ,$$

$$\psi_o, [\alpha_o]\varphi_1, r\geq dd \Rightarrow [dd:=dd*3]\psi_o \qquad (G112)\ ,$$

$$\psi_o, [\alpha_o]\varphi_1, r\geq dd \Rightarrow [dd:=dd*3][\alpha_o]\varphi_1 \qquad (G113)\ .$$

G111 and G112 are immediately reduced to

$$dd=d \Rightarrow \exists i.dd=d*3^i \quad \text{and}$$

$$\exists i.dd=d*3^i, r\geq dd \Rightarrow \exists i.(dd*3)=d*3^i$$

which are program-free. We use **R3** to reduce G12. Taking

$$\psi \equiv \psi_o \wedge [\alpha_o]\varphi_1 \wedge \neg r\geq dd$$

we get two new subgoals:

$$\psi_o, [\alpha_o]\varphi_1, \neg r\geq dd, \neg dd=d \Rightarrow [\beta_3](\psi_o \wedge [\alpha_o]\varphi_1 \wedge \neg r\geq dd) \qquad (G121)\ ,$$

$$\psi_o, [\alpha_o]\varphi_1, \neg r\geq dd, dd=d \Rightarrow \neg r\geq d \qquad (G122)\ .$$

As above we do not mention those subgoals which are already axioms. If we omit $[\alpha_o]\varphi_1$ from the antecedent, G122 is program-free.
Using **R2** and again **R3** (both with $\psi \equiv \psi_o \wedge [\alpha_o]\varphi_1$) we obtain from G121:

$$\psi_o, [\alpha_o]\varphi_1, \neg r\geq dd, \neg dd=d \Rightarrow [dd:=dd/3]\psi_o \qquad (G1211)\ ,$$

$$\psi_o, [\alpha_o]\varphi_1, \neg r\geq dd, \neg dd=d \Rightarrow [dd:=dd/3][\alpha_o]\varphi_1 \qquad (G1212)\ ,$$

$$\psi_o, [\alpha_o]\varphi_1, r\geq dd \Rightarrow [r:=r-dd]\psi_o \qquad (G1213)\ ,$$

$$\psi_o, [\alpha_o]\varphi_1, r\geq dd \Rightarrow [r:=r-dd][\alpha_o]\varphi_1 \qquad (G1214)\ .$$

G1211 can easily be reduced to a program-free sequent. We concentrate on G1214. **R4** yields:

$$\psi_o, \forall c.[\mathbf{loop}\ \ \mathbf{if}\ r\geq d\ \mathbf{then}\ r:=r-d\ \mathbf{fi}\ \ \mathbf{times}\ c][\alpha_o]\varphi_1, r\geq dd$$
$$\Rightarrow [r:=r-dd][\alpha_o]\varphi_1 \qquad (G12141)\ .$$

R1: $$\frac{\Gamma, [\mathbf{while}\ \varepsilon\ \mathbf{do}\ \alpha\ \mathbf{od}]\varphi \Rightarrow [\beta]([\mathbf{while}\ \varepsilon\ \mathbf{do}\ \alpha\ \mathbf{od}]\varphi \wedge \neg\varepsilon)}{\Gamma, [\mathbf{while}\ \varepsilon\ \mathbf{do}\ \alpha\ \mathbf{od}]\varphi \Rightarrow [\beta]\varphi}$$

R2: $$\frac{\Gamma \Rightarrow [\alpha]\psi \quad | \quad \psi \Rightarrow [\beta]\varphi}{\Gamma \Rightarrow [\alpha][\beta]\varphi}$$

R3: $$\frac{\Gamma \Rightarrow \psi \quad | \quad \psi, \varepsilon \Rightarrow [\alpha]\psi \quad | \quad \psi, \neg\varepsilon \Rightarrow \varphi}{\Gamma \Rightarrow [\mathbf{while}\ \varepsilon\ \mathbf{do}\ \alpha\ \mathbf{od}]\varphi}$$

R4: $$\frac{\Gamma, \forall c.[\mathbf{loop}\ \ \mathbf{if}\ \varepsilon\ \mathbf{then}\ \alpha\ \mathbf{fi}\ \ \mathbf{times}\ c][\mathbf{while}\ \varepsilon\ \mathbf{do}\ \alpha\ \mathbf{od}]\varphi \Rightarrow \psi}{\Gamma, [\mathbf{while}\ \varepsilon\ \mathbf{do}\ \alpha\ \mathbf{od}]\varphi \Rightarrow \psi}$$

R5: $$\frac{\Gamma \Rightarrow \varphi}{\Gamma \Rightarrow [x:=\tau]\varphi}$$

Figure 1: Some Derived Rules

This last goal can be achieved using a general lemma which we prove without specifying ψ:

$$dd \geq d, d \mid dd, r \geq dd, \forall c.[\textbf{loop if } r \geq d \textbf{ then } r:=r-d \textbf{ fi times } c]\psi \Rightarrow [r:=r-dd]\psi$$

We still have to deal with G113, G1212 and G1213. Using **R5** in all three cases we complete the proof of G. Because of the assignment dd:=o the validation is successful even without specifying φ_0.

4 THE KIV SYSTEM

In this chapter we give a brief description of the general system features followed by some technical remarks on how proofs can be performed in the system.

4.1 General System Features

First of all let us take a look at the representation of inference rules. In our system rules are represented as schemes. They are collected in a rule base containing the (fixed) set of rules required for the sequent calculus and a set of user-defined rules. Application is done by matching.

The system offers two modes for rule application: in the forward mode the premisses of a rule scheme are matched with theorems or axioms yielding a new theorem. In the backward mode, the conclusion of a rule scheme may be matched with a goal producing new subgoals to be proved.

As in the first-order sequent calculus [Praw65], [Ri78] some basic rules are associated with variable conditions that have to be checked before application.

Let us consider the concept of user-defined rules in more detail. Since the user may add new inference rules to the rule base, the problem of soundness arises. That means, there can be instantiations of rules such that the premisses are valid while the conclusion is not. Therefore the user has to provide a validation for each new rule which describes uniformly how to prove the conclusion from the premisses for all instantiations. Validations are expressed in terms of proof-constructing programs written in a simple functional programming language having rule applications as the only primitive operations. The execution of such a program results in a proof-tree having the conclusion as its root and the premisses as leaves.

However, we want to define rule schemes which do not preserve validity for all instantiations. Rule schemes of this type play an important role in proofs performed in our system. For example in rule scheme **R5** (see figure 1) x must not occur "free" in φ. This constraint is checked implicitly by performing the validation program. A sucessful execution yields a proof-tree containing only basic rule applications thereby demonstrating that the given instance of **R5** preserves validity. An instance like

$$\frac{y=o, z=o \Rightarrow y=z}{y=o, z=o \Rightarrow [y:=y+1]y=z}$$

obviously does not represent an admissible proof step. Thus the validation program will fail in this case.

It is possible to postpone the execution of validations until it is certain that the application of the corresponding rules leads to a proof.

Note that the validation can use basic as well as user-defined rules. It is even possible that the validation of a rule contains applications of itself, i.e. our proof-programming language allows recursion. Thus proof-constructing programs need not always terminate. Execution fails if for example a situation is reached where none of the rules proposed in the validation program is applicable. If on the contrary the execution of the program is successful the generated proof-tree always represents a proof of the conclusion from the premisses where only basic rules are used.

Validations often are quite complex. In order to achieve efficiency we simplify validation programs as far as possible by "symbolic execution". This means, if parts of a validation do not make use of the particular instantiations of the metavariables it is possible to execute certain proof steps independently of later instantiations. Thus these steps need not be carried out at every application of the corresponding rule, but only once when the rule is defined.

Our concept of user-defined rules is related to ideas incorporated in ML of Edinburgh LCF [GMW79], and generalized in [Sch84]. We name the notions of tacticals and tactics as well as the idea of ensuring soundness through validation.

4.2 Working with the System

Now, how is a proof in the KIV system actually carried out? Let us suppose we wanted to prove something, say, about programs working with trees.

First of all we have to define the theory of trees by specifying (i) an appropriate signature and (ii) additional signature-dependent rules and axioms, which are needed e.g. for termination proofs. Using the respective modules (see figure 2) these are integrated into the system kernel which initially only contains signature-independent rules and axioms. Thus we have created an incarnation of the system that is tailor-made for proofs about programs dealing with tree structures.

Now we are able to state goals we want to prove. We use the I/O-module for putting them into the goal base. Next we could try to do the proofs by using basic rules. But usually we will soon realize that this gets too complex. Therefore we must try to structure our proofs by defining appropriate derived rules.

Now we can select between two modes for carrying out the proofs: on the one hand we may proceed step by step, i.e. we use the system as a mere proof checker. On the other hand we recently started to implement elementary strategies which are executed by the supervisor module: most steps are performed automatically, only from time to time we are asked for additional informations, e.g. intermediate assertions (cf. chapter 3).

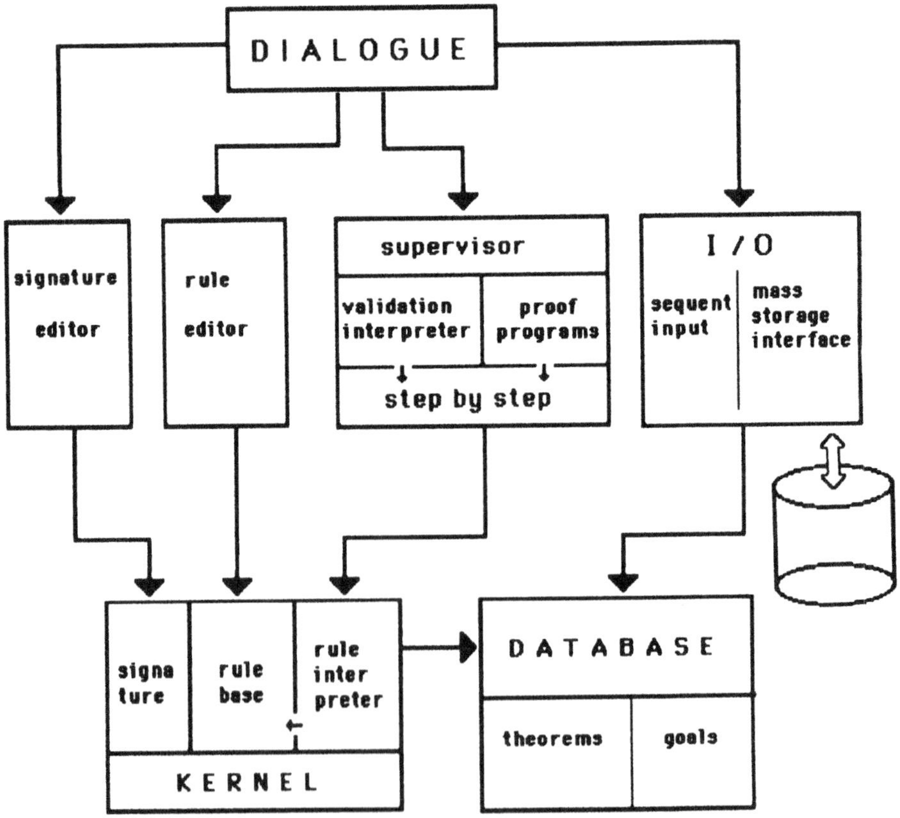

Figure 2: KIV System Configuration

5 CONCLUDING REMARKS

We have presented an interactive system for proofs in dynamic logic. The intention is to develop the system from a mere proof checker to a more and more complex and powerful tool. The concept of user-defined rules and some elementary strategies have already been implemented while the development of sophisticated proof strategies is subject of further research.

REFERENCES

[Di76] Dijkstra, E.W. A Discipline of Programming. Prentice-Hall (1976)

[Go82] Goldblatt, R. Axiomatising the Logic of Computer Programming. Springer-LNCS 130 (1982)

[GMW79] Gordon, M. & Milner, R. & Wadsworth, C. Edinburgh-LCF. Springer-LNCS 78 (1979)

[Haj81] Hajek, P. Making dynamic logic first order. Proc. Math. Foundations of Computer Science, Springer-LNCS 118 (1981), 287-295

[Har79] Harel, D. First Order Dynamic Logic. Springer-LNCS 68 (1979)

[Har84] Harel, D. Dynamic Logic. Handbook of Philosophical Logic, D. Gabbay and F. Guenther (eds.), Reidel (1984), vol. 2, 496 -604

[HHRS86] Hähnle, R. & Heisel, M. & Reif, W. & Stephan, W. The Karlsruhe Interactive Verifier - A Verification System based on Dynamic Logic. Interner Bericht 1/86, Fakultät für Informatik, Universität Karlsruhe (1986)

[Ho69] Hoare, C.A.R. An axiomatic basis for computer programming. C.A.C.M. 12 (1969), 576 - 580

[Prat76] Pratt, V.R. Semantical considerations on Floyd-Hoare logic. Proc. 17th Ann. I.E.E.E. Symp. on Foundations of Computer Science, 109-121

[Praw65] Prawitz, D. Natural Deduction. Stockholm Studies in Philosophy 3, Almquist & Wicksell, Stockholm (1965)

[Ri78] Richter, M.M. Logikkalküle. Teubner (1978)

[RS84] Reif, W. & Stephan, W. Vollständigkeit einer modifizierten Goldblatt-Logik und Approximation der Omegaregel durch Induktion. Diplomarbeit, Fakultät für Informatik, Universität Karlsruhe (1984)

[Sch84] Schmidt, D.A. A programming notation for tactical reasoning. Proc. 7th Int. Conf. on Automated Deduction, R.E. Shostak (ed.), Springer-LNCS 170 (1984),445 - 460

[St85] Stephan, W. A Logic for Recursive Programs. Interner Bericht 5/85, Fakultät für Informatik, Universität Karlsruhe (1985)

WHAT YOU ALWAYS WANTED TO KNOW ABOUT CLAUSE GRAPH RESOLUTION

Norbert Eisinger
Fachbereich Informatik, Universität Kaiserslautern
D-6750 Kaiserslautern, F.R.Germany

Abstract: Clause graph (or connection graph) resolution was invented by Robert Kowalski in 1975. Its behaviour differs significantly from that of traditional resolution in clause sets. Standard notions like completeness do not adequately cover the new phenomena introduced by clause graph resolution and standard proof techniques do not work for clause graphs, which are the major reasons why important questions have been open for years. This paper defines a series of relevant properties in precise terms and answers several of the open questions. The clause graph inference system is refutation complete and refutation confluent. Compared to clause set resolution, clause graph resolution does not increase the complexity of simplest refutations. Many well-known restriction strategies are refutation complete, but most are not refutation confluent for clause graph resolution, which renders them useless. Exhaustive ordering strategies do not exist and contrary to a wide-spread conjecture the plausible approximations of exhaustiveness do not ensure the detection of a refutation.

Key words: Clause Graphs, Connection Graphs, Resolution, Strategies, Completeness, Confluence

1. Introduction

A clause graph is essentially obtained from a set of clauses by inserting links between potentially resolvable literals. Kowalski adapted the resolution operation to the enriched structure of clause graphs [K75]. Among others, he proposed to obtain the links of new clauses by link inheritance, made link removal an integral part of resolution, and included a powerful purity removal rule with the potential to cause a chain reaction of removals. The present paper refers to an improvement of the original link inheritance mechanism based on internal links joining potentially resolvable literals of the same clause. This modification was proposed by Bruynooghe [B75] and was also described in [K79b]. In addition a subsumption operation is incorporated. All results also hold for improved treatments of the merging and factoring operations [E86], which are not discussed here. The reader should have some acquaintance with graph based resolution and should be familiar with the standard theoretical notions of the field.

Traditional clause set resolution can be regarded as an inference system whose states are finite clause sets and whose inference rules derive from a clause set S a clause set $S \cup \{C\}$, where C is a factor or resolvent of members of S. Any clause set may serve as an initial state. There are two classes of final states: refutation states contain the empty clause and affirmation states are closed under resolution and factoring but do not contain the empty clause.

Analogously, the states of clause graph resolution are clause graphs. Its inference rules derive from a clause graph G a clause graph G' by adding a resolvent or factor as indicated above, or by merging two identical literals of a clause, or by removing a pure clause, a tautology, or a subsumed clause. The only final affirmation state is the empty graph comprising neither clauses nor links, denoted by Ø. The final refutation states are the graphs containing the empty clause. For technical reasons all operations but subsumptions are prohibited on these graphs, such that □ subsumes any other clauses and a unique refutation state denoted by {□} is obtained. In contrast to clause set resolution only certain states may be initial, namely clause graphs in which no possible link is missing. The clause graph obtained from a clause set S by inserting all possible links is called the <u>initial clause graph for S</u>, denoted INIT(S).

The two inference systems differ in many aspects, but the most far-reaching consequences are due to the following: clause set resolution is commutative [N80], i.e. whenever two inference steps are possible in some state, each of them remains applicable after application of the other, and the resulting state is independent of the order in which the two steps are performed. Clause graph resolution, however, is not commutative.

1_1 Example: Resolution on the two links in G: [-Q1]—[+Q1|+P|+Q2]—[-Q2], without application of the purity or any other rule, results in either of the graphs

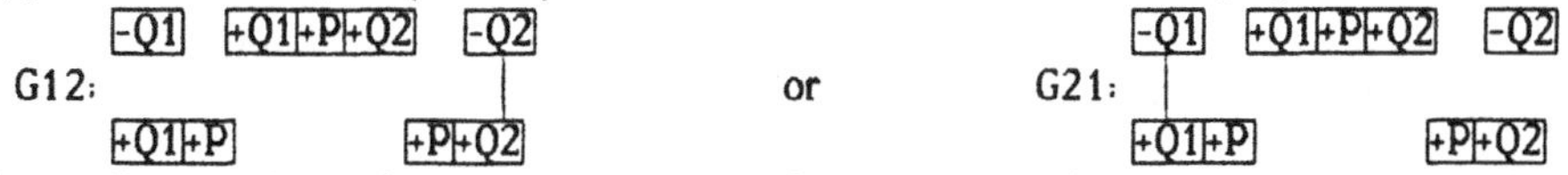

depending on the order of the two steps. Clause set resolution would produce the same state regardless of the order in which the two resolvents are created, but the two graphs have different structures. Thus with clause graph resolution even the resolution rule alone violates commutativity, let alone the combination of all inference rules. ■

The advantage of commutative inference systems lies in their automatic admittance of irrevocable control strategies [N80]. Otherwise a tentative control strategy might be necessary, which after each inference step has to make provisions for later reconsideration of alternatives. An irrevocable control stores just one state at a time, whereas a tentative control essentially requires the administration of a set of states and therefore tends to become unfeasible for inference systems with complex states, such as clause set resolution or clause graph resolution. Now the traditional conception of clause set resolution centers around just a single irrevocably modified clause set. The commutativity of this system is such a matter of course that nobody would seriously consider to undo the insertion of a resolvent in order to enable a continuation. Thus beyond questioning for clause set resolution, the appropriateness of irrevocable strategies does require reflection within the non-commutative clause graph resolution.

This is but one of the new problems which are considerable for clause graphs, but either vacuous or trivial in the context of clause set resolution. Such phenomena are the major reason why it is not even obvious what one actually wants to ascertain about clause graph resolution. In fact the history of clause graph resolution has been the history of identifying the difficulties, which may to some extent explain the leisurely pace of progress in this field.

2. Properties of the Inference System

The basic notions describing qualities of interest for logical inference systems are soundness and completeness, further Noetherianness and confluence [H80], which is a generalization of commutativity. Most of these are not directly properties of clause graph resolution. For instance, the purity rule leads to a satisfiable successor when applied to a graph consisting of two complementary unit clauses but no link, i.e. the system is not complete. Or consider the clause graph $\boxed{-P|-P}-\boxed{+P}$: depending on whether we start with a purity removal or a merging step, we can derive $\emptyset$ or $\{\square\}$. There is no common successor, so the system is not confluent.

The snag in these and similar easily contrived examples is that the derivations do not start from initial clause graphs. Were the missing links present in the example graphs, all derivations would ultimately arrive at $\{\square\}$. Now it might be that the given graphs can in fact be derived from some initial graphs, although this seems more than dubious. However, so far no criterion exists to test this reachability for arbitrary clause graphs, and even its decidability is open. Thus derivations from non-initial clause graphs only demonstrate properties of the inference system's unreachable parts, but convey no information about the crucial reachable subsystem. For clause set resolution this problem is void since initial states are not particularly distinguished.

Thus it appears useful to restrict the standard concepts to the reachable subsystem and even to subdivide the definitions according to the logical status of the initial state. Writing $G \rightarrow G'$ if G' is obtained from G by an inference step, and with the usual notation $\xrightarrow{+}$ and $\xrightarrow{*}$ for the transitive and the reflexive-transitive closure of $\rightarrow$ we obtain

2.1 Definition: Clause graph resolution is called

refutation sound	iff $\text{INIT}(S) \xrightarrow{*} \{\square\} \Rightarrow S$ unsatisfiable;
refutation complete	iff S unsatisfiable $\Rightarrow \text{INIT}(S) \xrightarrow{*} \{\square\}$;
refutation confluent	iff S unsatisfiable, $\text{INIT}(S) \xrightarrow{*} G_1$ and $\text{INIT}(S) \xrightarrow{*} G_2$ $\Rightarrow G_1 \xrightarrow{*} G'$ and $G_2 \xrightarrow{*} G'$ for some G';
affirmation sound	iff $\text{INIT}(S) \xrightarrow{*} \emptyset \Rightarrow S$ satisfiable;
affirmation complete	iff S satisfiable $\Rightarrow \text{INIT}(S) \xrightarrow{*} \emptyset$;
affirmation confluent	iff S satisfiable, $\text{INIT}(S) \xrightarrow{*} G_1$ and $\text{INIT}(S) \xrightarrow{*} G_2$ $\Rightarrow G_1 \xrightarrow{*} G'$ and $G_2 \xrightarrow{*} G'$ for some G'.

For clause set resolution these notions are defined correspondingly. ■

Note that for lack of a better name the definition associates the term "affirmation" with concepts pertaining to satisfiability, and not to validity as customary. But validity plays no rôle in the present context, which might justify the deviation from convention. Let us briefly discuss what these properties signify in general and how far they hold for the familiar clause set resolution in particular.

Refutation soundness means that deriving $\{\square\}$ implies the unsatisfiability of the initial state. The dual property affirmation soundness says that from reaching an affirmation state the satisfiability of the start state follows. Clause set resolution is both refutation and affirmation sound. As demonstrated above, affirmation soundness is by no means an obvious property of clause graph resolution. Yet to some extent it is more important for clause graph resolution because final affirmation states are derived in more cases than with clause set resolution. For instance, S = {+Pfa, -Pfx+Px, -Pb} has no finite closure under resolution and therefore no final state can be reached from S with clause set resolution. On the other hand, a resolution step on the link incident with +Pfa in INIT(S) enables a derivation of Ø with the purity rule. This "collapse" being a fairly frequent event for clause graph resolution, one would really like to know whether it stands for satisfiability, e.g. due to an insufficient axiomatization, or whether another control strategy might have found a refutation. Affirmation soundness would always ensure the former to be the case.

Refutation completeness expresses the existence of a refutation from each unsatisfiable initial state. Dually, affirmation completeness ensures the derivability of an affirmation state in satisfiable cases. Clause set resolution is refutation complete, but of course affirmation complete only for certain decidable classes, e.g. for the ground level. Even in these cases affirmation completeness is rarely exploited and affirmation states are usually arrived at only by accident when trying to refute an underaxiomatized problem.

The commutativity of an inference system represents confluence in its most elementary form, hence clause set resolution is both refutation confluent and affirmation confluent. Refutation confluence carrys weight in that it allows for irrevocable control strategies. Suppose some unsatisfiable S is refutable, i.e. INIT(S) $\xrightarrow{*}$ {□}, and we perform an arbitrary series of inference steps leading to a graph G, i.e. INIT(S) $\xrightarrow{*}$ G. Refutation confluence then ensures G $\xrightarrow{*}$ {□}, i.e. from G we can still somehow proceed and reach the empty clause. In other words, any derivation can be continued to a refutation, and there are no dead ends that require a backup to earlier states in order to derive {□}. Thus refutation confluence is absolutely indispensible for clause graph resolution to be of any practical value. Again, the dual property affirmation confluence lacks this significance as long as the system is primarily intended to refute unsatisfiable clause sets.

Noetherianness was not included in definition 2_1. Clause set resolution is Noetherian in the ground case because only finitely many different clauses exist with the given atoms. Clause graph resolution, however, is not:

2_2 Example: Infinite clause graph derivation

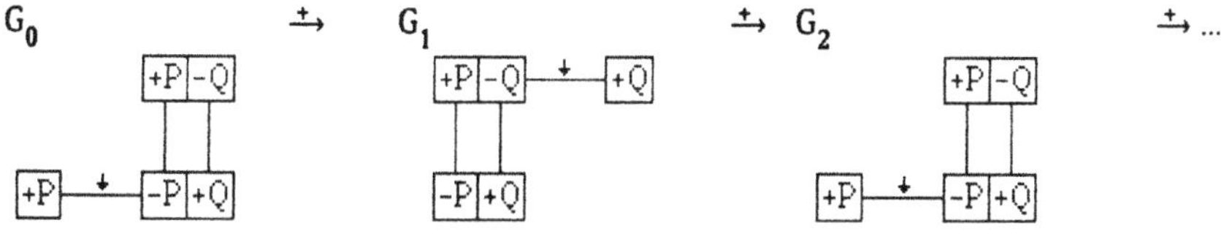

States directly preceding purity removals are not displayed, hence the $\xrightarrow{+}$ instead of $\rightarrow$. G_2 is isomorphic to G_0, and the corresponding inference steps can be repeated all over again. The graph's satisfiability is not the decisive point. Just add two unit clauses -R and +R to obtain an infinite derivation from an unsatisfiable initial graph. ■

At first sight the origin of the new phenomenon seems somewhat mysterious. After all, each clause graph derivation can directly be simulated by a clause set derivation, hence the same cyclicity ought to be immanent in clause set resolution. And indeed, it is, if we forgo the implicit fusion of identical clauses hidden in the set representation. Using states that are multisets rather than sets of clauses, ground clause set resolution admits of infinite derivations too. The usual definition based on sets of clauses happens to enforce a special form of subsumption. For clause set resolution circular derivations of the above kind are precluded by disallowing the inference of clauses that are subsumed by or at least variants of clauses already present in a state. Clause graph resolution faces the problem that the subsuming clause may no longer be present in the state because it was victimized by some previous purity removal.

Most of the results on properties defined in 2_1 derive from a fundamental lemma:

2_3 Lemma (Clause Graph Lemma): There is an invariance property $\mathfrak{R}$ (for refutable) such that for all clause graphs G, G' and clause sets S:

(i)	$\mathfrak{R}(G)$	$\Rightarrow$	G unsatisfiable;	
(ii)	S unsatisfiable	$\Rightarrow$	$\mathfrak{R}(\text{INIT}(S))$;	
(iii)	$\mathfrak{R}(G)$ and $G \rightarrow G'$	$\Rightarrow$	$\mathfrak{R}(G')$;	
(iv)	$\mathfrak{R}(G)$	$\Rightarrow$	$G \xrightarrow{*} \{\square\}$.	■

Once the clause graph lemma is available, most of the remaining proofs become straightforward (the labels (i)-(iv) refer to 2_3).

2_4 Theorem: Clause graph resolution is refutation sound.
Proof: trivial, since clause set resolution is sound. ■

2_5 Theorem: Clause graph resolution is refutation complete.
Proof: Let S be unsatisfiable

$\Rightarrow$	$\mathfrak{R}(\text{INIT}(S))$	by (ii)	
$\Rightarrow$	$\text{INIT}(S) \xrightarrow{*} \{\square\}$	by (iv)	■

2_6 Theorem: Clause graph resolution is refutation confluent.
Proof: Let S be unsatisfiable, $\text{INIT}(S) \xrightarrow{*} G_1$ and $\text{INIT}(S) \xrightarrow{*} G_2$

$\Rightarrow$	$\mathfrak{R}(\text{INIT}(S))$	by (ii)	
$\Rightarrow$	$\mathfrak{R}(G_1)$ and $\mathfrak{R}(G_2)$	by (iii)	
$\Rightarrow$	$G_1 \xrightarrow{*} \{\square\}$ and $G_2 \xrightarrow{*} \{\square\}$	by (iv)	■

2_7 Theorem: Clause graph resolution is affirmation sound.
Proof: By contradiction. Assume $\text{INIT}(S) \xrightarrow{*} \emptyset$ and S is unsatisfiable

$\Rightarrow$	$\mathfrak{R}(\text{INIT}(S))$	by (ii)	
$\Rightarrow$	$\mathfrak{R}(\emptyset)$	by (iii)	
$\Rightarrow$	$\emptyset$ unsatisfiable	by (i)	
$\lightning$	since $\emptyset$ is valid.		■

Alternatively, it is not difficult to see that refutation completeness and refutation confluence together imply affirmation soundness.

2_8 Theorem: Clause graph resolution is affirmation complete for the ground case.
Proof: Follows from the termination of the predicate cancellation strategy described in the next section. The applicability of this strategy to the ground case has several times been explicitly shown [SS76, SS80, B81b, B82b]. ∎

2_9 Theorem: Clause graph resolution is not affirmation confluent.
Proof: Let S = {-Px+Pfx, -Py+Pgy}, INIT(S) = G_0, and consider the derivation

$G_0 \quad \longrightarrow \quad G_1 \quad \longrightarrow \quad G_2$

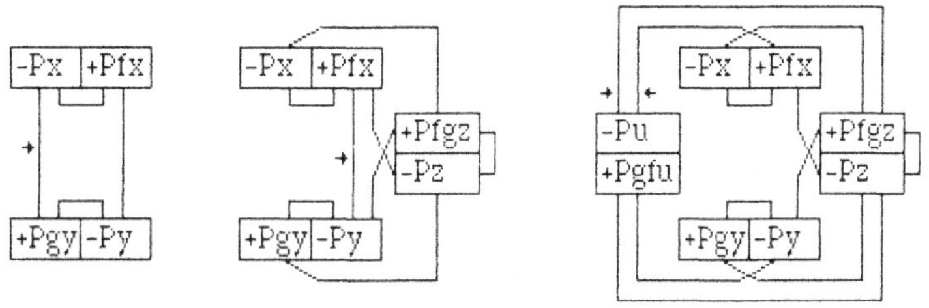

Let -Pu+Pgfu =: C and -Pz+Pfgz =: D. Next we resolve on the two indicated links enabling a purity removal of C. Let G be the resulting graph and note that G contains D but does not contain C. It can easily be shown that this holds for every graph derivable from G [E86]. Reversing the order of the first two steps essentially exchanges the rôles of C and D, i.e. from G_0 we can also derive a graph G', all of whose successors contain C but not D. Thus we have $G_0 \xrightarrow{*} G$ and $G_0 \xrightarrow{*} G'$, but there exists no common successor of G and G'. ∎

One thing about this example is conspicuous, though: there is no way whatsoever to cause a collapse, i.e. to derive Ø. One might suspect that each successor of a collapseable graph remains collapseable, and that therefore the derivability of Ø ensures confluence. A counterexample in [E86], requiring the precise definition of factoring, shows that this is not the case. Apparently it is exactly the critical subsystem on which the inference relation is confluent.

Most investigations of clause graph resolution published so far fit into the schema of the clause graph lemma. From this point of view Bibel in [B81b] proved parts (i), (ii), and (iii) for the ground case, $\mathfrak{R}(G)$ meaning "G is spanning". Other parts of his proof depend on a stronger $\mathfrak{R}'(G)$ meaning "G is spanning and contains only removable tautologies", for which he proved part (iv) while (i) and (ii) are obvious. However, part (iii) does not hold for $\mathfrak{R}'$ i.e. refutation confluence does not follow. In [B82b] Bibel used a lifting technique to show the completeness of first order clause set resolution, which can also be applied to clause graph resolution. Then the same parts hold for the first order level too. Confluence is presented as open in [B82b]. Smolka [S82a] used another $\mathfrak{R}(G)$ to prove all parts for the unit refutable class (the class of formulae for which unit resolution and input resolution are complete [CL73]). Proofs of all parts for the general case are contained in [E86], where $\mathfrak{R}(G)$ means "G contains the homomorphic image of a refutation graph". A refutation graph is a special clause graph which in a sense represents a derivation of {□}. This concept is due to Shostak [S76].

Bibel restricted the tautology and subsumption rule by certain link conditions, motivated by straightforward non-initial graphs. As pointed out above, such examples are not very informative. In the case of subsumption the restriction is in fact necessary. [E86] contains a ground example where Ø is derived from an unsatisfiable unit refutable initial graph using only resolution, purity removal, and forward subsumption with violations of the link condition. On the other hand Smolka showed that the link condition on tautologies can be abandoned for the unit refutable class. In the present context this means that all parts of the clause graph lemma remain true, which implies

2_10 Theorem: For the unit refutable class clause graph resolution with an unrestricted tautology rule is refutation complete, refutation confluent, and affirmation sound. ■

However, Smolka also found a ground but not unit refutable example, where Ø can be derived from an unsatisfiable initial graph using only resolution, purity, and unrestricted tautology removal. This directly contradicts affirmation soundness. In [E86] it is shown that for any unsatisfiable initial graph there exists a refutation producing no tautology. The combination of these results gives

2_11 Theorem: In general clause graph resolution with an unrestricted tautology rule is refutation complete, but neither refutation confluent nor affirmation sound. ■

Note that this represents the system as defined by Kowalski. Thus with the original version of clause graph resolution there exists a refutation for each unsatisfiable formula, but an attempt to find it can lead into dead ends from which it is no longer reachable. This result clearly demonstrates the uselessness of traditional completeness alone and ought to motivate the proposed notional framework.

3. Properties of Control Strategies

Since the underlying inference system is refutation complete, there are well-known standard control strategies [N80] ensuring the derivation of {□} from any unsatisfiable INIT(S) after finitely many steps. The most simple approach is a breadth-first search through the space of all states (i.e. all clause graphs) derivable from the given one. Unfortunately all of these standard strategies are of the tentative type and hence unfeasible. The complexity of the states necessitates irrevocable control strategies with the property above. The inference system's refutation confluence establishes the existence of such a strategy. So it seems as if in order to develop a serviceable control regime for clause graph resolution one simply has to scrutinize the confluence proof, all the more so because this proof even happens to be constructive.

The resulting approach amounts to finding within the given initial clause graph the homomorphic image of a refutation graph, whose structure is then used to control the application of the inference rules. However, finding such a subgraph is equivalent to proving the given formula's unsatisfiability. Many advanced proof methods using clause graphs are in fact based on this very idea of searching an initial graph for appropriate substructures [A81, AO83, B82b, CS79, Si76]. Thus the proposed strategy would actually solve the overall problem in another system and then just guide the clause graph resolution system to produce a rederivation of the solution found beforehand. Of course a strategy of this kind, albeit perfect in terms of goal-directedness, fails to comply with intuitive intentions (its remarkable kinship to a familiar way of working with theorem provers belongs to a different story).

By these considerations the questions of whether there exists an irrevocable control strategy guaranteed to find a refutation and whether it can be explicitely specified, are positively settled. But beyond that we expect a control strategy to base its decisions on some reasonably plain syntactic criteria that are extractable from the current clause graph in a direct and uncomplicated way. When selecting a step, it may well compute some information from the local vicinity of any object in the graph, and it may employ a number of easily updatable flags and markers, but it is not supposed to perform a global search of the entire clause graph. A strategy meeting these demands is called a pragmatic one. The requirements are admittedly hard to capture in a rigid definition, so typical control strategies for clause set resolution may serve as a guideline.

Most of the theorem proving literature further classify control strategies. In the terminology of [WOLB84] a restriction strategy constrains the application of inference rules, while an ordering strategy dictates where next to apply a rule. Restrictions decide about the admissibility of operations once and for all, whereas orderings only postpone operations relative to others, but do not totally prohibit any.

The most famous example of an ordering strategy is the unit preference strategy [WCR64]. From the multitude of restriction strategies we shall address the following ones: the unit and the input restriction, which are both sufficient for the unit refutable class [CL73], which is especially good-natured towards clause graph resolution; further the set-of-support restriction [WRC65], the merging restriction [A68], and the linear [L78], s-linear [L78], t-linear [KK71], and the SL restriction [KK71]; in addition, the half-factoring restriction [No80]. Finally, factoring can be restricted in a new way. Let "kindred literals" be such as have a common ancestor literal. For instance, resolution between C = +Px+Qz and D = -Qa-Qb produces R = +Py-Qb, which again resolves with C giving R' = +Px'+Py'. R' consists of kindred literals. The kindred factoring restriction confines factoring to kindred literals.

For a formal description of strategies let us for a moment consider the general level of an arbitrary inference system with a set **S** of states and an inference relation $\rightarrow$. Here the difference between restrictions and orderings disappears.

3_1 Definition: A filter for an inference system is a one-place predicate $\mathfrak{F}$ on the set of finite sequences of states. The notation $\underline{S_0 \xrightarrow{*} S_n \text{ with } \mathfrak{F}}$ stands for a derivation $S_0 \xrightarrow{*} S_n$ where $\mathfrak{F}(S_0 \dots S_n)$ holds. For an infinite derivation, $\underline{S_0 \rightarrow \dots \rightarrow S_n \rightarrow \dots \text{ with } \mathfrak{F}}$ means that $\mathfrak{F}(S_0 \dots S_n)$ holds for each n. ■

Intuitively, a filter $\mathfrak{F}$ conceals all derivations for which the predicate does not hold, i.e. $\mathfrak{F}$ retains a sub-portion of the inference system's original search space. While it is not impossible to apply some tentative control regime in the remaining search space, there is a canonical way to associate with $\mathfrak{F}$ a class of irrevocable strategies. Having derived from some initial state S_0 some non-final state S_n, such a strategy may choose as successor any state S with $S_n \to S$ and $\mathfrak{F}(S_0 \dots S_n S)$, thus pursuing one single derivation with $\mathfrak{F}$. The strategy may freely exploit any indeterminism left by the filter and one may not make any assumptions about its choices. Under these conditions the system's behaviour depends entirely on the properties of $\mathfrak{F}$.

In the present context the states are clause graphs and the filters have to be defined on sequences of clause graphs. Traditional strategies are often described by means of a structure called deduction tree [L78], which is an upward growing tree whose nodes are clauses and whose arcs connect resolvents or factors with their parent clauses (3_8 displays two specimens). One can think of deduction trees as being embedded in the clause graph resolution states in the following way: given $G_0 \xrightarrow{*} G_n$, associate with each clause C in G_n a deduction tree made up according to the steps of the derivation. Its nodes are clauses from the union of all G_i, with leaves from G_0 and root C. Loosely speaking, G_n contains a set of such deduction trees. An inference step $G_n \to G$ producing a new clause introduces a new deduction tree, which is contained in G and not contained in G_n, but its immediate subtrees are.

A restriction essentially depends upon a predicate that admits only certain deduction trees. The corresponding restriction filter $\mathfrak{F}$ is defined as $\mathfrak{F}(G_0 \dots G_n G)$ iff $G_n \to G$ and the deduction tree introduced by G is admitted by this predicate. An ordering relies on a merit ordering of deduction trees (or of linearizations thereof [K70]). Given a merit ordering one obtains the corresponding ordering filter $\mathfrak{F}$ qua $\mathfrak{F}(G_0 \dots G_n G)$ iff $G_n \to G$ and none of the deduction trees introduced by any other potential successor of G_n has better merit than the deduction tree introduced by G.

3_2 Example: Let $\mathfrak{F}_{INPUT}(S_0 \dots S_n S)$ iff the root of any deduction tree introduced by S is adjacent to a leaf from S_0. Further let a deduction tree T_1 have better merit than T_2, if the depth of T_1 is less than the depth of T_2, and let $\mathfrak{F}_{LEVEL}$ be the ordering filter corresponding to this merit ordering. For clause set resolution an irrevocable control strategy based on $\mathfrak{F}_{INPUT} \wedge \mathfrak{F}_{LEVEL}$ performs a level saturation derivation in the search space left by the input restriction. The remaining indeterminism leaves it up to the strategy to sequentialize at pleasure the generation of clauses of the same level. ■

Reduction steps can also be treated by both kinds of filters. An ordering filter might postpone the removal of subsumed clauses to certain resolution steps. A restriction filter might preclude backward subsumptions. Furnished with appropriate node attributes indicating if and why the clause is not present in the respective state, deduction trees may serve as a definitional aid for such cases too.

With this background the behaviour of control strategies can be described in terms of filters and their properties. There is a natural way to accomodate to filters the notions introduced for the inference system in definition 2_1:

3_3 Definition: A filter $\mathfrak{F}$ for clause graph resolution is called

<u>refutation sound</u>	iff INIT(S) $\xrightarrow{*}$ {□} with $\mathfrak{F}$ $\Rightarrow$ S unsatisfiable;
<u>refutation complete</u>	iff S unsatisfiable $\Rightarrow$ INIT(S) $\xrightarrow{*}$ {□} with $\mathfrak{F}$;
<u>refutation confluent</u>	iff S unsatisfiable, INIT(S) $\xrightarrow{*}$ G_1 with $\mathfrak{F}$ and INIT(S) $\xrightarrow{*}$ G_2 with $\mathfrak{F}$ $\Rightarrow$ INIT(S) $\xrightarrow{*}$ G_1 $\xrightarrow{*}$ G' with $\mathfrak{F}$ and INIT(S) $\xrightarrow{*}$ G_2 $\xrightarrow{*}$ G' with $\mathfrak{F}$ for some G';
<u>refutation Noetherian</u>	iff for no unsatisfiable S there exists an infinite derivation INIT(S) $\to G_1 \to G_2 \to \ldots \to G_n \to \ldots$ with $\mathfrak{F}$.

Note that $\to$ with $\mathfrak{F}$ need not be transitive, hence the special form of confluence. The dual concepts and the adaptation to clause set resolution read correspondingly. ■

The inference system's soundness properties are not affected by any strategy, i.e. each filter for clause graph resolution is refutation sound and affirmation sound.

Refutation completeness of a filter $\mathfrak{F}$ only signifies the existence of a refutation with $\mathfrak{F}$ from any unsatisfiable initial state. For clause set resolution this is the central and usually the only property of restrictions ever investigated. The afore-named restrictions are refutation complete for clause set resolution and most of them can even be combined without destroying refutation completeness. Of course it cannot be expected that such results carry over to clause graph resolution in some straightforward way.

As in the case of the inference system, refutation confluence of a filter is necessary to avoid the need for backups to earlier states of a derivation. Among others, the restrictions above are refutation confluent for clause set resolution, although no mention of this property seems to appear anywhere in the literature. One might suspect that this omission can be attributed to some automatic transfer of the (trivial) confluence of the inference system to any filter. But non-confluent filters for clause set resolution do exist:

3_4 Example: Let the "once-only" restriction ordain that no clause be used more than once as a parent clause in resolution steps. Consider the following derivations from S = {-Q, +Q+P+R, -P-R, +Q}, where clauses already used as parent clauses are underlined:

$$S \left\langle \begin{array}{l} \{\underline{-Q}, \underline{+Q+P+R}, -P-R, +Q, +P+R\} \to \{\underline{-Q}, \underline{+Q+P+R}, \underline{-P-R}, +Q, \underline{+P+R}, -R+R\} =: S_1 \\ \\ \{\underline{-Q}, +Q+P+R, -P-R, \underline{+Q}, \square\} \xrightarrow{+} \{\square\} \end{array} \right.$$

There is no continuation from S_1, i.e. no common successor state of S_1 and {□} can be reached with once-only. Hence this restriction is not refutation confluent for ground unit refutable clause sets, although its refutation completeness for this class can be shown. ■

Thus restriction filters for clause set resolution can indeed demand a tentative application in order to reach {□}. So why is it that this problem never arose for traditional strategies? The reason seems to be a tacit convention to consider only "context-free" restrictions, where the decision about the admission or rejection of a potential resolvent or factor remains constant and is never reconsidered throughout a derivation. Then a clause that may be added to some state may also be added to any of its successors, which implies confluence. The once-only restriction violates this arrangement. In the context of clause graph resolution the convention cannot be observed from the very outset.

Refutation completeness and confluence of a restriction filter do not exclude infinite branches in the remaining search space. In order to ensure that an existing refutation will actually be found by an irrevocable control regime, the filter has to be refutation Noetherian. None of the traditional restrictions comes anywhere near Noetherianness and termination is considered a problem to be handled by an ordering. The definition of an appropriate merit ordering yields an ordering filter. For each of them one has to ascertain refutation completeness, refutation confluence, and refutation Noetherianness, and ideally these qualities should be maintained in conjunction with any refutation complete and refutation confluent restriction filter.

Now in fact all orderings proposed for clause set resolution belong to the same class of exhaustive orderings. A merit ordering is exhaustive, if each derivation admitted by the corresponding filter potentially, i.e. if sufficiently continued to non-final states, reaches every deduction tree in the search space [K70]. Note the repugnance of this property with the very nature of restriction filters. Exhaustive ordering filters trivially enjoy all the properties we strive for (assuming the technicality that a derivation of $\{\square\}$ is enforced whenever S contains $\square$, but $S \neq \{\square\}$). Unfortunately exhaustiveness heavily depends on the inference system's commutativity. For the non-commutative clause graph resolution exhaustive ordering filters do in general not exist.

3_5 Example: From G in example 1_1 either of the following deduction trees can be obtained, but due to link removal no derivation reaches both of them:

```
      -Q1    +Q1+P+Q2                          +Q1+P+Q2    -Q2
        \   /                                          \   /
        +P+Q2    -Q2          and              -Q1    +Q1+P
              \  /                               \   /
(T12)          +P                                 +P        (T21)
```

■

Exhaustiveness being beyond attainment for clause graph resolution, we must attempt to capture its intention with a weaker notion. The central quality of an exhaustive ordering is fairness, which means that each possible operation has a finite chance to be performed and none is infinitely postponed. In a clause graph the possible operations are represented by the links. Hence the most obvious rephrasing of fairness consists in precluding the presence of any link in infinitely many states of a derivation.

3_6 Definition: An ordering filter $\mathfrak{F}$ for clause graph resolution is called covering, if the following hold: Let G_0 be an initial clause graph, let $G_0 \xrightarrow{*} G_n$ with $\mathfrak{F}$ be a derivation, and let λ be a link in G_n. Then there is a number $n(\lambda)$, such that for any derivation $G_0 \xrightarrow{*} G_n \xrightarrow{*} G$ with $\mathfrak{F}$ extending the given one by at least $n(\lambda)$ steps, λ is not in G. ■

With a covering filter no link may be infinitely delayed. Siekmann and Stephan [SS76, SS80] used the terminology coveringthree for this concept and further proposed two stronger properties. As an aid, imagine distinct "colours" associated with the links in the initial graph, and let each descendant of a coloured link inherit the ancestor's colour.

3_7 Definition: An ordering filter $\mathcal{F}$ for clause graph resolution is called coveringtwo, if it is covering and one link of each colour must have been operated upon after at most finitely many steps. $\mathcal{F}$ is coveringone, if each colour must have completely disappeared after at most finitely many steps. ■

Since links can vanish in consequence of a clause removal, a covering filter might admit derivations circumventing an essential operation by ridding of an initial link and of all its descendants that way. The additional coveringtwo condition enforces that some descendant is actually resolved upon. Coveringtwo corresponds to the notion of exhaustiveness as defined by Brown in [B76].

Each coveringone filter is evidently Noetherian, but as yet no ordering filter of this type has been known for the general case. Siekmann and Stephan discuss the uncomfortable stringency even of the coveringtwo class and argue that a slightly weaker condition ensues from covering anyway. Thus it seems plausible that in combination with a refutation complete and refutation confluent restriction filter (or at least with the null restriction) each covering ordering filter is refutation complete, refutation confluent, and refutation Noetherian. This presumption is a precise wording of what has been known as the strong completeness conjecture for clause graph resolution.

Attempts at proving this conjecture have been undertaken, but apart from such overall approaches refutation Noetherianness has not been definitely established for any particular ordering filter for the general case. A special covering ordering for clause graph resolution is founded on the idea to select some positive literal occurrence L = +P(...) and then to successively resolve on all links incident with L. If (a) L is not incident with an internal resolution link, it must eventually become pure. Provided that (b) no positive literal with predicate symbol P was introduced into any resolvent, the number of positive occurrences of P decreases with the subsequent purity removal. For classes of formulae where (a) and (b) can be satisfied, e.g. the ground case, a systematic elimination of all predicate symbols is thus possible. The resulting predicate cancellation filter is trivially Noetherian. In general, however, the conditions are not met and a predicate cancellation filter does not exist. Even if it does, it is incompatible with almost every restriction filter.

To sum up, the existence of tentative clause graph control strategies ferreting out the empty clause need not be troubled about, and neither does the existence of irrevocable strategies if they may exploit unlimited information. To establish a pragmatic irrevocable strategy for clause set resolution, the standard procedure has been as follows: define some restriction filter, prove its refutation completeness (i.e. the existence of some refutation with the filter from each unsatisfiable initial state), take confluence for granted, and implicitly assume an exhaustive ordering filter that will sure enough mind the irksome problem to actually find one of the existing refutations in the remaining search space. For clause graph resolution each of these steps occasions some idiosyncratic obstacles, whose difficulty increases in the given order. Now let us see how far they have at present been surmounted.

3_8 Theorem: The conjunction of the set-of-support and the merging restriction is not refutation complete for clause graph resolution (but it is for clause set resolution [A68]).

Proof: Let S = {+P, -P+Q, -Q+R, -Q-R} and let the support set be {+P}. It is easily verified that each deduction tree in the search space is a subtree of one of the following (note that all resolutions between the initial non-unit clauses are locked by the set-of-support restriction and that no merge can be reached):

```
    +P   -P+Q                          +P    -P+Q
      \  /                               \  /
       +Q    -Q+R                         +Q    -Q-R
         \  /                               \  /
    -Q-R  +R                          -Q+R  -R
       \  /                              \  /
  -P+Q    -Q                        -P+Q    -Q
     \  /                              \  /
   +P  -P                            +P  -P
    \  /                              \  /
     □                                 □
```

Thus each refutation starts with a resolution step using the input clause +P as parent and ends with a resolution step using the same parent. With clause graph resolution this clause inevitably becomes pure during the first step, and the final steps are impossible. ■

3_9 Theorem: The combination and thus each subset of the following restrictions is refutation complete for clause graph resolution: set-of-support, linear, s-linear, t-linear, SL, half-factoring, kindred factoring. Also the combination of merging, half-factoring, and kindred factoring is refutation complete. For the unit refutable class both the unit and the input restriction are refutation complete.

Proof outline: In [YRH70] an analogous completeness theorem treating a slightly different set of restrictions is shown for clause set resolution. The technique used there can be modified to apply to refutation graphs [S76, S79], and thus to clause graphs. Using this principle all of the results follow in a similar way. The proofs are carried out in [E86]. ■

Theorem 3_8 demonstrates that some of the popular restriction strategies do indeed loose their refutation completeness when applied to clause graph resolution. This means that each interesting combination has to be investigated anew. But 3_9 suggests that by and large the refutation completeness of restriction strategies is not extremely critical for clause graph resolution. The situation is entirely different for the next property.

3_10 Theorem: The linear, s-linear, t-linear, SL, and the merging restriction are not refutation confluent. Any combination of set-of-support, half-factoring, and kindred factoring is refutation confluent. For the unit refutable class the unit restriction is refutation confluent, but the input restriction is not.

Proof: (Of the negative statements only, the positive proofs appear in [E86].) Consider the unit refutable S = {+P, -P+Q, -Q+R, -Q-R} from 3_8 and the derivation

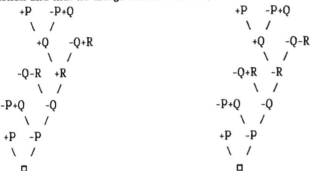

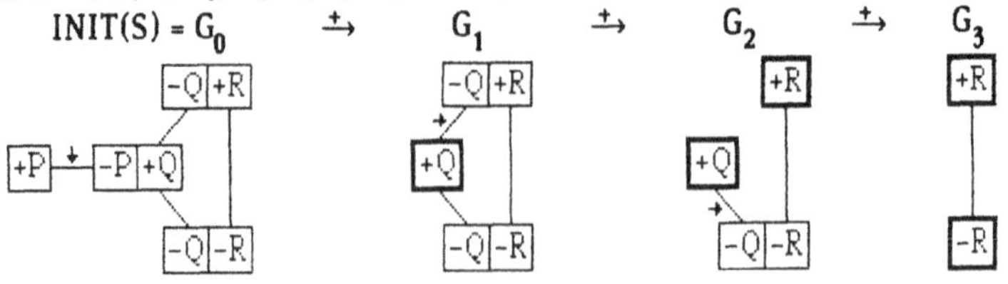

Bold boundary lines mark non-input clauses. Obviously G_3 was obtained with an input derivation, which by definition is also a derivation with merging. Since the clauses of G_3 are neither input clauses nor merges, the only possible continuation from this state is not admitted by either restriction.

Taking +P as the top clause, +P was used as a near parent of +Q, which in turn was used as a near parent of both +R and -R. No ancestry step is possible between these clauses. Since all of them are units, the condition to select a single most recently introduced literal trivially holds. Thus the input derivation is also a derivation with SL and consequently also a derivation with each of SL's successive generalizations t-linear, s-linear, and linear. Neither of the clauses of G_3 is an ancestor of the other, both are non-input, and again the restrictions allow no continuation.

By the above completeness results we have INIT(S) $\xrightarrow{*}$ {□} with each of the restrictions in question. The derivation of a common successor of {□} and G_3 is prevented by the restrictions. ■

Again, the uselessness of refutation completeness alone is evident. For instance, we know that there always exists a linear refutation from any unsatisfiable initial graph. However, when trying to find one, we can get lost in a state where the strategy does not permit any continuation. Note that due to the inference system's refutation confluence there always exists a continuation from such a state leading to the empty clause, but it requires steps precluded by the strategy. Refutation confluence appears to be a fairly rare property of restriction strategies for clause graph resolution. Remarkably, all the negative results also concern the unit refutable class.

There is a redundancy involved in the SL restriction (and others) due to the possibility of two or-branches in the search tree representing essentially opposite orders of the same sequence of input steps. If both have been halfway developed, the continuation of either repeats work already done in the other. Kowalski points out that such a repetition of steps is not possible with clause graph resolution and argues that a clause graph simulation of SL resolution overcomes the redundancy problem [K75]. From the above example the sequences of resolvents +Q, +R, -Q and +Q, -R, -Q represent two such branches, and indeed neither of them can be continued from G_3. Unfortunately the "shortcut" between the branches is no legal SL step. An appropriate extension of SL would have to cope with cases where each of the clauses contains further literals, leading to clauses $C_0C_1C_2$+R and C_0C_1-RC_3, for instance, with most recent parts to the right. In longer branches the continuation from the former may have proceeded before generation of the latter, and it is at least not obvious, when exactly such steps between different branches have to take place. In any case the extension would somewhat contravene the input-style flavour of SL resolution.

The overall expenditure of deriving □ is directly correlated to the complexity of deduction trees with root □. Kowalski and Kuehner [KK71] proposed to measure this complexity by any function on the so-called <u>rm-size</u> (r,m), where r is the number of resolution steps and m is the number of factoring or merging steps represented in a deduction tree. For instance, both of the deduction trees in 3_8 have rm-size (5,0) and hence have the same complexity. In comparing deduction trees with different rm-sizes, it is only assumed that the complexity is monotonously non-decreasing in r and m, and that a boost of m does not increase the complexity more than the same boost of r. If complexity is defined as a weighted sum of r and m, the latter condition means that the weight of m must not exceed the weight of r.

Unfortunately it is not uncommon for restrictions to reject deduction trees with minimal rm-size, and thus to pay for the lower branching rate of the search space with the increased complexity of refutations. For instance, from S in 3_8 the unrestricted clause set resolution system can reach deduction trees with root □ whose rm-sizes are (3,1) and (4,0), whereas all trees retained by a combination of the merging and the set-of-support restriction have rm-size (5,0). We have seen that clause graph resolution also rules out certain deduction trees reachable with clause set resolution, and the question suggests itself whether the same price has to be paid for this reduction of the search space. The negative answer is deserving of positive appreciation.

3_11 Theorem: If for some clause set S there exists a clause set refutation with rm-size (r,m) under the null, the set-of-support, or the unit restriction, then there exists a clause graph refutation with rm-size (r',m') under the same restriction, such that $r' \le r$ and $m' \le m$.
Proof outline: In [S79] Shostak showed how to translate any clause set refutation into a refutation graph, whose structure reflects the refutation's rm-size. The proof for part (iv) of the clause graph lemma consists in using the structure of a refutation graph to control the clause graph derivation. This might decrease, but never increases the rm-size. The combination of the two techniques requires only little technical adjustment. ■

This result means that the complexity of simplest refutations, no matter how defined, is not worse for clause graph resolution than for clause set resolution. Amiably clause graph resolution prunes sub-optimal branches of the search space only, and does not throw away garbage and gold alike as most restrictions do.

3_12 Theorem: For the unit refutable class the strong completeness conjecture is true, i.e. the conjunction of a covering ordering filter with any refutation complete and refutation confluent restriction filter is refutation complete, refutation confluent, and refutation Noetherian.
Proof: Smolka [S82a] established a complexity measure that decreases with any step belonging to some minimal refutation present in the graph. From this result the statements follow directly. ■

3_13 Theorem: In general the strong completeness conjecture is wrong even if no restriction filter is involved. The same holds for a strengthening of the conjecture based on the coveringtwo definition.

Proof: Let S = {+P+Q, -P+Q, -Q-R, +R+S, +R-S}. We perform a derivation from INIT(S) = G_0:

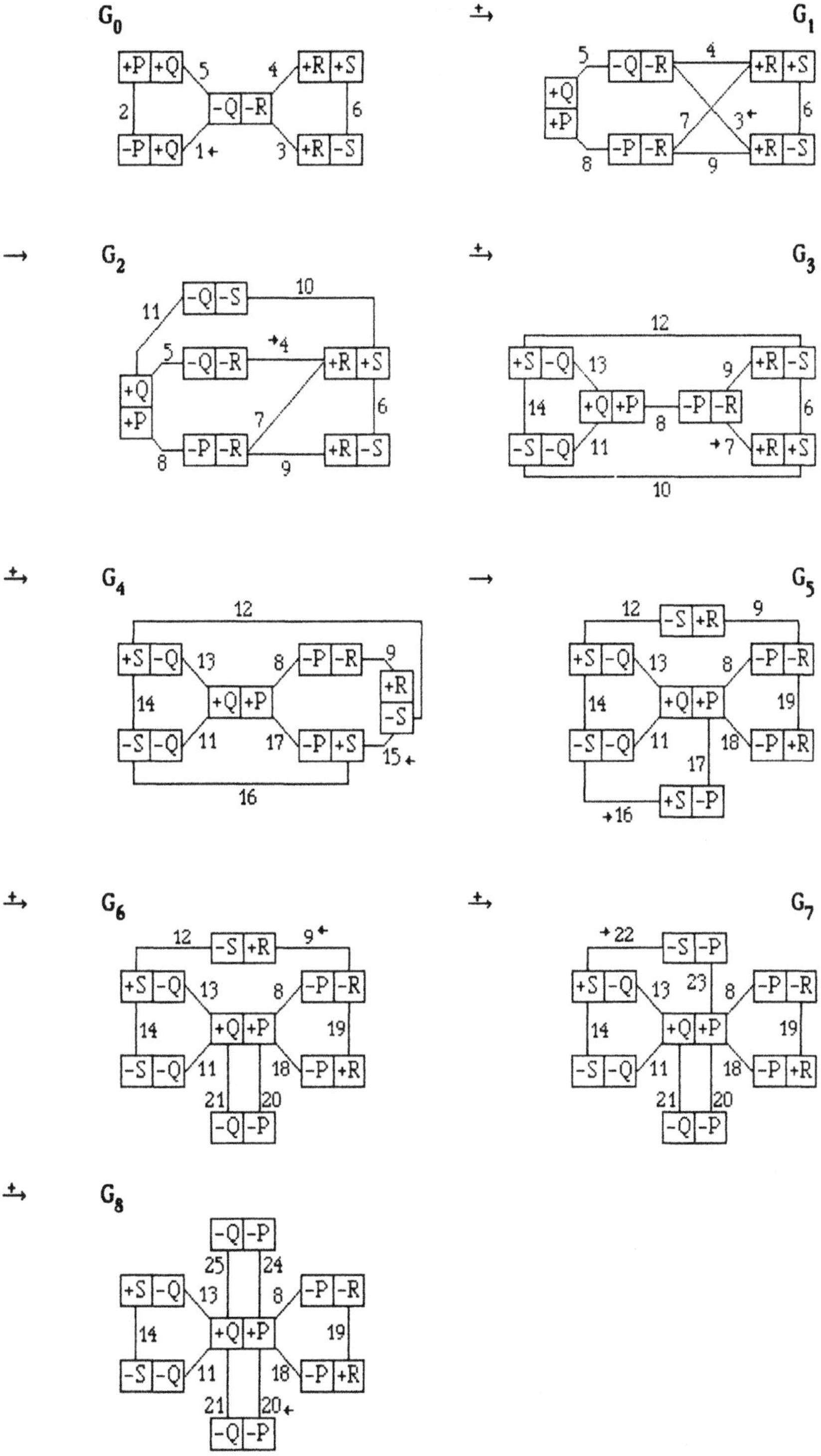

G_0
G_1
G_2
G_3
G_4
G_5
G_6
G_7
G_8

G_8 includes two copies of -Q-P, one of which might be removed by subsumption. Instead, we perform the corresponding resolution steps for both of them in succession.

$\xrightarrow{\pm}$ G_9

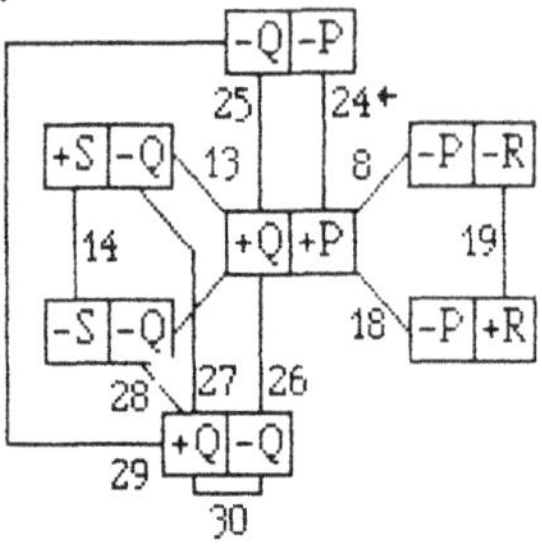

$\xrightarrow{\pm}$ G_{10}

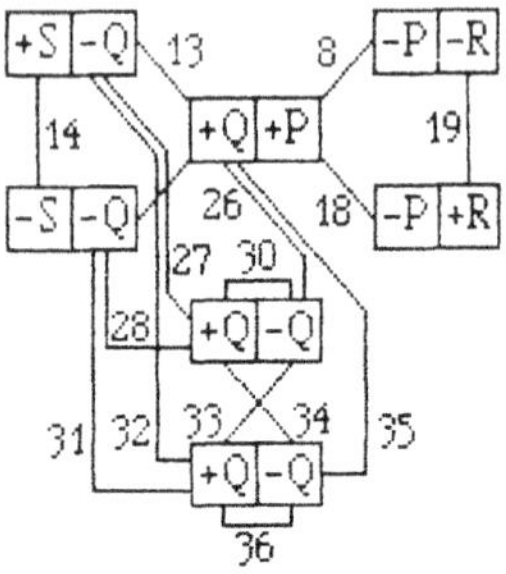

The last graph contains two tautologies and all links possible among its clauses. In other words, it is the initial clause graph of {+S-Q, -S-Q, +Q+P, -P-R, -P+R, +Q-Q, +Q-Q}. So far only resolution steps and purity removals were performed, now we apply two tautology removals.

$\xrightarrow{\pm}$ G_{11}

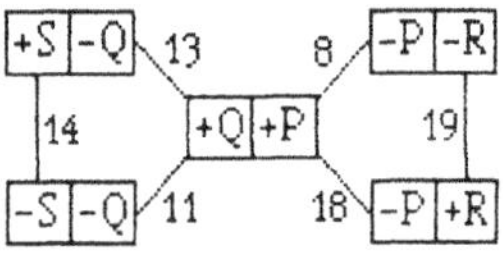

G_{11} has the same structure as G_0, from which it can be obtained by applying the literal permutation Π: $\pm Q \mapsto \mp Q$, $\pm P \mapsto \pm S \mapsto \mp R \mapsto \pm P$. Since $\Pi^6 = \mathrm{id}$, five more "rounds" with the analogous sequence of inference steps will reproduce G_0 as G_{66}.

The only object of G_0 still present in G_{11} is the clause +P+Q. In particular, all initial links disappeared during the derivation. Hence G_0 and G_{66} have no object in common, which implies that the derivation is covering. The following classes of link numbers represent the "colours" introduced for the coveringtwo definition 3_7, the numbers of links resolved upon are asterisked: {1*}, {2, 8, 17, 18, 20*, 23, 24*}, {3*, 9*, 19}, {4*, 7*}, {5, 11, 13, 21, 25, 26,...,36}, {6, 10, 12, 14, 15*, 16*, 22*}. Only the colour {5, 11,...} was never selected for resolution during the first round, and it just so happens that the second round starts with a resolution on link 11, which bears the critical colour. Hence the derivation also belongs to the coveringtwo class.

The ordering filter admitting exactly the steps of successive rounds is covering and coveringtwo, but neither refutation Noetherian nor refutation complete. Any modification of this filter permitting the same infinite derivation plus a second derivation from G_0 to {□} is not refutation confluent. ■

Note that the effect is not due to some fancy reduction rules. The tautology removal in initial graphs can hardly be considered the cause of the above cyclicity. Moreover, there is a refinement of the link inheritance rule [B82b, E86], which in both of the steps producing G_9 and G_{10} prevents the link between the tautology and its parent clause +Q+P, such that the tautologies can be removed by the purity rule. Further, in the example the problem can be overcome by requiring that resolutions producing a merge must be preferred to other steps. But then a slightly more complex graph serves as a counterexample, where additional two-literal clauses separate the respective parent clauses of merges.

This disconcerting state of affairs makes one appreciative of even modest affirmative results proving the desired properties for some special pragmatic ordering strategy. The only promising candidates for that purpose seem to be a level monotonicity strategy based on the ordering filter $\mathfrak{f}_{LEVEL}$ and an adaptation of the predicate cancellation filter to the first order level, neither of which is very useful for practical implementations.

4. Conclusions

Considering just the inference system per se, clause graph resolution possesses all the qualities one could reasonably ask for. However, as Kowalski points out, notions like the efficiency of proof procedures make sense only in the context of search strategies [K70]. The value of a logical inference system really ought to be judged in regard of the control strategies it admits. And as to this aspect the situation appears somewhat delicate.

While the occasional incompleteness of restriction strategies represents a mere nuisance, their frequent non-confluence potentially involves more disturbing impairments. Nevertheless, practical studies suggest that most of the traditional restrictions fail to significantly improve a system's performance anyway [WM76]. And those restrictions which do tend to yield profit, namely unit and set-of-support, also happen to maintain refutation confluence for clause graph resolution. Hence we need not be over-scrupulous about relinquishing the dubious benefits of the other restrictions, especially in the light of the fact that they increase the complexity of simplest refutations in return for their pruning, whereas clause graph resolution does not. On the whole one should be able to get along with the state of affairs concerning restriction strategies.

But then the results on the orderings leave a devastating impression. More than anything else it is the capacity to exploit domain-dependent and problem-specific heuristic knowledge that determines the power of an automated reasoning system. The smaller the class of control strategies admissible for an inference system, the tighter its limitations in taking full advantage of such knowledge. Clause set resolution requires only exhaustiveness, which can be achieved by virtually any control regime. This is the central pillar supporting all kinds of heuristic search strategies, and for clause graph resolution this foundation falls into ruin due to the absence of exhaustive ordering filters and the inadequacy of the plausible approximations thereof. The class of strategies based on the ordering filters that may turn out reliable, is extremely narrow and provides but a poor substitute, even when furnished with a series of more or less tricky refinements.

Remembering in addition all the unwieldy non-standard details of clause graph resolution, a pessimist can truly find ample reason to forget about this inference system.

On the other hand this insight does not necessarily cast a damper over the attachment of ardent adherents to clause graph resolution. Quite apart from the standard objections to the relevance of completeness or termination (and, come to that, of any theoretical investigation) for practical applications, proponents would plead as follows: the strong completeness conjecture holds for a significant subclass. Note that every refutation must eventually reach a state containing a unit refutable subset of clauses and that thereupon there is no getting around the empty clause for any covering ordering. Apparently the search spaces are very sparsely supplied with infinite branches of the new type. The counterexample demands utmost cunning to direct the derivation in such a way as to circumnavigate the ubiquitous "trap states", from which the empty clause can no longer be avoided. Small wonder that it took a decade of experience until such an example could even be construed. Moreover, actual implementations have demonstrated the competitiveness of clause graph resolution since its invention, but no failure to refute a clause set was ever traced back to the new cyclicity problem.

Which of these two extreme points of view comes closer to reality remains to be seen, but they might be reconciled by further research. A weakened form of clause graph resolution renouncing link removal faces none of the difficulties with control strategies, while maintaining the incontestable virtues of an explicit representation for operations and of link inheritance. But neither in the system including link removal does the class of covering ordering filters seem to surpass the desired class of terminating ones by much. If some simple amendment suffices to straighten things out, the benefits from link removal, such as the avoidance of redundancies and the power of reduction rules, can also be brought to bear. There is guarded hope that this is indeed possible.

A potential hint to that end may stem from the observation of a flaw in the so appealing intuitive justification for the purity rule. Allegedly a pure clause is useless for a deduction of the empty clause because there is no way ever to get rid of a pure literal or its descendants. This argument overlooks that a literal may be eliminated from a clause not only by resolution, but also by merging. As a matter of fact, its purities getting "merged away" a pure clause can indeed very well contribute to a refutation.

4_1 Example:

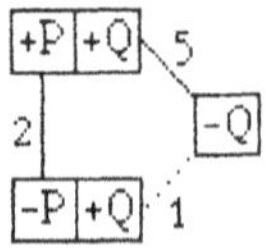

Suppose link 1 is missing and the clause -P+Q is pure. Resolution on link 2 and a subsequent merging operation result in the unit clause +Q with a descendant of link 5 joining the two complementary units. ■

Provided that the situation occurs in a graph derived from an initial one, the inference system's refutation confluence assures the existence of another sequence of steps removing the literal +P without recourse to the pure clause, be it ever so long a detour compared to the derivation above. The depicted graph is a fragment of the counterexample to the strong completeness conjecture. There the purity removal has the critical consequence to dispense the strategy from selecting link 2, which would produce a unit refutable clause set. If -P+Q may not be removed while opportunities persist to redress the purity by merging, a covering filter will encounter quite some difficulty when trying to evade this trap. The only alternative seems to lie in rendering pure all occurrences of Q, thus decreasing the number of predicate symbols incident with any links. In either case termination is impending.

Perhaps the purity rule ought to be appropriately weakened, e.g. such that a clause may be removed only if each of its literals is pure, or only if all potential merging partners of the pure literal are in turn pure. Admittedly this solution would to some extent deprive the system of an attractive feature.

Acknowledgement: The stimulating environment provided by the Markgraf Karl group has had more influence on this work than can be credited in detail. Special thanks go to Hans Jürgen Ohlbach for the profound discussions that helped clarify many a concept, and to Karl Hans Bläsius, Jörg Siekmann, and Graham Wrightson for thorough readings of earlier drafts that resulted in numerous improvements.

References

A68 Andrews, P. B.: *Resolution with Merging*, JACM, Vol. 15, No. 3 (1968), 367-381

A76 Andrews, P. B.: *Refutations by Matings*, IEEE Trans. Comp., Vol. C-25, No. 8 (1976), 801-807

A81 Andrews, P. B.: *Theorem Proving via General Matings*, JACM, Vol. 28, No. 2 (1981), 193-214

AO83 Antoniou, G., Ohlbach, H. J.: *Terminator*, Proc. 8th IJCAI, Karlsruhe (1983), 916-919

B75 Bruynooghe, M.: *The Inheritance of Links in a Connection Graph*, Report CW2, Applied Mathematics and Programming Division, Katholieke Universiteit Leuven (1975)

B76 Brown, F.: *Notes on Chains and Connection Graphs*, Personal Notes, Dept. of Computation and Logic, University of Edinburgh (1976)

B80 Bibel, W.: *A Strong Completeness Result for the Connection Graph Proof Procedure*, Bericht ATP-3-IV-80, Institut für Informatik, Technische Universität, München (1980)

B81a Bibel, W.: *On the Completeness of Connection Graph Resolution*, Proc. GWAI-81, Springer Informatik Fachberichte, Vol.47, (edited by Jörg H. Siekmann), Springer, Heidelberg (1981), 246-247

B81b Bibel, W.: *On Matrices with Connections*, JACM, Vol. 28, No. 4 (1981), 633-645

B82a Bibel, W.: *A Comparative Study of Several Proof Procedures*, Artificial Intelligence, Vol. 18, No. 3 (1982), 269-293

B82b Bibel, W.: *Automated Theorem Proving*, Vieweg, Wiesbaden (1982)

CL73 Chang, C.-L., Lee, R. C.-T.: *Symbolic Logic and Mechanical Theorem Proving*, Computer Science and Applied Mathematics Series (Editor Werner Rheinboldt), Academic Press, New York (1973)

CS79 Chang, C.-L., Slagle, J. R.: *Using Rewriting Rules for Connection Graphs to Prove Theorems*, Artificial Intelligence, Vol.12, No. 2 (1979), 159-178

E86 Eisinger, N.: *Completeness, Confluence, and Related Properties of Clause Graph Resolution*, Dissertation, Fachbereich Informatik, Universität Kaiserslautern (to appear1986)

H80 Huet, G.: *Confluent Reductions: Abstract Properties and Applications to Term Rewriting*, JACM, Vol. 27, No. 4 (1980), 797-821

K70 Kowalski, R.: *Search Strategies for Theorem-Proving*, Machine Intelligence (B. Meltzer and D. Michie, eds.), Vol. 5, Edinburgh University Press, Edinburgh (1970), 181-201

K75 Kowalski, R.: *A Proof Procedure Using Connection Graphs*, JACM, Vol. 22, No. 4 (1975), 572-595

K79b Kowalski, R.: *Logic for Problem Solving*, Artificial Intelligence Series (Nils J. Nilsson, Editor), Vol. 7, North-Holland, New York (1979)

KK71 Kowalski, R., Kuehner, D.: *Linear Resolution with Selection Function*, Artificial Intelligence, Vol. 2, No. 3-4 (1971), 227-260

L78 Loveland, D.: *Automated Theorem Proving: A Logical Basis*, Fundamental Studies in Computer Science, Vol. 6, North-Holland, New York (1978)

N80 Nilsson, N.: *Principles of Artificial Intelligence*, Tioga, Palo Alto, CA (1980)

No80 Noll, H.: *A Note on Resolution: How to Get Rid of Factoring without Losing Completeness*, Proc. 5th CADE, Springer Lecture Notes in Computer Science, Vol. 87 (edited by W. Bibel and R. Kowalski), Springer, Heidelberg (1980), 250-263

S76 Shostak, R. E.: *Refutation Graphs*, Artificial Intelligence, Vol. 7, No. 1 (1976), 51-64

S79 Shostak, R. E.: *A Graph-Theoretic View of Resolution Theorem-Proving*, Report SRI International, Menlo Park, CA (1979)

S82a Smolka, G.: *Einige Ergebnisse zur Vollständigkeit der Beweisprozedur von Kowalski*, Diplomarbeit, Fakultät Informatik, Universität Karlsruhe (1982)

S82b Smolka, G.: *Completeness of the Connection Graph Proof Procedure for Unit Refutable Clause Sets*, Proc. GWAI-82, Springer Informatik Fachberichte, Vol. 58 (1982), 191-204

Si76 Sickel, S.: *A Search Technique for Clause Interconnectivity Graphs*, IEEE Trans. Comp., Vol. C-25, No. 8 (1976), 823-835

SS76 Siekmann, J., Stephan, W.: *Completeness and Soundness of the Connection Graph Proof Procedure*, Bericht 7/76, Fakultät Informatik, Universität Karlsruhe (1976)

SS80 Siekmann, J., Stephan, W.: *Completeness and Consistency of the Connection Graph Proof Procedure*, Interner Bericht Institut I, Fakultät Informatik, Universität Karlsruhe (1980)

W84 Wrightson, G.: *An Approach to the Completeness of the Connection Graph Proof Procedure*, Personal Notes, Dept. of Comp. Sc., Victoria University of Wellington, NZ (1984)

WM76 Wilson, G. A., Minker, J.: *Resolution, Refinements and Search Strategies: A Comparative Study*, IEEE Trans. Comp., Vol. C-25, No. 8 (1976), 782-801

WOLB84 Wos, L., Overbeek, R., Lusk, E., Boyle, J.: *Automated Reasoning - Introduction and Applications*, Prentice-Hall, Englewood Cliffs, NJ (1984)

WCR64 Wos, L., Carson, D. F., Robinson, G. A.: *The Unit Preference Strategy in Theorem Proving*, Proc. AFIPS-26, Spartan Books, Washington, D.C. (1964), 615-621

WRC65 Wos, L., Robinson, G. A., Carson, D. F.: *Efficiency and Completeness of the Set of Support Strategy in Theorem Proving*, JACM, Vol. 12, No. 4 (1965), 536-541

YRH70 Yates, R. A., Raphael, B., Hart, T. P.: *Resolution Graphs*, Artificial Intelligence, Vol. 1, No. 3-4 (1970), 257-289

Parallel Theorem Proving with Connection Graphs

Rasiah Loganantharaj
Center for Advanced Computer Studies
USL, P.O. Box 44330
Lafayette, LA 70504-4330

Robert A. Mueller
Department of Computer Science
Colorado State University
Fort Collins, Colorado 80523

Abstract

In general, theorem provers are relatively slow. Speed up can be achieved by directing the search towards finding a proof and by using parallelism. The parallelisms identified in connection graph refutations are: *or* parallelism, *and* parallelism, and *dcdp* parallelism.

In *dcdp* parallelism, the links (edges) incident to distinct clauses and edge disjoint pairs are resolved in parallel. Optimally selecting potential parallel links is equivalent to solving the optimal graph coloring problem. Fortunately, however, optimal solutions to this *NP-hard* problem are not crucial. We describe a parallel solution of a sub-optimal graph coloring algorithm

In *and* parallelism, all literals in the *sun clause* are resolved concurrently and all the resolvents obtained. The resolvents, along with their inherited links, are inserted into the graph. The sun clause, and all the links connected to it, are removed from the graph. Because shared variables are restricted to have the same instantiation when there are shared variables in the sun clause, resolving literals concurrently becomes difficult. We discuss different approaches for performing this *and* parallelism.

1 Introduction and Background

Many applications such as expert problem solving, general problem solving, program verification, program synthesis, and robotics use theorem proving. Wos et al., [24] used it to solve some open questions in ternary Boolean algebra, finite semi-groups, equivalential calculus and the design of digital circuits. However, a major limitation of theorem provers is their relatively slow speed. Speed up can generally be achieved by both directing the search towards finding a proof and using parallelism. Generally, in the process of finding a proof, an inference rule will generate redundant well formed formulas (*wffs*). To reduce the search space, redundant wffs, tautologies and subsumed wffs, are removed. Several refinement strategies [3,10,15,20,25,26] have been suggested to restrict the wffs considered by the inference rules, thereby directing the proof procedure towards an empty wff (assuming a refutation proof procedure). Also, several ordering strategies have been suggested to select the clauses to apply inference rule. In addition, heuristic strategies have been suggested to assist the ordering and the search process. Yet, very little has been done to use parallelism in theorem proving. In this paper, we discuss parallel link resolution of connection graph refutation.

A ***connection graph*** is a schema for representing a first-order, clausal-form wff in a refutation proof [9]. For each literal in an input clause, a literal-node in the corresponding connection graph exists. Further, literals of a single clause are organized into a "group." Node refers to a clause in a connection graph. If two literals are ***potentially complementary*** in the unification sense, then a ***link*** (undirected edge) connects their corresponding literal-nodes in the connection graph. If a pair of complementary literals exist in a clause, then the clause may be either a tautology or a self-resolving clause. The ***pseudo-link*** connecting the self-resolving clauses's complementary literals stands for a link between literals in copies of the clause [1,8].

The basic operation in connection graph refutation is ***link resolution***, in which a link is selected and the corresponding terms of the pair of literal-nodes are unified. Then, the resolvent is computed and entered into the graph. The link is deleted, and appropriate links from the computed resolvent node are inserted. A ***pure*** clause (group node) occurs if no links join the literal-node it labels in the connection graph. As such, we can delete it and all other members of its group and all links incident to the other link-nodes. Similarly, if the resolvent clause is a ***tautology***, then the clause's corresponding node and its incident links are removed from the graph. In fact, when the resolvent is a tautology it need not be inserted into the graph.

A positive property of connection graph resolution is that it requires no search either to determine the pair of clauses to resolve or to determine the links to the computed resolvent. We can resolve any pair of clauses connected to a link since links connect complementary literals. A resolvent always inherits the links connecting the literal-nodes of the clauses incident to the resolved link. Further, once resolved, a link is removed from the graph, thus avoiding further redundant resolution between that pair of clauses (as might be the case in binary resolution). Other advantage of connection graph representation is the availability of the entire search space, (in contrast to (say) AND/OR proof trees [5], where the search space is built up as we resolve literals). A set of clauses corresponding to a connection graph is unsatisfiable if we can deduce the empty clause from the connection graph [9]. The links of the connection graph may be selected top-down, bottom-up, or by a mixture of both methods. Since the whole search space of the problem is known, heuristic information can be used to select the best links to resolve at each step [21,22]. Simplification strategies such as pure literal elimination, subsumption and tautology elimination have a snowball effect, reducing the connection graph further.

The presence of the complete search space suggests the opportunity to use parallel evaluation strategies to improve the efficiency of a generally very slow process. We can identify three different kinds of parallelism which are *or* parallelism, *dcdp* parallelism and *and* parallelism, depending on which links have been selected for parallel deduction. However, pseudo links are not considered for parallel link resolution because they represent different copies of a clause. The potential parallelism in a connection graph is explained with an example. In the connection graph shown in Figure 1, variable terms are shown in upper case letters and constant terms are shown in the lower case.

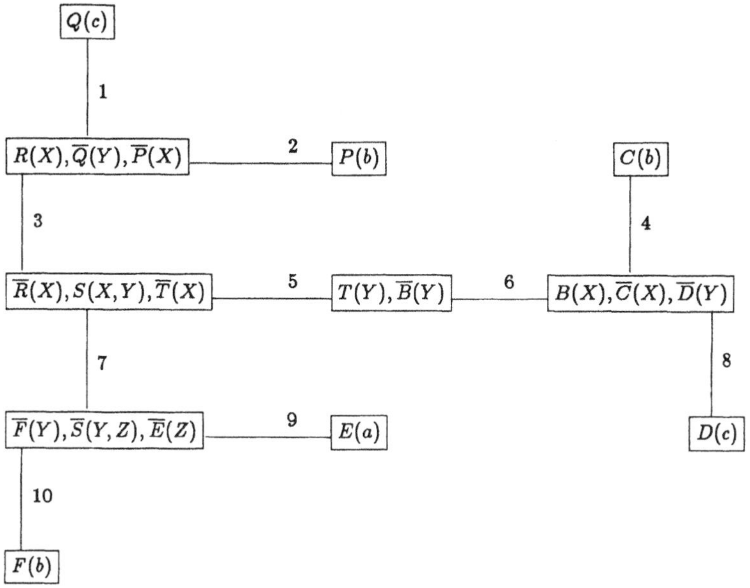

Figure 1

If we resolve links 1 and 7 we have *dc* parallelism (distinct clause parallelism) [7], where the clauses connected by the resolved links are distinct. Unrestricted *dc* parallelism will result in logical inconsistency [11,14]. This motivates the study of a restricted version of *dc* parallelism that maintains logical consistency. We call such parallelism *dcdp* parallelism for ***distinct clause***, edge ***disjoint pairs***. The set of links which connect to distinct clauses and edge disjoint pairs is called a ***parallel link set*** or a *dcdp* link set. In *dcdp* parallel link resolution, the *dcdp* links are resolved in parallel. In the above example, many combinations of links become candidates for the *dcdp* links. Some of them are {2,8,10}, {1,6,9}, {1,4,9} and {1,8,10}. We do not need to find all the sets of *dcdp* links. Only a set with a reasonable cardinality is sufficient. In section 2, we will show that the complexity of optimal selection of *dcdp* links is equivalent to an optimal graph coloring problem. Fortunately, we do not need optimal solution to this *NP*-hard problem. So, we can present a parallel solution to a sub-optimal graph coloring algorithm [2].

If we resolve links 1,2 and 3, we have *and* parallelism, where all the literals of the clause $R(X) \vee \overline{Q}(Y) \vee \overline{P}(X)$ (which we call a ***sun clause***) are resolved concurrently. We call the clause connected to the literal of the sun clause a ***satellite clause***. All the resolvents, along with the inherited links from the satellite clauses, are inserted into the graph; the sun clause, along with all the connected links, are removed. The *and* parallelism is similar to hyper-resolution [4] in resolving more then one pair of literals simultaneously. The difference is in obtaining the resolvents. Although *and* parallel link resolution is not a proof procedure by itself, it, used with link resolution, reduces the graph more rapidly. The effective use of *and* parallelism depends on the the selection of the proper sun clause and the efficiency of the algorithm obtaining the solution unifiers. We discuss details of *and* parallelism in section 4.

2 *dcdp* parallelism

Sequential refutation of links in connection graphs has been shown to be both sound and consistent [9,23]. By *sound* we mean that a refutation will succeed only if the set of wffs is unsatisfiable. By *consistent* we mean no set of wffs from which both the empty clause and the empty set can be derived exists. We will describe a refutation system as being *correct* if it is both sound and consistent.

In *dcdp* parallel link resolution, the set of links which connect to distinct clauses and edge disjoint pairs are resolved concurrently. We have shown in [11,14] that the *dcdp* parallel link resolution is correct. Assuming that there are neither clauses containing pure literals nor tautologies in the initial connection graph, we now describe the *dcdp* parallel link resolution. If there is a tautology, it is deleted along with all the links connected to it, and if there is an empty clause, then the connection graph is unsatisfiable, and no further processing is necessary.

While no terminating condition is true **do**
{
    Select a set of links from the *dcdp* link set.

    Resolve the links in parallel and obtain their resolvents.

    If any resolvent is empty, then terminate successfully, else insert the resolvents in the connection graph and continue.
}

Figure 2

The termination conditions are either: i) the presence of the empty clause or ii) the absence of any necessary condition for refutation. In the case where the empty clause is present, the procedure terminates successfully. The necessary conditions for refutation are: i) the graph should not be empty and ii) the set of clauses must have at least one positive and one negative clause [17]. If any one of these conditions is not satisfied, the procedure terminates unsuccessfully indicating that the original set of clauses is satisfiable. Note, however, the possibility that the above procedure will not terminate if the initial set of clauses are satisfiable.

Ideally, we would like to select a link which, when resolved, yields either the empty clause or a connection graph from which the empty clause could be derived in the fewest number of steps. Unfortunately no method is available to select such a link. However, heuristic strategies exist for making intelligent guesses. The heuristic strategies used by Siekmann et al. [21,22] assign two parameters to links: link-sum and priority. They defined *link-sum* as the total number of resolution links by which the graph is increased or decreased after resolution on this link. *Priority* is the value of a weighted polynomial representing various features of a resolution link. The features considered are i) age, ii) literal number, iii) term complexity, iv) factorization, v) parent isolation, and vi) resolvent isolation. The given formulae to calculate the numerical values of these features were found by computing a large number of examples and analyzing the proof protocols. Of all links lying inside the priority window, the link with smallest link-sum is selected as the *best* link.

Connection graph proof procedures may also use resolution refinement. The application of refinement strategies in connection graph refutation restricts the links that are eligible for resolution. One may also view this as rearranging the links for resolution consideration. The eligible links are those connected to the eligible clauses (nodes). For example, if we apply the *set of support* strategy to connection graph refutation, the links connected to the nodes corresponding to the clauses with support become the eligible links. When we resolve an eligible link,the links connected to the resolvent also become eligible links.

When considering a reasonable problem, the number of candidate links for *dcdp* parallelism is substantial, and we want to avoid or delay link resolution of "useless links" (i.e., those links that do not help directly to find an empty clause at the present state of the connection graph). Therefore, we use refinement strategies and ordering strategies to sort out the "useful links." The set of links selected by the refinement and ordering strategy is the potential link set. The *dcdp* links of the potential link set are the parallel link set. In sequential link resolution, the best link is selected from the potential link set by applying the heuristic "link-sum" and is resolved. Thus, we resolve links with a high potential for rapidly reducing the graph to the empty clause. However, in parallel link resolution more than one link is resolved at a time; therefore, we do not need to select the single best link. All links in the parallel link set can be resolved in parallel.

The basic functions of the proof procedure using *dcdp* parallelism are

1. Selecting a parallel link set from potential link set
2. For each link in the parallel link set, obtaining the most general unifier (m.g.u.) and calculating the resolvent
3. Inserting the resolvent to the graph, with all the inherited links.

For a pair of clauses, it is easy to find the most general unifier of a pair of complementary literals using the unification algorithm and to obtain the resolvent. Martelli [16] proposed an efficient unification algorithm. The difficulty is creating the potential parallel link set. If we can select *dcdp* link set from a graph, we can solve the problem of selecting *dcdp* links from the potential link set.

We consider the problem of finding the *dcdp* links of a connection graph. Assume the links have assigned labels. For each link L, a set of links cannot be *dcdp* linked with L. We call such a set a ***conflicting set*** for L. We call the set of links *dcdp* linked with L is the ***compatible*** link set of L. The following formulae give the conflicting and compatible link set of L. The adjacent links of L (AdjLink(L)) are the edges (excluding L) connected to the pair of clauses incident to L. Cnf(L) and Cmp(L) denote the conflicting and compatible link sets of L respectively.

$$\begin{aligned} Cnf(L) &= AdjLink(L) \cup \{\cup_{i \in AdjLink(L)} AdjLink(i)\} \\ Cmp(L) &= Lset - Cnf(L) \end{aligned} \tag{1}$$

For each link of a connection graph, we easily find the conflicting set of links using formula (1). We can create a color graph in which each node corresponds to a link of the connection graph, and the adjacent nodes of a node *b* are those of *Cnf(b)* in the connection graph. Finding the *dcdp* links in a connection graph is the the same problem as coloring the corresponding color graph. Garey [6] has shown the optimal graph coloring problem to be *NP-hard*. Instead of solving the graph coloring problem optimally, we have obtained a parallel version to the suboptimal graph coloring algorithm.

Chaitan et al. [2] presented a sub optimal graph coloring algorithm which selects and colors a node with the highest degree at each iteration. The node degree becomes the heuristic to select a node.

3 Parallel Graph Coloring Algorithm

Consider the problem of assigning colors from the color set *Cset* to the nodes in *Vset*. Let colorset(N) (initialized to *Cset*) be the assignable color set for node N. Select a node N with highest degree in the graph. Pick any color C from colorset(N), assign it to N, and remove C from the assignable color set of any node adjacent to N. The adjacent nodes of a node N (AdjNodes(N)) are connected to the edges incident to the node N. Then, remove N and all edges incident to N to obtain a new graph. When updating the neighbors, if any node becomes isolated (i.e., degree = 0), then the node may assume any color in its assignable color set. We delete it from the graph. Selecting and coloring nodes from *Vset* continues until all the nodes have been colored, i.e., until *Vset* is empty or we detect a failure. If any node's colorset of any node becomes empty, the procedure terminates unsuccessfully.

Before describing a parallel graph coloring algorithm, we consider a sequential version.

```
procedure Color-graph;
{ while (| Vset | > 0) do
    { Select a node N from Vset, having the greatest degree;
      Assign N the first color C from colorset(N);
      Vset := Vset - {N };
      for all X ∈ AdjNodes(N) do
      { Delete C from colorset(X) and delete edge(X,N);
        if colorset(X) = ∅ then terminate with failure
        elseif degree(X) = 0 then Vset := Vset - { X }
      }
    }
}
```

Figure 3

If the algorithm terminates successfully, we have colored all the nodes. To find the desired parallel link set, we must select a group with the maximum number of members having the same color. The number of groups will be same as | *Cset* |. The procedure in the Figure 4 groups all the nodes in the graph.

```
procedure Form-groups;
{ for i := 1 to | Cset | do
    { for all N ∈ Vset do
        { if Cset[i] ∈ colorset(N) then group[i] := group[i] ∪N }
    }
}
```

Figure 4

Once the grouping is done, we select the largest group. In the sequential algorithm, we can combine 1) assigning a color to the selected node; 2) updating its neighbors; and 3) deleting the node from the graph into a procedure Update(N), where parameter N is the selected node. Multiple instances of Update can run in parallel if we limit selecting a node for coloring to a group compatible with the ongoing updates. That is, none of N's neighbors can be candidates for coloring until we have updated all the nodes incident to N

and deleted N from the graph. Therefore, the candidates for coloring, are $Vset - AdjNodes(N) - \{ N \}$. We call the candidate vertex set *Comp. Vset* and the set of nodes being colored *Col. Vset.*

After a node N is colored and deleted from the graph, we can group it with other nodes of the same color. Multiple instances of a process Form-groups runs in parallel as long as we prevent simultaneous access to a particular color group. The multiple instances of Form-groups manipulate the variables *group*. The main procedure and the multiple instances of Update manipulate the variables *Vset* and *Comp. Vset.* Since different processes interact closely, Form-groups, the main procedure and Update communicate via shared variables.

Conventional semaphores S with operators **P**(S) and **V**(S) implement mutually exclusive access by processes to shared data. Frequently in this paper parallel algorithms expressed in pseudocode will use semaphores neither mentioned in the text nor declared or initialized. In addition, a "conditional critical section" allows a process to execute the guarded statements when the associated boolean expression is *true* and in mutual exclusion. The **await** block has the following syntax:

await <bool-expr> **then** <statements> **endawait**;

Concurrent processes execute one at a time among all **await** blocks. If the boolean expression is *false*, the executing process releases exclusive execution rights and waits for the boolean expression to become *true*. If the boolean expression is *true*, the process executes statements in the block and awakens all delayed processes to allow other processes to attempt to execute some **await** block. Conditional critical sections are efficiently implemented via semaphores; see [18].

In the parallel algorithm's description, the **fork** statement initiates processes and continues the current process.

fork <process> [(<parameters>)] [, <count>];

If a count is present, the operation increments its value indivisibly. A process terminates its own execution by executing the **quit** statement with syntax:

quit

Parallel-color-graph is the main procedure. Initially, the candidate vertex set *Comp. Vset* is *Vset.* Once we select a node N for coloring we remove N and the neighboring nodes of N from *Comp. Vset* and include the node N in *Col. Vset.* We fork off the process Update with node N as parameter. The shared variables updatecount and groupcount denote the number of current updating and grouping processes, respectively. The procedure loops until failure is detected or *Comp. Vset* becomes empty. Update processes remove nodes from *Vset*; the entire algorithm repeats until *Vset* becomes empty; then, when all forked processes have completed, the largest monochrome node group is selected. The pseudocode is given in Figure 5.

The process Update assigns node N a color C, deletes N from *Vset*, and forks the process Form-groups with parameter N. Then for each neighbor X of N, C is removed from colorset(X) and the edge(X,N) is deleted. (It is possible that X could be a neighbor of more than one node being colored; therefore each update of node X is protected in a critical section.) Once all the neighbors of node N are updated, we add to *Comp. Vset* all the neighbors of N except those that are also neighbors of nodes currently being colored. When updating the neighbors, if any node becomes isolated (i.e., degree = 0), then it may assume any color in its assignable color set. We can delete it from the graph. If any colorset(N) becomes empty before its node N obtains a color, the entire procedure terminates in failure. The pseudocode for the processes Update and Form-groups are given in Figure 6.

```
procedure Parallel-color-graph;
{  Comp.Vset := Vset; /* initialize compatible vertices set */
   Col.Vset := ∅; /* initialize the nodes being colored */
   updatecount := groupcount := 0; /* initialize the counters */

   while (| Vset | > 0) do
   { while (| Comp.Vset | > 0) do
      {  Select a node N from Comp.Vset having the greatest degree;
         P( S-Comp.Vset );
            Comp.Vset := Comp.Vset - AdjNodes(N) -N;
            Col.Vset := Col.Vset ∪ { N } ;
         V( S-Comp.Vset );
         fork Update(N), updatecount;
      }
   };
   await updatecount = groupcount = 0 then
      Select the group with greatest number of members
   endawait
}
```

Figure 5

```
process Update(N);
{  Assign N the first color C from colorset(N);
   P( S-Vset ); Vset := Vset - { N }; V( S-Vset );
   fork Form-groups(N), groupcount;
   for all X ∈ AdjNodes(N) do
   {  P(S-node-X);
         Delete C from colorset(X) and delete edge(X,N);
         if degree(X) = 0 then isolated := true;
         else isolated := false;
      V(S-node-X);
      if colorset(X) = ∅ then terminate with failure;
      if isolated then
      {  Assign to X any color in colorset(X);
         fork Form-groups(X), groupcount;
         P( S-Vset ); Vset := Vset - { X }; V( S-Vset )
      }
   };
   P( S-Comp.Vset );
      Col.Vset := Col.Vset - { N };
      Comp.Vset := Comp.Vset ∪ AdjNodes(V) - { { ∪_{i∈Col.Vset} AdjNodes(i)} ∩ AdjNodes(V)};
   V( S-Comp.Vset);
   await true then updatecount := updatecount - 1 endawait ;
   quit
}
```

```
process Form-groups(N);
{ for i := 1 to | Cset | do
   { if Cset[i] ∈ colorset(N) then
      { P(S-group-i); group[i] := group[i] ∪N; V(S-group-i) }
   };
   await true then groupcount := groupcount -1 endawait ;
   quit
}
```

Figure 6

We have described a parallel algorithm to find a *dcdp* link set from the potential link set of a graph. That is, if the color graph corresponding to the potential link set of a connection graph is given to this algorithm, it will find the parallel link set.

4 AND Parallelism

In *and* parallelism, all the links connecting literals in a sun clause are resolved concurrently. All the resolvents are inserted in the graph along with all the inherited links from the satellite clauses. The sun clause and the links connected to it are removed from the graph. Given the correctness property of the sequential connection graph refutation, we have shown the correctness of *and* parallelism in [11,13].

In link resolution, unification combines with the inference rule. If we separate these two (that is, if we can save all the m.g.u.'s), then we can save the time used to rediscover the unification. Saving of unification [19] improves backtracking.

With the m.g.u. separated from the inference rule, binding of shared variables with the application of the m.g.u.'s in different literals should be consistent. To formally express this, we introduce some notation.

$l_1 \vee l_2 \vee, ..., \vee l_n$: is the sun clause with l_i as the ith literal of the clause

$l_{ij} \vee C^{ij}_{rest}$: is the jth clause connected to the ith literal of the sun clause.

σ_{ij} : is the most general unifier between the terms of the pair of complementary literals l_{ij} of the satellite clause and l_i of the sun clause.

$Sv(X)$: if the variable X is shared in the predicates considered, it is *true* and is *false* otherwise.

$Const(y)$: if the argument y is constant then the predicate is *true* and *false* otherwise.

$Var(y)$: if the argument y is a variable then the predicate is *true* and *false* otherwise.

In sequential link resolution and general resolution, a most general unifier of a pair of complementary literals is found, and then it is applied to the remaining literals of the pair of resolved literal's clauses, to form the resolvent. By separating the unification from the inference rule, important information which otherwise would have been lost should be saved with the unifiers. Consider a unifier in which more than one sun clause variable is bound to the same variable of the satellite clause. Let $P(X, Y, Z)$ be a literal of a sun clause and $\overline{P}(XX, XX, c)$ be a literal of a satellite clause. Then the unifier of the pair of clauses is $\{X \leftarrow XX, Y \leftarrow XX, Z \leftarrow c\}$. In general resolution and sequential link resolution, this unifier is applied to the rest of the literals and, thereby, the variables X and Y of the sun clause are constrained to have the same instantiation. If we separate the unification from the inference rule, the information about X and Y must be saved. So, for each unifier, the sun clause variables that are bound to the same satellite clause variables are saved in a set called "var-of-same-inst" (which stands for variables of same instantiation). A unifier can have more than one set of var-of-same-inst.

First, we will consider a special case in which the sets var-of-same-inst are empty. The most general unifiers σ_{1j} and σ_{2k} of literals one and two in the sun clause are said to be consistent if the following statement is true:

$$\forall X(Sv(X) \wedge Const(X \cdot \sigma_{1j}) \wedge Const(X \cdot \sigma_{2k}) \wedge (X \cdot \sigma_{1j} = X \cdot \sigma_{2k}))$$
or
$$\forall X(Sv(X) \wedge Const(X \cdot \sigma_{1j}) \wedge Var(X \cdot \sigma_{2k}))$$
or
$$\forall X(Sv(X) \wedge Var(X \cdot \sigma_{1j}) \wedge Const(X \cdot \sigma_{2k}))$$
or
$$\forall X(Sv(X) \wedge Var(X \cdot \sigma_{1j}) \wedge Var(X \cdot \sigma_{2k}))$$

Figure 7

Note that in the statement of the consistent unifier, we did not consider the possibility of a variable binding to a function variable term because of the difficulty of determining the consistency. This restriction has an effect on the selection of a sun clause. With this restriction, there is no need for "occur check". If the combination of σ_{1j} and σ_{2k} is consistent, we can obtain the consistent unifier. A unifier is a set of tuples of the form $< X_i \leftarrow t_i >$ where X_i and t_i are a variable and a term, respectively. The set of tuples of the consistent unifier σ is obtained as follows.

$$\sigma_{sh} = \bigcup_{X_i \in shared\ variable\ set} < X_i \leftarrow t_i >$$

where

$$t_i = \begin{cases} X_i \cdot \sigma_{ik} & if\ \exists \sigma_{ik}\ (Const(X_i \cdot \sigma_{ik})) \\ X & otherwise \end{cases}$$

$$\sigma = \sigma_{sh} \cup \{\text{tuples of the non shared variables}\}$$

Figure 8

The resolvent is $C^{1j}_{rest} \cdot \sigma \vee C^{2k}_{rest} \cdot \sigma$.

We consider a general case in which the "sets var-of-same-inst" are not empty. We use the notation

var-of-same-inst[I,σ]

to refer to the Ith set of variables of the unifier σ, such that they must be bound to the same instantiation. Let group-1 and group-2 consist of all the sets of var-of-same-inst's of the unifiers σ_{1j} and σ_{2k}, respectively. Let us say $X_1 \in$ group-1 and $X_2 \in$ group-2. For any X_1 and X_2, if $X_1 \cap X_2 \neq \emptyset$, then there must be at least one variable common to X_1 and X_2. Therefore, X_1 and X_2 are combined to form a new var-of-same-inst of the consistent unifier for σ_{1j} and σ_{2k}. The algorithm in Figure 9 obtains the set of var-of-same-inst for the given pair of sun unifiers. In the algorithm, X_i stands for a set of var-of-same-inst (p. 352).

We illustrate the algorithm with an example. The sets of var-of-same-inst of the unifiers, and the group-1 and group-2 of the algorithm are shown in the tabulation of Figure 10.

var-of-same-inst[1,σ_{1j}] = $\{Y_1, Y_2\}$	var-of-same-inst[1,σ_{2k}] = $\{Y_1, Y_6\}$
var-of-same-inst[2,σ_{1j}] = $\{Y_3, Y_4\}$	var-of-same-inst[2,σ_{2k}] = $\{Y_2, Y_7\}$
var-of-same-inst[3,σ_{1j}] = $\{Y_5, Y_6\}$	
group-1 = $\{\{Y_1, Y_2\}, \{Y_3, Y_4\}, \{Y_5, Y_6\}\}$	group-2 = $\{\{Y_1, Y_6\}, \{Y_2, Y_7\}\}$

Figure 10

Both group-1 and group-2 are not null. Therefore, we start with $X_1 = \{Y_1, Y_2\}$ from group-1. Each set of group-2 has a common variable with group-1, and therefore is marked. The new-same-inst[1] =

$\{Y_1, Y_2, Y_6, Y_7\}$ is created, and the marked sets of group-2 are deleted. X_1 is deleted from group-1. The rest of the sets of group-1 are checked with new-same-inst[1] for common variables. The set $\{Y_5, Y_6\}$ has a common variable and is included in new-same-inst[1]. Then, the set $\{Y_5, Y_6\}$ is deleted from group-1. Since group-2 is null, the algorithm terminates with

$$\text{new-same-inst}[1] = \{Y_1, Y_2, Y_5, Y_6, Y_7\}$$
$$\text{new-same-inst}[2] = \{Y_3, Y_4\}.$$

We can obtain the sets of variables which must be bound to the same instances using the algorithm of Figure 9. The consistent binding of shared variables that are not in the sets of new-same-inst's for a given pair of unifiers can be obtained using the statements of Figures 7 and 8.

If all the variables in a new-same-inst set are bound to variable terms by σ_{1j} and σ_{2k}, they are consistent. If at least one variable in a new-same-inst set is bound to a constant term, the rest of the variables of the set must be bound either to the same constant term or to a variable term by the pair of unifiers. The algorithm of Figure 11 obtains the consistent unifiers for the variables in the sets of var-of-same-inst's.

```
procedure Find-const-unifier(σ1, σ2, set of all new-same-inst's);
{  K := cardinality of the sets of new-same-inst's;
   for I = 1 to K do
   {  found := false;
      for X ∈ new-same-inst[i] do
      {  if (found) then
         {  if (const(X · σ1) ∧ const(X · σ2) ∧ (X · σ1 = const-term)
                        ∧(X · σ2 = const-term )) or
            (const(X · σ1) ∧ var(X · σ2) ∧ (X · σ1 = const-term)) or
            (var(X · σ1) ∧ const(X · σ2) ∧ (X · σ2 = const-term)) or
            (var(X · σ1) ∧ var(X · σ2)) then terms are consistent;
            else exit with failure;
         }
         else
         {  if (const(X · σ1) ∧ const(X · σ2) ∧ (X · σ1 = X · σ2)) or
            (const(X · σ1) ∧ var(X · σ2)) or
            (var(X · σ1) ∧ const(X · σ2)) then
            {  if (const(X · σ1) then const-term := X · σ1;
               else const-term := X · σ2;
               found := true;
            }
            else if (var(X · σ1) ∧ var(X · σ2)) then ;
               else exit with failure;
         }
      }
      if (found) then
      {  for X ∈ new-same-inst[I] do
            < X ← const-term >;
         delete the new-same-inst[I]
      }
   }
}
```

Figure 11

We have described the condition under which two m.g.u.'s of two literals can be consistent and how to obtain the consistent unifier with the pair of consistent m.g.u.'s. This could be extended to any number of unifiers for any number of literals.

In *and* parallelism in terms of consistent unifiers, the m.g.u.'s of all the links connected to the sun clause's literals are obtained, and the combinations of the unifiers that result in a consistent binding of the shared variables are also found. First we obtain the consistent unifiers. For each solution unifier, we create the corresponding resolvents with all the inherited links of the satellite clause and insert them in the graph. We then remove the sun clause and the links connected to it.

The above definition of *and* parallelism has high potential for parallelism. Finding the m.g.u.'s of all the links can be done in parallel.

5 Algorithms to Obtain Solution Unifiers

In *and* parallelism, the most general unifiers to all the literals to a sun clause are obtained first. All the unification can be done in parallel. An efficient algorithm has been proposed in [16] to perform unification. Obtaining the resolvent using the solution unifier is merely a trivial substitution. The difficult step in *and* parallelism is to find all solution unifiers from the given set of m.g.u.'s. By partitioning the sun clause's literals into independent groups (based on shared variables), we can obtain solution unifiers for different groups in parallel.

A pair of most general unifiers is said to be *consistent* if it is not inconsistent. In an inconsistent pair of m.g.u.'s, there is at least a single shared variable to which each unifier unifies with a different unequal constant term. We will consider an algorithm to obtain a consistent unifier σ_k for literal 1,...,k given σ_{k-1} and a unifier of literal k, say σ_{Ik}. Since the unification is separated from the inference rule, the sets of var-of-same-inst, must be saved for each unifier. First, we consider a special case in which the sets of var-of-same-inst are empty. The algorithm (Figure 12), combine consistency checking and obtaining the consistent unifier. The variable temp-tuple saves the potential tuple when a pair of corresponding terms are consistent. Another variable saved-tuples collects the temp-tuple's. The variable common-variables denotes the set of variables common to the variables of literal k and the sequence of literals 1,...,k-1, that is

common-variable = (Variables of the literals 1,...,k-1) $\cap$ (Variables of literal k)

if (common-variables = $\emptyset$) **then**
{ σ_{k-1} and σ_{Ik} are consistent and the
consistent unifier $\sigma_k = \sigma_{k-1} \cup \sigma_{Ik}$ };
else
{ **for** ($X \in$ common-variables) **do**
{ consistent := *false*;
if ($Var(X \cdot \sigma_{k-1}) \wedge Var(X \cdot \sigma_{Ik})$ **then**
{ temp-tuple = $X \leftarrow$ VAR; consistent := *true*};
if ($Var(X \cdot \sigma_{k-1}) \wedge const(X \cdot \sigma_{Ik})$ **then**
{ temp-tuple = $X \leftarrow X \cdot \sigma_{Ik}$; consistent := *true*}
if ($const(X \cdot \sigma_{k-1}) \wedge Var(X \cdot \sigma_{Ik})$ **then**
{ temp-tuple = $X \leftarrow X \cdot \sigma_{k-1}$; consistent := *true*}
if ($const(X \cdot \sigma_{k-1}) \wedge cons(X \cdot \sigma_{Ik}) \wedge (X \cdot \sigma_{k-1} = X \cdot \sigma_{Ik})$) **then**
{ temp-tuple = $X \leftarrow X \cdot \sigma_{k-1}$; consistent := *true*}
if (not (consistent)) **exit**
saved-tuples := saved-tuples $\cup$ temp-tuple;
}
}
consistent-unifier := ($\cup$ tuples of non common variables) $\cup$ saved-tuples

Figure 12

The algorithms in Figures 9 and 11 found the consistent unifiers for the general case. This algorithm (Figure 12) checks the pair of terms corresponding to each of the common variables in sequence. This check could be improved if the comparison were done in parallel. Further, improvement of the algorithm is possible if we can compare all the unifiers of a literal, say l_k, to the consistent unifier of the previous literals i.e., with σ_{k-1}, in parallel rather than in sequence.

An improved algorithm based on the use of associative memory for storing unifiers and a heuristic cost function for ordering literals to be unified has been developed by the authors. The details of the algorithm are given in [12].

6 Summary

We have described *dcdp* and *and* parallel link resolutions in connection graph refutation proof procedure. The correctness of these parallel link resolutions are presented in [11,13,14].

The effectiveness of the *dcdp* parallelism depends on the following: links being resolved, in the overhead involved in creating the *dcdp* link set, and in the management of shared resources when running on a MIMD architecture. We have described a parallel solution to a sub optimal algorithm that selects a parallel link set (*dcdp* links). The parallel algorithm for *dcdp* link resolution can be mapped easily onto a MIMD architecture we have outlined such an implementation [11,14].

In *and* parallel link resolution, the m.g.u.'s of all links incident to a sun clause are obtained first. Thereafter, all the solution unifiers are found. Then, for each solution unifier, a resolvent is created and inserted into the graph. Finally, the sun clause is removed from the graph. We have considered a restricted *and* parallelism which does not consider pseudo links and function term bindings to a constant or a variable term. The correctness of *and* parallelism is shown in [11,13]. The difficult step in *and* parallelism is to obtain the solution unifiers from the set of m.g.u's.

We described different approaches to obtain the solution unifiers from the set of m.g.u.'s of the sun clause. An algorithm that uses associative memory(AM) and a sub optimal algorithm to order the unifiers is given in [12].

Although *and* parallel link resolution is not a refutation proof procedure by itself, it can be used with link resolution to reduce the graph more rapidly. But, the effective use of *and* parallelism depends on the selection of the proper sun clause. We have not investigated strategies to select a good candidate. However, investigation in this area would be interesting. More than one sun clause could be selected and the links resolved in parallel, provided the *dcdp* condition is satisfied. For example, the *and* parallelism could be applied to sun clauses S_1 and S_2, if each link in S_1 is *dcdp* linked with all the links in S_2. We do not yet know whether the overhead of checking all the links for *dcdp* condition will be outweighed by the advantages of an additional *and* parallelism.

References

[1] M. Bruynooghe. *The inheritance of links in connection graph and its relation to structure sharing.* Technical Report, Applied Mathematics and Programming Division, Katholieke Universiteit, Katholieke universiteit, Leuven, Belgium., 1977.

[2] G. J. Chaitin, M. A. Auslander, A. K. Chandra, J. Cocke, M. E. Hopkins, and P. W. Markstein. Register allocation via coloring. *Computer Languages*, 6:47–57, 1981.

[3] C. L. Chang. The unit proof and the input proof in theorem proving. *Journal of the ACM*, 17:698–707, October 1970.

[4] C. L. Chang and R. C. T. Lee. *Symbolic Logic and Mechnical Theorem Proving*. Academic Press, New York, 1973.

[5] J. S. Conery. *The AND/OR process model for parallel interpretation of logic programs*. PhD thesis, University of California, Irvine, June 1983.

[6] M. R. Garey and D. S. Johnson. *Computers and Intractability: A Guide to the Theory of NP-Completeness*. W. H. Freeman and Company, San Francisco, 1979.

[7] G. Hornung, A. Knapp, and U. Knapp. A parallel connection graph proof procedure. In *German Workshop on Artificial Intelligence. Lecture Notes in Computer Science*, pages 160–167, Berlin: Springer-Verlag, 1981.

[8] R. Kowalski. *Logic for Problem Solving*. Elsevier-North Holland, New York, 1979.

[9] R. Kowalski. A proof procedure using connection graphs. *Journal of the ACM*, 22(4):572–595, October 1975.

[10] R. Kowalski and D. Kuchner. Linear resolution with selection function. In *Artificial Intelligence*, pages 221–260, 1971.

[11] R. Loganantharaj. *Theoretical and implementational aspects of parallel link resolution in connection graphs*. PhD thesis, Department of Computer Science, Colorado State University, 1985.

[12] R. Loganantharaj and R. A. Mueller. Parallel algorithms for obtaining solution unifiers in connection graph theorem proving. To be submitted for publication, April 1986.

[13] R. Loganantharaj, R. A. Mueller, and R. R. Oldehoeft. *Connection graph refutation: aspects of AND-parallelism*. Technical Report CS-85-10, Department of Computer Science, Colorado State University, 1985.

[14] R. Loganantharaj, R. A. Mueller, and R. R. Oldehoeft. *Some theoretical and implementational aspects of Dcdp-parallelism in connection graph refutation*. Technical Report CS-85-9, Department of Computer Science, Colorado State University, 1985.

[15] D. W. Loveland. *Automated Theorem Proving: a Logical Basis*. North Holland Publishing Company, New York, 1978.

[16] A. Martelli and U. Montanari. An efficient unification algorithm. *TOPLAS*, 4:258–282, 1982.

[17] B. Meltzer. Theorem proving for computers: some results on resolutions and renaming. *Computer J.*, 8:341–343, January 1966.

[18] R. R. Oldehoeft and S. J. Allan. Execution support for hep sisal. In J. S. Kowalik, editor, *Parallel MIMD Computation: The HEP Supercomputer and its Applications*, MIT Press, Cambridge, MA, 1985.

[19] T. Pietrzykowski and S. Matwin. Exponential improvement of efficient backtracking. In *6 th Conference on Automated Deduction, Lecture Notes in Computer Science, 138*, pages 223–239, Springer-Verlag, 1982.

[20] J. A. Robinson. A machine oriented logic based on the resolution principle. *Journal of the ACM*, 12(1):23–41, January 1965.

[21] Seki-Projekt. *The Markgraf Carl refutation procedure.* Technical Report Memo-SEKI-MK-84-01, Institut fuer Informatik I, Universitat Karlsruhe, West Germany, 1984.

[22] Seki-Projekt. *The Markgraf Carl refutation procedure: the logic engine.* Technical Report Nr. 24/82, Institut fuer Informatik I, Universitat Karlsruhe, West Germany, 1982.

[23] J. Siekmann and W. Stephan. *Completeness and consistency of the connection graph proof procedure.* Technical Report Nr. 7/76, Institut fuer Informatik I, Universitat Karlsruhe, West Germany, 1976.

[24] L. Wos. Solving open questions with an automated theorem proving. In *Proc. 6th conference on Automated deduction, Lecture Notes in Computer Science 138*, pages 1–31, Springer-Verlag, New York, 1982.

[25] L. Wos, D. F. Carson, and G. A. Robinson. Efficiency and completeness of the set of support strategy in theorem proving. *Journal of the ACM*, 12:687–697, 1965.

[26] L. Wos, D. F. Carson, and G. A. Robinson. The unit preference strategy in theorem proving. In *Proceedings of the Fall Joint Computer Conference*, Thompson Book Company, New York, 1964.

```
procedure Find-new-same-inst (σ1, σ2);
{  group-1 := all the set var-of-same-inst of σ1
   group-2 := all the set var-of-same-inst of σ2
   I := 0; /* initialize I */
   if ( group-1 = ∅) then
      for all X2 ∈ group-2 do { I := I + 1; new-same-inst[I] := X; }
   elseif ( group-2 = ∅) then
      for all X1 ∈ group-1 do { I := I + 1; new-same-inst[I] := X; }
   else
   {  for X1 ∈ group-1 do
      {  something-changed := false;
         for X2 ∈ group-2 do
         {  if X1 ∩ X2 ≠ ∅ then mark X2; something-changed := true; }
         if (something-changed) then
         {  I := I +1;
            new-same-inst[I] := X1∪ (all the marked X2's);
            Remove all marked X2's from group-2;
            Remove X1 from group-1;
         }
         while (something-changed) do
         {  something-changed := false;
            for X3 ∈ group-1 do
            {  if X3∩ new-same-inst[I] ≠ ∅ then
               mark X3; something-changed := true; }
            if (something-changed) then
            {  something-changed := false;
               new-same-inst[I] := new-same-inst[I] ∪ (all the marked X3's);
               Remove all marked X3's from group-1;
               for X4 ∈ group-2 do
               {  if X4∩ new-same-inst[I] ≠ ∅ then
                  mark X4; something-changed := true; }
            }
            if (something-changed) then
            {  new-same-inst[I] := new-same-inst[I] ∪ (all the marked X4's);
               Remove all marked X4's from group-2;}
         }
      }
      for X1 ∈ the rest of group-1 or group-2 do
      {  I = I +1; new-same-inst[I] := X1; }
   }
}
```

Figure 9

Theory Links in Semantic Graphs

Neil V. Murray

State University of N.Y. at Albany
Department of Computer Science
Albany, NY 12222

Erik Rosenthal

Wellesley College
Department of Computer Science
Wellesley, MA 02181

ABSTRACT

Recently, Stickel developed *Theory Resolution*, a theorem proving technique in which inferences use an existing 'black box' to implement a theory. In this paper we examine the black box and expand his results. The analysis of the black box is accomplished with the introduction of a generalization of link which we call *theory link*. We demonstrate that theorem proving techniques developed for ordinary links are applicable to theory links.

1. Introduction

In [8,9,10,11] we developed a graphical representation of NNF quantifier-free predicate calculus formulas and a new rule of inference, path resolution, which employs this representation. In [19,20,21] Stickel introduced theory resolution in which inferences depend on the existence of a 'black box' to implement a theory. Stickel designed theory resolution to be "a method of incorporating specialized reasoning procedures in a resolution theorem prover so that the reasoning task will be effectively divided into two parts: special cases ... are handled efficiently by specialized reasoning procedures, while more generalized reasoning is handled by resolution."

Path resolution operations hinge on the discovery of subgraphs (called resolution chains) that have the special property that all their c-paths contain a link. Many results from path resolution go through when we consider a generalization of link that we call a *theory link*. Intuitively, an ordinary link is a set of two c-connected (conjoined) literals such that under **no** assignment can both be true; a theory link is a set of n c-connected literals such that under no T-assignment, i.e., an assignment satisfying the axioms of theory T, can all be true. These specialized theory links can then be used in resolution-like procedures.

A simple example illustrating theory links is as follows. Suppose the statements "elephants are mammals" and "mammals are animals" comprise a theory expressed as clauses. Each clause yields a theory link, and it is immediate from the sequel that "elephants are animals" yields one also. Sometimes the theory links may be incorporated directly into a formula in such a way that subsequently the theory need never be consulted.

A brief summary of required background material is presented in Section 2; for a more detailed exposition see [8,9]. Three equivalent formulations of theory link are presented in Section 3. In Section 4 we compare this work to Stickel's Theory Resolution. In section 5 we relate earlier results on the deletion of ordinary links to the deletion of theory links. In Section 6 we present sample deductions involving both ordinary and theory links. We make use of techniques originally developed for semantic graphs with ordinary links.

The symbol '•' is used to indicate the end of a proof.

2. Preliminaries

We briefly summarize semantic graphs and path resolution, including only those results necessary for the introduction of theory links.

A *semantic graph* is empty, a single node, or a triple (N, C, D) of *nodes, c-arcs,* and *d-arcs*, where a node is a literal occurrence, a c-arc is a conjunction of two non-empty semantic graphs, and a d-arc is a disjunction of two non-empty semantic graphs. We use the notation $(G, H)_c$ for the c-arc containing G and H, and, similarly, $(G, H)_d$ for the d-arc.

The construction of a graph may be thought of as a sequence of such arcs. There will always be exactly one arc (X, Y) with the property that every other arc is an arc in X or in Y. We call this arc the *final arc* of the graph, and we call X and Y the *final subgraphs*. Since this arc completely determines G, we frequently write $G = (X, Y)$. The notion of *fundamental subgraph* is often useful: if $G = (X, Y)_\alpha$ and the final arc of Y is **not** of type α, then Y is a fundamental subgraph of G; otherwise the fundamental subgraphs of Y are fundamental subgraphs of G.

A semantic graph may be thought of as a binary [n-ary] tree in which each node represents an explicit [fundamental] subgraph, and the children of a node are its final [fundamental] subgraphs. The root is of course the entire graph, and the leaves are the literals.

If $\mathbf{a} = (X, Y)$ is an arc in a graph, and if A and B are nodes in X and Y, respectively, then we say that $\mathbf{a}$ is the arc *connecting* A and B. If $\mathbf{a}$ is a c-arc, we say that A and B are *c-connected*, and if $\mathbf{a}$ is a d-arc, A and B are *d-connected.* A *c-path* is a maximal collection of c-connected nodes, and a *d-path* is a maximal collection of d-connected nodes. The semantics of a graph may be characterized by its paths: it is easy to verify that a c-path and a d-path have exactly one node in common, and that a graph is satisfied by an interpretation I iff I satisfies (every literal on) some c-path, and the graph is falsified iff some d-path is falsified by I.

The following example illustrates some of these notions. Consider the formula

$$(D \vee (A \Longleftrightarrow B)) \wedge ((C \wedge E) \vee \sim A \vee (\sim B \wedge P))$$

The corresponding graph is

$$
\begin{array}{ccc}
D & & C \to E \\
\downarrow & & \downarrow \\
\overline{A} \to \overline{B} & \to & \overline{A} \\
\downarrow & & \downarrow \\
B \to A & & \overline{B} \to P
\end{array}
$$

Examples of c-paths are $\{B, A, \overline{A}\}$ and $\{D, \overline{B}, P\}$, and some d-paths are $\{D, \overline{B}, A\}$ and $\{C, \overline{A}, P\}$.

In [8,9], the subgraph of a given graph with respect to a given set of nodes is precisely defined. Intuitively, it is a graph which contains the given nodes and those arcs associated with those nodes. Certain classes of subgraphs, the *blocks*, turn out to be especially important. There are three types of blocks: the *c-blocks*, the *d-blocks*, and the *full-blocks*.

A *c-block* C is a subgraph of a semantic graph with the property that any c-path p which includes at least one node from C must pass through C; that is, the subset of p consisting of the nodes which are in C must be a c-path through C. A *d-block* is similarly defined with d-paths, and a *full block* is a subgraph which is both a c-block and a d-block. We define *a strong c-block* in a semantic graph G to be a subgraph C of G with the property that every c-path through G contains a c-path through C. (If G is in CNF then C is one or more clauses.) A *strong d-block* is similarly defined.

Recall that a *link* in a formula in cnf is a pair of literals from different clauses that can be made complementary by an appropriate substitution. A link in a semantic graph is similarly defined for a pair of c-connected nodes. A formula in cnf is unsatisfiable iff every c-path contains a link; the same is true for a semantic graph. (A formula satisfying this condition is said to be *spanned* by its links.)

A *resolution subgraph* R in a semantic graph G is a subgraph with the property that every c-path through it contains a link. If a c-path **p** through the entire graph is satisfiable, it cannot possibly pass through the resolution subgraph. Thus **p** must miss part of R. The c-blocks of R are the parts of R with the property that one of them must be missed by **p**. Associated with each c-block is an *auxiliary subgraph*: that part of G that must be hit by a c-path that misses the c-block. Intuitively, the path resolvent of R in G contains the disjunction of the auxiliary subgraphs.

We define WS(H, G), the *weak split graph of H in G* as follows:

Let the fundamental subgraphs of G that meet H be $F_1, ..., F_k$, and let $F_{k+1}, ..., F_n$ be those that do not.

$$\mathrm{WS}(\emptyset, G) = G \quad \text{and} \quad \mathrm{WS}(G, G) = \emptyset.$$

$$\mathrm{WS}(H, G) = \mathrm{WS}(H_{F_1}, F_1) \vee \cdots \vee \mathrm{WS}(H_{F_n}, F_n)$$
$$\text{if the final arc of } G \text{ is a d-arc}$$

$$\mathrm{WS}(H, G) = \mathrm{WS}(H_{F_1}, F_1) \vee \cdots \vee \mathrm{WS}(H_{F_k}, F_k)$$
$$\text{if the final arc of } G \text{ is a c-arc}$$

The *strong split graph of H in G* is defined in a similar manner except that the last equation becomes

$$\mathrm{SS}(H, G) = \mathrm{SS}(H_{F_1}, F_1) \vee \cdots \vee \mathrm{SS}(H_{F_k}, F_k) \wedge F_{k+1} \wedge \cdots \wedge F_n$$

The weak or strong split graph of a resolution chain will be referred to as a *path resolvent*. When the graph is in CNF, both operations yield the same result. Intuitively, $WS(H,G)$ is the disjunction of the auxiliary subgraphs of the maximal c-blocks of H. (Certain redundancies are automatically removed by weak split. See [8], Theorems 4 and 5 for a precise statement.) On the other hand, strong split is essentially formed with the nodes lying on c-paths that miss the c-blocks of H. (It is surprising that these two notions in general lead to different inferences, although they are the same in CNF.) When we write $\mathrm{WS}(H_{F_1}, F_1)$ in the above definition, H_{F_1} denotes the subgraph of G relative to the nodes in H that are in F_1. Since the second argument F_1 determines the relevant nodes of H, we will often (in this and in similar situations) use the notation $\mathrm{WS}(H, F_1)$.

Considering the example above, the subgraph

$$\begin{array}{ccccc} & & & & \bar{A} \\ A & \rightarrow & B & \rightarrow & \downarrow \\ & & & & \bar{B} \end{array}$$

forms a resolution subgraph. Its path resolvent (in this case both weak and strong) is

$$\begin{array}{c} C \rightarrow E \\ \downarrow \\ D \\ \downarrow \\ \bar{A} \rightarrow \bar{B} \end{array}$$

3. Theory Links in the Ground Case

Suppose we express a propositional theory T as a semantic graph. Let H be a semantic graph with the property that $G = \mathrm{T} \rightarrow H$ is unsatisfiable. We assume that both T and H are satisfiable; since G is unsatisfiable, H is obviously T-unsatisfiable in the sense of Stickel [21]. Some c-paths through H may contain links, but there must be at least one linkless c-path p_H through H (since H is satisfiable). Yet the literals on such a c-path p_H must be T-unsatisfiable, and it is likely that some proper subset Q_{p_H} of those literals is T-unsatisfiable. One way to compute Q_{p_H} is to extend p_H to a c-path $p_T p_H$ in every possible way, i.e., form the c-path

$p_T p_H$ for every c-path p_T through T. By recording a minimal set of nodes on p_H, each member of which is linked to some linkless p_T, we can determine Q_{p_H}. If we now compute Q_{p_H} for each linkless p_H, then in some sense the Q's and the ordinary links of *H* are sufficient evidence that *H* is T-unsatisfiable.

Our intent then is to define, in a computationally feasible way, a minimal collection of such Q's so that any T-unsatisfiable semantic graph *H* is spanned by its links and the Q's. The Q's will generically be called theory links. Three characterizations of theory links are discussed and shown to be equivalent in power.

3.1. T-links

Assume the axioms of a theory T are expressed as the m clauses $C_1, C_2, ..., C_m$. (We assume only T to be in CNF.) Let R(T) be the union of T and all possible binary resolvents of clauses in T, and let $R^n(T) = R(R^{n-1}(T))$. Then T^* is the set of all clauses obtainable from T by (ordinary binary) resolution; i.e., $T^* = \bigcup_{j=1}^{\infty} R^j(T)$. Of course T^* is finite in the ground case.

A *T-link* is defined to be a set Q of c-connected nodes such that $Q = \overline{C}$, where C is a clause in T^*. In other words, Q is a set of c-connected nodes which are complements of the nodes in some clause in T^*. The following lemma is obvious.

Lemma 1. Any T-link Q is T-unsatisfiable. •

Theorem 1. Given a ground theory T defined by the m clauses $C_1, ..., C_m$ and a T-unsatisfiable semantic graph *H*, *H* is spanned by its links and T-links; i.e., every linkless c-path in *H* contains a T-link.

Proof: Let p_H be a linkless c-path through *H*. Since p_H is T-unsatisfiable, $\overline{p_H}$ is a logical consequence of T. Resolution is of course complete for consequence finding in the sense that if clause M′ is a logical consequence of T, then some M which subsumes M′ can be derived from T by resolution. Any such M for p_H does the trick: $Q = \overline{M}$ is the required T-link. •

The above proof is clear and concise, but it is not very constructive, and it relies crucially on the completeness of resolution for the derivation of logical consequences. We have developed an alternative proof that is somewhat more constructive, and that contains a proof of the completeness result required above. The proof is derived directly from the structure of the semantic graph. We do not include it here for lack of space.

3.2. Strong T-links and resolvent T-links

Path resolution and semantic graphs lead us to two other natural ways of characterizing sets of T-unsatisfiable nodes. Given sets L_1, L_2 of theory links, we say L_1 ***subsumes*** L_2 if every member of L_2 is a superset of some member of L_1. Recall that in CNF a strong c-block is a collection of clauses. Define a ***strong T-link*** to be a linkless set Q of c-connected nodes such that $S_T \rightarrow Q$ is a resolution chain, where S_T is a strong c-block in T. Let ^{S}T denote the set of all strong T-links.

Theorem 2. Given a ground theory T, ^{S}T is subsumed by T^*.

Proof: Let Q be a strong T-link with S_T its associated strong c-block in T. By the definition of strong c-block, no c-path through Q can be extended through T unless it passes through S_T, and hence through the resolution chain, guaranteeing the T-unsatisfiability of Q. By Theorem 1, Q must contain a T-link Q′ ; this is true for any strong T-link Q, and so ^{S}T is subsumed by T^*. •

We now give a third characterization of theory links. Define a ***resolvent T-link*** to be either the negation of a clause in T or a linkless set Q of c-connected nodes such that $\overline{Q}$ is a path resolvent of some resolution chain in T, and let ^{R}T denote the set of all resolvent T-links. (Since T is in CNF, it is irrelevant whether we use weak or strong split.)

Theorem 3. Given a ground theory T, ^{R}T is subsumed by ^{S}T.

Proof: Let R be a resolution chain whose path resolvent is $\overline{Q}$. Consider the strong c-block S_T consisting of the clauses in T that meet R. We know that the literals in $\overline{Q}$ are exactly the literals in S_T - R. We are done if we can show that $S_T \rightarrow Q$ is a resolution chain (making Q a strong T-link also), so consider a c-path p through $S_T \rightarrow Q$. If p goes through R it must have a link; if it does not, then p contains a node from $\overline{Q}$, and this node is linked to its complement on Q. •

We have now established that T^* subsumes ^{S}T, and that ^{S}T subsumes ^{R}T. Theorem 4 completes the cycle, assuring us that each characterization will give us all ***essential*** theory links: those that have no other theory links as subsets.

Theorem 4. Given a ground theory T, T^* is subsumed by ^{R}T.

Proof: Let Q be a T-link. The semantic graph $T \rightarrow Q$ is unsatisfiable. Define L to be the set of nodes in T that are linked to some node in Q. Consider first the case where L contains some clause C from T. Then $\overline{C} \subseteq Q$, and $\overline{C}$ is the desired resolvent T-link.

Otherwise, let *R* be the subgraph of T relative to the nodes in T – L. Since the above case does not apply, *R* meets each clause in T, so any c-path p_R through *R* is a c-path through T. By the definition of L, p_R Q cannot contain a link to Q. But it is a c-path through $T \rightarrow Q$ and must therefore contain some link. This link lies entirely within *R*. Therefore *R* is a resolution chain, its path resolvent is $\overline{L}$, and $L \subseteq Q$, so L is the desired resolvent T-link.

Theorems 2-4 imply that any minimal set of T-unsatisfiable nodes can be characterized in each of the three ways discussed. It thus seems natural to define a theory link to be a set of c-connected nodes satisfying any of the three conditions. A deductive system could be designed to recognize and/or record theory links based on whichever criterion is most convenient for the given application.

We now have an interesting result on consequence completeness for path resolution in the ground case, when the graph G is in CNF. ***Any*** (single clause) logical consequence of G is in G or is subsumed by a path resolvent of G. Looked at another way, Theorem 4 guarantees that for each $i \geq 0$, the clause obtained in the ith step of a resolution derivation is subsumed by a one-step path resolvent from the original graph.

3.3. First order theory links

Some of our ground level results lift directly into first order logic; others lift in somewhat modified form. In this brief section we restrict attention to the most straightforward cases.

Lemma 1 and Theorem 1 concerning T-links lift directly in much the same way that resolution does. In general T^* may be infinite, but a finite subset of T^* is always sufficient to demonstrate the T-unsatisfiability of a formula H. Incorporating sufficient T-links into H is semi-decidable; it becomes decidable in cases where T^* is known to be finite, such as when T is ground, or when, as in the second example of Section 7, $T = T^*$ because T is linkless. Furthermore, when T is function-free, its Herbrand universe is finite, and generating enough T-links is decidable whether or not (T^*) is finite.

Theorems 2 and 4 relate T-links to strong T-links and to resolvent T-links. They lift from *any one instance* of T: Suppose R is a resolution chain in $\{T\theta_1, T\theta_2, ..., T\theta_n\}$, a set of ground instances of a theory T. Unless $n = 1$, we cannot be sure that the chain lifts to T. Of course, R does lift if we have n copies of T. Thus the resolvent T-links do lift from any single instance of T, and hence, in view of Theorem 3, the same is true of strong T-links.

Theorem 3 remains true at the general level. But the reason is that the essential strong T-links lost in lifting are exactly the essential resolvent T-links similarly lost.

Unification becomes an issue upon consideration of first order theory links. Unifying an ordinary link is straightforward since the only atoms requiring inspection are present in the potential link; unifying a theory link may be more difficult since atoms in the theory may also require inspection. For example, if the literal A from a theory link {A, B, C} is inherited, it may be instantiated in such a way that B, C, and the inherited version of A cannot be unified to form a theory link. We have developed some results concerning the unification of first order theory links (and their descendants) [12], but those results are beyond the scope of this paper.

4. Theory Links and Theory Resolution

The work of Stickel [19,20,21] on theory resolution is closely related to our work on theory links. One important contrast in these approaches occurs in the assumptions made about how T-unsatisfiable sets of literals (or sets of clauses) are recognized. Theory resolution assumes a 'black box' for this recognition, and different categories of inference arise as a function of the power of the black box. We have instead assumed that the black box can be expressed as a formula in CNF. The consequences of the formula can then be incorporated into a knowledge base in the form of theory links, and used in the inference process.

Let us suppose that a T-decision procedure for theory resolution can be expressed as a set of clauses T. Let H be some knowledge base on which we wish to do deduction relative to T-interpretations. Path resolving on a single T-link in H corresponds to a *total narrow* theory resolution. Path resolving on a resolution chain built from several T-links (not all on the same c-path) corresponds to a *total wide* theory resolution.

Within our framework, *partial* theory resolution can be stated as the following theorem.

Theorem 5. Suppose H is a semantic graph containing a set l of c-connected nodes. Suppose further that L is a theory link, and that $l \subseteq$ L. If r is the clause consisting of the negations of the literals appearing in L $-$ l, then

$$\mathrm{WS}(l, H) \ \bigvee \ r$$

may be soundly (with respect to the theory) inferred.

The proof is straightforward and left to the reader. •

For well known theories (like equality) whose usefulness is almost universal, a highly developed and streamlined black box may be the best answer. But there may be situations in which a theory, representable as a set of clauses, is being learned or acquired. Adding new clauses to the theory-knowledge base would give rise to new theory links in the assertional knowledge, i.e., that part of the knowledge base on which a system performs deduction. It would be helpful to record the theory clauses on which a theory link is based. If a dynamic system should remove one such clause, the dependent theory link could then be immediately removed also.

5. Link Deletion

Deleting a link after activation reduces the size of the search space since not only will the given link never be used again, but also it will never be inherited. In [10] and [11] we developed several link deletion results for ordinary links. In this section, we briefly summarize those results.

Any subgraph R of a semantic graph G has the form:

$$C_1 \bigcup C_2 \bigcup \cdots \bigcup C_n \bigcup C_s$$

where C_s is a strong c-block, and the others are c-blocks that are *not* strong. (It is certainly possible that C_s is empty; moreover the C_i's need not be disjoint. The technical term for this decomposition is 'proper c-family' (see [8]).)

Suppose now that R is a resolution chain, and that it has the form:

$$C_1 \rightarrow C_2 \rightarrow \ldots \rightarrow C_n \rightarrow C_s$$

(In general, resolution chains need not have this form.) If R is activated, some of its links may be deletable. In [11] we proved the following result assuming that R has been activated using **weak** split.

Theorem: If n=2, and if C_1 and C_2 are strong with respect to c-connected full blocks M_1 and M_2 respectively, then every link that meets both M_1 and M_2 is deletable.

The following considerably stronger result holds if **strong** split is used.

Theorem: If n=2, then every link from C_1 to C_2 may be deleted when R is activated.

These two theorems apply at the ground level. Their proofs are purely structural and may be applied to (ground) theory links as well. The key to both proofs is the observation (first made by Bibel [2] for single (ordinary) link chains in CNF) that a c-path that contains a deleted link must go through both C_1 and C_2; the structure of the chain then guarantees that such a c-path will pick up an inherited link. The reason that n = 2 is necessary in these theorems is that the link must meet each non-strong c-block. The identical results apply to theory links that meet every non-strong c-block in the chain.

The two theorems lift (for theory links as well as for ordinary links) with additional restrictions on unifiers: A sufficient condition for a first order link to be deletable is that its unifier be identical to the simultaneous unifier of the chain being activated; i.e., the entire chain is present in any instance of the graph in which the link in question is present.

6. Examples

Consider the theory consisting of the clauses $C_1 = \{\overline{E(x)}, M(x)\}$ and $C_2 = \{\overline{M(x)}, A(x)\}$ representing "elephants are mammals" and "mammals are animals," respectively. The resolvent of C_1 and C_2 is $C_3 = \{\overline{E(x)}, A(x)\}$. This yields $T^* = \{C_1, C_2, C_3\}$.

Suppose we know that elephants like peanuts and that dumbo is an elephant. As a graph, we have:

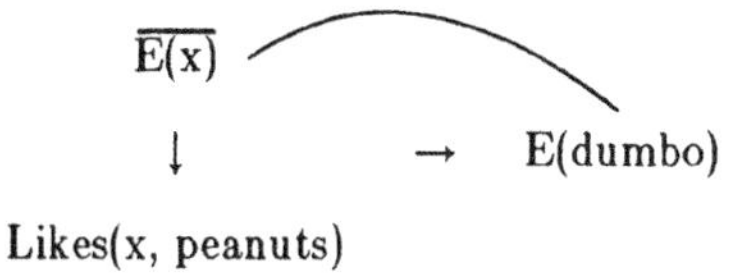

Note that E(dumbo) is a subset of an instance of the theory link $\overline{C_3}$. Theorem 6 applies, yielding partial theory resolvent A(dumbo). The ordinary resolvent of the link $\{\overline{E(x)}, E(dumbo)\}$ is Likes(dumbo, peanuts), and we have proven that dumbo is an animal that likes peanuts.

The next example has been discussed by Stickel [21]. His *theory resolution* inference rule has been used in the KLAUS [5] and KRYPTON [3,14] systems to generate refutations for this knowledge base. Suppose we have the following information about boys and girls: boys are persons whose sex is male, and girls are persons whose sex is female. We represent this as a theory T of four clauses.

$$T \;=\; \{\ \{\overline{B(x)}\ P(x)\}\ \ \{\overline{G(x)}\ P(x)\}\ \ \{\overline{S(x,y)}\ \overline{B(x)}\ M(y)\}\ \ \{\overline{S(x,y)}\ \overline{G(x)}\ F(y)\}\ \}$$

The semantic graph below defines **N**o**S**ons (as **P**ersons all of whose children are **G**irls) and **N**o**D**aughters; it also declares that every **P**erson has a **S**ex, **M**ales and **F**emales are disjoint, and that **c**hris has neither sons nor daughters and yet has a **C**hild. This semantic graph is T-unsatisfiable. The T-links are represented by the double curves, and the ordinary links by the single curves. There are only four T-links since $T^* = T$.

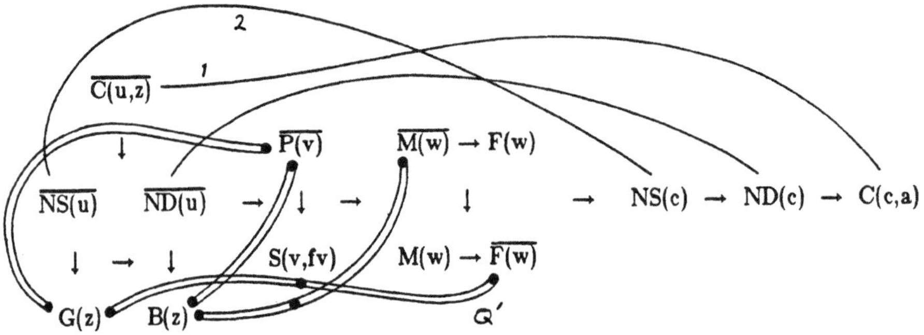

We now illustrate a path resolution refutation using a combination of ordinary links and theory links. The (ordinary) links numbered 1 and 2 in the diagram form a 'nice' resolution chain: It consists of two c-blocks, one of which is strong; each link meets both c-blocks; and the mgu of link 1 equals the mgsu of the chain, allowing deletion of the link. Had we used weak split, the path resolvent would be G(a); We instead used strong split, producing $G(a) \wedge (\overline{ND(c)} \vee B(a)))$ and allowing for the deletion of link 1. The Pure Lemma [10,11] then permits removal of the two fundamental subgraphs that meet link 1 and of their links. The resulting graph is shown below.

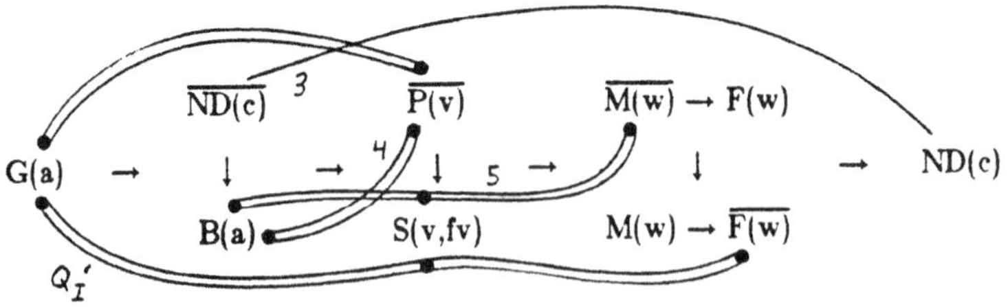

Notice that Q_I' (see the diagram above) is a potential theory link. Theorem 5 may be used to confirm that it is in fact a theory link and the same is true for the three other theory links in the diagram.

We now resolve on link 3 and theory link 4 (which comprise what might be called a 'hybrid' resolution chain). The path resolvent is S(a,fa). Link 4, the theory link, is deletable, and the resulting graph is shown below. We show only the inherited theory links since they are all that are required to complete the proof.

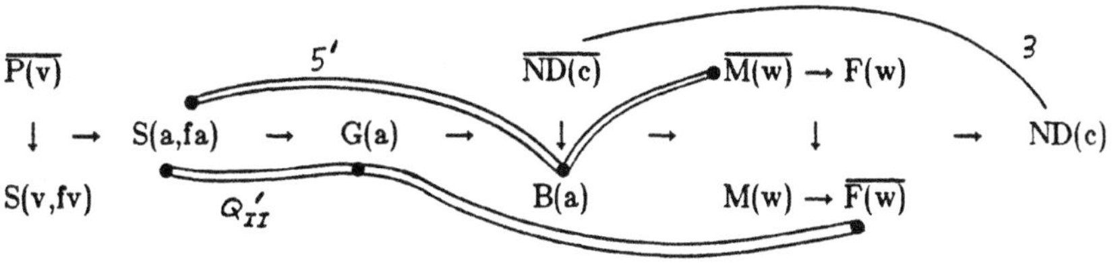

The inherited theory links (5′ and Q_{II}') form a five-node resolution chain with mgsTu {fa/w} (in which link 5′ is deletable). The path resolvent is $\overline{ND(c)}$, which inherits a link (to ND(c)) that spans the entire graph.

Note that the last link could have been added to the chain in the previous step producing a contradiction then. In fact, the whole proof could have been done in one step since one instance of this graph is contradictory.

A more difficult problem that has received considerable attention recently is Schubert's Steamroller. Although this problem confounded automated theorem provers for several years after its introduction, it can now be handled by a variety of techniques including Stickel's theory resolution [21], and the many-sorted provers of Cohn [4] and Walther [22]. We have constructed a 9-step proof that uses path resolution and theory links. The first twelve clauses (which assure the existence of, and describe the 'sorts' of foxes, birds, etc.) are treated as a theory, and only finitely many theory links are generated.

1. Summary

We have introduced the notion of theory link, and related it to Stickel's theory resolution. Many techniques and results from the study of semantic graphs and path resolution are directly applicable to performing inferences with theory links. In particular, the inferencing mechanisms designed for path resolution are completely adequate to handle resolution chains built in an arbitrary way. Link deletion results are also applicable. We have demonstrated that in the ground case any clause C that is a logical consequence of a set G of clauses can be essentially derived from G in one step by path resolution.

Acknowledgement

This research was partially supported by the Research Foundation of the State University of New York at Albany through Faculty Research Award #320-9788A.

References

1. Andrews, P.B. Theorem proving via general matings. JACM 28,2 (April 1981), 193-214.
2. Bibel, W. On matrices with connections. JACM 28,4 (Oct. 1981), 633-645.
3. Brachman, R.J., Gilbert, V. Pigman, and Levesque, H.J. An essential hybrid reasoning system: Knowledge and symbol level accounts of KRYPTON. Proceedings of The 9th International Joint Conference on Artificial Intelligence, Los Angeles, CA, August 18-24, 1985, 532-539.
4. Cohn, A.G. On the solution of Schubert's Steamroller in many sorted logic. Proceedings of the 9th International Joint Conference on Artificial Intelligence, Los Angeles, CA, August 1985, 1169-1174.
5. Haas, N. and Hendrix, G.G. An approach to acquiring and applying knowledge. Proceedings of the AAAI-80 National Conference on Artificial Intelligence, Stanford, CA, August 1980, 235-239.
6. Kowalski, R. A proof procedure using connection graphs. J.ACM 22,4 (Oct. 1975), 572-595.
7. Manna, Z. and Waldinger, R. Special relations in automated deduction. J.ACM 33,1 (Jan. 1986), 1-59.
8. Murray, N.V., and Rosenthal, E. Path resolution and semantic graphs. Proceedings of EUROCAL '85, Linz Austria, April 1-3, 1985. In Lecture Notes in Computer Science, Springer-Verlag, vol. 204, 50-63.
9. Murray, N.V., and Rosenthal, E. Inference with path resolution and semantic graphs. Submitted, June 1985.
10. Murray, N.V., and Rosenthal, E. On deleting links in semantic graphs. To appear in the proceedings of The Third International Conference on Applied Algebra, Algebraic algorithms, Symbolic Computation and Error Correcting Codes, Grenoble, France, July 15-19, 1985.
11. Murray, N.V., and Rosenthal, E. Improved link deletion and inference: proper path resolution. Submitted, August 1985.
12. Murray, N.V., and Rosenthal, E. Theory links: applications to automated theorem proving. Submitted, March 1986.
13. Nilsson, N.J. A production system for automatic deduction. Technical Note 148, SRI International, 1977.
14. Pigman, V. The interaction between assertional and terminological knowledge in KRYPTON. Proceedings of the IEEE Workshop on Principles of Knowledge-Based Systems, Denver, Colorado, December 1984.
15. Prawitz, D. An improved proof procedure. Theoria 26 (1960), 102-139.
16. Robinson, J.A. A machine oriented logic based on the resolution principle. J.ACM 12,1 (1965), 23-41.
17. Robinson, J.A. An overview of mechanical theorem proving. *Theoretical Approaches to Non-Numerical Problem Solving,* Springer-Verlag, New York, Inc., 1970, 2-20.
18. Shostak, R.E. Deciding combinations of theories. Proc. Sixth Conf. on Automated Deduction, New York, New York, June 1982, 209-222.
19. Stickel, M.E. Theory resolution: building in nonequational theories. Proc. of the Nat. Conf. on A.I., Washington, D.C., Aug. 1983.
20. Stickel, M.E. Automated deduction by theory resolution. Technical Note 340, SRI International, Oct. 1984.
21. Stickel, M.E. Automated deduction by theory resolution. Proc. of the 9th International Joint Conf. on Artificial Intelligence, Los Angeles, CA, August 18-24, 1985, 1181-1186.
22. Walther, C. A mechanical solution of Schubert's Steamroller by many-sorted resolution. Proceedings of the AAAI-84 National Conference on Artificial Intelligence, Austin, Texas, August 1984, 330-334.

ABSTRACTION USING GENERALIZATION FUNCTIONS

David A. Plaisted
Department of Computer Science
New West Hall 035A
University of North Carolina at Chapel Hill
Chapel Hill, North Carolina 27514

Abstract

We show how a generalization operator may be incorporated into resolution as a method of guiding the search for a proof. After each resolution, a "generalization operation" may be performed on the resulting clause. This leads to a more general proof than the usual resolution proof. These general proofs may then be used as guides in the search for an ordinary resolution proof. This method overcomes some of the limitations of the abstraction strategies with which the author has experimented for several years. Some of the results of these previous experiments and comparisons of the two approaches are given.

1. Introduction

The idea of using abstraction or analogy as a guide in the search for a proof has much intuitive appeal. The idea is this: When trying to obtain a proof of a theorem A, convert A to a simpler theorem B, try to prove B, and use the proofs of B as guides in searching for a proof of A. The idea can be extended to many levels, converting A to B_1 then to B_2 and so on, and the idea can also be extended to use more than one abstraction at the same time. This idea, in the context of resolution theorem proving (Robinson[65]), was presented in Plaisted[81], and some earlier work on the use of analogy was done by Kling[71]. For some general discussions of resolution, see Bundy[83], Chang and Lee[73], Loveland[84], Manna[74], and Wos et al[84]. The role of analogy is being recognized in artificial intelligence; see for example Winston[77]. However, despite the philosophical appeal of this concept, it has been difficult to translate it into the resolution framework in a useful way. We have done some experiments with a resolution theorem prover at the University of Illinois which has a variety of syntactic abstractions built in. There have been some successes. In some cases, the abstraction prover was able to quickly detect that a set of clauses is consistent (for example, if there is an error in the input) while the ordinary resolution prover would consume large amounts of time and space. There was a simple problem from Chang and Lee[73] which we could get using abstraction but could not get otherwise. However, we later modified our resolution prover and it was able to get this proof quickly without abstraction by using another syntactic strategy. We were able to get one moderately hard proof using abstraction that could not be obtained otherwise. This was reported in Plaisted and Greenbaum[84]. However, this used a "semantic abstraction" that was hard to devise. Our University of Illinois prover has the facility for performing many levels of abstraction, each level making use of the proofs found at the previous level.

One problem with our abstractions is that they only change the input clauses, not the resolution operation itself. We therefore refer to such abstractions as *input abstractions*, in contrast to the generalization abstractions defined below. Suppose S is the set of input clauses; these are modified to obtain an abstract set T of clauses by "throwing away" certain information. For example, certain arguments of certain predicates may be deleted. Then we look for refutations from T, and use these to guide the search for refutations from S. Note that this does not give any control over what happens during the search for a proof. There is no way, for example, to specify that any clause of depth greater than 3 will be replaced by a clause of depth 3. We will show later how this introduces large and unavoidable jumps between successive levels of abstraction, making an abstract search space a poor guide to the less abstract search space.

We propose to overcome this problem by abstracting the resolution operation itself, rather than the input clauses. That is, after each resolution, certain information is discarded from the resolvent. To be precise, each resolvent C is replaced by a more general clause D. This permits control of what happens during the proof itself. For example, it is possible to recognize and eliminate certain subterms in certain contexts, or to specify that clauses do not get deeper than a certain amount. It is necessary to restrict the generalization operation in such a way as to insure that the abstract search space can be used as a guide to the less abstract search space. The intuitive idea is that we are looking for a proof. The first idea we have of the proof is "blurry"; many details are missing. As we move from abstract levels of search to less abstract levels, the details of the proof come into clearer focus. Finally, at the input level, a complete proof is obtained, or else no proof exists within the specified bounds on the search space.

It may be that one of the reasons we had trouble with input abstraction is that resolution is a poor inference system for searching for proofs. Possibly the problems with resolution are so severe that input abstractions were not able to overcome them. However, the methods of this paper (as well as previous methods) may be applied to proof systems other than resolution. Furthermore, these new methods overcome some problems with previous abstractions, and may be helpful even for resolution.

2. Definitions

Definition. If C and D are clauses, then we write $C \leq D$ if there is a substitution Θ such that $C = D\Theta$, that is, C is an instance of D.

Definition. If S and T are sets of clauses, then we write $S \leq T$ if for all clauses C in S, there exists a clause D in T such that $C \leq D$.

Definition. If C and D are clauses, let Res(C, D) be the set of resolvents of C and D, as originally described in Robinson[65]. If S is a set of clauses, let R(S) be $\cup\{Res(C, D) : C, D \text{ in } S\}$.

Definition. If S is a set of clauses, let $P_n(S)$ be the clauses appearing in refutations from S of depth n or less. Formally, we have the following definition:

$$P_n(S) = \text{if } n > 0 \text{ then } P_{n-1}(S) \cup \{C \in S \;:\; \exists D \in S \; Res(C, D) \cap P_{n-1}(S \cup R(S)) \neq \emptyset\}$$
$$\text{else } \{NIL : NIL \in S\}$$

Definition. A *generalization function* g is a function from clauses to sets of clauses such that for all D in g(C), C $\leq$ D. We permit g(C) = $\emptyset$ for some C. We do not require that g be monotone. That is, we do not require that if $C_1 \leq C_2$ then $g(C_1)$ must in some sense be instances of $g(C_2)$. If g is a generalization function and S is a set of clauses, then g(S) is $\cup\{g(C) : C \text{ in } S\}$.

An example of a generalization function is the mapping g from clauses to sets of clauses such that g(C) = {D} where D is obtained from C by replacing the first argument of each predicate by a new variable. Another example is defined by g(C) = {D, E} where D is as before and E is obtained from C by replacing the last argument of each predicate by a new variable. For more examples, see sections 4 and 5.

Definition. A generalization function g *interpolates* for a set T of clauses if ($C \leq B$ and $B \in T$) implies there exists D in g(C) such that $C \leq D \leq B$, and if for all D in g(C), there exists B in T such that $D \leq B$.

Definition. We define clauses in proofs in which each resolution operation is followed by a generalization operation, as follows:

$$P_{g,\ n}(S) = \text{if } n > 0 \text{ then}$$
$$P_{g,\ n-1}(S) \cup \{C \in S \ :\ \exists D \in S\ g(Res(C,\ D)) \cap P_{g,\ n-1}(S \cup g(R(S))) \neq \emptyset\} \text{ else}$$
$$\{NIL : NIL \in S\}$$

Definition. Suppose g is a generalization operation. Then a clause E in g(Res(C, D)) is called a *g-resolvent* of clauses C and D. That is, E is in g(B) for some B in Res(C, D). A *g-deduction from S* is a sequence C_1, C_2, C_3, $\cdots$, of clauses together with partial functions l and r mapping integers to integers such that each C_i is either in S, and l(i) and r(i) are undefined, or C_i is a g-resolvent of clauses $C_{l(i)}$ and $C_{r(i)}$ where l(i) and r(i) are defined and l(i) < i and r(i) < i. If l(i) and r(i) are defined then clauses $C_{l(i)}$ and $C_{r(i)}$ are called the *parents* of clause C_i. The *depth* of a clause in a g-deduction is 0, if no parents l(i) and r(i) are defined, or else one plus the maximum depths of its parents. The depth of a g-deduction is the maximum depth of any clause in the g-deduction. A *g-refutation* is a g-deduction in which the last clause is NIL (the empty clause).

Proposition. If g is a generalization function then $P_{g,\ n}(S)$ is the set of clauses appearing in g-refutations of depth less than or equal to n.

It is useful to consider g-refutations for purposes of proving properties of strategies based on generalization functions, but the recursive definition of $P_{g,\ n}(S)$ is more convenient for computation purposes.

Definition. The *identity* generalization function id is the function such that id(C) = {C} for all C.

Definition. A refutation is a g-refutation where g is the identity function above; similarly a deduction is a g-deduction where g = id.

Definition. Suppose g and h are generalization functions. Suppose S and T are sets of clauses. We define an ordering $\leq_{P,\ n}$ on pairs (S, g) and (T, h) by (S, g) $\leq_{P,\ n}$ (T, h) if for all g-deductions C_1, C_2, C_3, $\cdots$, from S of depth n or less there is an h-deduction D_1, D_2, D_3, $\cdots$, from T such that $C_i \leq D_i$ for all i, and such that if clauses C_j and C_k are the parents of C_i then D_j and D_k are the parents of D_k. Intuitively, for every g-deduction from S of depth n or less there is an h-deduction from T of the same structure.

3. The search strategy

The following results are the basis of our method of using generalization functions as a guide in the search for a proof.

Theorem 1. Suppose h is a generalization function and for all clauses D, h(D) $\neq \emptyset$. Suppose S and T are sets of clauses and S $\leq$ T. Then (S, id) $\leq_{P,\ n}$ (T, h) for all n.

Theorem 2. Suppose h is a generalization function. Suppose for all sets S and T of clauses, if S $\leq$ T then (S, id) $\leq_{P,\ n}$ (T, h). Then for all sets S and T of clauses, if S $\leq$ T and g interpolates for $P_{h,\ n}$(T) then (S, id) $\leq_{P,\ n}$ (T, g) and $P_{g,\ n}(S) \leq P_{h,\ n}(T)$.

Theorem 3. Suppose g is a generalization function such that for all clauses C, (g(C) = $\emptyset$ or g(C) = {C}). Suppose S is a set of clauses and (S, id) $\leq_{P,\ n}$ (S, g). Then $P_n(S) = P_{g,\ n}(S)$.

We use Theorem 1 to start the search for a proof. Once h-refutations are found, then Theorem 2 may be used to obtain g-refutations from them, that throw away progressively less information. This may be done a number of times. Finally, theorem 3 may be used to find ordinary resolution refutations. For an example of a sequence of generalization functions which may be used in this way, see section 5.

Proposition. If g and h are generalization functions and (S, g) $\leq_{P,\ n}$ (S, h) then $P_{g,\ n}(S) \leq P_{h,\ n}(S)$.

This shows that h-refutations may be used as a filter for g-refutations. That is, in searching for g-refutations, we need only retain g-resolvents C if there exists D in some h-refutation such that C $\leq$ D. This same idea is implicit in the recursive definition of $P_{g,\ n}$(S) where g interpolates for the set $P_{h,\ n}$(S), since only such g-resolvents will be generated.

3.1 The overall method

The method for using a sequence of generalization functions to guide the search for a refutation of depth n or less from a set S of clauses is as follows:

Algorithm A

begin

Let g_0 be some generalization function such that for all clauses C, g(C) $\neq\emptyset$;

Compute $P_{g_0,\ n}(S)$;

For i = 1 to m do

let g_i be a generalization function that interpolates for $P_{g_{i-1},\ n}(S)$;

compute $P_{g_i,\ n}(S)$ od;

Let g be such that g(C) = {C} if there exists D in $P_{g_m,\ n}(S)$

such that C $\leq$ D, g(C) = $\emptyset$ otherwise;

Let U be the set of clauses $P_{g,\ n}(S)$;

Return U;

end;

Theorem 4. Algorithm A is complete, that is, the set U of clauses returned is $P_n(S)$, the clauses contributing to refutations from S of depth n or less.

Proof. By theorem 1, (S, id) $\leq_{P,\ n}$ (S, g_0). By theorem 2, if (S, id) $\leq_{P,\ n}$ (S, g_i) then (S, id) $\leq_{P,\ n}$ (S, g_{i+1}). By theorem 2, (S, id) $\leq_{P,\ n}$ (S, g) since g interpolates for $P_{g_m,\ n}(S)$. By theorem 3, $P_n(S) = P_{g,\ n}(S)$.

Algorithm A is still complete if tautologies are deleted, unlike our previous input abstraction strategies (Plaisted[81]). Also, it is not necessary to use "multiclauses", as in Plaisted[81]; we found that they could substantially increase the search space. It is possible to use Algorithm A together with any complete resolution strategy; to do this, Res(C, D) should be defined as the resolvents permitted by the strategy. For this, tautology deletion is only complete if it is compatible with the strategy; for example, tautology deletion is not compatible with locking resolution (Boyer[71]). Also, the strategy must satisfy the following *compatibility condition:*

$$\text{If } C_1 \leq D_1 \text{ and } C_2 \leq D_2 \text{ then } Res(C_1,\ C_2) \leq Res(D_1,\ D_2).$$

This condition is satisfied by unrestricted resolution. During the generation of g-refutations, it is permissible to delete a clause C if a clause D has been generated already and C $\leq$ D. However, the depth information associated with D will have to be updated if C is at a smaller depth than D. It is sometimes possible to terminate Algorithm A early. If a refutation is obtained, and for all clauses C in this refutation, $g_i(C)$ = {C}, then we know that S is inconsistent and it is not necessary to continue iterating through levels of abstraction. In fact, in this way Algorithm A may be used with infinitely many levels of abstraction, if it is known that for any specific proof eventually this condition on g_i will hold.

4. Examples of generalization functions

Having developed the theory of generalization functions, we now give some examples of such functions. First we define *selection functions.* These select a subset of the subterm occurrences of a clause. From these, generalization functions may be obtained; these replace the occurrences by distinct new variables. Given an arbitrary generalization function h and a set T of clauses, we show how another generalization function g may be obtained that interpolates for T.

Definition. A *selection function* σ is a function from clauses C to sets of occurrences of terms in C. We write $\sigma_1 \subset \sigma_2$ if for all clauses C, $\sigma_1(C) \subset \sigma_2(C)$.

Definition. If σ is a selection function then let $\hat{\sigma}$ be the function from clauses to clauses defined by $\hat{\sigma}(C) = D$ where D is C with all maximal term occurrences in $\sigma(C)$ replaced by distinct new variables. That is, if two subterms s and t are selected, and t occurs in the selected occurrence of s, then s is replaced by a variable. (We could also permit different occurrences of the same term to be replaced by the same variable, but have chosen not to do so.)

Proposition. If σ is a selection function then $\hat{\sigma}$ is a generalization function, and for all clauses C, $\hat{\sigma}(C) \neq \emptyset$. Also, if $\sigma_1 \subset \sigma_2$, then for all clauses C, $\hat{\sigma}_1(C) \leq \hat{\sigma}_2(C)$.

Proposition If g and h are generalization functions then so is their composition defined as the mapping from C to $\cup\{h(D) : D \text{ in } g(C)\}$.

Definition. If g is a generalization function and T is a set of clauses, let g^T be defined by $g^T(C) = \{E : E \text{ in } mgi(C', D), C' \text{ in } g(C), D \text{ in } T, C \leq D, C \leq E\}$. Here "mgi" is the "most general instance" operation, applied to clauses rather than literals. This produces a *set* of clauses.

Note that the computation of g^T given g and T requires the computation of most general instances of pairs of clauses. This is more complicated than computing the most general instances of two literals, but it is still computable (since an instance of a clause cannot contain more literals than the clause itself). However, there may be more than one most general instance of two clauses. For example, the clauses {P(a, x), P(b, y)} and {P(z, a), P(w, b)} have the following two most general instances:

{P(a, a), P(b, b)}, {P(a, b), P(b, a)}

Proposition. If g is a generalization function and T is a set of clauses, then g^T interpolates for T. Also, if g_1 and g_2 are generalization functions, and $T \leq U$, and for all C, $g_1(C) \leq g_2(C)$, then for all C, $g_1^T(C) \leq g_2^U(C)$.

To sum up, we can choose selection functions with great freedom. From a selection function σ, a generalization function $\hat{\sigma}$ may be obtained in a simple way. Finally, from a generalization function g, a generalization function h that interpolates for T may be obtained easily. Therefore, from a sequence of selection functions, we may directly obtain a sequence of generalization functions as required by algorithm A. Also, if σ_i is a sequence of selection functions such that $\sigma_{i+1} \subset \sigma_i$ for all appropriate i, then in this way we may obtain a sequence g_i of interpolation functions as required by algorithm A such that for all C, $g_{i+1}(C) \le g_i(C)$. That is, these interpolation functions are throwing away less and less information, until finally the last one does not throw away any information at all. Note that selection functions can be completely arbitrary; it is not necessary that $\sigma(C)$ and $\sigma(D)$ have any relation to each other for C and D such that $C \le D$.

4.1 Examples of selection functions

Here are some examples of selection functions.

1. Let $\sigma(C)$ be all subterms of C at depth d or greater.

2. Suppose term s is a subterm of term t. Let $\sigma(C)$ be the occurrences of s within t in C, $\emptyset$ if t does not occur in C. That is, σ only selects s if it occurs in the "context" t.

3. Let $\sigma(C)$ be the first arguments of all predicates in C.

4. Let $\sigma(C)$ be the second arguments of all occurrences of f in C.

5. Let $\sigma(C)$ be all subterm occurrences s of C that occur within a specified subterm t, such that s occurs at depth greater than d within t.

Note also that if σ_1 and σ_2 are selection functions, so is there union σ defined by $\sigma(C)=\sigma_1(C)\cup\sigma_2(C)$.

Also, in a literal P(a, a) we need not select both of the occurrences of a; we can select either one independently of the other. We can mimic the propositional abstraction of Plaisted[81] by selecting *all* subterms of a clause; the corresponding generalization function replaces all arguments of all predicates by distinct new variables. We can mimic the operator abstraction of Plaisted[81] by selecting all arguments of all function symbols; the corresponding generalization function replaces all arguments of all function symbols by distinct new variables. However, no input abstraction known to us can select all terms of greater than some specified depth, or all subterms in a given context, or all terms of greater than some specified depth within a given context, et cetera.

4.2 Interactive choice of the generalization functions

Suppose that in a proof about sets many terms of the form $s\cap(t\cap u)$ are generated, and the user does not think that these will contribute to a proof. Then the user can select all occurrences of terms of form $t\cap u$ in the context $s\cap(t\cap u)$; the corresponding generalization function would cause this context to be replaced by $s\cap z$ for some new variable z. This is not the same as deleting all clauses having subterms of the form $s\cap(t\cap u)$. In general, as the user sees the proof progressing, he or she can see what kinds of terms are being generated. In the author's experience, often a combinatorial explosion occurs within a particular kind of subterm structure. When the user sees this happening, he or she can select that structure to be replaced by a variable, reducing the search space. At later levels of generalization, hopefully the combinatorial explosion will not be so severe, since only the clauses contributing to proofs will be retained.

5. Comparison with input abstractions

We now show how this approach compares to the abstraction approach of Plaisted[81]. Suppose we have the following set S of clauses:

{P(a)}
{¬ P(x), P(f(x))}
{¬ P(f(f(f(f(x)))))}

If we select all subterms at depth greater than 1, we get the following proof:

1. P(a)	Input clause
2. ¬ P(x), P(f(y))	Generalized input clause
3. P(f(y))	1, 2
4. ¬ P(f(z))	Generalized input clause
5. NIL	3, 4

Let us call this generalization function g_1. If we select all subterms at depth greater than 2, we get the following proof:

1. P(a)	Input clause
2. ¬ P(x), P(f(x))	Input clause
3. P(f(a))	1, 2
4. P(f(f(x)))	2, 3 using resolution with generalization
5. ¬ P(f(f(z)))	Generalized input clause
6. NIL	4, 5

Let us call this generalization function g_2. If we select all subterms at depth greater than 3, we obtain the following proof:

1. P(a)	Input clause
2. ¬ P(x), P(f(x))	Input clause
3. P(f(a))	1, 2
4. P(f(f(a)))	2, 3
5. P(f(f(f(x))))	2, 4 using resolution with generalization
6. ¬ P(f(f(f(z))))	Generalized input clause
7. NIL	5, 6

Let us call this generalization function g_3. Note that in order to show that $(S, g_3) \le_{P,\ 4} (S, g_1)$ it is necessary to consider proofs involving g_1 resolution in which the same clause is derived more than once. Also, note how we are able to get progressively more and more of the actual proof with the successive generalization functions. However, with input abstraction, we have few choices. If we delete the argument of f, we get the following set of input clauses:

{P(a)}
{¬ P(x), P(f)}
{¬ P(f)}

This leads to a proof having two resolutions. If we retain the argument of f, then we get the original set of input clauses, and a full length proof having 5 resolutions. There is nothing in between. So, the distance between levels of abstraction is large, and the abstract space is not a good guide to finding proofs at the less abstract space. However, the use of generalization functions permits us to control what is happening in the proof itself, and thus gives a closer correspondence between the levels of abstraction.

Let us call a set of clauses *flat* if it contains no function symbols except zero-ary function symbols (constants). Let us call an abstraction flat if it produces a flat set of clauses. Now, a flat abstraction produces proofs in which all clauses are flat. An abstraction which is not flat produces proofs in which the term depth may increase with increasing depth in the proof. There is nothing in between. In contrast, generalization functions permit sets of input clauses that are not flat, but proofs in which the term depth is bounded. By increasing the bound on term depth little by little, we obtain a sequence of generalizations in which each level is close to adjacent levels of generalization. The situation is illustrated as follows:

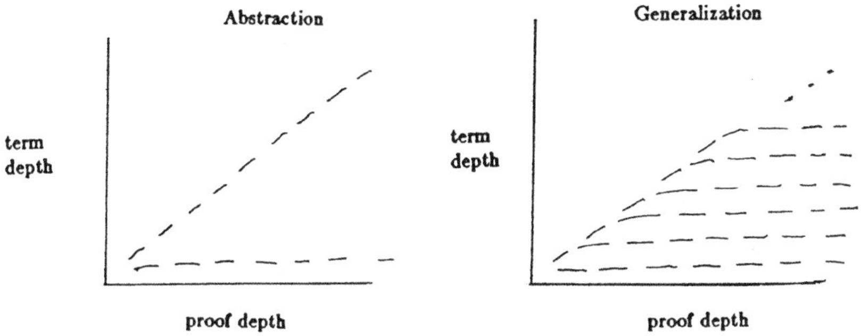

6. Use of instances in proofs

Another advantage of generalization functions is that it permits us to obtain the instances of the clauses actually used in the proofs. In the above example, the instances of the clauses actually used in the ordinary resolution proof are

{P(a)}
{¬ P(a), P(f(a))}
{¬ P(f(a)), P(f(f(a)))}
{¬ P(f(f(a))), P(f(f(f(a))))}
{¬ P(f(f(f(a)))), P(f(f(f(f(a)))))}

$\{\neg P(f(f(f(f(a)))))\}$

Without going into details, if g is a generalization function, then given a g-refutation from S it is possible to obtain the instances of S actually used in the g-refutation. Suppose T is the set of instances used in g-refutations from S of depth n or less, and (S, id) $\leq_{P, n}$ (S, g). Then one can show that T ≤ S, and if there is a refutation from S of depth n or less, there is a refutation from T of depth n or less. Therefore if the depth bound is n or less, it suffices to look for refutations from T instead of S. This can be helpful, since T will have more bindings of variables than S, and some resolutions may not be possible because of conflicting bindings. Similarly, we can obtain the instances of all lemmas in the proof, that are actually used in the proof. That is, if a clause D is not an input clause, but appears in a refutation, then some instance or instances $D\Theta$ of D will be actually used in the proof. Corresponding to the set $P_{g, n}(S)$ of clauses appearing in g-resolution proofs of depth n or less, we may define the set $I_{g, n}(S)$ of instances appearing in g-resolution proofs of depth n or less. Some experiments with a strategy implemented at the University of North Carolina have indicated that this idea may be helpful on some hard problems. Using this idea, we obtain the following proof procedure.

Algorithm B

begin

Let g_0 be some generalization function such that for all clauses C, g(C) ≠∅;

Compute $I_{g_0, n}(S)$;

Let T_0 be instances of S actually used in these proofs;

For i = 1 to m do

let g_i be a generalization function that interpolates for $I_{g_{i-1}, n}(T_{i-1})$;

compute $I_{g_i, n}(T_{i-1})$;

let T_i be instances of S actually used in these proofs od;

Let g be such that g(C) = {C} if there exists D in $I_{g_m, n}(T_m)$

such that C ≤ D, g(C) = ∅ otherwise;

Let U be the set of clauses $P_{g, n}(T_m)$;

Return U;

end;

7. Application to other inference systems

The same ideas apply to other proof systems based on skolemized formulae. The general idea is that after every inference, a generalization operator is applied. This gives "generalized proofs." By using a sequence of generalization operators, these generalized proofs may be gradually converted to non-generalized proofs. This may be done as in Algorithm A, by choosing each generalization operator to interpolate for the set of clauses (or corresponding structures) used in the previous level of abstraction. This requires that the inference system satisfy a compatibility condition similar to that given for resolution in section 3.1. Typical systems based on unification will satisfy such a condition. In this way, we believe that these techniques will apply to the Prolog technology theorem prover approach of Stickel[84], the simplified problem reduction format of Plaisted[82], and versions of connection graph proof procedures mentioned in Bibel[82]. The first two are methods for extending the back chaining search strategy of Prolog to non-Horn clauses. It is also possible that these techniques apply to the ''mating'' strategy of Andrews[81]. We do not yet have a technique for applying generalization operators to formulae containing explicit universal and existential quantifiers, although it seems likely that one exists.

8. Conclusions

Generalization abstractions hold promise of overcoming some of the problems with input abstractions and providing guidance in the search for refutations from sets of clauses. Note that the formalization of generalization abstractions given here is much more concise and elegant than that of abstractions given in Plaisted[81]. In addition to the potential for practical applications, the mathematical simplicity and elegance of these ideas is appealing in itself. These techniques may warrant a wider distribution because of their mathematical simplicity and generality and because of the insight they give into the structure of proofs. Our experience with input abstractions has shown some potential, and generalization abstractions should have more. It would be useful to have these in a theorem prover's ''tool kit'' for experimentation purposes. Also, because of their simplicity, implementation of these ideas should not be difficult.

9. References

Andrews, P.B., Theorem proving via general matings, J. ACM 28 (1981)193 - 214.

Bibel, W., Automated Theorem Proving, Vieweg, 1982.

Boyer, R., Locking, a restriction of resolution, Ph.D. thesis, University of Texas at Austin, TX (1971).

Bundy, A., The Computer Modelling of Mathematical Reasoning (Academic Press, New York, 1983).

Chang, C. and Lee, R., Symbolic Logic and Mechanical Theorem Proving (Academic Press, New York, 1973).

Kling, R., A paradigm for reasoning by analogy, Artificial Intelligence 2 (1971) 147-178.

Loveland, D., Automated theorem proving: a quarter century review, in Automated Theorem Proving: After 25 Years, W. Bledsoe and D. Loveland, eds., (American Mathematical Society, Providence, RI, 1984), pp. 1-45.

Manna, Z., Mathematical Theory of Computation (McGraw-Hill, New York, 1974).

Plaisted, D., and Greenbaum, S., Problem representations for back chaining and equality in resolution theorem proving, First Annual AI Applications Conference, Denver, Colorado, December, 1984.

Plaisted, D., Theorem proving with abstraction, Artificial Intelligence 16 (1981) 47 - 108.

Plaisted, D., A simplified problem reduction format, Artificial Intelligence 18 (1982) 227-261.

Robinson, J., A machine oriented logic based on the resolution principle, J. ACM 12 (1965) 23-41.

Stickel, M.E., A Prolog technology theorem prover, Proceedings of the 1984 International Symposium on Logic Programming, IEEE, Atlantic City, New Jersey, February, 1984, pp. 212-217.

Winston, P.H., Artificial Intelligence (Addison-Wesley, Reading, Mass., 1977).

Wos, L., Overbeek, R., Lusk, E., and Boyle, J., Automated Reasoning: Introduction and Applications (Prentice-Hall, Englewood Cliffs, NJ, 1984).

An Improvement of Deduction Plans:
Refutation Plans *)

Hans-Albert Schneider
Universität Kaiserslautern
Fachbereich Informatik
D-6750 Kaiserslautern

Abstract

A theorem proving method is described which takes the advantages of the methods of Cox and Pietrzykowski (called deduction plans) and of Shostak (called refutation graphs) and avoids some of their disadvantages. The method separates deduction and unification, thus allowing both AND- and OR-parallelism and directed backtracking. It even allows to drop the solution of a literal L without dropping it as a solution of literals factored to L. In addition, the unification algorithm used allows simple and directed backtracking of the unifiers, too.

I. Introduction

One of the problems automated theorem provers are concerned with is the handling of unification conflicts. This usually is done by revising the last decision in the search tree and trying another choice, if any. If there is none, the procedure applies to the previous one, and so on.

However, if the conflict is caused by a decision made many steps before, this results in a useless and exhaustive search before revising the decision which was wrong. If we knew the deduction steps responsible for the conflict, then we could direct backtracking to them and save a lot of time.

This idea was realized by Cox, Pietrzykowski and Matwin [COP 81,PIM 82]. They separated deduction and unification building an inference graph (the deduction plan) and a unification graph [MAP 82]. From the latter, a set of deduction steps can be computed the removal of which will also remove all unification conflicts.

---

*) Project supported by the Deutsche Forschungsgemeinschaft under grant number Ma 581/4-1

The deduction plan method is a refutation method. It includes five inference rules according to which the deduction plan is built up. Using the namings of [PIM 82] the substitution rule (it is similar to an input resolution step) and the reduction rule (similar to an ancestor resolution step) are a complete subset of them. The other rules allow factoring to unresolved and to resolved literals and the use of lemmata.

There is, however, a problem: if a proof of a literal L has to be removed due to unification conflicts, there might be literals factored to L, and the proof might be without conflicts for some of them. In this case, some arcs must be replaced by arcs pointing to another node (note that there may be chains of factoring arcs) and/or changed in their type, or this proof might have to be reconstructed later on. On the other hand, this problem would not occur in the refutation graphs of Shostak [SHO 76].

Combining these two approaches, the author has developed a theorem proving method which inherits their advantages stated above. It also allows parallel processing (which is not possible in refutation graphs) and even AND-parallelism. Together with the unification algorithm presented in [DIJ 83,DIJ 84] directed backtracking of the unification can be done in a simple way.
In the second section the inference graph (called refutation plan) is defined and then some results concerning its soundness and completeness are stated (section III). Finally, the fourth section gives an example.

II. Refutation Plans

Let A be an atom. If L = A or L = ¬A, then L is a literal. We define

$$|L| := A, \text{ and } sgn(L) = \begin{cases} +1 & \text{if } L = A \\ -1 & \text{if } L = \neg A. \end{cases}$$

Let S be the set of input clauses. Let

$$B(S) = \{P(x_1,\dots,x_n),\neg P(x_1,\dots,x_n) \mid P \text{ is a predicate symbol of some literal in } S,\ x_1,\dots,x_n \text{ are variables not occurring in } S\}.$$

Obviously, each literal in B(S) is unifiable with every literal that occurs in S and hs the same predicate symbol and sign. A deduction step will not insert an edge between two clauses but between a clause and a bridge (an element of B(S)). This allows better backtracking in the presence of factorization.

A refutation plan of S is a graph G = (V(G),B(G),I(G),R(G)), where V(G) consists of *nodes* (clauses), B(G) of *bridges*, I(G) of *labels*, and R(G) consists of *ramps* (edges). A label is a pair (t,c), where $t \in$ {SUB,RED,FACT} is the type of the arc, and c is a literal. V(G) ∪ B(G) are the vertices of G, R(G) its edges, $R(G) \subset V(G) \times I(G) \times B(G) \cup B(G) \times I(G) \times V(G)$. The edges are directed.
The refutation plan has to be built according to the following rules 0) - 3), which also define the set of all *open subproblems* (unresolved literals) os(G) and the *constraint set* C(G) (a constraint is a set of two terms which are to be unified).

0) G is a *base plan*, if

   $V(G) = \{\tau C\}$ for some $C \in S$ and a renaming τ,

   $B(G) = \emptyset$

   $R(G) = \emptyset$.

   For the base plan, $os(G) = \tau C$ and $C(G) = \emptyset$. τC is called the *base node*.

I.e., the base node is any input clause.

1) *Substitution Rule*

   Let G be a refutation plan, $C \in V(G)$, $c \in C \cap os(G)$, $D \in S$, $d \in D$, such that $sgn(c) = -sgn(d)$ and $|c| = P(s_1,\ldots,s_n)$, $|d| = P(t_1,\ldots,t_n)$ for some terms $s_1,\ldots,s_n$, $t_1,\ldots,t_n$. Let $K \in B(S)$ such that $|K| = P(x_1,\ldots,x_n)$ and $sgn(k) = sgn(c)$. Let τ be a renaming, such that V(G), τC, and τK have no variables in common.

   Define $V(G') := V(G) \cup \{\tau D\}$,

   $B(G') := B(G) \cup \{\tau K\}$,

   $R(G') := R(G) \cup \{(C,(SUB,c),\tau K),(\tau K,(SUB,\tau d),\tau D)\}$,

   $os(G') := (os(G) - \{c\}) \cup (\tau D - \{\tau d\})$,

   $C(G') := C(G) \cup \bigcup_{i=1}^{n} \{\{x_i,t_i\},\{x_i,s_i\}\}$.

   Then G' is a refutation plan.

I.e., choose an open subproblem c of the plan, and solve it using a variant of an input clause D with a literal d having the same predicate symbol and opposite sign. Note that no unifiability is required.

2) *Reduction Rule*

Let G be a refutation plan, C $\in$ V(G), c $\in$ C $\cap$ os(G) and K $\in$ B(G) such that sgn(c) = -sgn(K) and |c| = $P(t_1,\dots,t_n)$, |K| = $P(x_1,\dots,x_n)$. Let K be a strong ancestor of C, that is, every path from the base node to C contains K or it contains an arc labelled RED.

Define
$$V(G') := V(G),$$
$$B(G') := B(G),$$
$$R(G') := R(G) \cup \{(K,(RED,c),C)\},$$
$$os(G') := os(G) - \{c\},$$
$$\mathbf{C}(G') := \mathbf{C}(G) \cup \bigcup_{i=1}^{n} \{\{x_i,t_i\}\}.$$

Then G' is a refutation plan.

This rule is similar to ancestor resolution: there is a literal c which can be resolved with an ancestor. This ancestor, say a, does not occur here, but there is an arc (A,(SUB,a),K) to the bridge K.

3) *Factorization Rule*

Let G be a refutation plan, C $\in$ V(G), c $\in$ C $\cap$ os(G), K $\in$ B(G) such that sgn(c) = sgn(K) and |c| = $P(t_1,\dots,t_n)$, |K| = $P(x_1,\dots,x_n)$.

Define
$$V(G') := V(G),$$
$$B(G') := B(G),$$
$$R(G') := R(G) \cup \{(C,(FACT,c),K)\},$$
$$os(G') := os(G) - \{c\},$$
$$\mathbf{C}(G') := \mathbf{C}(G) \cup \bigcup_{i=1}^{n} \{\{x_i,t_i\}\}.$$

If for every D $\in$ V(G) and M $\in$ B(G), such that M is a strong ancestor of D in G, M also is a strong ancestor of D in G', then G' is a refutation plan.

I.e., c is factored to a literal a such that (A,(SUB,a),K) is an arc of G.

G is *closed*, if os(G) = $\emptyset$. G is *correct*, if **C**(G) is unifiable. In this case the most general unifier of **C**(G) is denoted θ(G). θ(G)os(G) is the *clause derived by* G.

The rules correspond to those of [COP 81] as follows: rule 1 corresponds to simple replacement, rule 2 corresponds to reduction, and rule 3 to backfactoring. They also correspond to those of [SHO 76]: rule 1 to extension, rule 2 to A-literal reduction, and rule 3 to C-literal reduction.

III. Some Results

It can be shown [SCH 86] that for every deduction plan $\tilde{G}$ of a set S of clauses there exists an acyclic refutation plan G such that $V(G) = V(\tilde{G})$, $os(G) = os(\tilde{G})$, C(G) is unifiable iff $C(\tilde{G})$ is unifiable. The converse also holds. The proof is constructive. (Note that $C(\tilde{G})$ has not been defined in this paper, but the definition is analogous to C(G). The definition of constraint is slightly different in [COP 81]: there, a constraint is an unordered pair of literals, here, it is an unordered pair of terms).
As a consequence, the following theorem holds:

Theorem
Let S be a set of clauses.
There is a closed correct acyclic refutation plan of S iff there is a closed correct deduction plan of S.

Together with the results of [COP 81], the refutation plan method is sound and complete:

Theorem
Let S be a set of clauses.
S is unsatisfiable iff there is a closed correct acyclic refutation plan of S.

IV. An Example

Consider the situation in figure 1.
There are several unification conflicts, e.g. u_2 has to be unified with $h(u_2)$ (by: $u_2 - x_1 - x_5 - y_1 - h(u_2)$, where $x_1 - x_5$ is due to arc 4), with $g(y_5)$ and $g(f(x_3,y_3))$. The conflicts can be removed by dropping edge 4, but L_5 is a solution for L_2 and L_4.

Figure 2 shows the corresponding part of a refutation plan. (The arcs are labelled with their type only, the literal of the label is the literal at the beginning or end of the respective arc). The unification conflicts, of course, are the same but we can remove them by deleting arcs 2 and 4a and thus save L_5 as a solution of L_2 and L_4. Because there is no SUB-arc ending in the bridge $L(x',y',z')$ we have to select one of its ingoing

FACT-arcs, say 3, and relabel it SUB. It is easy to see that the corresponding operation is more complex in Fig. 1.

Similarly removal of arc 4b can be repaired by relabelling arc 5 as SUB: this, however, involves the necessity to redirect all arcs labelled SUB along a path from the clause containing L_5 to L_6 which consists of SUB-arcs only (this could be done very similar in fig. 1).

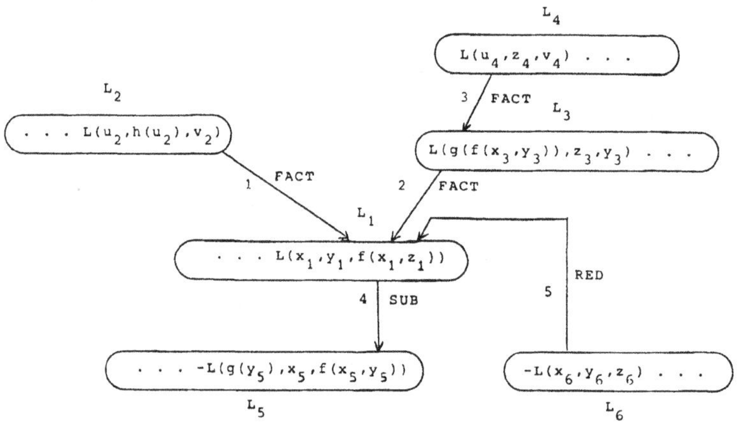

Fig. 1: A sample part of a deduction plan.

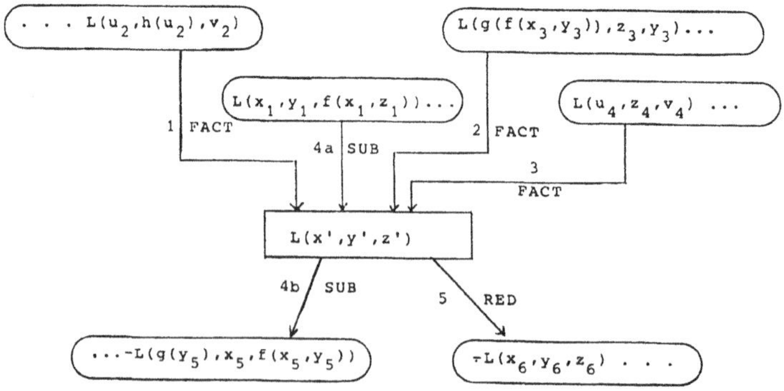

Fig. 2: The refutation plan corresponding to fig. 1.

V. Conclusion

A theorem proving method has been presented which allows directed backtracking. Due to the separation of deduction and unification it also provides the ability of AND-parallel processing.

This method and the unification algorithm we use are well-suited for parallel processing, especially for use with content addressable memories. Therefore, the model of an associative processor has been developed for it [DIS 85] and a simulation program is being implemented.

References

COP 81 Cox, P.T., Pietrzykowski, T.:
Deduction Plans: A Basis for Intelligent Backtracking
IEEE Trans. Pattern Analysis and Machine Intelligence, Vol., PAMI-3, Jan. 1981, pp. 52-65

DIJ 83 Dilger, W., Janson, A.:
Unifikationsgraphen für intelligentes Backtracking in Deduktionssystemen
Proc. GWAI-83, Informatik-Fachberichte 76, Springer Verlag Berlin 1983, pp. 189-196

DIJ 84 Dilger, W., Janson, A.:
A Unification Graph with Constraints for Intelligent Backtracking in Deduction Systems
Interner Bericht 100/84 Univesität Kaiserslautern, Fachbereich Informatik 1984

DIS 85 Dilger, W., Schneider, H.-A.:
A Theorem Proving Associative Processor
Interner Bericht 145/85, Fachbereich Informatik, Universität Kaiserslautern, 1985

PIM 82 Pietrzykowski, T., Matwin, S.:
Exponential Improvement of Efficient Backtracking: A Strategy for Plan-Based Deduction
Proc. 6th Conf. on Automated Deduction, Lecture Notes in Computer Science, New York 1982, pp. 223-239

SCH 86 Schneider, H.-A.:
Refutation Plans: Definition, Soundness, and Completeness.
Interner Bericht, Universität Kaiserslautern, Fachbereich Informatik, in preparation

SHO 76 Shostak, R.E.:
Refutation Graphs
Artificial Intelligence, Vol. 7 (1976), pp. 51-64.

Controlling deduction with proof condensation and heuristics

F. Oppacher and E. Suen
School of Computer Science
Carleton University
Ottawa, Ontario, Canada, K1S 5B6

Abstract

This paper presents a proof condensation (or redundancy elimination) procedure and heuristic rules that are used to enhance the tableau-based theorem prover HARP. The proof condensation procedure makes proofs easier to construct and more readable by excising redundancies from proof trees. Since the entire language of first-order logic is used without preprocessing, heuristics can be formulated to capture efficient and human-like deduction strategies. We also present evidence that the combination of these two techniques enable HARP to solve challenging problems such as Schubert's Steamroller.

1. Introduction

This paper describes a complete theorem prover for first-order predicate logic that combines the semantic tableau technique due to [Be65], [Hi55],and [Sm68] with declaratively expressed heuristic strategies . The resulting proof procedure HARP (Heuristics Augmented Refutation Procedure) is "natural" and easy to use for people (see Section 2) as well as computationally efficient (for example, as will be pointed out in section 2, HARP seems to be the only theorem prover that can solve the unmodified, nonclausal version of Schubert's Steamroller Problem [Wa85]).

Before discussing the underlying design goals we shall briefly describe the method of semantic tableaux.

HARP is based neither on variants of the resolution principle nor on natural deduction methods. Instead it implements a version of the semantic idea that proving a formula amounts to an unsuccessful attempt to construct a falsifying model for it. This approach to proving by model construction was first studied, although not in the context of automated theorem proving, by [Be65], [Hi55], and [Sm68]. The method provides a comprehensive test for consistency and, hence, for inconsistency, logical truth and logical entailment.

It is convenient to represent models by binary trees. To prove that $\Phi_1,\ldots,\Phi_n$ imply Φ, the set $\{\Phi_1,\ldots,\Phi_n, \neg\Phi\}$ is assumed to be satisfiable and put at the root of the tree. From the satisfiability assumption smaller and smaller (weak) subformulas are derived. If the assumption of semantic consistency is substantiated, i.e. if there is at least one truth-value assignment to all the atomic subformulas of $\Phi_1,\ldots,\Phi_n,\neg\Phi$ on which the former are simultaneously true, then a counterexample to the claim that $\Phi_1,\ldots,\Phi_n$ imply Φ can be directly retrieved from the constructed model. However, if the assumption of consistency turns out to be untenable, then the method will ascertain this in a finite number of steps. In other words, the method is complete.

The tree is grown by entering on its branches the results of applying heuristically selected decomposition rules to sentences previously entered. The decomposition rules (see below) are similar to natural deduction elimination rules. When such a rule is applied to any compound sentence that is not a universal quantification the compound is checked off to prevent further rule applications to it and its subformula or subformulas are attached to the end of each open, i.e. consistent branch which passes through the compound. As soon as a contradiction turns up on a branch, the latter is said to be closed and marked with a "*".

A tree is said to be closed if all its branches are closed, and is said to be open otherwise. A branch is said to be complete if it contains either no unchecked compounds or the universal quantifications on it have already been instantiated with respect to all individual parameters appearing anywhere on the branch. A tree is said to be complete if all its branches are complete.

In the tree calculus all sentences are classified as either elementary, i.e. atoms or negations of atoms, or as of type α, β, γ or δ. Sentences of type α, the conjunctive type, are those truth-functionally complex sentences from whose truth one can uniquely infer the truth values of their immediate subsentences. Sentences of type β, the disjunctive type, are all the other truth-functionally complex sentences. Sentences of type γ, the universal type, are those quantificationally complex sentences from whose truth one can uniquely infer the truth values of their instantiations. Sentences of type δ, the existential type, are all the other quantificationally complex sentences.

The semantic rationale for this classification lies in the fact - used in the completeness and correctness proofs - that for each truth value assignment a the following holds:

a satisfies $\alpha \Leftrightarrow a$ satisfies α_1 and a satisfies α_2;
a satisfies $\beta \Leftrightarrow a$ satisfies β_1 or a satisfies β_2;
a satisfies $\gamma \Leftrightarrow a$ satisfies $\gamma[\tau]$ for each parameter τ;
a satisfies $\delta \Leftrightarrow a$ satisfies $\delta[\tau]$ for some parameter τ.

It is perhaps the most remarkable property of the tableau approach that its syntactic decomposition rules precisely mirror the downward ($\Rightarrow$) direction of the above semantic evaluation rules. The natural style of proof construction in HARP is largely due to this property.

In the following,the usual predicate-logical definitions of terms and well-formed formulas are assumed. A "|" in a rule indicates a branch in a tree. We shall use "$\forall$" and "$\exists$" for the universal and existential quantifiers, respectively.

The rules are to be understood as follows: from a formula of the structure shown above the line, derive a formula (or formulas) of the structure shown below the line.

Type α:

$$\frac{\neg\neg\,\Phi}{\Phi} \qquad \frac{\Phi \wedge \Psi}{\begin{matrix}\Phi \\ \Psi\end{matrix}} \qquad \frac{\neg(\Phi \vee \Psi)}{\begin{matrix}\neg\Phi \\ \neg\Psi\end{matrix}} \qquad \frac{\neg(\Phi \Rightarrow \Psi)}{\begin{matrix}\Phi \\ \neg\Psi\end{matrix}}$$

Type β:

$$\frac{\Phi \vee \Psi}{\Phi \mid \Psi} \qquad \frac{\Phi \Rightarrow \Psi}{\neg\Phi \mid \Psi} \qquad \frac{\neg(\Phi \wedge \Psi)}{\neg\Phi \mid \neg\Psi} \qquad \frac{(\Phi \Leftrightarrow \Psi)}{\begin{matrix}\Phi \mid \neg\Phi \\ \Psi \mid \neg\Psi\end{matrix}} \qquad \frac{\neg(\Phi \Leftrightarrow \Psi)}{\begin{matrix}\Phi \mid \neg\Phi \\ \neg\Psi \mid \Psi\end{matrix}}$$

Type γ:

$$\frac{\forall\zeta\phi(\zeta)}{\phi(\tau)} \qquad \frac{\neg\exists\zeta\phi(\zeta)}{\neg\phi(\tau)} \qquad \text{with proviso (1).}$$

Type δ:

$$\frac{\exists\zeta\phi(\zeta)}{\phi(\tau)} \qquad \frac{\neg\forall\zeta\phi(\zeta)}{\neg\phi(\tau)} \qquad \text{with proviso (2).}$$

Proviso (1): Let $\phi(\zeta)$ be any schema and ζ any variable which may or may not occur free in $\phi(\zeta)$. Let $\phi(\tau)$ be the result of replacing all free occurrences, if any, of ζ in $\phi(\zeta)$ by the term τ. The rule of universal instantiation permits the inference from $\forall\zeta\phi(\zeta)$ to $\phi(\tau)$, provided that no free occurrence of ζ in $\phi(\zeta)$ is within the scope of an occurrence of $\forall\tau$ or $\exists\tau$.

Proviso (2): Let $\phi(\zeta)$ and $\phi(\tau)$ be as before.When an existential quantifier is dropped the instantiating term τ must be an individual parameter new to the entire branch. In the implementation, it is easier to use a constant foreign to the entire tree.

In principle, a universal quantification is instantiated with respect to every individual term anywhere on its branch unless such an instance is already on the branch. In practice, HARP relies on its full indexing scheme and heuristics to try first only those instantiations which are most likely to contribute to the early closing of a branch. Only if no individual parameter has appeared is a new one chosen for the instantiation. Unlike all other compounds, a universal quantification is not checked off and, thus, can be used repeatedly.

2. Design goals

Since HARP is intended to be used both interactively and as an inference engine for AI applications, its construction is influenced by the overall design goals of a) naturalness, b) efficiency, c) usefulness in an AI environment, d) partial specification of the control structure by heuristic rules.

a) The goal of naturalness dictates the use of the full language of first order logic and a natural style of proof construction. Our algorithm seems to be unique in that it requires no conversion to any canonical form and accepts its (fully parenthesized) input without any preprocessing. For example, Bibel's system [Bi82] presupposes that negations are driven inward until they apply only to atoms. Even nonclausal resolution [Mu82] which is claimed to be least restricted in truth-functional form, requires quantifier-free formulas.

The two major disadvantages of the conversion to clausal form required by many other theorem provers have been pointed out by various authors (see e.g. [Ni80]). First, by splitting a given premise into several, apparently unrelated clauses, conversion to clausal form may lead to a loss of potentially valuable control information and make it difficult to attach heuristic advice to individual premisses. Second, the use of clausal form leads to multiplicative growth of the number of literals. For example, converting the formula $(p \wedge q \wedge r) \vee (s \wedge t \wedge u)$ to clausal form yields 18 literals (9 clauses with 2 disjuncts each). HARP, on the other hand, will never generate more than 6 nodes for the literals present in the original formula (and since it decomposes nothing before embarking on a proof it might not even need to decompose the formula at all).

A natural style of proof should deploy principles of inference commonly used by people. Therefore we have opted for normal quantifier instantiation rules instead of unification and for natural-deduction-like rules instead of resolution or connection graph methods which are favored in current approaches[Lo78], [Bi82]. We feel that the method of analytic tableaux [Sm68] provides an extremely natural and elegant codification of logic. It also allows for a very lucid semantic adequacy proof [St84], [Fi83]. Many introductory logic texts also rely on tableaux as a didactic medium (eg. [Je67]). Even Robinson uses tableaux to introduce the idea of resolution [Ro8].

The natural flavor of the tableau calculus is due to the fact, mentioned in Section 1, that its rules are syntactic reformulations of semantic evaluation rules [St84]. In addition, the tableau calculus is a paradigm of a cut-free or analytic calculus, i.e. to prove a sentence p one only considers its weak subsentences where a weak subsentence of p is either a subsentence of p or the negation of such a subsentence. This contrast with resolution systems, in which the length of a derived clause may exceed that of either parent clause, is exploited to provide increased efficiency in our system. The analytic property also makes it possible to keep the basic control structure of the algorithm simple. By comparison, natural deduction systems would seem to face harder control problems due to the fact that they need to maintain some machinery for formally distinguishing between assumptions and assertions and for introducing and discharging assumptions. Furthermore, the tableau calculus can be easily extended to modal and other nonclassical logics [Fi83].

b) The addition of an indexing scheme and heuristics to the basic control structure of analytic tableaux [Sm68] results in a highly efficient proof procedure. HARP's empirical efficiency is demonstrated by the solution of accepted benchmark problems, in particular Schubert's Steamroller[Wa85].

HARP is currently implemented in approximately 3000 lines of ZetaLisp on a Symbolics 3600. It has solved many of the traditional test problems for theorem provers. These problems include the monkey and banana problem, the plane geometry problem [Lo78], problems from [Bl83], and Schubert's Steamroller.

Schubert's Steamroller is a challenge problem that had been unsolved for 6 years until recently in [Wa85]. The quickest solution time of 7.11 seconds [Wa85] was achieved by using a many-sorted clausal version of the problem (a further refinement of this approach, a polymorphic many-sorted logic, is described in [Co85]).

HARP is able to solve the original, nonclausal version of Schubert's Steamroller in 14 seconds on a Symbolics 3600. The resulting proof tree has 48 closure points and is built without any special purpose reasoning mechanisms and without any domain-dependent knowledge. It should be noted that the techniques used in converting traditional resolution theorem provers to many-sorted logic are equally applicable to tableaux. For example, the instantiation mechanism could be restricted by using a sort hierarchy of the universe. It is interesting to note that HARP has a solution time comparable to solutions using these techniques. Furthermore HARP seems to be the only theorem prover able to solve the original, nonclausal version of the problem.

Solutions not using a many-sorted version of the problem have also been reported [Wa85]. Without using special knowledge, Stickel's theorem prover has a reported solution time of 2 hours and 53 minutes [Co85]. The ITP system has a reported solution time of 660 seconds while using a limited many-sorted type logic[Wa85], [Co85].

c) Since HARP is intended to work in an AI environment, it has to cope with situations in which there are many irrelevant premisses and requests to prove nontheorems. An indexing scheme allows HARP to select relevant premisses and heuristics are used to recover quickly from many attempts to prove nontheorems. To date little work has been done on the use of heuristics to detect nontheorems. A brief discussion of this issue will be given in section 5.

d) Because of the experimental nature of the use of heuristics in theorem proving, a control structure modifiable by a set of explicitly represented heuristics is preferable to a rigid architecture where the heuristics are hardwired into the control structure. These heuristics can be classified into the following three groups (see Section 5): heuristics to enable an efficient and humanlike proof construction, heuristics to detect nontheorems, and domain dependent heuristics (e.g. heuristics for equality, commutativity, etc.). We also feel that this modifiable control structure is crucial to our current investigation of strategy - learning theorem provers (eg. [Co81]). However this issue will not be discussed in this paper.

3. The basic algorithm

The tree calculus described above is implemented with modifications as the basic loop of HARP :

```
Initialization
Loop until proved or nontheorem
        Select a node from the priority queue
        Decompose the node  -- prioritize the descendants using heuristics
        if a branch closure results
        then  condense the proof
              if there are more branches
              then  prepare for the next branch
              else  proved endif endif
        if current branch is complete, according to heuristics
        then  nontheorem endif
Endloop
```

In the preparatory stage each formula is internally represented as a node. All formulas are put into a completely parenthesized format by making the connective precedences explicit. This enables efficient, O(1) time, node decomposition. All decomposable nodes are kept on a priority queue. The priority of a node is determined via heuristics (see section 5 for examples). These particular heuristics use the information provided by the indexing scheme.

The indexing scheme is implemented by a connection graph-like mechanism . Links are kept between node pairs which consist of corresponding conjugate literals. Conjugate literal links are determined using a matching procedure which is a relaxed form of unification. Universally quantified variables match anything; constants match only the same constants; existentially quantified variables match only the same existentially quantified variables within the same scope. Any literals that are unifiable will also match using our procedure. However since we do not eliminate existential quantifiers, occasionally links will be made between nonunifiable literals.

Each link is labelled with a weight referred to as the link strength. The link strength is currently calculated by a heuristic using the formula $\lceil 32/2^n \rceil$ where n is the combined sum of the β split level of each literal. The β split level of a subformula is the number of β node decompositions necessary to reach the subformula from the top level formula. Since in a dyadic tree branching increases exponentially with the height of the tree, we have decided to express the strength of a link as inversely exponentially related to the number of β node decompositions necessary to reach the closure point specified by that link.

If one of the pair is a γ node then the link will supply a candidate parameter for universal instantiation when our matching procedure matches the universally quantified variable with the parameter. In contrast, [Wr84] colors the link with the most general unifier. The same heuristic is used to calculate the link strength. For example, given nodes **∀xFx, ¬Fb, Fa,** the γ node **∀xFx** forms only one link to **¬Fb. b** is also specified as a candidate parameter along with the link strength 32. No link to **Fa** is built because instantiation with **a** would not lead to closure.

The current heuristic which determines the priority of a node simply sums up the link weights of all the links associated with a node.

Decomposition of a node is performed by the rules in section 1. In the special case when an universal formula is decomposed, our universal instantiation procedure is called to produce a constant which is literally substituted (i.e. using a modified version of the Lisp Subst function) for occurrences of the universally quantified variable. Since the formulas are completely parenthesized, no unnecessary memory allocation or formula duplication is performed.

β node decompositions spawn a new branch of the tree. The ability to return to the state of computation at which the branch was created is provided by remembering the changes made since the most recent branch point. Upon return to this branch point, these changes are undone. Thus instead of saving the entire proof state, only the changes are noted.

If literals are produced by a decomposition then checking for a possible branch closure is done in O(1) time. Since each literal is treated as a symbol, checking for a closure simply amounts to checking for the existence of the conjugate symbol of each of the produced literals.

4. Proof condensation

In this section we will augment the basic tableau control structure with a proof condensation procedure. Like the heuristics described in the next section, this procedure makes proof construction more efficient and easier to follow.

The objective of the proof condensation procedure is to eliminate redundancies in the proof tree. It is called whenever a branch can be closed with a pair of conjugate literals. The algorithm searches for and eliminates unnecessary branch points. During the process of proof condensation, the proof tree is reshaped to highlight additional unnecessary branch points.

A branch point is a decomposed β node in the tableau tree. A node **a** is an ancestor of a node **b** if node **b** was produced by a series of decompositions from node **a**. A branch closure point (node) is created from two conjugate literals of the same branch. A node is an ancestor of a branch closure point if it is an ancestor of one of the conjugate literals of the branch closure point. In this case we say that the branch closure point depends on that node. For example in the first tree of Fig. 1 the beta node **f∨g** is an ancestor of the closure point (f, ¬f) in the leftmost branch. A branch point is necessary if and only if it is an ancestor of at least one branch closure from each of its two subtrees.

Because a decomposition operation may produce more than one literal and since also the same literal may be present several times on a single branch, several candidate closure points could exist for the same branch. If there are several competing candidates, then the closure node is chosen that depends on the least number of branch points which previously had no dependents. This strategy facilitates the effective operation of the proof condensation procedure. Other strategies are also considered (eg. strategies that favor closure points with dependencies closer to the tree root).

The algorithm has been implemented by using reference counts. Each branch point has an associated reference count that determines whether the node is necessary for completing the proof. The reference count denotes the number, in the current subtree, of dependent closure points for the branch point. If the reference count is zero after closing one of the subtrees of the branch point then that branch point can be eliminated. This algorithm exploits the ordered dyadic tree property of analytic tableaux (i.e. each branch point has a left subtree and a right subtree). Because of space limitations, we will not explain the mechanism of maintaining and updating reference counts but will present the basic outline of the algorithm along with some examples below (in Figures 1 - 3).

Descendant branches of a branch point are those created below that branch point. Two types of branch removals are possible: (A) removal of a branch point whose descendant branches are all closed and (B) removal of a branch point with some open descendant branches. Both these removals shorten the resulting proof and thus yield a more transparent output. Type A removals are initiated from the closing of the rightmost branch of the subtree and do not immediately decrease the number of open branches left to close. But they increase the chances of other removals (both type A and type B) since reference counts are updated when unnecessary subtrees are eliminated. Type B removals do immediately decrease the amount of work left to be done but they do not increase the chances of further removals higher in the tree.

A type A branch point removal is always initiated from the right subtree of that branch point. It is triggered from previous proof condensation, closure point construction, or a combination of both. Removals are performed by the retraction of the corresponding β node decomposition choice.

Type A removal

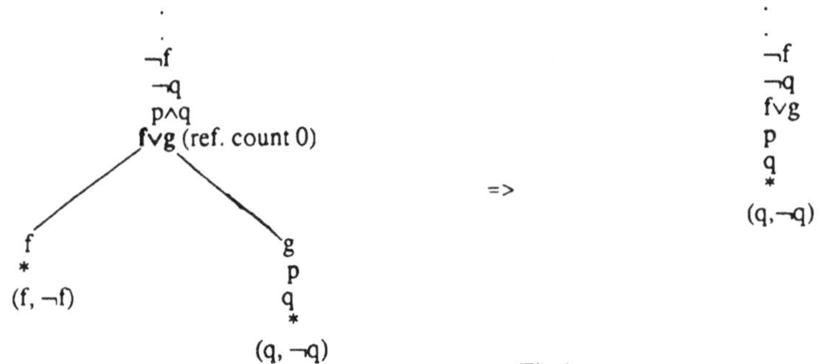

Fig.1

The branch closure point of (q,¬q) does not depend on the β node f∨g. Therefore this β node can be eliminated. Since the subtree spawned by this β node has already been fully closed, this removal does not immediately reduce the number of open branches left to close. Notice that proof condensation could not have been performed when the leftmost branch was closed because the branch closure point (f,¬f) was a descendant of the β node f∨g.

Type B removals are initiated from the left subtree by similar events (i.e. previous proof condensation, closure point construction, or a combination of both).

Type B removal

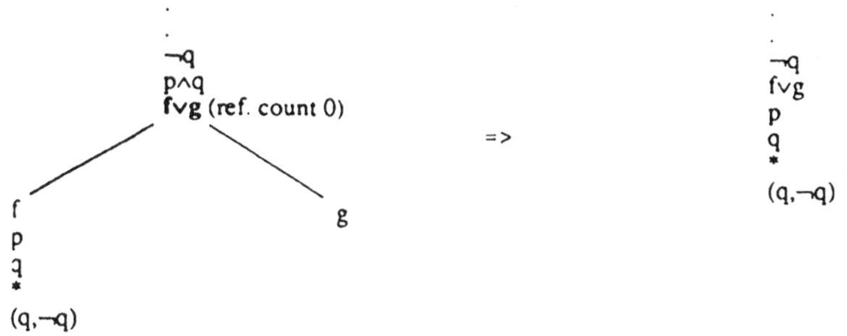

Fig. 2
Again the branch closure point of (q,¬q) does not depend on the β node f∨g. However in this case, removal of this branch point also causes the removal of an open branch which reduces the number of open branches left to close.

Whenever a type A removal is performed the closure points of the left subtree of that branch point are no longer needed. The following figure (Fig. 3) illustrates the retraction of dependencies emanating from closure points eliminated during a type A removal. These dependency retractions are reflected in reference count updates which could result in the identification of further redundancies (eg. see the subsequent type B removal shown in Fig.3).

Type A removal causing a Type B removal

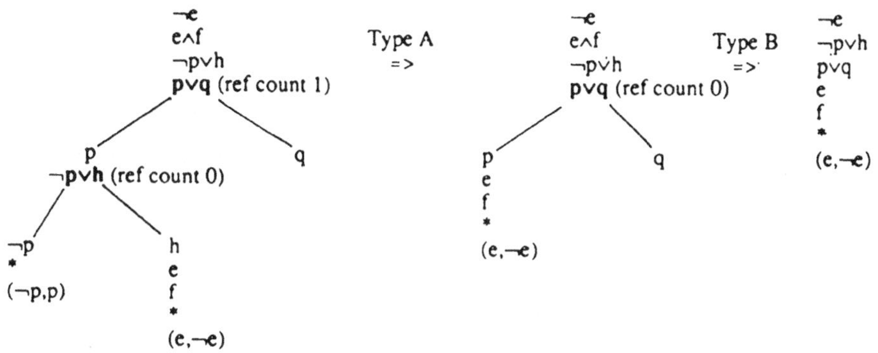

Fig. 3
The type A removal causes the descendants of the leftmost branch of the β node ¬p∨h to be eliminated. This elimination reduces the reference count of the β node p∨q which sets up the resulting type B removal.

Every subtree generated by the decomposition of a β node has a branch direction attribute which denotes the number of right branches immediately preceding the subtree. It indicates the number of subtrees which could be closed off by entirely closing this subtree. For example, the branch direction of a left subtree is 0; the branch direction of a right subtree of a left subtree is 1; the branch direction of a right subtree of a right subtree of a left subtree is 2, etc. This branch direction attribute is used by the proof condensation procedure to control the search for redundant branch points.

We will now present a general outline of the proof condensation procedure, **Condense**. It is called with two parameters: 1) branch-direction which is the branch direction of the current subtree, and 2) branch-points which is an ordered list of the branch points in the current branch.

```
Condense (branch-direction, branch-points)
    if branch-points is not an empty list {i.e. non-nil}
    then  let br-pt be the most recent branch point {i.e in lisp (car branch-points)}
          let new-branch-direction be the branch direction of br-pt
          if br-pt is redundant {i.e. if it has a reference count of 0}
          then  retract the branch point br-pt
                if type A removal {i.e. branch-direction > 0}
                then  remove the left subtree from the proof tree
                else  undo changes since last branch point endif
                Condense (new-branch-direction, rest(branch-points))
          else  if coming from the right {i.e. branch-direction > 0}
                then  Condense (new-branch-direction, rest(branch-points)) endif
          endif
    endif
End Condense
```

The proof condensation procedure is an efficient mechanism which compensates for some shortcomings in the set of heuristics. To illustrate this point clearly, the above examples do not apply the full set of heuristics that HARP normally uses. Indeed, the frequency of use of this condensation procedure is a measure of the pertinence of the heuristic advice. We rely on this measure to develop heuristics and to evaluate the performance of different sets of heuristics.

5. Heuristics

HARP's basic control structure, with the exception of proof condensation, is a modification of Smullyan's [Sm68] tree calculus and is thus amenable to a similar completeness proof. Although it can be proved that every inconsistent set has a closed tree it is clearly not the case that every tree for an inconsistent set closes. (For instance, a branch might grow forever simply because of the indefinite repeatability of universal instantiation. To prevent this, the algorithm closes a branch as soon as a contradiction appears on it.)

But the basic control structure will often build needlessly branching trees. Our heuristics are designed to prune trees without affecting HARP's completeness. (See the remarks below for some qualifications of this claim.) The heuristic rules prioritize tasks, i.e. tasks more likely to lead to early closure are attempted sooner while heuristically less likely appearing tasks are only delayed rather than abandoned.

We feel that a knowledge-based approach to heuristics, i.e. treating heuristics as explicitly represented, independently modifiable rules, has several advantages over hardwired heuristics. For example, it is easy to experiment with different control regimes and to study interactions among heuristics; it is also straightforward to reformulate them for certain applications so as not to reorder but abandon attempted steps, thereby giving up on the goal of completeness in the interests of greater efficiency.

Our heuristics fall into three groups: a) heuristics for efficient and human-like proof construction, b) heuristics for detecting nontheorems, i.e. for discovering as soon as possible when an open branch is complete, c) domain-specific heuristics.

Below we shall describe a few sample heuristics from groups a) and b). (We do not describe the rules in c) because we have not yet fully investigated them; moreover, no example mentioned in this paper uses them.) No rules in a) affect HARP's completeness property but some rules in b) do. The latter apply in situations in which a set is consistent but the algorithm could never establish that fact conclusively. As an example, suppose we try to prove the invalid formula $\forall x \exists y Hxy \Rightarrow \exists y \forall x Hxy$. The proof takes the form of the following, potentially infinite branch: $\forall x \exists y Hxy$, $\neg\exists y \forall x Hxy$, $\forall y \exists x \neg Hxy$, $\exists y Hxy$, Hxa, $\exists x \neg Hxa$, $\neg Hba$, $\exists x \neg Hxb$, $\neg Hcb$, etc. Heuristics in b) keep track of this type of loop and declare the above branch complete after a few cycles. This approach will, of course, not work in all cases but - in view of Church's theorem - no theorem prover can detect all nontheorems. It should be noted that successful proofs of theorems do not rely on any heuristics in b), i.e. the latter are only used to detect some types of nontheorems.

Heuristics are expressed in terms of the information provided by the indexing scheme. However, in order to improve readability (and also because of space limitations) we shall present the sample heuristics in English.

H1: Work on a compound until atoms are reached.

H1 implements a depth-first strategy which is appropriate given the goal of closing all branches.

H2: Favor rules that introduce as few new nodes as possible.

H2 obviously prevents needless ramification.

H3: Prefer existential to universal instantiation.

H3 reflects the fact that existential quantifiers, unlike universal ones, are dropped only once. H3 is overridden only when the available indexing information indicates that an universal instantiation would produce a contradiction immediately, as in the following example: Fa, ∀x¬Fx, ∃yGy.

H4: Favor compounds derived from the negation of the conclusion.

H4 corresponds to the set-of-support strategy and aids in selecting relevant premisses.

H5: Avoid clearly useless work.

H5 comprises several work reducing rules like the following two: if a propositional variable occurs only once in the premisses and not in the conclusion it is thrown away; if '¬' occurs nowhere in the root set then no proof is attempted because at least one branch is sure to be open.

H6: Choose carefully among branching compounds.

According to H6 a conditional is decomposed only if either its antecedent or the negation of its consequent (or both) appear as atoms higher up on the same branch. Similar rules apply to the other β formulas. The following example illustrates how selective H6 is. Suppose we wish to prove b ⇒ c, given these premisses: f ⇒ g, ¬b ⇒ d, ¬d ⇒ ¬c, a ⇒ d, a ⇒ ¬b, ¬c ⇒ a. Then,because of H4 and H6, HARP will build the tree below:

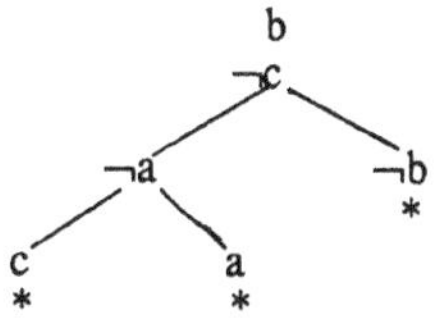

H7: Favor fresh universal quantifications.

For each universal quantifier a record is kept of how often it has been used and with what instantiating terms. In the absence of countervailing heuristic information, less frequently used quantifiers are preferred. H7 reflects the empirical observation that usually in AI applications each relevant premise makes a definite and unique contribution to a proof.

H8: Minimize the introduction of new parameters.

H8 suggests that universal instantiations be done only with respect to terms already on a branch unless there are none yet. Moreover, for the terms on the branch, only those instantiations are made that are sanctioned by the other heuristics.

H9: Identify complete open branches as quickly as possible.

H9 is intended to determine quickly whether the root set of an infinite tree is consistent. If an open branch contains no unchecked compounds except universal quantifications and if every such quantification is instantiated with respect to all and only the terms above it on that branch then the branch may be declared complete. For example, given ∃xFx, ∀x¬Gx, Fa, ¬Ga, it would be pointless to add further instantiations like ¬Gb, ¬Gc etc.

H10: Watch out for nonconverging ∀∃-patterns.

H10, like H9, attempts to identify nontheorems as quickly as possible. H10 consists of several rules from group b) that are invoked whenever a universal quantifier precedes an existential quantifier and universal instantiation is reapplied to new constants introduced by existential instantiation. The various rules of H10 recognize particular ∀∃-patterns that we have empirically identified as frequently indicating infinite branches that will never close. Whenever such a pattern is matched, its branch is immediately declared complete. The basic nonclosing ∀∃-pattern, of which all the other currently detected patterns are variants, is this: ∀x∃yFxy, ∃yFay, Fab, ∃yFby, Fbc, etc. At this point, H10 decides that it is pointless to continue the branch.

It should be noted that the rules in H10 do not, of course, declare a branch complete as soon as an ∀∃-pattern is found. Instead, when an ∀∃-pattern is found each rule of H10 begins to look in the current branch for a particular pattern of instantiations that is likely to continue indefinitely. If such a pattern occurs then the branch is declared complete, otherwise these rules have no effect on the proof. As an example, consider the proof of the following theorem:

$\exists x \forall y$ (Fy $\Rightarrow$ Fx).
$\neg\exists x \forall y$ (Fy $\Rightarrow$ Fx), $\forall x \exists y \neg$ (Fy $\Rightarrow$ Fx) (at this point, the rules in H10 are activated), $\exists y$ $\neg$ (Fy $\Rightarrow$ Fx), $\neg$ (Fa $\Rightarrow$ Fx), Fa , $\neg$ Fx, $\exists y \neg$ (Fy $\Rightarrow$ Fa), $\neg$ (Fb $\Rightarrow$ Fa), Fb, $\neg$ Fa (since the rules have not found a nonconverging pattern of instantiations, the branch closes without their interference).

The following example derivation (of "All tails of horses are tails of animals" from "All horses are animals") illustrates how our heuristics generate short and natural proofs.

1) $\forall x(Hx \Rightarrow Ax)$ *premise*
2) $\neg\forall x(\exists y(Hy \wedge Txy) \Rightarrow \exists y(Ay \wedge Txy))$ *negation of conclusion*
3) $\exists x\neg (\exists y(Hy \wedge Txy) \Rightarrow \exists y(Ay \wedge Txy))$ *quantifier negation from 2)*
4) $\neg (\exists y(Hy \wedge Tay) \Rightarrow \exists y(Ay \wedge Tay))$ *EI from 3), H2,H3,H4*
5) $\exists y(Hy \wedge Tay)$ *from 4), H2*
6) $\neg \exists y(Ay \wedge Tay)$ *from 4), H2*
7) $Hb \wedge Tab$ *EI from 5)*
8) Hb *from 7), H1*
9) Tab *from 7), H1*
10) $\forall y\neg (Ay \wedge Tay)$ *quantifier negation from 6)*
11) $\neg(Ab \wedge Tab)$ *UI from 10), H8. Note that* b *rather than* a *is chosen as an instantiating term because of a link from* Tay *in 10) to 9):* $\neg(Aa \wedge Taa)$ *could not produce a closure.*

12) $\neg$Ab

  14) Hb $\Rightarrow$ Ab *UI from 1), H7, H8. Note that* b *rather than* a *is chosen as an instantiating term because of links from* Hx *and* Ax *in 1) to 8) and 12), respectively:* Ha $\Rightarrow$ Aa *could not produce a closure.*

    15) $\neg$Hb *from 14)*
    *

    16) Ab *from 14)*
    * *from 8) and 12), respectively.*

13) $\neg$Tab *from 11)*
* *from 13) and 9)*

6. Conclusions

We have presented in this paper an efficient, human-oriented theorem prover as a potential tool in AI environments. To facilitate experimenting with different sets of heuristics - an ongoing concern of ours - we have pursued the design goal of separating the basic control structure from explicitly stated heuristics.

Analytic tableaux have been used to provide a natural control structure. It appears that the theorem proving community has generally ignored tableaux because of their presumed inefficiency (for a few exceptions see [Wr84], [Sc85]). On the other hand, many logicians find tableaux to yield an elegant and natural codification of logic (e.g [St84]). In this paper we have presented evidence that tableaux form a basis for combining both efficiency and naturalness.

To make HARP more efficient and natural, the basic tableau method was augmented with proof condensation and control heuristics supported by an indexing scheme. Since we intend HARP to be useful as an AI tool, it is especially important for it to rapidly detect nontheorems. To this end we have provided a few tentative heuristics.

References

[Be65] E.W.Beth. *The Foundations of Mathematics.* North-Holland, 1965.

[Bi82] W.Bibel. "*A Comparative Study of Several Proof Procedures* ". Artificial Intelligence (18, 1982), pp 269-293.

[Bl83] W.W.Bledsoe. *The UT Interactive Prover.* University of Texas at Austin, ATP-17B, 1983.

[Co81] D.N.Cohen. *Knowledge Based Theorem Proving and Learning.* UMI Research Press, Ann Arbor, Michigan, 1981.

[Co85] A.G.Cohn. "*On the solution of Schubert's Steamroller in Many Sorted Logic*".

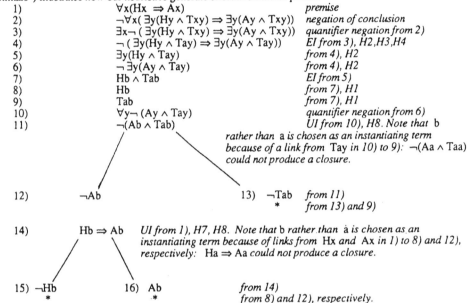

IJCAI85, pp 1169-1173.

[Fi83] M.C.Fitting. *Proof Methods for Modal and Intuitionistic Logics.* Reidel, Dordrecht, Holland, 1983.

[Hi55] J.Hintikka. *Form and Content in Quantification Theory.* Acta Philosophica Fennica 8, 1955.

[Je67] R.C.Jeffrey. *Formal Logic: Its Scope and Limits.* McGraw-Hill, New York, 1967.

[Lo78] D.Loveland. *Automated Theorem Proving:A Logical Basis .* North Holland, New York, 1978.

[Mu82] N.V.Murray. "*Completely Non-Clausal Theorem Proving*". Artificial Intelligence (18, 1982), pp 67-85.

[Ni80] N.Nilsson. *Principles of Artificial Intelligence.* Tioga, Palo Alto, 1980.

[Ro81] J.A.Robinson. *Logic: Form and Function.* North Holland, New York, 1981.

[Sc85] W. Schonfeld. "*Prolog extensions based on Tableau Calculus*". IJCAI 85, pp 730-732.

[Sm68] R.M.Smullyan. *First-Order Logic.* Springer Verlag, Berlin , 1968.

[St84] W.Stegmuller, M.V.v.Kibed. *Strukturtypen der Logik.* Springer-Verlag, New York, 1984.

[Wa83] C.Walther. "*A many-sorted calculus based on resolution and paramodulation*". IJCAI 83, Vol. 2, pp 882-891.

[Wa85] C.Walther. "*A mechanical Solution of Schubert's Steamroller by Many-Sorted Resolution*". Artificial Intelligence, (May 1985), pp 217-224.

[Wr84] G.Wrightson. *Semantic Tableaux, Unification and Links.* Technical Report CSD-ANZARP-84-001, 1984, Victoria University, Wellington, New Zealand.

NESTED RESOLUTION

Jonathan Traugott
Computer Science Department
Stanford University

ABSTRACT

In this paper we introduce nested resolution, a new nonclausal resolution rule which offers shorter proofs than are possible using conventional nonclausal resolution. Nested resolution also has a local flavor that makes it particularly suitable for parallel implementation. Completeness and soundness proofs are given.

INTRODUCTION

Resolution is a complete inference rule for first-order logic. It was proposed by Robinson in 1965 [Ro]. In order to apply Robinson's rule, a sentence must first be reduced to conjunctive normal form. Nonclausal resolution is a generalization of resolution that does not impose this restriction. Nonclausal resolution was developed independently by Manna and Waldinger [MW1] and Murray [Mu] in 1980. In this note we propose a new nonclausal resolution rule called *nested resolution* that differs substantially from the 1980 rule and offers shorter and more intuitive proofs. Nested resolution also has a local flavor that makes it particularly suitable for parallel implementation.

Before presenting the nested resolution rule we must define the notions of subexpression replacement and polarity.

Definition (Replacement)

Let $F[P]$ denote a (first-order logic) sentence containing one or more occurrences of the subsentence P. Then by $F[Q]$ we denote the new sentence obtained by replacing all occurrences of P in $F[P]$ with the sentence Q.

Definition (Polarity)

Here we give only an informal definition of polarity. For a more rigorous definition the reader is referred to [MW1].

We say that P has *positive polarity* in $F[P]$ iff P occurs within an even number of explicit or implicit negations. In this case we write $F[P^+]$. We say that P has *negative polarity* in $F[P]$ iff P occurs within an odd number of explicit or implicit negations. In this case we write $F[P^-]$. If P occurs within an equivalence connective or within the if clause of the if-then-else connective then P has no polarity.

From the above definition we may prove the following proposition

---

This research was supported in part by the National Science Foundation under grants DCR-8512356 and MCS-82-14523, by DARPA under Contract N00039-84-C-0211, by the United States Air Force Office of Scientific Research under Contract AFOSR-81-0014, by the Office of Naval Research under Contract N00014-84-C-0706, by the United States Army under contract DAJA-45-84-C-0040, and by a contract from the International Business Machines Corporation.

Proposition (Polarity)

For any ground sentences P, Q, $F[P^+]$ and $G[Q^-]$ the following sentences are valid:

$$(P \supset Q) \quad \supset \quad (F[P^+] \supset F[Q^+])$$

$$(P \supset Q) \quad \supset \quad (G[Q^-] \supset G[P^-]).$$

Example (Polarity)

B has negative polarity in $\neg(A \supset B)$ because it occurs within a single negation. On the other hand, A has positive polarity because it occurs within two negations, one explicit ($\neg$) and one implicit (the left-hand side of $\supset$). The polarity proposition allows us to infer

$$(A \supset C) \quad \supset \quad (\neg(A \supset B) \supset \neg(C \supset B))$$

$$(B \supset C) \quad \supset \quad (\neg(A \supset C) \supset \neg(A \supset B)).$$

Definition (Ground Resolution)

Let P, $F[P^+, P^-]$, and $G[P]$ be ground sentences. Then we may express the nested resolution rule as follows:

$$\begin{array}{ll} (1) & F[P^+, P^-] \\ (2) & G[P] \\ \hline (3) & F[G[\mathit{true}], \neg G[\mathit{false}]] \end{array}$$

More precisely, we take all occurrences of P in G and replace them by the propositional symbol *true* to obtain $G[\mathit{true}]$. All positive occurrences of P in F are then replaced with $G[\mathit{true}]$. In a like manner we then replace the remaining negative occcurrences of P with $\neg G[\mathit{false}]$. The resulting sentence (3) is the nested resolvent of $F[P]$ and $G[P]$. We may refer to (3) unambiguously by $\mathcal{R}_P(F, G)$.

This rule is an alternative to the conventional nonclausal resolution rule proposed by [Mu] and [MW1]:

$$\begin{array}{ll} (1') & F[P] \\ (2') & G[P] \\ \hline (3') & F[\mathit{true}] \;\vee\; G[\mathit{false}] \end{array}$$

It is easy to show that the nested resolution rule gives a stronger resolvent than the original rule. In other words, sentence 3 (strictly) implies $F[\mathit{true}, \mathit{true}] \;\vee\; G[\mathit{false}]$, i.e. the result of applying the conventional rule to sentences 1 and 2.

Soundness

We must show that (3) follows from (1) and (2).

Case: P is true

In this case (1) is equivalent to

$$F[\mathit{true}, \mathit{true}]$$

and (2) is equivalent to

$$G[\mathit{true}].$$

We may then infer (by replacement)

$$F[G[\mathit{true}], \mathit{true}^-].$$

Now

$$\neg G[false] \supset true$$

So by the polarity proposition we have

$$F[G[true], \neg G[false]].$$

Case: P is false

In this case (1) is equivalent to

$$F[false, false]$$

and (2) is equivalent to

$$G[false].$$

So we may infer

$$F[false^{+}, \neg G[false]].$$

But

$$false \supset G[true]$$

So again by the polarity proposition we have

$$F[G[true], \neg G[false]].$$

Transformation Rules

In order to simplify newly formed resolvents we use a set of transformation rules. These include the *true-false* removal rules such as

$$true \wedge P \longrightarrow P$$

$$false \wedge P \longrightarrow false$$

$$false \supset P \longrightarrow true$$

and the propositional simplification rules such as

$$P \vee P \longrightarrow P$$

$$P \supset P \longrightarrow true$$

$$\neg P \wedge P \longrightarrow false.$$

For completeness only the *true-false* rules are necessary.

Examples

The following examples illustrate how shorter proofs may be obtained using the new rule.

(1)	$A^{+} \wedge B$
(2)	$\neg A \vee \neg B$
(3)	$(\neg true \vee \neg B) \wedge B$

which simplifies to *false*, using the transformation rules described above. Resolving (1) and (2) using the original rule would have produced the weaker sentence $\neg B$. ∎

Suppose we want to show the transitivity of $\supset$. Then we must show the unsatisfiability of $\{A \supset B\ ,\ B \supset C\ ,\ \neg(A \supset C)\}$.

$$\begin{array}{ll} (1) & A \supset C^{+} \\ (2) & B \supset C \\ \hline (3) & A \supset C \end{array}$$

We may now resolve (3) with $\neg(A \supset C)$ to obtain a refutation:

$$\begin{array}{ll} (3) & (A \supset C)^{+} \\ (4) & \neg(A \supset C) \\ \hline (5) & \mathit{false} \end{array}$$

Using the original rule we would have obtained $\neg A \vee C$ at line 3. This is equivalent to $A \supset C$ but, because of its disjunctive form, it does not resolve completely with $\neg(A \supset C)$. Note that even if we had expressed $\neg(A \supset C)$ as $\neg(\neg A \vee C)$ (or alternatively as $A \wedge \neg C$) the nested refutation would still have required only two steps. ┛

Let us show the unsatisfiability of $\{B \equiv C \ , \ B \equiv \neg C\}$. Note that neither B nor C has any polarity in $B \equiv C$. This prohibits us from applying the rule directly. In order to proceed we must divide the first equivalence into separate implications.

$$\begin{array}{ll} (1) & B^{-} \supset C \ \wedge \ C \supset B^{+} \\ (2) & B \equiv \neg C \\ \hline (3) & \neg C \supset C \ \wedge \ C \supset \neg C \end{array}$$

Which simplifies to *false* using the propositional simplification rules. Thus the refution using nested resolution requires only a single step. Using the conventional rule requires a minimum of three steps. ┛

We will now prove a theorem that is slighly more complex; from $A \supset (B \wedge C), C \supset D$, and $(B \wedge D) \supset E$ we show $A \supset E$.

$$\begin{array}{ll} (1) & A \supset (B \wedge C^{+}) \\ (2) & C \supset D \\ \hline (3) & A \supset (B \wedge D) \end{array}$$

$$\begin{array}{ll} (3) & A \supset (B \wedge D)^{+} \\ (4) & (B \wedge D) \supset E \\ \hline (5) & A \supset E \end{array}$$

$$\begin{array}{ll} (5) & (A \supset E)^{+} \\ (6) & \neg(A \supset E) \\ \hline (7) & \mathit{false} \end{array}$$

To prove this theorem using the original rule would have required an additional two steps. ┛

Comparison with the Original Rule

The above examples illustrate two advantages of nested resolution. First it is stronger than the original rule and therefore moves us more quickly toward a refutation. Second it introduces no new binary connectives and so better preserves the syntactic character of resolved sentences. (Repeated use of the original rule tends to produce sentences in clause form.) This often allows us to resolve on bigger subsentences.

Another advantage of nested resolution is a local flavor that makes the rule particularly suitable for parallel implementation. Application of the ground rule may be looked upon as a local transformation of the first resolved sentence, a particular subexpression being replaced while the rest of the sentence remains unchanged. Thus several applications of the rule could be made in parallel, each resolving upon a different subsentence. [Ma] discusses use of nested resolution for parallel implementation of the logic programming language TABLOG.

A disadvantage of nested resolution is the sentence expansion that may result from replacing P with $G[\mathit{true}]$ (or $\neg G[\mathit{false}]$) in $F[P]$. Multiple replacements may result in unreasonably long sentences. However in our proofs we have found the need for multiple replacement to be extremely rare.

COMPLETENESS

In this section we will prove completeness for ground nested resolution. Here we assume that sentences are restricted to the connectives $\wedge,\vee,\supset$ and $\neg$.

Notation

$\mathcal{S}$: a finite set of ground sentences.

$\mathcal{S}^*$: the deductive closure of $\mathcal{S}$ under resolution.

π : a subexpression replacement. In the following proof we will assume that the domain of π is a set of ground atomic formulae and the range is a subset of $\{\mathit{true}, \mathit{false}\}$.

$\{P \leftarrow Q\}$: the replacement which replaces P by Q.

L : a finite set of ground atomic formulae.

L_i : the ith member of L.

$\mathbf{T}$: a binary tree.

$\mathbf{T_j}$: the set of nodes of $\mathbf{T}$ at level j.

$F \Rightarrow \mathit{false}$: F reduces to *false* by *true-false* transformation.

Definition (Replacement Tree)

A replacement tree $\mathbf{T}$ for L is an binary tree such that:

1. The root node of $\mathbf{T}$ is the empty replacement.

2. If π is a node of $\mathbf{T}$ at level $j-1$ then the left child of π is $\pi\{L_j \leftarrow \mathit{false}\}$ and the right child of π is $\pi\{L_j \leftarrow \mathit{true}\}$.

Definition (Atom Set)

The atom set of $\mathcal{S}$ is the set of all atomic formulae occurring in $\mathcal{S}$ other than the propositional symbols *true* and *false*.

Definition (Complete Replacement)

A complete replacement π for $\mathcal{S}$ is one whose domain contains the atom set of $\mathcal{S}$.

Lemma (Unsatisfiability)

If $\mathcal{S}$ is unsatisfiable then for every complete replacement π, there is an F in $\mathcal{S}$ such that $F\pi \Rightarrow$ *false*.

Proof

For a given π consider the interpretation I which for each L_i in the atom set of $\mathcal{S}$ falsifies(verifies) L_i just when $L_i\pi$ is *false* (*true*). Then for any sentence F in $\mathcal{S}$, $I(F)$ is $I(F\pi)$ (since all the atomic formulae in F and $F\pi$ have the same value under I).

Since $\mathcal{S}$ is unsatisfiable there is some F in $\mathcal{S}$ which is falsified by I. Hence $F\pi$ is also falsified by I. But, by completeness of π, $F\pi$ reduces to either *true* or *false* so it must reduce to *false*. ∎

Lemma (resolution)

Let $dom(\pi)$ include the atom set of $\{F, G\}$ minus P. If $F\pi\{P \leftarrow \mathit{false}\} \Rightarrow \mathit{false}$ and $G\pi\{P \leftarrow \mathit{true}\} \Rightarrow \mathit{false}$ then $\mathcal{R}_P(F,G)\pi \Rightarrow \mathit{false}$.

Proof

Since π is a complete replacement for $\mathcal{R}_P(F,G)$, $\mathcal{R}_P(F,G)\pi$ reduces to either *true* or *false*. Let us show that it implies *false* and so must also reduce to *false*.

$$\begin{aligned}
&\mathcal{R}_P(F,G)\pi \quad \equiv \\
&F[G[\mathit{true}], \neg G[\mathit{false}]^-]\pi \quad \text{(by definition)} \quad \equiv \\
&F[G[P]\pi\{P \leftarrow \mathit{true}\}, \neg G[\mathit{false}]^-]\pi \quad \text{(by properties of } \pi) \quad \equiv \\
&F[\mathit{false}, \neg G[\mathit{false}]^-]\pi \quad \text{(by assumption)}
\end{aligned}$$

We must now show that this last sentence implies *false*. First recall that from the polarity proposition we know that for any sentences Q, R, H:

$$(Q \supset R) \quad \supset \quad (H[R^-] \supset H[Q^-]).$$

In particular if Q is the propositional symbol *false* then this reduces to $H[R^-] \supset H[\mathit{false}^-]$. Taking $H[R^-]$ to be $F[\mathit{false}, \neg G[\mathit{false}]^-]\pi$ we obtain

$$\begin{aligned}
&F[\mathit{false}, \neg G[\mathit{false}]^-]\pi \quad \supset \\
&F[\mathit{false}, \mathit{false}^-]\pi \quad \text{(by the polarity proposition)} \quad \equiv \\
&F\pi\{P \leftarrow \mathit{false}\} \quad \text{(by properties of } \pi) \quad \equiv \\
&\mathit{false} \quad \text{(by assumption)}
\end{aligned}$$
∎

Definition ($D_{\mathcal{S}}$)

Let $L : \{L_1, ..., L_n\}$ be the atom set for $\mathcal{S}$. Let $\mathbf{T}$ be the replacement tree for L. Then for $j \in [0, n]$, $D_{\mathcal{S}}(j)$ iff for every π in $\mathbf{T_j}$, $F\pi \Rightarrow \mathit{false}$ for some F in $\mathcal{S}$, where the atom set for F is in $\{L_1, ..., L_j\}$.

Lemma ($D_{\mathcal{S}}$)

Let $L : \{L_1, ..., L_n\}$ be the atom set for $\mathcal{S}$. Then for all $j \in [1, n]$, $D_{\mathcal{S}^*}(j) \supset D_{\mathcal{S}^*}(j-1)$.

Proof

Assume $D_{\mathcal{S}^*}(j)$. Now consider an arbitrary π in $\mathbf{T_{j-1}}$. For $D_{\mathcal{S}^*}(j-1)$, we need to show that for some H in $\mathcal{S}^*$, $H\pi \Rightarrow \mathit{false}$ where the atom set of H is in $\{L_1, ..., L_{j-1}\}$.

Let α and β be π's left and right children respectively. Then since α and β belong to $\mathbf{T}_j$, and since $D_{S^*}(j)$ is true, we know there exist F and G in S^* such that $F\alpha \Rightarrow$ *false* and $G\beta \Rightarrow$ *false* and where the atom set of $\{F, G\}$ is in $\{L_1, ..., L_j\}$. But $dom(\pi) = \{L_1, ..., L_j\} - L_j$ so by the resolution lemma $\mathcal{R}_{L_j}(F, G)\pi \Rightarrow$ *false*.

We may take H to be $\mathcal{R}_{L_j}(F, G)$: We have just shown that $H\pi \rightarrow$ *false*. Further H is in S^* since it is the resolvent of two members of S^*. Finally the atom set of H is in $\{L_1, ..., L_{j-1}\}$ since resolution eliminates all occurrences of the resolved subexpression. ▟

Theorem (Completeness of ground resolution)

If S is unsatisifiable then S^* contains a sentence F such that $F \Rightarrow$ *false*.

Proof

Suppose S is unsatisfiable; we wish to show that S^* contains a sentence F that reduces to *false*. If we can show $D_{S^*}(0)$ then the theorem is proved since T_0 contains only the empty replacement. Note that by the unsatisfiability lemma we have $D_{S^*}(n)$ (where n is the size of S's atom set). Now consider the least number j for which $D_{S^*}(j)$. If j were different from 0 then by the D_S-lemma we would have $D_{S^*}(j-1)$ contradicting the leastness of j. Hence j is 0. ▟

NESTED RESOUTION: GENERAL VERSION

The general version of the rule allows us to instantiate the variables of the given sentences as necessary to create common subsentences. It is expressed as follows:

Definition (General Resolution)

For any (quantifier free) sentences $P, \widetilde{P}, F[P]$, and $G[\widetilde{P}]$ we have

$$\frac{\begin{array}{c} F[P^+, P^-] \\ G[\widetilde{P}] \end{array}}{F\theta[G\theta[\mathit{false}], \neg G\theta[\mathit{true}]]}$$

More precisely,

- The variables in F and G have been renamed so that the two sentences share no common variables.
- F has one or more subsentences $P_1, P_2, \ldots$.
- G has one or more subsentences $\widetilde{P}_1, \widetilde{P}_2, \ldots$.
- θ is a most general unifier of $P_1, P_2, \ldots$, and $\widetilde{P}_1, \widetilde{P}_2, \ldots$; hence $P_1\theta = P_2\theta = \ldots = \widetilde{P}\theta = \widetilde{P}_1\theta = \widetilde{P}_2\theta = \ldots$.
- The conclusion of the rule is obtained by replacing all occurrences of $\widetilde{P}\theta$ in $G\theta$ with *true* (*false*), obtaining $G\theta[\mathit{true}]$ ($G\theta[\mathit{false}]$), and replacing all positive (negative) occurrences of $P_1\theta$ in $F\theta$ with $G\theta[\mathit{true}]$ ($\neg G\theta[\mathit{false}]$).

In other words, we apply the ground version of the rule to $F\theta$ and $G\theta$, taking $P_1\theta$ as the common subsentence. ▟

COMPLETENESS

We have already established completeness for the ground version. To show that the general version is complete it suffices to show the semicommutivity of saturation and resolution. To the extent possible we repeat the argument given in [Ro] (and assume familiarity with the same terminology).

Lemma (lifting)

If $\mathcal{S}$ is any set of sentences and $\mathcal{P}$ is any subset of the Herbrand universe of $\mathcal{S}$, then $\mathfrak{R}(\mathcal{P}(\mathcal{S})) \subset \mathcal{P}(\mathfrak{R}(\mathcal{S}))$.

Proof

Assume that $H \in \mathfrak{R}(\mathcal{P}(\mathcal{S}))$. Then either $H \in \mathcal{P}(\mathcal{S})$, in which case $H \in \mathcal{P}(\mathfrak{R}(\mathcal{S}))$ since $\mathcal{S} \subset \mathfrak{R}(\mathcal{S})$; or H is $\mathcal{R}_P(F\alpha, G\beta)$. where $F, G \in \mathcal{S}$ and

$$\alpha = \{U_1 \leftarrow S_1, U_2 \leftarrow S_2 \ldots\}$$

$$\beta = \{V_1 \leftarrow T_1, V_2 \leftarrow T_2 \ldots\}$$

where $(U_1, U_2 \ldots)$ and $(V_1, V_2 \ldots)$ are the distinct variables occurring in F and G, and where the terms $(S_1, S_2, \ldots, T_1, T_2, \ldots)$ are in $\mathcal{P}$.

Let $P_1, P_2 \ldots$ be the atomic formula in F such that $P_i\alpha = P$. Let $\widetilde{P}_1, \widetilde{P}_2 \ldots$ be the atomic formula in G such that $\widetilde{P}_i\beta = P$. Finally let

$$\sigma \;=\; \{x_1 \leftarrow S_1, x_2 \leftarrow S_2, \ldots, y_1 \leftarrow T_1, y_2 \leftarrow T_2, \ldots\}$$

Then $F\alpha = F\sigma$, $G\beta = G\sigma$ and $P_i\sigma = \widetilde{P}_j\sigma = P$. Let θ be the most general unifier of the sentences $P_1, P_2, \ldots, \widetilde{P}_1, \widetilde{P}_2, \ldots$. Then there is a substitution λ over $\mathcal{P}$ such that $\sigma = \theta\lambda$. So $F\alpha = F\theta\lambda$ and $G\beta = G\theta\lambda$. Hence

$$\begin{aligned} &\mathcal{R}_P(F\alpha, G\beta) \;= \\ &F\theta\lambda[G\theta\lambda[\mathit{true}], \neg G\theta\lambda[\mathit{false}]] \;= \\ &F\theta[G\theta[\mathit{true}], \neg G\theta[\mathit{false}]]\lambda \end{aligned}$$

But this is just a ground instance over $\mathcal{P}$ of the general resolvent of F and G, and hence a member of $\mathcal{P}(\mathfrak{R}(\mathcal{S}))$. ∎

Corollary (lifting)

If $\mathcal{S}$ is any set of sentences and $\mathcal{P}$ is any subset of the Herbrand universe of $\mathcal{S}$, then $\mathfrak{R}^n(\mathcal{P}(\mathcal{S})) \subset \mathcal{P}(\mathfrak{R}^n(\mathcal{S}))$.

Proof

The proof is identical to that given in [Ro]. It is a trivial induction over the integers. ∎

Theorem (Completeness of General Resolution)

If $\mathcal{S}$ is any finite, unsatisfiable set of sentences then, for some n, $\mathfrak{R}^n(\mathcal{S})$ contains a sentence F such that $F \Rightarrow$ *false*.

Proof

Since $\mathcal{S}$ is unsatisfiable we know (by Herbrand's Theorem) that there is a finite subset $\mathcal{P}$ of the Herbrand universe of $\mathcal{S}$ such that $\mathcal{P}(\mathcal{S})$ is also unsatisfiable. Now from the completeness proof of ground resolution we know that $D_{\mathcal{P}(\mathcal{S})^*}(0)$: $\mathcal{P}(\mathcal{S})^*$ contains a sentence F such that $F \Rightarrow$ *false*, where the atom set of F is empty. Now because F is in $\mathcal{P}(\mathcal{S})^*$ it must be in $\mathfrak{R}^n(\mathcal{P}(\mathcal{S}))$ for some n, and by the lifting corollary also in $\mathcal{P}(\mathfrak{R}^n(\mathcal{S}))$. So F is a ground instance of some sentence in $\mathfrak{R}^n(\mathcal{S})$. But F contains no terms so it can only be an instance of itself. ∎

References

[MW1]: Manna and Waldinger. A deductive approach to program synthesis. *ACM Transactions on Programming Languages and Systems*, Vol. 2, No 1, 1980.

[MW2]: Manna and Waldinger, Special relations in automated deduction. Journal of the ACM., Vol 33, No 1, Jan 1986.

[Ma]: Malachi, Nonclausal logic programming. Ph.D. Thesis. Stanford University. March 1986.

[Mu]: Murray, Completely non-clausal theorem proving. Artificial Intelligence, Vol 18, 1982.

[Ro]: Robinson, A machine-oriented logic based on the resolution principle. Journal of the ACM., Vol. 12, No 1, Jan 1965

Mechanizing Constructive Proofs

Gérard Huet

INRIA

Domaine de Voluceau

78150 Rocquencourt

FRANCE

Abstract

This talk will give a short history on the evolution of constructive concepts in Mathematics. It will review the intuitionistic paradigm, and the most recent schools of thought in proof theory. More specifically, it will compare various proposals for higher order type theories, due to Russell, Church, Martin-Löf, Girard, de Bruijn, Constable and Coquand.

The formalist paradigm of mathematical foundations has been vigorously revived in the last decades with the introduction of computers. Three areas of Computer Science are developing formal systems which may be considered as candidates for constructive mathematics foundations: Automatic Theorem Proving, Algebraic Computation Systems, and Programming Language Theory. The talk will highlight a few especially striking achievements in these areas, and will discuss promising topics of research.

Implementing Number Theory: An Experiment with Nuprl *

Douglas J. Howe
Department of Computer Science
Cornell University, Ithaca, NY 14853, U.S.A.

Abstract

We describe the results of an experiment in which the Nuprl proof development system was used in conjunction with a collection of simple proof-assisting programs to constructively prove a substantial theorem of number theory. We believe that these results indicate the promise of an approach to reasoning about computationally meaningful mathematics by which both proof construction and the results of formal reasoning are mathematically comprehensible.

1 Introduction

In this paper we describe a step toward a high-level environment for formally proving theorems of constructive mathematics. In particular, we describe a formalized proof of the fundamental theorem of arithmetic. The aspects we will emphasize are not restricted in relevance to number theory but instead, we believe, indicate the promise of our approach for other branches of constructive mathematics. The basis for our work is the Nuprl proof development system [4], a system which combines a sequent calculus formulation of a higher-order constructive logic, with a proof system supported by a proof-editor, a definition facility, and a metalanguage in which one can write proof-assisting programs. We have constructed a set of such programs and used them to prove the well-known theorem that every positive integer has a unique factorization into a product of powers of primes. The resulting collection of definitions, theorems and proofs is mathematically understandable and very readable, and also implicitly defines correct programs which realize the constructive, or computational, content of the theorems proven.

We describe here the results of one of the first major experiments with the Nuprl system. In the process we will discuss some of the features of the system and refer to aspects of the Nuprl logic, but our main concern is to show the kind of formal reasoning that Nuprl makes possible. Central to the discussion in this paper is the structure of reasoning in Nuprl and the role played in it by *proof tactics*. The basic unit of inference in Nuprl is called a *refinement*; one constructs, via a proof editor, a tree-structured proof in a top-down style by successively refining a goal to

*This work was supported, in part, by NSF grant no. MCS-81-04018

produce subgoals. The "size" of the refinement steps is under the control of the user; one can write programs which can be used as new inference rules. If such a program can provide justification in terms of the primitive inference rules that the subgoals it generates from a given goal entail the goal, then the name of the program can appear in the proof as the rule used in the refinement step. It thus becomes possible to completely suppress much of the meticulous detail involved in a formal proof, and to construct a collection of programs which approximate the higher level steps of informal mathematics. Although the collection of tactics we used is a long way from meeting this goal, it is an indication of what can be done, and using it we were able to completely prove from scratch in a reasonable amount of time a substantial theorem of number theory.

There are many other systems for aiding in the construction of formal proofs; three such systems, AUTOMATH [3], LCF [8], and the system of Boyer and Moore [2], have been extensively used. None, however, combine the features which we believe make Nuprl a promising tool for developing formalized mathematics, although the project of Coquand and Huet [7] appears to be heading in a similar direction. One of the most important distinguishing features is the mechanism for constructing and manipulating proofs, i.e., the coupling of a high-level programming language serving as the formal system's metalanguage with a highly visual proof editor and definition mechanism. Another is the expressive power of the logic, which permits natural representations of concepts from computationally meaningful higher mathematics.

In the next section we give a short overview of some important aspects of Nuprl. That section describes work done by others at Cornell; the work described in the subsequent sections was done by the author. Section 3 contains a discussion of how the necessary concepts of number theory were formalized, and in section 4 we discuss some of the proof tactics employed and illustrate how they were used in the actual proofs. In the last section we draw conclusions from the experiment and discuss some future directions for our work.

2 Nuprl

Nuprl [4] is a proof development system developed at Cornell under the direction of Joseph Bates and Robert Constable. Its formal basis is similar to the constructive type theory of Martin-Löf [13] and is intended to be suited to the formalization of constructive mathematics. Both constructive logic and objects of constructive mathematics are represented naturally in the Nuprl theory. The implemented system provides a general definition facility so that mathematical formulas have a compact display form which approximates that of mathematics textbooks.

The basic objects of reasoning in the Nuprl theory are types and members of types. The rules of Nuprl deal with *sequents*, i.e., objects of the form

$$x_1 : H_1,\ x_2 : H_2,\ \ldots,\ x_n : H_n \ >>\ A$$

(sequents, in the context of a proof, are also called *goals*). To assert the truth of the sequent essentially means to assert that given members x_i of the types H_i, a member of the type A can be constructed. An important point about the Nuprl rules is that they allow one to construct a member in a top-down fashion. They allow one to *refine* a goal, obtaining subgoal sequents such that a construction for the goal can be computed from constructions for the subgoals.

Space limitations prevent us from giving details concerning the Nuprl type theory; however, a few general remarks should suffice for the purpose of this paper. Nuprl has a rich set of type-building operations. In addition to such conventional type constructors as cartesian product, there are constructors whose purpose is to represent fundamental notions of constructive logic, via the *propositions as types* correspondence. This correspondence gives a direct translation for the usual quantifiers and propositional connectives of logic. The basic idea behind this correspondence is that we can associate a type with each formula such that the formula is (constructively) true if and only if the type associated with it has a member. To prove a statement P of constructive mathematics, then, one first translates it into the Nuprl type theory, obtaining a type T, and then attempts to prove >> T by applying refinement rules until no more unproved subgoals exist. The system then can compute, or *extract*, a term t which is a member of the type T and which embodies the computational content of the theorem P. For example, if we prove in this way the formula

$$\forall x : int \;\; \exists y : int \;\; where \;\; x + y = 0$$

the term extracted from the proof will be a function which takes an integer x and produces an integer y such that $x + y = 0$. The existence of such a function is the meaning of constructive truth for the formula. An important point about this translation is that it is largely transparent to the Nuprl user; we will return to this point later. We have phrased the preceding discussion in terms of constructive mathematics. However, one can also view Nuprl as a system for program synthesis in the spirit of [12]; one proves theorems which are program specifications, and from the proofs the system can extract proven correct programs. Applications of the Nuprl methodology to program synthesis are discussed in [4]. The Nuprl system also provides a mechanism for evaluating the programs extracted from proofs.

Proofs in Nuprl are trees where each node has associated with it a sequent and a refinement rule. The children of a node are the subgoals which result from the application of the refinement rule of the node to the sequent. The refinement rule is either a primitive inference rule, or a program written in ML [8]. Such a program is called a *refinement tactic* (being similar to an LCF tactic [8]), and when given a sequent as input it applies primitive inference rules and other tactics to build a proof tree with the sequent as the root. This resulting proof tree is hidden except for its unproved leaves; these become the children of the input sequent, and the name of the tactic becomes its associated rule. Tactics, then, act as derived inference rules: the Nuprl display of a node of a proof tree can show as an individual step a tactic invocation; such a higher level step is correct because of the way the type structure of ML is used. For more on the Nuprl tactic mechanism, see [6]. There is one substantial decision procedure which is not a tactic, i.e., which is part of the Nuprl system and is invoked as a primitive inference rule. This procedure is called **arith**, and it proves subgoals which follow by certain simple kinds of reasoning about the primitive relations (equality and less-than) over the integers.

The basic component of a Nuprl session is the *library*, which contains a linearly ordered collection of definitions and theorems. Proofs are stored with the theorems. One interacts with Nuprl by creating, deleting, and manipulating objects using special purpose editors.

Some of the details regarding the components of Nuprl just discussed will be given as necessary in what follows. For a complete account of the system, we refer the reader to [4].

3 Representation

In this section we discuss how we represent some basic concepts of elementary number theory, and then we present the statements of the main theorems and of some of the important lemmas. An important point to note here is the conciseness and readability of the Nuprl mathematical definitions and statements we have written. In what follows, when we exhibit Nuprl objects what is presented is in exactly the same form (except for small differences in white space) as would appear on the screen during a Nuprl session.

The complete self-contained library constructed for the fundamental theorem of arithmetic contains 59 definitions. Of these, 36 are for generic objects, such as the logical connectives, which would be of use in any mathematical theory built in Nuprl. There are 15 general definitions dealing with basic list and integer relations and types, and only 8 which are in any way particular to the development of this theorem.

As indicated earlier, the logical connectives of predicate calculus have direct encodings in the Nuprl type theory. An important point is that one need not be aware of these representations. The definition mechanism can be used to suppress their display, and the Nuprl rules which apply to the representations are just what one might expect for the corresponding logical notions (interpreted constructively).

The definition mechanism of Nuprl is basically a macro facility. For the definitions for logic, we use this mechanism directly. For mathematical definitions, however, we employ a level of indirection in order to achieve a kind of abstraction. As a simple example we consider the definition of `N`, the natural numbers. This can be defined in terms of the (primitive) type of integers, `Int`, using the subtype constructor. To do this, we first prove a theorem (named `N_`) whose statement and first refinement rule is:

```
>> U1
BY explicit intro { n:Int | 0≤n }
```

`U1` is the type of all ("small") types, and we prove that it has a member by explicitly introducing the type we wish to define. The Nuprl definition for `N`, then, will reference this theorem instead of the actual term; this reference is made using the Nuprl term `term_of(`*theorem-name*`)`, which denotes the term extracted from named theorem. Thus `N` appearing in Nuprl text is just a display form for `term_of(N_)`, which in turn denotes `{n:Int|0≤n}`. More generally, we will use lambda abstraction; e.g., if we were to redefine integer addition in this way, we would prove a theorem `>> Int->Int->Int` by introducing `λx.λy.x+y`. This technique has several advantages, the most important of which is that it associates a type with each defined object. This makes it possible to construct an effective membership tactic (to be described in the next section) without which our proofs would have been unbearably tedious.

```
Fact:
U1
({2..} # N+) list

ordered:
Fact -> U1
λl. ∀ tails h·t of l. if hh = hd(t) then h.1 < hh.1

divides:
Int -> Int -> U1
λ i n. ∃ k:Int where i*k = n

prime:
Int -> U1
λ n. 1<n & ∀ i:{2..n-1}. ¬ i|n

all_prime:
Fact -> U1
λ l. ∀ tails h·t of l. h.1 prime

exp:
Int -> N -> Int
λ m n. ind(n; n,y.0; 1; n,y.m*y)

eval:
Fact -> Int
λ l. list_ind(l; 1; h,t,v. v * h.1↑h.2 )

PrimeFact:
U1
{l:Fact | l ordered & all factors in l prime }
```

Figure 1: The main definitions.

In figure 1 we show the 8 definitions particular to this theory. The name of each definition is shown followed by the statement of the theorem it is extracted from (which is the type of the object), followed by the actual object being defined. The type **Fact** has a definition which uses two previously defined types and the primitive type contructors **list** and **#** (the cartesian product constructor). Thus **Fact** is the type of all lists of pairs of integers where the first integer is at least 2 and the second is positive; these lists will represent the factorizations of integers into products of powers. The function **eval** is used to multiply out such factorizations; its definition uses the primitive Nuprl form **list_ind** for recursion over lists and also references the definitions for integer exponentiation **exp** (whose display form uses ↑), and for projection from pairs (denoted **.1** and **.2**). Informally, this definition of a recursive function can be read as follows. Given a list **l**, if **l** is **nil** then the value is **1**; otherwise, **l** is **h.t** (the dot serves as the Nuprl notation for *cons*) and so first compute the value **v** of the function on **t**, then multiply it by the result of applying **exp** to the two integers comprising the pair **h**. For example, the result of computing the expression **eval(<3,2>.<4,1>.nil)** would be 36. **prime** is defined in terms of **divides** (the display form for the divides relation uses "|", and a dot is used to separate the parts of the quantified formula). The definitions of **ordered** and **all_prime** use definitions which are based on **list_ind**. The definition

```
∀ tails h·t of l . P
```

is a recursively defined predicate which is true for a list l if P is true whenever the cons of **h** and **t** is a tail (or suffix) of the list **l**. **if h = hd(l) then** P is true if **l** is **nil** or if **l** is not **nil** and P is true for **h** the head of **l**. This kind of predicate is noteworthy, since it illustrates the expressive power of Nuprl's higher order logic; types are first-class citizens of the theory, and so the same form which is used to define recursive integer-valued functions over lists can also be used to recursively define type-valued functions, and hence predicates (since we represent propositions as types), over lists. Finally, the type **PrimeFact** of all prime factorizations can be defined as a subtype of **Fact** using previous definitions (**all factors in l prime** is the display form of the definition **all_prime**, etc.). For example, the prime factorization of 12 is **<2,2>.<3,1>.nil**.

Most of the lemmas proved in this theory concerned elementary properties of the defined objects, such as the fact that the value of **eval** was always at least 1. Many of these were discovered during the course of proving the major lemmas. These major lemmas, however, were straightforward expressions of lemmas used in the informal proof which the formal proof was based on.

Three of the major lemmas express familiar properties of the integers:

```
>> ∀ i,j,p:N+ where p prime & p|i*j. p|i ∨ p|j

>> ∀ i,n:Int. i|n ∨ ¬ i|n

>> ∀ a,b:N+ where ( ∀ d:{2..}. ¬(d|a & d|b) ).
            ∃ m,n:Int. m*a + n*b = 1
```

All three of these have interesting computational content. For example, the system can extract from the first theorem a (proven correct) program which takes three positive integers i, j and p, where p is a prime dividing the product ij, and returns a value which indicates whether the prime divides i or j, along with the appropriate factor. The program extracted from the third lemma takes two relatively prime positive integers as input and returns a pair of numbers which are the coefficients of a linear combination of the inputs which equals 1.

The existence part of the main theorem is

```
>> ∀ n:{2..}. ∃ l:PrimeFact where eval(l) = n
```

and its proof is just an application of the lemma

```
>> ∀ k:N. ∀ n,i:{2..} where i≤n & n-i≤k & ( ∀ d:{2..(i-1)}. ¬(d|n) ).
            ∃ l:PrimeFact where eval(l) = n
```

This lemma is a recasting of the existence part of the main theorem in a form which allows us to carry more information through the induction (the main step of the proof is to do induction on **k**). The **where** clause of this lemma can be viewed as a loop invariant.

Finally, we have the uniqueness half of the fundamental theorem preceded by two supporting lemmas.

```
>> ∀p1,p2,i:Int. p1 prime => p2 prime => p1<p2 => 0<i => ¬(p1|(p2↑i))
```

```
>> ∀l:PrimeFact. ∀ p:Int. p prime => (if h = hd(l) then p<h.1) => ¬ p|eval(l)
```

```
>> ∀ l1,l2:PrimeFact. eval(l1) = eval(l2) => l1 = l2 in Fact
```

In the immediately preceding theorem, **in Fact** is required in order to indicate that the equality relation is over the type **Fact**.

4 Proofs and Tactics

It is the combination of the proof editor, high level metalanguage, and definition mechanism which gives proofs in Nuprl their distinctive character. It provides a basis for the construction of understandable proofs where the component formulas have a form which makes their meaning apparent, and where the inference steps follow an understandable course. At the present, the construction of powerful tactics for use in formalizing mathematics is at a beginning stage, and so the degree of automation of the proving process is rather small in comparison to, say, the system of Boyer and Moore. In this section we attempt to convey the flavour of proofs in Nuprl by discussing some of the tactics used and by discussing the proof of an important lemma. First, however, we give some general information about the experiment.

All of the proofs constructed in this effort were done using only general purpose tactics. These tactics were designed beforehand, without number theory as a target, to be of use in most theorem proving efforts. The total time required to complete the library was under forty hours. This time includes all work relevant to the effort; in particular, it includes the time spent on entering definitions, on informal planning, on lemma discovery and aborted proof attempts, and on proving all the necessary results dealing with the basic arithmetic operators and relations. We estimate that at least half of this time is due to certain gross inefficiencies of the current implementation (having in part to do with the maintenance of display forms for definitions) that we believe are simple to correct and that should not be present in the next version of the system. It is interesting to constrast this figure with the approximately 8 weeks of effort required to prove the same theorem in the PLCV system [5], a natural deduction system for reasoning about PL/C programs which had powerful built-in support for arithmetical and propositional reasoning but which had no tactic mechanism or proof editor. Boyer and Moore also conducted a proof of the fundamental theorem of arithmetic, but they do not say how long the effort took (although they do say that it took only ninety seconds for the final sequence of definitions and lemmas to be checked). Thirty-four theorems were proved in our effort, most of which were simple properties of the defined objects.

Much of the work went into the uniqueness theorem and a major lemma for the existence theorem. The final collection of proofs contains 879 refinement steps, most of which were entered by the author (some were automatically applied by a very primitive analogy tactic). This number might seem somewhat large, but we believe that current tactic construction efforts will quickly add large increases in power to the system.

An important point concerns the relationship between the informal proof sketch and the formal Nuprl proof. The informal proof was two pages long and fairly detailed, but no consideration was given to the type theoretic encodings or to proof obligations arising specifically because of Nuprl. However, this sketch was able to serve as a guide; the progression of refinement for the main lemmas of the library for the most part followed the informal argument. Most of the more tedious steps due to formalization occurred near the leaves of the proof trees, and these lower level tactic-generated proof obligations often suggested useful lemmas.

The most important tactic was the *autotactic*. This tactic is generally automatically applied to any unproved subgoals that result when the user invokes a refinement tactic. Only the unproved subgoals generated by the autotactic appear on the screen as children of the refinement, and so the user need never be aware of the many details handled by it. This tactic will prove subgoals which follow by certain kinds of equality and arithmetic reasoning, and also has a component called **Member** which attempts to prove *membership* subgoals, i.e., subgoals requiring proof that some term is a member of a certain type. These subgoals arise because of the nature of Nuprl's type theory; for example, one must prove the well-formedness of all formulas introduced into the proof, and these well-formedness obligations have the form of membership subgoals. Because of the number of membership subgoals which arise, and because of the uninteresting nature of the vast majority of them, Nuprl would be unusable if it were not possible to handle automatically most of the work of proving them. There is no algorithm to prove all true subgoals of this form, since, for example, this is the form of a statement that a program meets its specification. Therefore, **Member** was designed with user participation as a primary concern. Such a concern is to some extent in conflict with the other main purpose of such a tactic, which is to automate as much of the proving process as possible. However, attempting to reduce a membership goal to the simplest possible subgoals will often result in some of those subgoals being false. At present, **Member** is rather conservative, stopping whenever the next step might create a false subgoal. For example, it makes no attempt to prove the validity of an application of a partial function (i.e., a function whose domain is a subtype), since this can involve proving an arbitrary proposition. Even so, **Member** was able to prove almost all of the membership goals which arose. Usually in the cases where it failed to complete a proof, it succeeded after one user-provided step. A crucial factor in the success of this tactic was the "**term_of**" style of definition, since it provided types for defined objects.

Most of the other tactics used are of a more familiar nature. We will briefly describe a few tactics which are somewhat representative of the collection used. Particularly important were the simplification tactics; e.g., **Normalize** was used to put sequents into a kind of normal form. Also important were tactics based on pattern matching. For example, the tactic **Lemma** takes as an argument the name of a lemma and attempts to find an instance of the lemma to apply to the goal, generating as subgoals the non-trivial hypotheses of the lemma instance. There is also **Backchain**, which applies simple backchaining from the conclusion of the goal via the hypotheses of the goal. Nuprl has only one integer induction rule, so several tactics were written to emulate several other common forms of integer induction. Recursive definitions were extensively used in our library, so fairly frequent use was made of tactics which performed unfolding of recursive definitions.

Also used were forms of the LCF tactic-combining functions, such as **THENW** which applies its first argument and then applies its second argument to any remaining subgoals which are not "well-formedness" ones (i.e., subgoals requiring one to show that some term is a well-formed type — these were all handled by the autotactic). An interesting kind of tactic is exemplified by the pair **SquashElim** and **SquashIntro**. They apply to a defined construct we call a "squash", which we denote $\downarrow P$ for P a proposition. The purpose of this operator has to do with information hiding, but the interesting point is that one can use it, reasoning about it via the pair of tactics, without knowing the details of its definition. We are currently working on a general framework for this kind of abstraction mechanism.

Space limitations prevent us from giving a thorough account of the proofs contructed in this effort. Instead we focus on the proof of one of the more important lemmas. This lemma, whose statement appears in the preceding section, is the one which states that any two relatively prime positive integers have a linear combination equal to one, and its proof is somewhat representative of the other proofs in the library. The first step in the proof is to assert a form of the lemma which will give a stronger induction hypothesis. This step gives two subgoals; the nontrivial one is to prove that the new form of the lemma is true. This subgoal, along with the rule applied to it, is shown in figure 2. What is shown is part of a snapshot of the screen of a Nuprl session; window borders have been removed, but otherwise the contents of the figure are what a Nuprl user would see. The first line contains the status of the proof ("*" means that the proof is complete) and an address of the current node within the proof tree. The four major components below the first line are, from top to bottom, the goal of the node, the rule applied to it, and two subgoals generated by the rule application. The numbered vertical lists of formulas in the subgoals are the hypotheses lists. The rule used here was actually a refinement tactic corresponding to the informal step "do induction on k". **THENW** was used to chain together a tactic which stripped off one universal quantifier and made a corresponding new hypothesis, and a tactic which performed induction with a specified base case (the arguments "`1 1 'j'`" are, respectively, the hypothesis number of the variable (k here) induction is being done on, the base of the induction, and a new identifier). This simple looking refinement step actually hides 58 primitive refinements. An interesting point that this snapshot illustrates is that although we are proving that there is a program that performs a certain task, we are not (explicitly) reasoning about computational objects but instead are dealing with something more like conventional mathematics.

To convey what the rest of the proof of the relative primes lemma is like, we give an informal description of some of the other proof steps. The entire proof of the lemma required 44 refinement steps. The main step is the induction step just described. The first step in the proof of its second

```
* top 1
>> ∀ k:{2..}. ∀ a,b: N+
       where a+b<k & ∀ d:{2..}. ¬(d|a&d|b).
     ∃ m,n:Int. m*a + n*b = 1

BY -- Intro THENW NonNegInductionUsing 1 1 'j'

1* 1. k:{2..}
   2. j:int
   3. j=1 in int
   >> ∀ a,b: N+
          where a+b<j & ∀ d:{2..}. ¬(d|a&d|b).
        ∃ m,n:Int. m*a + n*b = 1

2* 1. k:{2..}
   2. j:int
   3. 1<j
   4. ∀ a,b: N+
          where a+b<j-1 & ∀ d:{2..}. ¬(d|a&d|b).
        ∃ m,n:Int. m*a + n*b = 1
   >> ∀ a,b: N+
          where a+b<j & ∀ d:{2..}. ¬(d|a&d|b).
        ∃ m,n:Int. m*a + n*b = 1
```

Figure 2: A snapshot of the main induction step of the relative primes lemma.

subgoal (the inductive case) corresponds to the informal step "suppose **a** and **b** are such that ...", and generates one subgoal, with conclusion (the part of the sequent after the **>>**) the existential statement from the snapshot. The next step, resulting in three subgoals, is to do a case analysis on whether **a<b**, **a=b**, or **b<a**. The first step in the proof of the first case is to apply the inductive hypothesis (numbered 4 in the snapshot) to **a** and **b-a**; this step generates three subgoals. The first of the three is to show that **b-a** is positive; this is done in one further step. The second is to show that **a** and **b-a** satisfied the **where** clause of the induction hypothesis; this is done in seven additional steps. The third is to show that supposing there is a linear combination of **a** and **b-a** equal to 1, one for **a** and **b** can be found. This requires two extra steps. The rest of the proof of the lemma is at about the same level, except for several trivial steps where lemmas about the monotonicity of less-than with respect to addition had to be explicitly applied.

We end this section with a word about the extracted program. As we mentioned earlier, each Nuprl proof implicitly defines a correct program whose specification is the statement of the theorem. The program extracted from the existence part of the fundamental theorem of arithmetic takes as input a number greater than one, and returns its factorization into primes. The program is not hopelessly inefficient; using a fairly naive interpreter and no preprocessing, it took about ten seconds (on a Symbolics 3670) to factor 100!, (a number with 158 digits).

5 Conclusions, Directions For Further Work

We have written a collection of tactics and used it within the framework of the Nuprl proof development system to construct a highly readable and mathematically comprehensible formalization of a substantial theorem of number theory. The average size of a refinement step was rather low in the resulting proofs, however. More powerful tactics will be required (and are under construction) for the more ambitious projects in formalized mathematics that we believe are possible using Nuprl. One such project is the current work of the author, to formally prove a major theorem of constructive analysis. A significant problem we have encountered concerns the speed of refinement. There are certain gross inefficiencies present in the current system which seriously hamper theorem-proving activity. These inefficiencies can, for the most part, be corrected, but there is a more fundamental problem. As in LCF, tactics work by applying primitive inference rules, and so derived rules of inference that are encapsulated as tactics must be rejustified at each application. This is a significant obstacle to increasing tactic power; for example, term rewriting tactics are prohibitively slow if required to use the substitution rule for each individual application of a rewrite rule. A solution for this problem is to use a reflection technique, using the data types of the theory to represent classes of Nuprl terms, writing in Nuprl the desired term-rewriting programs, and providing a tactic which applies the results of these programs to the Nuprl terms being reasoned about. The approach to rewriting of Paulson [14] will be useful in this context. Also, the higher-order nature of the Nuprl logic allows kinds of abstraction which will greatly aid this approach. See [10] for more on this scheme. Also relevant here is [11], in which is described a technique for using Nuprl to reason about tactics in order to avoid, in many common cases, having to run them. Another problem we are currently addressing concerns the structure and use of developed theories. At the present, the collection of facts in a Nuprl library has little structure, and the user must invoke explicitly many of the lemmas that are required in a proof.

6 Acknowledgements

We wish to thank Rance Cleaveland, Bob Constable, Todd Knoblock, and Prakash Panangaden for their helpful suggestions concerning the presentation of this material.

References

[1] E. Bishop. *Foundations of Constructive Analysis.* McGraw–Hill, 1967.

[2] R. Boyer and J S. Moore. *A Computational Logic.* Academic Press, NY, 1979.

[3] N. G. deBruijn. A Survey of the Project AUTOMATH. In *Essays in Combinatory Logic, Lambda Calculus, and Formalism*, J. P. Seldin and J. R. Hindley, eds., pages 589–606. Academic Press, 1980.

[4] R. Constable, et al. *Implementing Mathematics with the Nuprl Proof Development System*, Prentice Hall, 1986.

[5] R. Constable, S. Johnson, and C. Eichenlaub. *Introduction to the PL/CV2 Programming Logic*, *Lecture Notes in Computer Science, vol. 135*, Springer–Verlag, Berlin, 1982.

[6] R. Constable, T. Knoblock, and J. Bates. Writing Programs that Construct Proofs. *Journal of Automated Reasoning* v.1 n.3, (1985), pages 285–326.

[7] T. Coquand & G. Huet. *Constructions: A Higher Order Proof System for Mechanizing Mathematics*. EUROCAL 85, Linz, Austria, April 1985.

[8] M. Gordon, R. Milner, and C. Wadsworth. *Edinburgh LCF: A Mechanized Logic of Computation. Lecture Notes in Computer Science, vol. 78*, Springer-Verlag, Berlin, 1979.

[9] D. Howe. *Implementing Analysis*. PhD Dissertation, Department of Computer Science, Cornell University, 1986 (expected).

[10] D. Howe *Reflected Term Rewriting In Type Theory*. Technical Report, Department of Computer Science, Cornell University, 1986 (*to appear*).

[11] T. Knoblock, R. Constable. *Formalized Metareasoning in Type Theory*. Technical Report TR 86-742, Department of Computer Science, Cornell University, 1986.

[12] Z. Manna, R. Waldinger. *A Deductive Approach to Program Synthesis. ACM Trans. on Prog. Lang. and Sys.* v.2 n.1 (January 1980), pp 90-121.

[13] Per Martin-Löf. Constructive Mathematics and Computer Programming. In *Sixth International Congress for Logic, Methodology, and Philosophy of Science*, North Holland, Amsterdam, 1982, pages 153–175.

[14] L. Paulson. A Higher-Order Implementation of Rewriting. *Science of Computer Programming* n. 3, 1983, pp 119-149.

PARALLEL ALGORITHMS FOR TERM MATCHING

Cynthia Dwork
IBM Research Lab, San Jose and MIT Lab for Computer Science

Paris Kanellakis[1]
Brown University and MIT Lab for Computer Science

Larry Stockmeyer
IBM Research Lab, San Jose

Abstract: We present a new randomized parallel algorithm for term matching. Let n be the number of nodes of the directed acyclic graphs (dags) representing the terms to be matched, then our algorithm uses $O(\log^2 n)$ parallel time and M(n) processors, where M(n) is the complexity of n by n matrix multiplication. The number of processors is a significant improvement over previously known bounds. Under various syntactic restrictions on the form of the input dags only $O(n^2)$ processors are required in order to achieve deterministic $O(\log^2 n)$ parallel time. Furthermore, we reduce directed graph reachability to term matching using constant parallel time and $O(n^2)$ processors. This is strong evidence that in practice, taking M(n) to be n^3, no deterministic algorithm can beat the processor bound of our randomized algorithm. We also improve the lower bound of [DKM] on the unification problem. We show that unification is logspace-complete in PTIME even if both input terms are linear, i.e., no variable appears more than once in each term.

1. INTRODUCTION

Unification of terms is an important step in resolution theorem proving [R], with applications to a variety of symbolic computation problems. In particular, unification is used in PROLOG interpreters [CM, Ko], type inference algorithms [M], and term rewriting systems [HO]. Informally, two symbolic terms s and t are unifiable if there exists a substitution of additional terms for variables in s and t such that under the substitution the two terms are syntactically identical. For example, the terms f(x,x) and f(g(y),g(g(z))) are unified by substituting g(z) for y and g(g(z)) for x.

Unification was defined in 1964 by Robinson in his seminal paper "A Machine Oriented Logic Based on the Resolution Principle" [R]. Robinson's unification algorithm required

[1]Research supported partly by an IBM Faculty Development Award, and partly by NSF grant MCS-8210830.

time exponential in the size of the terms. The following years saw a sequence of improved unification algorithms, culminating in 1976 with the linear algorithm of Paterson and Wegman [PW]. A general interest in parallel computing, together with specific interest in parallelizing PROLOG, led Dwork, Kanellakis and, Mitchell to search for a fast (time polynomial in log(n)) processor efficient (polynomially many processors) parallel unification algorithm [DKM]. Their results were negative: they proved that unification is complete for polynomial time, even if the input terms are represented as trees. A similar lower bound was independently derived in [Ya]. (The [Ya] bound is slightly weaker since it assumes that the input terms are encoded as directed acyclic graphs). Thus, the existence of a fast, efficient parallel algorithm is "popularly unlikely", in that it would contradict the popularly believed complexity theoretic conjecture that PTIME, the class of problems solvable sequentially in polynomial time, is not contained in NC, the class of problems solvable in polylog time using polynomially many processors. However, [DKM] found that term matching, a special case of unification in which one of the terms contains no variables, is in NC. Specifically, they obtained a matching algorithm requiring $O(\log^2 n)$ time and about $O(n^5)$ processors. Motivated by [DKM] some researchers interested in parallelizing PROLOG examined extant PROLOG programs to see whether in practice unification can be replaced by term matching. Preliminary results show that "often" the full power of general unification is not needed, and that term matching indeed suffices [MK].

Although the [DKM] algorithm is in NC, and is therefore "efficient" in a theoretical sense, the number n^5 of processors is unacceptable in practice. The current paper provides substantially improved upper bounds on processors for the term matching problem (i.e., $O(n^3)$ in practice), at no asymptotic cost in running time. The model of computation is the Concurrent Read Exclusive Write parallel RAM [FW], with word size $O(\log n)$ on inputs of length n.

Let s and t be two terms. We say that *s matches t* if there exists a substitution σ mapping the variables in s to terms, such that $\sigma(s)$, the term obtained by replacing each occurrence of each variable x in s by $\sigma(x)$, is syntactically equal to t. To state our results on matching it is necessary to describe the manner in which we represent the inputs to our algorithms. Terms are represented by labelled directed, acyclic, graphs (*dag*'s) as follows:

1. every vertex is labelled by either a variable or a k-ary function symbol ($k \geq 0$);

2. vertices labelled by variables have outdegree 0;

3. a vertex labelled by a k-ary function symbol has outdegree k and its outedges are labelled 1 through k; the term represented by the subdag rooted at the head of the ith edge is the ith argument of the function.

Although in practice, function symbols might have arities bounded above by some fixed constant, such a bound is not required by our algorithms; arities can be as large as n, the number of vertices in the dag. Figure 1 shows some examples of labelled dags and their corresponding terms. Note that if the term contains a repeated subexpression then its corresponding dag is not unique.

The input to our matching algorithm is a (not necessarily connected) labelled dag with two roots. The two terms s and t are the terms represented by the subdags rooted at the roots. Our principal result is a randomized algorithm for term matching requiring $O(\log^2 n)$ time and M(n) processors. This represents a substantial improvement over [DKM], which requires $M(n^2)$ processors, where $M(n) \approx n^{2.49}$ is the best known sequential bound for multiplying two $n \times n$ matrices [CW]. The theoretical improvement we get is therefore from about $O(n^5)$ processors to $O(n^{2.49})$ processors.

Although the Coppersmith and Winograd algorithm is fast asymptotically, in practice, it is less efficient than the obvious n^3 algorithm for "reasonable" values of n. In fact, despite the importance of matrix multiplication, we know of no case in which either [CW] or Strassen's [St] asymptotically fast algorithms are used in practice. Taking M(n) to be n^3 in practice yields an improvement over the [DKM] bound from $O(n^6)$ to only $O(n^3)$ processors.

Our notion of *randomized algorithm* is the one used by Rabin [Ra] (in the sequential case) and by Karp, Upfal and Wigderson [KUW] in the parallel case. Specifically, each processor can, at any step, flip an unbiased coin and use the outcome in its computation. The coin flips of different processors are independent. When applied to terms s and t which are unifiable, our algorithm always produces a unifying substitution. If the algorithm is applied to an s and t which are not unifiable, then with probability bounded above by a constant (say, 1/2) the algorithm might say that the terms are unifiable and produce a substitution which is not a unifying substitution. However, by repeating the algorithm c times (either sequentially or in parallel), the error probability decreases to at most 2^{-c}.

The study of parallel algorithms for fundamental problems is currently a very active area in theoretical computer science (a few recent papers are [A, Ga, GP, KUW, PR, Re, TV, Vi]). Randomization has also played an important role in some of this work. For example, [KUW] give a randomized polylog time algorithm using polynomially many processors for finding a maximum matching in a graph.[2] . Without randomization, an algorithm with this running time and number of processors is not known. Reif [Re] and Vishkin [Vi] have reduced the known processor requirements for certain problems by using randomization.

---

[2]Matching in graphs and matching of terms are two totally different problems

Vitter and Simons [VS] examine how limited parallelism can be used to give limited speed-ups for the general unification problem. (Because of the PTIME-completeness result of [DKM], limited speed-ups are all one should expect for unification in general.) Our approach for term matching is to aim for very small parallel running times (in particular, polynomial in log(n)), and then try to decrease the number of processors as much as possible. This approach has been taken for other problems, e.g., [KUW, GP, Re], and it is justified by the practical and fundamental importance of the term matching problem.

To explain some additional results, we need a few more definitions. Given a labelled dag G, let t_v denote the term represented by the subdag of G rooted at v, where v is any vertex of G. A dag is a *tree* if no vertex has indegree greater than 1. A dag is *compact* if for all pairs of vertices u and v, $t_u \neq t_v$. That is, a compact dag contains no repeated subdags.

In addition to our algorithm for matching with arbitrary dags, we obtain two stronger results for the cases in which t, the constant term to be matched, is represented by a tree or a compact dag. Both of these algorithms are deterministic and require only $O(n^2)$ processors, although they require the dags representing the two terms to be disjoint. (The disjointness assumption can be removed, causing the number of processors to increase to M(n) but retaining determinism.) The simplest algorithm is for the case in which t is represented by a compact dag. After a first stage common to all three algorithms it requires only one additional step. The algorithm for the case in which t is represented by a tree is more complicated, and relies on results of Tarjan and Vishkin [TV].

Although we have proven no lower bound for the term matching problem we have strong reason to believe that, in practice, no deterministic algorithm can beat the processor bound of our randomized algorithm. The evidence for this is as follows. Equivalence of terms is trivially reducible to term matching, since testing for equivalence is just matching in which neither term contains variables. Further, in this paper we prove that directed graph reachability is reducible to equivalence of terms within constant time using $O(n^2)$ processors. Thus, any algorithm for term matching solves the directed graph reachability problem. The parallel complexity of the reachability problem has been the subject of intensive research, but the best NC algorithm known requires matrix multiplication.

In addition to strengthening the known upper bounds on term matching, we strengthen the known lower bounds on unification by proving that unification is complete for PTIME even if the terms are represented by trees and are linear, where a term t is *linear* if no variable appears more than once in t. For example f(x,y) is a linear term and f(x,x) is non-linear. Restricting attention to linear terms is common in term rewriting systems. By insisting on terms represented by trees we make sure that our lower bounds do not depend on concise dag notation.

From the completeness proofs of [DKM] it follows that: unification of two trees is complete for PTIME even if each variable appears exactly in one tree (i.e., there is no sharing of variables) and there are at most two occurrences of the variable in that tree. In fact the reductions in [DKM] require that both terms be non-linear. If one of the terms to be unified is linear, and each variable appears exactly in one term (no sharing of variables), then it is not hard to see that unification belongs to NC. (We believe that our algorithms for term matching described above can be modified for this case). Thus the only remaining question, which is answered here is, the complexity of unifying linear terms- i.e., where each variable appears at most once in each term, but where there can be sharing of variables. The proof of completeness in this case is quite different and more intricate than that of [DKM], and in a sense provides the strongest PTIME-completeness result possible.

2. PARALLEL ALGORITHMS FOR TERM MATCHING

Recall that in the term matching problem we are given a dag s containing variables and a constant dag t, and we want either to find a substitution of terms for the variables of s to make s and t syntactically equivalent or to determine that no such substitution exists. As explained in the Introduction, we have somewhat different results depending on the form of t. However, the general outline of the algorithm is the same in all cases. For exposition it is convenient to break the algorithm into four main steps. Since s is a general dag in all cases, we can assume without loss of generality that each variable of s is the label of exactly one vertex of s (multiple copies of x can be merged into one copy).

M1. A spanning tree r of s is formed by having each vertex v of s arbitrarily choose one of the edges which are directed into v.

M2. Embed r in t. Formally an *embedding* of r in t is a mapping d from the vertices of r to the vertices of t (not necessarily a one-to-one mapping) such that (i) d maps the root of r to the root of t, (ii) for all vertices u of r, u and d(u) are labelled by the same function symbol, and (iii) for all vertices u and v of r, if there is an edge labelled i from u to v then there is an edge labelled i from d(u) to d(v). If there is no embedding then clearly s and t are not unifiable.

M3. The embedding d gives a substitution of terms for all the variables of s. Specifically, if x is a variable of s and v is the unique vertex of s labelled by x, then the term rooted at d(v) in t is substituted for x. The substitution is performed by replacing all edges of the form (u,v) by (u,d(v)). If s and t are unifiable, then clearly this substitution is the only one possible. However, this substitution is not necessarily a unifying substitution since the spanning tree r may not contain all the edges of the dag s. There is one more thing to check.

M4. Let c be the constant dag obtained from s by making the substitution defined in M3. Check whether the constant dags c and t are equivalent, i.e., whether they define the same term. Then s and t are unifiable iff c and t are equivalent.

We now sketch the methods used in the parallel algorithms for each of these steps.

Assuming that each vertex has an ordered list of its parents, step M1, finding a spanning tree of s, can clearly be done in constant time using O(n) processors.

For the embedding step M2 we use the following simple lemma, whose proof is omitted. It is based on the idea of doubling the length of the embedded path with each iteration.

> **Lemma 1:** Given a directed path p with m vertices whose vertices are labelled with function symbols and whose edges are labelled with integers, and given a dag t with n vertices ($n \geq m$) there is an algorithm which checks whether p is embeddable in t (and produces an embedding if there is one) within time O(log n) using O(mn) processors.

We shall at various times use the following results of Tarjan and Vishkin [TV].

> **Lemma 2:** (Tarjan, Vishkin). There are parallel algorithms which, when given a tree T, produce (1) the depth-first numbering of T, (2) for each vertex u of T, the number of vertices in the subtree rooted at u, and (3) for each vertex u, the distance of u from the root of T. The algorithms take time O(log m) using O(m) processors, where m is the number of vertices of T.

The embedding of a tree r in a dag t is done by a recursive divide-and-conquer approach. Say that r has m vertices and t has n vertices.

E1. Find a vertex v of r and a set C of children of v such that the subtrees rooted at vertices in C contain between 1/3 and 2/3 of the vertices of r. The existence of v and C is well-known. To find v and C fast in parallel (in time O(log m) using O(m) processors) we use Lemma 2(2).

E2. Let p be the path from the root of r to v. Use Lemma 1 to embed p in t. If there is no embedding, then r is not embeddable in t. So assume that an embedding d is found.

E3. For each $u \in C$ which is the i-child of v, let d(u) be the i-child of d(v). Let r' be the subtree of r obtained by deleting the vertices of C (together with all their descendants). We have now reduced the original problem to $|C|+1$ subproblems: embed r' in t; for each $u \in C$, embed the subtree of r rooted at u in the subtree of t rooted at d(u). Note that each vertex of r appears in exactly one of these subproblems.

Solving the straightforward recurrences for the running time and the number of processors establishes the following.

> **Lemma 3:** The problem of embedding a tree r having m vertices in a dag t having n vertices can be done in time $O(\log^2(mn))$ using O(mn) processors.

Step M3 of the general term matching algorithm is straightforward and can be done in constant time using O(n) processors.

The portion of the term matching algorithm described so far can be summarized in the following key fact, where we now let n denote the total number of vertices in disjoint dags s and t.

Theorem 4: The problem of performing term matching for a dag s and a constant dag t is reducible, in time $O(\log^2 n)$ using $O(n^2)$ processors, to the problem of checking equivalence of constant dags c and t, where c has at most n vertices.

In the remainder of Section 2, we describe algorithms for checking equivalence of constant dags c and t, (as described in Theorem 4). There are three cases depending on the form of t. We present them in increasing order of difficulty. If u and v are two vertices of a dag, we say that u and v are equivalent if they describe equivalent subterms of the dag, i.e., if $t_u = t_v$.

2.1. t Is a Compact Dag

As before we find a spanning tree r of c, and embed r in t. (When equivalence is being checked as step M4 of term matching, this has already been done.) Let d be the embedding of r in t. For each edge e = (u,v) which is in c but not in the spanning tree r, we must check the following: letting i be the label of e and letting v' be the i-child of d(u) in t, d(v) and v' must be equivalent. Since t is compact, this can be checked simply by checking that v' = d(v). This establishes our result for compact dags.

Theorem 5: Term matching for a general dag s and a constant compact dag t can be done in time $O(\log^2 n)$ using $O(n^2)$ processors.

2.2. t Is a Tree.

In order to be able to quickly do all the checking of the form v' = d(v) just described, we first determine, for all vertices u and v of t, whether u and v are equivalent. In effect, we turn the tree t into a compact dag. We first observe that, since each term is described by exactly one tree, u and v are equivalent iff the subtrees rooted at u and v are identical, taking the edge and vertex labels into account as well as the structure of the subtrees. The following uses the depth-first numbering produced by Lemma 2(1) and is easy to prove.

Lemma 6: Checking whether two trees are equivalent can be done in time $O(\log n)$ using $O(n)$ processors.

Our approach is to first eliminate many pairs of vertices from consideration since they are obviously not equivalent, and then apply Lemma 6 in parallel to the remaining pairs. By using the notion of level defined next, we can immediately determine that many vertices are not equivalent. The *level* of a vertex v in a dag, denoted level(v), is the length of a longest path from v to some leaf of the dag.

For trees, we use the following lemma to compute levels. The proof uses Lemma 2(3), but details are omitted.

Lemma 7: Given a tree t, the level numbers of all the vertices can be found in time O(log n) using $O(n^2)$ processors.

Using Lemma 2(2) we also compute size(u), the size of the subtree rooted at u, for all u in t. Clearly, vertices u and v can be equivalent only if level(u)=level(v) and size(u)=size(v), but the converse is false. However, this observation allows us to bound the total cost of checking the subtree equivalences to yield the following.

Lemma 8: Given a tree t, we can determine all equivalent vertices in time O(log n) using $O(n^2)$ processors.

Applying the algorithm of Lemma 8 and then using the approach of Section 2.1 proves that $O(\log^2 n)$ parallel time and $O(n^2)$ processors suffice for trees. In fact we can do a better in terms of time.

Theorem 9: Term matching for a general dag s and a constant tree t can be done in time O(log n) using $O(n^2)$ processors.

Proof: We only have to argue for O(log n) instead of $O(\log^2 n)$ parallel time.

In the case that t is a tree, the only part of the algorithm which uses time log-squared is the embedding of the spanning tree r in the tree t (Lemma 3). Since r and t are both trees, the embedding can be done in time O(log n) as follows.

Construct the product graph G whose vertices are all pairs (u,v) such that u is a vertex of r and v is a vertex of t. There is an edge labelled i directed from (u,v) to (u',v') if there is an edge labelled i from u to u' in r and an edge labelled i from v to v' in t. Since r and t are both rooted trees, it is easy to see that G is a forest of rooted trees. Letting s and s' be the root of r and t, respectively, we want to find all vertices of G which are reachable from (s,s'), since the embedding maps u to v iff (u,v) is reachable from (s,s'). This can be done in time O(log n) using $O(n^2)$ processors by the Euler tour technique of Tarjan and Vishkin [TV].

2.3. t Is a General Dag

If t is a general dag, we do not know how to determine all equivalent vertices in polylog time using even close to M(n) or n^3 processors. However, in order to perform step M4 of the general term matching algorithm and to prove the main result claimed in the Introduction, we just have to produce a randomized parallel algorithm which checks two constant dags for equivalence in time $O(\log^2 n)$ using M(n) processors. Briefly, our approach is to represent the given constant dags, c and t, by multivariate polynomials, P_c and P_t, in such a way that c and t are equivalent iff P_c and P_t are equivalent. We then use a randomized algorithm of Schwartz [S] to check whether P_c and P_t are equivalent. Since Schwartz was only interested in sequential algorithms, we have some additional work to prove that the algorithm can be done in time $O(\log^2 n)$ using M(n) processors.

We first present our proof for $O(n^3)$ processors. We will have to modify the argument to get M(n) processors, however, the basic (and practical) idea works for naive matrix multiplication.

We first show how to define P_t from t (the definition of P_c from c is identical). First add a new vertex z and add an edge labelled 1 from each leaf of t to z. For simplicity, let us call the new dag t also. Next divide the vertices of t into levels, where the level of a vertex v is the maximum distance from v to z, as in the preceding section. It is finding the levels of the vertices that seems to require $O(n^3)$ processors. We now label the edges of t with indeterminates of the polynomials as follows. If v is a vertex of t which is on level j and which is labelled by the function symbol f and if e is an edge directed out of v and labelled i, then e is labelled with the indeterminate $f_i^{(j)}$. For each path ρ directed from the root of t to z, let the monomial $m(\rho)$ be the product of all the indeterminates labelling edges of ρ. P_t is defined to be the sum of $m(\rho)$ over all paths ρ from the root to z. The following can be proved by induction on the depth of the dags.

Lemma 10: For any two constant dags c and t, c and t are syntactically equivalent iff P_c and P_t are equivalent polynomials.

Remark. If the indeterminates were not superscripted with level numbers, Lemma 10 would not hold. This is because the indeterminates f_1 and f_2 commute, i.e. $f_1f_2 = f_2f_1$, but the directed paths labelled f_1f_2 and f_2f_1 are in general not the same in the dag t.

Schwartz's randomized algorithm for checking equivalence of polynomials $P(y_1,...,y_m)$ and $Q(y_1,...,y_m)$ is to choose independent random numbers from some range [-N,N] to assign to the indeterminates $y_1,...,y_m$, choose a random prime p from some range [2,M], and evaluate P (mod p) and Q (mod p) for the chosen assignment. If the answer is "not equal", then we know that P and Q are not equivalent. If the answer is "equal", then we know that P and Q are equivalent with probability at least 1/2, provided that N and M are chosen large enough. It turns out that in our case, N is an O(log n)-bit integer and $M = O(n \log n)$. We sketch how the three key steps of the algorithm are performed.

K1. Computing level numbers. Let t be a dag and let A be the adjacency matrix of t ($a_{ij} = 1$ if there is an edge directed from vertex i to vertex j, or $a_{ij} = 0$ otherwise.) By defining matrix multiplication so that the inner operation is "+" and the outer operation is "max", the (i,j)-entry of A^n is the length of the longest path from i to j. By repeated squaring, A^n can be computed in O(log n) matrix multiplications. Each matrix multiplication takes time O(log n) using n^3 processors.

K2. Picking a random prime. Using Rabin's randomized algorithm for testing primality [Ra], for each $z \leq M$, a group of O(log n) processors checks whether z is prime. The size O(log n) of the groups guarantees that the probability that all checks are correct is sufficiently close to 1. By sorting, construct a table of the primes $\leq$ M. Picking a random prime is now easy. (Of course, in practice, the table of primes could be precomputed once and for all, given an upper bound on the size of the terms which the algorithm is expected to handle.)

K3. Computing P_c (mod p) and P_t (mod p) given a prime p and an assignment of integers to the indeterminates. Consider, for example, P_t. Let B be the matrix such that b_{ij} is the sum of the indeterminates on all edges directed from vertex i of t to vertex j of t; if there are no edges then $b_{ij} = 0$. Assuming that the root of t is numbered 1 and the new vertex z is numbered n, it is not hard to see that P_t is the (1,n)-entry of

$$I + B + B^2 + B^3 + \dots + B^n$$

$$= (I+B)(I+B^2)(I+B^4)\dots(I+B^{n'})$$

where n' is the smallest power of 2 which is $\geq$ n. This computation takes O(log n) matrix multiplications. Since all arithmetic can be done modulo p, word size O(log n) suffices.

Putting this all together proves our result for general dags using $O(n^3)$ processors.

Lemma 11: Term matching for a dag s and a constant dag t can be done in time $O(\log^2 n)$ using n^3 processors.

A modification of the above argument leads to an algorithm using M(n) processors. The technical reason is that the [CW] algorithm uses only addition, multiplication, and integer constants. The basic ideas, without formal proofs, are as follows.

Instead of computing the level number of each vertex, we compute for each vertex v the number of paths in the dag from v to the single leaf z; call this the "path number" of v. [We first add a new edge directed from each nonleaf of the original dag to the new vertex z; this ensures that if u is an ancestor of v then the path number of u is at least 1 more than the path number of v.] Path numbers can be computed using O(log n) multiplications of n X n matrices, where matrix multiplication is the usual definition with inner operation multiplication and outer operation addition, so that the fast matrix multiplication algorithms work.

Actually, we cannot compute the exact path numbers because they might be too large; a dag with n vertices could have path numbers as large as n^n. Instead, we pick a random prime p and compute path numbers modulo p. If p is chosen from a large enough range, such as from 2 to $O(n^4)$, then the probability is very high that no two different path numbers will become equal when computed mod p. We define a polynomial from a dag as described above except that instead of using the level numbers as superscripts on the indeterminates, we use the path numbers mod p. Assuming that we have not been unlucky in choosing p, i.e., assuming that different path numbers map to different numbers when reduced mod p, it can be shown that two constant dags are syntactically equivalent iff their associated polynomials are equivalent.

To compute path numbers mod p using a fast matrix multiplication algorithm which uses only addition, multiplication, and integer constants, we reduce mod p after every arithmetic operation in the algorithm.

We also need a new method of choosing a random prime. Since we are choosing a prime from a larger interval I, up to about n^4, we cannot afford to check every integer in this range for primality. Instead, we first pick a set C of d log n random numbers from the interval I (where d is a large enough constant), test each integer in C for primality, and pick a random prime from the numbers in C which turned out to be prime. If d is large enough, the probability is very high that C contains at least one prime. Thus,

Theorem 12: Term matching for a dag s and a constant dag t can be done in (randomized) parallel time $O(\log^2 n)$ using M(n) processors.

3. REDUCTIONS

Theorem 13: The directed graph reachability problem is reducible to testing equivalence of terms in constant time using $O(n^2)$ processors, where n is the number of vertices in the graph.

Proof: Given a dag D_1 with distinguished vertices s_1 and t_1 we can determine whether D_1 contains a path from s_1 to t_1 by constructing two term dags which will be equivalent if and only if there is no such path.

We begin by turning D_1 into a labelled dag. For each k, $0 \leq k \leq n$, and for each vertex v of D_1 with outdegree k, label v with the function symbol f^k. Label the outedges from 1 to k in arbitrary order.

Create a copy D_2 of D_1, identical to D_1 but with t_2, the D_2 copy of t_1, labelled with a new function symbol g. Let s_2 be the D_2 copy of s_1. There is a path from s_1 to t_1 if and only if s_1 and s_2 are inequivalent. However, we are not done, since the s vertices are internal, and we have been assuming that an equivalence algorithm takes as input two roots.

Create a new dag E_1 composed of D_1 and D_2 by creating a new root R_1 and making all the roots of D_1 and D_2 children of R_1. Let s_1 also be a child of R_1. Let k be the resulting outdegree of R_1. Label the edge from R_1 to s_1 by 1 and label the remaining outedges of R_1 from 2 to k in any order. Label R_1 with the function symbol f^k.

Create a copy E_2 of E_1, identical to E_1 but with the 1-edge of R_2, the root of E_2, pointing to the E_2 copy of s_2 (instead of to the copy of s_1). The construction is illustrated in Figure 2.

R_1 and R_2 are the inputs to the equivalence algorithm.

The preceding reduction provided a form of lower bound for the number of processors necessary to solve the matching problem. The following theorem says that unification is complete in PTIME even if both terms are linear, i.e., contain at most one occurrence of each variable.

Theorem 14: Unification is complete for PTIME even if both terms are represented by trees and are linear.

The proof is by reduction from the circuit value problem [Go], and somewhat more involved than the construction in [DKM]. What is important in this case is that the two linear terms are allowed to share variables, i.e., in the worst case a variable might appear once in each term. If there is no sharing of variables the linearity of one of the terms suffices for unification to be in NC. Thus the new bound together with the result of [DKM], (that unification of trees is complete in PTIME if there is no sharing of variables and each variable occurs at most twice), provide, in a sense, the strongest possible lower bound.

Proof: The proof has the same structure as that of [DKM] with one important difference. Given a boolean circuit each one of its wires will correspond to four nodes in the trees representing our terms. There will be two possible ways that these four nodes can be unified corresponding to the wire carrying a 1 or a 0, (see Figure 3a). The subterms for the boolean gates are in Figures 3b and 3c.

4. CONCLUSIONS

We have strengthened the lower bounds for the operation of unification of terms, to include linear terms. We have also provided better algorithms for an important special case of unification namely term matching. The proposed algorithms are practical and utilize at most M(n) processors, at no asymptotic cost in running time. We have also provided some arguments that no deterministic algorithm can significantly beat the processor bound of our randomized algorithm. The precise lower bounds here are still an open question.

5. REFERENCES

[A] Aggarwal, A., Chazelle, B., Guibas, L., O'Dunlaing, C., and Yap, C., "Parallel computational geometry", *Proc. 26th IEEE FOCS*, 1985, pp. 468-478.

[CM] Clocksin, W.F., Mellish, C.S., *Programming in Prolog*, Springer-Verlag, 1981.

[CW] Coppersmith, D., and Winograd, S., "On the asymptotic complexity of matrix multiplication", *SIAM J. Comput. 11* (1982), pp. 472-492.

[DKM] Dwork, C., Kanellakis, P., Mitchell, J., "On the Sequential Nature of Unification", *J. of Logic Programming 1*(1), pp.35-50.

[FW] Fortune, S., Wyllie, J., "Parallelism in Random Access Machines", *Proc. 10th ACM STOC*, 1978, pp. 114-118.

[Ga] Galil, Z., "Optimal parallel algorithms for string matching", *Proc. 16th ACM STOC*, 1984, pp. 240-248.

[GP] Galil, Z., and Pan., V., "Improved processor bounds for algebraic and combinatorial problems in RNC", *Proc. 26th IEEE FOCS*, 1985, pp. 490-496.

[Go] Goldschlager, L.M., "The Monotone and Planar Circuit Value Problems are Log Space Complete for P", *SIGACT News 9*(2), 1977, pp. 25-29.

[HO] G. Huet, D. Oppen, "Equations and Rewrite Rules: A Survey", in: *Formal Language Theory: Perspectives and Open Problems*, R.V. Book (ed), Academic Press, 1980.

[KUW] Karp, R. M., Upfal, E., and Wigderson, A., "Constructing a perfect matching is in random NC", *Proc. 17th ACM STOC*, 1985, pp. 22-32.

[Ko] Kowalski, R., "Predicate Logic as a Programming Language". *Proceedings IFIP 74*, 1974, pp. 569-574.

[MK] Maluszynski J., Komorowski H.J., "Unification-free Execution of Horn-clause Programs", *Proc. 2nd Logic Programming Symposium*, IEEE, July 1985.

[M] Milner, R., "A Theory of Type Polymorphism in Programming", *JCSS 17*, 1978, pp. 348-375.

[PR] Pan, V., and Reif, J., "Efficient parallel solution of linear systems", *Proc. 17th ACM STOC*, 1985, pp. 143-152.

[PW] Paterson, M.S., Wegman, M.N., "Linear Unification", *JCSS 16*, 1978, pp.158-167.

[Ra] Rabin, M.O., "Probabilistic algorithm for testing primality", *J. Number Theory 12* (1980), pp. 128-138.

[Re] Reif, J., "Optimal parallel algorithms for integer sorting and graph connectivity", *Proc. 26th IEEE FOCS*, 1985, pp.496-505.

[R] Robinson, J.A., "A Machine Oriented Logic Based on the Resolution Principle", *JACM 12*(1), 1965, pp. 23-41.

[S] Schwartz, J.T., "Fast Probabilistic Algorithms for Verification of Polynomial Identities", *JACM 27*(4), 1980, pp. 701-717.

[St] Strassen, V., "Gaussian elimination is not optimal", *Numerische Mathematik 13* (1969), pp. 354-356.

[TV] Tarjan, R.E., and Vishkin, U., "Finding biconnected components and computing tree functions in logarithmic parallel time", *Proc. 25th IEEE FOCS*, 1984, pp. 12-20.

[Vi] Vishkin, U., "Randomized speed-ups in parallel computation", *Proc. 16th ACM STOC*, 1984, pp. 230-239.

[VS] Vitter, J.S., and Simons, R., "New classes for parallel complexity: a study of unification and other complete problems for P", *IEEE Trans. on Computers*, to appear.

[Ya] Yasuura, H., "On the Parallel Computational Complexity of Unification", Yajima Lab., Research Report, ER 83-01, Oct. 1983.

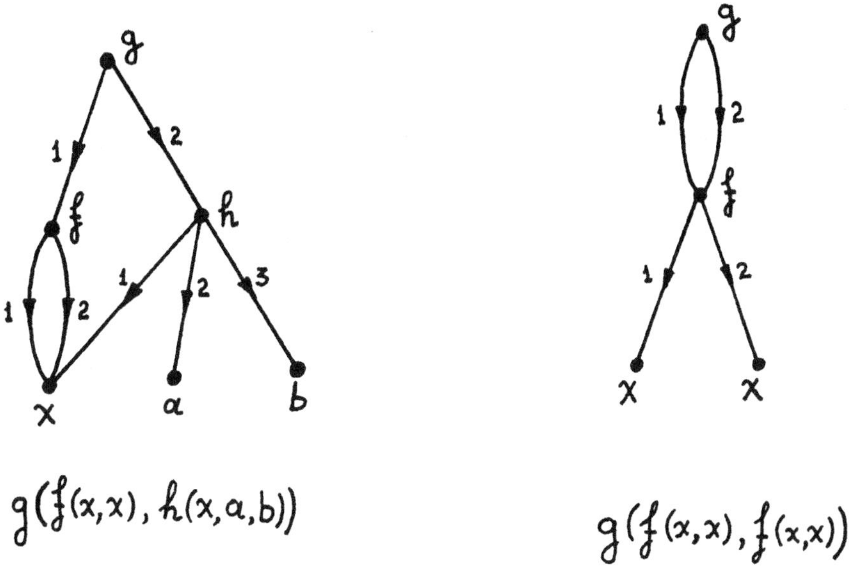

Figure 1

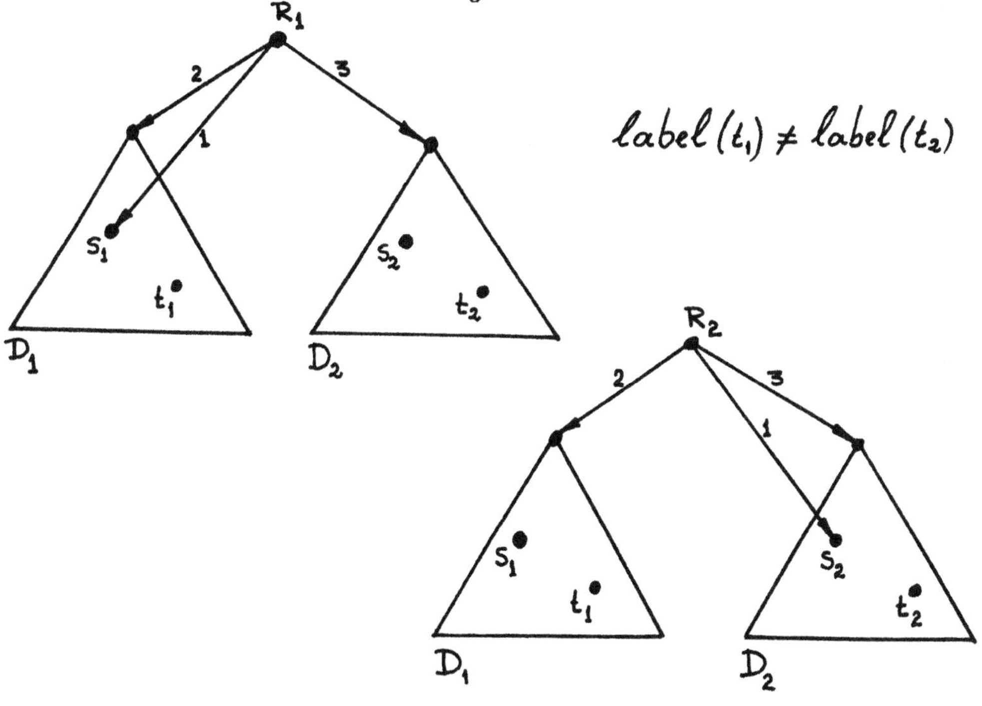

Figure 2

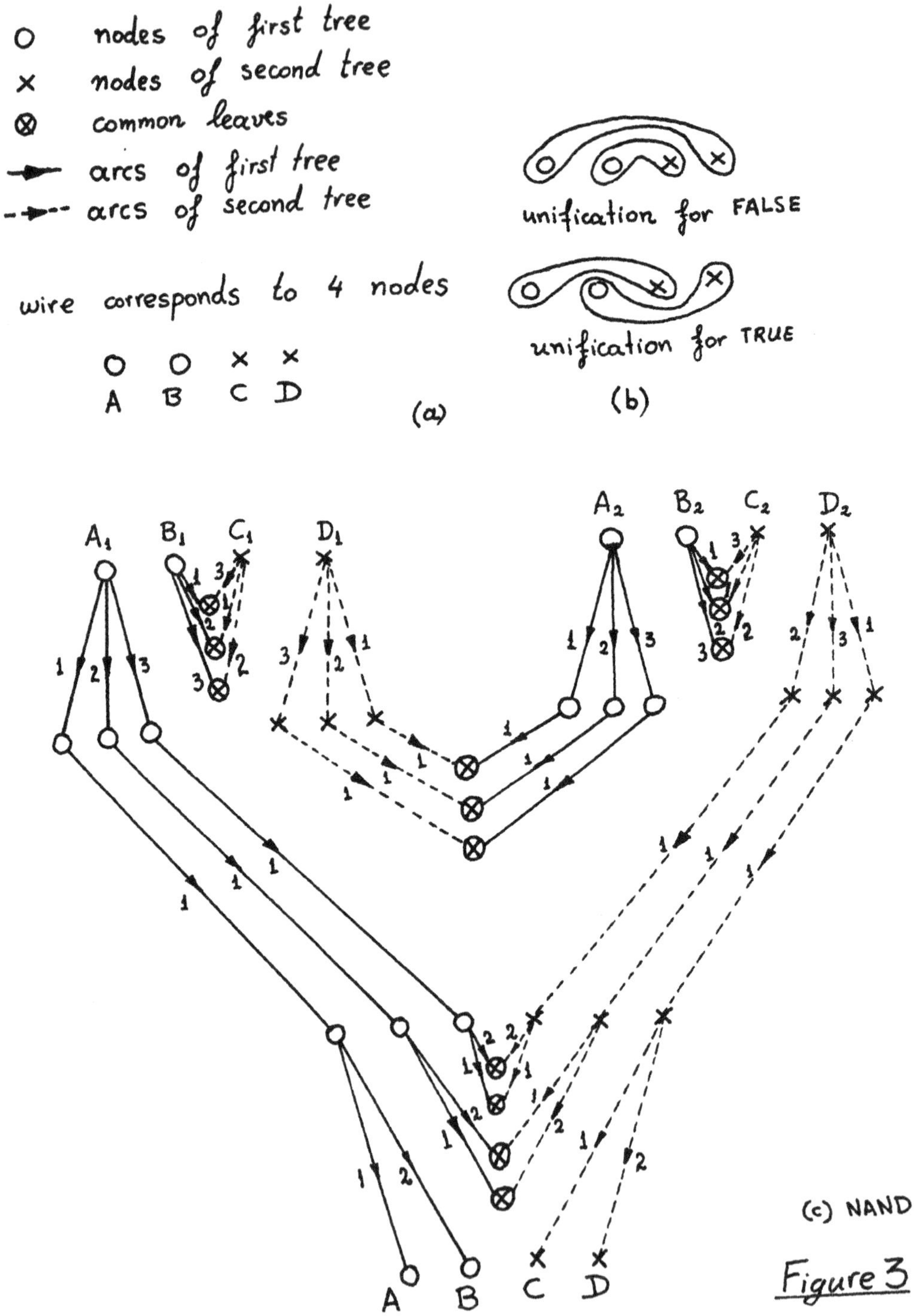
nodes of first tree
nodes of second tree
common leaves
arcs of first tree
arcs of second tree
wire corresponds to 4 nodes
A B C D
(a)
unification for FALSE
unification for TRUE
(b)
A_1 B_1 C_1 D_1
A_2 B_2 C_2 D_2
A B C D
(c) NAND

Figure 3

Unification in Combinations of Collapse-Free Theories with Disjoint Sets of Function Symbols

Erik Tidén

Royal Institute of Technology, Stockholm
Department of Computing Science *

Abstract

A unification algorithm for combinations of collapse-free equational theories with disjoint sets of function symbols is presented. The algorithm uses complete unification algorithms for the theories in the combination to compute complete sets of unifiers in the combined theory. It terminates if the algorithms for the theories in the combination terminate. The only restriction on the theories in the combination — apart from disjointness of function symbol sets — is that they must be collapse-free (i.e., they must not have axioms of the form $t = x$, where t is a non-variable term and x is a variable). This extends the class of equational theories for which the unification problem in combinations can be solved to collapse-free theories. The algorithm is based on a study of the properties of unifiers in combinations of non-regular collapse-free theories.

1. Introduction

Unification in equationally defined theories has a large number of applications and is also a fairly difficult problem in the sense that relatively few standard techniques are known and these do not seem to cover most cases. An important problem of a general nature is to characterize the unification properties of combinations of theories in terms of properties of their constituent theories. This is a largely unstudied problem, as is shown by the following quotation from Siekmann's state-of-the-art survey of unification theory [12]:

> *"Even the most trivial extension in this spirit, which is the extension of a known unification result to additional 'free' functions (i.e., the empty theory for every function symbol which is not part of the known unification result) as mentioned above is unsolved."*

An example of a unification procedure for a specific theory that was difficult to extend in this way is the procedure for associative-commutative functions [9,13]. The termination of these procedures was proved by Fages [3] after having been a well-known open problem for close to a decade. A more general problem than the extension of known unification procedures to additional 'free' functions is the problem of combining known unification procedures for equational theories with disjoint sets of function symbols. K. Yelick [17], C. Kirchner [6], and A. Herold [19], have recently each presented different algorithms that solve this problem essentially for the class of *regular* and *collapse-free* equational theories. (Here, a *regular* theory is a theory in which for every equation $s = t$ the two terms s and t contain the same set of variables. A *collapse-free* theory is a theory that contains no equation of the form $x = t$, where t is a non-variable term and x is a variable.) We extend these results to *non*-regular, but still collapse-free, theories. Our result is based on a study of the properties of unifiers in combinations of non-regular collapse-free theories which may

* Author's present address: Swedish Institute of Computer Science (SICS), Box 1193, S-163 62 Spånga, Sweden.

be interesting in its own right. Another apparently new result is a lemma that shows that the algorithm delivers complete sets of unifiers for some unification problems even in the presence of collapse axioms. This work was carried out independently of all three of the above authors, but we are indebted to both K. Yelick and C. Kirchner for pointing out an error in the termination proof in an early draft.

In the part that concerns the combination of regular theories, our algorithm can be seen as a generalization of Stickel's original method for dealing with associative-commutative unification (AC-unification) of general terms [13], i.e., terms that may contain arbitrary 'free' function symbols in addition to the AC-symbols. The termination proof for our algorithm is based on the technique developed by Fages [3]. Other sources of inspiration for this work are the solutions to the related problem of deciding formulas in combinations of unquantified first-order theories (*see, e.g.*, Nelson and Oppen [10], and Shostak [11]). Indeed, Shostak [11] sketches a procedure for solving a problem resembling the unification problem for combinations of certain unitary theories (i.e., theories in which two unifiable terms always have a single most general unifier).

We have not been able to construct a terminating algorithm for the case when one or both of the theories in the combination have collapse axioms. In conclusion, we offer a lemma which might be useful if this problem is to be solved. Note though that the solution of this problem is likely to be non-trivial, since it contains, as a special case, the unification problem for general terms in Abelian groups, which has been a known open problem for several years [8].

2. Equational theories and unification

We will use the standard concepts and notations of equational theories, unification, etc. (*see, e.g.*, Taylor [15], Huet and Oppen [5], Huet [4], and Lankford and Brady [7]), throughout this paper. However, for the sake of reference, some definitions and basic facts are repeated here.

Let $\mathcal{F}$ be a set of function symbols with arity $\alpha : \mathcal{F} \rightarrow N$ and $\mathcal{V}$ be a set of variables. Terms constructed with the symbols in $\mathcal{F}$ and the variables of $\mathcal{V}$ in the usual way are denoted by letters r, s, t; $\mathcal{T}$ is the set of such terms. For a term t, $\mathcal{V}(t)$ denotes the set of variables that occur in t and $\mathcal{H}(t)$ denotes the top symbol of t.

We allow ourselves one non-standard piece of notation in that we use *contexts* to describe terms. Informally, a context C is a term in which some subterms have been replaced by special symbols $\langle 1\rangle$, $\langle 2\rangle, \ldots$ The expression $C\langle t_1, t_2, \ldots, t_n\rangle$ denotes the term derived from the context C by replacing the special symbols $\langle 1\rangle$, $\langle 2\rangle, \ldots, \langle n\rangle$ in C by the terms $t_1, t_2, \ldots, t_n$. Thus, the phrase: "t can be written $t = C\langle t_1, \ldots, t_n\rangle$ for some context C", is just a graphically suggestive way of saying that the term t has the proper, non-overlapping subterms $t_1, \ldots, t_n$.

The greek letters $\eta, \rho, \sigma, \theta$, denote first-order substitutions, ϵ denotes the identity substitution $(\forall x \in \mathcal{V}[\epsilon(x) = x])$; $\mathcal{S}$ is the set of all first-order substitutions. Let

$$\mathcal{D}(\sigma) = \{x \mid x \in \mathcal{V} \quad \sigma(x) \neq x\} \qquad \text{and} \qquad \mathcal{I}(\sigma) = \bigcup_{x \in \mathcal{D}(\sigma)} \mathcal{V}(\sigma(x)).$$

The restriction of a substitution σ to a set of variables V is denoted $\sigma|_V$. If U is a set of terms then $\sigma(U)$ is the image of U under σ, i.e., a set containing $\sigma(t)$ for all terms t in U. We use $\circ$ to denote functional composition of substitutions in the ordinary way.

An equation is a pair of terms, written $s = t$. A set of equations $\boldsymbol{T}$ is an equational theory iff an equation $s = t$ is in $\boldsymbol{T}$ whenever $s = t$ is true in every model of $\boldsymbol{T}$ — written $\boldsymbol{T} \models s = t$ —, i.e., whenever $s = t$ is a *consequence* of $\boldsymbol{T}$.

A *set of axioms* T of an equational theory $\boldsymbol{T}$ is a set of equations such that $\boldsymbol{T}$ is the least equational theory that contains this set. There may be several different sets of axioms for a single equational theory — even sets of axioms with different sets of function symbols may in a precise sense (definitional equivalence) define the 'same' equational theory [15]. In the following we shall take note of this only when it is necessary. Otherwise we shall, for the sake of simplicity, speak of "an equational theory T", and then mean the least equational theory that contains the set T of axioms.

Birkhoff [2] axiomatized the relation "$s = t$ is a consequence of T" for equational theories T. One writes $T \vdash s = t$ if there is a derivation of $s = t$ with a finite number of steps in which the equation used in every step is in T. The fundamental result of equational logic is the well-known completeness theorem of Birkhoff:

Theorem 1. *(G. Birkhoff [2])* $T \models s = t$ *iff* $T \vdash s = t$.

As usual, $T \vdash s = t$ is written $s =_T t$. The relation $=_T$ is called T-equality. T-equality is extended to substitutions in the obvious way. The preorder $\leq_T$ on $\mathcal{T}$ is defined by:

$$s \leq_T t \iff \exists \sigma \in \mathcal{S} \qquad \sigma(s) =_T t.$$

For every set of variables V, a corresponding preorder on substitutions is defined by:

$$\sigma \leq_T \theta \ [\![V]\!] \iff \exists \rho \in \mathcal{S} \qquad \rho \circ \sigma|_V =_T \theta|_V.$$

A T-unifier of two terms s and t is a substitution σ such that $\sigma(s) =_T \sigma(t)$. $\mathcal{U}_T(s,t)$ denotes the set of such substitutions. We shall use the term 'solution of the equation $s =_T t$' as a synonym for 'T-unifier of s and t', and in general from now on use the term 'equation' as a synonym for 'unification problem'. A complete set of unifiers of s and t away from $V \supseteq L$, where $L = \mathcal{V}(s) \cup \mathcal{V}(t)$, is a set of substitutions Σ such that:

1. $\forall \sigma \in \Sigma \quad \mathcal{D}(\sigma) = L \quad \text{and} \quad \mathcal{I}(\sigma) \cap V = \emptyset$
2. $\Sigma \subseteq \mathcal{U}_T(s,t)$
3. $\forall \sigma \in \mathcal{U}_T(s,t) \quad \exists \theta \in \Sigma \quad \theta \leq_T \sigma \ [\![L]\!]$

$\mathcal{CSU}_T(s,t,V)$ denotes the set of all such Σ's. Note that the condition $\mathcal{D}(\sigma) = L$ implies that every unifier σ in Σ contains a complete renaming of the variables in L, so that, e.g., $\mathcal{V}(\sigma(s)) \cap \mathcal{V}(s) = \emptyset$.

An equational theory T is *unitary unifying* or just *unitary* if every equation $s =_T t$ has a complete set of unifiers with at most one element, it is *finitary unifying* if every equation has a finite complete set of unifiers, and it is *infinitary unifying* if it is not finitary unifying. (These infinitary theories comprise both what Siekmann and Szabó [12] [14] call infinitary theories, and what they call type-0 theories — a distinction we do not need in the following.)

A solution to a set or *system* of equations Γ is a substitution that simultaneously solves all the equations in Γ. The notations $\mathcal{V}(\Gamma)$, $\mathcal{U}_T(\Gamma)$, and $\mathcal{CSU}_T(\Gamma)$, all have their obvious meanings. $\mathcal{T}(\Gamma)$ denotes the set of all terms that appear as the left- or right-hand side of an equation in Γ. If σ is a substitution then $\sigma(\Gamma) = \{\sigma(s) = \sigma(t) \mid s = t \in \Gamma\}$.

A lot of notational hardship can be avoided if the following somewhat irregular, but common, convention is allowed: whenever the phrase 'new variable x' is used it is assumed that the variable x does not occur in any term except when this is explicitly or implicitly stated. We also use the following convention: *a complete renaming of a set V of variables* is a substitution that, restricted to variables, is a 1–1 mapping from V to a set of new variables.

3. Unifiers in combined theories

We shall study unification in an equational theory $T = \bigcup_{i=0}^{n-1} T_i$, where, for $0 \leq i \leq n-1$, the sets $\mathcal{F}_i$ of function symbols of the theories T_i are mutually disjoint. We shall assume that the theories T_i are consistent (otherwise our problem is trivial). $\mathcal{T}_i$, 'the T_i-terms', is the set of all terms that can be built using symbols in $\mathcal{F}_i$ and variables. The general case is trivially reducible to the case where $n = 2$, so, for convenience, we shall in the rest of this paper be concerned with a theory $T = T_0 \cup T_1$, where $\mathcal{F}_0 \cap \mathcal{F}_1 = \emptyset$. We shall call terms in $\mathcal{T}_0$ or $\mathcal{T}_1$ *pure* terms, all other terms are *mixed*. A substitution σ is pure if, for all variables x, the term $\sigma(x)$ is a pure term; it is *pure in* T_i if for all variables x $\sigma(x)$ is a term in $\mathcal{T}_i$. $\mathcal{HT}_i$ denotes the set of terms whose top symbols are in $\mathcal{F}_i$.

Example 1. Let I be the equational theory defined by the axiom $f(x, i(x)) = e$. This axiom can be thought of as saying: "the 'multiplication' f of x and its inverse $i(x)$ equals the unit element e", and appears, e.g., in groups. Let M be the '(endo-)morphism theory' defined by the single axiom $h(g(x,y)) = g(h(x), h(y))$. Taking $T_0 = I$ and $T_1 = M$ we have, e.g., $\mathcal{F}_0 = \{f, i, e\}$, $\mathcal{F}_1 = \{g, h\}$, $g(h(x), h(y))$ is a pure term, $g(h(x), h(e))$ is a mixed term, etc. The combination of I and M, which will be called IM, has the axioms $\{f(x, i(x)) = e,\ h(g(x,y)) = g(h(x), h(y))\}$. We shall use IM in all examples in the following. One good property of IM is that it has a unification procedure. This procedure is given by narrowing modulo equations, with I as the rewrite rule (in the obvious direction) and using Vogel's unification procedure for M [16] (M is unitary). This scant hint is enough to permit one to hand-check small unification problems in IM.

The purpose of this paper is to construct a complete unification algorithm for the theory T given that there are complete unification algorithms for the theories T_0 and T_1. We shall do this essentially by reducing the unification problem for mixed terms to the problem of unifying pure terms. The first task at hand is then to characterize the unifiers of pure terms in the combined theory in terms of the unifiers in the theories T_0 and T_1. The following lemma states that a complete set of unifiers for two pure terms in one of the theories T_0 and T_1 is also a complete set of unifiers for these terms in the combined theory:

Lemma 1. For $i = 0$ or 1 let s and t be two pure terms, $s, t \in \mathcal{T}_i$. A complete set of T_i-unifiers of s and t away from $V \supseteq (\mathcal{V}(s) \cup \mathcal{V}(t))$ is also a complete set of T-unifiers of s and t away from V.

Notice that this lemma does not require that T_0 and T_1 are collapse-free. This appears to constitute a new result, and makes the lemma less trivial than would otherwise be the case. One use of this lemma is that it makes it possible to show that the algorithm in the following section is complete for some unification problems even when T is not collapse-free. (T is collapse-free iff T_0 and T_1 are.)

We also need a lemma about the unifiers of equations between terms whose top symbols come from *different* sub-theories — if T is collapse-free there are none.

Lemma 2. Let $s \in \mathcal{HT}_0$ anf $t \in \mathcal{HT}_1$. If T is a collapse-free equational theory then the equation $s =_T t$ has no solution.

Proof. In the absence of collapse-axioms there are no derivation steps which can 'change' the top symbol of a term from a symbol in $\mathcal{F}_0$ to a symbol in $\mathcal{F}_1$.

H.-J. Bürckert and M. Schmidt-Schauß [18] have shown the following interesting result: There exists an infinitary non-regular equational theory such that unification in this theory is trivially decidable, but if the theory is enriched with a free constant, unification becomes undecidable. Hence, there exist equational theories T_0 and T_1 with disjoint sets of function symbols, such that unification is decidable in T_0 and T_1, but undecidable in $T = T_0 \cup T_1$. There is no contradiction between this result and what we attempt to achieve here. If T_0 or T_1 is infinitary, what we can do is to provide an algorithm that enumerates infinite complete sets of unifiers in T, which corresponds to having a *semi*-decision procedure for unification in T.

4. A unification algorithm

In this section we construct a unification algorithm for a theory T that is the combination of two collapse-free equational theories T_0 and T_1. The basic idea of the algorithm can be stated as follows:

Suppose that we want to unify the two terms s and t. If s and t are both pure terms we use Lemma 1 to obtain a complete set of unifiers for $s =_T t$. This requires of course that complete unification procedures are available for T_0 and T_1. If, on the other hand, either of s or t is mixed, we proceed as follows:

1. Replace the proper subterms r_i of s and t which cause s and t to be mixed with new variables x_i. This results in two pure terms, $\bar{s}$ and $\bar{t}$.
2. Unify the system of equations $\{\bar{s} =_T \bar{t}\} \cup \{x_i =_T r_i\}$, which has (essentially) the same set of unifiers as $s =_T t$.

Step 1 is usually called abstracting the subterms r_i — $\bar{t}$ is a 'variable abstraction' of t. The point is that $\bar{s} =_T \bar{t}$ is an equation between two pure terms, which we know how to solve — and the system of equations can be solved, e.g., by solving this equation first, substituting the solutions of it into the rest of the equations, and then solving them in turn in the same way. The following example should make clear how this functions.

Example 2. Let $s = f(z, h(y))$ and $t = f(e, g(h(z), h(z)))$. We wish to compute a complete set of unifiers of s and t.

- Abstract subterms:

$$\bar{s} = f(z, x_1) \quad , \quad x_1 = h(y)$$
$$\bar{t} = f(e, x_2) \quad , \quad x_2 = g(h(z), h(z))$$

  The system of equations that is to be solved is: $\Gamma = \{ f(z, x_1) =_{IM} f(e, x_2),\ x_1 =_{IM} h(y),\ x_2 =_{IM} g(h(z), h(z)) \}$
- Solve the first of these equations. This results in the two unifiers: $\sigma_1 = \{ z \to e, x_1 \to x_2 \}$ $\sigma_2 = \{ x_1 \to i(z), x_2 \to i(e) \}$. (In the examples we do not include the 'away-from' property of unifiers (condition 1. in the definition of CSU_T) hoping that this makes them easier to understand.) We now treat these two unifiers in turn.
  - Applying σ_1 to $\Gamma \setminus \{ f(z, x_1) =_{IM} f(e, x_2) \}$ gives: $\{ x_2 =_{IM} h(y),\ x_2 =_{IM} g(h(e), h(e)) \}$. These two equations can be solved by a recursive 'call' to the procedure we are describing. In fact, in this example, these equations are pure, so no more terms need be abstracted during this process. The single most general unifier is: $\sigma_3 = \{ x_2 \to h(g(e, e)),\ y \to g(e, e) \}$. Therefore, one unifier of s and t is: $\sigma_3 \circ \sigma_1|_{\{y,z\}} = \{ z \to e,\ y \to g(e, e) \}$.
  - Turning now to the other unifier of the first pure equation, we apply it to $\Gamma \setminus \{ f(z, x_1) =_{IM} f(e, x_2) \}$: $\{ i(z) =_{IM} h(y),\ i(e) =_{IM} g(h(z), h(z) \}$. These equations are equations between terms whose top symbols come from different sub-theories. Since there are no collapse axioms in T, by Lemma 2 there are no solutions to such equations. Consequently, σ_2 does not lead to any solutions of $s =_{IM} t$.

The described algorithm resembles Stickel's algorithm for unifying general terms in an associative-commutative theory [13], and is quite easy to understand. Essentially, Stickel's algorithm results if we let the theory T_0 be the theory of an associative-commutative function symbol and T_1 be the empty theory of every function symbol except this one.

There are of course several ways of solving a system of equations given a procedure that solves a single equation. The method chosen here is the most straightforward one; partly because the proofs of termination would otherwise become even more involved than they already are. Eder [2] has studied a more efficient way of solving systems of equations in the empty theory (i.e., Robinson unification of sets of equations), and a similar method could presumably be used for systems in a non-empty equational theory.

The informal description of the algorithm just given is unfortunately incomplete. As it stands, the algorithm fails to terminate in some important cases, namely certain equations of the form $x =_T t$ between a variable x, and a non-variable term t, as in the following example.

Example 3. Consider the unification problem $x =_{IM} t$, where $t = f(h(x), i(y))$. When we abstract subterms from this equation we get the system of equations $\Gamma = \{ x =_{IM} f(x_1, i(y)),\ x_1 =_{IM} h(x) \}$. Now the subterm containing x in the right-hand side of the original equation, $h(x)$, has been abstracted. Therefore, the first of the equations in the set only has the trivial solution $\{x \rightarrow f(x_1, i(y))\}$, and if one follows the algorithm a few steps further, one sees that it loops — the original equation appears again (with a change of variable names) as a subproblem to solve. Still, the equation does have a unifier. $\sigma = \{ y \rightarrow h(e), x \rightarrow e \}$ is a unifier, since: $\sigma(f(h(x), i(e))) = f(h(e), i(h(e))) =_{IM} e$, and $\sigma(x) =_{IM} e$.

What is wrong? The equation that causes the problem is of the general form "x equals term containing x", but the crucial information that the term in the right-hand side contains the variable in the left-hand side is, so to speak, lost when the subterms are abstracted. We shall call an equation of this troublesome form an ***equation with a hidden variable***.

In order to construct a terminating algorithm we must clearly solve the problem of unifying equations with hidden variables. This problem is trivial in regular equational theories. There, equations with hidden variables have no solution. The problem is more difficult in non-regular equational theories, where a method for finding complete sets of solutions to equations with hidden variables must be constructed. In the rest of this section we will first define and prove correct an algorithm for regular equational theories, and then extend it to non-regular theories.

4.1 Regular theories

As a preamble to a more formal definition of the unification algorithm we make some handy definitions.

Definition. The *pure part*, the *residue*, and the *residual equations* of a term t, denoted $\bar{t}$, res(t), and reseq(t) respectively, are defined as follows:

- If t is a pure term, then $\bar{t} = t$, res$(t) = \emptyset$, and reseq$(t) = \emptyset$.
- Otherwise, t can be written in the form: $t = C_t\langle t_1, \ldots, t_n \rangle$, where, if $t \in \mathcal{HT}_i$ ($i = 0$ or 1), then C_t is a context constructed entirely from symbols in $\mathcal{F}_i$, and the terms $t_1, \ldots, t_n$ are all in $\mathcal{HT}_{1-i}$. Then

$$\bar{t} = C_t\langle x_1, \ldots, x_n \rangle \quad \text{where} \quad x_1, \ldots, x_n \quad \text{are new variables,}$$
$$\text{res}(t) = \{ t_1, \ldots, t_n \},$$
$$\text{reseq}(t) = \{ x_1 =_T t_1, \ldots, x_n =_T t_n \}.$$

resV(t) denotes the set res$(t) \cup (\mathcal{V}(\bar{t}) \cap \mathcal{V}(t))$. We extend these notations to sets of terms U as follows: $\bar{U} = \{\bar{t} \mid t \in U\}$, res$(U) = \bigcup_{t \in U}$ res(t), and similarly for reseq(U) and resV(U).

Remark. This definition is informal because the same 'new' variables mysteriously appear in both $\bar{t}$ and reseq(t). A proper definition would require the variables actually used in $\bar{t}$ to be a parameter of reseq. However, that would be notationally inconvenient. We shall just adopt the convention that whenever $\bar{t}$ and reseq(t) are used in concert, the variables are the same just as in the definition.

Definition. The variable x is *hidden in the term t* if $\exists r \in \text{res}(t)[x \in \mathcal{V}(r)]$.

Lemma 3. Let T be a consistent regular and collapse-free equational theory. Then, equations of the form $x =_T t$, where x is hidden in t, have no solution.

Definition. The *pure part* of an equation $s =_T t$, denoted $\overline{s =_T t}$, is defined by: $\overline{s =_T t} = \bar{s} =_T \bar{t}$, the *residue* of $s =_T t$, denoted $\text{res}(s =_T t)$, is defined by: $\text{res}(s =_T t) = \text{res}(s) \cup \text{res}(t)$, and the *residual equations* of $s =_T t$, denoted $\text{reseq}(s =_T t)$, is defined by: $\text{reseq}(s =_T t) = \text{reseq}(s) \cup \text{reseq}(t)$.

Example 4. Let $s = f(z, h(y))$ and $t = f(e, g(h(z), h(z)))$. Then the pure part of $s =_T t$ is $\overline{s =_T t} = f(z, x_1) =_T f(e, x_2)$, and the residual equations of $s =_T t$ are $\text{reseq}(s =_T t) = \{x_1 =_T h(y), x_2 =_T g(h(z), h(z))\}$.

The following lemma justifies the process of abstracting subterms.

Lemma 4. Let $s =_T t$ be an equation, and let $\Gamma = \{\overline{s =_T t}\} \cup \text{reseq}(s =_T t)$. Let $V \supseteq \mathcal{V}(s) \cup \mathcal{V}(t)$, and let $V_\Gamma = V \cup \mathcal{V}(\Gamma)$. If Σ_Γ is a complete set of unifiers of Γ away from V_Γ, then the set $\Sigma = \{\sigma|_V \mid \sigma \in \Sigma_\Gamma\}$ is a complete set of unifiers of s and t away from V.

Figure 1 shows a definition of the unification algorithm. It is in the form of a procedure, UNIFY, that takes three arguments: two terms, s and t, and a set of variables, V. If $V \supseteq \mathcal{V}(s) \cup \mathcal{V}(t)$, then UNIFY returns a complete set of unifiers for s and t away from V. UNIFY calls the two auxiliary procedures UNIFY-T_0 and UNIFY-T_1 to obtain complete sets of unifiers in the sub-theories. Both of these procedures are assumed to have the same 'semantics' w.r.t. arguments as UNIFY. The 'meta-variable' i, as used, e.g., in expressions such as '$t \in \mathcal{H}T_i$', can take the values 0 or 1, just as in the rest of this paper.

The procedure UNIFYSET shown in Figure 2, which is used in UNIFY, is a procedure that solves a system of equations using the straightforward method. Γ is the set of equations that is to be solved, and V is the set of protected variables. UNIFYSET calls UNIFY recursively to obtain a complete set of unifiers for a single equation.

If one of the theories T_0 and T_1 is infinitary, then the set of unifiers Σ' returned by one of the calls to UNIFY-T_i may be infinite. In that case it is possible to re-organize the procedure so that it enumerates an infinite complete set of unifiers, e.g., by making the search for unifiers truly depth-first. However, since the set of all substitutions is trivially enumerable, a complete set of unifiers for two terms is enumerable in *any* theory T in which the equality $=_T$ is decidable. The result for infinitary theories is therefore not so interesting, and we shall not discuss the application of the algorithm to infinitary theories further.

We now prove, first that the procedure is partially correct, i.e., that it returns a complete set of unifiers if it terminates, then that it terminates if the procedures UNIFY-T_0 and UNIFY-T_1 do.

Theorem 2. Let T_0 and T_1 be two consistent, regular, and collapse-free equational theories, and assume that, for $i = 0$ or 1, the call UNIFY-$T_i(s, t, V)$ returns a complete (and finite) set of T_i-unifiers of the T_i-terms s and t away from $V \supseteq \mathcal{V}(s) \cup \mathcal{V}(t)$. If a call UNIFY$(s, t, V)$, where s

```
procedure UNIFY(s, t, V);                                            Justification/comment
begin
  if s ∈ 𝒱 and t ∈ 𝒱 then
    return { s → x, t → x }   (where x is a new variable)            trivial
  else if s ∈ 𝒱 and t ∉ 𝒱 and ∃r ∈ res(t)[s ∈ 𝒱(r)] then
    return ∅;                                                        Lemma 3 (Th. 4)
  else if t ∈ 𝒱 and s ∉ 𝒱 and ∃r ∈ res(s)[t ∈ 𝒱(r)] then
    return ∅;                                                        Lemma 3 (Th. 4)
  else if s ∈ ℋ𝒯_i and t ∈ ℋ𝒯_{1-i} then
    return ∅;                                                        Lemma 2 (Th. 4)
  else if s ∈ 𝒱 and s ∉ 𝒱(t) then
    return ρ ∘ {s → t}; (where ρ is a complete renaming of 𝒱(t))     straightforward
  else if t ∈ 𝒱 and t ∉ 𝒱(s) then
    return ρ ∘ {t → s}; (where ρ is a complete renaming of 𝒱(s))     straightforward
  else begin
    V_Γ := V ∪ 𝒱(s̄) ∪ 𝒱(t̄);
    Σ' := UNIFY-T_i(s̄, t̄, V_Γ);
    for each σ' ∈ Σ' do begin
      if ∃x ∈ (V_Γ \ V)[σ(x) ∈ (𝒯_i \ 𝒱)] then
        Σ''_σ' := ∅;                                                 Lemma 2 (Th. 4)
      else
        Σ''_σ' := UNIFYSET(σ'(reseq(s =_T t)), V_Γ ∪ I(σ'));
      end if;
    end for;
    return {σ'' ∘ σ'|_V | σ' ∈ Σ', σ'' ∈ Σ''_σ' };                   Lemma 4
  end if;
end
```

Figure 1. Unification procedure for the combination of two collapse-free and regular equational theories with disjoint sets of function symbols.

```
procedure UNIFYSET(Γ, V);
begin
  if Γ = ∅ then return { ε }
  else begin
    let s =_T t ∈ Γ;
    Σ' := UNIFY(s, t, V);
    for each σ' ∈ Σ' do
      Σ''_σ' := UNIFYSET(σ'(Γ \ { s =_T t }), V ∪ I(σ'));
    return {σ'' ∘ σ'|_V | σ' ∈ Σ', σ'' ∈ Σ''_σ' };
  end
end
```

Figure 2. Auxiliary procedure for solving a set of equations.

and t are terms and $V \supseteq (\mathcal{V}(s) \cup \mathcal{V}(t))$, terminates, then the set of unifiers returned is a complete set of T-unifiers of s and t away from V.

Proof. The lemmas justifying the statements in UNIFY are shown in the right column of Figure 2.

Theorem 3. If the procedures UNIFY-T_i terminate for every input, then every call UNIFY(s,t,V) (where $V \supseteq \mathcal{V}(s) \cup \mathcal{V}(t)$) terminates.

Proof. The proof is by well-founded, or Nœtherian, induction and uses a generalization of Fages' ordering for AC-unification [3].

Theorem 4. If none of the failure statements marked (Th. 4) in the right column of Figure 1 are reached, then the set returned by UNIFY is a complete set of unifiers even if T is non-regular and/or has collapse axioms.

Proof. Follows more or less immediately from Lemma 1.

How does the algorithm embodied in the procedure UNIFY compare to the algorithms of Yelick [17], Herold [19], and C. Kirchner [6]? Yelick's algorithm can also be seen as a generalization of Stickel's AC-unification procedure [13], and consequently UNIFY resembles her algorithm very closely. As Example 3 showed, generalizations of Stickel's algorithm such as UNIFY and Yelick's algorithm need the failure case corresponding to Lemma 3 in order to terminate. On this important point we are indebted to Yelick. Herold's algorithm can be seen as a generalization of Livesey and Siekmann's AC-unification algorithm [9], and, roughly speaking, it relates to UNIFY and Yelick's algorithm as Livesey and Siekmann's algorithm relates to Stickel's. Herold's algorithm is also restricted to regular and collapse-free theories. C. Kirchner's appealing algorithm, finally, is based on his interesting general method for constructing equational unification algorithms. As a consequence of this, the applicability of his algorithm is not so easy to describe in terms of simple properties of axioms such as regularity and freedom from collapse axioms. It appears that, with minor limitations, it too can handle combinations of regular collapse-free theories.

4.2 Non-regular theories

In this section, we extend the algorithm of the previous section to handle non-regular theories. We assume throughout that T_0 and T_1 (and hence T) are collapse-free. The main problem is to find complete sets of unifiers for equations with hidden variables. It turns out to be sufficient to study equations $x =_T t$ where, if t is in $\mathcal{HT}_i$, then x has at least one occurrence under a function symbol from $\mathcal{F}_{1-i}$ in t. We shall say that x is *well-hidden* in t if this condition is satisfied, and we call an equation $x =_T t$ in which x is well-hidden in t *an equation with a well-hidden variable*.

Definition. The *set of immediate operators of a term s in a term t*, denoted $\mathrm{op}(s,t)$, is defined by $\mathrm{op}(s,t) = \{f \mid f \in \mathcal{F} \wedge s \text{ occurs as an argument of } f \text{ in } t\}$.

Definition. For $i = 0$ or 1, the variable x is *well-hidden in the term $t \in \mathcal{HT}_i$* if $\mathrm{op}(x,t) \cap \mathcal{F}_{1-i} \neq \emptyset$.

The real reason why it is sufficient to study equations with well-hidden variables is buried in the proof of termination — the algorithm terminates even if equations $x =_T t$, where x is *not* well-hidden in t, are treated just as ordinary equations. To support this conjecture with some intuition, here is an example which shows how an equation $x =_T t$ where x is hidden, but not well-hidden, is transformed to an (essentially) equivalent system of equations with well-hidden variables by UNIFY.

Example 5. Consider the equation $x =_{IM} f(g(f(x,z)),i(y))$, in which x is hidden, but not well-hidden. After abstracting subterms we are left with $\{x =_{IM} f(x_1,i(y)),\ x_1 =_{IM} g(f(x,z))\}$. The first of these has the trivial solution $\sigma = \{x \to f(x_1,i(y))\}$, and if we substitute this into the second equation we get $\{x_1 =_{IM} g(f(f(x_1,i(y)),z))\}$, in which x_1 is well-hidden.

The following example shows why it is not trivial to compute complete sets of solutions to equations with well-hidden variables.

Example 6. Consider the equation $x =_{IM} f(g(x),i(y))$. This equation has the single most general unifier $\sigma = \{x \to e,\ y \to g(e)\}$. For the same reasons that made the algorithm fail in Example 3, this unifier cannot be found by the algorithm.

What are the characteristics of the unifier in Example 6? The answer is that part of the unifier makes the variable x 'disappear' from the term t. The unifier $\sigma = \{x \to e, y \to g(e)\}$ can be written $\sigma_1 \circ \sigma_2$, where $\sigma_2 = \{ y \to g(x) \}$, and $\sigma_1 = \{ x \to \sigma_2(t) \}$. In a sense the variable x does not appear in $\sigma_2(t)$: $\sigma_2(t) = f(g(x), i(g(x)))$, in which x certainly appears, but $f(g(x), i(g(x))) =_{IM} e$, in which x does not appear. It turns out that all solutions σ to an equation $x =_T t$, where x is well-hidden, have the property that $\sigma(x)$ 'disappears' — or *is eliminated in*, as we shall say — $\sigma(t)$. Before we make precise what it means that a set of terms is eliminated in a term, we need to define certain sets of subterms of a term. The *connected subterms* of a term are essentially those subterms whose top symbols come from a different sub-theory than the symbol under which they occur in the term.

Definition. The *non-variable connected subterms of a term t*, denoted $\mathcal{CS}(t)$, is a set of terms defined by:

$$\mathcal{CS}(t) = \begin{cases} \emptyset, & \text{if } t \in \mathcal{V}; \\ t \cup \bigcup_{r \in \text{res}(t)} \mathcal{CS}(r), & \text{otherwise.} \end{cases}$$

We also define the *proper* non-variable connected subterms of t, $\overline{\mathcal{CS}}(t) = \mathcal{CS}(t) \setminus \{t\}$, the *connected subterms* of t, $\mathcal{CSV}(t) = \mathcal{CS}(t) \cup \mathcal{V}(t)$, and finally, the *proper* connected subterms of t, $\overline{\mathcal{CSV}}(t) = \mathcal{CSV}(t) \setminus \{t\}$.

The connected subterms are, disregarding minor technicalities, what Fages [3] calls 'admissible' subterms, Yelick [17] 'significant' subterms, and Herold [19] 'alien' subterms.

Example 7. Let $t = f(g(x), i(y))$. Then $\mathcal{CS}(t) = \{f(g(x), i(y)), g(x)\}$, $\overline{\mathcal{CS}}(t) = \{g(x)\}$, $\mathcal{CSV}(t) = \{f(g(x), i(y)), g(x), x, y\}$, and $\overline{\mathcal{CSV}}(t) = \{g(x), x, y\}$.

Definition. Let S be a set of terms and let t be a term. S is said to be *T-eliminated in t*, abbreviated $\text{elim}_T(S, t)$, *iff* there exists a term r such that

1. $r =_T t$
2. $\neg\exists s_r \in \overline{\mathcal{CSV}}(r),\ s \in S \quad s_r =_T s$

S is *T-eliminated in the set of terms U*, abbreviated $\text{elim}_T(S, U)$ *iff* S is T-eliminated in each of the terms in U.

When S is a singleton, $S = \{s\}$, we say that s is T-eliminated in t or U if $\{s\}$ is, and whenever T can be understood from the context, we say just 'eliminated' instead of T-eliminated.

We now characterize solutions to equations $x =_T t$, with x well-hidden in t, by showing that if σ is a solution to such an equation, then $\sigma(x)$ is T-eliminated in t. We shall say that σ is a *T-eliminator* for x in t, in the sense of the following definition.

Definition. Let t be a term, let s be a proper connected subterm of t, and let σ be a substitution. σ is a *T-eliminator for s in t iff* $\sigma(s)$ is a proper connected subterm of $\sigma(t)$ and $\sigma(s)$ is T-eliminated in $\sigma(t)$. $\mathcal{E}_T(s, t)$ denotes the set of all T-eliminators for s in t.

When T is understood, we shall sometimes just say 'eliminator' instead of T-eliminator.

In order to prove that any solution σ to an equation $x =_T t$, with x well-hidden in t, is a T-eliminator for x in t, we introduce a function on terms that in a sense tells how mixed a term is. To be more precise, the *theory height* of a term t is the maximum number of times that one moves from a symbol in $\mathcal{F}_0$ to a symbol in $\mathcal{F}_1$, or conversely, from a symbol in $\mathcal{F}_1$ to a symbol in $\mathcal{F}_0$, when moving from the root of t to a leaf along any path.

Definition. *Theory height* is a mapping δ from terms to non-negative integers defined as follows: If t is a variable then $\delta(t) = 0$. Otherwise let C be a pure context such that t can be written

$t = C\langle t_1, \ldots, t_n\rangle$, where, for $1 \leq j \leq n$, t_j is either a variable or else the top symbol of t_j is from a different sub-theory than that of t. Then

$$\delta(t) = 1 + \max_{1 \leq j \leq n} \delta(t_j).$$

Remark. If s is a proper connected subterm of t then $\delta(t) > \delta(s)$.

We also need the *minimum theory height* of a term t, which is the least theory height of any term T-equal to t:

Definition. The *minimum theory height* $\delta_{\min}(t)$ of a term t is defined by

$$\delta_{\min}(t) = \min_{\{r \mid r =_T t\}} \delta(r).$$

Remark. If $s =_T t$ then $\delta_{\min}(s) =_T \delta_{\min}(t)$.

Lemma 5. If s is a proper connected subterm of the term t, and s is not T-eliminated in t, then $\delta_{\min}(t) > \delta_{\min}(s)$.

Proof. $\delta_{\min}(t) = \delta(r)$ for some term $r =_T t$. Since s is not T-eliminated in t there is a proper connected subterm s_r of r such that $s_r =_T s$. Hence,

$$\delta_{\min}(t) = \delta(r) > \delta(s_r) \geq \delta_{\min}(s_r) = \delta_{\min}(s).$$

Lemma 6. Let s be a proper connected subterm of a term t. If $s =_T t$ then s is eliminated in t.

Proof. $s =_T t$ implies that $\delta_{\min}(s) = \delta_{\min}(t)$. Assume that s is not eliminated in t. Then, by the preceding lemma, $\delta_{\min}(t) > \delta_{\min}(s)$, which is a contradiction.

We can now state and prove the characterization of solutions to equations with well-hidden variables. The following lemma will be our main tool for finding the solutions to these equations.

Lemma 7. Let t be a term and let x be a variable which is well-hidden in t. If σ is a solution to the equation $x =_T t$, then σ is a T-eliminator for x in t.

Proof. The theory T is collapse-free. Hence, any solution σ of the equation $x =_T t$ is such that if $t \in \mathcal{HT}_i$ ($i = 0$ or 1) then $\sigma(x) \in \mathcal{HT}_i$ — i.e., the top symbol of $\sigma(x)$ must come from the same sub-theory as the top symbol of t. It then follows from the fact that x is well-hidden in t that $\sigma(x)$ is a proper connected subterm of $\sigma(t)$. Since $\sigma(x) =_T \sigma(t)$ it follows from Lemma 6 that $\sigma(x)$ is eliminated in $\sigma(t)$.

Lemma 7 relates solutions to equations with well-hidden variables to eliminators for the variable. What we are after is *complete sets* of solutions to equations, however, and we shall relate these to complete sets of eliminators, which can be defined in analogy with complete sets of unifiers:

Definition. A complete set of T-eliminators for a term s in a term t away from $V \supseteq \mathcal{V}(t)$ is a set of substitutions Σ such that:

1. $\forall \sigma \in \Sigma \quad \mathcal{D}(\sigma) = \mathcal{V}(t) \quad \text{and} \quad \mathcal{I}(\sigma) \cap V = \emptyset$
2. $\Sigma \subseteq \mathcal{E}_T(s, t)$
3. $\forall \sigma \in \mathcal{E}_T(s, t) \quad \exists \theta \in \Sigma \quad \theta \leq_T \sigma \; [\![\mathcal{V}(t)]\!]$

$\mathcal{CSE}_T(s,t,V)$ denotes the set of all such Σ's.

Lemma 8. Let $\Sigma_\mathcal{E}$ be a complete set of T-eliminators for x in t, where x is well-hidden in t, away from $V \supseteq \mathcal{V}(t)$. For each $\sigma_\mathcal{E}$ in $\Sigma_\mathcal{E}$, let

$$\Sigma_{\sigma_\mathcal{E}} \in \mathcal{CSU}_T(\sigma_\mathcal{E}(x), \sigma_\mathcal{E}(t), V \cup \mathcal{I}(\sigma_\mathcal{E})),$$

and define a set of substitutions Σ by:

$$\Sigma = \{\, \sigma_{\sigma_\mathcal{E}} \circ \sigma_\mathcal{E}|_{\mathcal{V}(t)} \mid \sigma_\mathcal{E} \in \Sigma_\mathcal{E}, \sigma_{\sigma_\mathcal{E}} \in \Sigma_{\sigma_\mathcal{E}} \,\}.$$

Then Σ is a complete set of unifiers of x and t away from V.

Proof. *Correctness:* Immediate. *Completeness:* Let σ be an arbitrary unifier of x and t. By Lemma 7, σ is an eliminator for x in t. Since $\Sigma_\mathcal{E}$ is a complete set of eliminators for x in t, there is a substitution $\sigma_\mathcal{E} \in \Sigma_\mathcal{E}$ such that $\sigma_\mathcal{E} \leq_T \sigma\ [\![\mathcal{V}(t)]\!]$. Let θ be a substitution such that $\theta \circ \sigma_\mathcal{E} =_T \sigma\ [\![\mathcal{V}(t)]\!]$. θ is then a T-unifier of $\sigma_\mathcal{E}(x)$ and $\sigma_\mathcal{E}(t)$, and since $\Sigma_{\sigma_\mathcal{E}}$ is a complete set of unifiers of $\sigma_\mathcal{E}(x)$ and $\sigma_\mathcal{E}(t)$, there is a unifier $\sigma_{\sigma_\mathcal{E}}$ in $\Sigma_{\sigma_\mathcal{E}}$ such that $\sigma_{\sigma_\mathcal{E}} \leq_T \theta\ [\![\mathcal{I}(\sigma_\mathcal{E})]\!]$. It follows that $\sigma_{\sigma_\mathcal{E}} \circ \sigma_\mathcal{E} \leq_T \sigma\ [\![\mathcal{V}(t)]\!]$. $\sigma_{\sigma_\mathcal{E}} \circ \sigma_\mathcal{E}|_{\mathcal{V}(t)}$ is in Σ, so this shows that σ is complete.

The use in Lemma 8 of a complete set of T-unifiers of $\sigma_\mathcal{E}(x)$ and $\sigma_\mathcal{E}(t)$ for every eliminator $\sigma_\mathcal{E}$ for x in t will correspond to recursive calls to the procedure UNIFY. If $\sigma_\mathcal{E}(x)$ happens to be a variable, this will immediately lead to a loop. Therefore, we need the following lemma, which says that if x is eliminated in t, then, even though $x \in \mathcal{V}(t)$, the trivial complete set of unifiers we would have if $x \notin \mathcal{V}(t)$ is a complete set of unifiers of x and t.

Lemma 9. Let x be a variable, t a term, V a set of variables, $V \supseteq \mathcal{V}(t)$, and let ρ be a complete renaming of the variables in t away from V (i.e., ρ is a 1–1 mapping from $\mathcal{V}(t)$ to a set of new variables). If x is T-eliminated in the term t, then $\{\, \rho \circ \{x \to t\} \,\}$ is a complete set of unifiers of x and t away from V.

Proof. Since x is T-eliminated in t there exists a term r such that $r =_T t$ and $x \notin \mathcal{V}(r)$. Now $\{\, \rho' \circ \{x \to t\}|_{\mathcal{V}(r)} \,\}$, where $\rho' = \rho$ on $\mathcal{V}(t)$ and contains a complete renaming of the variables in r away from V, is a complete set of unifiers of x and r, and the lemma follows.

We can already see what the modifications that allow UNIFY to handle equations with hidden variables are. Assume that CSE(x,t,V) is a procedure that computes a complete set of eliminators for x in t away from $V \supseteq \mathcal{V}(t)$. UNIFY should then be modified as follows: We replace the two failure cases

else if $s \in \mathcal{V}$ **and** $t \notin \mathcal{V}$ **and** $\exists r \in \text{res}(t)[s \in \mathcal{V}(r)]$ **then**
  return $\emptyset$;
else if $t \in \mathcal{V}$ **and** $s \notin \mathcal{V}$ **and** $\exists r \in \text{res}(s)[t \in \mathcal{V}(r)]$ **then**
  return $\emptyset$;

with the code shown in Figure 3.

Under the assumption that we have the procedure CSE we have the following two theorems about the modified procedure UNIFY:

Theorem 5. Let T_0 and T_1 be two consistent and collapse-free equational theories, and assume that, for $i = 0$ or 1, the call UNIFY-$T_i(s,t,V)$ returns a complete (and finite) set of T_i-unifiers of the T_i-terms s and t away from $V \supseteq \mathcal{V}(s) \cup \mathcal{V}(t)$. If a call UNIFY$(s,t,V)$, where s and t are terms and $V \supseteq (\mathcal{V}(s) \cup \mathcal{V}(t))$, terminates, then the set of unifiers returned is a complete set of T-unifiers of s and t away from V.

else if $s \in \mathcal{V}$ **and** $t \notin \mathcal{V}$ **and** $\exists r \in \text{res}(t)[s \in \mathcal{V}(r)]$ **and**
$\text{op}(s,t) \cap \mathcal{F}_{1-i} \neq \emptyset$ **then begin** — s well-hidden in t
  $\Sigma_\mathcal{E} := \text{CSE}(s,t,V)$;
  for each $\sigma_\mathcal{E}$ **in** $\Sigma_\mathcal{E}$ **do begin**
    if $\sigma_\mathcal{E}(s) \in \mathcal{V}$ **then**
      $\Sigma_{\sigma_\mathcal{E}} := \{\rho \circ \{x \to \sigma_\mathcal{E}(t)\}\}$ — Lemma 9
      (where ρ is a complete renaming of $I(\sigma_\mathcal{E})$ away from $V \cup I(\sigma_\mathcal{E})$)
    else
      $\Sigma_{\sigma_\mathcal{E}} := \text{UNIFY}(\sigma_\mathcal{E}(s), \sigma_\mathcal{E}(t), V \cup I(\sigma_\mathcal{E}))$;
  end for;
  return $\{\sigma_{\sigma_\mathcal{E}} \circ \sigma_\mathcal{E}|_{\mathcal{V}(t)} \mid \sigma_\mathcal{E} \in \Sigma_\mathcal{E}, \sigma_{\sigma_\mathcal{E}} \in \Sigma_{\sigma_\mathcal{E}}\}$; — Lemma 8
end
else if $t \in \mathcal{V}$ **and** $s \notin \mathcal{V}$ **and** $\exists r \in \text{res}(s)[t \in \mathcal{V}(r)]$ **and**
$\text{op}(t,s) \cap \mathcal{F}_{1-i} \neq \emptyset$ **then begin** — t well-hidden in s
  $\Sigma_\mathcal{E} := \text{CSE}(t,s,V)$;
  for each $\sigma_\mathcal{E}$ **in** $\Sigma_\mathcal{E}$ **do begin**
    if $\sigma_\mathcal{E}(t) \in \mathcal{V}$ **then**
      $\Sigma_{\sigma_\mathcal{E}} := \{\rho \circ \{x \to \sigma_\mathcal{E}(s)\}\}$ — Lemma 9
      (where ρ is a complete renaming of $I(\sigma_\mathcal{E})$ away from $V \cup I(\sigma_\mathcal{E})$)
    else
      $\Sigma_{\sigma_\mathcal{E}} := \text{UNIFY}(\sigma_\mathcal{E}(t), \sigma_\mathcal{E}(s), V \cup I(\sigma_\mathcal{E}))$;
  end for;
  return $\{\sigma_{\sigma_\mathcal{E}} \circ \sigma_\mathcal{E}|_{\mathcal{V}(s)} \mid \sigma_\mathcal{E} \in \Sigma_\mathcal{E}, \sigma_{\sigma_\mathcal{E}} \in \Sigma_{\sigma_\mathcal{E}}\}$; — Lemma 8
end

Figure 3. Modification for UNIFY that handles equations with a hidden variable.

Proof. This essentially follows from Lemmas 8 and 9, and the correctness of the procedure CSE, which is indicated in section 4.2.1 below.

Theorem 6. If the procedures UNIFY-T_i terminate for every input, then every call UNIFY(s,t,V) (where $V \supseteq \mathcal{V}(s) \cup \mathcal{V}(t)$) terminates.

Proof. This proof is similar to the proof of Theorem 3. The new recursive calls to UNIFY in the modified code in Figure 3 can be shown to terminate by refining the ordering used to prove Theorem 3 so that an equation $s =_T t$, where s and t are non-variable terms, is *less* than an equation $x =_T r$, where x is a variable, if $s =_T t$ and $x =_T r$ were similar in the old ordering. The necessary termination proof for the procedure CSE uses similar techniques.

A stronger version of Theorem 4 is not possible since the technique developed in this section is not applicable to equational theories that have collapse axioms.

The following is an example of how UNIFY now computes the solutions to equations with well-hidden variables.

Example 8. Consider the equation $x =_{IM} t$, where t is the term

$$t = h(\ f(\ f(x, i(h(y,z))),\ v\ ),\ f(z, i(w))\ ).$$

A $\mathcal{CSE}_{IM}(x, t, \mathcal{V}(t))$ contains the two substitutions σ_1 and σ_2:

$$\sigma_1 = \{\, v \to i(f(x, i(h(y,z))))\,\},$$
$$\sigma_2 = \{\, x \to h(y,z)\,\}.$$

(Recall that we do not include the 'away from' or idempotency properties of substitutions in the examples.) We have:

$$\sigma_1(t) = h(\ f(\ f(x, i(h(y,z))),\ i(f(x, i(h(y,z))))\ ),\ f(z, i(w))\ ) =_{IM} h(e, f(z, i(w))),$$
$$\sigma_2(t) = h(\ f(\ f(h(y,z), i(h(y,z))),\ v\ ),\ f(z, i(w))\ ) =_{IM} h(f(e,v), f(z, i(w))),$$

which shows that σ_1 and σ_2 are IM-eliminators for x in t.

- $\sigma_1(x) = x \in \mathcal{V}$, so σ_1 corresponds to the case treated in Lemma 9, and the *IM*-unifier of x and t corresponding to σ_1 is:
$$\{x \rightarrow \sigma_1(t)\} \circ \sigma_1 =_{IM} \{x \rightarrow h(e, f(z, i(w))),\ v \rightarrow i(f(x, i(h(y, z))))\}.$$
- $\sigma_2(x) = h(y, z) \notin \mathcal{V}$, so for σ_2 the recursive call UNIFY($\sigma_2(x), \sigma_2(t), \ldots$) is made. There is only one unifier of $\sigma_2(x)$ and $\sigma_2(t)$:
$$\sigma_{\sigma_2} = \{y \rightarrow f(e, v), w \rightarrow e, z \rightarrow e\},$$
and the *IM*-unifier of x and t corresponding to σ_2 is:
$$\sigma_{\sigma_2} \circ \sigma_2 = \{\, y \rightarrow f(e, v), w \rightarrow e, z \rightarrow e, x \rightarrow h(f(e, v), e)\,\}.$$

In the following section, we turn to the task of computing complete sets of eliminators.

4.2.1 Computing complete sets of eliminators

We have not found any direct, 'goal-oriented', way of computing complete sets of eliminators. Instead, in order to compute a complete set of eliminators, we first compute a sort of set that we know must *contain* a complete set of eliminators and filter out a complete set of eliminators from this set afterwards. We shall call those larger sets *total complete sets of eliminators*. Before they can be defined, we need an order relation on substitutions which is a refinement of the usual subsumption order on substitutions.

Definition. For every set of terms U and set of variables V, the *eliminator ordering for* U is a relation on substitutions defined by:

$$\begin{array}{l} \sigma \leq_T \eta \ [\![V, U]\!] \ \textit{iff} \\ \forall S_\eta \quad \mathrm{elim}_T(S_\eta, \eta(U)) \implies \exists \theta \quad \theta \circ \sigma =_T \eta \ [\![V]\!] \text{ and } \mathrm{elim}_T(S, \sigma(U)), \\ \text{where } S \text{ is the largest subset of } \overline{\mathcal{CSV}}(\sigma(U)) \text{ such that } \theta(S) \subseteq_T S_\eta. \end{array}$$

(By $\theta(S) \subseteq_T S_\eta$ we mean $\theta(S)/{=_T} \subseteq S_\eta/{=_T}$.)

So $\sigma \leq_T \eta\ [\![V, U]\!]$ means that for every set S_η that is eliminated in $\eta(U)$ there is a substitution θ such that $\theta \circ \sigma =_T \eta\ [\![V]\!]$ — hence $\sigma \leq_T \eta\ [\![V]\!]$ — and the set S of all proper connected subterms of $\sigma(U)$ such that $\theta(S) \subseteq_T S_\eta$ is T-eliminated in $\sigma(U)$. Informally, σ is more general than η, and 'anything' that is eliminated in $\eta(U)$ is 'already' eliminated in $\sigma(U)$.

Definition. Let U be a set of terms and let V be a set of variables, $V \supseteq \mathcal{V}(U)$. A *total complete set of T-eliminators for U away from V* is a set Σ of substitutions such that:

$$\begin{array}{ll} 1. & \forall \sigma \in \Sigma \quad \mathcal{D}(\sigma) = \mathcal{V}(U) \quad \text{and} \quad \mathcal{I}(\sigma) \cap V = \emptyset \\ 2. & \forall \eta \quad \exists \sigma \in \Sigma \quad \sigma \leq_T \eta\ [\![\mathcal{V}(U), U]\!] \end{array}$$

We denote the set of all such Σ's by $\mathcal{TCSE}_T(U, V)$. If U is a singleton, $U = \{t\}$, we leave out the braces and write just $\mathcal{TCSE}_T(t, V)$.

There are two reasons why we are interested in total complete sets of eliminators. The first reason — which also motivates the somewhat formidable name of these sets — is that a total complete set of eliminators for a term t *contains* a complete set of eliminators for every term s in t.

Lemma 10. Let t be a term. If Σ is a total complete set of T-eliminators for t away from $V \supseteq \mathcal{V}(t)$, then for every term s, there is a subset Σ' of Σ such that $\Sigma' \in \mathcal{CSE}_T(s, t, V)$.

Proof. What we have to prove is that for any eliminator η for s in t, there is an eliminator σ for s in t in Σ such that $\sigma \leq_T \eta$ $[\![\mathcal{V}(t)]\!]$. Assume (without loss of generality) that η is idempotent. $\eta(s)$ is T-eliminated in $\eta(t)$. Since $\Sigma \in \mathcal{TCSE}_T(t,V)$, there is a substitution $\sigma \in \Sigma$ such that $\sigma \leq_T \eta$ $[\![\mathcal{V}(t),\ \{t\}]\!]$. From the definition of the eliminator ordering it is immediate that $\sigma \leq_T \eta$ $[\![\mathcal{V}(t)]\!]$. Is σ a T-eliminator for s in t? We first verify that $\sigma(s)$ is a proper connected subterm of $\sigma(t)$. We know that s is a proper connected subterm of t and that $\eta(s)$ is a proper connected subterm of $\eta(t)$. Since $\sigma \leq_T \eta$ $[\![\mathcal{V}(t)]\!]$, $\sigma(s)$ is either a variable, in which case $\sigma(s)$ is a proper connected subterm of $\sigma(t)$ because s is a proper connected subterm of t, or else the top symbol of $\sigma(s)$ comes from the same sub-theory as the top symbol of $\eta(s)$, in which case $\sigma(s)$ is a proper connected subterm of $\sigma(t)$ since $\eta(s)$ is a proper connected subterm of $\eta(t)$. Finally, we must verify that $\sigma(s)$ is eliminated in $\sigma(t)$. For every substitution θ such that $\theta \circ \sigma =_T \eta$ $[\![\mathcal{V}(t)]\!]$, we have $\theta(\sigma(s)) =_T \eta(s)$. This means that $\sigma(s) \in S$, where S is the largest subset of $\overline{\mathcal{CSV}}(\sigma(t))$ such that $\theta(S) \subseteq_T \{\eta(s)\}$, and, by the definition of the eliminator ordering, S, and hence $\sigma(s)$, is T-eliminated in t.

The second reason for our interest in total complete sets of eliminators is that, for pure terms, we know how to compute them:

Lemma 11. For $i = 0$ or 1 let U be a set of pure terms in $\mathcal{T}_i$, and let ρ be a complete renaming of the variables in U (i.e., ρ is a 1–1 mapping from $\mathcal{V}(U)$ to a set of new variables). If σ is a complete set of T_i-unifiers of the set of equations $\{t =_T \rho(t) \mid t \in U\}$ away from $V \supseteq (\mathcal{V}(U) \cup \mathcal{I}(\rho))$, then the set

$$\{\sigma|_{\mathcal{V}(t)} \mid \sigma \in \Sigma\}$$

is a total complete set of T_i-eliminators for U away from V.

Proof. A complete proof of this lemma is too long to be presented here. We try to make some of the intuition behind the lemma clear with an example. Let U consist of the single term $t = f(x,y)$, and let $\rho(t) = f(x',y')$. Consider the unifiers of t and $\rho(t)$. One unifier is obviously the trivial one: $\{x' \to x,\ y' \to y\}$. But there is another unifier, which is *not* subsumed by the trivial one. Let $\sigma = \{y \to i(x)\}$. Then $\sigma(t) = f(x, i(x)) =_I e$, so x is I-eliminated in $\sigma(t)$. We can now construct a unifier σ' of t and $\rho(t)$ as follows: $\sigma' = \sigma \cup \{y' \to i(x')\}$. We have $\sigma'(t) = \sigma'(\rho(t)) =_I e$. σ' is *not* subsumed by the trivial unifier because σ' does not contain the component $\{x' \to x\}$ — to be very informal one could say that σ' does not *need* to contain this component because σ' eliminates x in t and x' in $\rho(t)$. In essence, a complete set of unifiers of t and $\rho(t)$ contains the trivial unifier and σ', and, in fact, $\{\sigma'|_{\mathcal{V}(t)}\} = \{\sigma\}$ is a total complete set of eliminators for x in t (disregarding idempotency and protection of variables).

It would be well if there were some direct way of computing a complete set of eliminators for a single variable in a pure term — not just a total complete set of eliminators as in the above lemma. Alas, we have found no such way. One, perhaps natural, conjecture, is that if ρ in the lemma were restricted to just renaming one single variable in t, a complete set of unifiers of t and $\rho(t)$ would be a complete set of eliminators for this single variable. However, this is not true: $f(x,y)$ and $f(x',y)$ have only the trivial unifier $\{x' \to x\}$.

Lemma 11 relates total complete sets of eliminators for pure terms to certain complete sets of unifiers. It now remains to extend this knowledge to mixed terms in some way. We have the following lemma:

Lemma 12. For $i = 0$ or 1 let U be a set of terms in $\mathcal{HT}_i$ and let $V \supseteq \mathcal{V}(U)$. We define a set Σ of substitutions via a series of sets as follows:

$$\Sigma^{(1)} \in \mathcal{TCSE}_{T_i}(\bar{U}, V)$$
$$\forall \sigma_1 \in \Sigma^{(1)} \quad \Sigma^{(2)}_{\sigma_1} \in \mathcal{CSU}_T(\sigma_1(\text{reseq}(U)), V \cup \mathcal{V}(\sigma_1(\text{reseq}(U))))$$
$$\Sigma^{(12)} = \{\sigma_2 \circ \sigma_1 \mid \sigma_1 \in \Sigma^{(1)},\ \sigma_2 \in \Sigma^{(2)}_{\sigma_1}\}$$
$$\forall \sigma_{12} \in \Sigma^{(12)}\ \forall U' \subseteq \text{resV}(\sigma_{12}(U)) \quad \Sigma^{(3)}_{\sigma_{12},U'} \in \mathcal{TCSE}_T(U', V \cup \mathcal{V}(U'))$$
$$\Sigma = \{\sigma_{3,U'} \circ \sigma_{12}|_{\mathcal{V}(U)} \mid \sigma_{12} \in \Sigma^{(12)}, \sigma_{3,U'} \in \Sigma^{(3)}_{\sigma_{12},U'}\}$$

Σ is a total complete set of eliminators for t away from V. (By $\bar{U}$ we mean $\{\bar{t} \mid t \in U\}$.)

Proof. The proof is too long to be presented here. It consists essentially of showing that the 'equality propagation' that goes on between all the sets in the lemma gives the correct result.

On the basis of Lemmas 11 and 12, we can construct the procedure shown in Figure 4 for computing total complete sets of eliminators. The procedures UNIFYSET-T_0 and UNIFYSET-T_1 are assumed to be just like UNIFYSET, except that they solve sets of equations in T_0 and T_1 respectively, instead of sets of equations in T.

procedure TCSE(U, V);	Justification/comment
begin	
$\Gamma_{\bar{U}} := \{\bar{t} =_{T_i} \rho(\bar{t}) \mid t \in U\}$;	
(where ρ is a complete renaming of $\mathcal{V}(\bar{U})$)	
$\Sigma^{(1)} :=$ UNIFYSET-T_i($\Gamma_{\bar{U}}, V \cup \mathcal{V}(\Gamma_{\bar{U}})$);	Lemma 11
for each $\sigma_1 \in \Sigma^{(1)}$ **do begin**	
$\Sigma^{(2)}_{\sigma_1} :=$ UNIFYSET($\sigma_1(\text{reseq}(U)), V \cup \mathcal{V}(\sigma_1(\text{reseq}(U)))$);	
end for;	
$\Sigma^{(12)} := \{\sigma_2 \circ \sigma_1 \mid \sigma_1 \in \Sigma^{(1)}, \sigma_2 \in \Sigma^{(2)}_{\sigma_1}\}$;	
for each $\sigma_{12} \in \Sigma^{(12)}$ **do**	
for each $U' \subseteq \text{resV}(\sigma_{12}(U))$ **do begin**	
$\Sigma^{(3)}_{\sigma_{12},U'} :=$ TCSE($U', V \cup \mathcal{V}(U')$);	
end for;	
return $\{\sigma_{3,U'} \circ \sigma_{12}|_{\mathcal{V}(U)} \mid \sigma_{12} \in \Sigma^{(12)}, \sigma_{3,U'} \in \Sigma^{(3)}_{\sigma_{12},U'}\}$;	Lemma 12
end	

Figure 4. A procedure that computes a $\mathcal{TCSE}$ for a set of terms.

Lemma 13. For every set of terms U and set of variables $V \supseteq \mathcal{V}(U)$, TCSE($U, V$) terminates and returns a total complete set of eliminators for U away from V.

Proof. Correctness follows from Lemmas 11 and 12. Termination is proved by the use of the same type of ordering as for UNIFY.

Here is an example of the workings of TCSE.

Example 9. We consider again the term $t = h(\ f(\ f(x, i(h(y, z))), v\ ), f(z, i(w))\ )$ from Example 8. It is impossible to go through all of the steps of the procedure call TCSE($\{t\}, \mathcal{V}(t)$) here. Instead we concentrate on how the eliminator $\{x \to h(y, z)\}$ is found by the procedure. This eliminator is found by the recursive call TCSE($\{f(x, i(h(y, z)))\}, \ldots$), which will eventually be made. In this call of TCSE we have:

$$U = \{\ f(x, i(h(y, z)))\ \}$$
$$\Gamma_{\bar{U}} = \{\ f(x, i(v_1)) =_I f(x', i(v_1'))\ \}$$
$$\Sigma^{(1)} = \{\ \{x' \to x, v_1' \to v_1\}, \{v_1 \to x, v_1' \to x'\}\ \}$$

The first of the unifiers in $\Sigma^{(1)}$ is the trivial one, and is uninteresting. Let $\sigma_1 = \{v_1 \rightarrow x, v_1' \rightarrow x'\}$. Then

$$\sigma_1(\text{reseq}(U)) = \{x =_{IM} h(y,z)\}.$$

This equation has the trivial unifier $\sigma_2 = \{x \rightarrow h(y,z)\}$, and we get

$$\Sigma^{(12)} = \{\ldots, \sigma_2 \circ \sigma_1\} = \{\ldots, \{\ldots, x \rightarrow h(x,y)\}\}.$$

$\text{resV}(\sigma_2 \circ \sigma_1(U)) = \{h(y,z)\}$, so one of the final recursive calls to TCSE will be TCSE($\{h(x,y)\}, \ldots$), which will return $\{\epsilon\}$ as a total complete set of eliminators for $h(y,z)$. Thus, one of the unifiers returned will be

$$\epsilon \circ \sigma_2 \circ \sigma_1|_{\mathcal{V}(f(x,i(h(y,z))))} = \{x \rightarrow h(y,z)\},$$

precisely the eliminator we sought.

The call TCSE($\{t\}, \mathcal{V}(t)$) will return a set that contains the eliminator found above, the other eliminator in Example 8, *and* a lot of 'garbage' — the identity substitution, the useless eliminator $\{w \rightarrow z\}$, which eliminates z in the connected subterm $f(z, i(w))$ of t, etc. All of this garbage is filtered out by the procedure CSE below.

What we need now in order to compute a complete set of eliminators for a term s in a term t, is a way of testing the substitutions in a total complete set of eliminators for t to see which ones really eliminate s in t. The following lemma shows how it can be tested for a pure term whether a set of variables is eliminated in the term or not.

Lemma 14. For $i = 0$ or 1, let t be a pure term in $\mathcal{T}_i$, let ρ be a complete renaming of the variables in t, and let $V \subseteq \mathcal{V}(t)$. V is T_i-eliminated in t *iff* $\rho|_{\mathcal{V}(t)\setminus V}$ is a T_i-unifier of t and $\rho(t)$.

Proof. ρ is idempotent, i.e., $\rho(t) = \rho(\rho(t))$, so ρ is a T_i-unifier of t and $\rho(t)$. If V is T_i-eliminated in t, then there exists a term r such that $r =_{T_i} t$ and $V \cap \mathcal{V}(r) = \emptyset$. We have $\rho(r) =_{T_i} \rho(t)$, and $(\rho|_{\mathcal{V}(t)\setminus V})(r) = \rho(r)$. Trivially $(\rho|_{\mathcal{V}(t)\setminus V})(\rho(t)) = \rho(t)$, so $\rho|_{\mathcal{V}(t)\setminus V}$ is a T_i-unifier of r and $\rho(t)$, and hence of t and $\rho(t)$. For the only-if part, assume that $\rho|_{\mathcal{V}(t)\setminus V}$ is a T_i-unifier of t and $\rho(t)$, i.e., $(\rho|_{\mathcal{V}(t)\setminus V})(t) =_{T_i} (\rho|_{\mathcal{V}(t)\setminus V})(\rho(t)) = \rho(t)$. $(\rho|_{\mathcal{V}(t)\setminus V})(t)$ does not contain any of the variables in $\rho(V)$, so $\rho(V)$ is T_i-eliminated in $\rho(t)$. Since ρ is a bijection $\rho : \mathcal{V}(t) \rightarrow \mathcal{V}(\rho(t))$, this implies that V is T-eliminated in t.

To save space, we just outline briefly how Lemma 14 can be used to check if a term s is T-eliminated in a mixed term t. Start by computing a variable abstraction $\tilde{t}$ of t, which is similar to $\bar{t}$ except that in $\tilde{t}$ T-*equal* terms in res(t) are replaced with *equal* variables. Then use Lemma 14 to decide which subsets of $\mathcal{V}(\tilde{t})$ are T_i-eliminated in $\tilde{t}$. If a variable x in $\tilde{t}$ is T_i-eliminated in $\tilde{t}$, then the corresponding term r in res(t) and all its connected subterms are T-eliminated in t, *unless* r (or some of its connected subterms) is *not* T-eliminated in some other term in res(t). The latter condition can be checked by recursive calls to the procedure we are describing. The procedure CSE can now be written as in Figure 5.

```
procedure CSE(s, t, V);
begin
  if s ∉ CSV‾(t) then return ∅
  else begin
    Σ := TCSE({t}, V);
    return {σ | σ ∈ Σ, σ(s) ∈ CSV‾(σ(t)) ∧ elim_T(σ(s), σ(t)) };
  end
end
```

Figure 5. A procedure that computes a complete set of eliminators.

Lemma 15. For all terms s, t and sets of variables $V \supseteq \mathcal{V}(t)$, CSE$(s, t, V)$ terminates and returns a total complete set of eliminators for U away from V.

Proof. Correctness and termination follow from Lemmas 10 and 13.

5. Concluding remarks

To err is human. If there are any serious errors in this work, they are probably to be found in the termination proofs or in the proof of Lemma 12, neither of which is included here, since their inclusion would have approximately doubled the size of the paper.

In order to end on a positive note, we conclude with a lemma that might be of use if the problem of unification in combinations of theories with collapse-axioms is to be solved. However, as remarked in the introduction, this is likely to be a difficult problem since its solution would, as a special case, solve the well-known open problem of unification of general terms in Abelian groups [8]. One problem if T_0 or T_1 contain collapse axioms is that Lemma 2 is then invalid — there *are* solutions to equations between terms whose top-symbols come from different sub-theories. The result is the following: Every unifier of $s =_T t$, where, say, $s \in \mathcal{T}_0$ and $t \in \mathcal{T}_1$, can be obtained by first matching s (or t) with a new variable to obtain a matcher σ, and then unifying $\sigma(t)$ $(\sigma(s))$ with the variable. (A T-matcher of s and t is a substitution σ such that $\sigma(s) =_T t$. Complete sets of matchers are defined in analogy with complete sets of substitutions [5].) The result has a certain formal likeness to Craig's interpolation lemma in first-order logic with equality — the new variable here corresponds to what is usually called a Craig interpolant.

Lemma 16. Let s and t be two terms, $s \in \mathcal{T}_0$ and $t \in \mathcal{T}_1$, let x_0, x_1 be two new variables, and let $V \supseteq L = \mathcal{V}(s) \cup \mathcal{V}(t)$. Let also the following sets of matchers and unifiers be given:

$$\Sigma'_0 \in \mathcal{CSM}_{T_0}(s, x_0, V) \quad \text{and for each } \sigma \in \Sigma'_0 \quad \Sigma''_{0,\sigma} \in \mathcal{CSU}_T(\sigma(t), x_0, V \cup \mathcal{I}(\sigma))$$
$$\Sigma'_1 \in \mathcal{CSM}_{T_1}(t, x_1, V) \quad \text{and for each } \sigma \in \Sigma'_1 \quad \Sigma''_{1,\sigma} \in \mathcal{CSU}_T(\sigma(s), x_1, V \cup \mathcal{I}(\sigma)),$$

Then

$$\Sigma = \{\sigma'' \circ \sigma'|_L \mid \sigma' \in \Sigma'_0, \sigma'' \in \Sigma''_{0,\sigma'}\} \cup \{\sigma'' \circ \sigma'|_L \mid \sigma' \in \Sigma'_1, \sigma'' \in \Sigma''_{1,\sigma'}\}$$

is a complete set of unifiers of s and t away from V.

References.

1. Birkhoff, G., 'On the structure of abstract algebras', *Proc. Cambridge Philos. Soc.* **31**, 433–454 (1935).
2. Eder, E., 'Properties of Substitutions and Unifications', *Journal of Symbolic Computation* **1**, (1985).
3. Fages, F., 'Associative-Commutative Unification', in *7th Int. Conf. on Automated Deduction* (ed. R. E. Shostak), Springer-Verlag, LNCS 170, 194–208 (1984).
4. Huet, G., 'Confluent Reductions: Abstract Properties and Applications to Term Rewriting Systems', *JACM* **27**, 1–12 (1980).
5. Huet, G. and Oppen, D.C., 'Equations and Rewrite Rules: A Survey', in *Formal Languages: Perspectives and Open Problems* (ed. R. Book), Academic Press (1980).
6. Kirchner, C., 'Méthodes et outils de conception systematique d'algorithmes d'unification dans les théories equationelles', *Thèse de doctorat d'état*, Université Nancy I (1985).
7. Lankford, D. and Brady, B., 'On the Foundations of Applied Equational Logic', *Department of Mathematics and Statistics*, Louisiana Tech University (1984).
8. Lankford, D., Butler, G. and Brady, B., 'Abelian Group Unification Algorithms for Elementary Terms', in *Proc. of an NSF Workshop on the Rewrite Rule Laboratory* (ed. J.V. Guttag, D. Kapur, and D.R. Musser), Report 84GEN008, General Electric Company, Schenectady, New-York, 101–108 (1984).

9. Livesey, M. and Siekmann, J., 'Unification of A + C-terms (bags) and A + C + I - terms (sets)', *Intern Bericht Nr. 5/76*, Institut für Informatik I, Universität Karlsruhe (1976).
10. Nelson, G. and Oppen, D.C., 'Simplification by cooperating decision procedures', *ACM Trans. on Progr. Lang. and Syst.* **1**, 245–257 (1979).
11. Shostak, R.E., 'Deciding Combinations of Theories', *JACM* **31**, 1–12 (1984).
12. Siekmann, J., 'Universal Unification', in *7th Int. Conf. on Automated Deduction* (ed. R. E. Shostak), Springer-Verlag, LNCS 170, 1–42 (1984).
13. Stickel, M.E. A Unification Algorithm for Associative—Commutative Functions, *JACM* **28**, 423–434 (1981).
14. Szabó, P., 'Unifikationstheorie erster Ordnung', *Thesis*, Univ. Karlsruhe (1982).
15. Taylor, W., 'Equational Logic', *Houston Journal of Mathematics* **5** (1979).
16. Vogel, E., 'Morphismenunifikation', *Diplomarbeit*, Univ. Karlsruhe (1978).
17. Yelick, K., 'Combining unification algorithms for confined equational theories', *Proc. First Int. Conf. on Rewriting Techn. and Appl.* (ed. J-P. Jouannaud), Springer-Verlag (1985).
18. Bürckert, H.-J., 'Some Relationships between Unification, Restricted Unification, and Matching', *SEKI-Report*, Universität Kaiserslautern (1986).
19. Herold, A., 'Combination of Unification Algorithms', *MEMO SEKI-85-VIII-KL*, Universität Kaiserslautern (1986).

Acknowledgements.

I wish to thank Stefan Arnborg for suggesting the problem studied here to me, for his help and encouragement. Jörg Siekmann, Alexander Herold and Hans-Jürgen Bürckert received me in Kaiserslautern, read an early draft and had much help and useful comment to offer. This work was supported by the Swedish Board for Technical Development (STU).

COMBINATION OF UNIFICATION ALGORITHMS

Alexander Herold
Fachbereich Informatik, Universität Kaiserslautern
Postfach 3049
D-6750 Kaiserslautern, F.R.Germany
(electronic mail: UUCP: ...!seismo!mcvax!unido!uklirb!herold)

ABSTRACT: Unification in equational theories, i.e. solving equations in varieties, is a basic operation in many applications of Computer Science, particularly in Automated Deduction [Si 84]. A combination of unification algorithms for regular finitary collapse free equational theories with disjoint function symbols is presented. The idea is first to replace certain subterms by constants and to unify this constant abstraction and then to handle the replaced subterms in a recursive step. Total correctness is shown, i.e. the algorithm terminates and yields a correct and complete set of unifiers provided the special algorithms do.

1. INTRODUCTION

Unifcation theory is concerned with problems of the following kind: given two terms built from function symbols, constants and variables, do there exist terms that can be substituted for the variables such that the two terms thus obtained become equal? Robinson [Ro 65] was the first to give an algorithm to find such a substitution with the additional property that the returned 'unifier' is most general (or is an mgu for short), i.e. all other substitutions 'unifying' the two terms can be computed from that substitution. From an algebraic point of view unification is solving equations and an mgu is a 'basis' of the whole set of unifiers.

Equational unification extends the classical unification problem to solving equations in equationally defined theories. But then there may not exist one single mgu. Depending on the equational theory there are finite or infinite sets of mgu's and in some cases the set of mgu's does not even exist. The equational theories can therefore be classified into unitary, finitary and infinitary theories and the class of nullary theories. In the literature there are many unification algorithms for special equational theories, but they only solve problems with input terms built from the function symbols defining the equational theories, arbitrary constants and variables. For a detailed bibliography we refer to the-state-of-the-art survey of J. Siekmann [Si 84]. In that article the problem of extending these algorithms to handle terms with additional 'free' function symbols (i.e. there is no equational theory defined for that function symbol) is mentioned. M. Stickel was the first who proposed a solution to that problem for the equational theory defined by an associative and commutative (AC) function symbol [St 81] and F. Fages showed the termination of Stickel's algorithm. Building upon this work K. Yelick [Ye 85] and E. Tidén [Ti 85] independently gave algorithms for combining finitary theories by abstracting those subterms to variables that do not belong to the theory of the top function symbol. Yelick restricts the problem to regular finitary collapse free theories (she calls them confined) whereas Tidén gives a proof of the completeness of his algorithm for the whole class of finitary theories. But in that general framework termination fails. Another approach is given by C. Kirchner [Ki 85] who tackles the problem by a decomposition of the terms to be unified. His algorithm only admits a more restrictive class of equational theories than the regular finitary collapse free theories.

This paper presents a different combination of unification algorithms for regular finitary collapse free equational theories based on an extension of the AC-unification algorithm of Livesey and Siekmann [LS 76][HS 85]. The total correctness of the algorithm is shown, i.e. the algorithm terminates with a complete and correct set of unifiers. The essential idea of the algorithm is as follows: for the given terms the subterms not starting with a function symbol from the same equational theory as the original terms are temporarily replaced by special constants not occurring in the whole problem, thus reducing the case at hand to a problem that can be solved by special unification algorithms. The replaced subterms are then taken care of in a recursive call of the same process.

After some definitions and notation we shall present our algorithm, demonstrate its working by an example, and prove its total correctness. Finally we shall compare our algorithm with those of Yelick and Tidén.

2. DEFINITIONS AND NOTATIONS

2.1 Terms and Substitutions

Unification theory rests upon the usual algebraic notions (see e.g. [Gr 79], [BS 81]) with the familiar concept of an algebra $\mathcal{A} = (\mathbf{A}, \mathbf{F})$ where $\mathbf{A}$ is the **carrier** and $\mathbf{F}$ is a family of **operators** given with their arities. For a given **congruence relation** ϱ the quotient algebra modulo ϱ is written as $\mathcal{A}_{/\varrho} = (\mathbf{A}_{/\varrho}, \mathbf{F})$.

Assuming that there is at least one constant (operator of arity 0) in $\mathbf{F}$ and a denumerable set of variables $\mathbf{V}$, we define $\mathbf{T}$, the set of **first order terms**, over $\mathbf{F}$ and $\mathbf{V}$, as the least set with (i) $\mathbf{V} \subseteq \mathbf{T}$, and if arity(f) = 0 for $f \in \mathbf{F}$ then $f \in \mathbf{T}$, and (ii) if $t_1, \dots, t_n \in \mathbf{T}$ and arity(f) = n then $f(t_1 \dots t_n) \in \mathbf{T}$. For a given term $t = g(t_1 \dots t_n)$ the term $t_k, 1 \le k \le n$, is called an **immediate subterm** of t and t is the **immediate superterm** of t_k; the **leading function symbol** of t is g denoted as hd(t) = g. If t is a constant or a variable then hd(t) = t.

Let $\mathbf{V}(s)$ be the set of variables occuring in a term s; a term s is **ground** if $\mathbf{V}(s) = \emptyset$.

As usual $\mathcal{T}$ denotes the algebra with carrier $\mathbf{T}$ and with operators namely the term constructors corresponding to each operator of $\mathbf{F}$. $\mathcal{T}$ is called the absolutely free (term) algebra i.e. it just gives an algebraic structure to $\mathbf{T}$. If the carrier is the set of ground terms it is called the initial algebra [GT 78] or Herbrand Universe [Lo 78].

A **substitution** σ: $\mathbf{T} \to \mathbf{T}$ is an endomorphism on the term algebra $\mathcal{T}$ which is identical almost everywhere on $\mathbf{V}$ and can be represented as a finite set of pairs: $\sigma = \{ x_1 \leftarrow t_1, \dots, x_n \leftarrow t_n \}$. The restriction $\sigma|_V$ of a substitution σ to a set of variables V is defined as $\sigma|_V x = \sigma x$ if $x \in V$ and $\sigma|_V x = x$ else.

$\mathbf{\Sigma}$ is the set of substitutions on $\mathcal{T}$ and ε the identity. The application of a substitution σ to a term $t \in \mathbf{T}$ is written as σt. The composition of substitutions is defined as the usual composition of mappings: $(\sigma\tau)t = \sigma(\tau t)$ for $t \in \mathbf{T}$.

Let $\mathrm{DOM}\sigma = \{ x \in \mathbf{V} \mid \sigma x \neq x \}$ (domain of σ)

$\mathrm{COD}\sigma = \{ \sigma x \mid x \in \mathrm{DOM}\sigma \}$ (codomain of σ)

$\mathrm{VCOD}\sigma = \mathbf{V}(\mathrm{COD}\sigma)$ (variables in codomain of σ)

If $\mathrm{VCOD}\sigma = \emptyset$ then σ is a **ground substitution**.

A set of substitutions $\Sigma \subseteq \mathbf{\Sigma}$ is said to be **based** on a set of variables W away from some set of variables $Z \supseteq W$ iff the following two conditions are satisfied

(i) $\mathrm{DOM}\sigma = W$ for all $\sigma \in \Sigma$

(ii) $\mathrm{VCOD}\sigma \cap Z = \emptyset$ for all $\sigma \in \Sigma$

In particular for substitutions based on some W we have $\mathrm{DOM}\sigma \cap \mathrm{VCOD}\sigma = \emptyset$ which is equivalent to the idempotence of σ, i.e. $\sigma\sigma = \sigma$. We shall use this property in the proofs later on.

2.2 Equational Logic and Unification

An **equation** $s = t$ is a pair of terms. A set of equations $\mathbf{T}$ is called an **equational theory** iff an equation e is in $\mathbf{T}$ whenever e is true in every model of $\mathbf{T}$ i.e. e is a consequence of $\mathbf{T}$ (or for short: $e \in \mathbf{T}$ whenever $\mathbf{T} \models e$). A set of axioms T of an equational theory $\mathbf{T}$ is a set of equations such that $\mathbf{T}$ is the least equational theory containing this set T. We sometimes say that the equational theory $\mathbf{T}$ is presented by T. For simplicity we do not distinguish between the equational theory and its presentation.

The equality $=_T$ generated by a set of equations T is the finest congruence over $\mathbf{T}$ containing all pairs $\sigma s = \sigma t$ for $s = t \in T$ and $\sigma \in \mathbf{\Sigma}$. (i.e. the $\mathbf{\Sigma}$-invariant congruence relation generated by T). The following is Birkhoffs well-known completeness theorem of equational logic [Bi 35]

Theorem 2.1: $\mathbf{T} \models s = t$ iff $s =_T t$.

We shall sometimes use another derivation system for equational logic which has been useful in induction proofs (see e.g. McNulty [Mc 76]):

$$s \longrightarrow_{\sigma,e} t$$

iff there exists an equation e of the form $l = r$ in T and a substitution $\sigma \in \mathbf{\Sigma}$ such that t results from s by replacing a subterm of s equal to σl by σr. By a derivation of $s =_T t$ we mean a finite sequence of steps $s_{i-1} \longrightarrow_{\sigma i,ei} s_i$

$$s = s_0 \longrightarrow_{\sigma 1,e1} s_1 \longrightarrow_{\sigma 2,e2} s_2 \longrightarrow_{\sigma 3,e3} \cdots \longrightarrow_{\sigma n,en} s_n = t$$

where $\sigma_i \in \mathbf{\Sigma}$ and $e_i \in T$. If they are clear from the context we omit the indices σ and e. If we consider T as a directed rewrite system, we have $=_T \; = \; \overset{*}{\longleftrightarrow}$ where $\overset{*}{\longleftrightarrow}$ is the reflexive, symmetric and transitive closure of $\longrightarrow$. Our definitions and notations are consistent with [Gr 79][HO 80][Mc 76] and [Ta 79].

We extend T-equality in $\mathbf{T}$ to the set of substitutions $\mathbf{\Sigma}$ by:

$$\sigma =_T \tau \quad \text{iff} \quad \forall x \in \mathbf{V} \quad \sigma x =_T \tau x .$$

If T-equality of substitutions is restricted to a set of variables W we write

$$\sigma =_T \tau\ [W] \quad \text{iff} \quad \forall x \in W \quad \sigma x =_T \tau x$$

and say σ and τ are **T-equal on W**.

A substitution τ is **more general than** σ **on W** (or σ is **a T-instance of** τ **on W**):

$$\sigma \leq_T \tau\ [W] \quad \text{iff} \quad \exists \lambda \in \Sigma \quad \sigma =_T \lambda\tau\ [W].$$

Two substitutions σ,τ are called **T-equivalent on W**

$$\sigma \equiv_T \tau\ [W] \quad \text{iff} \quad \sigma \leq_T \tau\ [W] \text{ and } \tau \leq_T \sigma\ [W].$$

Given two terms s, t and an equational theory T, a unification problem for T is denoted as

$$\langle s = t \rangle_T$$

We say $\sigma \in \Sigma$ is a solution of $\langle s = t \rangle_T$ (or σ is a T-unifier of s and t) iff $\sigma s =_T \sigma t$. For the set of all T-unifiers of s and t we write $U\Sigma_T(s, t)$. Without loss of generality we can assume that the unifiers of s and t are idempotent (if not, one can find an equivalent unifier that is idempotent). For a given unification problem $\langle s = t \rangle_T$, it is not necessary to compute the whole set of unifiers $U\Sigma_T(s, t)$, which is always recursively enumerable for a decidable theory T, but instead a smaller set useful in representing $U\Sigma_T$. Therefore we define $cU\Sigma_T(s, t)$, a **complete set of unifiers of s and t** on $W = \mathbf{V}(s, t)$ as:

(i)	$cU\Sigma_T \subseteq U\Sigma_T$	(correctness)
(ii)	$\forall \delta \in U\Sigma_T \quad \exists \sigma \in cU\Sigma_T: \quad \delta \leq_T \sigma\ [W]$	(completeness)

A **set of most general unifiers** $\mu U\Sigma_T(s, t)$ is a complete set with

(iii)	$\forall \sigma, \tau \in \mu U\Sigma_T: \quad \sigma \leq_T \tau\ [W]$ implies $\sigma = \tau$	(minimality).

For technical reasons it turned out to be useful to have the following requirement:

(iv) $\mu U\Sigma_T(s, t)$ (resp. $cU\Sigma_T(s, t)$) is based on W away from Z (protection of Z)

If conditions (i) - (iv) are fulfilled we say $\mu U\Sigma_T$ is a **set of most general unifiers away from Z** (resp. $cU\Sigma_T(s, t)$ is a **complete set of unifiers away from Z**) [PL72].

The set $\mu U\Sigma_T$ does not always exist [FH 83][Sc 86]; if it does then it is unique up to the equivalence $\equiv_T [W]$ (see [Hu 76][FH 83]). For that reason it is sufficient to generate just one $\mu U\Sigma_T$ as some representative of the equivalence class $[\mu U\Sigma_T]_{\equiv T}$.

Depending on the cardinality of the set of most general unifiers we can classify the equational theories into the following subclasses:

- a theory is **unitary** iff $\mu U\Sigma_T$ exists and $|\mu U\Sigma_T(s, t)| = 1$ for all s and t
- a theory is **finitary** iff $\mu U\Sigma_T$ exists and $|\mu U\Sigma_T(s, t)| < \infty$ for all s and t
- a theory is **infinitary** iff $\mu U\Sigma_T$ exists and $|\mu U\Sigma_T(s, t)| = \infty$ for some s and t
- a theory is **nullary** iff $\mu U\Sigma_T$ does not exist for some s and t.

Sometimes it turned out to be useful to change the relation $\leq_T [W]$ used in the definition of completeness and minimality to $\leq_T [X]$ with $W \subseteq X \subseteq Z$. This procedure is justified by the so called "Fortsetzungslemma" as follows:

Lemma 2.1: For two idempotent substitutions θ_1, θ_2 and the sets of variables $U \subseteq V$ with $DOM\theta_2 \subseteq U$ and $VCOD\theta_2 \cap V = \emptyset$:

$$\theta_1 \leq_T \theta_2 [\![U]\!] \text{ iff } \theta_1 \leq_T \theta_2 [\![V]\!].$$

Another technical lemma, which is useful for later proofs, is the following:

Lemma 2.2: For idempotent substitutions δ, σ, τ and a set of variables V with $DOM\tau = \mathbf{V}(\sigma(V))$ and $VCOD\tau \cap (VCOD\sigma \cup V) = \emptyset$:

(i) $DOM\tau\sigma = V \cup DOM\sigma$

(ii) if $\delta \leq_T \tau [\![\mathbf{V}(\sigma(V))]\!]$ then $\delta\sigma \leq_T \tau\sigma [\![DOM\tau\sigma]\!]$.

A proof of both lemmas can be found in [He 85].
A unification algorithm is called complete (and minimal) if it returns a correct and complete (and minimal) set of unifiers for any pair of terms.

2.3 Combination of Equational Theories

In this section we shall describe the equational theories, for which we shall give a unification algorithm.
An equation $l = r$ is called **regular** iff $\mathbf{V}(l) = \mathbf{V}(r)$. It is called a **collapse axiom** iff it is of the form $x = t$ where t is a non-variable term. A set of equations is called **regular** iff all equations are regular, and **collapse free** iff it does not contain any collapse axioms. In [Ye 85] collapse free theories are called confined. A theory T is **consistent** iff the equation $x =_T y$ is not deducible in T.

Lemma 2.3: (i) A theory T is regular iff some presentation of T is regular.

(ii) A theory T is collapse free iff some presentation of T is collapse free.

Let T be a presentation of an equational theory then $\mathbf{F}(T)$ is the set of function symbols and constants occurring in T. We sometimes call them interpreted function symbols or interpreted constants to distinguish them from the set $\mathbf{F}_\emptyset$ of function symbols and $\mathbf{C}_\emptyset$ of constants for which no equational theory is defined. We say these function symbols belong to the empty theory $\emptyset$ or are uninterpreted. A term t is **constrained** by a theory T iff $hd(t) \in \mathbf{F}(T)$ and we write $TH(t) = T$.
In the sequel we shall assume that T is the union of a set of presentations T_i whose set of function symbols are mutually disjoint, i.e. $T = \bigcup\{ T_i \mid 1 \leq i \leq n \}$ and $\mathbf{F}(T_i) \cap \mathbf{F}(T_j) = \emptyset$ for $1 \leq i \neq j \leq n$.
We say a subterm r is **alien** in s if it is not an uninterpreted constant or a variable, i.e. $r \notin \mathbf{C}_\emptyset \cup \mathbf{V}$, and if it is constrained by another theory than its immediate superterm. In other words r is an immediate subterm of some subterm r' of s and $TH(r) \neq TH(r')$. By abuse of notation t is an alien subterm of t if $t \notin \mathbf{C}_\emptyset \cup \mathbf{V}$. For a set S of terms we denote as **ALIEN(S)** a set of representatives of the T-equivalence classes of all alien subterms of S.

Hence $s =_T t$ iff $s = t$ for all $s, t \in$ ALIEN(S).

We have to impose some restrictions on the subtheories T_i of T in order for the algorithm to work:

(i) each equational theory T_i must be regular;

(ii) each equational theory T_i must not contain any collapse axioms;

(iii) each equational theory T_i must be consistent;

(iv) the wordproblem in the equational theory T must be decidable, i.e. it must be decidable whether $s =_T t$ for every $s, t \in \mathbf{T}$;

(v) each equational theory T_i must be unitary or finitary;

(vi) for each equational theory there must exist a complete unification algorithm for uninterpreted constants and variables, i.e. we can solve unification problems $\langle s = t \rangle_{T_i}$ where $s, t \in \mathbf{T}(\mathbf{F}(T_i) \cup \mathbf{C}_\emptyset, \mathbf{V})$ and $\mathbf{C}_\emptyset$ is a denumerable set of uninterpreted constants.

The first restriction is needed for the completeness and termination of the main algorithm. The second restriction is necessary in order to know which special unification algorithm is to be called. Regularity and the absence of collapse axioms is inherited to the whole equational theory T by Lemma 2.3. The third restriction is obvious since otherwise the unification problem would be trivial. The next condition is heavily relied upon in our proofs and without this restriction any unification problem is senseless. If one of the theories is infinitary or nullary, but there exists a special unification algorithm that enumerates a complete set of unifiers one can reorganize the combination algorithm to enumerate also a complete set. The last restriction is necessary since the main idea of our algorithm is to abstract subterms constrained by a different theory than the original terms to uninterpreted constants.

We want to compare these restrictions with those imposed by K. Yelick [Ye 85] and E. Tidén [Ti 85]: Tidén shows that the regularity requirement can be dropped, but Yelick [Ye 85] requires the same five conditions (i) - (v) as stated above. Concerning the last we would need special algorithms to handle free constants, but this is no restriction since theoretically we can always find such an algorithm (e.g. using the results of Yelick and Tidén) and practically nearly all known special unification algorithms work for variables and constants.

We shall now formalize this abstraction process and give some useful lemmata about T-equal terms.

Given a set of constants $\mathbf{C}_{=T} = \{ c_{[t]} \mid [t] \in \mathbf{T}(\mathbf{F}(T), \mathbf{V})_{/=T} \} \subseteq \mathbf{C}_\emptyset$ indexed by the equivalence classes modulo $=_T$, the replacement of the subterms of a term t not constrained by the theory of t can be described by the following recursive function

$$\text{C-abstract}_i\colon\ \mathbf{T}(\mathbf{F}(T), \mathbf{V}) \longrightarrow \mathbf{T}(\mathbf{F}(T_i) \cup \mathbf{C}_\emptyset, \mathbf{V})$$

with C-abstract$_i(t) = t$ if $t = x$ with $x \in \mathbf{V}$ or $t = c$ with $c \in \mathbf{C}_\emptyset$ and C-abstract$_i(t) = f($C-abstract$_i(t_1) \dots$ C-abstract$_i(t_n))$ if $t = f(t_1 \dots t_n)$ and $f \in \mathbf{F}(T_i)$ and C-abstract$_i(t) = c_{[t]}$ if $t = f(t_1 \dots t_n)$ and $f \notin \mathbf{F}(T_i)$. We omit the indices of C-abstract$_i$ if it is clear to which subalgebra we abstract. Note that we replace T-equal subterms by the same constant.

Let $TH(t) = T_i$ then we denote the abstracted term by $\underline{t}$ = C-abstract$_i$(t). The set of **immediate alien** subterms is defined as the set I-ALIEN(t) = $\{t_1, ..., t_m\} \subseteq$ ALIEN(t) of immediate subterms constrained by another theory than t and replaced by some constants $c_{[t1]}, ..., c_{[tm]}$. Note that the set I-ALIEN is again a set of representatives of the T-equivalence classes. The **subterm replacement** is denoted as $\mathfrak{a} = [t_1 \Leftarrow c_{[t1]}, ..., t_m \Leftarrow c_{[tm]}]$ with $\underline{t} = \mathfrak{a}t$. Now consider the inverse subterm replacement $\mathfrak{a}^{-1} = [c_{[t1]} \Leftarrow t_1, ..., c_{[tm]} \Leftarrow t_m]$. If we treat the constants $c_{[t1]}, ..., c_{[tm]}$ in $\mathfrak{a}^{-1}$ as '**special variables**' there is no need to formally distinguish between the subterm replacement $\mathfrak{a}^{-1}$ and the substitution $\alpha = \{c_{[t1]} \leftarrow t_1, ..., c_{[tm]} \leftarrow t_m\}$. We then have $\alpha\underline{s} = \alpha\mathfrak{a}s =_T s$.

We define the **theory height** of a term t as the maximal number of theory changes in that term:

$$h_T(t) = \begin{cases} 1 + \max\{ h_T(s) \mid s \in \text{I-ALIEN}(t) \} & \text{if I-ALIEN}(t) \neq \emptyset \text{ and} \\ 1 & \text{else} \end{cases}$$

and call terms whose theory height equals 1 **T-pure** terms (or **pure** for short). Note that for pure terms t = C-abstract(t) and I-ALIEN(t) = Ø.

Next we collect some lemmata which are needed later on. All lemmas apply to equational theories T which satisfy the above restrictions. The proofs are always induction proofs on the length of a derivation of $s \xrightarrow{*} t$ and are carried out in [He 85].

Lemma 2.4: If $s =_T t$ then TH(s) = TH(t).

As an immediate consequence we have the following fact which is useful as an extended 'clash criterium' for the equational theories under consideration:

Corollary 2.1: If TH(s) ≠ TH(t) then s and t are not T-unifiable.

Lemma 2.5: If $s =_T t$ then $h_T(s) = h_T(t)$.

We finally want to show that a complete set of T_i-unifiers for T_i-pure terms is a complete set of T-unifiers. The idea of the proof is to abstract the non-pure subterms in the codomain of an arbitrary unifier by constants to get a pure unifier which is more general than the original unifier. E. Tidén [Ti 85] showed the lemma for the more general case of nonregular theories with collapse axioms. A proof for regular collapse free theories can be found in [He 85].

Lemma 2.6: Let s and t be pure terms with $TH(s) = TH(t) = T_i$. Then every complete set $cU\Sigma_{Ti}(s, t)$ of T_i-unifiers is a complete set of T-unifiers.

Finally we want to introduce a new class of theories called **simple theories**: a theory T is simple iff $\mu U\Sigma_T(x, t) = \emptyset$ for all terms t and all variables x occurring in t. This is equivalent to the fact that a term is never T-equal to one of its subterms. It is easy to see that theories with finite congruence classes are always simple. The converse does not hold. For simple theories some of the proofs are easier and especially in the algorithms a lot of recursive calls can be dropped.

3. THE ALGORTIHM

Before we state the algorithm we need some notation: given two arbitrary terms s and t we defined the set I-ALIEN(s, t) of toplevel subterms that are constrained by another theory than s resp. t. We then need the set of subproblems of s and t

$$SP(s, t) := \{ (s', t') \mid s', t' \in \text{I-ALIEN}(s, t) \text{ and } TH(s') = TH(t') \}$$

i.e. those pairs which are potentially unifiable.

We do not explicitly consider the details of basing the unifier on $\mathbf{V}(s, t)$ away from some set of variables Z containing $\mathbf{V}(s, t)$ since it would only complicate the notation. The proofs demonstrating that the unifiers are based on $\mathbf{V}(s, t)$ away from Z are not difficult. The special algorithms for the particular theories T_i are denoted by T_i-UNIFY.

In our main algorithm we shall use an operation called the merge $\sigma * \alpha$ of two substitutions σ and α. Essentially the merge is the set of most general instances of the two substitutions and is defined in chapter 4 along with some properties. For a set Σ of substitutions we abbreviate $\{ \sigma\delta \mid \sigma \in \Sigma \}$ by $\Sigma \cdot \delta$ and $\bigcup \{ \sigma * \delta \mid \sigma \in \Sigma \}$ by $\Sigma * \delta$.

The main idea of the unification algorithm is first to unify the constant abstractions of the original terms. In order to obtain finally the unifiers for the original terms we have to merge the unifiers of the abstracted terms with the substitution reversing the abstraction, where the newly introduced constants are now regarded as variables (STEP 5). Once we have solved the unification problem for the constant abstraction we have to apply it recursively to all subterms that have been 'abstracted away' (STEP 6).

```
FUNCTION   UNIFY
   INPUT:    Two arbitrary terms s and t

   STEP 1:   if       s ∈ V  or  t ∈ V      then  Σ_T(s, t) := VARIABLE-UNIFY(s, t)
   STEP 2:   elseif   s ∈ C  or  t ∈ C      then  Σ_T(s, t) := CONSTANT-UNIFY(s, t)
   STEP 3:   elseif   TH(s) ≠ TH(t)         then  Σ_T(s, t) := Ø
   STEP 4:   elseif   I-ALIEN(s, t) = Ø     then  Σ_T(s, t) := TH(s)-UNIFY(s, t)
             else     let  s, t  be the constant abstraction and α be the
                      corresponding substitution reversing the abstraction
   STEP 5:            in  Σ_T(s, t) := TH(s)-UNIFY(s, t) * α
   STEP 6:            forall (s', t') ∈ SP(s, t)  do
                         forall σ' ∈ UNIFY (s', t')  do
                               Σ_T(s, t) := Σ_T(s, t) ∪ UNIFY(σ's, σ't)•σ'
                         od
                      od

   OUTPUT:   The set of unifiers Σ_T(s, t) away from Z ⊇ V(s, t)
ENDOF        UNIFY
```

There remain the cases where at least one term is either a variable or a constant.

FUNCTION CONSTANT-UNIFY

INPUT: Two non-variable terms s and t where at least one is a constant

STEP 1: if $s \in \mathbf{C}_{\emptyset}$ and $t \in \mathbf{C}_{\emptyset}$
then if $s = t$ then $\Sigma_T(s, t) := \{\varepsilon\}$ else $\Sigma_T(s, t) := \emptyset$

STEP 2: elseif $s \in \mathbf{C}_{\emptyset}$ then $\Sigma_T(s, t) := \emptyset$

STEP 3: elseif $t \in \mathbf{C}_{\emptyset}$ then $\Sigma_T(s, t) := \emptyset$

STEP 4: elseif $TH(s) \neq TH(t)$ or $h_T(s) \neq h_T(t)$
then $\Sigma_T(s, t) := \emptyset$ else $\Sigma_T(s, t) :=$ TH(s)-UNIFY(s, t)

OUTPUT: The set of unifiers $\Sigma_T(s, t)$ away from $Z \supseteq \mathbf{V}(s, t)$

ENDOF CONSTANT-UNIFY

Provided the T_i-unification algorithms terminate, the termination of CONSTANT-UNIFY is is easy to see.

Lemma 3.1: For a constant and a non-variable term CONSTANT-UNIFY returns a correct and complete set of unifiers.

The lemma follows from Corollary 2.1, Lemma 2.5 and Lemma 2.6. The variable case is more complicated if the considered theories are not simple.

FUNCTION VARIABLE-UNIFY

INPUT: Two terms s and t where at least one is a variable

STEP 1: if $s \in \mathbf{V}$ and $t \in \mathbf{V}$
then if $s = t$ then $\Sigma_T(s, t) := \{\varepsilon\}$
else $\Sigma_T(s, t) := \{\{s \leftarrow t\}\}$

STEP 2: elseif $s \in \mathbf{V}$ then $\Sigma_T(s, t) :=$ VARIABLE-TERM-UNIFY(s, t)

STEP 3: elseif $t \in \mathbf{V}$ then $\Sigma_T(s, t) :=$ VARIABLE-TERM-UNIFY(t, s)

OUTPUT: The set of unifiers $\Sigma_T(s, t)$ away from $Z \supseteq \mathbf{V}(s, t)$

ENDOF VARIABLE-UNIFY

The following example shows the difficulty in the variable-term-case: let $T_1 = \{ f(f(x, y), y) = f(x, y) \}$, $T_2 = \emptyset$ with $g \in \mathbf{F}(T_2)$ and $T = T_1 \cup T_2$ be the equational theories and $\langle u = f(u, g(v)) \rangle_T$ the unification problem. Then there exists a unifier $\sigma = \{ u \leftarrow f(u', g(v)) \}$ which will be computed by the following algorithm in STEP 5.

FUNCTION VARIABLE-TERM-UNIFY

INPUT: A variable x and a non-variable term t

```
STEP 1:  if      x ∉ V(t)             then  Σ_T(x, t) := {{ x ← t }}
STEP 2:  elseif  TH(t) is simple      then  Σ_T(x, t) := Ø
STEP 3   elseif  I-ALIEN(t) = Ø       then  Σ_T(x, t) := T_i-UNIFY(x, t)
         else let t̲ be the constant abstraction and α be the
              corresponding substitution reversing the abstraction
              in
STEP 4:         if   x ∈ VCODα        then  Σ_T(x, t) := Ø
                else
STEP 5:           Σ_T(x, t) := T_i-UNIFY(x, t̲) * α
STEP 6:           forall (s', t') ∈ SP(t) do
                    forall σ' ∈ UNIFY (s', t') do
                      Σ_T(x, t) := Σ_T(x, t) ∪ UNIFY(σ'x, σ't)∘σ'
                    od
                  od
```

OUTPUT: The set of unifiers $\Sigma_T(x, t)$ away from $Z \supseteq \mathbf{V}(s, t)$

ENDOF VARIABLE-TERM-UNIFY

In STEP 1 and STEP 2 termination, correctness and completeness are trivial. By Lemma 2.6 we know that in STEP 3 a correct and complete set of unifiers is returned, whereas termination is inherited from termination of T_i-UNIFY. In STEP 4 correctness and completeness follows from the next lemma and termination is trivial. For the other steps termination, correctness and completeness is shown as for the main algorithm.

Lemma 3.2: If $x \in \mathbf{V}(t)$ and x occurs in some immediate alien subterm of t, i.e. $x \in VCOD\alpha$, then $U\Sigma_T(x, t) = \emptyset$.

Proof: Suppose there exists a σ such that $\sigma x =_T \sigma t$. Since $x \in VCOD\alpha$ there exists a subterm s in t that is constrained by another theory as t and $x \in \mathbf{V}(s)$. Hence σx is a subterm of σs and therefore $h_T(\sigma x) \le h_T(\sigma s) < h_T(\sigma t)$ which is a contradiction to $\sigma x =_T \sigma t$ and Lemma 2.5. ■

The lemma is a generalized 'occur-check' for non-simple regular collapse free theories. With the theories T_1 and T_2 of the above example the unification problem $\langle u = f(v\ g(u)) \rangle_T$ has no solution since u occurs in the immediate alien subterm $g(u)$.

4. THE MERGE OF SUBSTITUTIONS

The algorithm of the previous section used an operation called the merge of substitutions or unification of substitutions. Given two substitutions σ and τ we say σ and τ are T-unifiable iff there exists λ such that $\lambda\sigma =_T \lambda\tau$. Then λ is called a T-unifier of σ and τ. The sets $U\Sigma_T(\sigma, \tau)$, $cU\Sigma_T(\sigma, \tau)$, $\mu U\Sigma_T(\sigma, \tau)$ are defined accordingly. If $\mu U\Sigma_T(\sigma, \tau)$ exists then $\sigma * \tau := \{ \lambda\sigma \mid \lambda \in \mu U\Sigma_T(\sigma, \tau) \}$ is called a merge of σ and τ.

In the special situation of the previous algorithms there are certain constraints on the two substitutions which we want to exploit in the computation of the merge: first $\tau = \{x_1 \leftarrow t_1, ..., x_m \leftarrow t_m\}$ is a unifier of the constant abstractions which are pure terms and $\alpha = \{c_1 \leftarrow r_1, ..., c_n \leftarrow r_n\}$ reverses a constant abstraction. Note that the newly introduced constants are now considered as special variables. Hence we have

(i) $\quad DOM\tau \cap DOM\alpha = \emptyset$

(ii) $\quad VCOD\tau \cap VCOD\alpha = \emptyset$.

The last equation holds since $COD\tau$ only contains the special variables and new variables not occurring in s and t whereas $VCOD\alpha \subseteq \mathbf{V}(s, t)$.

In the following lemmata we shall always assume that τ and α satisfy these conditions. We show that unifying two substitutions is the same as unifying two termlists (σ unifies $(s_1, ..., s_n)$ and $(t_1, ..., t_n)$ iff $\sigma s_i =_T \sigma t_i$ for $1 \le i \le n$, or equivalently σ unifies the set of termpairs $\{ (s_i, t_i) \mid 1 \le i \le n \}$). All the proofs of this chapter are omitted and can be found in [He 85].

Lemma 4.1: For τ and α as above:

$$U\Sigma_T(\tau, \alpha) = U\Sigma_T((x_1, ..., x_m, c_1, ..., c_n), (t_1, ..., t_m, r_1, ..., r_n))$$

Since for simple equational theories $\mu U\Sigma_T(x, t) = \emptyset$ if $x \in \mathbf{V}(t)$ and $\mu U\Sigma_T(x, t) = \{\{x \leftarrow t\}\}$ otherwise the most general unifier (if it exists) of the termlists in Lemma 4.1 is just the composition of $\{x_i \leftarrow t_i\}$ and $\{c_j \leftarrow r_j\}$. Hence we define $\tau_0 = \tau$ and $\tau_j = \sigma_j \tau_{j-1}$ and $\sigma_j = \{c_j \leftarrow \tau_{j-1} r_j\}$ for $1 \le j \le n$.

Lemma 4.2: In a simple theory T the termlists $(x_1, ..., x_m, c_1, ..., c_n)$ and $(t_1, ..., t_m, r_1, ..., r_n)$ are T-unifiable iff τ_n is the most general unifier.

Corollary 4.1: Let T be a simple theory. If the substitutions τ and α are T-unifiable then τ_n is a single most general unifier of τ and α and $\tau * \alpha = \{ \tau_n \}$ with $\tau_n = \tau_n \alpha = \tau_n \tau$ is the merge of τ and α. If the substitutions τ and α are not T-unifiable then there exists j, $1 \le j \le n$ with $c_j \in \mathbf{V}(\tau_{j-1} r_j)$.

For non-simple theories the unification problem $\langle c_j = \tau_{j-1} r_j \rangle_T$ is not trivial as we have seen in chapter 3. It can produce a set of substitutions different from $\{c_j \leftarrow \tau_{j-1} r_j\}$. Hence we cannot directly state the set of most general unifiers of τ and α. But we shall show that it is sufficient to compute the set of most general unifiers of two smaller termlists.

We later use these termlists to show the termination of our main algorithm.

Lemma 4.3: (i) If τ and α are T-unifiable then the termlists $(\tau c_1, ..., \tau c_n)$ and $(\tau r_1, ..., \tau r_n)$ are T-unifiable and vice versa.
(ii) For $\lambda \in \mu U\Sigma_T((\tau c_1, ..., \tau c_n), (\tau r_1, ..., \tau r_n))$ there exists $\theta \in \mu U\Sigma_T(\tau, \alpha)$ with $\lambda\tau =_T \theta\, [\![\mathbf{V}(\tau) \cup \mathbf{V}(\alpha)]\!]$ and vice versa.

We shall now describe an algorithm that computes a complete set $cU\Sigma_T((s_1, ..., s_n), (t_1, ..., t_n))$ of unifiers of two termlists. We assume the existence of our main algorithm UNIFY since LIST-UNIFY and UNIFY are mutually recursive.

FUNCTION LIST-UNIFY
INPUT: Two arbitrary termlists $(s_1, ..., s_n)$ and $(t_1, ..., t_n)$ of length n

$\Sigma_0 := \{\varepsilon\}$
for $i := 1, ..., n$ do
$\Sigma_i := \{ \tau_i \sigma_{i-1} \mid \sigma_{i-1} \in \Sigma_{i-1}$ and $\tau_i \in$ UNIFY$(\sigma_{i-1}s_i, \sigma_{i-1}t_i) \}$
od

OUTPUT: The set of unifiers Σ_n away from $Z \supseteq \mathbf{V}((s_1, ..., s_n), (t_1, ..., t_n))$
ENDOF LIST-UNIFY

Provided that our main algorithm UNIFY terminates and returns a correct and complete set of unifiers it is shown in [He 85] that LIST-UNIFY returns a correct and complete set of unifiers. Termination of our main algorithm and hence of LIST-UNIFY is shown in section 6. In order to compute the merge it is sufficient to take a complete (not necessarily minimal) set of unifiers of τ and α. We then have some redundant unifiers which can be eliminated in a minimizing step.

5. AN EXAMPLE

Given the unification problem

$$\langle f(x\, f(x\, f(y\, g(x\, u)))) = f(z\, f(g(a\, b)\, g(a\, b))) \rangle_T$$

where $T = T_1 \cup T_2$ with T_1 the theory of associativity and commutativity (AC) for the function symbol f denoted as AC_f and T_2 the empty theory $\emptyset$ for g; a, b are two uninterpreted constants. To ease the notation we drop the unnecessary function symbols f and represent the terms as abelian strings. The immediate alien subterms for $s = f(x^2\, y\, g(x\, u))$ and $t = f(z\, g(a\, b)\, g(a\, b))$ are I-ALIEN(s, t) = { g(x u), g(a b) }. The only subproblem is therefore (s', t') = (g(x u), g(a b)) with the most general unifier $\sigma' = \{x \leftarrow a, u \leftarrow b\}$. The constant abstractions of s and t are $\underline{s} = f(x^2\, y\, c_1)$ and $\underline{t} = f(z\, c_2^2)$ with $\alpha = \{c_1 \leftarrow g(x\, u), c_2 \leftarrow g(a\, b)\}$. The set of most general unifiers for $\underline{s}$ and $\underline{t}$ is (see [St 81][HS 85]):

$$\begin{aligned}\mu U\Sigma_T(\underline{s}, \underline{t}) = \{ & \{x \leftarrow f(z_1\ c_2),\ y \leftarrow z_2,\ z \leftarrow f(z_1\ z_2^2\ c_1)\},\\ & \{x \leftarrow c_2,\ y \leftarrow u_2,\ z \leftarrow f(u_2^2\ c_1)\},\\ & \{x \leftarrow v_1,\ y \leftarrow f(v_2 c_2^2),\ z \leftarrow f(v_1 v_2^2 c_1 c_2)\},\\ & \{x \leftarrow w_1,\ y \leftarrow f(c_2^2),\ z \leftarrow f(w_1 c_1 c_2)\}\ \}.\end{aligned}$$

Merging these unifiers with α we get:

$$\begin{aligned}\mu U\Sigma_T(\underline{s}, \underline{t}) * \alpha = \{ & \{ x \leftarrow f(z_1\ g(a\ b)),\ y \leftarrow z_2,\ z \leftarrow f(z_1\ z_2^2\ g(\ f(z_1\ g(a\ b))\ z_3\ )),\ u \leftarrow z_3 \},\\ & \{ x \leftarrow g(a\ b),\ y \leftarrow u_2,\ z \leftarrow f(u_2^2\ g(\ g(a\ b)\ u_3)),\ u \leftarrow u_3 \},\\ & \{ x \leftarrow v_1,\ y \leftarrow f(v_2\ g(a\ b)\ g(a\ b)),\ z \leftarrow f(v_1\ v_2^2\ g(a\ b)\ g(v_1 v_3)),\ u \leftarrow v_3 \},\\ & \{ x \leftarrow w_1,\ y \leftarrow f(g(a\ b)\ g(a\ b)),\ z \leftarrow f(w_1\ g(a\ b)\ g(w_1\ w_3),\ u \leftarrow w_3 \}\ \}.\end{aligned}$$

The only unifier of (s', t') is $\sigma = \{x \leftarrow a, u \leftarrow b\}$ and hence $\sigma s = f(a^2\ y\ g(a\ b))$ and $\sigma t = f(z\ g(a\ b)\ g(a\ b))$ have the set of most general T-unifiers $\mu U\Sigma_T(\sigma s, \sigma t) = \{\tau_1, \tau_2\}$ with $\tau_1 = \{y \leftarrow f(x_1\ g(a\ b)),\ z \leftarrow f(x_1\ a^2)\}$ and $\tau_2 = \{y \leftarrow g(a\ b),\ z \leftarrow f(a^2)\}$. Hence

$$\tau_1\sigma = \{x \leftarrow a,\ y \leftarrow f(x_1\ g(a\ b)),\ z \leftarrow f(x_1\ a^2),\ u \leftarrow b\}$$
$$\tau_2\sigma = \{x \leftarrow a,\ y \leftarrow g(a\ b),\ z \leftarrow f(\ a^2),\ u \leftarrow b\}$$

are two more most general T-unifier of s and t. So finally we have

$$\mu U\Sigma_T(s, t) = \mu U\Sigma_T(\underline{s}, \underline{t}) * \alpha \cup \{\tau_1\sigma, \tau_2\sigma\}.$$

6. TERMINATION

In order to prove the termination of Stickel's AC-unification algorithm, F. Fages [Fa 84] gave a complexity measure for two terms which can be used in the more general case [Ye 85][Ti 85]. We shall use a slightly modified version of that measure for showing that our algorithm terminates. The following are prerequisites for the definition of that measure.

We define the immediate function symbols of a term r in a term s by

$$Op(r, s) = \{ hd(t) \mid t \text{ is an immediate superterm of r and a subterm of s} \}$$

We write $Op(r, S)$ for $\cup\{ Op(r, s) \mid s \in S \}$ and omit the parenthesis in $Op(r, \{s, t\})$ and write $Op(r; s, t)$. The set of theories by which r is constrained in s is denoted as

$$T\text{-}Op(r, s) = \{ T \mid f \in \mathbf{F}(T) \text{ and } f \in Op(r, s) \}$$

and $T\text{-}Op(r, S)$ as $\cup\{ T\text{-}Op(r, s) \mid s \in S \}$. The set of shared variables of a set of terms S is then defined as the set of those variables, which are constraint by at least two different theories, or which are an element of S and constrained by one theory:

$$\mathbf{V}_s(S) = \{ x \in \mathbf{V}(S) \mid |T\text{-}Op(x, S)| > 1 \text{ or } x \in S \text{ and } |T\text{-}Op(x, S)| = 1 \}.$$

The complexity of a pair of terms s and t, which we shall use to show the termination of our algorithm, is:

$$C(s, t) = (\nu, \tau) \quad \text{where} \quad \nu = |\mathbf{V}_s(s, t)| \quad \text{and} \quad \tau = |ALIEN(s, t)|.$$

where the set ALIEN(s, t) of alien subterms is as defined in section 2.3. To illustrate the definitions we take the example of the previous chapter: let $s = f(x\ f(x\ f(y\ g(x\ u))))$ and $t = f(z\ f(g(a\ b)\ g(a\ b)))$ then $Op(x; s, t) = \{f, g\}$ and $T\text{-}Op(x; s, t) = \{AC_f, \emptyset\}$. Since the other

variables only occur immediately under one function symbol we have $\mathbf{V}_s(s,t) = \{x\}$. The set of alien subterms of s and t is ALIEN(s, t) = {s, t, g(x u), g(a b)}. Note that an uninterpreted constant is not an alien subterm in our definition whereas in the definitions of Fages, Tiden and Yelick it is.
Taking the lexicographic order on the complexities we obtain a Noetherian order. For this section we always assume that one of the given terms s and t is not a variable or an uninterpreted constant, the other cases were treated in section 3. The following lemma states that the complexity of alien subterms is less than the complexity of the terms itself and is shown in [He 85].

Lemma 6.1: Given two terms s and t. If $s', t' \in$ ALIEN(s, t) are proper subterms of s or t or $t' \in$ ALIEN(s, t) is a proper subterm of s or t and $s' = x \in \mathbf{V}_s(s, t)$ is a shared variable of s and t then

$$\mathbf{C}(s', t') < \mathbf{C}(s, t).$$

To show the termination of our main algorithm UNIFY we have to show $\mathbf{C}(\sigma s, \sigma t) < \mathbf{C}(s, t)$ if σ unifies some immediate alien subterms of s and t, i.e. we have to show that unifiers produced by the algorithm decrease the complexity of the original terms.
We say a substitution σ is **monotone for s and t** iff $\mathbf{C}(\sigma s, \sigma t) \le \mathbf{C}(s, t)$ and **strictly monotone for s and t** iff $\mathbf{C}(\sigma s, \sigma t) < \mathbf{C}(s, t)$. In the following lemmata we show the monotony of certain substitutions. The proofs appear in [He 85]. We call a substitution σ **alien for s and t** iff $\sigma = \{x \leftarrow r\}$ with $x \in \mathbf{V}(s, t)$, $r \in$ ALIEN(s, t) and $x \notin \mathbf{V}(r)$.

Lemma 6.2: If a substitution σ is alien for two terms s and t then σ is monotone for s and t.

A substitution σ is called **T-pure for s and t** iff $\mathrm{DOM}\sigma \subseteq \mathbf{V}(s, t)$, $\mathrm{VCOD}\sigma \cap \mathbf{V}(s, t) = \emptyset$ and the following two conditions are satisfied

- $T \in$ T-Op(x; s, t) for all $x \in \mathrm{DOM}\sigma$ and
- σx is a T-pure term (i.e. $\mathrm{COD}\sigma \subseteq \mathbf{T}(\mathbf{F}(T) \cup \mathbf{C}_{\emptyset}, \mathbf{V})$).

We asssume that the algorithms T-UNIFY only generate T-pure substitutions for T-pure terms; we simply speak of pure substitutions and pure terms if there are no ambiguities.

Lemma 6.3: If a substitution σ is T-pure for two terms s and t then σ is monotone for s and t.

In the termination proof we shall often use the fact that a substitution that is pure or alien for s' and t' is pure or alien for s and t as well if s' and t' are alien subterms of s and t:

Lemma 6.4: Let s and t be two terms and let $s', t' \in$ ALIEN(s, t) $\cup\ \mathbf{V}(s, t)$ be proper subterms of s or t and not both variables. If σ is alien for s' and t' then σ is alien for s and t. If σ is a pure substitution for s' and t' and $\mathrm{VCOD}(\sigma) \cap \mathbf{V}(s, t) = \emptyset$ then σ is pure for s and t.

The next lemma is the key for the termination proof and shown in [He 85].

Lemma 6.5: Let σ be a pure or alien substitution for s and t and let $s', t' \in$ ALIEN(s, t) be proper and distinct subterms of s or t. If σ unifies s' and t' then σ is strictly monotone.

Since in the algorithm the substitutions are built up by composition we say a substitution σ is **elementary for the problem** $\langle s = t \rangle_T$ (or short for s and t) iff it is a composition of pure or alien substitutions, i.e. $\sigma = \sigma_n \sigma_{n-1} \ldots \sigma_1$ where σ_i is pure or alien for $\sigma_{i-1} \ldots \sigma_1 s$ and $\sigma_{i-1} \ldots \sigma_1 t$ for $2 \le i \le n$ and σ_1 is pure or alien for s and t. By an induction argument we have:

Lemma 6.6: (i) If σ is an elementary substitution for s and t then σ is monotone for s and t.

(ii) If in addition σ unifies two distinct and proper subterms $s', t' \in$ ALIEN(s, t) then σ is strictly monotone for s and t.

(iii) Let $s', t' \in$ ALIEN(s, t) $\cup$ **V**(s, t) be proper subterms of s or t and not both are variables. If σ is elementary for s' and t' then σ is elementary for s and t provided the newly introduced variables are away from **V**(s, t).

To summarize: first we introduced monotone substitutions. Then we have shown that alien and pure substitutions (the elements of the generated unifiers) and their compositions are monotone.
The main termination proof is by Noetherian induction on the complexities of the input terms. We show UNIFY(s, t) terminates and generates substitutions elementary for s and t. Therefore it is sufficient:

> for two terms s and t the complexity of the input terms s' and t' in every recursive call of UNIFY in UNIFY(s, t) is smaller than the complexity of the original terms, i.e. $\mathcal{C}(s', t') < \mathcal{C}(s, t)$ (hence we can apply the induction hypothesis) and the substitutions generated by UNIFY(s, t) are elementary for s and t.

First we prove that the merge operation terminates by showing that every call of UNIFY in LIST-UNIFY($(c_1, \ldots, c_n), (\tau r_1, \ldots, \tau r_n)$) terminates and the generated substitutions are elementary for s and t.

Theorem 6.1: Given s and t with the corresponding abstractions $\underline{s}$ resp. $\underline{t}$, the substitutions $\alpha = \{c_1 \leftarrow r_1, \ldots, c_n \leftarrow r_n\}$ reversing the abstraction and $\tau \in$ TH(s)-UNIFY($\underline{s}, \underline{t}$) unifying the abstractions. Then the merge operation $\tau * \alpha$ terminates and the merges are elementary for s and t.

Proof: We show by induction on n that for $i = 1, \ldots, n$ $\mathcal{C}(\tau_{i-1} c_i, \tau_{i-1} r_i) < \mathcal{C}(s, t)$ and τ_i is elementary for s and t with $\tau_0 = \tau$, $\tau_i = \sigma_i \tau_{i-1}$ and $\sigma_i \in$ UNIFY$(\tau_{i-1} c_i, \tau_{i-1} r_i)$. Since $\tau \in$ TH(s)-UNIFY($\underline{s}, \underline{t}$) $\tau_0 = \tau$ is pure and hence elementary for s and t, w.l.o.g. we can assume VCOD$\tau \supseteq \{c_1, \ldots, c_n\}$.

Base step: first we have $\tau_0 c_1 = c_1 \in \mathbf{V}(\tau_0 s, \tau_0 t)$. Since $r_1 \in$ I-ALIEN(s, t) it is $\tau_0 r_1 \in$ I-ALIEN($\tau_0 s, \tau_0 t$). To show that τ_1 is elementary for s and t we distinguish two cases:

CASE 1: $c_1 \in \mathbf{V}(\tau_0 r_1)$: We want to apply the second part of Lemma 6.1 and have therefore to show that c_1 is a shared variable in $\tau_0 s$ and $\tau_0 t$. Let $V_1 := \mathbf{V}(r_1) \cap \mathrm{DOM}\tau_0$. Suppose for all $x \in V_1$ $\tau_0 x \neq c_1$ then $\mathrm{Op}(c_1, \tau_0 x) = \mathrm{TH}(s)$, since τ_0 is TH(s)-pure, and hence $\mathrm{Op}(c_1, \tau_0 r_1) = \mathrm{TH}(s) \neq \mathrm{TH}(r_1)$, i.e. c_1 and $\tau_0 r_1$ are not unifiable by Corollary 2.1. Now if $\{ x \in V_1 \mid \tau_0 x = c_1 \} \cap \mathbf{V}_s(s, t) = \emptyset$ then again $\mathrm{Op}(c_1, \tau_0 r_1) = \mathrm{TH}(s) \neq \mathrm{TH}(r_1)$ by the TH(s)-purity of τ_0, which again contradicts the unifiability of c_1 and $\tau_0 r_1$. Hence $\tau_0 c_1 = c_1 \in \mathbf{V}_s(\tau_0 s, \tau_0 t)$ and with the second part of Lemma 6.1 we have $\mathcal{C}(\tau_0 c_1, \tau_0 r_1) < \mathcal{C}(\tau_0 s, \tau_0 t) \leq \mathcal{C}(s, t)$. Let $\sigma_1 \in$ UNIFY($\tau_0 c_1, \tau_0 r_1$) then by the main Noetherian induction hypothesis σ_1 is elementary for $\tau_0 c_1$ and $\tau_0 r_1$ and by Lemma 6.6 (iii) σ_1 is elementary for $\tau_0 s$ and $\tau_0 t$.

CASE 2: $c_1 \notin \mathbf{V}(\tau_0 r_1)$: with $\sigma_1 = \{c_1 \leftarrow \tau_0 r_1\}$ we have $\{\sigma_1\} =$ UNIFY($\tau_0 c_1, \tau_0 r_1$) and with Lemma 6.4 σ_1 is alien for $\tau_0 s$ and $\tau_0 t$.

Summarizing $\tau_1 = \sigma_1 \tau_0$ is elementary for s and t.

Induction step: If $\tau_{n-1} c_n = c_n$ then the proof is analogue to the base step. Now let $\tau_{n-1} c_n \neq c_n$. If $\mathrm{TH}(\tau_{n-1} c_n) = \mathrm{TH}(s)$ then by Corollary 2.1 $\tau_{n-1} c_n$ and $\tau_{n-1} r_n$ are not unifiable since $\mathrm{TH}(\tau_{n-1} r_n) \neq \mathrm{TH}(\tau_{n-1} c_n)$. Now if $\mathrm{TH}(\tau_{n-1} c_n) \neq \mathrm{TH}(s)$ $\tau_{n-1} c_n$ is in ALIEN($\tau_{n-1} s, \tau_{n-1} t$) by $\mathrm{TH}(s) \in$ T-Op($c_i; \tau_0 s, \tau_0 t$). By the same argument as above $\tau_{n-1} r_n \in$ ALIEN($\tau_{n-1} s, \tau_{n-1} t$). Hence with Lemma 6.1 and the induction hypothesis (τ_{n-1} is elementary for s and t) we get that $\mathcal{C}(\tau_{n-1} c_n, \tau_{n-1} r_n) < \mathcal{C}(\tau_{n-1} s, \tau_{n-1} t) \leq \mathcal{C}(s, t)$. Hence by the main Noetherian induction $\sigma_n \in$ UNIFY($\tau_{n-1} c_n, \tau_{n-1} r_n$) is elementary for $\tau_{n-1} c_n$ and $\tau_{n-1} r_n$ and with Lemma 6.6 (iii) for $\tau_{n-1} s$ and $\tau_{n-1} t$. Finally we have $\tau_n = \sigma_n \tau_{n-1}$ is elementary for s and t. ∎

We now state the two main theorems:

Theorem 6.2: For a variable x and a term t VARIABLE-TERM-UNIFY(x, t) terminates and generates substitutions which are elementary for x and t.

The proof is analogue to that of the next theorem since termination is obvious for the steps in TERM-VARIABLE-UNIFY that are different from UNIFY and the substitutions generated in these steps are elementary by definition.

Theorem 6.3: For two terms s and t at least one of which is not a variable UNIFY(s, t) terminates and generates substitutions which are elementary for s and t.

Proof: In STEP 1 termination is established by Theorem 6.2 as well as the property of the generated substitutions being elementary.

For STEP 2 termination of CONSTANT-UNIFY is obvious. The empty substitution and unifiers of pure terms are elementary for s and t.

For STEP 3 we are done.

In STEP 4 termination follows from termination of TH(s)-UNIFY and the generated substitutions are elementary since they are T-pure by the remark before Lemma 6.3.

As in STEP 4 TH(s)-UNIFY($\underline{s}$, $\underline{t}$) terminates. By Theorem 6.1 the merge operation terminates and the merges are elementary for s and t. So the theorem is shown for STEP 5.

In STEP 6 let $(s', t') \in SP(s,t)$ then by Lemma 6.1 $C(s', t') < C(s, t)$ and hence UNIFY(s', t') terminates and the substitutions $\sigma' \in$ UNIFY(s', t') are elementary for s' and t' by the main induction hypothesis. With Lemma 6.6 (iii) σ' is elementary for s and t since $s', t' \in$ ALIEN(s, t). Since σ' unifies s' and t' we know by Lemma 6.6 (ii) that σ' is strictly monotone for s and t, i.e. $C(\sigma's, \sigma't) < C(s, t)$. Hence by induction hypothesis UNIFY(σ's, σ't) terminates and produces substitutions σ'' which are elementary for σ's and σ't, i.e. $\sigma = \sigma''\sigma'$ is elementary for s and t. ∎

7. CORRECTNESS AND COMPLETENESS

All the proofs of this chapter are by induction on the recursion depth of the term pair which is a Noetherian order by the last chapter. The set of substitutions returned by the unification algorithm is a correct set of unifiers of s and t :

Theorem 7.1: For $s, t \in \mathbf{T}$ UNIFY(s, t) returns a correct set of unifiers.

Proof: Consider each step in UNIFY in succession:

STEP 1: The theorem follows from the theorem below (correctness for the variable-term-case).

STEP 2: Correctness is obvious (confer chapter 3).

STEP 3: Nothing is to show.

STEP 4: By assumption T_i-UNIFY is correct.

STEP 5: Let $\underline{s}$, $\underline{t}$ be the constant abstractions of s and t with $\alpha\underline{s} =_T s$ and $\alpha\underline{t} =_T t$. By induction hypothesis let τ be a correct T-unifier of $\underline{s}$ and $\underline{t}$. Since for $\theta \in \tau * \alpha$ $\theta = \lambda\alpha =_T \lambda\tau$ for some λ we have (using the idempotence of τ and α):

$$\theta\tau =_T \lambda\tau\tau = \lambda\tau =_T \lambda\alpha = \lambda\alpha\alpha =_T \theta\alpha$$

Hence
$$\begin{aligned} \theta s &=_T \theta\alpha\underline{s} \\ &=_T \theta\tau\underline{s} \\ &=_T \theta\tau\underline{t} \qquad \text{by assumption} \\ &=_T \theta\alpha\underline{t} \\ &=_T \theta t. \end{aligned}$$

STEP6: Let (s', t') be a subproblem of s and t. By induction hypothesis let σ' be a correct unifier of s' and t' and σ'' be a correct T-unifier of σ's and σ't. Then for $\sigma = \sigma'' \circ \sigma' \in \Sigma_T$ we have

$$\sigma s = \sigma''(\sigma' s) =_T \sigma''(\sigma' t) = \sigma t$$ ∎

Theorem 7.2: Let $x \in \mathbf{V}$ and $t \in \mathbf{T}$ then VARIABLE-TERM-UNIFY(x, t) returns a correct and complete set of unifiers.

The correctness proof is analogous to the one above, completeness is shown as below. The following theorem shows that the main algorithm returns a complete set of unifiers. The technical details can be found in [He 85].

Theorem 7.3: Let s, t be terms and let θ be a T-unifier of s and t. Then there exists $\sigma \in \Sigma_T(s, t)$ (returned by UNIFY(s, t)) such that

$$\theta \leq_T \sigma \ [\![V]\!] \quad \text{with } V = \mathbf{V}(s, t).$$

Proof: Again we consider each step in turn.
STEP 1: The theorem follows from the next theorem (completeness for the variable-term-case).
STEP 2: By Lemma 3.1.
STEP 3: By Corollary 2.1.
STEP 4: By Lemma 2.6.
STEP 5: Now I-ALIEN(s, t) ≠ Ø and assume for all $(s', t') \in SP(s, t)$ it is $\theta s' \neq_T \theta t'$ (else STEP 6 applies). Then there exists $\underline{\theta}$ with $\underline{\theta}\underline{s} =_T \underline{\theta}\underline{t}$ and $\theta^* \in \underline{\theta} * \alpha$ with $\theta \leq_T \theta^* \ [\![V]\!]$, where $\underline{s} = \mathbf{a}s$ and $\underline{t} = \mathbf{a}t$ are the constant abstractions of s and t and α the substitution reversing the constant abstraction **a**. By Lemma 2.6 there exists $\theta' \in \Sigma_T(\underline{s}, \underline{t})$ (returned by TH($\underline{s}$)-UNIFY($\underline{s}$, $\underline{t}$)) such that $\underline{\theta} \leq_T \theta' \ [\![\underline{V}]\!]$ where $\underline{V} = \mathbf{V}(\underline{s}, \underline{t})$. Then θ' and α are T-unifiable and there exists $\sigma \in \theta' * \alpha \subseteq \Sigma_T(s, t)$ (returned by UNIFY(s, t)) with

$$\theta \leq_T \sigma \ [\![V]\!].$$

STEP 6: In this step the subproblems are considered and moreover there exists $(s', t') \in SP(s, t)$ with $\theta s' =_T \theta t'$. By induction there exists $\sigma' \in \Sigma_T(s', t')$ (returned by UNIFY(s', t')) such that $\theta \leq_T \sigma' \ [\![V']\!]$ with $V' = \mathbf{V}(s', t')$. In other words there exists λ with $\theta =_T \lambda\sigma' \ [\![V]\!]$ using Lemma 2.1. But then λ is a unifier of σ's and σ't. By induction there exists $\sigma'' \in \Sigma_T(\sigma's, \sigma't)$ (returned by UNIFY(σ's, σ't)) with $\lambda \leq_T \sigma'' \ [\![V'']\!]$ with $V'' = \mathbf{V}(\sigma's, \sigma't)$ and by Lemma 2.1 and 2.2 we obtain $\lambda\sigma' \leq_T \sigma''\sigma' \ [\![V]\!]$. Hence with $\sigma := \sigma''\sigma'$ there exists $\sigma \in \Sigma_T(s, t)$ (returned by UNIFY(s, t)) such that

$$\theta \leq_T \sigma \ [\![V]\!].$$ ■

8. CONCLUSION

We presented a general unification algorithm that combines unification algorithms for regular finitary collapse free equational theories. Correctness, completeness, and termination are shown, i.e. the combination of regular finitary collapse free equational theories is again finitary. The algorithm is not minimal, but the redundant unifiers can be eliminated in a minimizing step. Our method does not apply to theories with collapse axioms: for example given an idempotent function symbol f, i.e. the equational theory

$I = \{ f(x\ x) = x \}$, an uninterpreted function symbol g, and the problem $\langle g(x, f(x, y)) = g(a, a) \rangle_{I \cup \emptyset,}$ our algorithm would not find a unifier since the constant abstraction $\langle g(x, c) = g(a, a) \rangle_{I \cup \emptyset}$ is not unifiable and there does not exist any further subproblem. But the original problem is solvable by the substitution $\sigma = \{x \leftarrow a, y \leftarrow a\}$. The reason is that in equational theories with collpase axioms terms can collapse to variables by instantiation. It is an open problem to find a terminating unification algorithm for the whole class of finitary theories.

Given a special unification algorithm for a regular collapse free theory we can extend this algorithm at once to handle uninterpreted function symbols by our method. To get an efficient implementation however we are not forced to compute the whole set of subproblems as defined in the abstract algorithm. Depending on the theory and the variables in the problem under consideration, the algorithm can be improved by taking only a subset of SP(s, t) in the iterative step. For common equational theories it is an open problem to find such sets.

The combinations of unification algorithms for regular collpsefree theories as proposed by E. Tidén and K. Yelick are based on the same method: both abstract every immmediate alien subterms to a new variable, unify the variable-pure abstracted terms and then merge the resulting unifiers with the substitution reversing the abstraction. Our algorithm however uses a different approach. We abstract all T-equal immediate alien subterms to a new constant, unify the variable - constant equation and handle the unification of the subterms in a recursive step. This implies checking T-equality of subterms, but abstracting T-equal terms to different constants is not necessary since T-equal terms can never become different by applying a substitution. This checking process - if it is not done before - must be done during the unification process (and then by unifying T-equal terms which is certainly more expensive). Moreover replacing T-equal subterms by the same constant bounds the number of unifiers of the abstractions as compared to replacing T-equal subterms by different constants. Just as Yelick's algorithm was motivated by the AC-unification algorithm of Stickel [St 81] and its extension by Fages [Fa 84] we started our work on extending the AC-unification algorithm of Livesey and Siekmann [LS 76] resulting in a unification algorithm for AC-function symbols and uninterpreted function symbols [HS 85]. Variable-pure terms are almost always unifiable and the number of most general unifiers increases with the number of variables in the terms to be unified. Hence the main objection to variable abstraction is that generating all unifiers of the abstracted terms is too inefficient since most of these unifiers are not unifiable with the substitution reintroducing the original subterms. For certain equational theories it is perhaps possible to constrain the generation of unifiers to eliminate the formation of most the unifiers that will not unifiy with the substitution reversing the abstraction as proposed for AC-unification [St 81]. But discovering an algorithm for applying constraints in the variable abstraction case is as difficult as finding improvements for the set of subproblems.

Acknowledgements: I would like to express my gratitude to Hans-Jürgen Bürckert and Jörg Siekmann for their patience in endless discussions. Their support and constructive criticism have contributed much to the present form of this paper. I am also grateful to Manfred Schmidt-Schauß for a thorough reading of an earlier draft of this paper.

REFERENCES:

[Bi 35] Birkhoff, G., 'On the Structure of Abstract Algebra', Proc. Cambrigde Phil. Soc., Vol. 31, 433-454, (1935)

[BS 81] Burris, S. and Sankappanavar, H.P., 'A Course in Universal Algebra', Springer-Verlag, (1981)

[Fa 84] Fages, F., 'Associative-Commutative Unification', in Proc. of 7th CADE (ed. R.E.Shostak), Springer-Verlag, LNCS 170, 194-208, (1984)

[FH 83] Fages, F. and Huet, G., 'Unification and Matching in Equational Theories', Proc. of CAAP'83 (ed. G. Ausiello and M. Protasi), Springer-Verlag, LNCS 159, 205-220, (1983)

[GT 78] Goguen, J. A., Thatcher, J.W. and Wagner, E. G., 'An Initial Algebra Approach to the Specification, Correctness and Implementation of Abstract Data Types', in 'Current Trends in Programming Methodology, Vol.4, Data Structuring' (ed. R. T. Yeh), Prentice Hall, (1978)

[Gr 79] Grätzer, G., 'Universal Algebra', Springer-Verlag, (1979)

[He 85] Herold, A., 'A Combination of Unification Algorithms', MEMO SEKI-85-VIII-KL, Universität Kaiserslautern, (1985)

[HO 80] Huet, G. and Oppen, D. C., 'Equations and Rewrite Rules: A Survey', in 'Formal Languages: Perspectives and Open Problems (ed R. Book), Academic Press, (1980)

[HS85] Herold, A. and Siekmann, J., 'Unification in Abelian Semigroups', MEMO SEKI-85-III-KL, Universität Kaiserslautern, (1985)

[Hu 76] Huet, G., 'Résolution d'équations dans des langages d'ordre 1, 2 ... ω', Thèse de doctorat d'état, Université Paris VII, (1976)

[Ki 85] Kirchner, C., 'Methodes et outils de conception systematique d'algorithmes d'unification dans les théories équationelles', Thèse de doctorat d'état, Université de Nancy 1, (1985)

[Lo 78] Loveland, D., 'Automated Theorem Proving', North-Holland, (1978)

[LS 76] Livesey, M. and Siekmann, J. , 'Unification of Sets and Multisets', Universität Karlsruhe, Techn. Report, (1976)

[Mc 76] McNulty, G., 'The Decision Problem for Equational Bases of Algebras', Annals of Mathematical Logic 10, 193-259, (1976)

[Ro 65] Robinson, J. A., 'A Machine-Oriented Logic Based on the Resolution Principle', JACM 12, Nº. 1, 23-41, (1965)

[Sc 86] Schmidt-Schauß, M., 'Unification under Associativity and Idempotence is of Type Nullary', MEMO SEKI, Universität Kaiserslautern, (1986), (submitted to JAR)

[Si 84] Siekmann, J., 'Universal Unification', in Proc. of 7th CADE (ed R.E. Shostak), Springer-Verlag, LNCS 170, 1-42, (1984)

[St 81] Stickel, M.E., 'A Unification Algorithm for Associative-Commutative Functions', JACM 28, Nº. 3, 423-434, (1981)

[Ta 79] Taylor, W., 'Equational Logic', Houston Journal of Mathematics 5, (1979)

[Ti 85] Tidén, E., 'Unification in Combinations of Theories with Disjoint Sets of Function Symbols', Royal Institute of Technology, Department of Computing Science, S-100 44 Stockholm, Sweden, (1985)

[Ye 85] Yelick, K., 'Combining Unification Algorithms for Confined Regular Equational Theories', in Proc. of 'Rewriting Techiques and Applications' (ed J.-P. Jouannaud), Springer-Verlag, LNCS 202, 365-380, (1985)

Unification in the Data Structure Sets

Wolfram Büttner
Siemens AG
Corporate Laboratories for
Information Technology

D-8000 Munich 83

Acknowledgements

I began and completed this work during a stay as a visiting researcher at the "Sonderforschungsbereich Künstliche Intelligenz" at Kaiserslautern University during the springterm 1985.

For helpful discussions and suggestions I am particularly grateful to A. Herold and J. Siekmann. I further gratefully acknowledge Prof. Dr. J. Siekmann from the University of Kaiserslautern and Dr. H. Schwärtzel from the Siemens Corporate Laboratories for Information Technology for making this stay possible.

Unification in the Data Structure Sets

Abstract:

Sets are a basic data structure in artificial intelligence. Embedding this datastructure into deductive processes based on resolution requires inventing a suitable unification algorithm (ACI-unification algorithm) respecting the particular properties of the constructor function union.
We shall present such an algorithm based on a careful analysis of the underlying mathematical structures. It is the surprising simplicity of this algorithm which suggests an embedding of ACI-unification into deductive processes.

Introduction

An extension to terms with additional function symbols [HS 85] and optimization of the restricted AC-unification [BU 85] have made AC-unification a theoretically mature unification process. Applications in automatic theorem proving, term rewrite systems, program verification, and logic programming have now attained practical relevance. One of the first papers on AC-unification [LS 76] also touches the ACI-unification problem and suggests a treatment which parallels the work on AC-unification.

The notion of ACI-unification and unification of sets refer to equivalent structures. Namely, in a free term algebra built of variables, constants, and a two-place function symbol we consider the following congruence relation: terms which differ in a number of associative and commutative and idempotent manipulations, are collected in a congruence class. The set defined by the (different) leaves of a term can be identified with the congruence class containing this term. Since sets are particular multisets, the unification of two sets can be considered as a special AC-unification problem. One requires the main result of AC-unification to show that a most general AC-unifier is also a most general ACI-unifier. Additional unifiers exist however, which do not arise this way. The authors first attempt to solve the ACI-unification problem via AC-unification turned out to be far less economical than a direct approach. Since it is an important unification process, we believe that ACI-unification deserves a treatment on its own right exploiting the particular underlying structures.

This paper will provide the theoretical basis of the ACI-unification process as well as an algorithmic treatment. In comparison to AC-unification, the decision problem and the enumeration of all most general unifiers turns out to be very simple. Despite the enormous proliferation of unifiers, it is this simplicity which suggests an embedding of ACI-unification into deducive processes.

The paper has been divided into three chapters. The first chapter presents the basics about free commutative bands and establishes a useful transfer from free commutative semigroups to free commutative bands. ACI-unification can be discussed most properly by employing the structures defined in this chapter.
Chapter 2 introduces the ACI-unification problem and transfers this problem into the algebraic language developed in the previous chapter. In chapter 3 the set of most general unifiers for a given ACI-problem will be computed.
The terminology we use is the usual one. Some basic semigroup theory and linear algebra are the only mathematical prerequisites [CP 61]. The reader is assumed to be familiar with the concept of unification under theories as presented in [SS 82].

1. Free Abelian Idempotent Monoids and some Representations

For countable sets X (variables) and C (constants) with $X \cap C = \emptyset$, we consider the free abelian idempotent monoid $\mathfrak{J}$ generated by $X \cup C$ [1]. We denote semigroup multiplication by $\vee$ and abbreviate $(...(f_1 \vee f_2) \vee) ... f_n)$ $(f_i \in \mathfrak{J})$ by $\bigvee f_i$. 0 denotes the identity of $\mathfrak{J}$, which is assumed to be an element of C. The theory [CP 61] ensures the existence of $\mathfrak{J}$ and shows that $\mathfrak{J}$ is determined by $X \cup C$ up to isomorphisms. $\mathfrak{J}$ is often called a free abelian band.

Finite free abelian idempotent monoids will be particularly important for our purposes. We shall provide basic facts about these structures and agree for the rest of this chapter that $X \cup C$ is finite. Given a linear ordering of X and C, we have an obvious choice for a basis of $\mathfrak{J}$. The free abelian band $\mathfrak{J}_1$, generated by one element, can be constructed over the set $\{0,1\}$ by putting $a \dot{\vee} b := \max\{a, b\}$ $(a, b \in \{0,1\})$ [2]. We construct the free abelian band generated by $X \cup C$ similarly.

For elements x, c in X, C respectively, let $\mathfrak{J}_x$, $\mathfrak{J}_c$ denote distinguished copies of $\mathfrak{J}_1$. The operation $\dot{\vee}$ extends naturally to tuples and turns $((\times \mathfrak{J}_x) \times (\times \mathfrak{J}_c) , \dot{\vee})$ into an abelian band.

As semigroups, $(\mathfrak{J} , \vee)$ and $((\times \mathfrak{J}_x) \times (\times \mathfrak{J}_c) , \dot{\vee})$ are isomorphic. Once a linear ordering has been put upon the sets X, C, an explicit isomorphism f is given by:

$$\Big(\bigvee_{v=1}^{n} x_{i_r}\Big) \vee \Big(\bigvee_{s=1}^{m} c_{j_s}\Big) \xrightarrow{\ f\ } (a_1 , \dots , a_{|X|} , a_{|X|+1} , \dots , a_{|X|+|C|}) \in \{0, 1\}^{|X \cup C|}$$

with

$$a_e = \begin{cases} 1 & \text{for } e = i_r \ (1 \le r \le n) \ \text{ or } \ e = |X|+j_s \ (1 \le s \le m) \\ 0 & \text{otherwise.} \end{cases}$$

Another representation of the free abelian band generated by $X \cup C$ will be used:
With respect to union of sets as a semigroup operation, the powerset of $X \cup C$ - $\mathbb{P}(X \cup C)$ - forms an abelian band. $(\mathfrak{J} , \vee)$ and $(\mathbb{P}(X \cup C) , \cup)$ are isomorphic semigroups, an explicit isomorphism g being given by

$$\Big(\bigvee x_{i_r}\Big) \vee \Big(\bigvee c_{j_s}\Big) \xrightarrow{\ g\ } \{ x_{i_r} \mid 1 \le r \le n \} \ \cup \ \{ c_{j_s} \mid 1 \le s \le m \} \in \mathbb{P}(X \cup C).$$

Soon the advantage of having different "descriptions" of $(\mathfrak{J} , \vee)$ at our disposal will become apparent. For the unification problem we are aiming at, we have to

---

1) There is no need at this point to split the generator set of $\mathfrak{J}$ into disjoint subsets. Chapter 2 will explain the meaning of these distinguished subsets.

2) We reserve $\vee$ for the multiplication in the abstract finite abelian band and we use different symbols for multiplication in certain representations of this band. If the band is represented as a set of n-tuples we use $\dot{\vee}$ as a multiplication symbol.

study homomorphisms between (finite) free abelian bands $\mathfrak{J}$, $\mathfrak{J}'$ [3]. Representing $\mathfrak{J}$ and $\mathfrak{J}'$ as sets of tuples - as on the previous page - we may describe homomorphisms between $\mathfrak{J}$ and $\mathfrak{J}'$ by matrices.

Let $(\mathfrak{F}, +)$, $(\mathfrak{F}', +)$, $((\mathfrak{J}, \vee)$, $(\mathfrak{J}', \vee))$ be free abelian (free abelian idempotent) monoids with generating sets $X \cup C := \{x_i \mid 1 \le i \le p\} \cup \{c_j \mid 1 \le j \le q\}$, $X' \cup C := \{u_r \mid 1 \le r \le m\} \cup \{c_j \mid 1 \le j \le q\}$ respectively.
We denote the map from $\mathbb{N}_0$ to $\{0, 1\}$ given by $\overline{0} = 0$ and $\overline{a} = 1$ for $a > 0$ by $\bar{\ }$. Using $\bar{\ }$, we obtain a homomorphism from $\mathfrak{F}$ onto $\mathfrak{J}$ by

$$\overline{\sum a_i x_i + \sum u_j c_j} := (\bigvee \overline{a}_i x_i) \vee (\bigvee \overline{u}_j c_j).$$

Now let ϱ be a homomorphism from $\mathfrak{F}$ to $\mathfrak{F}'$. Then ϱ uniquely defines a homomorphism $\overline{\varrho}: \mathfrak{J} \to \mathfrak{J}'$ by $\overline{\varrho}((\bigvee a_i x_i) \vee (\bigvee u_j c_j)) := \overline{\varrho(\sum a_i x_i + \sum u_j c_j)}$.
Conversely, every homomorphism from $\mathfrak{J}$ to $\mathfrak{J}'$ arises from some homomorphism mapping $\mathfrak{F}$ to $\mathfrak{F}'$ this way. Finally, by connecting commuting diagrams

$$\begin{array}{ccccc} \mathfrak{F} & \xrightarrow{\varrho_1} & \mathfrak{F}' & \xrightarrow{\varrho_2} & \mathfrak{F}'' \\ \downarrow & & \downarrow & & \downarrow \\ \mathfrak{J} & \xrightarrow{\overline{\varrho_1}} & \mathfrak{J}' & \xrightarrow{\overline{\varrho_2}} & \mathfrak{J}'' \end{array}$$

we derive $\overline{\varrho_2 \circ \varrho_1} = \overline{\varrho}_2 \circ \overline{\varrho}_1$.

Shortly speaking we have:

<u>1.1 Lemma:</u> Let $\bar{\ }: \mathbb{N}_0 \to \{0, 1\}$ be given by $\overline{0} = 0$, $\overline{a} = 1$ for $a > 0$. Then $\bar{\ }$ induces a functor from the full category of free finitely generated abelian monoids to the full category of free finite abelian bands. □

Of course, the mapping $\varrho \to \overline{\varrho}$ is not injective. By restricting $\bar{\ }$ to those homomorphisms which map the generators of $\mathfrak{F}$ onto linear combinations of generators of $\mathfrak{F}'$ involving only coefficients from $\{0, 1\}$ we obtain an injective map.
The linear order on the generating sets of $\mathfrak{F}$ defines a basis of this semigroup. Therefore we may identify $\mathfrak{F}$ with $\mathbb{N}_0^{|X \cup C|}$ equipped with vector addition. Similarly we may identify the semigroups $\mathfrak{F}'$ and $\mathbb{N}_0^{|X' \cup C|}$. Given this identification, the effect of ϱ can be described by some $|X' \cup C| \times |X \cup C|$-matrix $R(\varrho) := (a_{ij})$ $(a_{ij} \in \mathbb{N}_0)$.

3) For the sake of simplicity we write $\vee$ for the semigroup operation in $\mathfrak{J}$ and $\mathfrak{J}'$

<u>1.2 Remark:</u> i) If ϱ fixes the set C pointwise we have

$$R(\varrho) = \begin{bmatrix} R_X(\varrho) & 0 \\ R_C(\varrho) & I \end{bmatrix} \begin{matrix} \}\ |X'| \\ \}\ |C| \end{matrix}$$

$$\underbrace{\quad}_{|X|} \quad \underbrace{\quad}_{|C|}$$

with a $|C|\times|C|$-identity matrix I, an $|X'|\times|C|$-all-zero matrix 0, an $|X'|\times|X|$-matrix $R_X(\varrho)$ and a $|C|\times|X|$-matrix $R_C(\varrho)$.

ii) Conversely any matrix $M = \begin{bmatrix} R_1 & 0 \\ R_2 & I \end{bmatrix} \begin{matrix} \}\ |X'| \\ \}\ |C| \end{matrix}$

$$\underbrace{\quad}_{|X|} \quad \underbrace{\quad}_{|C|}$$

defines a homomorphism $\varrho: \mathfrak{F} \longrightarrow \mathfrak{F}'$ which fixes C pointwise.

□

The map $\overline{\varrho}$ defined by ϱ can be expressed in terms of $R(\varrho)$ as follows:

<u>1.3 Remark:</u> If ϱ has been associated with the matrix $R(\varrho)$, the homomorphism $\overline{\varrho}$ between $\mathfrak{J}$ and $\mathfrak{J}'$ can be calculated by

$$\overline{\varrho}((\vee a_i\, x_i) \vee (\vee u_j\, c_j)) = \overline{R(\varrho)\,(a_1, \dots, a_p, n_1, \dots, n_q)^T}$$ [1].

1.1 - 1.3 are basic for the computation of most general unifiers.

□

For a matrix (a_{ij}) ($a_{ij} \in \mathbb{N}$) we define $\overline{(a_{ij})} := (\overline{a_{ij}})$.

1) T denotes the transposition of matrices.

2. The ACI-Unification Problem

Let $\mathfrak{J}$ be a (possibly infinite), free abelian band generated by $X \cup C$. Since $\mathfrak{J}$ is free on $X \cup C$, any mapping from $X \cup C$ to $\mathfrak{J}$ can be uniquely extended to a semigroup endomorphism of $\mathfrak{J}$. Conversely, any semigroup endomorphism of $\mathfrak{J}$ is determined by its effect on $X \cup C$. Certain endomorphisms of $\mathfrak{J}$ are particularly important for unification theory.

<u>2.1 Definition:</u> A substitution of $\mathfrak{J}$ is an endomorphism of $\mathfrak{J}$ which is different from the identity morphism only on a finite subset of X.

□

Elements f, f' of $\mathfrak{J}$ may be equal after application of suitable substitutions.

<u>2.2 Definition:</u> For elements f, f' of $\mathfrak{J}$, a unifier of {f, f'} is a substitution σ with $\sigma(f) = \sigma(f')$.

□

We can compose unifiers σ, τ of {f, f'} by setting $\sigma \le \tau$ iff there exists a substitution λ such that the equality $\sigma = \lambda\tau$ holds for all variables occurring in f and f'.
The set $U\Sigma$(f, f') of all unifiers of {f, f'} carries an equivalence relation $\approx$ given by $\sigma \approx \tau$ iff $\sigma \le \tau$ and $\tau \le \sigma$. The relation $\le$ induces in $U\Sigma(f, f')/_{\approx}$ a partial ordering and here we can speak of maximal elements.

<u>2.3 Definition:</u> i) A most general unifier σ of {f, f'} is a unifier which induces in $U\Sigma(\{f, f'\})/_{\approx}$ a maximal element.

ii) $\mu U\Sigma(\{f, f'\})$ is a minimal set of unifiers of {f, f'} which induces in $U\Sigma(\{f, f'\})/_{\approx}$ all maximal elements.

□

Note, that the axiomatic definition of standard unification theory as in [PL 72], [SI 84] produces the same set of most general unifiers. $\mu U\Sigma(\{f, f'\})$ is uniquely determined up to $\approx$.

Now the ACI-unification problem poses three questions:

A: Given elements f, f' in $\mathfrak{J}$, can the existence of a unifier of {f, f'} be decided upon?
B: Can the cardinality of $\mu U\Sigma(\{f, f'\})$ be calculated?
C: Can $\mu U\Sigma(\{f, f'\})$ be computed?

A positive answer to these questions implies, among other things (refutation)-completeness of a theorem prover with built-in ACI-unification.
ACI-unification is a local problem in the following sense:

Given elements f, f' in $\mathfrak{J}$ and a unifier σ of {f, f'} then f, f' lie in a finite subsemigroup $\mathfrak{J}_0$ and σ maps $\mathfrak{J}_0$ to some subsemigroup $\mathfrak{J}_0'$ thereby making f and f' equal.
Precisely speaking: Let X_0 be the set of variables occurring in f, f' and put X_1 := set of variables of X_0 occurring exclusively in f, X_2 := set of variables of X_0 occurring exclusively in f', X_3 := set of variables of X_0 occurring in f and f'. Similarly we split C_0, the set of all constants occurring in f, f', into subsets C_i $(1 \le i \le 3)$. Finally we write $W(\sigma)$ for the (finite) set of all variables occurring in $\sigma(x)$ for all x in X_0.
$\mathfrak{J}_0$ will now be taken as the subsemigroup generated by $X_0 \cup C_0$ and $\mathfrak{J}_0(\sigma)$ denotes the subsemigroup generated by $W(\sigma) \cup C_0$ [1].
As we shall explain in 2.5 since we are mainly interested in most general unifiers we may assume without loss of generality that σ does not "introduce new constants" when applied to elements of $\mathfrak{J}_0$.
Hence $\sigma|\mathfrak{J}_0$ is a homomorphism between the finite commutative bands $\mathfrak{J}_0$ and $\mathfrak{J}_0(\sigma)$. Our results from chapter 1 apply to this situation.

We now address the decision problem A:
Given elements f, f' in $\mathfrak{J}$ we ask for conditions guaranteeing the existence of a substitution σ such that $\sigma|\mathfrak{J}_0$ maps $\mathfrak{J}_0$ to $\mathfrak{J}_0(\sigma)$ and "makes f and f' equal". As mentioned in chapter 1, we may identify $\mathfrak{J}_0$ and $\mathfrak{J}_0'$ with $(\mathbb{P}(X_0 \cup C_0), \cup)$ and $(\mathbb{P}(W \cup C_0), \cup)$. The following example shows phenomena which may occur.

Let f := {x, a, b, c}, f' :={d} $(a, b, c, d \in C_0, x \in X_0)$ be elements of $\mathfrak{J}_0$. f and f' cannot be unified; for any substitution applied to f yields a set containing at least three elements, whereas f' remains unchanged. The situation changes entirely if f' is taken to be {y, d} with an arbitrary variable y (which may equal x).
Now a unifier can be obtained substituting x by d and y by {a, b, c} (for $x \neq y$) and substituting x by {a, b, c, d} otherwise.
It is easy to generalize these observations in the following statement, the proof of which is immediate:

<u>2.4 Lemma:</u> Let $\mathfrak{J}_0$ be identified with $(P(X_0 \cup C_0), \cup)$; for subsets f, f' of $X_0 \cup C_0$ we have:

i) If f and f' contain variables, they can be unified.
ii) If f contains variables and f' is variable-free, the two sets can be unified iff $f \cap C_0$ is a subset of f'.
iii) If f and f' are variable-free, the two sets can be unified iff they are equal. □

---

1) Given elements f, f' and a unifier σ of {f, f'}, the meaning of Xi, Ci $(0 \le i \le 3)$, $W(\sigma)$, $\mathfrak{J}_0$, $\mathfrak{J}_0(\sigma)$ will always be as above.

By a result of G. Plotkin [PL 72], the existence of unifiers of $\{f, f'\}$ implies the existence of most general unifiers of $\{f, f'\}$. Lemma 2.4 thus reduces the ACI-decision problem to a containment check.

We still have to explain why we may confine ourselves to unifiers which do not introduce new constants.

<u>2.5 Remark</u>:

i) Let σ be a unifier of subsets f, f' of $X_0 \cup C_0$. Then there exists a unifier σ_1 of $\{f, f'\}$ such that $\sigma = \lambda_1 \circ \sigma_1$ for a suitable substitution λ_1 and all constants in $\sigma_1(f)$ occur in f or f'.

ii) Let σ be a most general unifier of $\{f, f'\}$. Then all constants in $\sigma_1(f)$ occur in f or f'.

The proof of 2.5 is easy and will be omitted.

As we have seen in the examples, "cancellation" of variables yields different unification problems. This phenomenon, contrasting the corresponding situation in the AC-case, can partly be explained as follows: any free abelian semigroup is cancellative (since it can be embedded into a group). A semigroup, however, containing nontrivial idempotent elements, can never be embedded into a group.
Given f, f' and a unifier of $\{f, f'\}$ such that $\sigma(f)$ contains only constants occurring in f or f' remark 1.2 applies if we put $\sigma|\mathfrak{J}_0 =: \overline{\rho(\sigma)}$ and $\rho(\sigma)$ is the corresponding homomorphism between the free abelian semigroups $\mathfrak{F}_0, \mathfrak{F}_0(\sigma)$

with $\overline{\mathfrak{F}_0} = \mathfrak{J}_0$, $\overline{\mathfrak{F}_0(\sigma)} = \mathfrak{J}_0(\sigma)$.
$R_X(\sigma)$ has a quite natural meaning. Let $\mathfrak{J}_X, \mathfrak{J}_C$ be the subsemigroups of $\mathfrak{J}$ generated by X and C, respectively. $\tilde{\mathfrak{J}} := \mathfrak{J}/\mathfrak{J}_C$ is isomorphic to $\mathfrak{J}_X$, i.e. $\tilde{f} = f \setminus f \cap C$ for all sets f in $\mathfrak{J}$. Any substitution σ fixes C pointwise. Therefore setting $\tilde{\sigma}(\tilde{f}) := \sigma(f)$ we obtain a substitution $\tilde{\sigma}$ of $\mathfrak{J}_X$. Furthermore we set $\widetilde{\mu U\Sigma(\{f, f'\})} := \{\tilde{\sigma} \mid \sigma \in \mu U\Sigma(\{f, f'\})\}$. Now replace f, f' by $\tilde{f}, \tilde{f}'$. We have:

<u>2.6 Remark</u>:

i) $\widetilde{\mu U\Sigma(\{f, f'\})} \subseteq \mu U\Sigma(\{\tilde{f}, \tilde{f}'\})$

ii) If -according to remark 1.2- σ has been associated with the matrix $R(\rho(\sigma))$, $\tilde{\sigma}$ will be associated with the matrix $R_X(\rho(\sigma))$.

□

The meaning of $R_C(\sigma)$ is less obvious and since we do not need it we omit any explanation.

3. The Computation of $\mu U\Sigma(\{f, f'\})$

This chapter provides the computation of $\mu U\Sigma(\{f, f'\})$ as its main result. We shall need some abbreviations for certain matrices which will occur frequently. By I_n we denote the n×n identity matrix, $O_{n,m}$ stands for the n×m all-zero matrix and, finally, for a subset A of $X_0 = \{x_1, \dots, x_{|X_0|}\}$ we let v(A) denote the $|X_0|$-tuple with 1 as j-th component iff $x_j \in A$, and zero otherwise.
Let f, f' be elements of $\mathfrak{J}$ which can be unified. According to 1.2, any unifier σ of {f, f'} determines a matrix $R(\varrho(\sigma))$ where $\varrho(\sigma) = \sigma | \mathfrak{J}_0$ is considered to be a mapping from $\mathfrak{J}_0$ to $\mathfrak{J}_0'$.
The following lemma translates the order ≤ on unifiers to their associated matrices. Recall that $\overline{(a_{ij})} := (\bar{a}_{ij})$ for any matrix (a_{ij}) with nonnegative integers a_{ij}.

<u>3.1 Lemma:</u> Let σ, τ be unifiers of {f, f'} and let λ be a substitution such that $\sigma = \lambda \cdot \tau$ and consider $\lambda | \mathfrak{J}_0(\tau)$ as a homomorphism from $\mathfrak{J}_0(\tau)$ to $\mathfrak{J}_0(\sigma)$. Then we have:

i) $\sigma | \mathfrak{J}_0 = (\lambda | \mathfrak{J}_0(\tau)) \circ (\tau | \mathfrak{J}_0)$

ii) $\overline{R(\varrho(\sigma))} = \overline{R(\varrho(\lambda)) \cdot R(\varrho(\tau))}$

iii) $\overline{R_X(\varrho(\sigma))} = \overline{R_X(\varrho(\lambda)) \cdot R_X(\varrho(\tau))}$ and

$$\overline{R_C(\varrho(\lambda)) \cdot R_X(\varrho(\tau)) + R_C(\varrho(\tau))} = \overline{R_C(\varrho(\sigma))}$$

Proof: i) and ii) are immediate. For iii) we multiply the matrices $R(\varrho(\lambda))$, $R(\varrho(\tau))$ blockwise.

□

It remains to characterize a unifier σ in terms of the associated matrix $R(\varrho(\sigma))$. Using the decomposition of the sets f, f' as in chapter 2, we have:

<u>3.2 Lemma:</u> Let f, f' be elements of $\mathfrak{J}$ which can be unified. Assume f, f' to be sets. Let σ be a unifier of {f, f'} and let $R(\varrho(\sigma))$ be its associated matrix. Then we have:

i) A row of $R_X(\varrho(\sigma))$ is a vector $v_{|X_0|}(A)$ (for some subset A of X_0) such that $\overline{|A \cap (X_1 \cup X_2)|} = \overline{|A \cap (X_2 \cup X_3)|}$.[1)]

---

1) By |M| we denote the cardinality of the set M. This is a natural number, hence ‾ may be applied.

ii) A row of $R_C(\varrho(\sigma))$ corresponding to $c \in C_1$ is a vector

$v_{|X0|}(A)$ $(A \subseteq X_0)$ such that $\overline{|A \cap (X_2 \cup X_3)|} = 1$.

iii) A row of $R_C(\varrho(\sigma))$ corresponding to $c \in C_2$ is a vector

$v_{|X0|}(A)$ $(A \subseteq X_0)$ such that $\overline{|A \cap (X_1 \cup X_3)|} = 1$.

iv) A row of $R_C(\varrho(\sigma))$ corresponding to $c \in C_3$ is a vector $v_{|X0|}(A)$ for an arbitrary subset A of X_0.

v) Conversely, any $|W \cup C_0| \times |X_0 \cup C_0|$ - matrix

$$M := \begin{bmatrix} R_1 & O_{|W|,|C0|} \\ & \\ \underbrace{R_2}_{|X_0|} & \underbrace{I_{|C0|}}_{|C_0|} \end{bmatrix} \begin{matrix} \}\,|W| \\ \\ \}\,|C_0| \end{matrix}$$

defines a unifier σ of $\{f, f'\}$ with $M = R(\varrho(\sigma))$, $R_1 = R_X(\varrho(\sigma))$, $R_2 = R_C(\varrho(\sigma))$ iff the rows of R_1 satisfy i) and the rows of R_2 satisfy ii), iii), iv).

Proof: By remark 1.2 we have $R(\varrho(\sigma)) = \begin{bmatrix} R_X(\varrho(\sigma)) & 0 \\ & \\ \underbrace{R_C(\varrho(\sigma))}_{|X_0|} & \underbrace{I}_{|C_0|} \end{bmatrix} \begin{matrix} \}\,|W(\sigma)| \\ \\ \}\,|C_0| \end{matrix}$

Furthermore $\sigma(f) = \sigma(f')$ iff

$$\overline{R(\varrho(\sigma))\ (\underbrace{1,1,\dots,1}_{X_1},\underbrace{0,\dots,0}_{X_2},\underbrace{1,\dots,1}_{X_3},\underbrace{1,\dots,1}_{C_1},\underbrace{0,\dots,0}_{C_2},\underbrace{1,\dots,1}_{C_3})^T} =$$

$$\overline{R(\varrho(\sigma))\ (\underbrace{0,\dots,0}_{X_1},\underbrace{1,\dots,1}_{X_2},\underbrace{1,\dots,1}_{X_3},\underbrace{0,\dots,0}_{C_0},\underbrace{1,\dots,1}_{C_1},\underbrace{1,\dots,1}_{C_2})^T} \quad \text{iff}$$

$$\overline{R_X(\varrho(\sigma))\;(\underbrace{1,\dots,1}_{X_1},\underbrace{0,\dots,0}_{X_2},\underbrace{1,\dots,1}_{X_3})^T} = \overline{R_X(\varrho(\sigma))\;(\underbrace{0,\dots,0}_{X_1},\underbrace{1,\dots,1}_{X_1},\underbrace{1,\dots,1}_{X_3})^T}$$

and

$$\overline{R_C(\varrho(\sigma))\;(\underbrace{1,\dots,1}_{X_1},\underbrace{0,\dots,0}_{X_2},\underbrace{1,\dots,1}_{X_3})^T} \cup \overline{I_{|C0|}\;(\underbrace{1,\dots,1}_{C_1},\underbrace{0,\dots,0}_{C_1},\underbrace{1,\dots,1}_{C_3})^T} =$$

$$\overline{R_C(\varrho(\sigma))\;(\underbrace{0,\dots,0}_{X_1},\underbrace{1,\dots,1}_{X_2},\underbrace{1,\dots,1}_{X_3})^T} \cup \overline{I_{|C0|}\;(\underbrace{0,\dots,0}_{C_1},\underbrace{1,\dots,1}_{C_1},\underbrace{1,\dots,1}_{C_3})^T}$$

We now give an explicit proof for i), the statements ii), iii), iv) can be proved analogously.

The i-th row of $R_X(\varrho(\sigma))$ is some $|X_0|$-tuple and can be written as $v_{|X0|}(A_i)$ for some subset A_i of X_0. Evaluating the last of the above equivalences we have (here $\cdot$ denotes scalarproduct):

$$\Big(\overline{v_{|X0|}(A_i)\cdot(\underbrace{1,\dots,1}_{X_1},\underbrace{0,\dots,0}_{X_2},\underbrace{1,\dots,1}_{X_3})^T}, \dots, \overline{v_{|X0|}(A_{|W|})\cdot(\underbrace{1,\dots,1}_{X_1},\underbrace{0,\dots,0}_{X_2},\underbrace{1,\dots,1}_{X_3})^T}\Big) =$$

$$\Big(\overline{v_{|X0|}(A_i)\cdot(\overbrace{0,\dots,0},\overbrace{1,\dots,1},\overbrace{1,\dots,1})^T}, \dots, \overline{v_{|X0|}(A_{|W|})\cdot(\overbrace{0,\dots,0},\overbrace{1,\dots,1},\overbrace{1,\dots,1})^T}\Big).$$

i) is an immediate consequence of this. Using the above equivalences in reversed order we obtain a proof for v).

□

By specializing we can easily construct unifiers of (f, f') from lemma 3.2.

3.3 Lemma: i) Let f, f' be sets such that $|X_1| + |X_2| \neq 0$ or $|X_1| = |X_2| = |C_1| = |C_2| = 0$

$$\text{Any matrix} \quad M := \left[\begin{array}{cc} R_1 & O_{|W|,|C_0|} \\ & \\ \underbrace{R_2}_{|X_0|} & \underbrace{I_{|C_0|}}_{|C_0|} \end{array}\right] \begin{array}{l} \}\,|W| \\ \\ \}\,|C_0| \end{array}$$

with $|W| = |X_1| \cdot |X_2| + |X_3|$, where

$$R_1 = \left[\begin{array}{c|c|c}
\begin{array}{cccc} 1 & 0 & \dots & 0 \\ 1 & 0 & \dots & 0 \\ & & \cdots & \\ 1 & 0 & \dots & 0 \end{array} & I_{|X_2|} & O_{|X_2|,\,|X_3|} \\
\hdashline
\begin{array}{cccc} 0 & 1 & \dots & 0 \\ & & \cdots & \\ 0 & 1 & \dots & 0 \end{array} & I_{|X_2|} & O_{|X_2|,\,|X_3|} \\
\hdashline
\begin{array}{cccc} 0 & 0 & \dots & 0\ 1 \\ & & \cdots & \\ 0 & 0 & \dots & 0\ 1 \end{array} & I_{|X_2|} & O_{|X_2|,\,|X_3|} \\
\hdashline
O_{|X_3|,\,|X_1|} & O_{|X_3|,\,|X_2|} & I_{|X_3|}
\end{array}\right]$$

$$\qquad\qquad \underbrace{\qquad\quad}_{X_1} \qquad \underbrace{\qquad\quad}_{X_2} \qquad \underbrace{\qquad\quad}_{X_3}$$

$$
\text{and} \qquad R_2 = \begin{array}{ccc} \quad X_1 \quad & \quad X_2 \quad & \quad X_3 \quad \end{array}
$$

$$
\left[\begin{array}{c|c|c}
O_{|C1|,\,|X2|} & \begin{matrix} v_{|X2|}(A_1) \\ \cdots \\ v_{|X2|}(A_{|C1|}) \end{matrix} & O_{|C1|,\,|X3|} \\ \hline
\begin{matrix} v_{|X1|}(B_1) \\ \cdots \\ v_{|X1|}(B_{|C2|}) \end{matrix} & O_{|C2|,\,|X2|} & O_{|C2|,\,|X2|} \\ \hline
\multicolumn{2}{c|}{\begin{matrix} v_{|X1 \cup X2|}(D_1) \\ \cdots \\ v_{|X1 \cup X2|}(D_{|C3|}) \end{matrix}} & O_{|C3|,\,|X3|}
\end{array}\right]
$$

(Here for $1 \le i \le |C_1|$ A_i denotes an arbitary not empty subset of X_2,
for $1 \le j \le |C_2|$ B_j denotes an arbitary not empty subset of X_1
and for $1 \le k \le |C_3|$ D_k is an arbitary subset of X_1 or X_2.)

defines a unifier σ of $\{f, f'\}$ such that $M = R(\varrho(\sigma))$.

ii) For $|X_1| = |X_2| = 0$ and $C_1 \cup C_2 \neq \emptyset$, we modify M, eliminating R_1, and replacing R_2 by a matrix

$$
\left[\begin{array}{c}
\begin{matrix} v_{|X3|}(F_1) \\ \cdots \\ v_{|X3|}(F_{|C1|}) \end{matrix} \\ \hline
\begin{matrix} v_{|X3|}(F_{|C1|+1}) \\ \cdots \\ v_{|X3|}(F_{|C1|+|C2|}) \end{matrix} \\ \hline
O_{|C3|,\,|X3|}
\end{array}\right]
$$

where the sets F_i are singletons from X_3.
The proof is immediate from 3.2 .

□

Let Σ_i, Σ_{ii} be the set of all unifiers of {f, f'} defined by the matrices M which have been constructed in lemma 3.3 i), ii), respectively. We can now formulate our main result:

<u>3.4 Theorem:</u> Let f, f' be elements of the free abelian band $\mathfrak{J}$ which can be unified. According to the structure of f, f', let Σ be one of the sets Σ_i, Σ_{ii}. Then we have:

i) Any most general unifier of {f, f'} introduces $|X_1| \cdot |X_2| + |X_3|$ variables.
ii) Up to equivalence, $\mu U\Sigma(\{f, f'\})$ and Σ are equal.
iii) $|\Sigma| = (2^{|X_1|}-1)^{|C_2|} \cdot (2^{|X_2|}-1)^{|C_1|} \cdot (2^{|X_1|}+2^{|X_2|}-1)^{|C_3|}$ for $\Sigma = \Sigma_{ii}$.
$|\Sigma| = |X_3|^{|C_1 \cup C_2|}$ for $\Sigma = \Sigma_{ii}$.

Proof: We will assume $\Sigma = \Sigma_i$. The proof for $\Sigma = \Sigma_{ii}$ can be easily derived from this.

i) Let σ be a most general unifier of {f, f'}. Then $\tilde{\sigma}$ (see 2.6) is a most general unifier of the associated constant-free unification problem $\{\tilde{f}, \tilde{f'}\}$. Let W be the set of variables introduced by σ, let $R(\varrho(\tilde{\sigma}))$ be the matrix associated with $\tilde{\sigma}$ and let a_i $(1 \leq i \leq |W|)$ be the rows of $R(\varrho(\tilde{\sigma}))$. Note that $|W| = |W(\tilde{\sigma})|$.
Finally, let b_j $(1 \leq j \leq |X_1| \cdot |X_2| + |X_3|)$ be the rows of R_1 (see 3.3). The vectors a_i are restricted by 3.2 i). All of these vectors can be obtained by adding up suitable rows of R, and then performing the — operation i.e.

$$a_i = \overline{\sum_{j=1}^{|x_1| \cdot |x_2| + |x_3|}} a_{ij}\, b_j \,.$$

The coefficients a_{ij} determine a $|W| \times |X_1| \cdot |X_2| + |X_3|$ -matrix (a_{ij}) such that $R(\varrho(\tilde{\sigma})) = (a_{ij}) \cdot R_1$. Let $\tilde{\lambda}$ be the substitution with $R(\varrho(\tilde{\lambda})) = (a_{ij})$. Then $\tilde{\sigma} = \tilde{\lambda} \cdot \tilde{\sigma}_0$, where $\tilde{\sigma}_0$ is the unifier of $\{f, f'\}$ defined by R_1. Since $\tilde{\sigma}$ was assumed to be most general, $\tilde{\sigma}_0$ will also be a most general unifier of $\{f, f'\}$. Hence, up to equivalence, a constant-free ACI-unification problem possesses exactly one most general unifier. This unifier may then be taken as the substitution defined by R_1. Since R_1 has $|X_1| \cdot |X_2| + |X_3|$ rows we have i). Furthermore, by 2.6, passing to an equivalent unifier if necessary, we may assume that $R_X(\varrho(\sigma))$ equals R_1.

ii) We first show that Σ contains an element which is equivalent to the most general unifier σ from i.). If $R_C(\varrho(\sigma))$ is one of the matrices R_2 of 3.3, there is nothing to show. Assume that the first row of $R_C(\varrho(\sigma))$ is different from the first row of any of the matrices R_2.
Consider a substitution σ_1 where $R_X(\varrho(\sigma_1)) = R_X(\varrho(\sigma)) = R_1$, all rows of $R_C(\varrho(\sigma_1))$ and $R_C(\varrho(\sigma))$ -except the first- agree and the first rows z_1, z of $R_C(\varrho(\sigma_1))$, $R_C(\varrho(\sigma))$, respectively, agree in all positions corresponding to X_2. z can be obtained from z_1 by adding suitable rows of R_1 to z_1 and then performing the -operation i.e.

$$z = \overline{z_1 + \sum_{j-1}^{|x_1| \cdot |x_2| + |x_3|} c_j b_j} .$$

Let M be the $|C_0| \times (|X_1| \cdot |X_2| + |X_3|)$ -matrix

$$\begin{bmatrix} c_1 \; c_2 \; \dots \; c_{|X1| \cdot |X2| + |X3|} \\ \\ O_{|C0|-1,\; |X1| \cdot |X2| + |X3|} \end{bmatrix}$$

We compute $\overline{M \cdot R_1 + R_C(\varrho(\sigma_1))} = \overline{R_C(\varrho(\sigma_1))}$.

Now let λ be the substitution defined by the matrix

$$\begin{array}{c} \left[\begin{array}{cc} I & 0 \\ \underbrace{M}_{|X_1|+|X_2|+|X_3|} & \underbrace{I}_{|C_0|} \end{array}\right] \begin{array}{l} \} \ |X_1| \cdot |X_2| + |X_3| \\ \} \ |C_0| \end{array} \end{array}$$

Then we have $R(\varrho(\sigma)) = R(\varrho(\lambda)) \cdot R(\varrho(\sigma))$ i.e. $\sigma = \lambda \cdot \sigma_1$.
Hence, σ and σ_1 are equivalent and we may replace σ by σ_1. Iterating this process with appropriate changes, we deduce that σ is equivalent to some element in Σ. It remains to be shown that the elements of Σ are pairwise inequivalent, most general unifiers.

Let τ_1, τ_2 be elements of Σ and assume $\tau_1 = \lambda \cdot \tau_2$. By lemma 3.1 iii) this is equivalent to

$$\overline{R_X(\varrho(\tau_1))} = \overline{R_X(\varrho(\lambda)) \cdot R_X(\varrho(\tau_2))} \qquad \text{and}$$

$$\overline{R_C(\varrho(\lambda)) \cdot R_X(\varrho(\tau_2)) + R_C(\varrho(\tau_2))} = \overline{R_C(\varrho(\tau_1))}.$$

The first of these equations can easily be satisfied (for instance by setting $R_X(\varrho(\lambda)) = I_{|X1| \cdot |X2| + |X3|}$). The second equation defines equalities for the j-th rows of the matrices

$\overline{R_C(\varrho(\lambda)) \cdot R_X(\varrho(\tau_2)) + R_C(\varrho(\tau_2))}$, $R_C(\varrho(\tau_1))$ respectively. As an example we work out the case $j = 1$.

The first row $(\underbrace{0, \dots, 0}_{X_1}, \underbrace{v_{|X2|}(A_1)}_{X_2}, \underbrace{0, \dots, 0}_{X_3})$

of $\overline{R_C(\varrho(\tau_1))}$ is obtained from the first row
$(0, \dots, 0, v_{|X2|}(A_1'), 0, \dots, 0)$ of $R_C(\varrho(\tau_2))$ by adding

suitable rows of $R_X(g(\tau_2))$ and then performing the — operation. Since τ_2 is an element of Σ, we have $R_X(g(\tau_2)) = R_1$. By adding a row of R_1 to $(0, \dots, 0, v_{|X2|}(A_1'), 0, \dots, 0)$, however, we necessarily place a one at some position corresponding to X_1 or X_3.

$$\text{However, } (\underbrace{0, \dots, 0}_{X_1}, \underbrace{v_{|X2|}(A_1)}_{X_2}, \underbrace{0, \dots, 0}_{X_3})$$

does not carry one at these positions. Hence $A_1 = A_1'$. Analogously we obtain $A_j = A_j'$ (for $1 \le j \le |C_0|$) and therefore $R_C(g(\tau_2)) = R_C(g(\tau_1))$. Consequently, $\tau_1 = \tau_2$ and ii) has been proved.

iii) The cardinality of Σ can easily be computed from the choices for R_2 in 3.3 .

□

The following example demonstrates how to use the theorem in order to compute the set of most general unifiers for a given ACI-problem.

Let $f := \{x, y, d\}$ and $f' := \{a, b, w\}$. By 2.4 i) these sets can be unified. We have $X_1 = \{x, y\}$, $X_2 = \{w\}$, $X_3 = \emptyset$, $C_1 = \{d\}$, $C_2 = \{a, b\}$, $C_3 = \emptyset$. Let σ be a most general unifier of $\{f, f'\}$. Without loss of generality we may identify $R_X(g(\sigma))$ with R_1 and $R_C(g(\sigma))$ with some matrix R_2 as in 3.3 .

$$\begin{array}{c} \\ \\ \end{array}\begin{bmatrix} R_1 \\ \\ R_2 \end{bmatrix} = \begin{array}{c} \;\; x \;\; y \;\; w \\ \begin{bmatrix} 1 & 0 & 1 \\ 0 & 1 & 1 \\ 0 & 0 & 1 \\ \beta & \gamma & 0 \\ \delta & \varepsilon & 0 \end{bmatrix} \end{array} \begin{array}{c} \\ u \\ v \\ d \\ a \\ b \end{array}$$

with $\beta, \gamma, \delta, \varepsilon \in \{0, 1\}$ and $(\beta, \gamma) \neq (0, 0), (\delta, e) \neq (0, 0)$.

Hence we have 9 most general unifiers:

σ_i	$\sigma_i(x)$	$\sigma_i(y)$	$\sigma_i(w)$
σ_1	{u}	{v, a, b}	{u, v, d}
σ_2	{u, a}	{v, b}	{u, v, d}
σ_3	{u, a}	{v, a, b}	{u, v, d}
σ_4	{u, b}	{v, a}	{u, v, d}
σ_5	{u, b}	{v, a, b}	{u, v, d}
σ_6	{u, a, b}	{v}	{u, v, d}
σ_7	{u, a, b}	{v, a, }	{u, v, d}
σ_8	{u, a, b}	{v, b}	{u, v, d}
σ_9	{u, a, b}	{v, a, b}	{u, v, d}

4. References

[BU 85] W. Büttner
"Unification in the Datastructure Multisets"
To appear in JAR, 1986

[CP 61] A. Clifford, G. Preston
"The Algebraic Theory of Semigroups"
Math. Surveys, vol. 1, no. 7, 1981

[FA 84] F. Fages
"Associative-Commutative Unification"
CADE-7, LNCS 170, 1984

[FT 85] A. Fortenbacher
"An Algebraic Approach to Unification under Associativity and Commutativity"
CRTA-85, Dijon, 1985

[HS 85] A. Herold, J. Siekmann
"Unification in Abelian Semigroups"
Seki-Research Report, Univ. of Kaiserslautern, 1985

[LS 76] M. Liversey, J. Siekmann
"Unification of Sets and Multisets"
Univ. of Karlsruhe, Techn. Report, 1976

[PL 72] G. Plotkin
"Building in Equational Theories"
Machine Intelligence, vol.7, 1972

[SI 84] J. Siekmann
"Universal Unification"
LNCS 170, 1984

[SS 82] J. Siekmann, P. Szabo
"Universal Unification"
Informatik-Fachberichte, no. 58, 1982

[ST 81] M. Stickel
"A Unification Algorithm for Assoc. Comm. Functions"
JACM 28, no.3, 1981

NP-COMPLETENESS OF THE SET UNIFICATION AND MATCHING PROBLEMS

Deepak Kapur† *and Paliath Narendran*†

Computer Science Branch
Corporate Research and Development
General Electric Company
Schenectady, NY

ABSTRACT

The set-unification and set-matching problems, which are very restricted cases of the associative-commutative-idempotent unification and matching problems, respectively, are shown to be NP-complete. The NP-completeness of the subsumption check in first-order resolution follows from these results. It is also shown that commutative-idempotent matching and associative-idempotent matching are NP-hard, thus implying that the idempotency of a function does not help in reducing the complexity of matching and unification problems.

1. Introduction

In an earlier paper [1], we studied the complexity of term matching problems when certain operators in terms satisfy properties such as associativity and commutativity. It was shown that the associative-commutative matching problem is NP-complete and thus associative-commutative unification problem is NP-hard. This paper is a continuation of our work in analyzing the complexity of basic primitive operations involved in theorem proving, especially an approach towards theorem proving based on term rewriting which has drawn considerable attention in the past few years. In this paper, we concentrate on unification and matching problems involving operators that are *idempotent* (besides being associative and/or commutative); we define set-unification and set-matching problems, which, respectively, appear to be the simplest special cases of associative-commutative-idempotent unification and matching problems. We show that both of these problems are NP-complete. Furthermore, the construction employed in the proof gives us an alternative proof of the NP-completeness of associative-commutative matching.

---

† Partially supported by the National Science Foundation Grant no. DCR-8408461.

Examples of associative, commutative and idempotent operators are the *and* and *or* operations in propositional calculus. Set-unification and set-matching are two primitive operations in theorem proving methods developed in [4, 5] based on the rewriting approach when considered in full generality ; superpositions and critical pairs are generated using set unification, and set-matching is used for rewriting.[1] In resolution-based theorem provers also, set-unification (actually, set-matching) comes into play when a "redundant" clause (i.e., which contains an instance of some other clause) is to be deleted; this operation is also called the subsumption check (cf. [2], p.95). Given that the set-unification and set-matching problems are NP-complete and that associative-commutative matching is also NP-complete, these results cast a doubt about whether it is possible to get fast methods for generating canonical bases, that extensively use set-unification, set-matching, associative-commutative matching, or associative-commutative unification.

Since the set-unification problem is a special case of the associative-commutative-idempotent unification problem, it also follows from the results of this paper that associative-commutative-idempotent unification problem is NP-hard. Further, it is shown that the idempotency of an associative-commutative (or simply associative or commutative) function does not help in bringing the complexity of the corresponding unification and matching problems down into polynomial time.

2. Preliminaries

Let F be a finite set of function symbols (operators) and V a denumerable set of variables. Let $T(F, V)$ stand for the set of all possible terms that can be constructed using F and V. For a term t, $Var(t)$ denotes the set of all variables that occur in t. For example, $Var(f(x, y, g(y))) = \{x, y\}$. The *size* of a term s is the number of occurrences of function and variable symbols in s and is denoted by $|s|$.

A function f is *associative* if and only if it satisfies the following axiom:

$$f(f(x, y), z) = f(x, f(y, z)). \tag{A}$$

A function f is *commutative* if and only if it satisfies

$$f(x, y) = f(y, x). \tag{C}$$

A function f is *idempotent* if and only if it satisfies

$$f(x, x) = x. \tag{I}$$

A *substitution* is a mapping θ from variable symbols to terms such that $\theta(v) = v$ for all but a finite number of variables symbols. It can be denoted by $\{v_1 \leftarrow t_1, \cdots, v_k \leftarrow t_k\}$, where the $k \geq 0$ variable symbols $v_1, \cdots, v_k$ are distinct. (The case $k = 0$ is the identity substitution.) By the size of a substitution θ denoted in this way we shall mean $k + \sum_{i=1}^{k} |t_i|$. We shall denote the size of θ by $|\theta|$. The domain of a substitution θ is ex-

1. It is possible to avoid using set-unification for generating new rules in the method proposed in [5]; it appears that this may be the case for the method in [4] also.

tended to the set of all terms by inductively defining $\theta(f\ (t_1, \ldots, t_n))$ to be $f\ (\theta(t_1), \cdots, \theta(t_n))$.

A substitution θ is said to *match* a term s with a term t if and only if $t = \theta(s)$. A substitution θ is said to *unify* terms s and t if and only if $\theta(s) = \theta(t)$; we say "s and t are *unifiable*" if there is a substitution that unifies them. (Matching is thus one-way unification.)

Two terms s and t are said to be *E-equivalent*, where E is a non-empty subset of $\{A, C, I\}$, expressed as $s \underset{E}{=} t$, if and only if they are equivalent under the equational theory of the axioms in E. For example, if f is associative, commutative and idempotent, then $f\ (f\ (a, b), f\ (a, c)) \underset{ACI}{=} f\ (c, f\ (b, a))$.

3. Problem Definitions

The following is a list of problems that will be referred to in this paper.

1. **Associative-Commutative-Idempotent Equivalence** (referred to as **ACIEQ**)

   Instance: A set of variable symbols V, a set of function symbols F some of which may be associative, commutative and idempotent, and terms t_1, t_2 from $T(F, V)$.

   Question: Is $t_1 \underset{ACI}{=} t_2$?

2. **Associative-Commutative-Idempotent Matching** (referred to as **ACIM**)

   Instance: A set of variable symbols V, a set of function symbols F some of which may be associative, commutative and idempotent, and terms t_1, t_2 from $T(F, V)$.

   Question: Does there exist a θ such that $\theta(t_1) \underset{ACI}{=} t_2$?

3. **Associative-Commutative-Idempotent Unification** (referred to as **ACIU**)

   Instance: A set of variable symbols V, a set of function symbols F some of which may be associative, commutative and idempotent, and terms t_1, t_2 from $T(F, V)$.

   Question: Does there exist a θ such that $\theta(t_1) \underset{ACI}{=} \theta(t_2)$?

4. **Associative-Idempotent Matching** (referred to as **AIM**)

   Instance: A set of variable symbols V, a set of function symbols F some of which may be associative and idempotent, and terms t_1, t_2 from $T(F, V)$.

   Question: Does there exist a θ such that $\theta(t_1) \underset{AI}{=} t_2$?

5. **Commutative-Idempotent Matching** (referred to as **CIM**)

Instance: A set of variable symbols V, a set of function symbols F some of which may be commutative and idempotent, and terms t_1, t_2 from $T(F,V)$.

Question: Does there exist a θ such that $\theta(t_1) \underset{CI}{=} t_2$?

Let $S = \{s_1, s_2, \cdots s_m\}$ be a set of terms. By $\theta(S)$ we denote the **set** formed by the terms $\theta(s_1)$, $\theta(s_2)$, ... , $\theta(s_m)$. Note that for all θ, $card(S) \geq card(\theta(S))$.

Example 1: Let $S = \{g(x,a), g(a,x), f(x)\}$ and $\theta = \{x \leftarrow a\}$. Then $\theta(S) = \{g(a,a), f(a)\}$ whose cardinality is lower than that of S.

Let S_1 and S_2 be two sets of terms. Then θ is said to unify them if and only if $\theta(S_1) = \theta(S_2)$ and here we say that the two sets are *unifiable.*

Example 2: Let $S_1 = \{g(x,a), g(a,x), f(x)\}$ and $S_2 = \{g(y,y), f(y)\}$. They can be unified by the substitution $\{x \leftarrow a, y \leftarrow a\}$.

The **set-unification problem** (**SUP**) is

Instance: A set of variable symbols V, a set of function symbols F and two sets S_1 and S_2 of terms from $T(F,V)$.
Question: Are S_1 and S_2 unifiable ?

A substitution θ *matches* S_1 to S_2 if $\theta(S_1) = S_2$.

Example 3: Let $S_1 = \{g(x,a), g(a,x), f(x)\}$ and $S_2 = \{g(a,a), f(y)\}$. The substitution $\{x \leftarrow a, y \leftarrow a\}$ matches S_1 to S_2.

Thus the **set-matching problem** (**SMP**) is

Instance: A set of variable symbols V, a set of function symbols F and two sets S_1 and S_2 of terms from $T(F,V)$.
Question: Does S_1 match to S_2 ?

4. Main Results

Theorem 1: SUP is in NP.

Proof: Let $S_1 = \{s_1, s_2, \cdots, s_m\}$ and $S_2 = \{t_1, t_2, \cdots, t_n\}$ be two sets of terms. Without loss of generality assume $m \geq n$. If θ unifies the two sets then it is clear that for every s_i in S_1 there must be some t_j in S_2 (and vice versa too) such that $\theta(s_i) = \theta(t_j)$. We can view this correspondence between the s_i's and the t_j's as a relation defined as

$$\{(i,j) \mid \theta(s_i) = \theta(t_j)\}.$$

Thus all that we have to do is to "choose" this relation correctly (note that there are at the most $m \cdot n$ elements in this relation) and then use any polynomial-time algorithm (such as the one in [6]) to unify terms with uninterpreted function symbols. □

Theorem 2: SUP is NP-hard.

Proof: We reduce 3SAT to SUP by the following construction.

Let $C = \{c_1, c_2, \cdots c_m\}$ be an instance of 3SAT over the boolean variables $x_1, \ldots,$ x_n. For each clause c_i, we have a distinct ternary function symbol g_i. Let 0 and 1 be nullary (constant) symbols. (Thus $card(F) = m + 2$.)

Let $V = \{x_1, \ldots, x_n\} \cup \{u_{ij} \mid 1 \le i \le m, 1 \le j \le 6\}$. The truth and falsity of a boolean variable x_i is simulated by the conditions $x_i = 1$ and $x_i = 0$ respectively.

For each clause c_j, we do the following: let x_1, x_2 and x_3 be the variables in c_j. There are exactly 7 sets of truth-value-assignments that make the clause c_j true. Define 7 distinct terms $q_{j1}, \ldots, q_{j7}$ as follows: $q_{ji} = g_j(b_1, b_2, b_3)$ where $b_i \in \{1,0\}$ and the assignment (b_1, b_2, b_3) satisfies c_j. Thus, for instance, if $c_1 = \{x_1, x_2, x_3\}$, then the 7 terms are $g_1(0, 0, 1)$, $g_1(0, 1, 1)$, $g_1(1, 1, 1)$, $g_1(0, 1, 0)$, $g_1(1, 1, 0)$, $g_1(1, 0, 0)$ and $g_1(1, 0, 1)$.
Let

$$S_j = \{g_j(x_1, x_2, x_3)\} \cup \{u_{ji} \mid 1 \le i \le 6\} \textit{ and}$$

$$\overline{S_j} = \{q_{j1}, \cdots, q_{j7}\}.$$

Note that these two sets are unifiable if and only if the clause c_j is satisfiable. Finally form S which is the union of all the S_i s and $\overline{S}$ which is the union of all the $\overline{S_i}$ s. It is not hard to show that S and $\overline{S}$ are unifiable if and only if C is satisfiable. □

Since the set-unification problem is a special case of the associative-commutative-idempotent unification problem, it follows that associative-commutative-idempotent unification problem is NP-hard.

Note that in the above construction for showing that SUP is NP-hard, the terms in the second set ($\overline{S}$) do not have any variables. Thus we have also shown

Theorem 3: SMP is NP-complete.

Using the above results, we show below that in resolution based theorem-proving, the check whether a clause subsumes another clause is NP-complete.

A clause is a set of literals, where a literal is an atomic formula or its negation. A clause C is said to ***subsume*** another clause D iff there is a substitution σ such that $\sigma(C) \subseteq D$. The ***subsumption-check*** problem is to determine, given two clauses C and D, whether C subsumes D.

Theorem 4: Subsumption-check is NP-complete.

Proof: Membership in NP is immediate, since one can non-deterministically choose the subset of D that C can be matched with and then proceed as in the set-matching problem.

NP-hardness can be shown by reducing SMP to subsumption-check as follows: a set S_1 of terms matches another set S_2 of terms iff S_1 when viewed as a clause subsumes S_2 when viewed as a clause; if the outermost function symbol of a term in S_1 and S_2 is not a predicate, we use a unary predicate P different from any predicate symbols appearing in S_1 and S_2 to make terms in S_1 and S_2 into literals. □

5. Matching for Idempotent Functions

Using the construction given in [1], it can be shown that the ACIEQ is also in polynomial time. One just has to get rid of identical arguments of an idempotent function after flattening terms.

Theorem 5: ACIEQ can be done in polynomial time.

Corollary 6: ACIM is NP-hard.

Proof: Let $S_1 = \{s_1, s_2, \cdots, s_m\}$ and $S_2 = \{t_1, t_2, \cdots, t_n\}$ constitute an instance of SMP, the question being whether S_1 can be matched with S_2. Let g be an uninterpreted monadic function symbol which does not appear in any of the terms in S_1 or S_2 and let f be an associative-commutative-idempotent operator. Form the terms $\mathbf{s} =$ $f(g(s_1), g(s_2), \cdots, g(s_m))$ and $\mathbf{t} = f(g(t_1), g(t_2), \cdots, g(t_n))$. The reader can easily see that $\mathbf{s}$ can be "ACI-matched" with $\mathbf{t}$ if and only if S_1 can be "set-matched" with S_2. □

In fact, we can also show that having the idempotency property in addition does not help even when a function is just associative or commutative.

Theorem 7: AIM is NP-hard.

Proof: We reduce 3SAT to AIM by the following construction, similar to the one used in showing the NP-hardness of SUP.

Let $C = \{c_1, c_2, \cdots c_m\}$ be an instance of 3SAT over the boolean variables $x_1, \ldots,$ x_n. Take $F = \{h, f, g, a, 0, 1\}$, where f is an associative and idempotent function, g is a ternary function, a, 0, and 1 are nullary constants, and h is a m-ary function. Let $V = \{x_1, \ldots, x_n\} \cup \{u_i, v_i \mid 1 \leq i \leq m\}$; truth and falsity of a boolean variable x_i are simulated by the conditions $x_i = 1$ and $x_i = 0$ respectively.

For each clause c_j, we do the following: let x_1, x_2 and x_3 be the variables in c_j. There are exactly 7 sets of truth-value-assignments that make the clause c_j true. Define 7 distinct terms $q_{j1}, \ldots, q_{j7}$ as follows: $q_{ji} = g(b_1, b_2, b_3)$ where $b_k \in \{1,0\}$ and the assignment (b_1, b_2, b_3) satisfies c_j (see the proof of Theorem 2).

Let p_j be the term $f(u_j, g(x_1, x_2, x_3), v_j)$, and q_j be the term $f(a, q_{j1}, q_{j2}, q_{j3}, q_{j4}, q_{j5}, q_{j6}, q_{j7}, a)$. Note that p_j matches to q_j if and only if there is a substitution for u_j, x_1, x_2, x_3 and v_j which satisfies the clause c_j.

Finally let $s = h(p_1, \ldots, p_m)$ and $t = h(q_1, \ldots, q_m)$. The reader can easily verify that s can be "associative-idempotent matched" with t if and only if C is satisfiable. □

The construction used in [1] to show the NP-hardness of CM (commutative matching) can also be used to show the NP-hardness of CIM. Hence

Theorem 8: CIM is NP-hard.

6. REFERENCES

[1] Benanav, D., Kapur, D., and Narendran, P., "Complexity of Matching Problems," Proc. of the *First International Conf. on Rewrite Techniques and Applications*, Dijon, France, May 1985.

[2] Chang, C.-L., and Lee, R. C.-T., *Symbolic Logic and Mechanical Theorem Proving*. Academic Press, New York, 1973.

[3] Garey, M.R., and Johnson, D.S., *Computers and Intractability*, W.H. Freeman, 1979.

[4] Hsiang, J., "Refutational Theorem Proving Using Term Rewriting Systems," *Artificial Intelligence* 25 (3), pp. 255-300, March 1985.

[5] Kapur, D., and Narendran, P., "An Equational Approach to Theorem Proving in First-Order Predicate Calculus," Proc. of the *Ninth International Joint Conf. on Artificial Intelligence*, Los Angeles, August 1985.

[6] Paterson, M.S., and Wegman, M., "Linear Unification," *Journal of Computer and System Sciences* 16, pp. 158-167, 1978.

Matching with distributivity

Jalel Mzali

Centre de Recherche en Informatique de Nancy
Campus Scientifique B.P. 239
54506 Vandoeuvre Les Nancy Cedex
and
Greco de Programmation
(C.N.R.S)

ABSTRACT

We study matching problems for one-sided distributivity. A general method is presented to build up a matching algorithm in an equational theory, and illustrated by the case of one-sided distributivity. Using this algorithm, (right and left) distributivity matching is shown decidable and a method to compute distributive matches is given.
Key words: matching, distributivity, equational theory, decomposition, merging, mutation.

1. Introduction

Solving an equation $t_1 == t_2$ modulo a set of axioms E (or unifying t_1 and t_2 in the equational theory E) consists of finding a substitution σ such that $\sigma(t_1) =_E \sigma(t_2)$.

In many cases, such as program transformation [Der84] or Knuth-Bendix completion [JoK84], we are interested in a matching problem, i.e. a substitution σ is needed such that $\sigma(t_1) = t_2$. Matching is also the basic mechanism of languages based on rewriting such as OBJ2 [FGJ85]. Of course if a unification algorithm is known, it can be applied for computing matches. However, efficiency is crucial in applications such as Knuth-Bendix completion or rewriting-based languages. Therefore, it is still advantageous to find matching algorithms.

The axiom of distributivity (left and/or right) is especially important in theorem proving applications [KaN85] [Hsi85] and in other kind of applications as well [BeK84]. Decidability of unification under distributivity (left and right) is open [Sza82], but one sided-distributivity (left or right) has been shown decidable by Arborg & Tiden who gave a complete algorithm [ArT85]. On the other hand matching under distributivity has not yet been studied. Starting from a new method of constructing matching algorithms, we develop an efficient algorithm for the left (or right) distributivity. Furthermore, we give a simple method to decide the equality problem modulo left (or right) distributivity. Finally we solve the open problem of matching under left and right distributivity. However, our algorithm is derived from simple general consideration upon finite congruence classes, which leads to a somewhat inefficient algorithm.

2. Substitutions and Matches

We assume the reader to be familiar with the basic notions of term rewriting systems [HuO80], but recall some definitions related to our problem.

Definition 2.1 — Let F be a graded set of **function symbols**, and X a set of **variables**. M(F,X) is the free algebra over X. Elements of M(F,X) are called **terms**.

Definition 2.2 — **Substitutions** σ of M(F,X) are defined to be endomorphisms of M(F,X) equal to the identity on X except on a finite subset of it, called the **domain** $D(\sigma)$ of the substitution. $I(\sigma)$ denotes the set of variables occurring in the terms $\sigma(x)$ for all variables x in the domain of σ. $\sigma|W$ is the restriction of σ on the subset W of $D(\sigma)$. Composition of substitutions σ and ρ is written $\rho\sigma$. We say that ρ factors σ, if there is ξ such that $\sigma = \rho\xi$.

Definition 2.3 — We call an **axiom** any pair of terms $\{t_1, t_2\}$ written $t_1 = t_2$. For a finite set of axioms E, the **E-equality** is the smallest congruence on M(F,X) containing all pairs $\{\sigma(t_1), \sigma(t_2)\}$ where $t_1 = t_2$ is any axiom of E and σ any substitution. The E-equality is written $=_E$.

Definition 2.4 — We denotes by $\lesssim_E$ the quasi-ordering on M(F,X) defined by $t_1 \lesssim_E t_2$ iff $t_2 =_E \sigma(t_1)$, where σ called an **E-match**, is a substitution from t_1 to t_2. Let $V \subseteq X$, we say that a substitution σ factors σ' modulo E on V, written as $\sigma \lesssim_E \sigma'$ [V], iff there exists a substitution σ'' such that $\forall x \in V$, $\sigma'(x) =_E \sigma''(\sigma(x))$.
A substitution σ is written $\{(x_1 \leftarrow t_1), \ldots (x_n \leftarrow t_n)\}$. When $E = \emptyset$, $\lesssim_E$ is written as $\lesssim$.

Unlike the standard case of the empty theory, there may be (infinitely) many independent matches from t_1 to t_2 in general. We are therefore interested in finding a basis of the set of matches whenever one exists [FaH83]. We now give the definition of a complete set of E-matches.

Definition 2.5 — [FaH83]— Let M and N be two terms, V(M) the set of variables of M. The set of the E-matches from M to N is written $\{M{=}{=}N\}_E$. Let W be a set of variables containing V(M). A **complete set** of E-matches from M to N outside W is a set of substitutions denoted $C\{M{=}{=}N\}_E$ such that

* $\forall \sigma \in C\{M{=}{=}N\}_E$ $\quad D(\sigma) \subseteq V(M)$ and $I(\sigma) \cap (W\text{-}V(N)) = \emptyset$ (Protection of variables)
* $C\{M{=}{=}N\}_E \subseteq \{M{=}{=}N\}_E$ (correctness)
* $\forall \sigma \in \{M{=}{=}N\}_E$, there exists $\rho \in C\{M{=}{=}N\}_E$ so that $\rho \lesssim_E \sigma$ [V] (completeness)

Moreover $C\{M{=}{=}N\}_E$ is minimal if

* $\forall \sigma, \sigma' \in C\{M{=}{=}N\}_E \quad \sigma \lesssim_E \sigma' \Rightarrow \sigma = \sigma'$ (minimality)

3. Transformation of Equations

We now introduce equations and systems of equations that constitute the basis of our framework.

Definition 3.1 — We call an **equation** any pair of terms written M==N. A **trivial equation** is an equation whose first term is a variable. An **E-solution** (or E-match) of an equation M==N is any E-match from M to N.
A **system** is a set of equations $U = \{M_i {=}{=} N_i \mid i \in I\}$. An E-solution of U is any substitution σ such that σ is an E-solution of each equation $M_i {=}{=} N_i$. We denote by $\{M_i {=}{=} N_i \mid i \in I\}_E$ the set of all solutions.

Definition 3.2 — An **environment** is a system of trivial equations $\{x_i {=}{=} t_i \mid i \in I\}$ where x_i appears at most once, as the left part of an equation. The Domain of an environment e, written D(e) is the set of variables having an occurrence as left part of an equation.
Given an environment $e = \{x_i {=}{=} t_i \ \ i \in I\}$, we define an associated substitution σ_e of domain D(e) such that

$\sigma_e(x_i)$ = if t_i is a variable and $t_i \notin D(e)$ then t_i
else $\sigma_e(t_i)$

Remark 3.1 — σ_e is well defined, if there is no cycle in the relation $(x, \sigma_e(x))$ This is the case in our study because:

1) each added variables is a new variable, one which, does not occur anywhere else in the set of equations,
2) the right hand side of an equation can be considered as a ground term. As a consequence, environments like {(x==x)} or {(x==y),(y==x)} are not possible.

In some cases like distributivity matching, we need to introduce new variables during the matching process. This provides environments such as (x==y,y==z) where y is the new variable. This is why we need a recursive definition of σ_e as given above.

Example 3.1 — for e = {(x==a), (y==v), (v==f(a,b))},

$$\sigma_e(x) = a,$$
$$\sigma_e(z) = z,$$
$$\sigma_e(y) = f(a,b) \text{ and}$$
$$\sigma_e(a) = a.$$

Our matching method amounts to transforming any system of equations until it contains only trivial equations. Correctness and completeness of our transformation are consequences of using an equivalence relation among sets of equations.

Definition 3.3 — We say that systems S = $\{M_i == N_i \mid i \in I\}$ and S' = $\{M_j' == N_j' \mid j \in J\}$ are **E-equivalent** for the matching problem, and we write S $\Leftrightarrow_E$ S, iff $\{M_i == N_i \mid i \in I\}_E = \{M_j' == N_j' \mid j \in J\}_E$.

A system can contain trivial equations, and in that case the system is divided into two parts called **trivial part** (containing only trivial equations), and **non-trivial part** (containing the others).

The base of our method is to increase the trivial part and decrease the non-trivial one while respecting the equivalence of equations. This is done by looking at head symbols in the equations, as explained now.

Some symbols of F may behave, with respect to simplification of equations, as if E is an empty theory. Such symbols are called decomposable [Kir85]:

Definition 3.4 — Given a set of axioms, the set Fd(E) of **E-decomposable** symbols for the matching problem, is the largest subset of F such that

$\forall$ f, g $\in$ Fd $\quad t = f(t_1,\ldots, t_k), \quad t' = g(t_1',\ldots, t_k')$

$$f = g \Rightarrow [\{t = t'\} \Leftrightarrow_E \{(t_i = t_i')_{i=1..k}\}].$$

Similarly, some symbols of F may behave, for the problem of non existence of solutions, as if E is non empty:

Definition 3.5 — Given a set of axioms, the **E-exclusive** set Fex(E) is the largest subset of F×F such that for two terms $f(t_1,..,t_k)$ and $g(t_1',..,t_{k'})$, (f,g) $\in$ Fex:

$$f \neq g \Rightarrow \{f(t_1,..,t_k) == g(t_1',..,t_k')\}_E = \emptyset.$$

4. A General Scheme for Matching Modulo an Arbitrary Set of Equations E.

Computing a set of matches of two terms M and N in the equational theory defined by E, is done by transformation of the set of equations $EQ_1 = \{M_i == N_i \mid i \in I\}$ into a set of equations EQ_n E-equivalent to EQ_1, and which contains only trivial equations. This transformation consists of applying repeatedly the three following steps: merging, decomposition and mutation. In what follows we briefly present these three steps which constitute the main phases of our matching algorithm in an equational theory E.

A more detailed presentation of the method in the general case of unification is given by C. Kirchner [Kir85]. A study of the same method for associative-commutative and/or idempotent matching can be found in [Mza85].

Merging step:

During this step we group together trivial equations, to form the trivial part of the system. If two trivial equations have the same left part like $(x == t_1)$ and $(x == t_2)$, then, $(x == t_1)$ stays in the trivial part while $t_1 == t_2$ goes to the non-trivial part.

$$\left\{\begin{array}{l}(x+y)==(a+b)\\ x==a\\ y==b\\ x==t_1\end{array}\right. \rightarrow \left\{\begin{array}{l}(x+y)==(a+b)\\ a==t_1\\ \hline x\leftarrow a\\ y\leftarrow b\end{array}\right.$$

Decomposition step:

Non-trivial equations that have the same decomposable top symbol are transformed into a systems of simpler equations. During this step, only the decomposable and/or exclusive symbols are used, which insures the equivalence of the starting system and of the decomposed system. If for example f belongs to Fd and + does not, the system:

$$\{ (f(x,y)==f(a,b)), ((x+y)==(a+b)) \}$$

will be decomposed into the E-equivalent system

$$\{x==a, y==b, (x+y)==(a+b)\}$$

On the other hand, if we have an equation $f(t_1,..,t_k)==g(t_1',..,t_k')$ with $(f,g) \in$ Fex, we conclude that the starting system has no solution.

Mutation step:

The non-trivial part of the system only deals with the last two steps. The mutation step consists of transforming equations that cannot be transformed by the previous steps because their top symbols are not decomposable or do not belong to Fex. These transformations are specific to the theory E, unlike the two previous steps which are purely syntactic. Of course, they must transform a set of equations into an E-equivalent one.

For example, let us start from the system obtained at the previous step be $\{(x+y)==(a+b), x\leftarrow a, y\leftarrow b\}$, and let us assume that the symbol "+" is commutative. The mutation of this system yields a disjunction of two systems:

$$\left\{\begin{array}{l}(x+y)==(a+b)\\ \hline x\leftarrow a\\ y\leftarrow b\end{array}\right. \rightarrow \begin{array}{l}\left\{\begin{array}{l}x==a\\ y==b\\ \hline x\leftarrow a\\ y\leftarrow b\end{array}\right.\\ \text{or}\\ \left\{\begin{array}{l}x==b\\ y==a\\ \hline x\leftarrow a\\ y\leftarrow b\end{array}\right.\end{array}$$

Our matching algorithm modulo E is a repetitive application of the three previous steps, as long as the non-trivial part of the system is not empty. A sufficient condition for this algorithm to *stop* is that the mutation step introduces equations of a size strictly smaller than the size of the previous equations. The *correctness* of the algorithm follows from the proof that each transformation step maintains the equivalence of the systems.

5. Case of Left Distributivity

$$Dl:\ (x * y) + (x * z) = x * (y + z)$$

We only consider terms which are canonical with respect to the canonical system DL = $\{(x*y)+(x*z) \rightarrow x*(y+z)\}$. Non-canonical terms are reduced in their canonical form before the mutation process starts.

5.1. Merging

As described above, this step is completely independent from the Axiom Dl.

5.2. Decomposition

We first characterize Fd and Fex:

Proposition 5.1 — The symbols + and * are decomposable, thus

$$\{(t_1+t_2==t_3+t_4)\} \Leftrightarrow_{Dl} \{(t_1==t_3), (t_2==t_4)\}$$
$$\{(t_1*t_2==t_3*t_4)\} \Leftrightarrow_{Dl} \{(t_1==t_3), (t_2==t_4)\}$$

Proof :

See Arnborg and Tiden [ArT85]. []

Proposition 5.2 — Let Fr be the subset of F which contains all symbols not involved in the distributive axiom, then

$$Fex = (Fr \times F) \cup (F \times Fr)$$

Proof :

It is easy to see that $(Fr \times F) \cup (F \times Fr) \subseteq Fex$. The reversed inclusion follows from the fact that equations of the form +(..)==*(..) or *(..)==+(..) need a mutation process as detailed in subsection 5.3. []

5.3. Mutation Phase

We now have a system of the form

$$\begin{cases} (M==N) \\ (M_1==N_1) \\ \hline (x \leftarrow M_2) \\ (y \leftarrow N_2) \end{cases}$$

where the non-trivial equations M==N are not decomposable. More precisely, these equations are of the form

$$t_1+t_2 == s_1*s_2$$
or
$$t_1*t_2 == s_1+s_2$$

We study in this part the transformation of these different equations into simpler Dl-equivalent systems.

First we give a result which allows us to reduce the general problem to the case where only variables occur under the top symbol left hand side of the equations, as in $S=\{x_1+x_2==t_1'*t_2'\}$. The general case is dealt with by applying an appropriate transformation called **generalisation**: For example, the equation

$t_1+t_2 == t_1'*t_2'$, is solved by considering the variable case $x_1+x_2 == t_1'*t_2'$ first, then by instantiating the system obtained from S by the substitution $\sigma=\{x_1 \leftarrow t_1, x_2 \leftarrow t_2\}$.

Lemma 5.1 — Let S_1 and S_2 be two systems of equations and σ a substitution. if $S_1 \Leftrightarrow_E S_2$ then $\sigma(S_1) \Leftrightarrow_E \sigma(S_2)$.

Proof :

Straightforward. []

Now here is the mutation process for the variable case:

Lemma 5.2 —

$$\{(x+y)==(t_1*t_2)\} \Leftrightarrow_{DI} \begin{cases} (x_2+y_2)==t_2 \\ x==(t_1*x_2) \\ y==(t_1*y_2) \end{cases}$$

Proof :

Assume that σ is a solution to the equation on the left, we show that it is also a solution to the system on the right. Because of the decomposability of $*$, x must be of the form $x = x_1*x_2$ and $y = x_1*y_2$, thus $x+y = (x_1*x_2)+(x_1*y_2) =_{DI} x_1*(x_2+y_2)$. Hence $\sigma(x+y) =_{DI} \sigma(x_1*(x_2+y_2))$. According to the decomposability of $*$, we have $\sigma(x_1) =_{DI} t_1$ and $\sigma((x_2+y_2)) =_{DI} t_2$.

Conversely $\sigma(x+y) = \sigma(x)+\sigma(y) = ((\sigma(x_1)*\sigma(x_2)) + (\sigma(y_1)*\sigma(y_2)) = ((t_1*\sigma(x_2)) + (t_1*\sigma(y_2)) = (t_1 * (\sigma(x_2) + \sigma(y_2)) = (t_1*t_2)$. []

We omit the proof of the next lemma, which is similar.

Lemma 5.3 —

$$\{(x*y)==(t_1+t_2)\} \Leftrightarrow_{DI} \begin{cases} (x*y_1)==t_1 \\ (x*y_2)==t_2 \\ y==(y_1+y_2) \end{cases}$$

5.4. The Algorithm for DI-Matching

By repeating the three steps we have a matching algorithm modulo the axiom DI. The algorithm must terminate because equations introduced during the mutation process have a size smaller than that of the equations before the mutation process.

Note, however, that this property is not true for *unification* modulo DI. In this case the mutation process constructs equations of the same type (for example $\{y*z==w+z\} \Leftrightarrow \{v_1+v_2==y*v_1,\ w==y*v_2,\ w==y*v_1\}$). In the DI matching problem, this cannot occur even if we introduce new variables during the process. The equations containing variables in their right part are equations of the trivial part and are not equations to be transformed any more.

We now give an example which illustrates the different steps of this algorithm.

Example 5.1 — Let M = (((x*y)+(z*u))+x) and N = (((a+b)*(c+d))+(a+b)). The decomposition and the merging of {M==N} gives:

$$\begin{cases} (x*y)+(z*u)==(a+b)*(c+d) \\ \hline x==(a+b) \end{cases}$$

mutation:

$$\left\{\begin{array}{l} x_2+y_2==c+d \\ x*y==(a+b)*x_2 \\ z*u==(a+b)*y_2 \\ \hline x==(a+b) \end{array}\right.$$

decomposition:

$$\left\{\begin{array}{l} x_2==c \\ y_2==d \\ x==(a+b) \\ y==x_2 \\ z==(a+b) \\ u==y_2 \\ \hline x==(a+b) \end{array}\right.$$

merging:

$$\left\{\begin{array}{l} a+b==a+b \\ \hline x_2==c \\ y_2==d \\ y==x_2 \\ z==(a+b) \\ u==y_2 \\ x==(a+b) \end{array}\right.$$

decomposition:

$$\left\{\begin{array}{l} \hline x_2==c \\ y_2==d \\ y==x_2 \\ z==(a+b) \\ u==y_2 \\ x==(a+b) \end{array}\right.$$

the solution is then: $\{x \leftarrow (a+b),\ y \leftarrow c,\ z \leftarrow (a+b),\ u \leftarrow d\}$.

6. A Technique for Partly Eliminating Mutations.

The mutation step sometimes introduces inefficiency. We consider here a method for eliminating mutation steps by introducing a new symbol (Δ). The idea is to characterize the mutations that eventually fail because of some clash of symbols in Fex, or yield a further decomposition step.

The new symbol Δ is defined as follows:

$$x*(y+z)=\Delta(x,y,x,z)$$
$$(x*y)+(z*u)=\Delta(x,y,z,u)$$

The Knuth-Bendix completion procedure run on distributivity and these two rules returns the following canonical system ΔL:

$$x*(y+z) \rightarrow \Delta(x,y,x,z)$$
$$(x*y)+(z*u) \rightarrow \Delta(x,y,z,u)$$
$$x*\Delta(x_1,y,z,u) \rightarrow \Delta(x,x_1*y,x,z*u)$$
$$\Delta(x,y,x,z)+(z_1*u) \rightarrow \Delta(x,y+z,z_1,u)$$
$$(x*y_1)+\Delta(z_1,y,z_1,z) \rightarrow \Delta(x,y_1,z_1,y+z)$$
$$\Delta(x,y_1,x,z_2)+\Delta(z_1,y,z_1,z) \rightarrow \Delta(x,y_1+z_2,z_1,y+z)$$

Using the decomposability of + and *, it is easy to see that Δ in decomposable.

Terms are always considered in DL normal form. Every term of the forms $t_1*(t_2+t_3)$ or $(t_1*t_2)+(t_3*t_4)$ is now transformed into $\Delta(t_1',t_2',t_3',t_4')$, using ΔL, ($t_1'=t_3'=t_1$ in the first case and $t_i'=t_i$ in the second). Since Δ is decomposable, equations containing such terms are then transformed by using the decomposition process instead of a mutation process, and this is more efficient.

The mutation process is used only for those equations containing terms of the form (t_1+t_2) where t_1 and t_2 are ΔL-irreducible, or of the form (t_1*t_2) where the top symbol of t_2 is not "+".

Consequently, these transformations reduce the number of transformation steps by eliminating many mutations.

Example 6.1 — We consider the terms M and N of Example 5.1. They are transformed respectively into Δ(x,y,z,u)+x and Δ(a+b,c,a+b,d)+(a+b). Decomposition (of Δ and +) and merging steps are only used here to find solutions:
decomposition of (+) and merging:

$$\left\{\begin{array}{l} \Delta(x,y,z,u) == \Delta(a+b,c,a+b,d) \\ \hline x == a+b \end{array}\right.$$

decomposition of (Δ) and merging:

$$\left\{\begin{array}{l} \hline x == a+b \\ y == c \\ z == a+b \\ u == d \end{array}\right.$$

The introduction of a new symbol Δ also allows us to reduce the equality modulo Dl, to the equality of ΔL normal forms, which decreases the number of equality tests of the subterms of a term. For example for a term such as (x+y)*(z+u), the number of tests is 7 (explicitly the top symbol, +, x, y, +, z and u) whereas for its ΔL normal form Δ(x,y,z,u), the number of tests is only 5. We have thus transformed the two tests for * and + into a unique one for Δ.

Example 6.2 — Consider the following terms

$$M = (x * ((y_1 + y_2) + z))$$
$$N = (((x * y_1) + (x * y_2)) + (x * z))$$
$$M' = (x * ((y_1 + y_2) \# z))$$

The ΔL normal form of M is equal to Δ(x, (y_1+y_2), x, z) which is equal to the ΔL normal form of N. We conclude that M $=_{Dl}$ N. On the other hand, M' is ΔL-irreducible and the top symbols of ΔL-normal form of M and M' are not equal. This is enough to conclude that M and M' are not equal modulo Dl.

7. D = Dl ∪ Dr

The method used for Dl to characterize the set of term modulo Dl cannot be used for D = Dl ∪ Dr, because all systems considered until now are not convergent (the Knuth-Bendix completion runs forever). However, matching modulo D is decidable, and we have this result:

We note $[M]_E$ the set of all terms equal to M modulo E, and $\{M==[N]_E\}$ the union $\bigcup_{Ni \in [N]_E} \{M==Ni\}_E$

Proposition 7.1 — Let A and B be two sets of axioms s.t. $=_A$ commutes with $=_B$, then we have:

$$\{M==N\}_{A \cup B} = \{M==[N]_A\}_B = \{M==[N]_B\}_A$$

Proof :
Using a commutation of $=_A$ and $=_B$ we have this equality. []

Since D is a permutative theory (admits finite equivalence classes), we can apply this result to prove decidability of D matching, and we have an algorithm to compute all matches under the axiom of distributivity.

Corollary 7.1 —

$$\{M==N\}_D = \{M==[N]_D\}_\emptyset$$

Example 7.1 —

$$\{(x+x)*y==a*(b+b)\}_D = \{(x+x)*y==[a*(b+b)]_D\}_\emptyset$$

witch is equal to:

$$\{(x+x)*y==(a*b)+(a*b)\}_\emptyset \cup \{(x+x)*y==(a+a)*b\}_\emptyset$$

The first system has an empty set of $\emptyset$-solutions (clash * and + in the empty theory), the second system is $\emptyset$-equivalent to {x←a, y←b} witch is the unique solution.

8. Conclusion

The approach used here to describe an algorithm modulo Dl is based on a standard mechanism for the construction of matching algorithms, applicable to other equational theories. This mechanism is based on the following fundamental points:

- Decomposition of equations.
- Merging the constraints in order to build substitutions.
- Mutation of non-decomposable equations into decomposable equations.

In the particular case of Dl, this study is simplified by introducing a new function symbol Δ with appropriate equations to define it. Δ allows us to simplify the algorithm by transforming mutation steps into decomposition steps. In the case of left distributivity, it also provides with an efficient clash for Dl-equation.

9. Acknowledgement

I would like to thank the members of EURECA at CRIN. Special thanks are due to Jean-Pierre Jouannaud for many helpful suggestions. I also thank Claude Kirchner, Pierre Lescanne and Jean-Luc Remy for detailed comments and discussions during this work. The final draft of the paper is done when the author is visiting the State University of New York at Stony Brook sponsored by NCF grant DCR-8401624.

10. References

[ArT85] S. Arnborg and E. Tiden, "Unification problems with one-sided distributivity", in *Proc. 1st Conf. on Rewriting Techniques and Applications*, vol. 202, Springer Verlag, Dijon (France), 1985, 398-406.

[BeK84] J. A. Bergstra and J. W. Klop, "The Algebra of Recursively defined Processes and the Algebra of Regular Processes", in *ICALP 84*, vol. 172, 1984, 82-94.

[Der84] N. Dershowitz, "Computing With Term Rewriting Systems", *Procedings of An NSF Workshop On The Rewrite Rule Laboratory*, April 1984.

[FaH83] F. Fages and G. Huet, "Unification and Matching in Equational Theories", *Proceedings of CAAP 83*, **159**, (1983), 205-220, Springer Verlag.

[FGJ85] K. Futatsugi, J. A. Goguen, J. P. Jouannaud and J. Meseguer, "Principles of OBJ2", in *Proceedings, 12th ACM Symposium on Principles of Programming Languages Conference*, 1985.

[Hsi85] J. Hsiang, "Two Results in Term Rewriting Theorem Proving", in *Proc. 1st Conf. on Rewriting Techniques and Applications*, vol. 202, Springer Verlag, Dijon (France), 1985, 301-324.

[HuO80] G. Huet and D. Oppen, "Equations and Rewrite Rules: A Survey", in *Formal Languages: Perspectives And Open Problems*, B. R., (ed.), Academic Press, 1980.

[JoK84] J. P. Jouannaud and H. Kirchner, "Completion of a set of rules modulo a set of equations", *Proceedings 11th ACM Conference of Principles of Programming Languages*, Salt Lake City (Utah, USA), 1984.

[KaN85] D. Kapur and P. Narendran, "An Equational Approach to Theorem Proving in First-Order Predicate Calculus", in *Proc. Int. Joint Conf. on Artificial Intelligence*, Los Angeles, 1985.

[Kir85] C. Kirchner, "Méthodes et outils de conception systématique d'algorithmes d'unification dans les théories équationnelles", Thése de doctorat d'Etat, Université de Nancy I, 1985.

[Mza85] J. Mzali, "Filtrage associatif, commutatif ou idempotent", in *Proceedings of the conference Materiels et logiciels pour la 5ieme generation*, AFCET, Paris (France), 1985, 243-258.

[Sza82] P. Szabo, "Unificationtheorie erster Ordnung", Doktorarbeit, Universitat Karlsruhe, 1982.

Unification in Boolean Rings

Ursula Martin and Tobias Nipkow
Department of Computer Science
The University
Manchester M13 9PL

Abstract

A simple unification algorithm for terms containing variables, constants and the set operators intersection and symmetric difference is presented. The solution is straightforward because the algebraic structure under consideration is a boolean ring. The main part of the algorithm is finding a particular solution which is then substituted into a general formula to yield a single most general unifier. The combination with other equational theories is briefly considered but even for simple cases the extension seems non-trivial.

1 Introduction

The relevance of unification algorithms for particular theories in many areas of computer science and mathematics is well known, see for example Siekmann [Si 84]. In particular algebraic manipulation of formulae, automated theorem proving and some programming languages make heavy use of unification as a basic inference mechanism.

Over the years a number of theories have received special attention, among them the theories of associative, commutative operators and associative, commutative and idempotent operators [LiSi 76], [St 81], [Fa 84]. The latter is of particular interest because the set operators union and intersection have exactly these three properties. Yet none of the known solutions treat the full set theory, because they can only handle one of the operators, but not formulae containing both of them. Besides, even union and intersection are not sufficient to express additional operators like set difference.

The purpose of this paper is to describe a simple solution to the problem of unification in set theory, expressed in terms of the two operators symmetric difference (+) and intersection (*). This is no restriction since all other customary set operators like union and difference can be expressed in terms of + and *. The terms under consideration contain only variables $(x, y, \ldots)$ and constants $(a, b, \ldots)$. For ease of readability intersection will often be denoted by concatenation. We use the standard terminology of unification theory as described for example in [Si 84].

2 The Solution

Suppose two terms $s = s(x_1, \ldots, x_n)$ and $t = t(x_1, \ldots, x_n)$ are unifiable. Then we shall prove that the substitution

$$x_i \to x_i + (s+t)(x_i + w_i)$$

where $x_i \to w_i$ is any solution to $s = t$, is a most general unifier (mgu) for s and t. A simple algorithm determining whether or not s and t are unifiable, and returning a solution $x_i \to w_i$ to $s = t$ if they are, is given below.

A first example: Let $s = ax + by$, $t = a$. One solution to $s = t$ is

$$x \to a,\ y \to 0.$$

Thus the equation $ax + by = a$ has the mgu

$$\begin{aligned} x &\to x' + (a + ax' + by')(x' + a) \\ y &\to y' + (a + ax' + by')(y' + 0) \end{aligned}$$

which simplifies to

$$\begin{aligned} x &\to x' + ax' + bx'y' + aby' + a \\ y &\to y' + ay' + by' + ax'y'. \end{aligned}$$

Another solution to $s = t$ is given by $x \to a + ab$, $y \to ab$, which is obtained by substituting $x' = 0$, $y' = a$ into our mgu.

Our unification procedure takes this simple form because the operators $+$ and $*$ turn $P(S)$, the set of subsets of S, into a boolean ring, that is a ring in which every element is an idempotent. Union and difference can be expressed as

$$\begin{aligned} x \cup y &= x + y + xy \\ x \backslash y &= x + xy. \end{aligned}$$

Hence we can deal with terms containing union, intersection, difference and symmetric difference by translating them into a normal form containing only $+$ and $*$. This approach to boolean terms goes back to Stone [St 36]. It has been used as the basis for decision procedures for the propositional calculus by Watts and Cohen [WaCo 80] and Hsiang [Hs 85]. It is also important in the study of logic circuits where $+$ and $*$ become "exclusive or" and "and".

3 The Most General Unifier

Our proof depends upon the fact that the power set $P(S)$ of a set S forms a boolean ring under the operations $+$ and $*$. A set A is a *boolean ring* under these operations if for all $a, b, c \in A$ we have

$$\left.\begin{aligned} a + b &= b + a \\ (a + b) + c &= a + (b + c) \\ a + 0 &= a \\ a + (-a) &= 0 \\ (a * b) * c &= a * (b * c) \\ a * (b + c) &= a * b + a * c \\ (a + b) * c &= a * c + b * c \\ a * a &= a \end{aligned}\right\} E$$

where 0 is the zero element and $-a$ is the additive inverse of a. It then follows that $*$ is commutative and every element is its own additive inverse, that is

$$\begin{aligned} a * b &= b * a \\ a + a &= 0. \end{aligned}$$

An element $1 \in A$ with the property that

$$1 * a = a * 1 = a$$

for all $a \in A$ is called an identity element. In the case of $A = P(S)$ we have $1 = S$ and $0 = \{\}$.

In the sequel we will repeatedly make use of the identities $a * (1 + a) = 0$, and $a^k = a$ for $k \neq 0$.

Now suppose that we have two disjoint sets C of constants and V of variables and that A and B are the free boolean rings on C and $C \cup V$ respectively; that is $A = T(C, \Sigma)_E$, $B = T(C \cup V, \Sigma)_E$ where $\Sigma = \{0, 1, +, *\}$. Here $T(X, S)_E$ denotes the term algebra over the constants X and the signature S, factorized by the congruence induced by a set of equations E.

Any element $u \in B$ can be expressed (by repeatedly applying the distributive law) as

$$u = a_0 + a_1 v_1 + \cdots + a_n v_n \tag{1}$$

where $a_i \in A$ and each v_i is a product of elements of V_u, the set of variables occuring in u. We have

$$v_i = \prod_{v \in V_u} v^{e_v}$$

where each $e_v \in \{0, 1\}$ and not all the e_v are 0. All the v_i are distinct, i.e. $v_i \neq v_j$ for $i \neq j$. Since each v_i contains each variable at most once, it can be treated as a subset of V. We make use of this view below. We call the above form (1) *normalized.* If $a_0 = 0$ then u is said to be *homogeneous.*

We can consider u as a map from B^k to B and write

$$u = u(x_1, \ldots, x_k) = u(\underline{x})$$

where $\underline{x} = (x_1, \ldots, x_k)$ is in B^k. Similarly $u(a\underline{x} + b\underline{y})$ denotes $u(ax_1 + by_1, \ldots, ax_k + by_k)$ where a, b, x_i, y_i are in B. In this notation, $u(\underline{x})$ is homogeneous if and only if $u(\underline{0}) = 0$.

The following lemma gives a useful property of homogeneous terms which we shall use frequently.

Lemma Let $u(\underline{x})$ be homogeneous. Then

1 if b is in B then $bu(\underline{x}) = u(b\underline{x})$

2 if $\underline{x} = b_1\underline{x}_1 + \cdots + b_n\underline{x}_n$ and $b_i b_j = 0$ for $i \neq j$ then

$$u(\underline{x}) = b_1 u(\underline{x}_1) + \cdots + b_n u(\underline{x}_n)$$

Proof

1 We have

$$u(\underline{x}) = \sum_{i=1}^{n} a_i \prod_{x \in V} x^{e_{ix}}$$

where for each i not all the e_{ix} are 0. Then

$$\begin{aligned} bu(\underline{x}) &= \sum_{i=1}^{n} a_i b \prod_{x \in V} x^{e_{ix}} \\ &= \sum_{i=1}^{n} a_i \prod_{x \in V} (bx)^{e_{ix}} \\ &= u(b\underline{x}) \end{aligned}$$

since for each value of i we have $b = b^{k_i}$ where $k_i = \sum_{x \in V} e_{ix}$.

2 Observe that $b_i\underline{x} = b_i\underline{x}_i$ for each i. Then

$$\begin{aligned} & b_1 u(\underline{x}_1) + \cdots + b_n u(\underline{x}_n) \\ =\ & u(b_1\underline{x}_1) + \cdots + u(b_n\underline{x}_n) \\ =\ & u(b_1\underline{x}) + \cdots + u(b_n\underline{x}) \\ =\ & (b_1 + \cdots + b_n) u(\underline{x}) \\ =\ & u((b_1 + \cdots + b_n)\underline{x}) \\ =\ & u(b_1\underline{x}_1 + \cdots + b_n\underline{x}_n). \quad \square \end{aligned}$$

We can now prove the main result.

Theorem Let $s(\underline{x}), t(\underline{x}) \in B, \underline{w} \in B^n$ such that $s(\underline{w}) = t(\underline{w})$. Then the substitution

$$\underline{y} = \underline{x} + (s(\underline{x}) + t(\underline{x}))(\underline{x} + \underline{w})$$

is a mgu of the two terms $s(\underline{x})$ and $t(\underline{x})$.

Proof First we show that the above substitution is indeed a unifier. We can write $s(\underline{x}) + t(\underline{x}) = u(\underline{x}) + a$ where $u(\underline{x})$ is homogeneous and $a \in A$.

$$\begin{aligned} s(\underline{y}) + t(\underline{y}) &= u(\underline{y}) + a = u(\underline{x} + (u(\underline{x}) + a)(\underline{x} + \underline{w})) + a \\ &= u((1 + u(\underline{x}) + a)\underline{x} + (u(\underline{x}) + a)\underline{w}) + a \\ &= (1 + u(\underline{x}) + a)u(\underline{x}) + (u(\underline{x}) + a)u(\underline{w}) + a && \text{by the lemma} \\ &= 0 && \text{since } s(\underline{w}) = t(\underline{w}) \Rightarrow u(\underline{w}) + a = 0 \end{aligned}$$

It follows that $s(\underline{y}) = t(\underline{y})$.

Now suppose there exists some solution $\underline{z}$, i.e. $s(\underline{z}) = t(\underline{z})$. We need to show that $\underline{z}$ is an instantiation of $\underline{y}$. Fortunately $\underline{x} = \underline{z}$ will do: $\underline{z} + (s(\underline{z}) + t(\underline{z}))(\underline{z} + \underline{w}) = \underline{z}$ because $s(\underline{z}) + t(\underline{z}) = 0$. Therefore $\underline{y}$ is indeed a most general solution. □

Remark

The fact that we do not introduce any new variables in a unification step is for notational convenience only. Since the mgu substitutes all variables present in the original two terms, it is invariant under any bijective renaming of variables in the solution.

4 Finding Particular Solutions

In this section we describe how to test whether or not two terms are unifiable, and how to find a particular solution if they are.

There is a trivial solution to this problem since A is finite and enumerable. Given a finite set of constants C, $A = T(C, \Sigma)_E$ contains $2^{(2^{|C|})}$ elements. All one needs to do is to test all possible valuations. The complexity of this algorithm is $O(|A|^{|V|}) = O(2^{|V|2^{|C|}})$. The problem itself is NP-complete since it also covers the special case where $C = \{\}$, i.e. propositional formulae: finding a particular solution to $p = 1$ is equivalent to determining the satisfiability of p. Therefore it is unlikely that we can find a sub-exponential solution. However our method is significantly better than this.

We begin with an example: $u(\underline{x}) = axy + by = a$ where $\underline{x} = (x, y)$.

We want to find a solution in $A = T(\{a, b\}, \Sigma)_E$. Observe that this is a vector space over the field of two elements $F_2 = \{0, 1\}$ with basis $D = \{ab = d_1, a\bar{b} = d_2, \bar{a}b = d_3, \bar{a}\bar{b} = d_4\}$ where $\bar{c} = (1 + c)$. This means that each $c \in A$ can be expressed uniquely as

$$c = \sum_{i=1}^{4} c_i d_i \quad \text{with } c_i \in F_2$$

Furthermore we have $d_i d_j = 0$ for $i \neq j$ and hence $cd_i = c_i d_i \in \{d_i, 0\}$ for any $c \in A$.
Now suppose that

$$\begin{aligned} x &= x_1 d_1 + x_2 d_2 + x_3 d_3 + x_4 d_4 && \text{and} \\ y &= y_1 d_1 + y_2 d_2 + y_3 d_3 + y_4 d_4 && \text{and thus} \\ \underline{x} &= d_1 \underline{x}_1 + d_2 \underline{x}_2 + d_3 \underline{x}_3 + d_4 \underline{x}_4 && \text{where } \underline{x}_i = (x_i, y_i). \end{aligned}$$

We have

$$\begin{aligned}
u(\underline{x}) &= u(d_1\underline{x}_1 + d_2\underline{x}_2 + d_3\underline{x}_3 + d_4\underline{x}_4) \\
&= d_1u(\underline{x}_1) + d_2u(\underline{x}_2) + d_3u(\underline{x}_3) + d_4u(\underline{x}_4) && \text{by the lemma} \\
&= d_1(ax_1y_1 + by_1) + d_2(ax_2y_2 + by_2) + d_3(ax_3y_3 + by_3) + d_4(ax_4y_4 + by_4) \\
&= d_1(x_1y_1 + y_1) + d_2(x_2y_2) + d_3(y_3) + d_4(0)
\end{aligned}$$

So $u(\underline{x}) = d_1u_1(\underline{x}_1) + \cdots + d_4u_4(\underline{x}_4)$ where $u_i(\underline{x}_i)$ is a polynomial in x_i, y_i with coefficients in F_2.

Now since $a = d_1 + d_2$, we may equate coefficients of the d_i in $u(\underline{x}) = a$ to deduce that this has a solution in A if and only if the four equations

$$\begin{aligned}
u_1(\underline{x}_1) &= x_1y_1 + y_1 &= 1 \\
u_2(\underline{x}_2) &= x_2y_2 &= 1 \\
u_3(\underline{x}_3) &= y_1 &= 0 \\
u_4(\underline{x}_4) &= 0 &= 0
\end{aligned} \tag{2}$$

have a solution. It is easy to find solutions to these equations. For the third and fourth one, just set

$$x_3 = y_3 = x_4 = y_4 = 0.$$

For the first, find the shortest word on the left hand side, y_1, and set the variables appearing in it to 1 and the other variables to 0, to get the solution

$$x_1 = 0, y_1 = 1.$$

Similarly for the second: $x_2 = y_2 = 1$.

Thus we have a solution

$$\begin{aligned}
x &= d_2 &= a\bar{b} &= a + ab \\
y &= d_1 + d_2 &= ab + a\bar{b} &= a.
\end{aligned}$$

The equations (2) were so easy to solve because they were independent, i.e. each variable appeared only in one of them. This is not an accident - it always happens, as can be seen by generalizing the argument of the above example. We do this below.

However first we consider an example of two terms which cannot be unified. Let $u(\underline{x}) = ax = b$. We have $x = d_1x_1 + d_2x_2 + d_3x_3 + d_4x_4$, hence $ad_1x_1 + ad_2x_2 + ad_3x_3 + ad_4x_4 = d_1 + d_3$, that is $x_1d_1 + x_2d_2 = d_1 + d_3$, which gives, on equating coefficients,

$$x_1 = 1,\ x_2 = 0,\ 0 = 1,\ 0 = 0$$

which clearly has no solution. Thus ax and b cannot be unified.

In general then , this is our algorithm to determine if $s(\underline{x})$ and $t(\underline{x})$ are unifiable and to find a solution if they are.

We have $A = T(C,\Sigma)_E$ where $C = \{c_1,\ldots,c_k\}$. The elements of $D = \{\tilde{c_1} * \cdots * \tilde{c_k} \mid \tilde{c_i} \in \{c_i, \overline{c_i}\}\} = \{d_1,\ldots,d_m\}$, where $m = 2^k$, form a basis for A as a vector space over F_2, with the property that if $a \in A$, then a can be written uniquely as

$$a = \sum_{i=1}^{m} a_id_i \text{ with } a_i \in F_2.$$

We also have $d_id_j = 0$ for $i \neq j$ and so $ad_i = a_id_i \in \{d_i, 0\}$.

The algorithm to compute a particular solution for the equation $s(\underline{x}) = t(\underline{x})$ can be broken up as follows.

1. Normalize $s(\underline{x}) + t(\underline{x})$ as $u(\underline{x}) + a_0$ where $u(\underline{x}) = a_1 v_1(\underline{x}) + \cdots + a_r v_r(\underline{x})$ is homogeneous, $a_i \in A$ and the $v_i(\underline{x})$ are pairwise distinct strings of variables in $\underline{x}$ as described above.

2. For each x_i in $\underline{x} = (x_1, \ldots, x_n)$ write $x_i = d_1 x_{i1} + \cdots + d_m x_{im}$ where the x_{ij} lie in F_2. Substituting this back into u we get

$$u(\underline{x}) = u(d_1\underline{x}_1 + \cdots + d_m\underline{x}_m) = d_1 u(\underline{x}_1) + \cdots + d_m u(\underline{x}_m)$$

where $\underline{x}_j = (x_{1j}, \ldots, x_{nj})$. For each j we have

$$d_j u(\underline{x}_j) = d_j \sum_{i=1}^{r} a_i v_i(\underline{x}_j) = \sum_{i=1}^{r} d_j a_i v_i(\underline{x}_j) = d_j \sum_{i \in N_j} v_i(\underline{x}_j) = d_j u_j$$

where $N_j = \{i \in \{1..r\} \mid d_j a_i \neq 0\}$ and each $v_i(\underline{x}_j)$ is just $v_i(\underline{x})$ with x_{ij} substituted for x_i. Thus u_j is a homogeneous polynomial in the variables $x_{1j}, \ldots, x_{nj}$.

3. Express a_0 in terms of the d_i and equate coefficients:

   Let $a_0 = \sum_{i=1}^{m} p_i d_i$ where $p_i \in F_2$. Now $\sum_{j=1}^{m} d_j u_j = u(\underline{x}) = a_0 = \sum_{j=1}^{m} p_j d_j$ has a solution iff each equation $u_j = \sum_{i \in N_j} v_i(\underline{x}_j) = p_j$ has a solution. There are three possibilities for each equation:

   3.1. $p_j = 0$: then $\underline{x}_j = \underline{0}$, i.e. $x_{ij} = 0$ for all i, is a solution.

   3.2. $p_j = 1$:

   3.2.1. $N_j = \{\}$: then there is no solution because $1 = \sum_{i \in \{\}} \cdot = 0$ has no solution. Hence s and t are not unifiable.

   3.2.2. $N_j \neq \{\}$: the equation always has a solution. From among the $v_i(\underline{x}_j)$ select one with non-zero coefficient, i.e. $i \in N_j$, such that there is no smaller set of variables in u_j, i.e. there is no $k \in N_j$ with $v_k(\underline{x}_j)$ containing fewer variables than $v_i(\underline{x}_j)$. Set all x_{lj} in $v_i(\underline{x}_j)$ to 1 and all other x_{lj} to 0. Then $v_i(\underline{x}_j)$ is 1 and all other $v_k(\underline{x}_j)$ are either larger or of the same size but different from $v_i(\underline{x}_j)$. In both cases they must contain some x_{lj} not in $v_i(\underline{x}_j)$ which means they evaluate to 0.

The complete algorithm for finding a special solution is given below in a more formal and concise notation. Since all sets involved are finite, even the quantified expressions are in principle executable. The nondeterministic choice of shortest strings of variables is embodied in the **let** $min \in \ldots$ construct.

sol(s,t) =
  let $a_0 + a_1 v_1 + \cdots + a_r v_r = s + t$ **in**
  let $p_1 d_1 + \cdots + p_m d_m = a_0$ **in**
  let $A = \{i \in \{1..m\} \mid p_i = 1\}$ **in**
  let $N(i \in A) = \{j \in \{1..r\} \mid d_i a_j \neq 0\}$ **in**
    if $\exists i \in A : N(i) = \{\}$ **then fail**
    else let $min \in \{f : A \to \{1..r\} \mid f(i) \in N(i) \wedge \forall j \in N(i) : |v_{f(i)}| \leq |v_j|\}$ **in**
      let $I(i \in \{1..n\}) = \{j \in A \mid x_i \in v_{min(j)}\}$ **in**
        $\{x_i \to \sum_{j \in I(i)} d_i \mid i \in \{1..n\}\}$

Computing a special solution is the only algorithmic part in our unification algorithm. Once a particular solution has been derived, it just has to be substituted into the formula for the general solution.

The above algorithm solves $|D| = 2^{|C|}$ equations with at most $2^{|V|}$ variables in each of them. Thus the overall complexity is of the order of $O(2^{|V||C|})$.

5 Remarks and Extensions

Our mgu is not always "optimal" in the sense that it may use more variables than strictly necessary. Take the example $ax + y = 0$. Using the special solution $x = y = 0$ we get the mgu

$$\begin{aligned} x &\rightarrow x + (ax + y + 0)(x + 0) &= x + ax + xy \\ y &\rightarrow y + (ax + y + 0)(y + 0) &= axy \end{aligned}$$

However, since $ax + y = 0$ is equivalent to $ax = y$ we also have a mgu

$$x \rightarrow x, \; y \rightarrow ax$$

which is an instance of our solution obtained by substituting a for y, but a "simpler" one.

Our method can easily be used to solve a system of simultaneous equations. This is because the pair of equations $f = 0, g = 0$ is equivalent to the single equation $f + g + fg = 0$. For a set of n equations $f_1 = 0, \ldots, f_n = 0$ we need to solve the single equation $\sum_{I \subseteq \{1..n\}} \prod_{i \in I} f_i = 0$. On the other hand, it is not clear whether this is any more efficient than building up the solution one equation at a time.

The introduction of the universal set **1** into our calculus may be undesirable for particular applications. Fortunately **1** will not occur in the general solution if it was not present in the terms to be unified in the first place: a **1** can occur in $\underline{x} + (s(\underline{x}) + t(\underline{x}))(\underline{x} + \underline{w})$ only if it occurs already in s or t; if s and t do not contain **1**, one can see that a) the special solution $\underline{w}$ does not have to contain **1** either, and b) even if it did, the multiplication with $(s(\underline{x}) + t(\underline{x}))$ would get rid of it.

So far we have only treated variables and constants, but no additional operators. Yet one would like to unify terms which also contain function symbols of different theories. The two main obstacles are exactly the ones that prohibit the application of Yelick's solution in [Ye 85]. Her algorithm works only for equational theories that obey the following restrictions: they must not contain an equation $s = t$ where s is a variable and t a proper term, or where the set of variables contained in s is not the same as the set of variables contained in t. The set equations $x = x * x$ and $0 = 0 * x$ violate those restrictions.

As an example, take the equation $x = f(x * y)$ where f is a new uninterpreted function symbol from the empty theory. A unifier is $\{x \rightarrow f(0), y \rightarrow (1 + f(0))t\}$. Another example is $f(x) * f(y) = f(a)$. Here we get the unifier $\{x \rightarrow a, y \rightarrow a\}$. The first example demonstrates that a variable may be unifiable with a term that contains it, and the second that two terms with root symbols from different theories can be unifiable. As a third example consider $f(x) * f(y) = f(a) * f(b)$. Two unifiers are $\{x \rightarrow a, y \rightarrow b\}$ and $\{x \rightarrow b, y \rightarrow a\}$. This is of particular interest because there does not seem to exist a single mgu. This would be an example of a combination of two unitary theories (see [Si 84]) resulting in a non-unitary theory.

6 Relation to Other Work

As far as we are aware the unification problem for the complete class of set operators has not been treated in the literature. In [KRKaNa 85] the complexity of the unification problem for boolean rings is briefly considered but no explicit algorithm is given.

The exact relationship between our algorithm and the ones for commutative, associative and idempotent unification as in [LiSi 76] or [St 81] is not clear. We solve a different and more general problem but even in the case where the original equation contains only one set operator, e.g. union, our solution is still in terms of "$*$" and "$+$". We do not know whether one can derive the complete set of unifiers in the restricted theory from our solution.

Some Relationships between Unification, Restricted Unification, and Matching

Hans-Jürgen Bürckert
Universität Kaiserslautern, Fachbereich Informatik
Postfach 3049, D-6750 Kaiserslautern, W. Germany

Abstract:
We present restricted **T**-unification that is unification of terms under a given equational theory **T** with the restriction that not all variables are allowed to be substituted. Some relationships between restricted **T**-unification, unrestricted **T**-unification and **T**-matching (one-sided **T**-unification) are established. Our main result is that, in the case of an almost collapse free equational theory the most general restricted unifiers and for certain termpairs the most general matchers are also most general unrestricted unifiers; this does not hold for more general theories. Almost collapse free theories are theories, where only terms starting with projection symbols may collapse (i.e to be **T**-equal) to variables.

1 Introduction

Restricted **T**-unification is unification of terms under a given equational theory **T**, where only some of their variables are allowed to be substituted. Szabo mentioned this notion to be necessary for writing down some of his proofs more exactly (he used the notation 'partial **T**-unification' [Szabo82]). But he gave no definitions and he did not investigate the relationships of this notion with common ones. He described how to get unrestricted unification and matching as special cases. We present the necessary definitions and show some of the relationships between restricted **T**-unification, unrestricted **T**-unification and **T**-matching (one-sided **T**-unification). For that purpose we introduce the class of 'almost collapse free' equational theories - a slight generalization of the wellknown collapse free theories [Szabo82], [Yelick85], [Tiden85], [Herold85]. Collapse free theories have the property that no non-variable term is **T**-equal ('collapses') to a variable. In almost collapse free theories special terms starting with a projection symbol - a function symbol f for which the **T**-equation $f(v_1,...,v_i,...,v_n) =_T v_i$ holds - may be **T**-equal to variables. For those theories we show that every most general restricted **T**-unifier, and for term pairs, where one term is ground, also every most general **T**-matcher, is a most general unrestricted **T**-unifier. Additionally we present an algorithm to compute the most general restricted **T**-unifiers from the unrestricted ones. Some examples demonstrate that these results will not hold, if we admit arbitrary collapsing terms.
The results are of high significance in the area of theory classification known under the notion 'unification hierarchy' [Siekmann84+86]. Equational theories are classified by the cardinality of the most general unifier/matcher sets into unitary, finitary, infinitary and nullary unifying/matching theories. The result above shows that in the case of almost collapse free theories every infinitary (nullary) matching theory is also infinitary (nullary) unifying and every finitary (unitary) unifying theory is also finitary (unitary) matching. Some examples demonstrate that the matching classification can not be

References

[Fa 84] F. Fages: "Associative-Commutative Unification", Proc. 7th Int. Conf. on Automated Deduction, 1984, LNCS 170

[Hs 85] J. Hsiang: "Refutational Theorem Proving using Term-Rewriting Systems", Artificial Intelligence 25, 1985

[KRKaNa 85] A. Kandri-Rody, D. Kapur, P. Narendran: "An Ideal-theoretic Approach to Word Problems and Unification Problems over Finitely Presented Commutative Algebras"

[LiSi 76] M. Livesey, J. Siekmann: "Unification of Sets and Multisets", Memo Seki-76-II, Univ. Karlsruhe, 1976

[Si 84] J. H. Siekmann: "Universal Unification", Proc. 7th Int. Conf. on Automated Deduction, 1984, LNCS 170

[St 36] M. Stone: " The Theory of Representations for Boolean Algebra", Trans. AMS 40, 1936

[St 81] M.E. Stickel: "A Complete Unification Algorithm for Associative- Commutative Functions", JACM 28, No 3, 1981

[WaCo 80] D.E. Watts, J.K. Cohen: " Computer-Implemented Set Theory", Amer.Math.Month. 87, 1980

[Ye 85] K. Yelick: "Combining Unification Algorithms for Confined Regular Equational Theories", Proc. Rewriting Techniques and Applications, 1985, LNCS 202

obtained from the unification classification by replacing variables of one side of a unification problem by free constants. There might exist no free constants and addition of free constants might destroy decidability of the unification problems (see appendix).
There is some closer kinship between almost collapse free and collapse free theories: Every almost collapse free theory can be transformed into a collapse free theory without losing any information about the unification problems. This is a consequence of a result holding for arbitrary theories: All projection symbols can be removed by recursively replacing terms starting with such function symbols by the argument they are projecting to. In this way we can map every unification problem of a given theory to a problem of a theory containing no projections with essentially the same solution set as the original problem.
These theoretical investigations were triggered by some applications in automatic theorem proving [Loveland78]: Computation of matcher sets from unifier sets are used in several reduction mechanisms of clause graph theorem provers (for example subsumption and replacement factoring in the MKRP-system [Raph84]).
This paper is a draft of [Bürckert86], which contains all proofs.

2 Terms, Substitutions, Equations

In this section we summarize the common algebraic terminology [Grätzer79], [Burris&Sankappanavar83] used in unification theory [Siekmann84+86].
Given a signature $\mathbf{F}=\bigcup\mathbf{F}_n$ of finite sets $\mathbf{F}_n$ of *n-ary function* symbols ($n\geq 0$), and a denumerable set $\mathbf{V}$ of *variable* symbols, the *term algebra* $\mathbf{T}:=\mathbf{T}(\mathbf{F},\mathbf{V})$ is the least set with (i) $\mathbf{V}$, $\mathbf{F}_0\subseteq\mathbf{T}$ and (ii) $f\in\mathbf{F}_n$ and $t_1,\dots,t_n\in\mathbf{T} \Rightarrow f(t_1,\dots,t_n)\in\mathbf{T}$, together with the usual operations induced by the function symbols. $\mathbf{V}(O)$ denotes the set of variable symbols of an object O.
The *substitution set* $\mathbf{\Sigma}:=\mathbf{\Sigma}(\mathbf{F},\mathbf{V})$ of a term algebra is the set of endomorphisms σ on the term algebra with finite *domain* $\mathrm{Dom}\sigma:=\{v\in\mathbf{V}: \sigma v\neq v\}$. As usual they are represented by sets of variable-term pairs $\{v\leftarrow\sigma v: v\in\mathrm{Dom}\sigma\}$. $\mathrm{Cod}\sigma:=\{\sigma v\in\mathbf{T}: v\in\mathrm{Dom}\sigma\}$ is the *codomain* of the substitution σ, and $\mathrm{VCod}\sigma:=\mathbf{V}(\mathrm{Cod}\sigma)$ is the set of variables introduced by σ. For $V\subseteq\mathbf{V}$ the *V-restriction* $\sigma|_V$ of $\sigma\in\mathbf{\Sigma}$ is defined by $\sigma|_V v:=\sigma v$ ($v\in V$) and $\sigma|_V v:=v$ ($v\notin V$); $\mathbf{\Sigma}|_V$ is the set of all V-restrictions. A substitution $\varrho\in\mathbf{\Sigma}$ is a *V-renaming*, iff $\mathrm{Dom}\varrho=V$, $\mathrm{Cod}\varrho\subseteq\mathbf{V}\backslash V$ and $v\neq w \Rightarrow \varrho v\neq\varrho w$ ($\forall v,w\in V$). The *converse* ϱ^c of a V-renaming ϱ is defined by: $\varrho^c v=w$, if $v=\varrho w$ and $w\in\mathrm{Dom}\varrho$, and $\varrho^c v=v$ otherwise. The set of all V-renamings is Ren(V) [Herold83].

A finite set $\mathbf{T}:=\{(s\ ,\ t): s,t\in\mathbf{T}\}$ of term pairs is an *axiomatization* of an (equational) theory, the elements are the *axioms*. The *equational theory* is the least congruence relation $=_T$ on $\mathbf{T}$ containing the set $\{(\sigma s\ ,\ \sigma t): (s\ ,\ t)\in\mathbf{T},\ \sigma\in\mathbf{\Sigma}\}$. We only consider *consistent* theories, that are theories that do not collapse into a single equivalence class: $\forall v,w\in\mathbf{V}:\ v=_T w \Rightarrow v=w$.
We frequently call the axiomatization $\mathbf{T}$ itself a theory, and we write $s = t$ instead of $(s\ ,\ t)$ for the axioms, if no confusion is possible.
This algebraic notion of equational theories is equivalent to the common logical one: $s=_T t$ is denoted in first order (equational) logics by $\mathbf{T}\vdash s=t$, i.e. 's=t is *deducible* from $\mathbf{T}$'.

An *equation* $s=_T t$ is *regular*, iff $\mathbf{V}(s)=\mathbf{V}(t)$. Equations $t=_T v$ with $v\in\mathbf{V}$, $t\notin\mathbf{V}$ are called *collapse* equations. A theory is called *regular*, iff all equations are regular. A theory without collapse equations is called *collapse free*. Both properties are inherited from the axiomatization to the equational theory [Plonka69], [Yelick85], [Tiden85], [Herold86].

We need the following relations on terms s,t and substitutions σ,τ ($V\subseteq\mathbf{V}$): s is a $\mathbf{T}$-*instance* of t ($s\leq_T t$), iff $\exists\lambda\in\mathbf{\Sigma}$ with $s=_T\lambda t$ and s,t are $\mathbf{T}$-*equivalent* ($s\equiv_T t$), iff $s\leq_T t$ and $s\geq_T t$. Substitutions σ,τ are $\mathbf{T}$-*equal on V* ($\sigma=_T\tau$ [V]), iff $\sigma v=_T\tau v$ $\forall v\in V$. σ is a $\mathbf{T}$-*instance* of τ *on V* ($\sigma\leq_T\tau$ [V]), iff $\exists\lambda\in\mathbf{\Sigma}$ with $\sigma=_T\lambda\tau$ [V] and finally σ,τ are $\mathbf{T}$-*equivalent on V* ($\sigma\equiv_T\tau$ [V]), iff $\sigma\leq_T\tau$ [V] and $\sigma\geq_T\tau$ [V]. If $V=\mathbf{V}$, we drop the suffix [V]. The 'instance' relations are reflexive and transitive, and the 'equivalent' relations are in addition symmetric.

3 Unification, Restricted Unification, Matching

Unification theory is the general theory of solving equations [Huet76], [Szabo82], [Siekmann84+86] and as usual we are interested in computing a base of the solution space. Therefore we introduce the notion of the base of a substitution set as the set of its most general elements, that is, we admit only the $\leq_T$-maximal elements.

Let $\Sigma\subseteq\mathbf{\Sigma}$ be any set of substitutions and $W\subseteq\mathbf{V}$. Then $\mu\Sigma$ is a set of *most general substitutions on W* of Σ (or *base* or *μ-set on W*), iff the following conditions hold:

(B1) $\mu\Sigma\subseteq\Sigma$ (correctness)

(B2) $\forall\delta\in\Sigma\ \exists\sigma\in\mu\Sigma$ with $\delta\leq_T\sigma$ [W] (completeness)

(B3) $\forall\sigma,\tau\in\mu\Sigma$: $\sigma\leq_T\tau$ [W] $\Rightarrow\sigma=\tau$ (minimality)

The set $\mu\Sigma$ may not exist and if it exists, it is not unique. However, it is unique modulo the equivalence relation $\equiv_T$[W] [Huet76], [Fages&Huet83].

<u>3.Definition1:</u> For a pair of terms $s,t\in\mathbf{T}$ and a theory $\mathbf{T}$, the set $U\Sigma[s=_T t]:=\{\sigma\in\mathbf{\Sigma}: \sigma s=_T\sigma t\}$ is called the set of $\mathbf{T}$-*unifiers* of s and t or the *solution set* of the *unrestricted unification problem* $\langle s =_T t\rangle$. The set of *most general* $\mathbf{T}$-unifiers $\mu U\Sigma[s=_T t]$ is a base on $W:=\mathbf{V}(s,t)$ of $U\Sigma[s=_T t]$. We always choose a base $\mu U\Sigma[s=_T t]$ with the additional technical property:

(*) $\forall\sigma\in\mu U\Sigma[s=_T t]$: $Dom\sigma=\mathbf{V}(s,t)$ and $VCod\sigma\cap\mathbf{V}(s,t)=\emptyset$.

Note, that this is always fulfilled by an appropriate μ-set [Huet76], [Fages&Huet83]. ∎

Sometimes one is interested in substituting only into the variables of one side of an equation in order to solve it; for example, to find out, whether a term is an instance of another one. This is called matching.

<u>3.Definition2:</u> For $s,t\in\mathbf{T}$ and a theory $\mathbf{T}$ let $M\Sigma[s\leq_T t]:=\{\sigma\in\mathbf{\Sigma}: s=_T\sigma t\}$ be the set of $\mathbf{T}$-*matchers* of the *matching problem* $\langle s\leq_T t\rangle$. Its base on $W:=\mathbf{V}(t)$, the set of *most general* $\mathbf{T}$-matchers, is $\mu M\Sigma[s\leq_T t]$. The technical restriction is:

(*) $\forall\mu\in\mu M\Sigma[s\leq_T t]$: $Dom\mu\subseteq\mathbf{V}(t)$ and $VCod\mu\cap(\mathbf{V}(t)\setminus\mathbf{V}(s))=\emptyset$. ∎

Notice, that in general a matcher is not a special unifier. For example, the substitution $\mu=\{x\leftarrow f(x)\}$ is a matcher of the problem $\langle f(x)\leq_\emptyset x\rangle$, but not a unifier of the corresponding unification problem $\langle f(x) =_{\emptyset} x\rangle$: $\mu\in M\Sigma[f(x)\leq_{\emptyset}x]$ but $\mu\notin U\Sigma[f(x)=_{\emptyset}x]$ ($\emptyset$ is the empty theory).

Matching is closely related to semi-unification [Huet76], where the domain of a unifier is not allowed to contain the variables of one of the two terms. This kinship is particularly close for equations, where both sides have disjoint variable sets - especially if one side contains no variables (see 3.Proposition4(iii)). In certain applications it is necessary to generalize this concept by restricting the domain of the unifiers to arbitrary subsets V of the variables of the unification problem [Szabo82]. We call this V-restricted unification. It is somewhat similar to regarding the blocked variables of the problem as constants, a view often proposed in applications. However, this amounts to a change of the signature, which is not convenient and - as our definition shows - not necessary from a theoretical point of view. Moreover, such an extension of the signature might destroy the decidability of the solvability problem of **T**-unification (see appendix).

<u>3.Definition3:</u> For a *V-restricted unification problem* $\langle s =_T t , V\rangle$ the *V-restricted solution set*, or the set of *V-restricted* **T**-*unifiers* of $s,t\in\mathbf{T}$ is $U\Sigma|_V[s=_T t]:=\{\sigma\in\mathbf{\Sigma}: \sigma s=_T\sigma t, \mathrm{Dom}\sigma\subseteq V\}$ $(V\subseteq\mathbf{V}(s,t))$. Its base on $W:=\mathbf{V}(s,t)$ is $\mu U\Sigma|_V[s=_T t]$, the set of *most general* V-restricted **T**-unifiers of s and t. A special case is the *semi-unification problem* $\langle s =_T t , \mathbf{V}(t)\setminus\mathbf{V}(s)\rangle$.
Again we require a technical property:
(*) $\forall\sigma\in \mu U\Sigma|_V[s=_T t]$: $\mathrm{Dom}\sigma=V$ and $V\mathrm{Cod}\sigma\cap V=\emptyset$. ■

<u>Remarks:</u> 1. $U\Sigma|_V[s=_T t]\neq\emptyset$ is the *solvability problem* of V-restricted **T**-unification and $U\Sigma|_{\emptyset}[s=_T t]\neq\emptyset$ is just the *word problem* of the equational theory **T**.
2. Every semi-unifier is a matcher, but not conversely (see above).
3. Sometimes semi-unification is called matching and the more general definition of matching is not used. This seems to be allright for practical applications, because the instance problem will arise only for variable disjoint terms. ■

The following proposition shows some relationships between these three kinds of solving equations; especially that $\mathbf{V}(s,t)$-restricted unification and unrestricted unification are essentially the same and that semi-unification and matching (for variable disjoint terms) are closely related, but notice 5.Example5.

<u>3.Proposition4:</u>
(i) For $V\subseteq W\subseteq\mathbf{V}(s,t)$ we have $U\Sigma|_V[s=_T t]\subseteq U\Sigma|_W[s=_T t]\subseteq U\Sigma[s=_T t]$.
(ii) $U\Sigma|_{\mathbf{V}(s,t)}[s=_T t]$ is a complete subset of $U\Sigma[s=_T t]$, and thus
$\mu U\Sigma|_{\mathbf{V}(s,t)}[s=_T t]$ is a base of the unrestricted unification problem.
(iii) If $\mathbf{V}(s)\cap\mathbf{V}(t)=\emptyset$, then $\mu U\Sigma|_{\mathbf{V}(t)\setminus\mathbf{V}(s)}[s=_T t]$ is a complete subset of $M\Sigma[s\leq_T t]$.
(iv) If $\mathbf{V}(s)=\emptyset$, then $\mu U\Sigma[s=_T t]$ is a base of $M\Sigma[s\leq_T t]$ on $\mathbf{V}(t)=\mathbf{V}(s,t)$. ■

Note, that (iii) is less trivial than (ii), because of the different variable sets, the instance relations for the μ-sets are based on.

4 Almost Collapse Free Theories

We introduce the notion of projection equations and of almost collapse free theories, where projection equations are essentially the only collapse equations. We give some useful technical characterization of these theories, and we show that projections are superfluous in unification theory.

4.Definition1: Equations $p^{(i)}(v_1,\dots,v_i,\dots,v_n)=_T v_i$ (for some i with $1\le i\le n$) with pairwise different $v_1,\dots,v_n \in \mathbf{V}$ are called *projection* equations, $p^{(i)}\in \mathbf{F}_n$ is called a *projection* symbol, *projecting* to the i-th argument. The set of projection symbols induced by $\mathbf{T}$ is $\mathbf{P}_T$.
A theory is called *almost collapse free*, iff the leading function symbol of every collapse equation is a projection symbol. ■
Of course every collapse free theory is almost collapse free.
4.Example2: The theory $\mathbf{T}:=\{f(g(x)) = g(x),\ g(x) = x,\ h(f(x)) = x\}$ is almost collapse free. The function symbols are projection symbols: the equations $f(x)=_T x$, $g(x)=_T x$ and $h(x)=_T x$ hold. ■

Almost collapse free theories are characterized by some useful properties of terms and substitutions: Every term that is $\mathbf{T}$-equivalent to a variable is $\mathbf{T}$-equal to a variable, and every substitution that is $\mathbf{T}$-equivalent to a renaming is $\mathbf{T}$-equal to a renaming (note, that on their domain renamings are always $\mathbf{T}$-equivalent to the identity).
4.Lemma3: Let $\mathbf{T}$ be a theory. The following three statements are equivalent:
(i) $\mathbf{T}$ is almost collapse free.
(ii) $\forall t\in\mathbf{T}$: $t\equiv_T v$ for some $v\in\mathbf{V} \Rightarrow t=_T v'$ for some $v'\in\mathbf{V}$
(iii) $\forall \sigma\in\mathbf{\Sigma}$ with $\mathrm{Dom}\sigma\cap \mathrm{VCod}\sigma=\emptyset$ $\forall V\subseteq\mathbf{V}$: $\sigma\equiv_T \varepsilon\ [V] \Rightarrow \exists\varrho\in\mathrm{Ren}(W)$: $\sigma=_T\varrho\ [W]$, $W:=V\cap\mathrm{Dom}\sigma$. ■

These theories are also in some unification theoretical sense 'almost' collapse free: Every almost collapse free theory can be transformed (by a computable mapping on the terms) into a collapse free theory, such that the transformed unification problems have essentially the same bases as the original problems.
Given a theory $\mathbf{T}$ with the set $\mathbf{P}_T$ of its projection symbols, then we define a mapping $'$: $\mathbf{T}\to\mathbf{T}$, $t\mapsto t'$ recursively by
(i) $\forall v\in\mathbf{V}$: $v':=v$ and $\forall c\in\mathbf{F}_0$: $c':=c$
(ii) $\forall f\in\mathbf{F}_n\setminus\mathbf{P}_T$, $\forall t_1,\dots,t_n\in\mathbf{T}$: $(f(t_1,\dots,t_n))':=f(t_1',\dots,t_n')$ and
$\forall p^{(i)}\in\mathbf{F}_n\cap\mathbf{P}_T$, $\forall t_1,\dots,t_n\in\mathbf{T}$: $(p^{(i)}(t_1,\dots,t_n))':=t_i'$ $(n\ge1)$
and we extend this mapping to substitutions by $\forall\sigma\in\mathbf{\Sigma}$, $x\in\mathrm{Dom}\sigma$: $\sigma'x:=(\sigma x)'$.
The images of $\mathbf{T}$ and $\mathbf{\Sigma}$ are denoted $\mathbf{T}'$ and $\mathbf{\Sigma}'$. We apply this mapping to the axiomatization T and we get the axiomatization $T':=\{(l', r'): (l, r)\in T\}$. Thus this mapping removes all projection symbols and we may regard $\mathbf{T}'$ as term algebra with reduced signature $\mathbf{F}\setminus\mathbf{P}_T$. Then $\mathbf{\Sigma}'$ is the substitution set and T' is an axiomatization of an equational theory of this reduced term algebra. Both axiomatizations T and T' are equivalent on $\mathbf{T}'$, that is, they induce the same equational theory on this reduced term algebra:

<u>4.Lemma4:</u>

(i) $\forall s',t' \in \mathbf{T}': s' =_{T'} t' \Rightarrow s' =_{T} t'$ (i.e. $\mathbf{T}' \vdash s'=t' \Rightarrow \mathbf{T} \vdash s'=t'$).

(ii) $\forall s,t \in \mathbf{T}: s =_{T} t \Rightarrow s' =_{T'} t'$ (i.e. $\mathbf{T} \vdash s=t \Rightarrow \mathbf{T}' \vdash s'=t'$). ■

Now we can show that deletion of all projection symbols in this way will not affect the bases of the unifier sets of the unification problems, hence projection symbols are superfluous in a unification theoretical sense.

<u>4.Theorem5:</u> With the above notations we obtain:

(i) $U\Sigma'[s'=_{T'}t'] := \{\sigma' \in \mathbf{\Sigma}' : \sigma' s' =_{T'} \sigma' t'\}$ is a complete subset of $U\Sigma[s=_{T}t]$.

(ii) The base $\mu U\Sigma'[s'=_{T'}t']$ of $U\Sigma'[s'=_{T'}t']$ is also a base of $U\Sigma[s=_{T}t]$. ■

<u>4.Corollary6:</u> For each almost collapse free theory $\mathbf{T}$ there is a collapse free theory $\mathbf{T}'$, such that $\mathbf{T}$-unification and $\mathbf{T}'$-unification are related as above, that is (i) and (ii) of 4.Theorem5 hold. ■

The theorem is a generalization of a result in [Szabo82]:

> Collapse equations of the form $f(v) =_{T} v$ with $f \in \mathbf{F}_1$ are called *monadic*. They are superfluous in a model theoretic sense: For every theory $\mathbf{T}$ there is a theory $\mathbf{T}'$ without monadic collapse equations, but with essentially the same models as $\mathbf{T}$, that is they are 'definition-equivalent' (see also [Taylor79] for a more detailed definition of and some literature about definition-equivalence). It is easy to see that 4.Theorem5 holds for definition-equivalent theories $\mathbf{T}$ and $\mathbf{T}'$.

The theory of 4.Example2 is a regular theory with monadic collapse equations and it is definition-equivalent to the theory induced by the empty axiomatization. Thus unification in this theory is the same as syntactic unification.

These results can also be described with the notions of term rewriting [Huet&Oppen80]: The set of projection equations of a theory can be regarded as a canonical terms rewriting system. Then t' denotes just the normal form of a term t.

5 Relationhips Between μ-Sets

In this section we show the main result: In almost collapse free theories the most general V-restricted unifiers are most general unrestricted unifiers. Throughout this section we denote the blocked variables of a V-restricted unification problem by $V^c := \mathbf{V}(s,t) \setminus V$.

First we show, that a solvable unification problem has V-restricted $\mathbf{T}$-unifiers, iff there are some most general unrestricted unifiers being $\mathbf{T}$-equivalent on the blocked variables to the identity, in other words, the substitution of the blocked variables is not essential to solve the problem. We collect these unifiers in the set $U_V := \{\sigma \in \mu U\Sigma[s=_{T}t] : \sigma \equiv_{T} \varepsilon\ [V^c]\}$, for $s,t \in \mathbf{T}$ with existing $\mu U\Sigma[s=_{T}t]$ and for $V \subseteq \mathbf{V}(s,t)$.

<u>5.Lemma1:</u> Let $s,t \in \mathbf{T}$ with existing $\mu U\Sigma[s=_{T}t]$ and $V \subseteq \mathbf{V}(s,t)$. Then: $U_V \neq \emptyset \Leftrightarrow U\Sigma|_V[s=_{T}t] \neq \emptyset$. ■

In almost collapse free theories the elements of U_V are $\mathbf{T}$-equal to renamings (4.Lemma3), hence $U_V = \{\sigma \in \mu U\Sigma[s=_{T}t] : \sigma =_{T} \varrho_\sigma\ [V^c],\ \varrho_\sigma \in Ren(V^c)\}$. In this case we obtain a base of V-restricted unifiers by composing the converse of these renamings with the corresponding unifiers, hence the set $U^c:=\{(\varrho^c\sigma)|_V: \sigma\in U_V, \varrho=\varrho_\sigma\}$ is a base. This means that in almost collapse free theories the most general unrestricted unifiers differ from the restricted ones only by a renaming of the blocked variables.

5.Theorem2: Let $\mathbf{T}$ be almost collapse free and let $s,t\in\mathbf{T}$ with existing $\mu U\Sigma[s=_T t]$.

Then $U_V\neq\emptyset$ for $V\subseteq\mathbf{V}(s,t)$ implies U^c is a base of $U\Sigma|_V[s=_T t]$. ∎

Notice, that this can be generalized: Most general V-restricted unifiers are most general W-restricted unifiers, if $V\subseteq W\subseteq\mathbf{V}(s,t)$. In the lemma and theorem we can replace $\mu U\Sigma[s=_T t]$ by $\mu U\Sigma|_W[s=_T t]$.

The following corollary is an immediate consequence of 3.Proposition4(iii). It shows that the results of the lemma and the theorem also hold for matching problems with variable disjoint terms.

5.Corollary3: Let $s,t\in\mathbf{T}$ with existing $\mu U\Sigma[s=_T t]$. If $\mathbf{V}(s)\cap\mathbf{V}(t)=\emptyset$, then we obtain:

$U_V\neq\emptyset \Leftrightarrow M\Sigma[s\leq_T t]\neq\emptyset$ (with $V=\mathbf{V}(t)$ and $V^c=\mathbf{V}(s)$).

If in addition $\mathbf{T}$ is almost collapse free, then:

$U_V\neq\emptyset \Rightarrow U^c:=\{(\varrho^c\sigma)|_{\mathbf{V}(t)}: \sigma\in U_V\}$ is a complete subset of $M\Sigma[s\leq_T t]$. ∎

Remarks: 1. The results can be abbreviated by ('$\subseteq_T$' means 'subset modulo $=_T$'):

$V\subseteq W\subseteq\mathbf{V}(s,t) \Rightarrow \mu U\Sigma|_V[s=_T t]\subseteq_T\mu U\Sigma|_W[s=_T t]\subseteq_T\mu U\Sigma[s=_T t]$ (5.Theorem2)

$\mathbf{V}(s)\cap\mathbf{V}(t)=\emptyset \Rightarrow \mu M\Sigma[s\leq_T t]\subseteq_T\mu U\Sigma[s=_T t]$ (5.Corollary3).

2. If $\mathbf{T}$ is collapse free, then $U_V=\sigma\in\mu U\Sigma[s=_T t]: \sigma=\varrho_\sigma\ [V^c], \varrho_\sigma\in Ren(V^c)\}$, since being $\mathbf{T}$-equal to a renaming enforces being identical to a renaming. ∎

Let us illustrate these results by an example.

5.Example5: Let $\mathbf{T}:=\{fgx = fx\}$ with $f,g\in\mathbf{F}_1$; the theory is collapse free (for ease of notation we drop the parantheses for unary function symbols and abbreviate multiple nestings of the same function symbol by exponents: $f^0x:=x$, $f^{n+1}x:=f(f^nx)$ for $f\in\mathbf{F}_1$). We consider the terms $s:=fx$ and $t:=fy$.

1. The unification problem $\langle s =_T t\rangle$ has a base $\mu U\Sigma[s=_T t]=\{\sigma_{00}\}\cup\{\sigma_{nm}: n,m>0, n\neq m\}$ with $\sigma_{nm}:=\{x\leftarrow g^n v_{nm}, y\leftarrow g^m v_{nm}\}$, where $v_{nm}\in\mathbf{V}\setminus\{x,y\}$ are pairwise different $(n,m\geq 0)$.

2. For the semi-unification problem $\langle s =_T t, \{y\}\rangle$ we obtain $U_{\{y\}}=\{\sigma_{0m}: m\geq 0\}$, and a base is $U^c=\{\tau_m: m\geq 0\}$ with $\tau_m=(\varrho_m{}^c\sigma_{0m})|_{\{y\}}=\{y\leftarrow g^m x\}$, where $\varrho_m=\{x\leftarrow v_{0m}\}$ $(m\geq 0)$.

3. The matching problem $\langle s \leq_T t\rangle$ has the same sets $U_{\{y\}}$ and U^c, but it has a base with a single matcher: $\mu M\Sigma[s\leq_T t]=\{\mu\}$ with $\mu=\tau_0=\{y\leftarrow x\}\in U^c$. Every other element of U^c is a $\mathbf{T}$-instance of μ on $\mathbf{V}(t)=\{y\}$: $\tau_m=_T\lambda\mu\ [\{y\}]$ with $\lambda=\{x\leftarrow g^m x\}$ $(m\geq 1)$.

This also demonstrates that we cannot get minimality for U^c in 5.Corollary3. ∎

If we drop the 'almost collapse free' requirement, the following example is a semi-unification problem contradicting both 5.Corollary3 and 5.Theorem2: the most general unrestricted unifiers differ (in general) from the restricted ones not only by a renaming of the blocked variables.

5.Example6: Let $\mathbf{T}:=\{f(x,x) = x\}$ and consider the terms $s:=x$ and $t:=f(y,z)$ and the substitutions $\sigma:=\{x\leftarrow f(u,v), y\leftarrow u, z\leftarrow v\}$ and $\mu:=\{y\leftarrow x , z\leftarrow x\}$. Then $\sigma\in\mu U\Sigma[s=_T t]$, and $\mu\in\mu M\Sigma[s\leq_T t]$, but $\mu\leq_T\sigma$ $[\mathbf{V}(s,t)]$ with $\lambda=\{u\leftarrow x , v\leftarrow x\}$ and $\mu\neq_T\sigma$ $[\mathbf{V}(s,t)]$, i.e. the most general matcher (semi-unifier) is a proper instance of the most general unifier. Note, that 5.Lemma1 holds, since $\sigma\in U_{\{y,z\}}$. However, σ is not $\mathbf{T}$-equal to a renaming on $V^c:=\{x\}$. ∎

The next example shows, that 5.Corollary3 does not hold for arbitrary matching problems. There may exist more most general matchers than most general unifiers and in addition the latter may be proper instances of the former, if the terms have common variables. Note that the theory is collapse free.

5.Example7: Let $\mathbf{T}:=\{f(x,y) = f(y,x)\}$, with the terms $s:=f(g(x),y)$ and $t:=f(x,z)$ and with the substitutions $\sigma:=\{x\leftarrow u , y\leftarrow u , z\leftarrow g(u)\}$, $\mu_1:=\{x\leftarrow g(x) , z\leftarrow y\}$ and $\mu_2:=\{x\leftarrow y , z\leftarrow g(x)\}$ we obtain: $\sigma\in\mu U\Sigma[s=_T t]$ and $\mu_1,\mu_2\in\mu M\Sigma[s\leq_T t]$ and $\sigma\leq_T\mu_2$ $[\mathbf{V}(s,t)]$ with $\lambda=\{x\leftarrow u , y\leftarrow u\}$, but not conversely. On the other hand σ and μ_1 are not comparable in $\leq_T$ and σ is the only most general unifier. Note, that the corresponding semi-unification problem $\langle s =_T t , \mathbf{V}(t)\backslash\mathbf{V}(s)\rangle$ has no solution. ∎

6 Consequences and Applications

Depending on the cardinality of the μ-sets we classify the unification problems and the theories. This is known as *unification hierarchy* [Siekmann84+86].

6.Definition1:

(i) A solvable unification problem is called *nullary*, iff the μ-set does not exist (the cardinality is null). It is called *unitary/finitary/infinitary*, iff the μ-set exists and its cardinality is one/finite/infinite.

(ii) A theory is *unitary/finitary unifying*, iff every solvable unification problem is unitary/finitary. It is called *nullary/infinitary unifying*, iff at least one solvable unification problem is nullary/infinitary. ∎

Analogously we define this for matching and restricted unification.

The following theorem describes the relationship between the hierarchy classes of restricted and unrestricted unification problems.

6.Theorem2:

Let $\mathbf{T}$ be an almost collapse free theory.

(i) If $\mathbf{T}$ is nullary restricted unifying, then $\mathbf{T}$ is nullary unifying.

If in addition the μ-set exists for each problem (that is $\mathbf{T}$ is not nullary), then we have the following hierarchy results:

(ii) If $\mathbf{T}$ is unitary/finitary unifying, then $\mathbf{T}$ is unitary/finitary restricted unifying.

(iii) If $\mathbf{T}$ is infinitary restricted unifying, then $\mathbf{T}$ is infinitary unifying. ∎

Notice, that these results hold especially for semi-unification.

Remarks: 1. The converses of 6.Corollary3 do not hold in general. Szabo gives an infinitary unifying, but unitary semi-unifying theory [Szabo82].

2. The results for semi-unification are covered by some results in [Szabo82].
   - Every unitary unifying theory is unitary semi-unifying.
   - Every nullary semi-unifying theory is nullary unifying.

Here the theories need not to be almost collapse free! However, Szabo's proof of the nullary case is incomplete, and it is based upon the idea of replacing the blocked variables by ground terms, which is not possible in general (see 4. below and the appendix example)

3. We cannot get the nullary hierarchy result for matching, since we obtain only completeness of the matcher sets constructed by our method. But of course we have the other results provided all the bases exist.

4. Occasionally there has been the suggestion to simply replace the blocked variables by some ground terms in order to prove such hierarchy results. That this does not work in general is shown by the following counterexample (see also the example in the appendix):

   Let **F** consist of a single constant c, unary functions f,g,h, and let **F** contain no further functions at all (we again drop the parantheses). Consider the variable disjoint terms $s:=ghy$ and $t:=gx$ and the collapse free theory $\mathbf{T}:=\{gfx = gx, fc = c, hc = c\}$. Then the semi-unification problem $\langle s =_T t\ ,\ \{x\}\rangle$ has an infinite base $\mu U\Sigma|_{\{x\}}[s=_T t]:=\{\sigma_n : n\geq 0\}$ with $\sigma_n:=\{x\leftarrow f^n hy\}$ $(n\geq 0)$, that is, the theory **T** is infinitary semi-unifying. In order to show with the above idea that **T** is also infinitary unifying, we must replace the variable of s by a ground term to get an infinitary unification problem. But with the only existing ground substitution $\gamma=\{y\leftarrow c\}$ (all ground terms are **T**-equal to c) we get: $\mu U\Sigma[\gamma s=_T t]=\mu U\Sigma[ghc=_T gx]=\{\{x\leftarrow c\}\}$. Hence the corresponding unification problem is unitary. However, the theory is of course infinitary unifying by 6.Theorem2(iii). An infinitary unification problem will be given by the original terms s and t. ■

From the main theorem of the last section we infer an algorithm to compute most general restricted unifiers from most general unifiers. This implies, that for every almost collapse free theory with an existing minimal unification algorithm (that is an algorithm computing a base for every solvable unification problem) there is also a minimal restricted unification algorithm.

Algorithm Unifier_to_Restricted_Unifier

<u>Input:</u> a (finite) base of unifiers of a unification problem $<s =_T t>$ and a subset V of the variables of s and t

<u>Output:</u> a base of V-restricted unifiers of $<s =_T t\ ,\ V>$, if the problem is solvable, and FAILURE, otherwise

- <u>If</u> there is no most general unifier with a V-renaming part, <u>then</u> <u>return</u> FAILURE.
- <u>Else</u> <u>for</u> each most general unifier with a V-renaming part <u>do</u>:
  remove the V-renaming part, and apply the converse of the V-renaming to the codomain of the rest.
- <u>Return</u> all changed unifiers.

This algorithm is particularily useful for clause graph theorem proving procedures [Kowalski75] like the MKRP-system [Raph84] at Kaiserslautern:

In clause graph procedures the clause sets are transformed into graphs with
- nodes labelled with the literals of the clauses
- arcs between nodes labelled by unifiable literals (with opposite sign) of different clauses (*resolution links*)
- arcs between unifiable literals (with same sign) of different clauses (*subsumption links*)

The resolution links are labelled by $\mu U\Sigma$-sets and refer to resolution possibilities. The subsumption links support application of the subsumption rule [Loveland78]: If there are two clauses C,D and a substitution μ with $\mu C \subseteq D$, then the clause D can be removed. This can be extended to clause graphs using the above subsumption links [Eisinger81]. Therefore these links should be labelled by semi-unification bases, but since the direction of the semi-unfication problem is not known in advance, the links are also labelled by $\mu U\Sigma$-sets and the semi-unification bases are computed dynamically. This computation can be done with the above algorithm.

7 Conclusions

We have seen that for almost collapse free theories the most general restricted unifiers can be computed from a set of most general unrestricted unifiers. An open question is, whether this can be done in the general case. By 5.Lemma1 we can decide the restricted unification problem, if the minimal solution set of the unrestricted unification problem exists and is finite (if the unification problem is infinitary, we obtain at least a semi-decision procedure for the restricted problem). A proof of this lemma gives some hints for computing restricted unifiers from those most general unifiers that are equivalent to the identity on the blocked variables: we have to instantiate them and restrict the instances on the unblocked variables. Finding out the appropriate instantiations - for almost collapse free theories, these are the renamings - might also lead to a minimal solution set. Solving this problem then would also yield the still missing hierarchy results for theories with arbitrary collapse equations.
Our result on the unification theoretical relationship between almost collapse free and collapse free theories (4.Corollary6) affects also the problem of combining unification algorithm of theories with disjoint function sets [Yelick85], [Tiden85], [Herold86]. This is still only solved for collapse free theories and our result yields, that the collapse free requirement can be weakened by admitting projection equations.
Acknowledgements: I want to thank my colleagues A. Herold and M. Schmidt-Schauß for their support and helpful discussion during the preparation of this work. M. Schmidt-Schauß had the basic idea for the example presented in the appendix. Special thanks go to my supervisor J. Siekmann, who read two drafts of this paper. His helpful criticisms and hints contributed much to its present form.
This research was supported by the Bundesministerium für Forschung und Technologie of FR Germany and by the NIXDORF Computer AG within the framework of the Joint Research Project ITR-8501 A.

References

Burris S.& Sankappanavar H.P. [1983]: A Course in Universal Algebra; Springer, New York - Heidelberg - Berlin
Bürckert H.-J. [1986]: Some Relationships Between Unification, Restricted Unification, and Matching; SEKI-Report, Universität Kaiserslautern
Eisinger N. [1981]: Subsumption and Connection Graphs; Proc. IJCAI-81, Vancouver
Fages F. & Huet G. [1983]: Complete Sets of Unifiers and Matchers in Equational Theories; Proc. Caap-83, Springer Lec.Notes Comp. Sci. 159
Grätzer G. [1979]: Universal Algebra; Springer, New York - Heidelberg - Berlin

Herold A. [1983]: Some Basic Notions of First-Order Unification Theory, SEKI-Memo, Univ. Karlsruhe
Herold A. [1985]: A Combination of Unification Algorithms; SEKI-Memo, Univ. Kaiserslautern
Huet G. [1976]: Resolution d'Equations dans les Languages d'Ordre 1,2,...,ω; These d'Etat, Univ. Paris VIII
Huet G. & Oppen D.C. [1980]: Equations and Rewrite Rules (Survey); Technical Report, SRI International
Kowalski R. [1975]: A Proof Procedure Using Connection Graphs; J. ACM 22,4
Loveland D. [1978]: Automated Theorem Proving; North-Holland, Amsterdam - New York - Oxford
Plonka J. [1969]: On Equational Classes of Abstract Algebras Defined by Regular Equations; Fund. Math. LXIV
Raph K.M.G. [1984]: The Markgraf Karl Refutation Procedure (MKRP); SEKI-Memo, Univ. Kaiserslautern
Siekmann J. [1984]: Universal Unification (Survey); Proc. 7th CADE, Springer Lec.Notes Comp. Sci. 170
Siekmann J. [1986]: Unification Theory; Proc. 7th ECAI, Brighton, to appear
Szabo P. [1982]: Unifikationstheorie Erster Ordnung (in german); Dissertation, Univ. Karlsruhe
Taylor W. [1979]: Equational Logic (Survey); Houston J. Math. 5
Tiden E. [1985]: Unification in Combination of Theories with Disjoint Sets of Function Symbols; Royal Inst. Techn., Dep. Comp. Sci., Stockholm
Yelick K. [1985]: Combining Unification Algorithms for Confined Regular Equational Theories; Proc. Rewriting Technics and Applications, Springer Lec.Notes Comp. Sci. 202

Appendix

We consider unification under distributivity and associativity and some extension to obtain an interesting undecidability result.
Given the signature $\mathbf{F}'=\mathbf{F}'_0\cup\mathbf{F}'_2$ with $\mathbf{F}'_2=\{+,\times\}$ and $\mathbf{F}'_0=\{c\}$ (+ and × are written in infix notation), then we define the distributivity laws and the associativity law to be the theories:

$\mathbf{D} := \{ x\times(y+z) = (x\times y)+(x\times z),\ (x+y)\times z = (x\times z)+(y\times z) \}$ (distributivity)

$\mathbf{A} := \{ x+(y+z) = (x+y)+z \}$ (associativity)

The solvability problem of unification of $\mathbf{F}'$-terms in the theory $\mathbf{DA}:=\mathbf{D}\cup\mathbf{A}$ of distributivity and associativity is known to be undecidable [Szabo82].
Now given another signature $\mathbf{F}=\mathbf{F}_0\cup\mathbf{F}_3$ with $\mathbf{F}_0=\{a\}$ and $\mathbf{F}_3=\{f,g\}$, we define some generalizations of the distributivity and the associativity axioms for binary function symbols to ternary function symbols:

$\mathbf{D3} := \{ f(g(x,y,v),z,v) = g(f(x,z,v),f(y,z,v),v),\ f(x,g(y,z,v),v) = g(f(x,y,v),f(x,z,v),v) \}$

$\mathbf{A3} := \{ g(g(x,y,v),z,v) = g(x,g(y,z,v),v) \}$

Then we consider the collapse free theory $\mathbf{T} := \mathbf{D3}\cup\mathbf{A3}\cup\mathbf{Ta}$ with

$\mathbf{Ta} := \{ f(x,y,a) = a,\ f(x,y,f(u,v,w)) = a,\ f(x,y,g(u,v,w)) = a,$
$g(x,y,a) = a,\ g(x,y,f(u,v,w)) = a,\ g(x,y,g(u,v,w)) = a \}.$

In this theory every term starting with a ternary function symbol (a *complex* term) is **T**-equal to the constant a, if its third top argument is a non-variable. Every **T**-unification problem constructed with the signature **F** is solvable: We substitute all variables of the problem by the constant a. Then both terms will become **T**-equal to the constant a by the subtheory **Ta**. Hence unification under this theory is decidable within the given signature.
But, if we introduce a new free constant, say b, unification in this theory will become undecidable. We can reduce a subset of the unification problems to the unfication under distributivity and associativity of binary function symbols introduced above. Therefore we consider the subset $\mathbf{T}_b$ of the term algebra $\mathbf{T}(\mathbf{F}\cup\{b\},\mathbf{V})$ with the extended signature $\mathbf{F}\cup\{b\}$, defined by:

(i) $a,b\in\mathbf{T}_b$ and $\mathbf{V}\subseteq\mathbf{T}_b$

(ii) $t_1,t_2\in\mathbf{T}_b \Rightarrow f(t_1,t_2,b),g(t_1,t_2,b)\in\mathbf{T}_b$.

That is, $\mathbf{T}_b$ is the subset of $\mathbf{F}\cup\{b\}$-terms, where the third argument of every complex term is only allowed to be the constant b. Every **T**-unification problem $\langle s =_T t\rangle$ built up by those terms is solvable, iff the **DA**-unification problem $\langle s' =_{DA} t'\rangle$ is solvable. Here we obtain s′ and t′ from s and t by the following mapping:

(i) $a \mapsto c,\ b \mapsto c$ and $v \mapsto v\ \forall v\in\mathbf{V}$

(ii) $f(t_1,t_2,b) \mapsto t_1\times t_2,\ g(t_1,t_2,b) \mapsto t_1+t_2\ \forall t_1,t_2\in\mathbf{T}_b$.

Thus in this theory unification will become undecidable, if we introduce new constants.
An analogous reduction will demonstrate that restricted unification is not necessarily decidable, when unrestricted unification is.

A CLASSIFICATION OF MANY-SORTED UNIFICATION PROBLEMS

Christoph Walther, Institut für Informatik I
Universität Karlsruhe, Karlsruhe, W. Germany

Abstract. *Many-sorted unification is considered, i.e. unification in the many-sorted free algebras of terms, where variables as well as the domains and ranges of functions are restricted to certain subsets of the universe, given as a potentially infinite hierarchy of sorts. Many-sorted unification is the same as solving an equation in the corresponding heterogeneous 'order-sorted' algebra /7/ rather than in a homogeneous algebra. It is proved that many-sorted unification can be classified completely by conditions imposed on the structure of the sort hierarchy. It is shown that complete and minimal sets of unifiers may not always exist for many-sorted unification. Conditions for sort hierarchies which are equivalent to the existence of these sets with one, finitely many or infinitely many elements are presented. It is also proved that being a forest-structured sort hierarchy is a necessary and sufficient criterion for the Robinson Unification Theorem to hold for many-sorted unification, i.e. a criterion for the existence of a most general unifier without 'auxiliary variables'. An algorithm for many-sorted unification is given. This paper generalizes and extends the results presented in /22/. It is a shortened version of the technical report /24/, to which the reader is referred for any omitted proofs.*

1. Formal Preliminaries

Given pairwise disjoint alphabets, the infinite set of **variable symbols** $\mathcal{V}$ and the set of **function symbols** $\mathcal{F}$ together with an arity-function for them, we let $\mathcal{T}$ denote the set all **well formed terms** over $\mathcal{V}$ and $\mathcal{F}$. For a term t, $\mathcal{V}(t)$ denotes the set of all variable symbols in t, and for a set D of terms, $\mathcal{V}(D)$ is defined as $\cup_{t\in D}\mathcal{V}(t)$.

A mapping σ from $\mathcal{V}$ to $\mathcal{T}$ with $\sigma x=x$ almost everywhere is called a **substitution**. Substitutions are extended as **endomorphisms** to mappings from $\mathcal{T}$ to $\mathcal{T}$. ε denotes the **identity** substitution, i.e. $\varepsilon x=x$ for all $x\in\mathcal{V}$. We define $\sigma D=\{\sigma t \mid t\in D\}$ for subsets D of $\mathcal{T}$. The **domain** of σ, denoted $DOM(\sigma)$, is given by $\{x\in\mathcal{V} \mid \sigma x\neq x\}$ and the **codomain** $COD(\sigma)$ of σ is defined as $\sigma DOM(\sigma)$. Substitutions will be sometimes denoted by a finite set of variable-term pairs $\{x_1\leftarrow t_1,\ldots,x_n\leftarrow t_n\}$. SUB is the set of all substitutions. A **separation** of σ is a sequence $\sigma_1,\ldots,\sigma_n$ of substitutions with pairwise disjoint domains and codomains satisfying (1) $\sigma=\sigma_1\circ\ldots\circ\sigma_n$, (2) $COD(\sigma_i)=\{y_i\}\subset\mathcal{V}$ for each i with $1\leq i<n$, and (3) $COD(\sigma_n)\cap\mathcal{V}=\emptyset$ (where $\circ$ denotes functional composition).

A subset V of $\mathcal{V}$ induces an equivalence relation $=[\![V]\!]$ on SUB by $\sigma=\rho[\![V]\!]$ iff $\sigma x=\rho x$ for all $x\in V$. The quasi-ordering $\leq[\![V]\!]$ of **subsumption** in SUB over some $V\subset\mathcal{V}$ is given as $\sigma\leq\theta[\![V]\!]$ iff $\lambda\circ\sigma=\theta[\![V]\!]$ for some $\lambda\in SUB$. $\leq[\![V]\!]$ induces an equivalence relation $\equiv[\![V]\!]$ on SUB by $\sigma\equiv\rho[\![V]\!]$ iff $\sigma\leq\rho[\![V]\!]$ and $\rho\leq\sigma[\![V]\!]$. $[\![V]\!]$ is omitted in the above definitions if $V=\mathcal{V}$.

Given a finite subset $D=\{t_1,\ldots,t_n\}$ of $\mathcal{T}$, a substitution θ **unifies** D iff $|\theta D|=1$, i.e. $\theta t_1=\ldots=\theta t_n$. In this case we say D is **unifiable** and θ is a **unifier** of D. $U(D)$ is the set of all unifiers of D. A unifier σ of D is called a **most general unifier** (or **mgu** for short) iff $\sigma\leq\theta$ for each $\theta\in U(D)$. For any mgu σ of D we have (1) $\sigma\circ\sigma=\sigma$, (2) $DOM(\sigma)\cup\mathcal{V}(COD(\sigma))\subset\mathcal{V}(D)$, (3) $\theta=\theta\circ\sigma$ for each $\theta\in U(D)$, and (4) $\sigma\equiv\tau$ iff τ is also an mgu of D, cf. [4,5,9]. Each unifiable subset D of $\mathcal{T}$ induces an equivalence relation

$\underset{D}{\sim}$ on $\mathcal{T}\times\mathcal{T}$ by $q\underset{D}{\sim}r$ iff $\sigma q=\sigma r$ for some mgu σ of D. By the most generality of σ, this definition is independent of σ. $\mathcal{T}/\underset{D}{\sim}$ denotes the quotient set of $\mathcal{T}$ modulo $\underset{D}{\sim}$. The following lemma is used subsequently:

Lemma 1.1 Let $D\subset\mathcal{T}$ be unifiable and $M\in\mathcal{T}/\underset{D}{\sim}$. Then (1) $|M|<\infty$ and (2) $U(D)\subset U(M)$.

Proof (1) For each $t\in\mathcal{T}$ we let $c(t)$ denote the term complexity of t, i.e. the number of nodes of the tree associated with t. For each mgu σ of D, we have $|\sigma M|=1$, hence there is some $n\in\mathbb{N}$ such that $c(\sigma t)\leq n$ for all $t\in M$, and with $c(t)\leq c(\sigma t)$, $c(t)\leq n$ for all $t\in M$. Now assume $|M|=\infty$ for some $M\in\mathcal{T}/\underset{D}{\sim}$ such that $n\in\mathbb{N}$ with $c(t)\leq n$ (for all $t\in M$) is minimal and let $M'=M\setminus DOM(\sigma)$. Obviously $|M'|=\infty$, hence $a\notin M'$ for all constant- and variable symbols a, because with $\sigma a=a$ we would obtain $|M'|=1$ otherwise. But then $M'=\{f(t_1^i\ldots t_k^i)\mid i\in\mathbb{N}\}$ for some $f\in\mathcal{F}$. With $|M'|=\infty$, $T_j=\{t_j^i\mid i\in\mathbb{N}\}$ must be infinite for some $j\in\{1,\ldots,k\}$. With $M'\subset M\in\mathcal{T}/\underset{D}{\sim}$, we obtain $T_j\in\mathcal{T}/\underset{D}{\sim}$. But then $|T_j|<\infty$, because $c(t_j^i)<n$ and n was choosen minimal. ▽☒ (2) Obvious. ☒☒

A sort hierarchy is a pair $(\mathcal{S},<_{\mathcal{S}})$ such that $\mathcal{S}$ is a non-empty set partially ordered by $<_{\mathcal{S}}$. The members of $\mathcal{S}$ are called sort symbols and $<_{\mathcal{S}}$ is the subsort order of $\mathcal{S}$. If $s_1<_{\mathcal{S}}s_1$ or $s_1=s_1$, one writes $s_1\leq_{\mathcal{S}}s_2$, and says s_1 is a subsort of s_2. If $\mathcal{S}$ is known from the context we shall omit the indices, e.g. we write $<$ for $<_{\mathcal{S}}$.

A sort hierarchy $(\mathcal{S},<_{\mathcal{S}})$ is a forest structure iff $s_1\geq s\leq s_2$ implies $s_1\geq s_2$ or $s_1\leq s_2$ for all $s_1,s_2,s\in\mathcal{S}$. Following the standard terminology of lattice theory, e.g.[1], we define for each $s_0\in\mathcal{S}$, each $S\subset\mathcal{S}$ and each $s\in S$: s_0 is the least element of S iff $s_0\in S$ and $s_0\leq s$. s_0 is a maximal element of S iff $s_0\in S$ and $s_0\nleq s$. s_0 is a lower bound of S iff $s_0\leq s$. s_0 is the greatest lower bound or infimum of S, denoted $\sqcap S$, if additionally $s'\leq s_0$ for each lower bound s' of S.

The set of all maximal elements in S will be denoted max(S) and lbs(S) is the set of all lower bounds of S in $\mathcal{S}$, i.e. $max(S)=\{s_0\in S\mid s_0\nleq s\ \forall s\in S\}$ and $lbs(S)=\{s_0\in\mathcal{S}\mid s_0\leq s\ \forall s\in S\}$. S is a chain in $\mathcal{S}$ iff S is totally ordered by $<$, i.e. $s_1\leq s_2$ or $s_1\geq s_2$ for all $s_1,s_2\in S$. $(\mathcal{S},<_{\mathcal{S}})$ is called a meet-semilattice iff each pair of sort symbols s_1 and s_2 has an infimum.

For a sort hierarchy $(\mathcal{S},<)$ let $\mathcal{S}^*$ denote the set of all finite strings from $\mathcal{S}$ including the empty string e. For each $s\in\mathcal{S}$ and each $w\in\mathcal{S}^*$ let $\mathcal{V}_s$ be an infinite set of variable symbols and let $\mathcal{F}_{w,s}$ be a set of function symbols such that all these sets are pairwise disjoint. An $\mathcal{S}$-sorted signature Σ is a family $\Sigma_{w,s}$ of sets such that $\Sigma_{w,s}=\mathcal{V}_s\cup\mathcal{F}_{w,s}$. Setting $\mathcal{V}=\cup\mathcal{V}_s$ and $\mathcal{F}=\cup\mathcal{F}_{w,s}$ for each $s\in\mathcal{S}$ and each $w\in\mathcal{S}^*$ we define terms as in the unsorted case, where the arity of each $f\in\mathcal{F}_{w,s}$ is given by the length of the string w.

For a function symbol $f\in\mathcal{F}_{s(1)\ldots s(k),s}$, the ith domainsort of f, denoted $[f]_i$, is $s(i)$, provided $1\leq i\leq k$. The sort $[t]$ of a term t is s iff $t\in\mathcal{V}_s$ or $t=f(\ldots)$ and $f\in\mathcal{F}_{w,s}$ for some $w\in\mathcal{S}^*$. For $D\subset\mathcal{T}$, $[D]$ is the sortal image of D defined as $\{[t]\in\mathcal{S}\mid t\in D\}$. Since the sort of a term is determined only by the outermost symbol of t, we have $[\sigma t]=[t]$

and $[\sigma D]=[D]$ for each $\sigma\in SUB$, $t\in\mathcal{T}\setminus\mathcal{V}$ and $D\subset\mathcal{T}\setminus\mathcal{V}$.

For an $\mathcal{S}$-sorted signature Σ, the set $\mathcal{T}_\Sigma$ of all **well-sorted terms** (or **Σ-terms** for short) is the smallest subset of $\mathcal{T}$ satisfying (1) $\mathcal{V}_s\cup\mathcal{F}_{e,s}\subset\mathcal{T}_\Sigma$ for each $s\in\mathcal{S}$ and (2) $f(q_1...q_k)\in\mathcal{T}_\Sigma$, if $q_i\in\mathcal{T}_\Sigma$ and $[q_i]\leq[f]_i$ for each i with $1\leq i\leq k$. These notions are consistent with the standard terminology: for $<_\mathcal{S}=\emptyset$, i.e. $s_1\leq_\mathcal{S}s_2$ iff $s_1=s_2$, $\mathcal{T}_\Sigma$ is the **free Σ-algebra generated by** $\mathcal{V}$, denoted $\mathcal{T}(\Sigma\cup\mathcal{V})$ in [10] or $\mathcal{T}_\Sigma(\mathcal{V})$ in [8]. $\mathcal{T}_\Sigma$ is the **Σ-word algebra on** $\mathcal{V}$, denoted $W_\Sigma(\mathcal{V})$ in [2], if $\mathcal{S}$ is a singleton.

A substitution σ satisfying $\sigma\mathcal{T}_\Sigma\subset\mathcal{T}_\Sigma$ is called **well-sorted** or a **Σ-substitution**. A more useful characterization is $[\sigma x]\leq[x]$ and $\sigma x\in\mathcal{T}_\Sigma$ for all $x\in\mathcal{V}$ and it can be proved that both definitions are equivalent. SUB_Σ is the set of all Σ-substitutions. Again these notions are in accordance with the many-sorted unordered case, where a substitution is defined as a **sort-preserving** Σ-endomorphism from $\mathcal{T}_\Sigma$ to $\mathcal{T}_\Sigma$ [10], thus guaranteeing $[\sigma x]=[x]$ and $\sigma x\in\mathcal{T}_\Sigma$ for each $x\in\mathcal{V}$.

The quasi-ordering $\leq_\Sigma[V]$ of **Σ-subsumption** in SUB_Σ over some $V\subset\mathcal{V}$ is given as $\sigma\leq_\Sigma\theta[V]$ iff $\lambda\circ\sigma=\theta[V]$ for some $\lambda\in SUB_\Sigma$ and the corresponding equivalence relation $\equiv_\Sigma[V]$ is defined by $\sigma\equiv_\Sigma\rho[V]$ iff $\sigma\leq_\Sigma\rho[V]$ and $\rho\leq_\Sigma\sigma[V]$.

Given a finite subset D of $\mathcal{T}_\Sigma$ and a substitution θ, θ is a **Σ-unifier** of D and D is **Σ-unifiable** by θ iff θ is well-sorted and unifies D. $U_\Sigma(D)=U(D)\cap SUB_\Sigma$ is the set of all Σ-unifiers of D. σ is a **well-sorted most general unifier** or a **Σ-mgu** of D iff $\sigma\in SUB_\Sigma$ is an mgu of D, which obviously is equivalent with $\sigma\leq_\Sigma\theta$ for all $\theta\in U_\Sigma(D)$.

But unfortunately the notion of a most general unifier is too strong to provide us with all the necessary concepts when unification is generalized to the many-sorted case (with $<_\mathcal{S}\neq\emptyset$). The situation is similar (but not identical) to that of unification under equational theories, where the notion of an mgu is replaced by the concept of complete and minimal sets of unifiers [5,13,17]: $U\subset U_\Sigma(D)$ is a **complete** set of Σ-unifiers of some $D\subset\mathcal{T}_\Sigma$ iff for each $\theta\in U_\Sigma(D)$ there is some $\sigma\in U$ such that $\sigma\leq_\Sigma\theta[\mathcal{V}(D)]$. $U\subset U_\Sigma(D)$ is **minimal** iff $\sigma\leq_\Sigma\rho[\mathcal{V}(D)]$ implies $\sigma=\rho$ for all $\sigma,\rho\in U$.

2. Some Many-Sorted Unification Problems

The following examples should serve to demonstrate the problems associated with many-sorted unification and also to illustrate the results presented in the subsequent sections. We show how the existence and cardinality of complete and minimal sets of Σ-unifiers depend on the structure of a sort hierarchy.

<u>Example 2.1</u> Let $(\mathcal{S},<_\mathcal{S})=(\mathbb{N},<_\mathbb{N})$ and $D=\{x,y\}$ with $x\in\mathcal{V}_n$ and $y\in\mathcal{V}_m$, where $\mathbb{N}$ denotes the set of natural numbers ordered by $<_\mathbb{N}$. Then $\sigma_1=\{x\leftarrow y\}$ and $\sigma_2=\{y\leftarrow x\}$ are the only mgu's of D. For $n\geq_\mathbb{N}m$, $\sigma_1\in SUB_\Sigma$, and for $m\geq_\mathbb{N}n$, $\sigma_2\in SUB_\Sigma$. Hence D possesses a Σ-mgu in either case. ∎

Later we will prove that each Σ-unifiable Σ-termset has a Σ-mgu, provided $(\mathcal{S},<_\mathcal{S})$ is a forest. The reason is that in a forest-structured sort hierarchy each Σ-unifiable pair of variables, as x,y above, can be arranged in an mgu, i.e. $x\leftarrow y$ or $y\leftarrow x$, so that the resulting substitution is well-sorted.

We will also prove that being a forest structure is also a **necessary** condition for the Robinson Unification Theorem [14] to hold true for many-sorted unification. As a consequence we have to use **auxiliary variables** in order to obtain complete and minimal sets of Σ-unifiers if non forest-structured sort hierarchies are used.

<u>Example 2.2</u> Let $\mathcal{S}=S\cup S_k$ with $S=\{a,b\}$ and $S_k=\{n\in\mathbb{N}\mid n<_{\mathbb{N}}k\}$ where $k\in\mathbb{N}^+\cup\{\infty\}$ and let $<_{\mathcal{S}}=S_k\times S$. Figure 2.1 (i) is a diagram of the sort hierarchy under consideration, which is obviously not a forest structure. Furthermore let $D=\{x,y\}$ with $x\in V_a$ and $y\in V_b$, let $\sigma_1=\{x\leftarrow y\}$, $\sigma_2=\{y\leftarrow x\}$ and let $\theta_h=\{x\leftarrow z_h,\ y\leftarrow z_h\}$ where $z_h\in V_h$ for any $h\in S_k$. Since neither $a\not<_{\mathcal{S}}b$ nor $b\not<_{\mathcal{S}}a$, neither σ_1 nor σ_2 are Σ-substitutions, i.e. D has no Σ-mgu. Now consider $U_k=\{\theta_h\in SUB\mid h\in S_k\}$. Obviously $U_k\subset U_\Sigma(D)$ and we can prove that U_k is complete and minimal [24].∎

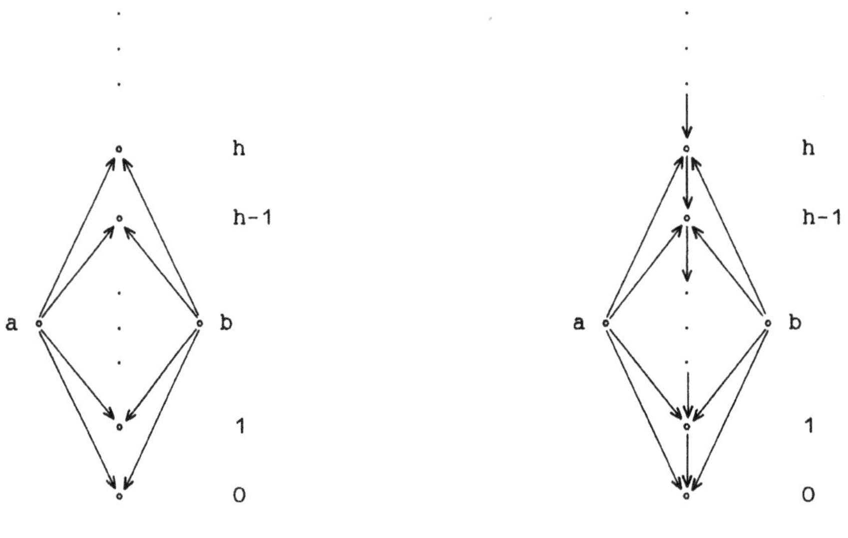

(i) (ii)

<u>Figure 2.1</u> The sort hierarchies of Examples 2.2 and 2.3

We shall also present necessary and sufficient conditions for $(\mathcal{S},<_{\mathcal{S}})$, cf. Theorem 6.2, such that each complete and minimal set of Σ-unifiers of some Σ-termset

(1) is a singleton (as U_k for k=1 in Example 2.2 above),

(2) is finite (as U_k for $k\in\mathbb{N}^+$ in Example 2.2 above), and

(2) is infinite (as U_k for $k=\infty$ in Example 2.2 above).

Unfortunately, however, the existence of complete and minimal sets of Σ-unifiers is not guaranteed at all, i.e. there are cases where a Σ-unifiable Σ-termset has no complete and minimal set of Σ-unifiers:

Example 2.3 Let all notions be given as in Example 2.2, except that we define $<_{\mathcal{S}}=S_k\times S \cup <_N$. Figure 2.1 (ii) is a diagram of this sort hierarchy. We can prove that $\{\theta_{k-1}\}$ is a complete and minimal set of Σ-unifiers of D if $k\in N^+$. But D has no complete and minimal set of Σ-unifiers for $k=\infty$, cf. [24]. ∎

Further on we shall formulate a condition for sort hierarchies, called the **maximal-sorts condition**, and we shall prove that this condition is equivalent for the existence of a complete and minimal set of Σ-unifiers, cf. Theorem 6.1.

3. Most General Weakening Sets

Throughout the following three sections we let D be any finite and Σ-unifiable subset of $\mathcal{T}_\Sigma$ for which we are seeking a complete and minimal set of Σ-unifiers. As we have seen in Example 2.2, auxiliary variables can become indispensable in order to obtain such a set. These new variables are formally introduced by certain Σ-substitutions, called **weakening substitutions**, which replace certain variables in the set D. The application of a weakening substitution can be thought of as a kind of **coercion process** which is necessary in programming languages with **types**, e.g. PASCAL or ADA, whenever the types of a variable and an expression do not agree in an assignment. Since there are several distinct variable replacements we have also several distinct weakening substitutions, which are collected in a so-called **most general weakening set** $\mathcal{W}(D)$ of D.

Let us look closer at the introduction of new variables before we go into formal details: Suppose that during the computation of a complete and minimal set of Σ-unifiers of D, as a subproblem some variables $\{x_1,\dots,x_n\}=V\subset\mathcal{V}(D)$ have to be Σ-unified 'as generally as possible'. If $\sqcap[V]\in[V]$, then $\sqcap[V]=[x]$ for some $x\in V$ and $\lambda_x=\{x_1\leftarrow x,\dots,x_n\leftarrow x\}$ is obviously a Σ-mgu of V, i.e. we do not need auxiliary variables in this case. But for $\sqcap[V]\notin[V]$, λ_x cannot be a Σ-substitution for any $x\in V$. We need a new variable, say y, and $\lambda_y=\{x_1\leftarrow y,\dots,x_k\leftarrow y\}$ unifies V.

But which sort do we stipulate for y? We have to guarantee $[y]\in lbs[V]$, because otherwise we cannot expect the result to be well-sorted (which is a matter of **correctness**). As we shall see in subsequent sections $[y]\in max(lbs[V])$ is a necessary condition to obtain a **minimal** set of Σ-unifiers. But $max(lbs[V])$ may have more than one element and as a matter of **completeness** we need a new variable $y(s)$ for **each** $s\in max(lbs[V])$. Hence $U=\{\lambda_{y(s)}\in SUB_\Sigma \mid s\in max(lbs[V])\}$ is a minimal subset of $U_\Sigma(V)$ but U is only complete if in addition for each $s\in lbs[V]$ some $s_m\in max(lbs[V])$ exists such that $s\leq s_m$.

To define the weakening substitutions for a given $D\subset\mathcal{T}_\Sigma$, we have to identify the subsets V of $\mathcal{V}(D)$ (as the set V above), which necessitate a variable replacement. The following technical lemma shows that these sets are given as those equivalence classes of $\underset{D}{\sim}$ of which the sortal image has no least element.

__Lemma 3.1__ For each $M\in\mathcal{T}/\underset{D}{\sim}$, $M\subset\mathcal{V}(D)$ or $\sqcap[M]\in[M]$.

Note that by this lemma, $\mathcal{M}=\{M\in\mathcal{T}/\underset{D}{\sim}\mid\sqcap[M]\notin[M]\}$ is a finite set of finite subsets of $\mathcal{V}(D)$. This fact guarantees that the following definition, which renders our idea of how new variables have to be introduced precise, is meaningful.

__Definition 3.1__ Let $\mathcal{M}=\{M\in\mathcal{T}/\underset{D}{\sim}\mid\sqcap[M]\notin[M]\}$. If $\mathcal{M}\neq\emptyset$, then by Lemma 3.1 $\mathcal{M}=\{V_1,\ldots,V_n\}$ for some $V_1,\ldots,V_n\subset\mathcal{V}(D)$ and we define for each i with $1\le i\le n$: (1) $Z_i\subset\mathcal{V}\setminus\mathcal{V}(D)$ such that $\max(\mathrm{lbs}[V_i])\subset[Z_i]$ and $Z_i\cap Z_h=\emptyset$ for any h with $1\le h\le n$ and $h\neq i$, (2) var_i is a mapping from $\max(\mathrm{lbs}[V_i])$ to Z_i such that $[\mathrm{var}_i(s)]=s$ for all $s\in\max(\mathrm{lbs}[V_i])$, and (3) $W_i=\{\mu_i^s\in SUB\mid\mu_i^s=\{x_i\leftarrow\mathrm{var}_i(s)\}$ for some $s\in\max(\mathrm{lbs}[V_i])\}$, where $x_i\in V_i$ is arbitrary but fixed. A **most general weakening set** or **mgw-set** $W(D)$ of D is a subset of SUB satisfying (4) $W(D)=\{\mu\in SUB\mid\mu=\mu_1\circ\ldots\circ\mu_n$, where $\mu_i\in W_i\}$, where each element of $W(D)$ is called a **weakening substitution**.

Finally, the **sortal degree** ΔD of D is defined as (5) $\Delta D=|\max(\mathrm{lbs}[V_1])\times\ldots\times\max(\mathrm{lbs}[V_n])|$ and D satisfies the **maximal-sorts condition** iff (6) $\forall s\in\mathrm{lbs}[V_i]$ $\exists s_m\in\max(\mathrm{lbs}[V_i])$ such that $s\le s_m$ holds true for each i. For $\mathcal{M}=\emptyset$ we define $W(D)=\{\varepsilon\}$, $\Delta D=1$ and D satisfies the maximal-sorts condition. ⊠

Mgw-sets do exist for each Σ-unifiable Σ-termset D. They consist of Σ-substitutions and their cardinality is given by the sortal degree ΔD of D. For instance n=1, $V_1=D$, $Z_1=\{z_h\mid h\in S_k\}$ and $W_1=W(D)=\{\{x\leftarrow z_h\}\mid h\in S_k\}$ is an mgw-set of D as given in Example 2.2, where $\Delta D=k$. As we have seen at the beginning of this section, $\sqcap[V]\in[V]$ is a sufficient condition that $V\subset\mathcal{V}(D)$ has a Σ-mgu. The introduction of auxiliary variables by a weakening substitution μ guarantees that this condition is satisfied for all equivalence classes of $\underset{\mu D}{\sim}$:

__Lemma 3.2__ Let μ be a member of an mgw-set of D. Then $\sqcap[N]\in[N]$ for each $N\in\mathcal{T}/\underset{\mu D}{\sim}$.

By this lemma there is no more need of 'coercions' after an application of a weakening substitution, i.e all weakening substitutions $\mu\in W(D)$ share the interesting property that each μD has a Σ-mgu. This result will provide the key for the main theorems about many-sorted unification in the subsequent sections.

__Theorem 3.3__ Let μ be a member of an mgw-set of D. Then μD has a Σ-mgu.

Proof For each $\theta\in U_\Sigma(D)$ some $\tau,\lambda\in SUB$ exists such that (1) τ is an mgu of μD, and (2) $\lambda\circ\tau=\theta[\![\mathcal{V}(D)\cap\mathcal{V}(\mu D)]\!]$. Let $\tau_1,\ldots\tau_n$ be a separation of τ, let $\{y_i\}=COD(\tau_i)$ and let $V_i=DOM(\tau_i)\cup\{y_i\}$ for each i with $1\le i<n$. Then $V_i\in\mathcal{T}/\underset{\mu D}{\sim}$ and by Lemma 3.2 each $[V_i]$ has a least element, i.e. $[x_i]=\sqcap[V_i]$ for some $x_i\in V_i$.

Let $\sigma_1,\ldots,\sigma_{n-1},\sigma\in SUB$ be defined by (3) $\sigma_i=\{y_i\leftarrow x_i\}\circ\tau_i$, for each i with $1\le i<n$, and (4) $\sigma=\sigma_1\circ\ldots\circ\sigma_{n-1}$. Since we obtain σ_i from τ_i by a variable renaming, we know

$\sigma\circ\tau_n \equiv \tau$, and it is easily proved that $[\sigma_i x] \leq [x]$ and $[\sigma_i x] \in \mathcal{V}$ for all $x \in \mathcal{V}$. Hence (5) $\sigma\circ\tau_n$ is an mgu of μD and (6) $\sigma \in SUB_\Sigma$.

We prove that $\sigma\circ\tau_n \in SUB_\Sigma$, i.e. we have to show for each $x \in \mathcal{V}$ (7) $[\sigma\tau_n x] \leq [x]$ and (8) $\sigma\tau_n x \in \mathcal{T}_\Sigma$. To prove (7) suppose $\tau_n x \in \mathcal{V}$ for some $x \in \mathcal{V}$. With $COD(\tau_n) \cap \mathcal{V} = \emptyset$ we obtain $\tau_n x = x$, i.e. $[\sigma\tau_n x] = [\sigma x] \leq [x]$ by (6). If $\tau_n x \notin \mathcal{V}$ for some $x \in \mathcal{V}$, then $x \in \mathcal{V}(D) \cap \mathcal{V}(\mu D)$ and we obtain $[\sigma\tau_n x] = [\tau_n x] = [\tau x] = [\lambda\tau x] = [\theta x] \leq [x]$ by (2) and $\theta \in SUB_\Sigma$. Hence (7) holds in either case.

For $x \notin DOM(\tau_n)$ we obtain $\sigma\tau_n x = \sigma x \in \mathcal{V}$ and (8) holds for this case. Now let $x \in DOM(\tau_n)$ and $t = \tau_n x$. We assume by way of contradiction that $\sigma t \notin \mathcal{T}_\Sigma$. Then t has a subterm $r = f(r_1 \ldots r_k)$ such that (9) $[\sigma r_j] \nleq [f]_j$ for some j with $1 \leq j \leq k$. From $x \in DOM(\tau_n)$ we obtain $x \in \mathcal{V}(D) \cap \mathcal{V}(\mu D)$, hence by (2) $\lambda t = \theta x$ and with $\theta \in SUB_\Sigma$, $\lambda t \in \mathcal{T}_\Sigma$. Therefore $\lambda r \in \mathcal{T}_\Sigma$ and in particular $[\lambda r_j] \leq [f]_j$.

Now if $r_j \notin \mathcal{V}$ then $[\sigma r_j] = [r_j] = [\lambda r_j] \leq [f]_j$, contradicting (9), hence (10) $r_j \in \mathcal{V}$. Since r is a subterm of some member t of $COD(\tau)$ and τ is an mgu of μD, a subterm $q = f(q_1 \ldots q_k)$ of some term in μD exists such that (11) $q \in \mathcal{T}_\Sigma$, and $\tau q = r$. Hence $\tau q_j = r_j$ and therefore (12) $[\sigma r_j] = [\sigma\tau q_j]$.

By (10) we obtain $\tau q_j \in \mathcal{V}$, hence $q_j \in \mathcal{V}$ and $q_j \notin DOM(\tau_n)$ because $COD(\tau_n) \cap \mathcal{V} = \emptyset$. If $q_j \notin \mathcal{V}_i$ for each i then $\sigma\tau q_j = \sigma q_j = q_j$ and if $q_j \in \mathcal{V}_i$ for some i, then $\sigma\tau q_j = \sigma\tau_i q_j = \sigma y_i = x_i$. With $[x_i] = \sqcap[\mathcal{V}_i]$ we have $[\sigma\tau q_j] = [x_i] \leq [q_j]$, hence in either case (13) $[\sigma\tau q_j] \leq [q_j]$. From (11) we infer $[q_j] \leq [f]_j$, hence by (12) and (13) $[\sigma r_j] \leq [f]_j$ contradicting (9). ∇

With (7) and (8) we obtain $\sigma\circ\tau_n \in SUB_\Sigma$, hence by (5) $\sigma\circ\tau_n$ is a Σ-mgu of μD. ▣

4. Many-Sorted Unification with Forest-Structured Sort Hierarchies

We consider here the special case where $(\mathcal{S}, <_\mathcal{S})$ is a forest structure, i.e. a (finite or infinite) collection of (finite or infinite) trees. This special case deserves our interest because we can prove that being a forest structure is a necessary and sufficient condition for a sort hierarchy such that the Robinson Unification Theorem [14] holds for many-sorted unification.

<u>Theorem 4.1</u> $(\mathcal{S}, <)$ is a forest-structured sort hierarchy iff each Σ-unifiable $D \subset T_\Sigma$ has a Σ-mgu.

Proof '→' Let $\theta \in U_\Sigma(D)$ and let $M \in \mathcal{T}/\underset{D}{\sim}$. Then θ unifies M by Lemma 1.1(2), i.e. $\theta q = r$ for some $r \in \mathcal{T}$ and each $q \in M$. With $\theta \in SUB_\Sigma$ we obtain $[r] = [\theta q] \leq [q]$ for each $q \in M$, hence $lbs([M]) \neq \emptyset$. But then [M] is a chain in $\mathcal{S}$ because $(\mathcal{S}, <)$ is a forest. Since D is unifiable, M is finite by Lemma 1.1(1), hence [M] is finite and since [M] is a chain in $\mathcal{S}$, [M] has a least element, i.e. $\sqcap[M] \in [M]$. Hence $W = \{\varepsilon\}$ is the only mgw-set of D and by Theorem 3.3 $\varepsilon D = D$ has a Σ-mgu. ▣

'←' Let $s(1), s(2), s \in \mathcal{S}$ such that (*) $s(1) \geq s \leq s(2)$ and let $x_1 \in \mathcal{V}_{s(1)}$, $x_2 \in \mathcal{V}_{s(2)}$ and $y \in \mathcal{V}_s$. Then $\theta = \{x_1 \leftarrow y, x_2 \leftarrow y\}$ unifies $D = \{x_1, x_2\}$ and with (*) $\theta \in SUB_\Sigma$. By the hypothesis D has a Σ-mgu, i.e. $\sigma_1 = \{x_1 \leftarrow x_2\}$ or $\sigma_2 = \{x_2 \leftarrow x_1\}$ is in SUB_Σ. Hence $s(1) \geq s(2)$ or $s(1) \leq s(2)$, i.e. $(\mathcal{S}, <_\mathcal{S})$ is a forest structure. ▣▣

5. Many-Sorted Unification with General Sort Hierarchies

In this section we prove the existence of complete and minimal sets of Σ-unifiers for Σ-unifiable Σ-termsets which satisfy the maximal-sorts condition: we define so called most general unifier sets and prove that each such set consists of Σ-unifiers, is complete and is also minimal.

Definition 5.1 Let W be some mgw-set of D. Then U(W) is a most general unifier set or mgu-set of D (relative to W) iff U(W) is a minimal subset of SUB satisfying: $\forall\mu\in W$ $\exists\sigma\in SUB$ such that (1) $\sigma\circ\mu\in U(W)$ and (2) σ is a Σ-mgu of μD. ▣

The existence of an mgu-set is guaranteed for each Σ-unifiable Σ-termset D by Theorem 3.3. It consists of Σ-unifiers and its cardinality is given by the sortal degree ΔD of D. For instance $\{\{y\leftarrow z_h\}\circ\{x\leftarrow z_h\} \mid h\in S_k\}=\{\theta_h \mid h\in S_k\}$ is an mgu-set of D as given in Example 2.2. To prove the completeness of an mgu-set we use the fact that each Σ-unifier is Σ-subsumed by some weakening substitution, provided the maximal-sorts condition is satisfied.

Lemma 5.1 Let $\theta\in U_\Sigma(D)$ and let W be an mgw-set of D. If D satisfies the maximal-sorts condition then $\mu\leq_\Sigma\theta[V(D)]$ for some $\mu\in W$.

Theorem 5.2 (Completeness Theorem) Let U be an mgu-set of D. Then for each $\theta\in U_\Sigma(D)$ some $\rho\in U$ exists such that $\rho\leq_\Sigma\theta[V(D)]$, provided D satisfies the maximal-sorts condition.

Proof Let U=U(W) be an mgu-set relative to some mgw-set W of D. By the hypothesis D satisfies the maximal-sorts condition, hence with Lemma 5.1 $\lambda\circ\mu=\theta[V(D)]$ for some $\mu\in W$ and some $\lambda\in SUB_\Sigma$. By Definition 5.1 $\sigma\circ\mu\in U$ for some Σ-mgu σ of μD. With $\theta\in U_\Sigma(D)$ we have $\lambda\in U_\Sigma(\mu D)$, hence $\lambda=\lambda\circ\sigma$. But then $\lambda\circ\sigma\circ\mu=\lambda\circ\mu=\theta[V(D)]$, i.e. $\sigma\circ\mu\leq_\Sigma\theta[V(D)]$. ▣

Note that the completeness of an mgu-set is guaranteed only if the Σ-unifiable Σ-termset under consideration satisfies the maximal-sorts condition. To prove the minimality no such proviso is required. Again we need a technical lemma before proving the main theorem.

Lemma 5.3 Let μ be a member of an mgw-set of D and let σ be a Σ-mgu of μD. Then $[\sigma\mu x]=[\mu x]$ for each $x\in DOM(\mu)$.

Theorem 5.4 (Minimality Theorem) Let U be an mgu-set of D and let $\rho_1,\rho_2\in U$. Then $\rho_1=\rho_2$, if $\rho_1\leq_\Sigma\rho_2[V(D)]$.

Proof Let us assume by way of contradiction that $\lambda\circ\rho_1=\rho_2[V(D)]$ for some $\lambda\in SUB_\Sigma$, but $\rho_1\neq\rho_2$. By Definition 5.1 there is an mgw-set W of D, some $\mu_1,\mu_2\in W$ and some σ_1, $\sigma_2\in SUB_\Sigma$ such that for j=1,2 (1) $\rho_j=\sigma_j\circ\mu_j$ and (2) σ_j is a Σ-mgu of μ_jD. For $\mu_1=\mu_2$,

$U\setminus\{\rho_j\}$ (for any $j\in\{1,2\}$) is also an mgu-set of D by Definition 5.1. But this is impossible by the minimality of U, hence $\mu_1\neq\mu_2$.

Let all notions be given as in Definition 3.1. From $\mu_1\neq\mu_2$ we obtain $\mu_1x_i\neq\mu_2x_i$ for some $x_i\in DOM(\mu_1)=DOM(\mu_2)$, hence (3) $[\mu_1x_i]\neq[\mu_2x_i]$, by Definition 3.1, (4) $\{[\mu_1x_i],[\mu_2x_i]\}\subset max(lbs[V_i])$, by Definition 3.1, and (5) $[\mu_2x_i]\leq[\mu_1x_i]$, because

$[\mu_2x_i]=[\sigma_2\mu_2x_i]$, by Lemma 5.3 using (2), ... $=[\rho_2x_i]$, by (1), ... $=[\lambda\rho_1x_i]$, by assumption, ... $\leq[\rho_1x_i]$, since $\lambda\in SUB_\Sigma$, ... $=[\sigma_1\mu_1x_i]$, by (1), ... $=[\mu_1x_i]$, by Lemma 5.3 using (2).

From (4) and (5) we obtain $[\mu_1x_i]=[\mu_2x_i]$ which contradicts (3). ∇■

6. A Classification of Sort Hierarchies

In section 3 we defined the **maximal-sorts condition** for Σ-unifiable Σ-termsets. Here we extend this notion to **sort hierarchies** and we will prove that satisfying this con dition is a necessary and sufficient criterion for a sort hierarchy such that a complete and minimal set of Σ-unifiers exists for each Σ-unifiable Σ-termset D. We also define properties of sort hierarchies which are equivalent in having one, finitely many or infinitely many members for a complete and minimal set of Σ-unifiers of D.

Definition 6.1 A sort hierarchy $(\mathcal{S},<)$ satisfies the **maximal-sorts condition** iff for each non-empty and finite $S\subset\mathcal{S}$: $\forall s\in lbs(S)\ \exists s_m\in max(lbs(S))\ s\leq s_m$. $(\mathcal{S},<)$ has **degree 1** iff $|max(lbs(S))|\leq1$, and has **degree** ω iff $|max(lbs(S))|<\infty$ for each non-empty and finite $S\subset\mathcal{S}$. $(\mathcal{S},<)$ has **degree** ∞ iff $(\mathcal{S},<)$ has not degree ω. ■

Theorem 6.1 A sort hierarchy $(\mathcal{S},<)$ satisfies the maximal-sorts condition iff each Σ-unifiable $D\subset\mathcal{T}_\Sigma$ has a complete and minimal set of Σ-unifiers U(D) with $|U(D)|=\Delta D$.

Proof '←' We prove for any finite and non-empty subset S of $\mathcal{S}$: (1) $\forall s\in lbs(S)\ \exists s_m\in max(lbs(S))\ s\leq s_m$. Let $S=\{s_1,\ldots,s_k\}$ and let $D=\{x_1,\ldots,x_k\}\subset\mathcal{V}$ with $[x_i]=s_i$. For $lbs(S)=\emptyset$ (1) holds trivially, so let us assume $lbs(S)\neq\emptyset$. Then we can define $\theta_s=\{x_1\leftarrow x_s,\ldots,x_k\leftarrow x_s\}$ where $x_s\in\mathcal{V}_s$ for each $s\in lbs(S)$. Obviously each $\theta_s\in U_\Sigma(D)$ and by the hypothesis U(D) exists. Let $U=\{\sigma\in U(D)\mid\sigma\leq_\Sigma\theta_s[\![\mathcal{V}(D)]\!]$ for some $s\in lbs(S)\}$. Since $\theta_s\in U_\Sigma(D)$ and U(D) is complete we know that $U\neq\emptyset$. Now for each $\sigma\in U$ there is some $y_\sigma\in\mathcal{V}$ such that $\sigma D=\{y_\sigma\}$ and $[y_\sigma]\leq s_i$ for each i because $\sigma\leq_\Sigma\theta_s[\![\mathcal{V}(D)]\!]$.

Let $S_U=\{[y_\sigma]\in\mathcal{S}\mid\sigma\in U\}$. With $[y_\sigma]\leq s_i$ for each i we have $[y_\sigma]\in lbs(S)$, i.e. (2) $S_U\subset lbs(S)$. By the completeness of U(D), for each $s\in lbs(S)$ there is some $\sigma\in U$ such that $\sigma\leq_\Sigma\theta_s[\![\mathcal{V}(D)]\!]$, hence $s\leq[y_\sigma]\in S_U$, i.e. (3) $\forall s\in lbs(S)\ \exists s_m\in S_U.\ s\leq s_m$. Now suppose that for some $\sigma\in U$ $[y_\sigma]$ is not maximal in lbs(S), i.e. $[y_\sigma]<s'$ for some $s'\in lbs(S)$. Then by (3), $s'\leq[y_\rho]$ for some $\rho\in U$. Hence $[y_\sigma]<[y_\rho]$, i.e. $\rho\leq_\Sigma\sigma[\![\mathcal{V}(D)]\!]$ and with the minimality of U(D), $\rho=\sigma$ contradicting $[y_\sigma]<[y_\rho]$.∇ Hence with (2) $S_U\subset max(lbs(S))$ and with (3), (1) is proved. ■

'→' Obvious with Theorems 5.2 and 5.4, because D satisfies the maximal-sorts condition if $(\mathcal{S},<)$ does. ■■

Having proved the existence of complete and minimal sets of Σ-unifiers the cardinality of these sets can be characterized completely as a function of the structure of a particular sort hierarchy.

Theorem 6.2 Let $(\mathcal{S},<)$ be a sort hierarchy satisfying the maximal-sorts condition. Then (1) $(\mathcal{S},<)$ has degree 1 iff $|U(D)|=1$ for each Σ-unifiable $D\subset\mathcal{T}_\Sigma$,

(2) $(\mathcal{S},<)$ has degree ω iff $|U(D)|<\infty$ for each Σ-unifiable $D\subset\mathcal{T}_\Sigma$, and

(3) $(\mathcal{S},<)$ has degree ∞ iff $|U(D)|=\infty$ for some Σ-unifiable $D\subset\mathcal{T}_\Sigma$,

where $U(D)$ denotes a complete and minimal subset of $U_\Sigma(D)$.

Proof Obvious consequence of Theorem 6.1. ☒

Theorems 6.1 and 6.2 are the central results for the characterization of sort hierarchies w.r.t. the existence and the cardinality of complete and minimal sets of Σ-unifiers. The following two lemmata deal with cases which are of interest in practical applications, viz. many-sorted unification for sort hierarchies with finite $\mathcal{S}$ and for those where $(\mathcal{S},<)$ forms a meet-semilattice.

Corollary 6.3 Let $(\mathcal{S},<)$ be a sort hierarchy with finite $\mathcal{S}$. Then each Σ-unifiable $D\subset\mathcal{T}_\Sigma$ possesses a finite, complete and minimal set of Σ-unifiers.

Corollary 6.4 Let $(\mathcal{S},<)$ be a meet-semilattice. Then each Σ-unifiable $D\subset\mathcal{T}_\Sigma$ has a complete and minimal set of Σ-unifiers which is a singleton.

The proofs are immediate consequences of Theorems 6.1 and 6.2 because both hypotheses imply that $(\mathcal{S},<)$ satisfies the maximal-sorts condition and has degree ω or 1 respectively.

7. A Many-Sorted Unification Algorithm

At the very heart of each (Robinson) unification algorithm variable symbols x have to be unified with terms t. The resulting substitution, represented by $\{x\leftarrow t\}$, is composed with other substitutions of this kind finally yielding an mgu for the set of terms initially given to the unification algorithm (provided the set is unifiable). Hence each unification algorithm contains a sequence of statements like the following:

```
(1) if x=t then return ({})
(2) if x∈V(t) then stop/failure
(3) return ({x←t})
```

Figure 7.1 Unification of variables and terms

On many-sorted unification a complete and minimal set of Σ-unifiers is obtained by introduction of new variable symbols using most general weakening substitutions, cf. Definition 3.1. We modify the unification algorithm to obtain a **many-sorted unification algorithm** by replacing statement (3) in Figure 7.1 by the following sequence of statements:

(3.1) if $[t]\leq_{\varphi}[x]$ then return $(\{\{x\leftarrow t\}\})$
(3.2) if $t\notin\mathcal{V}$ or $lbs\{[t],[x]\}=\emptyset$ then stop/failure
(3.3) if $[x]<_{\varphi}[t]$ then return $(\{\{t\leftarrow x\}\})$
(3.4) let $\{s_1,\ldots,s_k\}=max(lbs\{[t],[x]\})$
(3.5) let $\{z_1,\ldots,z_k\}\subset\mathcal{V}$ such that no z_i is used before and $[z_i]=s_i$
(3.6) return $(\{\{x\leftarrow z_1, t\leftarrow z_1\},\ldots,\{x\leftarrow z_k, t\leftarrow z_k\}\})$

Figure 7.2 Many-sorted unification of variables and terms

It can be verified using Theorems 6.1 and 6.2 that, for each finite $D\subset\mathcal{T}_\Sigma$ given as input, the many-sorted unification algorithm terminates with a failure indication if D is not Σ-unifiable. Otherwise it terminates with a finite, complete and minimal set U(D) of Σ-unifiers of D, provided $(\mathcal{S},<)$ satisfies the maximal-sorts condition and also has degree ω. If $(\mathcal{S},<)$ is a forest structure then $U(D)=\{\sigma\}$, where σ is a Σ-mgu of D, because in this case at least one of the conditions in the above statements (3.1)-(3.3) is always satisfied.

8. Applications, Related Work and Concluding Remarks

Starting with the Robinson Unification Problem [14], i.e. unification in the empty equational theory or the free algebra of terms respectively, several aspects of unification, e.g. unification under special equational theories or universal unification, have been investigated and unification theory has become an important subfield of artificial intelligence and computer science, cf. [17].

In particular unification is a central notion in resolution and paramodulation based theorem proving [14,25]. In [21] a many-sorted version of a resolution calculus with paramodulation is proposed in which many-sorted unification can lead to a drastic reduction of the search space (cf. [20,23] for a case study), and it is argued in [19] that many-sorted unification implements a special procedure for **taxonomic reasoning**, which can be viewed (at least in particular cases) as an instance of the **theory resolution** principle.

In [15] this calculus is extended with **polymorphic** function symbols (where the sort of a term $f(t_1\ldots t_n)$ can vary depending on the sorts of its arguments $t_1,\ldots,t_n$) and the existence of complete and minimal sets of Σ-unifiers with finite cardinality is proved under the proviso that there are finitely many sorts.

Unification is also one of the basic mechanisms in the Knuth-Bendix completion algorithm [10,12]. In this approach **conditional** equations are difficult to treat and therefore the use of sorts (and subsorts) can be advantageous because conditions can sometimes be expressed by sortal constraints.

A further advantage in this context is pointed out in [3]: sorts can be used to deal economically with partial functions without leaving the scope of equational reasoning or introducing excessive formal clutter by error elements. A unification algorithm is presented and it is shown that complete and minimal sets of Σ-unifiers can have more than one member, if $(\mathcal{S},<_{\mathcal{S}})$ is a lattice and polymorphic functions are admitted. An algorithm for Σ-matching is given which returns at most one substitution .

[7] presents rewrite rule-based operational semantics for many-sorted algebra with subsorts and polymorphism and discusses pattern matching as well as term rewriting in this context.

A short survey of advantages which sortal inferences by many-sorted unification may yield in logic programming, in parsing with Horne-clause grammars and in knowledge retrieval is presented in [6].

In all these cases of application only those sort hierarchies which avoid complete and minimal sets of Σ-unifiers with infinite cardinality are of interest. Corollary 6.3 provides a simple sufficient condition for this. Additionally, backtracking problems and the entering of deadends are avoided, if each many-sorted unification problem has at most one most general solution. Hence Theorem 4.1 and Corollary 6.4 are of practical importance because each sort hierarchy can be embedded into a meet-semilattice by invention of additional sorts (cf. also [11]).

However, non-polymorphism is an essential prerequisite for the results presented in the preceding sections to hold true. This work has treated unification in many-sorted empty equational theories. The investigation of many-sorted non-empty equational theories (with and without polymorphism) is an interesting field for future research, cf. [16].

ACKNOWLEDGEMENTS I am indebted to A. Herold, J.-P. Jouannaud, J. Siekmann and one of the referees for their helpful comments and suggestions on earlier drafts of this paper.

REFERENCES

1. **BIRKHOFF, G.** Lattice Theory. American Mathematical Society Colloquium Publications, vol xxv, 1967.
2. **COHN, P.M.** Universal Algebra. Reidel, Dordrecht, 1981.
3. **CUNNINGHAM, R.J. and DICK, A.J.J.** Rewrite systems on a lattice of types. Acta Informatica 22, 2 (1985), pp. 149-169.
4. **EDER, E.** Properties of substitutions and unifications. J. Symbolic Computation 1, 1 (1985), pp. 31-46.
5. **FAGES, F. and HUET, D.** Complete sets of unifiers and matchers in equational theories. In Proceedings Eight CAAP, Lecture Notes in Computer Science 159 (L'Aquila 1983), pp. 205-220.
6. **FRISCH, A.M.** An investigation into inference with restricted quantification and a taxonomic representation. SIGART Newsletter 91 (1985), pp. 28-31.

7. **GOGUEN, J.A., JOUANNAUD, J.-P., and MESEGUER, J.** Operational semantics for order-sorted algebra. In Proceedings Twelfth ICALP, Lecture Notes in Computer Science 194 (Nafplion 1985), pp. 221-231.

8. **GOGUEN, J.A., THATCHER, J.W. and WAGNER, E.G.** An initial algebra approach to the specification, correctness and implementation of abstract data types. In Current Trends in Programming Methodology 4, R.T. Yeh, Ed., Prentice Hall 1978, pp. 80-321.

9. **HEROLD, A.** Some basic notions of first-order unification theory. Interner Bericht 15/83, Universität Karlsruhe, Inst. f. Informatik I (1983), pp. 1-45.

10. **HUET, G. and OPPEN, D.C.** Equations and rewrite rules: a survey. In Technical Report CSL-111,SRI International (1980) also in Formal Languages Theory: Perspectives and Open Problems, R. Book, Ed., Academic Press, 1980, pp. 349-405.

11. **IRANI, K.B. and SHIN, D.G.** A many-sorted resolution based on an extension of a first-order language. In Proceedings Ninth IJCAI (Los Angeles 1985), pp. 1175-1177.

12. **KNUTH, D. and BENDIX, P.** Simple word problems in universal algebras. In Computational Problems in Abstract Algebra, J. Leech, Ed., 1970.

13. **PLOTKIN, G.** Building-in equational theories. Machine Intelligence 7 (1972), pp. 73-90.

14. **ROBINSON, J.A.** A machine-oriented logic based on the resolution principle, J.ACM 12, 1 (1965), pp. 23-41, also in [18].

15. **SCHMIDT-SCHAUSS, M.** A many-sorted calculus with polymorphic functions based on resolution and paramodulation. In Proceedings Ninth IJCAI (Los Angeles 1985), pp. 1162-1168.

16. **SCHMIDT-SCHAUSS, M.** Unification in Many-Sorted Equational Theories. In Proceedings Eight CADE (Oxford 1986), this volume.

17. **SIEKMANN, J.** Universal unification. In Proceedings Seventh CADE (Napa 1984), Lecture Notes in Computer Science 170, pp. 1-42.

18. **SIEKMANN, J. and WRIGHTSON, G.**, Eds., Automation of Reasoning-Classical Papers on Computational Logic 1, Springer, 1983.

19. **STICKEL, M. E.** Automated deduction by theory resolution. In Proceedings Ninth IJCAI (Los Angeles 1985), pp. 1181-1186, revised version in J. Automated Reasoning 1, 4 (1985), pp. 333-355.

20. **STICKEL, M. E.** Schubert's Steamroller Problem: Formulations and Solutions. J. Automated Reasoning 2, 1 (1986), pp. 89-101.

21. **WALTHER C.** A many-sorted calculus based on resolution and paramodulation. In Proceedings Eight IJCAI (Karlsruhe 1983), pp. 882-891.

22. **WALTHER, C.** Unification in many-sorted theories. In Advances in Artificial Intelligence - Proceedings Sixth ECAI (Pisa 1984), T. O'Shea, Ed., North-Holland 1985, pp. 383-392.

23. **WALTHER, C.** A mechanical solution of Schubert's Steamroller by many-sorted resolution. In Proceedings Fourth AAAI (Austin 1984), pp. 330-334, revised version in Artificial Intelligence 26, 2 (1985), pp. 217-224.

24. **WALTHER, C.** A classification of unification problems in many-sorted theories. Interner Bericht 10/85, Universität Karlsruhe, Inst. f. Informatik I (1985), pp. 1-43.

25. **WOS, L. and ROBINSON, G.** Maximal models and refutation completeness: semidecision procedures in automatic theorem proving. In Wordproblems, Boone, W.W., Cannonito, F.B. Lyndon, R.C., Eds., North-Holland, 1973, also in [18].

Unification in Many-Sorted Equational Theories.

ManfredSchmidt-Schauss
Fachbereich Informatik, Universität Kaiserslautern
6750 Kaiserslautern, F.R.G.
(UUCP: seismo!mcvax!unido!uklirb!schauss)

Abstract.

Unification in many-sorted equational theories is of special interest in automated deduction. It combines the work done on unification in unsorted equational theories and unification in many-sorted signatures without equational theories. The sort structure considered in this paper may be an arbitrary finite partially ordered set, the equational theory may be arbitrary. Combining them, however, requires some natural restrictions, one of which is that equal terms have equal sorts.

Given a unification algorithm A for an unsorted equational theory the corresponding unification algorithm for the many-sorted equational theory is obtained by simply postprocessing the substitutions generated by A with a uniform algorithm to be presented in this paper.

The results obtained concern the decidability of unification as well as the existence and the cardinality of complete and minimal sets of unifiers. The most useful results are obtained for a combination of finitary matching theories with a polymorphic signature.

Key words: Unification, Many-Sorted Logics, Heterogeneous Algebras, Equational Theories

Introduction.

The advantages of sorts in automated reasoning systems are well-known, see e.g. [Ob62, Hay71, Hen72, Wa83, Co83, CD83, GM85, Sch85a].

Building in equational theories into the unification procedure [Pl72, Si84, Sz82] is now a commonly accepted way to handle special equational theories, for a survey see [Si84]. However, all known unification procedures for equational theories are designed for unsorted signatures, whereas the work on unification in many-sorted signatures [Wa83, Sch85a, CD83] does not treat equational theories. An algorithm for solving unification problems in a many-sorted equational theory, however, may be a very useful tool in all fields of automated deduction, in which unification is of interest.

In [GM85] it is proposed to encode the sortal information into functions and equations and then use narrowing [Hl80] to solve unification problems.

In this paper we describe a method to construct a unification algorithm for a wide class of many-sorted theories, thus solving the problem of extending unification results to sorted terms. This is mentioned as an open problem in [Si84] : Given a unification algorithm A for an unsorted equational theory the corresponding unification algorithm for the many-sorted equational theory is obtained by simply postprocessing the substitutions generated by A with a uniform algorithm. The latter algorithm does not depend on the equational theory. It only exploits the given sort structure.

The following example illustrates the problem:

Example. Let + be the addition on integers (INT) considered as a polymorphic and associative function and let O and

---

This work was supported by the Deutsche Forschungsgemeinschaft, SFB 314, "Künstliche Intelligenz".

E be subsorts of the integers denoting odd and even integers.

Consider the unification problem $< x_O + y_O = x'_E + y'_E >$. Let us try an ad hoc method for solving this problem. First we ignore the sorts and compute a minimal set of solutions:

1) $\{x_O \leftarrow z_1 ;\ y_O \leftarrow z_2 ;\ x'_E \leftarrow z_1 ;\ y'_E \leftarrow z_2 \}$

2) $\{x_O \leftarrow z_3 + z_4 ;\ y_O \leftarrow z_5 ;\ x'_E \leftarrow z_3 ;\ y'_E \leftarrow z_4 + z_5 \}$

3) $\{x_O \leftarrow z_6 ;\ y_O \leftarrow z_7 + z_8 ;\ x'_E \leftarrow z_6 + z_7 ;\ y'_E \leftarrow z_8 \}$

Next we try to make the substitutions well-sorted resulting in:

2') $\{x_O \leftarrow z_{1,E} + z_{2,O} ;\ y_O \leftarrow z_{3,O} ;\ x'_E \leftarrow z_{1,E} ;\ y'_E \leftarrow z_{2,O} + z_{3,O} \}$

3') $\{x_O \leftarrow z_{4,O} ;\ y_O \leftarrow z_{5,O} + z_{6,E} ;\ x'_E \leftarrow z_{4,O} + z_{5,O} ;\ y'_E \leftarrow z_{6,E} \}$

The same procedure applied for example to the unification problem $<x_O + y_O = x'_O + y'_E>$ yields no well-sorted solution.

Two question arise:

* Is the set thus obtained a set of most general unifiers?
* How general is this procedure, i.e. to which class of many-sorted equational theories does it apply?

We show, that this procedure always generates a complete set of unifiers and in the case of sort-stable (polymorphic) signatures a minimal set can be obtained.

This paper is a shortened version of [Sch86], which contains all proofs.

1. Basic Notions

We assume a familiarity with the basic notions of unification theory [Si84, He83] and fix the following symbols.

F	(finite) set of function symbols.	elements: f,g,h
F_n	set of functions of arity n	
C	(finite) set of constants i. e. $= F_0$.	a,b,c
V	set of variables	x,y,z
T	set of terms (i.e. $T = T(F, V)$)	s,t

Ground objects and sets of ground objects are marked with the suffix "gr".

A substitution is an endomorphism on the term algebra associated with F that is identical almost everywhere on V.

Σ denotes the set of all substitutions

$V(o_1,\dots,o_n)$ denotes the set of variables occuring in the objects $o_1,\dots,o_n$.

$DOM(\sigma) := \{x \in V \mid \sigma x \neq x\}$; $COD(\sigma) := \sigma DOM(\sigma)$; $VCOD(\sigma) := V(\sigma DOM(\sigma))$

$V(\sigma) := DOM(\sigma) \cup VCOD(\sigma)$.

A substitution σ can be represented as a finite set of variable-term pairs:

$\{x_1 \leftarrow t_1 ,\dots, x_n \leftarrow t_n\}$, where $DOM(\sigma) = \{x_1 ,\dots, x_n\}$.

$\sigma_{|W}$ denotes a substitution (the restriction of σ to $W \subseteq V$) such that

$\sigma_{|W} x = \sigma x$ for $x \in W$ and $\sigma_{|W} x = x$ otherwise.

$\Sigma_{|W}$ $\{\sigma \in \Sigma \mid DOM(\sigma) \subseteq W\}$

ε denotes the identical substitution.

Occurrences or positions [HO80] are vectors of natural numbers which are used to select subterms of terms. Let q be such a vector, then

$t_{|q}$ denotes the subterm of t at position q. For example $f(x, y)_{|(2)} = y$

$t[q \leftarrow s]$ denotes the term, which is constructed from t by replacing the subterm at position q with the term s.

subterms(t) denotes the set of all subterms of the term t (including t).

A set of terms $\mathbf{T}_0$ is subterm-closed , iff $\mathbf{T}_0 = \cup\{\text{subterms}(t) \mid t \in \mathbf{T}_0\}$.

DEPTH (t) the depth of t (i.e. the maximal length of an occurrence in t)

DEPTH(σ) the maximal depth of terms in COD(σ).

A relation $\leq$ is a quasi-ordering, iff it is reflexive and transitive.

EQR($\leq$) the equivalence relation corresponding to a quasi-ordering $\leq$.

I.e. a EQR($\leq$) b $\Leftrightarrow$ $a \leq b$ and $b \leq a$

A subset SUB of a set U ordered by a quasi-ordering $\leq$ is called a lower segment of U, iff

$$\forall u \in SUB\ \forall v \in U: \ v \leq u \Rightarrow v \in SUB.$$

We denote with **S** the finite set of sorts, partially ordered by $\sqsubseteq$.

Let SORT: **T** $\rightarrow$ **S** be a partial, computable function on the set of terms. The domain of SORT is called the set of well-sorted terms **WST** and the set $\mathbf{W\Sigma} := \{\sigma \in \Sigma \mid \forall x \in V: \sigma x \in \mathbf{WST} \text{ and } SORT(\sigma x) \sqsubseteq SORT(x)\}$ is called the set of well-sorted substitutions. To avoid difficulties with empty sorts, we assume that for every sort S there exists a ground terms t_S with SORT(t_S) $\sqsubseteq$ S.

The following definition generalizes the definition of sorts in [HO80, Wa84, Sch85a] and is the basis for the subsequent chapters.

1.1 Definition. The function SORT is called a sort-assignment , iff the following conditions hold:

i) **WST** is a subterm-closed subset of **T** with V, C $\subseteq$ **WST**.

ii) For every S $\in$ **S**, there exist infinitely many variables x with SORT(x) = S.

iii) **WSΣ** is a monoid and **WSΣ** (**WST**) = **WST** ■

The unsorted case provides a trivial sort-assignment, where **S** is a singleton, **WST** = **T**, **WSΣ** = **Σ** and SORT is constant on **T**.

The sorts of terms defined for many-sorted signatures in [Wa84, Sch85a, HO80] are examples for nontrivial sort-assignments.

In the following we fix **WST** and **WSΣ** and always assume, that SORT is a sort-assignment.

As a consequence of definition 1.1 we have that well-sorted substitutions are compatible with the function SORT:

1.2 Lemma. $\forall t \in \mathbf{WST}\ \forall \sigma \in \mathbf{W\Sigma}$: SORT ($\sigma t$) $\sqsubseteq$ SORT(t). ■

Lemma 1.2 may be false, if the preconditions of definition 1.1 are not satisfied:

Example. Let S := {A,B} with A $\sqsupseteq$ B . Let f be a unary function and c be a constant. If we define SORT such that: i) there are infinitely many variables for sort A and B, ii) SORT(f(c)) := A , iii) SORT(t) := B for all other terms t, then $\{x_A \leftarrow c\} \in \mathbf{W\Sigma}$, SORT(f($x_A$)) = B, but $\{x_A \leftarrow c\}$ f(x_A) = f(c), which has sort A. The reason is that **WSΣ** is not a monoid ■

We denote with f: $S_1 \times \ldots \times S_n \rightarrow S_{n+1}$ the fact, that for every n-tuple $(t_1,\ldots,t_n)$ of well-sorted terms with SORT(t_i) $\sqsubseteq$ S_i the term $f(t_1,\ldots,t_n)$ is again well-sorted and has sort less than or equal to S_{n+1}.

The sort-assignments in [Wa84, Sch85a, HO80] have the property that SORT($f(t_1,\ldots,t_n)$) is a function of f and

$SORT(t_i)$ alone; i.e. it does not depend on the structure of the t_i's .
We call such sort-assignments sort-stable.
In other words, sort-stable sort-assignments characterize exactly those signatures, for which the sort-assignment SORT is completly determined by a finite number of assignments $f: S_1 \times ... \times S_n \to S_{n+1}$.
Note that for sort-stable sort-assignments there exists a minimal set of assignments $f: S_1 \times ... \times S_n \to S_{n+1}$ such that for every f there exists a monotone function $f_S : S \times ... \times S \to S$ with $SORT(f(t_1,...,t_n)) = f_S(SORT(t_1), ..., SORT(t_n))$ for every well-sorted term $f(t_1,...,t_n)$.
In [Wa84, HO80] the sort of a term is determined by its toplevel function symbol. In [CD83, Co83, Ob62, Sch85a] the sort of a term depends on the toplevel function symbol and the sort of its direct subterms. The sort-assignments in [GM85, Sch85b] may be not sort-stable.

We give an example for a non sort-stable sort-structure:
Example. Let $S := \{A,B\}$ with $A \sqsupseteq B$. Let f be a unary function and c be a constant. If we define SORT such that: i) there are infinitely many variables for sorts A and B, ii) SORT(f(c)) := B, iii) SORT(t) := A for all other terms, then the sort of f(t) depends on the structure of t.

If SORT is sort-stable then the sort of a term remains unchanged if an arbitrary subterm is replaced by a term of the same sort.

In the following we adapt some well-known notions to sorted terms. For the unsorted case see for example [He83] . We often use in parallel notations for the sorted and the unsorted case.The symbols denoting well-sorted objects are in general marked with the suffix -WS .

We define matching and subsumption for the unsorted and sorted case:
1.3 Definition. Let $s,t \in \mathbf{T}$. Then
i) $s \le t :\Leftrightarrow \exists \lambda \in \mathbf{\Sigma}: s = \lambda t.$ (t matches s; t is more general than s)
ii) $s \le_{WS} t :\Leftrightarrow \exists \lambda \in \mathbf{WS\Sigma}: s = \lambda t.$
iii) $\equiv := EQR(\le)$ and $\equiv_{WS} := EQR(\le_{WS})$ ■

1.4 Definition. Let $\sigma,\tau \in \mathbf{\Sigma}$ and $W \subseteq V$.
i) $\sigma = \tau\, [\![W]\!] :\Leftrightarrow \sigma x = \tau x$ for all $x \in W$.
ii) $\sigma \le \tau\, [\![W]\!] :\Leftrightarrow \exists\lambda \in \mathbf{\Sigma}: \sigma = \lambda\circ\tau\, [\![W]\!].$ (τ subsumes σ modulo W)
iii) $\sigma \le_{WS} \tau\, [\![W]\!] :\Leftrightarrow \exists\lambda \in \mathbf{WS\Sigma}: \sigma = \lambda\circ\tau\, [\![W]\!].$
iv) $\equiv [\![W]\!] := EQR(\le [\![W]\!])$ and $\equiv_{WS} [\![W]\!] := EQR(\le_{WS} [\![W]\!])$ ■

Note that the relations $\le$, $\le_{WS}$, $\le [\![W]\!]$ and $\le_{WS} [\![W]\!]$ are quasi-orderings.

The substitution $\rho \in \mathbf{\Sigma}$ is called a renaming [He83], iff ρ maps variables into variables, $DOM(\rho) \cap VCOD(\rho) = \emptyset$ and ρ is injective on $DOM(\rho)$.
Renamings with respect to $\mathbf{WS\Sigma}$ are defined like well-sorted renamings [Wa84, Sch85a]:

1.5 Definition. $\rho \in \mathbf{WS\Sigma}$ is called a $\mathbf{WS\Sigma}$-renaming, iff ρ is a renaming which satisfies in addition $SORT(\rho x) = SORT(x)$ for all variables x. ■

Note that $\mathbf{WS\Sigma}$-renamings are well-sorted renamings, but the converse does not hold.

1.6 Lemma. Let $W \subseteq Z \subseteq V$ be finite sets of variables. Then there exists a $\mathbf{WS\Sigma}$ -renaming $\rho \in \mathbf{WS\Sigma}$ such that $DOM(\rho) = W$ and $VCOD(\rho) \cap Z = \emptyset$. ■

1.7 Definition. Let ρ be a $\mathbf{WS\Sigma}$ -renaming.
Then we define the converse ρ^c of a $\mathbf{WS\Sigma}$ -renaming as follows [He83]:
$\rho^c x :=$ y, if there exists a variable $y \in DOM(\rho)$ such that $\rho y = x$ and
$\rho^c x :=$ x, otherwise. ■

1.8 Lemma. Let ρ be a $\mathbf{WS\Sigma}$ -renaming. Then:

i) ρ^c is a $\mathbf{WS\Sigma}$ -renaming

ii) $DOM(\rho) = COD(\rho^c)$

iii) $DOM(\rho^c) = COD(\rho)$

iv) $(\rho^c)^c = \rho$

v) $\rho^c \circ \rho = \varepsilon [\![DOM(\rho)]\!]$ ■

1.9 Definition. Let $U \subseteq \Sigma$ and let $W \subseteq Z \subseteq V$. Then U is said to be based on W away from Z, iff the following holds:
$\forall \sigma \in U$: $DOM(\sigma) = W$ and $VCOD(\sigma) \cap Z = \emptyset$ ■.

1.10 Lemma. [Sch85a] Let $W \subseteq V$ and let $\tau \in \Sigma$. Let ρ be a $\mathbf{WS\Sigma}$ -renaming with $DOM(\rho) = V(\tau W)$.
Then $\tau \equiv_{WS} \rho \circ \tau [\![W]\!]$. ■

The next proposition is trivial for the unsorted case and in the many-sorted case it is a consequence of the finiteness of the set of sorts S.

1.11 Proposition. Let W be a finite set of variables and let n be a natural number. Then the set $\{\sigma \mid \sigma \in \Sigma \text{ and } DEPTH(\sigma) \leq n\}$ contains a finite number of $\equiv_{WS}[\![W]\!]$ - congruence classes. ■

2. Congruences on Well-Sorted Terms.

Usually, congruences (equational theories) are generated by a set T of equations.
The case of arbitrary equations and sort-structure is beyond the reach of this paper. A treatment of general congruences can be found in [Sch86].
We confine ourselves to equational theories that satisfy the condition that a well-sorted term remains well-sorted after replacing a subterm by an equal term. This is enforced by the condition 2.1 ii)

We define the corresponding congruences for many-sorted signatures similar to strong congruences for partial algebras [Gr79] :

2.1 Definition. We say $=_{T,WS}$ is a WSΣ-invariant congruence on WST, iff the following conditions hold:

i) $=_{T,WS}$ is an equivalence relation on **WST**.

ii) if $s_i =_{T,WS} t_i$ and if $f(s_1,...,s_n) \in \mathbf{WST}$
then $f(t_1,...,t_n) \in \mathbf{WST}$ and $f(s_1,...,s_n) =_{T,WS} f(t_1,...,t_n)$.

iii) $\forall \sigma \in \mathbf{WS\Sigma}: \forall s,t \in \mathbf{WST}$: $s =_{T,WS} t \Rightarrow \sigma s =_{T,WS} \sigma t$ ($\mathbf{WS\Sigma}$ -invariance) ■

We write congruence for WSΣ-invariant congruence, if it is not ambiguous.
The congruence $=_{T,WS}$ is generated by T, iff $=_{T,WS}$ is the finest relation satisfying Definition 2.1 and the following:

$(s, t) \in T \Rightarrow s =_{T,WS} t$.
In the case there exists a finite set T generating $=_{T,WS}$, we say $=_{T,WS}$ is finitely generated, otherwise infinitely generated.

2.2 Definition. The congruence $=_{T,WS}$ on **WST** is called sort-compatible, iff
$\forall s,t \in$ **WST**: $s =_{T,WS} t \Rightarrow SORT(s) = SORT(t)$. ■

If the sort of a term depends solely on the toplevel function symbol, then a lot of interesting theories T generate non sort-compatible congruences or even worse a congruence in the sense of Definiton 2.1 does not exist.
For example let f be an idempotent function with range sort A defined on the sort A that has a subsort B. The term $f(x_B\ x_B)$ has sort A, whereas x_B has sort B. The generated congruence is not sort-compatible.
Suppose there is another (unary) function g defined on B, then $g(x_B)$ is well-sorted, but $g(f(x_B\ x_B))$ is not, hence there exists no congruence in the sense of Definition 2.1 in this case.

The following definition extends $=_{T,WS}$ to all terms and yields a congruence in the usual sense:

2.3 Definition. The congruence $=_T$ is defined as the finest relation on **T** satisfying the following conditions:

i) for all $s,t \in$ **WST**: $s =_{T,WS} t \Rightarrow s =_T t$
ii) $=_T$ is an equivalence relation on **T**.
iii) if $s_i =_T t_i$, then $f(s_1,\ldots,s_n) =_T f(t_1,\ldots,t_n)$ for all n and all $f \in \mathbf{F}_n$.
iv) $\forall \sigma \in \mathbf{\Sigma}$: $\forall s,t \in$ **T**: $s =_T t \Rightarrow \sigma s =_T \sigma t$. ■

For the related notions of algebras and E-models we refer to [Sch86].

For sort-stable SORT there is a criterion for recognizing theories that do not generate sort-compatible congruences, since in this case the set of sorts S itself is the carrier set of an E-model of this congruence.
For example let SORT be sort-stable, let f be an associative function defined everywhere, and let the generated congruence be sort-compatible. Then we can define f_S on the set of sorts S as follows:
$f_S(R\ S) := SORT(f(x_R\ x_S))$ for all sorts R,S. This definiton makes S together with f_S a semigroup with the additional property: $\forall R_1,R_2,S_1,S_2 \in S$: $R_1 \sqsubseteq S_1$ and $R_2 \sqsubseteq S_2 \Rightarrow f_S(R_1,R_2) \sqsubseteq f_S(S_1,S_2)$.
If we apply the above consideration to the group axioms, it can be shown that the only possibility is a set S of disjoint sorts and every sort corresponds to an element of the factor group with respect to a normal subgroup.

2.6 Definition. The congruence $=_{T,WS}$ on **WST** is called closed, iff
$\forall s,t \in$ **WST** $s =_{T,WS} t \Leftrightarrow s =_T t$ ■

In a closed congruence ill-sorted deductions do not yield new relations on the set of well-sorted terms. It may be more convenient to localize this condition. If for example the interesting part of the congruence is defined on a sort A and there exists a sort B that is separated from A and an equation $x_B = y_B$ is true on B, then ill-sorted deduction yields that all terms are equal and the congruence is not closed. However, ignoring this equation may yield a closed congruence.

2.7 Example.

a) Let $S := \{A,B\}$ with $A \sqsupseteq B$. Further let $\mathbf{F} := \{f\}$ such that $f: A \times A \to A$. Let $=_{T,WS}$ be generated by $T := \{f(x_B, x_B) = x_B\}$. Then $=_{T,WS}$ is neither closed nor sort-compatible: We have $f(x_A, x_A) =_T x_A$, but not $f(x_A, x_A) =_{T,WS} x_A$. Furthermore $SORT(f(x_B, x_B)) = A$, whereas $SORT(x_B) = B$.

b) Let $S := \{A,B\}$ with $A \sqsupseteq B$. Further let $F := \{f\}$ such that $f: A \times A \to A$ and $f: B \times B \to B$
Let $=_{T,WS}$ be generated by $T := \{f(x_B, x_B) = x_B\}$. Then $=_{T,WS}$ is sort-compatible, but not closed. ∎

We extend the definitions of subsumption to equational theories:

2.8 Definition. Let $\sigma,\tau \in \Sigma$ and let $W \subseteq V$. Then

$$\sigma =_{T,WS} \tau \, [\![W]\!] \quad :\Leftrightarrow \forall x \in W: \sigma x =_{T,WS} \tau x$$
$$\sigma =_{T} \tau \, [\![W]\!] \quad :\Leftrightarrow \forall x \in W: \sigma x =_{T} \tau x \quad ∎$$

2.9 Definition. Let $s,t \in \mathbf{T}$. Then

i) $s \leq_T t \quad :\Leftrightarrow \exists \lambda \in \Sigma: s =_T \lambda t.$

i) $s \leq_{T,WS} t \quad :\Leftrightarrow \exists \lambda \in \mathbf{WS\Sigma}: s =_{T,WS} \lambda t.$

iii) $\equiv_T := EQR(\leq_T)$ and $\equiv_{T,WS} := EQR(\leq_{T,WS})$ ∎

2.10 Definition. Let $\sigma,\tau \in \Sigma$ and $W \subseteq V$.

i) $\sigma \leq_T \tau \, [\![W]\!] \quad :\Leftrightarrow \exists \lambda \in \Sigma: \sigma =_T \lambda \circ \tau \, [\![W]\!].$

ii) $\sigma \leq_{T,WS} \tau \, [\![W]\!] \quad :\Leftrightarrow \exists \lambda \in \mathbf{WS\Sigma}: \sigma =_{T,WS} \lambda \circ \tau \, [\![W]\!].$

iii) $\equiv_T [\![W]\!] := EQR(\leq_T [\![W]\!])$ and $\equiv_{T,WS} [\![W]\!] := EQR(\leq_{T,WS} [\![W]\!])$ ∎

Note that the relations $\leq_T$, $\leq_{T,WS}$, $\leq_T [\![W]\!]$ and $\leq_{T,WS} [\![W]\!]$ are quasi-orderings.

The next theorem shows, how a sort-structure on terms may be changed to obtain a sort-compatible congruence.
We make some preliminary definitions:
Let $=_{T,WS}$ be a $\mathbf{WS\Sigma}$-invariant congruence on $\mathbf{WST}$. For $t \in \mathbf{WST}$ let $USORT(t) := \{SORT(s) \mid s =_{T,WS} t\}$.
If for all $t \in \mathbf{WST}$ the set USORT(t) has a unique minimal element (sort) and for variables this minimal element is equal to the sort of the variable then we say $=_{T,WS}$ generates a unique sort.
If $=_{T,WS}$ generates a unique sort we define $SORT_N(t)$ to be the unique minimal element of USORT(t).
Let $\mathbf{WST}_N := \mathbf{WST}$ and $\mathbf{WS\Sigma}_N := \{\sigma \in \Sigma \mid \forall x \in \mathbf{V}: \sigma x \in \mathbf{WST}$ and $SORT_N(\sigma x) \sqsubseteq SORT(x)\}$.

2.11 Theorem. Let $=_{T,WS}$ be a $\mathbf{WS\Sigma}$-invariant congruence on $\mathbf{WST}$ that generates a unique sort.
Then:
1) $SORT_N$ is a sort-assignment
2) $\mathbf{WS\Sigma} \subseteq \mathbf{WS\Sigma}_N$.
3) The generated $\mathbf{WS\Sigma}_N$-invariant congruence $=_{T,WS,N}$ on $\mathbf{WST}$ is the same relation as $=_{T,WS}$.
4) $=_{T,WS,N}$ is sort-compatible with respect to $SORT_N$.
5) The set of E-models of $=_{T,WS,N}$ and $=_{T,WS}$ is the same. ∎

In the case that USORT(t) has no unique minimal element, it is not possible to construct a sort-compatible congruence that satisfies 1)-5) of Theorem 2.11.
This theorem is useful in the case, where a sort of terms is given and the equational theory is defined by a canonical set of rewrite rules $l_i \to r_i$ and $SORT(l_i) \sqsupseteq SORT(r_i)$. In this case we can define a sort-assignment SORT' by computing SORT'(t) as the sort of its normalform, such that the congruence is sort-compatible.

If $=_{T,WS}$ is generated by a set T of equations, then we can characterize closed and sort-compatible congruences:

2.12 Proposition. Let T be a set of equations that generate the congruence $=_{T,WS}$. Then

a) $=_{T,WS}$ is closed, iff for all generating equations s = t:

$\forall \sigma \in \Sigma$: ($\sigma s \in$ **WST** or $\sigma t \in$ **WST**) $\Rightarrow$ $\sigma s, \sigma t \in$ **WST** and $\sigma s =_{T,WS} \sigma t$

b) If SORT is sort-stable, then $=_{T,WS}$ is closed, iff for all generating equations s = t:

For all renamings ρ: $\rho s \in$ **WST** or $\rho t \in$ **WST** $\Rightarrow$ $\rho s, \rho t \in$ **WST** and $\rho s =_{T,WS} \rho t$. ∎

2.13 Proposition. Let SORT be sort-stable. Let T be a set of equations, which generate the closed congruence $=_{T,WS}$

Further let the following condition be satisfied:

For all renamings ρ and all s=t $\in$ T: $\rho s \in$ **WST** $\Rightarrow$ SORT(ρs) =SORT(ρt).

Then $=_{T,WS}$ is sort-compatible. ∎

The following lemmas state some useful properties of sort-compatible, closed congruences, which are not true in the general case.

2.14 Lemma. Let $=_{T,WS}$ be a sort-compatible, closed congruence on **WST**.

Then: $\forall \sigma \in \mathbf{WS\Sigma}\ \forall \tau \in \Sigma : \sigma =_T \tau \Rightarrow (\sigma =_{T,WS} \tau$ and $\tau \in \mathbf{WS\Sigma})$. ∎

2.15 Lemma. Let $=_{T,WS}$ be a sort-compatible, closed congruence on **WST** and let W be a finite set of variables.

Then: $\forall \sigma \in \mathbf{WS\Sigma}|_W\ \forall \tau \in \Sigma|_W$: $\tau \le_{T,WS} \sigma\ [\![W]\!] \Rightarrow \tau \in \mathbf{WS\Sigma}$. ∎

That means $\mathbf{WS\Sigma}|_W$ is a lower segment in $\Sigma|_W$ w. r. t. $\le_{T,WS}\ [\![W]\!]$.

2.16 Lemma. Let $=_{T,WS}$ be a sort-compatible, closed congruence on (**WST**, **WSΣ**).

Let $\rho_1,\rho_2 \in$ **WSΣ** be renamings such that $\{\rho_1,\rho_2\}$ is based on W.

Then $\rho_1 \le_{T,WS} \rho_2\ [\![W]\!] \Rightarrow \rho_1 \le_{WS} \rho_2\ [\![W]\!]$. ∎

3. Computing Minimal and Complete Sets of Unifiers.

Assumptions: In the following it is always assumed, that the congruence $=_{T,WS}$ is decidable, i.e. for every term pair s,t $\in$ **WST** it is decidable whether s $=_{T,WS}$ t or not.

We also assume throughout this chapter, that $=_{T,WS}$ is sort-compatible and closed.

The following problems and their sets of solution are denoted as:

$\langle s =_{T,WS} t\rangle$	Find a substitution $\sigma \in$ **WSΣ** (a unifier) such that for s,t $\in$ **T** : $\sigma s =_{T,WS} \sigma t$. The set of all solutions (unifiers) is denoted as $U\Sigma_{T,WS}$ (s,t).
$\langle s =_T t\rangle$	Find a substitution $\sigma \in \Sigma$ such that for s,t $\in$ **T** : $\sigma s =_T \sigma t$. The set of all solutions is denoted as $U\Sigma_T$ (s,t).
$\langle s \le_{T,WS} t\rangle$	Find a substitution $\sigma \in$ **WSΣ** (a matcher) such that s $=_{T,WS} \sigma t$ and DOM(σ) $\cap$ V(s) = Ø, where s,t $\in$ **T**. The set of all solutions (matchers) is denoted as $M\Sigma_{T,WS}$ (s,t). We make use of the fact that matching problems are equivalent to unification problems in a signature with variables in s "constantified".
$\langle \tau \in \mathbf{WS\Sigma}\rangle$	Find a substitution $\sigma \in$ **WSΣ** such that $\sigma \circ \tau \in$ **WSΣ**, where $\tau \in \Sigma$. The set of all solutions (weakenings) is denoted as $W\Sigma_{WS}(\tau)$.

We say $=_{T,WS}$ is a <u>permutative congruence</u> [Sz82] , iff for every term $t \in$ **WST**, the equivalence class of t with respect to $=_{T,WS}$ is finite and effectively computable.
A binary , associative function symbol f may serve as an example for a permutative congruence. Other examples can be found in[Sz82].

We use the notation from [Sz82] and denote with $\mathcal{M}_{1\omega}$ the class of finitely generated congruences for which a terminating matching algorithm exists. In other words: the set of most general solutions for every matching problem is finite and can be computed by a terminating algorithm.
We give some definitions concerning relations.

Let U be a set ordered by a quasi-ordering $\leq$.
A set $C_{\leq}(U)$ is defined as a subset of U satisfying the following conditions:

i) $C_{\leq}(U) \subseteq U$ (<u>correctness</u>)

ii) $\forall u \in U\ \exists v \in C_{\leq}(U):\ u \leq v.$ (<u>completeness</u>)

A set $\mu_{\leq}(U)$ is a correct and complete subset of U, which is minimal in addition. I.e. satisfies i) and ii) and the following

iii) $\forall u,v \in \mu_{\leq}(U): u \leq v \Rightarrow u = v$ (<u>minimality</u>)

$\mu_{\leq}(U)$ is a set of EQR($\leq$)-representatives of the maximal elements of U.

A set $C_{\leq}(U)$ always exists (for example U). A set $\mu_{\leq}(U)$ does not exist in general, however if the set U is finite or if all increasing chains in U have an upper bound then $\mu_{\leq}(U)$ exists. A set $C_{\leq}(U)$ which has no proper complete subset, is minimal (i.e. = $\mu_{\leq}(U)$).
Note that $\mu_{\leq}(U)$ is unique up to EQR($\leq$) [Hu76, Fa83].

In the following we deal with a set $U \subseteq \Sigma$ (or $\subseteq$ **WS**Σ) ordered by a relation $\leq$.
Usually $\leq \in \{ \leq[\![W]\!], \leq_{WS}[\![W]\!], \leq_{T}[\![W]\!], \leq_{T,WS}[\![W]\!] \}$ (where W is a finite set of variables). We use the abbreviation $\mu_W(U)$, $\mu_{WS,W}(U)$, $\mu_{T,W}(U)$, $\mu_{T,WS,W}(U)$, respectively, for the corresponding complete, minimal subsets. To denote complete subsets, we use the abbreviations $C_W(U)$, $C_{WS,W}(U)$, $C_{T,W}(U)$, $C_{T,WS,W}(U)$, respectively.

We assume that every set $\mu_{\leq}(U)$ is based on the corresponding set of variables.
We use the following abbreviations, which are consistent with [Si84, Sz82]:

$\mu W\Sigma_{WS}(\sigma) := \mu_{WS,VCOD(\sigma)}(W\Sigma_{WS}(\sigma))$; $CW\Sigma_{WS}(\sigma) := C_{WS,VCOD(\sigma)}(W\Sigma_{WS}(\sigma))$

$\mu U\Sigma_T(s,t) := \mu_{T,V(s,t)}(U\Sigma_T(s,t))$; $CU\Sigma_T(s,t) := C_{T,V(s,t)}(U\Sigma_T(s,t))$

$\mu U\Sigma_{T,WS}(s,t) := \mu_{T,WS,V(s,t)}(U\Sigma_{T,WS}(s,t))$; $CU\Sigma_{T,WS}(s,t) := C_{T,WS,V(s,t)}(U\Sigma_{T,WS}(s,t))$

<u>3.1 Proposition.</u> Let $s,t \in$ **T** , $\sigma,\tau \in \Sigma$. Then:

i) $M\Sigma_{WS}(s,t) \neq \emptyset$ is decidable. ii) $M\Sigma_{WS}(\sigma,\tau) \neq \emptyset$ is decidable. ■

<u>3.2 Proposition.</u> Let $U \subseteq \Sigma$ and let W be a finite set of variables. Then $\mu_{WS,W}(U)$ exists.
<u>Sketch of Proof.</u> Proposition 1.11 implies that there exists no infinite chain
$\sigma_1 <_{WS} \sigma_2 <_{WS} \ldots$, where $\sigma_i \in \Sigma$. Hence $\mu_{WS,W}(U)$ exists. ■

<u>3.3 Theorem</u>. Let $\sigma \in \Sigma$. Then $\mu W\Sigma_{WS}(\sigma)$ exists and is recursively enumerable.

3.4 Theorem. Let SORT be sort-stable. Let $\sigma \in \Sigma$.

Then a minimal set of solutions $\mu W\Sigma_{WS}(\sigma)$ (which is based on W) consists of renamings.

Furthermore this set is finite and effectively computable. ■

The set $\mu W\Sigma_{WS}(\{x_1 \leftarrow t_1, \ldots, x_n \leftarrow t_n\})$ can be computed by a Robinson-style algorithm, namely successively computing $U_i := \cup \{ \mu W\Sigma_{WS}(\{x_i \leftarrow \sigma_{i-1} t_i\}) \mid \sigma_{i-1} \in U_{i-1} \}$, starting with $U_0 = \{ \varepsilon \}$.

For sort-stable sort-assignments $\mu W\Sigma_{WS}(\{x \leftarrow t\})$ is finite and consists of renamings that are well-sorted substitutions.

An ad hoc algorithm for computing $\mu W\Sigma_{WS}(\{x \leftarrow t\})$ works as follows:

Generate the set U_r of all renamings of V(t) that are well-sorted and then select the substitutions σ that satisfy $SORT(\sigma t) \sqsubseteq SORT(x)$.

Another possibility is the following recursive algorithm:

i) If t is a constant, then $\mu W\Sigma_{WS}(\{x \leftarrow t\})$ is either $\{\varepsilon\}$ if $SORT(t) \sqsubseteq SORT(x)$ or empty otherwise.

ii) If t is a variable, then $\mu W\Sigma_{WS}(\{x \leftarrow t\}) = \{ \{t \leftarrow z\} \mid SORT(z)$ is maximal in the set of all sorts that are common subsorts of SORT(x) and SORT(t)$\}$

iii) Let $t = f(t_1,\ldots,t_n)$. For all maximal assignments in the set $\{ f: S_1 \times \ldots \times S_n \rightarrow S_{n+1} \mid$ with $S_{n+1} \sqsubseteq SORT(x)$ and S_i and $SORT(t_i)$ have a common subsort$\}$ compute $\mu W\Sigma_{WS}(\{z_1 \leftarrow t_1, \ldots, z_n \leftarrow t_n\})$, where z_i are variables of sort S_i.
After that join these sets and remove instances.

Both algorithms have their advantages and their drawbacks. The number of renamings that have to be generated by the first algorithm may be far too large. However, if this number is small then the first algorithm is advantageous
The second algorithm is able to recognize that $\mu W\Sigma_{WS}(\{x \leftarrow t\})$ is empty without inspecting the variables in t. A drawback of this procedure is that substitutions may be computed more than once.

In general both algorithms are at least exponential, since the number of substitutions in $\mu W\Sigma_{WS}(\{x \leftarrow t\})$ may grow exponentially (cf. [Sch85a]).

3.5 Theorem. [Sch85b].

i) There exists no algorithm which decides for all sort-structures (with a finite description) the problem: $\mu W\Sigma_{WS}(\sigma) \neq \emptyset$.

ii) There exists a sort-structure (with a finite description) such that there exists no algorithm which decides the problem: $\mu W\Sigma_{WS}(\sigma) \neq \emptyset$. ■

The unification of a variable and a term not containing this variable is fairly simple:

3.6 Theorem. Let x be a variable and t a term with $x \notin V(t)$. Then

$\mu U\Sigma_{T,WS}(x,t) = \{ \sigma \circ \{x \leftarrow t\} \mid \sigma \in \mu W\Sigma_{T,WS}(\{x \leftarrow t\}) \}$ ■

It is always possible to add a greatest sort to S, if S has no greatest element.This completion does not affect the properties of sets of unifiers. Now we can assume that all variables in the codomain of substitutions in $C_T(U)$ and $\mu_T(U)$ have maximal sort.

The next theorem shows how to compute complete sets of unifiers in an equational theory with a many-sorted signature, namely compute first the unifiers without sorts, and then instantiate the unifiers into well-sorted ones.

In the case of sort-stable SORT, a minimal set of solutions can be obtained by minimizing finite subsets of the computed complete unifier set. In general only a complete set of solutions can be obtained.

3.7 Main Theorem. Let $=_{T,WS}$ be a congruence. Let W be a finite set of variables and let $U \subseteq \Sigma$ be a lower segment w.r.t $\leq_T[\![W]\!]$. Then:

i) $C_{T,WS,W}(U \cap WS\Sigma) = \{ \omega \circ \tau \mid \tau \in C_{T,W}(U), \omega \in \mu W\Sigma_{WS}(\tau) \}$

ii) If SORT is sort-stable and $\mu_{T,W}(U)$ exists, then

$$\mu_{T,WS,W}(U \cap WS\Sigma) = \cup \{ \mu_{T,WS,W}\{\omega \circ \tau \mid \omega \in \mu W\Sigma_{WS}(\tau)\} \mid \tau \in \mu_{T,W}(U) \}$$

Proof.

i) We show that the $\omega \circ \tau$ -set is a correct, complete set for $U \cap WS\Sigma$:

a) Correctness follows trivially, since U is a lower segment w.r.t. $\leq_T[\![W]\!]$.

b) Completeness: Let $\theta \in U \cap WS\Sigma$. Then there exists a $\lambda_1 \in \Sigma$ and a $\tau \in C_{T,W}(U)$, such that $\theta =_T \lambda_1 \circ \tau$ $[\![W]\!]$. Since we have assumed that all variables in VCOD(τ) have maximal sort, we have $\lambda_1 \in WS\Sigma$.

There exists an $\omega \in \mu W\Sigma_{WS}(\tau)$ and a $\lambda_2 \in WS\Sigma$, such that $\lambda_1 = \lambda_2 \circ \omega$ $[\![V(\tau(W))]\!]$.

Now $\theta =_T \lambda_2 \circ \omega \circ \tau$ $[\![W]\!]$ implies $\theta =_{T,WS} \lambda_2 \circ \omega \circ \tau$ $[\![W]\!]$, since $=_{T,WS}$ is closed; hence $\theta \leq_{T,WS} \omega \circ \tau$ $[\![W]\!]$.

ii) The set $\{\omega \circ \tau \mid \omega \in \mu W\Sigma_{WS}(\tau)\}$ is finite due to Theorem 3.4, hence a minimal subset exists. We show that it suffices to minimize the finite subsets $\{\omega \circ \tau \mid \omega \in \mu W\Sigma_{WS}(\tau)\}$ for a fixed τ:

Let $\tau_i \in \mu_{T,W}(U)$ and $\omega_i \in \mu W\Sigma_{WS}(\tau_i)$, i=1,2.

Assume, that $\omega_1 \circ \tau_1 \leq_{T,WS} \omega_2 \circ \tau_2$ $[\![W]\!]$. Then there exists a $\lambda \in WS\Sigma$, such that $\omega_1 \circ \tau_1 =_T \lambda \circ \omega_2 \circ \tau_2$ $[\![W]\!]$. By theorem 3.4 ω_1 is a renaming on VCOD(τ_1), hence by applying $\omega_1{}^c$, we obtain: $\tau_1 =_T \omega_1{}^c \circ \lambda \circ \omega_2 \circ \tau_2$ $[\![W]\!]$. The minimality of $\mu_{T,W}(U)$ implies $\tau_1 = \tau_2$. ∎

For applying this main theorem to unification, note that the set $U\Sigma_{T,WS}(s,t)$ is a lower segment in Σ w.r.t. $\leq_{T,WS}[\![W]\!]$ (Lemma 2.15). For the unification problem $\langle s =_{T,WS} t \rangle$, Theorem 3.7 i) yields

$CU\Sigma_{T,WS}(s,t) = \{\omega \circ \tau \mid \tau \in CU\Sigma_T(s,t), \ \omega \in \mu W\Sigma_{WS}(\tau) \}$.

In the case, where SORT is sort-stable, 3.7 ii) yields:

$\mu U\Sigma_{T,WS}(s,t) \subseteq \{ \omega \circ \tau \mid \omega \in \mu W\Sigma_{WS}(\tau), \tau \in \mu U\Sigma_T(s,t) \}$. Furthermore the test for instances is only necessary on the finite subsets $\{\omega \circ \tau \mid \omega \in \mu W\Sigma_{WS}(t)\}$.

We apply this theorem to matching problems in regular equational theories , i.e. for every equation $s =_T t$ we have $V(s) = V(t)$:

It is known that in regular theories every matcher is minimal [Sz82, Si84], in other words:

$\mu M\Sigma_T(s,t) = M\Sigma_T(s,t)_{|W}$ for all terms s,t and $W = V(t) \setminus V(s)$. In many-sorted regular theories we obtain a set $\mu M\Sigma_{T,WS}(s,t)$ by deleting all ill-sorted substitutions from $\mu M\Sigma_T(s,t)$: $\mu M\Sigma_{T,WS}(s,t) = \mu M\Sigma_T(s,t) \cap WS\Sigma$.

For regular, finitary matching and sort-compatible congruences there exists an algorithm, which decides $\sigma \leq_{T,WS} \tau$ for arbitrary substitutions: First compute the (finite) set of matchers of σ and τ and then test this set for a well-sorted substitution.

If in addition SORT is sort-stable the set $\mu U\Sigma_{T,WS}(s,t)$ is effectively computable for arbitrary terms s,t.

If SORT is not sort-stable but the set $CU\Sigma_{T,WS}(s,t)$ in Theorem 3.7 i) is finite the set $\mu U\Sigma_{T,WS}(s,t)$ is effectively computable.

3.8 Example. The minimizing step for sort-stable sort structures (part ii of 3.7) may be necessary to obtain a minimal set of solutions:

Let A,B be sorts with $A \sqsupseteq B$. Let $F := \{f,g,h\}$ with f: $A \times A \to A$; $A \times B \to B$; $B \times A \to B$; $B \times B \to B$.
and g: $B \to B$; h: $B \to B$.
Let $T := \{ g(f(x_A, y_B)) = h(f(x_A, y_B)),\ f(x_A, y_A) = f(y_A, x_A) \}$.
The generated congruence is sort-compatible, closed and sort-stable due to propositions 2.12 and 2.13.
Consider the unification problem $<g(z_B) =_{T,WS} h(z_B)>$.
The unsorted solution is: $\{ z_B \leftarrow f(x_1\ y_1) \}$, where x_1, y_1 are of maximal sort.
We have $\mu W\Sigma_{WS}(z_B \leftarrow f(x_1\ y_1)) = \{(x_1 \leftarrow x_{1,A}, y_1 \leftarrow x_{1,B}) ;(x_1 \leftarrow x'_{1,B}, y_1 \leftarrow x'_{1,A}) \}$.
The combination, however, can be minimized and the final set of solutions is:
$\{ z_B \leftarrow f(x_{1,A}\ y_{1,B}) \}$. ■

The results for permutative theories in [Sz82] can be extended to many-sorted term structures, due to Proposition 1.11: Permutative theories are in the class $\mathcal{M}_{1\omega}$, furthermore most general unifier sets always exist and are recursively enumerable.

The following table summarizes results, which can be obtained from the main theorem and results about permutative theories. The column "unsorted" contains the assumptions and the column "sort-stable" and "general" the conclusion. Note that every line of this table is a theorem.

	unsorted	sort-stable	general (not sort-stable)
$\lvert\mu U\Sigma\rvert$	1 n ∞	n n´ ∞	∞ ∞ ∞
	$T \in \mathcal{M}_{1\omega}$ and $\lvert\mu U\Sigma_T(s,t)\rvert < \infty$	$\mu U\Sigma_T \neq \emptyset$ decidable	$\mu US_T \neq \emptyset$ not decidable
	$\lvert\mu U\Sigma_T(s,t)\rvert = \infty$, $\mu U\Sigma_T \neq \emptyset$ decidable	$\mu U\Sigma_{T,WS} \neq \emptyset$ not decidable	$\mu U\Sigma_{T,WS} \neq \emptyset$ not decidable
$\mu U\Sigma_T(s,t)$	exists	$\mu U\Sigma_{T,WS}(s,t)$ exists	does not exist
	rec. enumerable and $T \in \mathcal{M}_{1\omega}$	rec. enumerable	open problem
	T permutative	$\mu U\Sigma_{T,WS}(s,t)$ exists, rec. enumerable	$\mu U\Sigma_{T,WS}(s,t)$ exists and is rec. enumerable

4. Example: Unification of Sets and Multisets.

4.1 Unification of Sets.

Unification of sets (where sets are not allowed as elements of sets) [LS76, Bu86b] can be modelled as unification of terms built from variables, constants and a binary ACI-function symbol f, i.e. f is associative, commutative and idempotent. For convenience we denote terms as sets of the occurring variables and constants. For example f(a x) is denoted as {a ,x}. Furthermore {a, a} = a . It follows from the axioms that it is allowed to remove duplicates.

It is well-known that ACI-unification is of type finitary and that ACI-matching is also finitary [LS76, Bu86b].
In unsorted ACI-unification problems it is not possible to make a difference between element variables and set-variables. However, there are applications, where sets have to be unified and some variables denote elements and not sets. Such a unification problem can easily be described using the following sort structure:

Let S := {SET,SINGLETON} with SET $\sqsupseteq$ SINGLETON.
Let f be defined everywhere with range sort SET. Let all constants be of sort SINGLETON.

Now SORT is a sort-assignment and all terms are well-sorted. However, SORT is not sort-compatible, since $SORT(f(x_{SINGLETON}, x_{SINGLETON})) = SET$ and $SORT(x_{SINGLETON}) = SINGLETON$.
We can use Theorem 2.11 to obtain a sort-compatible sort-assignment SORT':

SORT'(t) := SINGLETON, if t can be reduced to a variable or constant of sort SINGLETON
SORT(t) , otherwise.

From now on we use SORT' as sort-assignment. Note that this sort-assignment is not sort-stable.
Theorem 2.11 yields that SORT' is sort-compatible. Furthermore $=_{ACI,WS}$ is closed as can easily be verified.
An interesting observation is that $\mu W\Sigma_{WS}(\{x \leftarrow t\})$ is a singleton and that all terms in the codomain of substitutions in $\mu W\Sigma_{WS}(\{x \leftarrow t\})$ are variables or constants:

We have the cases:

i) x is of sort SET, then $\mu W\Sigma_{WS}(\{x \leftarrow t\}) = \{\varepsilon\}$
ii) x of sort SINGLETON, t contains more than one constant, then $\mu W\Sigma_{WS}(\{x \leftarrow t\}) = \emptyset$
iii) x of sort SINGLETON,t contains exactly one constant a,
then $\mu W\Sigma_{WS}(\{x \leftarrow t\}) = \{\mu\}$ with ,where μ maps all variables in t to a.
iv) x of sort SINGLETON , t contains only variables, then $\mu W\Sigma_{WS}(\{x \leftarrow t\}) = \{\mu\}$,
where μ maps all variables in t to a new variable of sort SINGLETON.

Theorem 3.7 i) now yields a finite set $CU\Sigma_{ACI,WS}(s,t)$ for arbitrary terms s,t.
Since ACI is finitary matching, we can remove instances and obtain $\mu U\Sigma_{ACI,WS}(s,t)$ as a subset of $CU\Sigma_{ACI,WS}(s,t)$.
Hence $\mu U\Sigma_{ACI,WS}(s,t)$ exists and is finite for all terms s and t.

The following example which is taken from [Bu86b] demonstrates how to compute the set $\mu U\Sigma_{ACI,WS}(s,t)$.
Let $s = \{x_{SET}, y_{SIN}, d\}$ and $t = \{a, b, w_{SET}\}$
The set of most general unifiers for the unsorted problem consists of 9 unifiers:

$\sigma_1 = \{x \leftarrow u,\quad y \leftarrow \{v,a,b\},\quad w \leftarrow \{u,v,d\}\}$
$\sigma_2 = \{x \leftarrow \{u,a\},\quad y \leftarrow \{v,b\},\quad w \leftarrow \{u,v,d\}\}$
$\sigma_3 = \{x \leftarrow \{u,a\},\quad y \leftarrow \{v,a,b\},\quad w \leftarrow \{u,v,d\}\}$
$\sigma_4 = \{x \leftarrow \{u,b\},\quad y \leftarrow \{v,a\},\quad w \leftarrow \{u,v,d\}\}$

$\sigma_5 = \{x \leftarrow \{u,b\},\quad y \leftarrow \{v,a,b\},\quad w \leftarrow \{u,v,d\}\}$
$\sigma_6 = \{x \leftarrow \{u,a,b\},\quad y \leftarrow v \quad w \leftarrow \{u,v,d\}\}$
$\sigma_7 = \{x \leftarrow \{u,a,b\},\quad y \leftarrow \{v,a\} \quad w \leftarrow \{u,v,d\}\}$
$\sigma_8 = \{x \leftarrow \{u,a,b\},\quad y \leftarrow \{v,b\} \quad w \leftarrow \{u,v,d\}\}$
$\sigma_9 = \{x \leftarrow \{u,a,b\},\quad y \leftarrow \{v,a,b\} \quad w \leftarrow \{u,v,d\}\}.$

We can assume that the variables u,v are of sort SET.
The next computation step is to examine each of these unifiers and instantiate them into well-sorted ones.
This is not possible for the unifiers σ_1, σ_3, σ_5 and σ_9, since y_{SIN} is of sort SINGLETON. Hence we obtain a complete set $CU\Sigma_{ACI,WS}(s,t)$ consisting of 5 unifiers:

$\sigma_2' = \{x \leftarrow \{u,a\},\quad y \leftarrow b,\quad w \leftarrow \{u,b,d\}\}$
$\sigma_4' = \{x \leftarrow \{u,b\},\quad y \leftarrow a,\quad w \leftarrow \{u,a,d\}\}$
$\sigma_6' = \{x \leftarrow \{u,a,b\},\quad y \leftarrow v',\quad w \leftarrow \{u,v',d\}\}$ (v' of sort SINGLETON)
$\sigma_7' = \{x \leftarrow \{u,a,b\},\quad y \leftarrow a,\quad w \leftarrow \{u,a,d\}\}$
$\sigma_8' = \{x \leftarrow \{u,a,b\},\quad y \leftarrow b,\quad w \leftarrow \{u,b,d\}\}$

Now we can remove σ_7' and σ_8', since they are instances of the unifier σ_6'. The set $\mu U\Sigma_{ACI,WS}(s,t)$ consists of the three unifiers σ_2', σ_4', and σ_6'. It may be the case that a modification of the unification algorithm in [Bu86b] yields a more efficient unification algorithm for the theory ACI together with the sort structure described above.

If a sort structure is used, where SINGLETON has subsorts, and these subsorts form a semi lattice, then there are only minor changes to the above procedure. The set $\mu W\Sigma_{WS}(\{x \leftarrow t\})$ is a singleton in this case, too.

4.2.Unification of Multisets.

The structure multisets [LS76, St81, HS85, Bu86a] is equivalent to the structure of AC-terms, generated by an associative and commutative function symbol, variables and constants.
The theory of AC-terms is regular and permutative, furthermore it is well-known that most general unifier sets are finite [LS76]. Hence for every sort-structure that makes the congruence closed and sort-compatible, the set of most general unifiers $\mu U\Sigma_{AC,WS}(s,t)$ exists for every term pair s,t.
The sort-structure on multisets, however, is in general sort-stable, i.e. it can be described by a finite set of range sort assignments to argument sorts, as is the case if we consider the addition + as an AC-symbol on the same sort-structure S = {INT,O,E} (cf.our introductory example).
In this case,we have:

i) $\mu W\Sigma_{WS}(\{x \leftarrow t\})$ is finite and consists of renamings.
ii) $\mu U\Sigma_{AC,WS}(s,t)$ exists and is finite

Conclusion.

The procedure described in this paper can easily be implemented and improves the efficiency of an automated deduction system. An advantage of the described procedure is the separation of the computation in an equational and a sort handling part. For sort-stable sort-assignments it is never necessary to compute most general weakening sets respecting equational theories.
Possibly some theories admit algorithms where equational steps and weakening are intermixed. If such an algorithm

exists, it may be more efficient, but it may be a hard task to find one for a given theory and it may be even harder to prove its completeness and minimality.

Acknowledgements.

I would like to thank: my colleagues H.J. Ohlbach, A. Herold and H.J. Buerckert for their support and for their helpful discussions during the preparation of this paper; special thanks go to my supervisor J. Siekmann.

References

Bu86a Buettner,W., "Unification in the datastructure multiset", *Journal of Automated Reasoning*, vol.2, pp. 75-88, (1986)

Bu86b Buettner,W., "Unification in the datastructure set", (to appear in this Proc. of 8th CADE)

CD83 Cunningham, R.J., Dick, A.J.J., "Rewrite Systems on a Lattice of Types", Rep. No. DOC 83/7, Imperial College, London SW7 (1983)

Co83 Cohn, A.G. "Improving the Expressiveness of Many- sorted Logic.", *AAAI-83*, Washington (1983)

Fa83 Fages, F. "Formes canonique dans les algèbres booleènnes et application à la dèmonstration automatique en logique de premier ordre." Thèse du 3ème cycle, Paris, (1983)

GM84 Goguen, J.A., Meseguer, J., "Equality, Types, Modules and Generics for Logic Programming", *Journal of Logic Programming*, (1984)

GM85 Goguen, J.A., Meseguer, J., "Order Sorted Algebra I. Partial and Overloaded Operators, Errors and Inheritance.", SRI Report (1985)

Gr79 Grätzer, G., "Universal algebra", Springer Verlag, (1979)

Hay71 Hayes, P., "A Logic of Actions.", *Machine Intelligence 6*, Metamathematics Unit, University of Edinburgh (1971)

He83 Herold, A. , "Some basic notions of first order unification theory.", Internal report 15/83 , Univ. Karlsruhe, (1983)

HS85 Herold, A.,Siekmann, J., "Unification in abelian semigroups", technical report, Universität Kaiserslautern, Memo SEKI-85-III, (1985)

Hen72 Henschen, L.J., "N-Sorted Logic for Automated Theorem Proving in Higher-Order Logic.", *Proc. ACM Conference Boston* , (1972)

Hu76 Huet, G., "Resolution d´equations dans des languages d´ordere 1,2,..., ω;", These d´Etat, Univ. de Paris, VII, (1976)

HO80 Huet, G., Oppen, D.C., "Equations and Rewrite Rules", SRI Technical Report CSL-111, (1980)

Hl80 Hullot, J.M., "Canonical forms and unification," *Proc of the 5th workshop on automated deduction*, Springer Lecture Notes, vol. 87, pp. 318-334, (1980)

KM84 Karl Mark G. Raph, "The Markgraf Karl Refutation Procedure", Technical report, Univ. Kaiserslautern, Memo SEKI-84-03, (1984)

Lo78 Loveland, D., "Automated Theorem Proving", North Holland, (1980)

LS76 Livesey,M., Siekmann, J., "Unification of sets and multisets," Technical report, Unive Kaiserslautern, Memo-SEKI-76-01 , (1976)

Ob62 Oberschelp, A., "Untersuchungen zur mehrsortigen Quantorenlogik.", *Mathematische Annalen 145* (1962)

Pl72 Plotkin, G., "Building in equational theories.", *Machine Intelligence, vol. 7*, 1972

Ro65 Robinson, J.A., "A Machine-Oriented Logic Based on the Resolution Principle." *JACM 12* (1965)

Sch85a Schmidt-Schauss, M., "A many-sorted calculus with polymorphic functions based on resolution and paramodulation.", *Proc 9th IJCAI*, Los Angeles ,(1985)

Sch85b Schmidt-Schauss, M., "Unification in a many-sorted calculus with declarations.", *Proc. of GWAI*, (1985)

Sch86 Schmidt-Schauss, M., "Unification in many-sorted equational theories.", Internal report. Institut für Informatik, Kaiserslautern (forthcoming)

Si84 Siekmann, J. H., "Universal Unification", *Proc. of the 7th CADE*, Napa California, (1984)

St81 Stickel, M.E.,"A unification algorithm for Associative-Commutative Functions", *JACM 28*,pp. 423–434, (1981)

Sz82 Szabo, P. , "Theory of First order Unification.", (in German, thesis) Univ. Karlsruhe, 1982

Wa83 Walther, C., "A Many-Sorted Calculus Based on Resolution and Paramodulation.", *Proc. of the 8th I JCAI*, Karlsruhe, (1983)

Wa84 Walther, C., "Unification in Many-Sorted Theories.", *Proc. of the 6th ECAI*, PISA, (1984)

Classes of First Order Formulas
Under Various Satisfiability Definitions

H. Kleine Büning
Th. Lettmann

Institut für Angewandte Informatik und Formale Beschreibungsverfahren
Universität Karlsruhe (TH), 7500 Karlsruhe

Abstract

In this paper we consider some special satisfiability problems of first order logic. We study effects of a unique name assumption and a domain closure assumption on complexity of satisfiability tests for certain classes of formulas interesting in logic programming or relational database theory. It is shown that the last assumption simplifies the satisfiability problem for first order logic. However for classes of formulas with lower complexity of the unrestricted satisfiability problem no general reduction of complexity can be determined.

1. Introduction

In relational database theory some problems of first order predicate calculus occur. For example in order to check the consistency of a formula–set describing the database we have to test the satisfiablity of this set. In this case we meet several additional assumptions as the unique name assumption, i.e. constants with distinct names are distinct, and the domain closure assumption, i.e. satisfiability must be tested over a fixed domain corresponding to the set of constants occuring in the database description [7]. Also studying query and update problems for databases we meet satisfiability problems of this restricted type. On the other hand we might consider a database or a knowledge based system as a conjunction of nonquantified clauses (facts) and quantified clauses (rules). Testing only the restricted satisfiability seems to be more natural then testing the general case, if we look at the understanding of formulas in everyday life. Looking for an individual which must have a certain property we try to select one of the known individuals. This behaviour we find also in the PROLOG resolution mechanism. Only those facts are considered to be true that are explicitly given in the database.

Our question is however, whether it is an advantage to restrict the satisfiability in this manner. Clearly the undecidability of predicate calculus we be reduced to a combinatorial problem, but what do these restrictions mean for the more special classes of formulas occurring for example in PROLOG programs.

After some basic observations we restrict ourselves to the satisfiability problem for cer-

tain classes of formulas as described with clauses of special form or with fixed maximal arity of the predicates or with restrictions for the use of the identity symbol.
The above mentioned assumptions can be formulated in first order logic terminology with respect to the satisfiability problem as follows:

Unique Name Assumption:

A first order formula α is satisfiable under the unique name assumption – $\mathrm{SAT}_{\mathrm{diff}}(\alpha)$ – iff α is satisfiable and α has a model with the restriction that constants with different names are different individuals.*)

Domain Closure Assumption:

A first order formula α is satisfiable under the domain closure assumption – $\mathrm{SAT}_{\mathrm{dom}}(\alpha)$ – iff α is satisfiable and α has a model with a subset of the set of the constants of α as domain.*)

Combined Assumption:

A formula α is satisfiable under the unique name assumption and the domain closure assumption – $\mathrm{SAT}_{\mathrm{diff,dom}}(\alpha)$ – iff α is satisfiable over a domain consisting exactly of the (names of the) constants.*)

*) We take free variables as constants. Therefore variables occurring free and bounded must be renamed.

In the following we assume, that each formula contains at least one constant and doesn't contain function symbols. Further we use some special notations and abbreviations.

α is a formula in conjunctive quantificational form – CQF-formula –, if α is a first order formula consisting of a conjunction of quantified prenex disjunctions (clauses), e.g.

$$\alpha \equiv \bigwedge_i \forall x \forall y (P_i(x) \rightarrow Q_i(y)).$$

The form of CQF–formulas naturally corresponds to descriptions in logic, because in many cases information is given as a collection of clauses and not as a prenex formula.

A n–clause is a (quantified) disjunction of at most n literals, i.e. negated or nonnegated primeformulas (predicates or equations); a n–Horn–clause is a n–clause with at most one nonnegated primeformula; a n–Prolog–clause is a n–Horn–clause with exactly one nonnegated primeformula.

The last clauses are called Prolog–clauses, because a CQF–formula without existential variables of this type represents a Prolog program, if we look upon each clause as an implication of primeformulas.

CQF(P,q,n–type,m) denotes the class of CQF–formulas – a so called CQF–class – with clause prefixes of type P, e.g. $\forall^*\exists^*, \exists^*\forall\forall\exists^*$, and prefix length of at most q, with n–type clauses and with at most m–ary predicate symbols. Sometimes we omit arguments or use an asterix "*" to indicate that item to be unrestricted.

2. Undecidable Classes of Formulas

The satisfiability problem for predicate calculus without function symbols is known to be undecidable, even under the unique name assumption, since formulas describing the behaviour of Turing machines can be given without using the identity and in the implied Herbrand universe all constants are interpreted to be different. On the other hand the domain closure assumption forces the decidability of the satisfiability problem, because we only need to test all possible assignments of constants to existential variables and all truth assignments.

Theorem 1:
The domain-restricted satisfiability problem for first order predicate calculus is at least PSPACE–complete and requires at most nondeterministic exponential time in length of the formula. If the arity of the predicates and functions or the number of universal quantifiers is bounded by a fixed k, the problem is PSPACE–complete resp. NP–complete.

First-order predicate calculus

	SAT, SAT_{diff}	SAT_{dom}, $SAT_{dom,\ diff}$
no restrictions	undecidable	≥ PSPACE–complete ≤ NEXPTIME
bounded arity of predicates ≥ 2	undecidable	PSPACE–complete
bounded number of universal quantifiers	undecidable	NP–complete

Proof:
Let F be a first order formula of length n. Without loss of generality we have the following restrictions:
F contains no function symbols. (Replace a function symbol f by a predicate P_f describing its graph.) All quantified variables and constants in F have different names. F doesn't contain the identity symbol. (Replace " = " by a predicate I and several postulates.) F is prenex.
A transformation of a formula to one with the above restrictions requires only polynomial space and time.
F may contain k constants, $k \leq O(n/log n)$. (Existential variables which don't lie in the scope of an universal quantifier or free variables can be replaced by nondeterministically chosen constants from the set of all constants of the formula.)
Ad bounded arity:
The number of arguments of predicates may be bounded by $m \geq 2$. A nondeterministic algorithm for testing the satisfiability of F can be described as follows:
For each predicate P and each constant c in F choose nondeterministically a truth assignment.

Let be $x_1, ..., x_r$, $r \leq n$, the universal variables ordered according to the prefix; then any existential variable y lies in the scope of $x_1, ..., x_{s(y)}$, $1 \leq s(y) \leq r$. If we replace the universal variables by constants in all possible ways according to the sequence $c_1, ..., c_1, c_1; c_1, ..., c_1, c_2; ...; c_k, ..., c_k, c_{k-1}; c_k, ..., c_k, c_k$, we nondeterministically choose a new constant of $\{c_1, ... c_k\}$ for substitution of an existential variables, only if the sequence of constants for the reigning universal variables has changed. Since in the mentioned sequence of substitutions we never meet twice an arrangement of leading constants with that arrangement altered meanwhile, we are shure not to choose different values for an existential variable in a Herbrand instance of F, if the reigning universal variables are instantiated with the same constants.
Now test the Herbrand instances of F in the described order with the predicate values chosen before.
The first step must be performed only once and takes polynomial time; the second step must be performed $r!$–times (number of different instantiations for universal variables, $r! \leq O(e^{pol(k)})$) and each step takes polynomial time. Therefore the satisfiability of F can be tested in exponential time using polynomial space.
Ad unrestricted case:
The prefix of F may contain r universal quantifiers. In this case we build the conjunction of all Herbrand instances of F with constants replacing universal variables getting a formula F' of length $\leq O(k^r * n)$. By nondeterministically choosing values for the functions replacing the existential variables in the Herbrand instances we obtain a propositional formula, which can be tested in nondeterministic polynomial time in length of the formula. So we have an nondeterministic algorithm needing at most $O(exp(polynom(n)))$ time.
Restricting the number of universal quantifiers to be limited by a fixed m implies the length of the above conjunction of Herbrand instances to be limited by a polynomial in n. In this case the satisfiability problem is in NP. On the other hand the satisfiability problem for first order theory of equations is PSPACE–complete [5] and SAT is NP–complete. This completes our proof.
q.e.d.

Corollary:
The satisfiability problems for the in general case undecidable classes of formulas with the additional restrictions have the following complexities:

	SAT_{diff}	SAT_{dom}, $SAT_{dom,\ diff}$
CQF($\forall^*\exists^*$, 3, 3–clauses)	undecidable	NP–complete
CQF($\forall^*\exists^*$, $*$, $*$–clauses) at most k predicates	undecidable	PSPACE–complete

For undecidability proofs see [3] and [4].

3. Formulas without variables

Formulas in conjunctive normal form (CNF) without bounded variables, i.e. without quantifiers, are the most simple form of describing finite relations and connections between finite relations. Classes of such formulas not containing the identity symbol correspond directly to propositional calculus and they are independent of our different definitions of satisfiability. For classes with identity symbol the domain closure assumptions has no effect to the complexity of satisfiability tests.

Theorem 2:
The complexity of the satisfiability problem for the class of formulas without variables in conjunctive normal form and with identitiy symbol and the following restrictions is

	$\mathrm{SAT}_{\mathrm{dom}}$	$\mathrm{SAT}_{\mathrm{diff}}, \mathrm{SAT}_{\mathrm{diff,\ dom}}$
Prolog–clauses	consistent	$O(n^2)$
3–Horn–clauses	$O(n^2)$	$O(n^2)$
2–clauses	NP–complete	$O(n^2)$
3–clauses	NP–complete	NP–complete

Proof:
Ad SAT, $\mathrm{SAT}_{\mathrm{dom}}$:
Formulas with Prolog–clauses are obviously consistent. For 3–clauses we immediately have a transformation to 3SAT [2]. For 2–clauses a proof can be found in [4] dealing with identity only. (The proof is based on a reduction to the "One–in–three– 3SAT"–Problem [2].)
In case of 3–Horn–clauses we obtain the lower bound $O(n^2)$, because propositional 3–Horn formulas can be described. The upper bound $O(n^2)$ can be shown analogue to propositional 3–Horn formulas by identifying the constants a and b, if a clause $a = b$ occurs during the execution.
Ad $\mathrm{SAT}_{\mathrm{diff}}, \mathrm{SAT}_{\mathrm{dom,\ diff}}$:
In case of formulas with Prolog–clauses we can associate to each 3–Horn formula α a Prolog–formula σ in linear time, such that $\mathrm{SAT}_{\mathrm{diff}\ (,\mathrm{dom})}(\alpha)$ holds iff $\mathrm{SAT}_{\mathrm{diff}\ (,\mathrm{dom})}(\sigma)$ holds. Let be $\alpha_i \equiv \neg\beta_{i_1} \vee \neg\beta_{i_2} \vee \neg\beta_{i_3}$ a 3–Horn–clause which isn't a Prolog–clause. Then we write $\sigma_i \equiv a = b \vee \neg\beta_{i_1} \vee \neg\beta_{i_2} \vee \neg\beta_{i_3}$ for different constants a and b. Under the unique name assumption it must hold $a \neq b$ and therefore we have α_i in this case equivalent to σ_i.
For 3–Horn–clauses, 2–clauses and 3–clauses we can reduce the formula in linear time by considering the terms $a = b$ and $a \neq b$.
As a result we get a contradiction or a formula without identity symbol, which is equivalent to propositional calculus (i.e. for $P_i(a, b, a)$ we have a propositional variable $c_{P_i}(a, b, a)$). Therefore we have for CQF–classes with 3–Horn–clauses, 2–clauses and 3–clauses the complexity of the corresponding classes of propositional formulas and for Prolog–clauses the Horn–complexity of propositional calculus.

4.Monadic classes with and without identity symbol

At first we take a closer look at formulas with identity as only predicate symbol. As we will see, the identity often is essential for the complexity of the specified satisfiability problems. So the next theorem may be considered as a preassumption for later propositions.

Theorem 3:
The satisfiability problem for CQF-classes with identity symbol only and the following restrictions is:

	SAT, SAT_{dom}	SAT_{diff}, $SAT_{diff,\ dom}$
CQF(3–Prolog)	consistent	$O(n)$
CQF(3–Horn)	$\leq O(n^2)$	$O(n)$
CQF(2–cl)	NP–complete	$O(n)$
CQF(3–cl)	NP–complete	$O(n)$

Proof:
Ad CQF(3–Prolog) $\cap$ SAT(SAT_{dom}):
It is a trivial observation, that each formula is satisfiable over a domain consisting of one element, i.e. each formula is satisfiable under the domain closure assumption.
Ad SAT_{diff}, $SAT_{dom,\ diff}$:
There exist formulas $\alpha \in$ CQF(3–Prolog), for which $SAT_{diff}(\alpha)$ and $SAT_{diff,\ dom}(\alpha)$ do not hold, for example $\alpha \equiv c = d$ for distinct constants c and d. So we must perform at least one run along the formula to look after such clauses, which requires linear time. Now it suffices to show, that CQF(3–cl) $\cap$ SAT_{diff}($SAT_{diff,\ dom}$) has complexity $O(n)$.
Let be $\alpha \in$ CQF(3–cl); then the formula α has the form $\bigwedge_i Q_i\alpha_i$, where Q_i is a sequence of quantifiers and α_i is a 3–clause. Each quantified clause is a proposition over the size of the domain, is true or is false (in general). We can associate in constant time to each quantified clause independent of the rest of the formula an information about the size of the domain as $D \leq (\leq, =, \geq, \geq)i$ for $i \in \{1, ..., 6\}$.
After these computations we decide whether the propositions over the domain together with the number of constants leads to a contradiction or not. Because of $D \leq (...)i (i \leq 6)$ this can be done in linear time. (We don't count the total number of distinct constants, but need only to test, whether we have j constants, $j \in \{1, ..., 6\}$, or more and in the first case this number matches with the domain propositions.)
Ad CQF(k–cl) $\cap$ SAT(SAT_{dom}) for $k = 2, 3$:
The NP–completeness can be found in, resp. followed from [4]. The proof is based on a reduction to the "One–in–three– 3SAT"– Problem [2].
Ad CQF(3–Horn) $\cap$ SAT(SAT_{dom}):
At first we test, whether the formula is satisfiable over a one–elementary domain. This

requires $O(n)$ time. If this is false, we reduce in linear time the formula to a formula without variables as follows (with the assumption, that the domain is greater than one):

1-clauses:

$\forall x \forall y(x = y) : \#$	$\forall x \forall y(x \neq y) : \#$	$\forall x(x = a) : \#$
$\forall x \exists y(x = y)$: delete	$\forall x \exists y(x \neq y)$: delete	$\exists x(x = a)$: delete
$\exists x \forall y(x = y) : \#$	$\exists x \forall y(x \neq y) : \#$	$\forall x(x \neq a) : \#$
$\exists x \exists y(x = y)$: delete	$\exists x \exists y(x \neq y)$: delete	$\exists x(x \neq a)$: delete

(# stands for contradiction)

2- or 3-clauses (contain a negated term):

$..\forall x..\forall y..(\sigma(x,y) \vee x \neq y) : ..\forall x..\sigma(x,x)$	$..\forall x..(\sigma(x) \vee x \neq a) : \sigma(a)$
$..\forall x..\exists y..(\sigma(x,y) \vee x \neq y)$: delete	$..\exists x..(\sigma(x) \vee x \neq a)$: delete
$..\exists x..\forall y..(\sigma(x,y) \vee x \neq y) : ..\exists x..\sigma(x,x)$	

After at most two applications of the above reduction the clauses don't contain terms $x \neq y$ or $x \neq a$.

Now we consider clauses with $x = y$ or $x = a$. Because we are dealing with Horn-clauses, only the following cases may occur (σ is a variable-free):

$\forall x \forall y(\sigma \vee x = y) : \sigma$	$\exists x \forall y(\sigma \vee x = y) : \sigma$	$\forall x(\sigma \vee x = a) : \sigma$
$\forall x \exists y(\sigma \vee x = y)$: delete	$\exists x \exists y(\sigma \vee x = y)$: delete	$\exists x(\sigma \vee x = a)$: delete

The resulting formula contains only equations like $a = b$ or $a \neq b$ and is a conjunction of j-Horn-clauses ($j \leq 3$).

The $O(n^2)$ time algorithm for propositional Horn formulas can be applied on the above formula identifying two constants $a = b$ during the decision procedure, if necessary.
q.e.d.

Now we investigate formulas without identity symbol, but with monadic (unary) predicate symbols and then monadic formulas with identity symbol.

Theorem 4:

The satisfiability problem for classes of monadic CQF-formulas without identity symbol and with the following restriction is:

	SAT, SAT_{diff}	$SAT_{dom}, SAT_{dom,\ diff}$
CQF(3-Prolog)	consistent	consistent
CQF(2-Horn)	$\leq O(n^4)$	NP-complete
CQF(3-cl)	NP-complete	NP-complete

Proof:
Ad CQF(3–Prolog):
For $\alpha \in$ CQF(3–Prolog) each quantified clause contains a positive primeformula $P(z)$ (z is a variable or a constant). Taking the set of constants as domain and interpreting those predicates $P(c)$ as true for any constant c gives us at once a model, which suffices the different satisfiability restrictions.
Ad CQF(2–Horn) $\cap$ SAT(SAT$_{\text{diff}}$):
Without loss of generality we have only clause prefixes of the types $\forall$ and $\forall\forall$, since the sequence of the quantifiers can be changed and existential variables can be taken as new distinct constants. Testing the Herbrand instances requires a total amount of at most $O(n^4)$ time.
Ad CQF(2–Horn) $\cap$ SAT$_{\text{dom}}$(SAT$_{\text{dom, diff}}$), CQF(3–cl) $\cap$ SAT$_{\text{dom}}$(SAT$_{\text{dom, diff}}$):
We associate to each 3–clause propositional formula α in polynomial time a formula $\sigma \in$ CQF(*,*,2–Horn,1) with SAT(α) iff SAT$_{\text{dom}}(\sigma)$, as follows: Let be $\alpha \in$ 3–CNF, $\alpha \equiv \bigwedge_i \alpha_i$, where $\alpha_i \equiv (\neg)a_{i_1} \vee \neg a_{i_2} \vee \neg a_{i_3}$ and the set of variables is $\{a_{i_j} | 1 \leq i \leq m, 1 \leq j \leq 3\}$.
To each propositional clause we associate 2–Horn–clauses:

$$a_{i_1} \cong Q(a_{i_1})$$

$$\neg a_{i_1} \cong \neg Q(a_{i_1})$$

$$a_{i_1} \vee a_{i_2} \cong \exists y P_i(y) \wedge \bigwedge_{j \neq i_1, i_2} \neg P_i(a_j) \wedge \forall x (P_i(x) \rightarrow Q(x))$$

$$a_{i_1} \vee \neg a_{i_2} \cong Q(a_{i_1}) \vee \neg Q(a_{i_2})$$

$$\neg a_{i_1} \vee \neg a_{i_2} \cong \neg Q(a_{i_1}) \vee \neg Q(a_{i_2})$$

$$a_{i_1} \vee a_{i_2} \vee a_{i_3} \cong \exists y P_i(y) \wedge \bigwedge_{j \neq i_1, i_2, i_3} \neg P_i(a_j) \wedge \forall x (P_i(x) \rightarrow Q(x))$$

$$a_{i_1} \vee a_{i_2} \vee \neg a_{i_3} \cong \exists y (P_i(y) \vee \neg Q(a_{i_3})) \wedge \bigwedge_{j \neq i_1, i_2} \neg P_i(a_j) \wedge \forall x (P_i(x) \rightarrow Q(x))$$

$$a_{i_1} \vee \neg a_{i_2} \vee \neg a_{i_3} \cong \exists y (\neg P_i(y) \vee Q(a_{i_1})) \wedge \bigwedge_{j \neq i_1, i_2} P_i(a_j) \wedge \forall x (\neg P_i(x) \rightarrow \neg Q(x))$$

$$\neg a_{i_1} \vee \neg a_{i_2} \vee \neg a_{i_3} \cong \exists y (\neg P_i(y) \vee \neg Q(a_{i_3})) \wedge \bigwedge_{j \neq i_1, i_2} P_i(a_j) \wedge \forall x (\neg P_i(x) \rightarrow \neg Q(x))$$

The conjunction of the right side formulas is called σ.
If α is satisfiable, then there exists a truth assignment B with $B(\alpha) = \mathit{true}$. Now we define for each variable a:

$$Q(a) \text{ iff } B(a) = \mathit{true},$$

and for each clause containing the predicate P:

$$P_i(a_{i_j}) \text{ iff } B(a_{i_j}) = \mathit{true} \text{ for } j = 1, 2(, 3).$$

Then it follows that SAT$_{\text{dom}}(\sigma)$ holds with the above interpretation.

For the other direction we define a truth assignment B as

$$B(a) = true \text{ iff } Q(a) \text{ for } a \in \{a_1, .., a_m\}.$$

Since σ is satisfiable under the domain closure assumptions, for each clause of α there exists a literal a with $Q(a)$ and therefore $B(\alpha) = true$ holds.
The rest of the proof, i.e. to show that CQF(3–cl) $\cap$ SAT_{dom} belongs to NP, is obvious.
q.e.d.

Theorem 5:
The satisfiability problem for the class of CQF–formulas with monadic predicate symbols, identity symbol and the following restrictions is

	SAT, SAT_{dom}	SAT_{diff}	$\text{SAT}_{\text{diff, dom}}$
CQF(2–Prolog)	consistent	P	NP–complete
CQF(3–Prolog)	consistent	P	NP–complete
CQF(3–Horn)	P	P	NP–complete
CQF(2–cl)	NP–complete	$\leq$ NP	NP–complete
CQF(3–cl)	NP–complete	NP–complete	NP–complete

Proof:
Ad NP–complete cases:
At first we will show the NP–completeness of CQF(2–Prolog) $\cap$ $\text{SAT}_{\text{diff, dom}}$. The NP–completeness of the other cases follows immediately or from the previous theorems. That the classes belong to NP is obvious.
Ad CQF(2–Prolog) $\cap$ $\text{SAT}_{\text{diff, dom}}$:
We associate to each propositional formula $\alpha \in$ 3–CNF a formula $\sigma \in$ CQF(2–Prolog), such that SAT(α) iff $\text{SAT}_{\text{diff, dom}}(\sigma)$ as follows:

$$a_{i_1} \vee a_{i_2} \vee a_{i_3} \cong \exists y P_i(y) \wedge \bigwedge_{j \neq i_1, i_2, i_3} (P_i(a_j) \rightarrow a_1 = a_2) \wedge \forall x (P_i(x) \rightarrow Q(x))$$

(first clauses describe $\neg P_i(a_j)$)

$$a_{i_1} \vee a_{i_2} \vee \neg a_{i_3} \cong \exists y (P_i(y) \vee \neg Q(a_{i_3})) \wedge \bigwedge_{j \neq i_1, i_2} (P_i(a_j) \rightarrow a_1 = a_2)$$
$$\wedge \forall x (P_i(x) \rightarrow Q(x))$$

$$a_{i_1} \vee \neg a_{i_2} \vee \neg a_{i_3} \cong \exists y (Q(a_{i_1}) \vee \neg P_i(y)) \wedge \bigwedge_{j \neq i_1, i_2} P_i(a_j) \wedge \forall x (\neg P_i(x) \rightarrow \neg Q(x))$$

$$\neg a_{i_1} \vee \neg a_{i_2} \vee \neg a_{i_3} \cong \exists x \forall y (P_i(x) \rightarrow x = y) \wedge \bigwedge_{j \neq i_1, i_2, i_3} P_i(a_j)$$
$$\wedge (Q(a_{i_3}) \rightarrow P_i(a_{i_3})) \wedge \forall x (\neg P_i(x) \rightarrow \neg Q(x))$$

(first two clause describes $\exists x \; \neg P_i(x)$)

The proof of the equivalence is similar to the proof of theorem 4.
Ad CQF(k–Prolog) $\cap$ SAT_{diff} for $k = 2, 3$, CQF(3–Horn) $\cap$ SAT (SAT_{diff}):
All classes can be reduced to CQF(3–Horn) $\cap$ SAT as follows:

Let be $\alpha \in$ CQF(3–Horn) $\cap$ SAT$_{\text{diff}}$ with constants $a_1, .., a_m$. We can add to α the clauses $a_i \neq a_j (1 \leq i \neq j \leq n)$. Then the resulting formula called α' is satisfiable iff SAT$_{\text{diff}}(\alpha)$. Let be given $\alpha \in$ CQF(3–Horn). We assume that α isn't satisfiable over a domain with exactly one element.

α has the form $\alpha \equiv \bigwedge_i Q_i \alpha_i$, where Q_i is a sequence of quantifiers and α_i is a Horn–clause.

In a first step we can rearrange the quantifiers or the quantified clause is true in any model and can be deleted, such that $\alpha' \equiv \bigwedge_j Q'_j \alpha_j$, Q'_j has the form $\exists^* \forall^*$.

E.g. $\forall x \exists y (x = y ...)$ is true and

$\forall x \exists y (P(x) \vee \neg Q(y))$ can be transformed to $\exists y \forall x (P(x) \vee \neg Q(y))$.

Now we can omit all existential quantifiers by taking the existential variables as pairwise different new constants and obtain a formula $\alpha'' \equiv Q''_i \alpha''_i$ with clause prefix Q''_i of type $\forall^*$.

Without loss of generality we assume, that the formula α'' contains at least one constant.

Now it holds:

SAT(α) iff SAT(α'') over a subset of the set of constants $\{c_1, .., c_m\}$.

We have obtained a CQF–formula with universal quantifiers only, but still with identity symbol.

Since the domain of a model must be a subset of $\{c_1, .., c_m\}$, we can introduce monadic identity descriptions as follows:

For all $1 \leq i, j, k \leq m$ we add clauses

$$\forall x : P_{c_i,c_i}(x) \qquad \forall x \forall y \forall z : (P_{c_i,c_j}(x) \wedge P_{c_j,c_k}(y) \rightarrow P_{c_i,c_k}(z))$$
$$\forall x \forall y : (P_{c_i,c_j}(x) \rightarrow P_{c_j,c_i}(y)) \qquad \forall x \forall y : (P_{c_i,c_j}(x) \rightarrow P_{c_i,c_j}(y))$$

and for each predicate symbol T of α and all $1 \leq i, j \leq m$

$$\forall x : (P_{c_i,c_j}(x) \wedge T(c_j) \rightarrow T(c_i)))$$
$$\forall x : (P_{c_i,c_j}(x) \wedge \neg T(c_j) \rightarrow \neg T(c_i))).$$

We add these quantified Horn–clauses to α''. The resulting formula is called α'''. Again since the domain of a model of α''' is a subset of $\{c_1, .., c_m\}$, we replace each quantified clause by clauses without quantifiers (variables), building the conjunction of all Herbrand instances, e.g.

1) $\forall x \forall y (R(x) \vee \neg Q(y))$ is replaced by $\bigwedge_{1 \leq i,j \leq m} (R(c_i) \vee \neg Q(c_i))$
2) $\forall x (x = c_i \vee \neg P(x))$ is replaced by $\bigwedge_{1 \leq i \leq m} (c_i = c_1 \vee \neg P(c_i))$.

We obtain a formula – called σ – without quantifiers, but with identity terms $c_i = c_j$ and $c_i \neq c_j$. Finally we replace equations $c_i = c_j$ resp. $c_i \neq c_j$ by $P_{c_i,c_j}(c_1)$ resp. $\neg P_{c_i,c_j}(c_1)$ and get a CQF(3–Horn) formula – called σ' – without identity symbol and without quantifiers, such that SAT(σ) iff SAT(σ'). All the above transformations are

polynomial. Applying theorem 2 with σ', we see that CQF(3–Horn) $\cap$ SAT is decidable in polynomial time.
q.e.d.

As we saw the two assumptions have different effects on the complexity of the satisfiability problems for the considered classes of formulas. For example the unique name assumption obviously has no effect for classes of formulas without identity symbol, but satisfiability is harder to prove for quantified equality clauses and easier for variable free equations. So these assumptions don't reduce the complexity of such problems in general, for some classes a surprising result.

References

[1] E. Börger: Berechenbarkeit, Komplexität, Logik, Vieweg, 1985

[2] M. Garey, D.S. Johnson: Computers and Intractability, Freeman and Company, 1979

[3] H. Kleine Büning: Complexity Results for Classes of First-Order Formulas with Identity and Conjunctive Quantificational Form, Forschungsberichte der Universität Karlsruhe, 1985

[4] H. Kleine Büning, Th. Lettmann: First-Order Formulas in Conjunctive Quantificational Form, Forschungsberichte der Universität Karlsruhe, 1985

[5] A.R. Meyer, L.J. Stockmeyer: Word Problems Requiring Exponential Time, Proc. 5th Ann. ACM Symp. on Theory of Computing, 1973

[6] D.A. Plaisted: Complete Problems in the First-Order Predicate Calculus, Journal of Computer and System Sciences, 1984

[7] R. Reiter: Towards a Logical Reconstrustion of Relational Database Theory, in On Conceptual Modelling, Ed. Brodie, Mylopoulos, Schmidt Springer Verlag

[8] T. J. Schäfer: The Complexity of Satisfiability Problems, Proc. 10th Ann. ACM Symp. on Theory of Computing, 1978

DIAMOND FORMULAS IN THE DYNAMIC LOGIC OF RECURSIVELY ENUMERABLE PROGRAMS

Volker Weispfenning
Mathematisches Institut
Universität Heidelberg
D-6900 Heidelberg, FRG

ABSTRACT. Dynamic logic QDL as presented in |3| provides a comprehensive logical framework for the study of the before-after behaviour of deterministic and non-deterministic programs. While the set of all valid QDL-formulas is highly complex (Π^1_1-complete) and hence not axiomatizable, the subset of valid termination assertions was shown to be axiomatizable in |5|. In |8| , this result was generalized to the effect that the much larger QDL-fragment of diamond formulas is still axiomatizable and satisfies a compactness theorem. The proofs were based on a rather delicate proof-theoretical treatment of consistency properties. We show how results of this kind can be obtained in the general framework of recursively enumerable dynamic logic by a very flexible approach that uses only the compactness and completeness of first-order logic and saturated structures. The method is also applicable to the dynamic logic involving undeclared global procedures and recursive procedure calls studied in |6|.

INTRODUCTION.

Dynamic logic - QDL and its variants (see |3|) - is an extension of classical first-order logic by modal operators associated with programs. It has proved to be a convenient and comprehensive framework for the study of the before-after behaviour of deterministic and non-deterministic sequential programs, incorporating in particular Hoare logic. The set of all QDL-assertions valid in all interpretations has turned out to be highly complex (Π^1_1-complete); in fact even the modest subset of universally valid partial correctness assertions $\phi \rightarrow [\alpha]\psi$ is not recursively enumerable. So it is not axiomatizable, i.e. it cannot be generated by a finitary system of axioms and rules (see |3|). By way of contrast, Meyer and Halpern showed in |5| that the set of termination assertions $\phi \rightarrow \langle\alpha\rangle \psi$ valid in all interpretations can be axiomatized, i.e. satisfies a completeness theorem. In |8| ,Schmitt generalized this result by proving a completeness and compactness theorem for the much more comprehensive set of diamond formulas in regular and context-free dynamic logic. The proof uses the technique of consistency properties for infinitary logic, which involves rather delicate syntactical considerations, whereas the proof in |5| is based only on the compactness of first-order logic. In |6| , Meyer and Mitchell extended the result in |5| to a completeness theorem for termination assertions about programs involving recursive procedure calls and undeclared global procedures. Again the proof employs the compactness of an extension of first-order logic by global procedure calls, which in turn is established via a completeness theorem proved by the Henkin method.

The purpose of the present note is to show that a compactness theorem and an extended completeness theorem can be proved for diamond formulas in the general setting of the dynamic logic of arbitrary recursively enumerable programs, using only a few basic facts on first-order logic: Completeness, compactness, existence of ω_1-saturated structures. In particular, our complete system of axioms and rules is quite flexible with respect to modifications for various subsystems of recursively enumaerable dynamic logic. It also incorporates random assignments and array assignments. Moreover, the method covers diamond formulas in the extended language of |6| , including local variable declarations, undeclared global procedures and non-deterministic recursive procedures with call-by-address and call-by-value parameters.

1. COMPACTNESS.

Let L be a recursively enumerable (r.e.) first-order language, let Φ_1 be the set of all first-order L-formulas, and let Π_0 be an r.e. set of symbols $a_0, a_1, \ldots$ called atomic programs. Then we define the set Σ of sequents, the set Π of programs, and the set Φ of formulas, together with the rank of sequents, programs and formulas inductively as follows:

1.1 DEFINITION. (1) If $a \in \Pi_0$, then $a \in \Sigma$ and $r(a) = 1$;
(2) If $p \in \Phi$, then $p? \in \Sigma$ and $r(p?) = r(p) + 1$;

(3) If $\alpha,\beta \in \Sigma$, then $\alpha;\beta \in \Sigma$ and $r(\alpha;\beta) = r(\alpha) + r(\beta) + 1$;
(4) $\Sigma \subseteq \Pi$;
(5) If W_e is an r.e. subset of Σ with r.e. index e , then $e \in \Pi$ and $r(e) = \sup\{ r(\alpha) + 1 : \alpha \in W_e \}$;
(6) If p is an atomic formula in Φ_1 , then $p \in \Phi$ and $r(p) = \text{length}(p)$;
(7) If $p, q \in \Phi$ and x is a variable of L, then $(p \wedge q)$, $(p \vee q)$, $\exists x\ p$, $\forall x\ p \in \Phi$, and $r(p \wedge q) = r(p \vee q) = r(p) + r(q) + 1$, $r(\exists x\ p) = r(\forall x\ p) = r(p) + 1$;
(8) If $p \in \Phi$, then $\neg p \in \Phi$ and $r(\neg p) = r(p) + 1$;
(9) If $p \in \Phi$, $\alpha \in \Pi$, then $\langle\alpha\rangle p \in \Phi$ and $r(\langle\alpha\rangle p) = \omega \cdot r(\alpha) + r(p)$.

So in each case, the rank r is a countable ordinal.

The semantics of programs and formulas in a first-order structure $\underline{A}$ are defined as usual (see |3|,|4|). If V is the set of variables of L and $s:V \longrightarrow A$ is a state (over $\underline{A}$), then we write $(\underline{A},s) \models p$ for "p <u>holds in</u> $\underline{A}$ <u>at state</u> s . $\underline{A} \models p$ means that $(A,s) \models p$ for all states $s:V \longrightarrow A$. $\models p$ means that p is universally valid, i.e. that $\underline{A} \models p$ for all L-structures $\underline{A}$. If $\Gamma \subseteq \Phi$, $p \in \Phi$, then $\Gamma \models p$ means that p is a semantical consequence of Γ , i.e. that for all L-structures $\underline{A}$ and states $s:V \longrightarrow A$, $(\underline{A},s) \models \Gamma$ implies $(\underline{A},s) \models p$ (,where $(\underline{A},s) \models \Gamma$ stands for $(\underline{A},s) \models q$ for all $q \in \Gamma$). In all these definitions, atomic programs may be interpreted by arbitrary binary relations on $\underline{A}$ subject only to the following restriction: We require that there is a recursive map assigning to every pair (a,p) with $a \in \Pi_o$, $p \in \Phi_1$, a formula $p_a \in \Phi_1$ such that $\models \langle a\rangle p \longleftrightarrow p_a$.

1.2 <u>EXAMPLES</u> (1) a is an assignment $x := \tau$, where τ is an L-term ; then p_a may be taken as the formula $p(x/\tau) \in \Phi_1$, obtained from p by substituting τ for each free occurence of x in p .
(2) a is a random assignment x := ? ; then p_a may be taken as $\exists x\ p$.
(3) a is an array assignment $g(x_1,\ldots,x_n) := \tau$; then p_a may be constructed as in (|4|, p.287).

The set $\Diamond$ of <u>diamond programs</u> and the set D of <u>diamond formulas</u> are defined like Π and Φ , except that 1.1(2) is taken only for $p \in D$, 1.1(8) is replaced by $\Phi_1 \subseteq D$, and 1.1(9) is taken for $\alpha \in \Diamond$ only.

Next we associate (by induction on the rank) with every diamond formula p a countable disjunction $p' = \bigvee\{ p_i : i \in I_p\}$ of first-order formulas $p_i \in \Phi_1$:

1.3 DEFINITION (1) If $p \in \Phi_1$, then $p' = p$;
(2) If $p' = \bigvee\{p_i : i \in I_p\}$, then $(\langle\alpha\rangle p)' = \bigvee\{(p_i)_a : i \in I_p\}$;
(3) $(\langle q?\rangle p)' = (q \wedge p)'$;
(4) $(\langle\alpha;\beta\rangle p)' = (\langle\alpha\rangle\langle\beta\rangle p)'$;
(5) If W_e is an r.e. subset of Σ with r.e. index e, then
$(\langle e\rangle p)' = \bigvee\{(\langle\alpha\rangle p)' : \alpha \in W_e\}$;
Next, let $p' = \bigvee\{p_i : i \in I_p\}$, $q' = \bigvee\{q_j : j \in I_q\}$; then
(6) $(p \diamond q)' = \bigvee\{p_i \diamond q_j : i \in I_p,\ j \in I_q\}$;
(7) $(\exists x\, p)' = \bigvee\{\exists x\, p_i : i \in I_p\}$;
(8) $(\forall x\, p)' = \bigvee\{\forall x(\bigvee\{p_i : i \in F\}) : F \text{ finite}, F \subseteq I_p\}$.

REMARK. With appropriate coding of the subscripts $i \in I_p$ as natural numbers, the map $(p,i) \longmapsto p_i$ is (partial) recursive. In particular, $\{p_i : i \in I_p\}$ is a recursive sequence for fixed p.

Recall (|7|,|9|) that an L-structure $\underline{A}$ is ω_1-saturated if for every set S of first-order formulas with countably many common free variables, every variable x, and every state $s:V \longrightarrow A$,
$(\underline{A},s) \models \bigwedge\{\exists x(\bigwedge S') : S' \text{ finite}, S' \subseteq S\} \longrightarrow \exists x(\bigwedge S)$.

1.4 LEMMA. (i) $\models p' \longrightarrow p$.
(ii) If $\underline{A}$ is ω_1-saturated, then $\underline{A} \models p \longleftrightarrow p'$.
(iii) Every L-structure $\underline{A}$ has an ω_1-saturated elementary extension $\underline{B}$.

The proof of (i) and (ii) is by induction on the rank of p. (i) is straightforward using the equivalence $\langle a\rangle p_i \longleftrightarrow (p_i)_a$ for $a \in \Pi_o$. The only case of interest in (ii) is (8), where one uses the ω_1-saturation of $\underline{A}$. (iii) is a well-known model-theoretic fact (see |7|).

For any set Ψ of formulas, we let $\neg\Psi = \{\neg p : p \in \Psi\}$.

1.5 COMPACTNESS THEOREM FOR DIAMOND FORMULAS.
Let $\Gamma \subseteq \neg D$, $p \in D$, $p' = \bigvee\{p_i : i \in I_p\}$, and suppose $\Gamma \models p$. Then there exist finite subsets Γ' of Γ and I'_p of I_p such that $\Gamma' \models \bigvee\{p_i : i \in I'_p\}$ and hence $\Gamma' \models p$.

PROOF. For $q \in D$, let $q' = \bigvee\{q_i : i \in I_q\}$. Let $\underline{A}$ be an L-structure, $s:V \longrightarrow A$ a state, and let $\underline{B}$ be an ω_1-saturated elementary extension of $\underline{A}$. Then $(\underline{B},s) \models p \vee \bigvee\{q : \neg q \in \Gamma\}$, and so by 1.4(ii), $(\underline{B},s) \models \bigvee\{p_i : i \in I_p\} \vee \bigvee\{q_i : \neg q \in \Gamma, i \in I_q\}$, and so $(\underline{A},s) \models \bigvee\{p_i : i \in I_p\} \vee \bigvee\{q_i : \neg q \in \Gamma, i \in I_q\}$. Consequently, this infinite disjunction is valid in all L-structures.

By the compactness of first-order logic, there exist finite subsets Γ' of Γ, I_p' of I_p such that $\models \bigvee\{p_i : i \in I_p'\} \vee \bigvee\{q_i : \neg q \in \Gamma', i \in I_q\}$, and hence $\Gamma' \models \bigvee\{p_i : i \in I_p'\}$, and by 1.4(i) $\models \bigvee\{p_i : i \in I_p'\} \longrightarrow p$.

2. COMPLETENESS.

Let Ax and Rℓ be an r.e. set of axioms and rules that are sound and complete for the first-order logic of L-formulas. Add the following axiom schemata and rules :

(A1) $p_a \longrightarrow \langle a\rangle p$ for $a \in \Pi_o$, $p \in \Phi_1$.

(A2) $q \wedge p \longrightarrow \langle q?\rangle p$ for $p, q \in D$.

(A3) $\langle\alpha\rangle\langle\beta\rangle p \longrightarrow \langle\alpha;\beta\rangle p$ for $\alpha, \beta \in \Sigma$, $p \in D$.

(A4) $\langle\alpha\rangle p \longrightarrow \langle e\rangle p$ for $e \in \Pi$ an r.e. index of an r.e. set W_e of sequents in $\Diamond$, $\alpha \in W_e$.

(R1) $p \vdash p$, if q is a propositional consequence of p, $p,q \in \Phi$.

(R2) $p \longrightarrow q \vdash \langle\alpha\rangle p \longrightarrow \langle\alpha\rangle q$ for $p,q \in D$, $\alpha \in \Diamond$.

(R3) $p \longrightarrow q \vdash \exists x\, p \longrightarrow \exists x\, q$ for $p,q \in D$.

(R4) $p \longrightarrow q \mid \forall x\, p \longrightarrow \forall x\, q$ for $p,q \in D$.

We write $\vdash p$ for "p is provable in this system". If $\Gamma \subseteq \neg D$, then provability relative to Γ is defined as follows : $\Gamma \vdash p$, if there exist $\neg q_1, \ldots \neg q_n \in \Gamma$ such that $\vdash q_1 \vee \ldots \vee q_n \vee p$. The following is obvious :

2.1 <u>SOUNDNESS THEOREM</u>.
Let $\Gamma \subseteq \neg D$, $p \in D$. Then $\Gamma \vdash p$ implies $\Gamma \models p$.

The important fact is :

2.2 <u>EXTENDED COMPLETENESS THEOREM FOR DIAMOND FORMULAS</u>.
Let $\Gamma \subseteq \neg D$, $p \in D$. Then $\Gamma \models p$ implies $\Gamma \vdash p$.

Notice that the set of axioms and rules of our system is r.e.. So we may conclude :

2.3 <u>COROLLARY</u>. If $\Gamma \subseteq \neg D$ is r.e., then the set $\{p \in D : \Gamma \models p\}$ is r.e..

PROOF of 2.2 : It suffices to show :

CLAIM 2.4. Let $p' = \bigvee\{p_i : i \in I_p\}$. Then $\vdash p_i \longrightarrow p$ for all $i \in I_p$.

Granted the claim, we may argue as follows : Suppose to begin with that $\Gamma = \emptyset$. If $\models p$, then by 1.5, $\models \bigvee\{p_i : i \in I'_p\}$ for some finite subset I'_p of I_p . So by the completeness of our system for first-order logic, $\vdash \bigvee\{p_i : i \in I'_p\}$, and by 2.4 , $\vdash \bigvee\{p_i : i \in I'_p\} \longrightarrow p$, which shows $\vdash p$. Next let Γ be arbitrary and assume $\Gamma \models p$. By 1.5, there exists a finite subset Γ' of Γ such that $\Gamma' \models p$, and so
$\models p \vee \bigvee\{q : \neg q \in \Gamma'\}$, and so by the first case
$\vdash p \vee \bigvee\{q : \neg q \in \Gamma'\}$, and so $\Gamma' \vdash p$, and so $\Gamma \vdash p$.

The proof of 2.4 by induction on rank(p) is straightforward.

2.5 REMARK. Our axioms and rules contain formulas that are not diamond formulas. This can be avoided by replacing the implications between diamond formulas by suitable rules similar to those used in |8|. Claim 2.4 must then be rephrased in the form $\{r \vee p_i\} \vdash r \vee p$ for $r \in D$.

3. VARIANTS AND EXTENSIONS.

The compactness theorem 1.5 is evidently valid for diamond formulas in all subsystems of r.e. dynamic logic. To get complete axiomatizations for the set of diamond formulas in such subsystems, the very abstract axiom schema (A4) has to be modified according to the program constructions admitted in the specific system. To be more specific, let us consider the following 3 subsystems : Regular QDL , strict (or deterministic) QDL, and context-free QDL (see |3|,|2|).

In regular QDL (or QDL for short) programs are built up inductively according to 1.1 (1),(2),(3) (with Σ replaced by Π), and in addition:

3.1 (4) If $\alpha, \beta \in \Pi$, then $(\alpha \cup \beta)$ and $r(\alpha \cup \beta) = r(\alpha) + r(\beta) + 1$;
(5) If $\alpha \in \Pi$, then $\alpha^* \in \Pi$ and $r(\alpha^*) = \omega \cdot r(\alpha)$.

Axiom schema (A4) is now replaced by

(A4.1) $\langle\alpha\rangle p \vee \langle\beta\rangle p \longrightarrow \langle\alpha \cup \beta\rangle p$ for $\alpha, \beta \in \Diamond$, $p \in D$;
(A4.2) $\langle\alpha^n\rangle p \longrightarrow \langle\alpha^*\rangle p$ for $\alpha \in \Diamond$, $p \in D$, $n < \omega$.

When regular programs are identified in a natural manner with certain r.e. sets of sequents, regular QDL becomes a subsystem of r.e.QDL, in which the new axioms prove (A4) .

Strict (or deterministic) QDL is a subsystem SQDL of QDL, in which the use of $\cup$ and * in programs is limited to the deterministic constructs ' if p then α else β'($(p?;\alpha) \cup ((\neg p)?;\beta)$) and 'while p do α ' ($(p?;\alpha)^* \cup (\neg p)?$). Accordingly (A4.1) and (A4.2) are replaced by the Hoare-like axiom schemata

(A4.3) $(p \wedge \langle\alpha\rangle q) \vee (\neg p \wedge \langle\beta\rangle q) \longrightarrow \langle$if p then α else $\beta\rangle q$ for $\alpha,\beta \in \Diamond$ $p,q \in D$;

(A4.4) $\langle(p?;\alpha)^n\rangle(\neg p \wedge q) \longrightarrow \langle$while p do $\alpha\rangle q$ for $\alpha \in \Diamond$, $p,q \in D$, $n < \omega$.

Context-free QDL (CFQDL) is an extension of QDL, in which the iteration operator * is replaced by the recursion operator $\mu X\tau(X)$ in the formation of programs (see |2| for more details). Accordingly, the iteration schema (A4.2) has to be replaced by the recursion schema

(A4.5) $\langle\tau^n(\text{false}?)\rangle p \longrightarrow \langle\mu X\tau(X)\rangle p$ for $\tau \in \Diamond$, $p \in D$, $n < \omega$ ' .

(compare also |8| , § 4). This yields a complete axiomatization for universally valid diamond formulas in CFQDL.

Finally, we consider the dynamic logic QDL' based on the program constructs of Meyer and Mitchell |6|. We indicate the changes necessary to obtain compactness and completeness for diamond formulas in QDL' : The rôle of the set Φ_1 of first-order formulas is taken over by the set Φ_1' of first-order formulas about global procedure calls; states with values in L-structures are replaced by states with procedure environment with values in the corresponding expanded L-structures (see |6|). By (|6|, lemma 2) there is a complete axiomatization for this logic, and so this logic satisfies the compactness theorem. The definition of the map $p \longmapsto p' = \bigvee\{p_i : i \in I_p\}$ (with $p_i \in \Phi_1'$) is essentially as before (compare |6|, lemma 5). Since the logic of first--order formulas with global procedure calls is compact, lemma 1.4 applies with a corresponding definition of ω_1-saturation for the expanded L-structures. To obtain a complete axiomatization of the universally valid diamond formulas in QDL', the axiom schemata (A1)-(A3) and the rules (R1)-(R4) of section 2 have to be augmented by the axiom schemata (P1)-(P8) of (|6|, lemma 2) and the axiom schemata and rules of (|6|, theorem 1). This will guarantee the claim 2.4 is valid, and so prove the extended completeness theorem 2.2 .

We close with a remark on our use of ω_1-saturated structures : Both in regular QDL and context-free QDL, the validity of a formula at a state s with values in an L-structure $\underline{A}$ depends only on the restriction of s to finitely many variables. So the use of ω_1-saturated structures in section 1 can be replaced by recursively saturated structures in these cases . This tool has a more constructive flavor (see |1|).

REFERENCES.

|1| J.Barwise, J.Schlipf, An introduction to recursively saturated and resplendent models, J.Symb.Logic 41 (1976), 531-536 .

|2| D.Harel, First-order dynamic logic, Springer LNCS vol.68, 1979.

|3| ---"---, Dynamic logic, in The Handbook of Philosophical Logic, vol.II, Reidel, Dordrecht, 1984 .

|4| A.R.Meyer, R.Parikh, Definability in dynamic logic, J.Comp.Syst.Sci. 23 (1981), 279-298 .

|5| A.R.Meyer, J.Y.Halpern, Axiomatic definitions of programming languages: A theoretical assessment, J. ACM 29 (1982), 555-576 .

|6| A.R.Meyer, J.C.Mitchell, Termination assertions for recursive programs: Completeness and axiomatic definability, Inf. & Control 56 (1983), 112-138 .

|7| G.Sacks, Saturated Model Theory, Benjamin, Reading, Mass.,1972 .

|8| P.H.Schmitt, Diamond formulas: A fragment of dynamic logic with recursively enumerable validity problem, Inf. & Control 61 (1984), 147-158 .

|9| V.Weispfenning, Infinitary model theoretic properties of κ--saturated structures, Z. math. Logik u. G. M. , 19 (1973), 97-109 .

A PROLOG Machine

D.Warren
Dept. of Computer Science
University of Manchester
Manchester M13 9PL

ENGLAND

A Prolog Technology Theorem Prover:

Implementation by an Extended Prolog Compiler [1]

Mark E. Stickel

Artificial Intelligence Center

SRI International

Menlo Park, California 94025

Abstract

A Prolog technology theorem prover (PTTP) is an extension of Prolog that is complete for the full first-order predicate calculus. It differs from Prolog in its use of unification with the occurs check for soundness, the model-elimination reduction rule that is added to Prolog inferences to make the inference system complete, and consecutively bounded depth-first search instead of unbounded depth-first search to make the search strategy complete. A Prolog technology theorem prover has been implemented by an extended Prolog-to-Lisp compiler that supports these additional features. It is capable of proving theorems in the full first-order predicate calculus at a rate of thousands of inferences per second.

1 PTTP Concept and Implementation

Despite Prolog's logic and theorem-proving heritage, and despite its use of theorem-proving unification and resolution operations, Prolog still fails to qualify as a full general-purpose theorem-proving system. There are three reasons for this:

- For the sake of efficiency, many Prolog systems use an unsound unification algorithm.
- Prolog's inference system is not complete for non-Horn clauses.
- Prolog's unbounded depth-first search strategy is incomplete.

[1]Preparation of this paper was supported by the Defense Advanced Research Projects Agency under Contract N00039-84-K-0078 with the Naval Electronic Systems Command. The views and conclusions contained herein are those of the author and should not be interpreted as necessarily representing the official policies, either expressed or implied, of the Defense Advanced Research Projects Agency or the United States government. Approved for public release. Distribution unlimited.

Nevertheless, Prolog is quite interesting from a theorem-proving standpoint because of its very high speed as compared with conventional theorem-proving programs. The objective of a *Prolog technology theorem prover (PTTP)* is to remedy the above deficiencies while retaining to the fullest extent possible the high performance of well-engineered Prolog systems.

Our current effort to secure the advantages of Prolog for general-purpose theorem proving is the construction of an extended Prolog compiler. Written in Common Lisp for the Symbolics 3600 Lisp machine, it translates Prolog procedures into Common Lisp functions that are then compiled by the Lisp compiler. This process yields performance on the standard Prolog `reverse` benchmark of about 6.7K lips, and a few thousand inferences per second for general-purpose theorem proving. Modifying a Prolog virtual machine [21], such as the one being developed at Argonne National Laboratory, should yield better performance and faster compilation with smaller object code.

Each Prolog procedure (a list of clauses with the same predicate in the head literal) is translated into a single Lisp function to which is passed the procedure's arguments and a continuation. The procedure

```
p(args1).
p(args2) :- q(...).
p(args3) :- r(...), s(...).
```

is translated into something like

```
function p (args,cont);
   begin
   if unify(args,args1) then cont;
   undo-unify;
   if unify(args,args2) then q(...,cont);
   undo-unify;
   if unify(args,args3) then r(...,s(...,cont));
   undo-unify
   end
```

This approach to Prolog compilation is also described in Cohen [3].

We discuss below Prolog's deficiencies for general-purpose theorem proving and examine the manner in which they are dealt with by the current PTTP implementation.

1.1 Unification

The first obstacle to general-purpose theorem proving that must be overcome is Prolog's use of unification without the occurs check. For efficiency, many implementations of Prolog do not check whether a variable is being bound to a term that contains that same variable. This can result in unsound or even nonterminating unification. A dramatic example of unsound deduction resulting from unification without the occurs check is the "proof" by the following program [16] that three is less than two:

```
X<(X+1).
3<2 :- (Y+1)<Y.
?- 3<2.
```

The invalid result relies upon the successful unification of the Prolog terms `X<(X+1)` and `(Y+1)<Y`, which creates circular bindings for `X` and `Y`. Unification of the values of `X` and `Y` will not terminate unless a unification algorithm capable of handling infinite terms is used [4].

Although applying the occurs check in logic programming can be quite costly, it is less likely to be too expensive in theorem proving, since the huge terms sometimes generated in logic programming are less likely to appear in theorem proving.

Accordingly, we simply compile in occurs checks except when they are obviously unnecessary. The unification of the actual and formal arguments, which are initially variable-disjoint, the first binding of a variable is guaranteed not to need the occurs check; only when a second occurrence of a variable is seen does it become necessary to start compiling in occurs checks.

An alternative approach is to allow creation of circular bindings by a unification algorithm that terminates even when applied to infinite terms and verify at the completion of a possible proof that no bindings used were circular. This may substantially reduce the cost of unification at the risk of allowing many inferences to be drawn after an infinite term is created that could have been cut off immediately by using the occurs check. We have no data on the trade-off between the cost of the occurs check and the amount of search saved.

1.2 Inference System

Prolog's inference system is complete for Horn sets of clauses, i.e., sets of clauses such that no clause has more than a single positive literal. In developing a Prolog technology theorem prover, it is necessary to extend the inference system so that it is complete for non-Horn sets of clauses as well.

However, one should consider only those means for extending Prolog's inference system that permit highly efficient Prolog implementation techniques. Some of the most important factors contributing to the high speed of well-engineered Prolog implementations are compilation and efficient representations for variable substitutions.

First, let us consider Prolog's representation for variable substitutions. This representation is made possible by the depth-first search strategy and by Prolog's use of a form of input resolution as its inference procedure.

Two methods for handling substitutions are used in conventional resolution theorem proving. The simple method is to form resolvents fully by applying the unifying substitution to the parent clauses. This is far more expensive in both time and space than Prolog inference.

The second method involves *structure sharing* [1], in which a resolvent is represented by the parents plus the unifying substitution. Whenever the resolvent must be examined (e.g., for printing or resolution with another clause), it is traversed, with variables being implicitly replaced by their substitution values. But this is much less efficient than the highly optimized form of structure sharing employed in Prolog in which only a single derived clause exists at a time and variables do not have multiple values.

The use of input resolution also facilitates the compilation of Prolog programs. In input resolution, there is a given set of input clauses such that (ignoring run-time assertions) these clauses are always used as one of the two inputs to each resolution operation. It is thus quite natural and effective to compile this given set of input clauses. It is more difficult and expensive to use compilation in more general forms of resolution, since derived clauses can be resolved with one another and there is thus no fixed set of clauses to compile.

All this suggests that a good approach to building a PTTP is to employ a complete inference system that is also an input procedure. Probably the simplest is [an affirmative form of] the *model elimination (ME)* procedure [8,9,10].

The ME procedure requires only the addition of the following inference operation to Prolog to constitute a complete inference system for the first-order predicate calculus:

> If the current goal matches the complement of one of its ancestor goals, then apply the matching substitution and treat the current goal as if it were solved.

This added inference operation is the ME *reduction* operation. The normal Prolog inference operation is the ME *extension* operation. The two together (without any need for factoring) comprise a complete inference system for the full first-order predicate calculus.

The reduction operation is a form of reasoning by contradiction. If, in trying to prove P, we discover that P is true if Q is true (i.e., $Q \supset P$) and also that Q is true if $\neg P$ is true (i.e., $\neg P \supset Q$), then P must be true. The rationale is that P is either true or false; if we assume that P is false, then Q must be true and hence P must also be true, which is a contradiction; therefore, the hypothesis that P is false must be wrong and P must be true.

In Prolog, when a goal is entered, a choice point is established at which the alternatives are matching the goal with the heads of all the clauses and executing the body of the clause if the match is successful. In this extension of Prolog, it is also necessary to consider the additional alternatives of matching the entered goal with the complements of each of its ancestor goals. For each such successful match, we proceed in the same manner as if we had matched the goal with the head of a unit clause.

In Prolog, when a goal is exited, the goal, instantiated by the current substitution, has been proved. In this extension of Prolog, when a goal is exited, all that has been proved is its instantiation disjoined with all the ancestor goals used in reduction operations in the process of "proving" the goal. Thus, in the example of proving P from $Q \supset P$ and $\neg P \supset Q$, expressed in Prolog by

```
p :- q.
q :- ¬p.
?- p.
```

when goal q is exited, $P \vee Q$, but not Q, has been proved. The top goal p, when exited, has been proved; there are no ancestor goals whose negation could have been assumed in trying to prove the top goal.

The reduction rule is implemented by maintaining, during execution, lists of the current ancestor goals. There is one such list for each predicate symbol or negated predicate symbol. The compiled code for a procedure then includes code that maintains this list (by pushing the current goal before and popping it after execution of the body of nonunit clauses) and unifies the incoming arguments with elements of the list of ancestor goals whose predicate is the complement of the procedure's predicate and executing the continuation for each successful match.

There are two additional prerequisites for using this inference system. First, contrapositives of the assertions must be furnished. For each assertion with n literals, n Prolog assertions must be provided so that each literal is the head of one of the Prolog assertions. The order of the literals in the clause body can be freely specified by the user, as is the case for ordinary Prolog assertions.

The second additional prerequisite relates to a feature of theorem proving that is absent in Prolog deduction: indefinite answers. Prolog, when provided with the goal $P(x)$, will attempt to generate all terms t such that $P(t)$ is definitely known to be true. In non-Horn clause theorem proving, however, there may be indefinite answers.

For example, consider proving $\exists x\ P(x)$ from $P(a) \vee P(b)$. In our extension to Prolog, this can be expressed as

```
p(a) :- ¬p(b).
p(b) :- ¬p(a).
?- p(X).
```

This set of assertions and the described inference procedure are still insufficient to solve the problem because there is no term t for which it is definitely known that $P(t)$ is true. To solve problems with indefinite answers, it is necessary to add the negation of the query as another assertion (n assertions if the query has n literals).

In this example, addition of the Prolog assertion ¬p(Y) results in the finding of two proofs (one in which p(X) is matched with p(a) and ¬p(Y) is matched with ¬p(b), one in which p(X) matched with p(b) and ¬p(Y) is matched with ¬p(a)). The answer to the query is thus $P(a) \vee P(b)$, i.e., either $P(a)$ or $P(b)$ (or both) is true, but neither $P(a)$ nor $P(b)$ has been proved. In general, indefinite answers are disjunctions of instances of the query. One instance of the query is included for each use of the query in the deduction (the use of the query as the initial list of goals and each use of the negation of the query).

Unfortunately, because the derivation of indefinite answers requires the inclusion of the conclusion's negation among the axioms, an otherwise static assertional database may need to be modified and expensively recompiled when indefinite answers are sought.

1.3 Search Strategy

Even if the problems of unification without the occurs check and an incomplete inference system are solved, Prolog is still unsatisfactory as a theorem prover because of its unbounded depth-first search strategy.

Consider the familiar problem of proving that a group is commutative if the square of every element is the identity element.[2] Using the common convention that $P(x, y, z)$ denotes $x \circ y = z$, the problem can be expressed in Prolog by the following assertions and goal:

```
p(X,e,X).
p(e,X,X).
p(X,X,e).
p(a,b,c).
p(U,Z,W) :- p(X,Y,U), p(Y,Z,V), p(X,V,W).
p(X,V,W) :- p(X,Y,U), p(Y,Z,V), p(U,Z,W).
?- p(b,a,c).
```

Prolog will fail to solve this problem because its unbounded depth-first search strategy will cause infinite recursion. Only the first of the two clauses for associativity will ever be used.

Obviously, to solve this problem, Prolog's unbounded depth-first search strategy must be replaced by some complete search strategy, such as breadth-first search or the A* algorithm [15]. However, arbitrarily choosing a complete search strategy may result in the loss of much of the efficiency of Prolog implementations. In particular, adopting breadth-first search or the A* algorithm would make it necessary for Prolog to represent and retain more than one derived clause at once. Moreover, such strategies would substantially increase memory requirements.

A simple solution to this problem is to replace Prolog's unbounded depth-first search strategy with a bounded one. Backtracking when reaching the depth bound would cause the entire search space, up to a specified depth, to be searched completely.

It is then necessary to determine the depth bound. Because the size of the search space grows exponentially as the depth bound increases, assigning too large a depth bound for a particular problem may result in an enormous amount of wasted effort. Furthermore, the amount of effort expended before discovering a proof will be highly dependent on the specified depth bound. An obvious solution to these problems is to run with increasing depth bounds—first one tries to find a proof with depth 1, then depth 2, and so on, until a proof is found. This is called *consecutively bounded depth-first search* [20] or *depth-first iterative deepening* [6]. The effect is similar to breadth-first search except that results from earlier levels are recomputed rather than

[2]This is Problem 9 for which performance results are given in Section 2. Problem 14 and Chang&Lee Problem 2 are the same except for order of the clauses. Problem 35 adds clauses for uniqueness, totality, and inverse of $\circ$, equality, and substitutivity. The two 4-literal clauses for associativity can be collapsed into the single Prolog assertion `p(X,V,W) :- p(U,Z,W), ((p(X,Y,U), p(Y,Z,V)); ((p(U,Y,X), p(Y,V,Z))` after renaming variables and reordering literals. This produces a smaller search space, though the technique was not used for our performance results.

stored. Thus, when searching is done to depth n, level $n - 1$ results are being computed for the second time, level $n - 2$ results are being computed for the third time, and results at level 1 are being computed for the nth time.

Because of the exponential growth in the size of the search space as the depth bound is increased, the number of recomputed results is small in comparison with the size of the search space. In particular, analysis shows that consecutively bounded depth-first search performs only about $\frac{b}{b-1}$ times as many operations as breadth-first search, where b is the branching factor [20].

Consecutively bounded depth-first search can also make use of heuristic information, in contrast to unbounded breadth- and depth-first search; the latter are uninformed search strategies that do not take into account heuristic estimates of the remaining distance to a solution. Informed search strategies such as the A* algorithm [15] utilize such information to order the search space. Consecutively bounded depth-first search does not do that, but can use an estimate of the minimum number of remaining steps to a solution to perform cutoffs if the estimated number exceeds the number of levels left before the depth bound is reached. These cutoffs result in lower effective branching factors for consecutively bounded depth-first search than for breadth-first search. If these estimates uniformly exceed the number of remaining levels by more than one, then one or more levels can be skipped when the next depth bound is set. This test can also be used to determine when a finite search space has been fully explored. As with the A* algorithm, admissibility—the guarantee of finding a shortest solution path first—is preserved, provided that the heuristic estimate never exceeds the actual number of remaining steps to a solution.

The consecutively bounded depth-first search strategy is implemented by using the new metalevel predicate `search`. The execution of `search(Goals,Max,Min,Inc)` attempts to solve `Goals` by a sequence of bounded depth-first searches that allow at least `Min` and at most `Max` subgoals, incrementing by at least `Inc` between searches. The last one, two, or three arguments of `search` can be omitted with default values of infinity, zero, and one provided for `Max`, `Min`, and `Inc`, respectively.

The `search` predicate succeeds for each solution it discovers. Backtracking into `search` continues the search for additional solutions. When, as in theorem proving, only a single solution (proof) is needed, the `search` call can be followed by a cut operation (as in the top-level goal `?- search(p(b,a,c)), !, write("proved"))` to terminate further attempts to find a solution. Although `search` does not check whether the solution it found is the same as a previously discovered one, it will avoid succeeding a second time with solutions previously found during a level m search that are rediscovered in the course of a later level n search.

At the beginning of each search, a depth bound representing the number of allowable subgoals is established by **search**. The compiled code for each nonunit clause decrements [undoably upon backtracking] this bound by the number of literals in its body. If the resulting bound is negative, resolution with the clause fails and backtracking occurs. The code also keeps track of the minimum amount by which the depth bound is exceeded; this is used to increment the depth bound for the next search.

This process merely counts subgoals to estimate the number of steps remaining to a solution and, as is required for admissibility, never overestimates their number, since each subgoal will require at least one inference step for its solution. It would be desirable to have a better estimator; this is difficult to achieve, however, because subgoals can often be removed in a single step, by resolution with a unit clause or by reduction. This makes an estimate of at least one step the only obvious, easily computed, admissible estimator of the number of steps remaining to a solution.

1.4 Refinements

The changes made in unification, the inference system, and the search strategy are all sufficient to create a Prolog technology theorem prover that is complete for the full first-order predicate calculus. It is of course possible to refine this system by adding restrictions on the current inference operations or by introducing entirely new inference operations.

The ME procedure justifies the completeness of our Prolog extension even if some goal states are disallowed.

For example, our extension of Prolog remains complete even if we cause the the current goal to fail under any of the following circumstances:

- A goal is identical to one of its ancestor goals (it is unnecessary to attempt to solve a goal while in the process of attempting to solve that same goal).
- An extended-upon goal is complementary to one of its ancestor goals (it is unnecessary to attempt to solve a goal that is complementary to an ancestor goal by any means other than the reduction operation).
- A goal with subgoals is an instance of a unit clause (it is unnecessary to solve a goal that is an instance of a unit clause by any means other than extension by the unit clause).

Some benefits of using such restrictions are that the use of commutativity assertions, such as `p(X,Y,Z) :- p(Y,X,Z)`, does not result by itself in an infinite search space and that propositional calculus problems can be solved safely and completely without any depth bound, since any state in which a goal is either identical to or complementary (unless removed by reduction) to an ancestor goal is rejected.

Because the search space in theorem proving is generally exponential, it is always worth considering criteria for failing goals, so that the exponentially many derivative deductions can be eliminated. However, the desire to cut off deductions must be balanced against the cost of checking to determine whether the present deduction is acceptable according to the criteria. Because consecutively bounded depth-first search entails minimal memory requirements, there is no point in reducing the number of inferences at the expense of overall increased running time. In contrast to other theorem-proving systems, it seems that the only reasonable measure of performance for a PTTP is the execution time for a proof.

After experimentation with various alternatives, the current implementation employs the following more limited forms of the restrictions:

- If the current goal (before unification with any clause in the procedure) is identical to one of its ancestor goals then fail.
- If the current goal is exactly complementary to one of its ancestor goals then perform the reduction operation and cut (disallow any other inferences on this goal).
- If the current goal is an instance of a unit clause then perform the extension operation and cut (disallow any other inferences on this goal).

Thus, these tests do some immediate checking, but will not detect when a substitution later causes one of the conditions to be violated. Relatively sophisticated code for detecting such conditions, depending on demons associated with individual variables that check for identity of goals when the variable is instantiated, has so far cost more in time than is saved by the diminished size of the search space.

Two possible solutions to this problem are to develop yet more efficient means for checking these conditions or to perform the checks less often. An effective means of reducing the frequency of the checks while maintaining most of their value is to restrict them to the earlier levels of the search. Given the exponential search space, cutoffs by earlier-level checks reduce the overall search-space size and running time more than do checks that are near the depth bound.

It is valuable to investigate other search-space pruning restrictions. Methods for "intelligent backtracking" in Prolog systems would also be beneficial for a PTTP.

The current implementation lacks either the model-elimination *lemma* facility [8,9] or the similar graph-construction procedure *C-reduction* operation [18] that can be used to shorten deductions by recognizing a goal as having been previously solved.

More exotic extensions are also worth considering. Among these are special unification for algebraic properties such as associativity and commutativity, or for sorted logic or types, and support for equality reasoning. An obvious first step toward including equality reasoning is the addition of a demodulation (equality rewriting) facility that simplifies goals to an irreducible form before attempting to solve them. Essentially the same compilation methods as are used for Prolog clauses can be applied to demodulators, making this equality simplification process quite rapid.

2 PTTP Performance

It is never an easy task to find a large number of problems with suitable accessibility, variety, and difficulty. We used the Wilson and Minker study [22] as a source of problems (the technical-report version of this article contains listings of the problems). Problems 1–9 were taken from Reboh et al. [17]; problems 10–19 were taken from Michie et al. [14]; problems 20–25 were taken from Fleisig et al. [5]; problems 26–58 were taken from Wos [23]; problems 59–86 were taken from Lawrence and Starkey [7].

We have so far solved about 70% of these problems, usually in well under a second (not counting compilation time). Wilson and Minker reported that their system MRPPS, with any one proof procedure, solved only about 55% of the 152 problems that include the above 86, their fully factored versions, and five problems devised by modifying and/or combining problems taken from the above 86; more than 15% of the problems could not be solved by any proof procedure (in particular, the Fleisig et al. and many of the Wos examples were especially difficult for Wilson and Minker as well as for us).

The following table presents the results for solved problems. The depth of proof is expressed as $m+n$, where m is the number of initial goals (i.e., the length of the top clause of the derivation) and n is the number of subgoals. At the end of the table, we also provide results for the more accessible but often easier problems that appear in Chang and Lee [2], pp. 298–305.

Problem	Number of Clauses	Depth of Proof	Number of Inferences	Run Time (sec)
1. BURSTALL	19	1+11	690	0.16
2. SHORTBURST	11	1+ 5	35	0.02
3. PRIM	9	2+11	2,271	0.60
4. HAS-PARTS1	8	1+ 9	84	0.05
5. HAS-PARTS2	8	1+23	3,948	3.25
6. ANCES2	7	2+17	957	0.08
7. NUM1	7	1+ 5	25	0.01
8. GROUP1	6	1+ 3	12	0.01
9. GROUP2	7	1+ 9	1,589	1.18
10. EW1	6	1+ 6	37	0.01
11. EW2	5	2+ 5	35	0.00
12. EW3	9	3+15	1,423	0.12
13. ROB1	3	4+ 4	15	0.00
14. ROB2	7	1+ 9	1,402	0.53
15. DM	4	1+ 3	5	0.00
16. QW	3	3+ 9	1,406	0.47
17. MQW	5	2+ 4	181	0.09
18. DBABHP	14	1+10	1,175	0.56
19. APABHP	18	1+13	1,707,214	1,353.67
26. WOS1	17	1+ 9	59,526	28.44
27. WOS2	16	1+ 9	26,570	11.54
28. WOS3	20	1+ 2	8	0.01
29. WOS4	23	1+12	8,233,689	4,403.61
30. WOS5	16	1+ 6	748	0.30
31. WOS6	20	1+ 8	18,631	9.97
32. WOS7	19	1+ 6	575	0.27
33. WOS8	18	1+ 5	223	0.10
34. WOS9	20	1+ 6	1,089	0.54
35. WOS10	20	1+ 9	61,330	31.37
36. WOS11	22	1+ 7	17,643	10.44
37. WOS12	21	1+ 3	11	0.02
38. WOS13	22	3+ 3	55	0.03
39. WOS14	21	1+ 6	374	0.16
41. WOS16	27	1+ 6	1,583	0.78
42. WOS17	30	1+ 7	25,063	14.63
43. WOS18	25	1+ 4	74	0.04
44. WOS19	33	1+ 7	8,450	4.21
59. LS5	4	2+ 4	31	0.01
60. LS17	12	3+ 6	175	0.05
61. LS23	6	1+ 6	268	0.11
62. LS26	10	1+ 6	9	0.02
63. LS28	13	1+ 6	954	0.99
64. LS29	13	1+ 6	679	0.72
65. LS35	6	1+12	8,224	6.32
66. LS36	20	1+11	2,689,558	1,467.88
68. LS41	11	1+ 2	8	0.01

Problem	Number of Clauses	Depth of Proof	Number of Inferences	Run Time (sec)
69. LS55	13	1+ 3	58	0.05
70. LS65	20	1+ 8	16,674	9.02
71. LS68	15	1+ 1	2	0.00
72. LS75	16	1+ 6	3,365	1.86
73. LS76?	17	1+ 2	4	0.02
74. LS86?	18	1+ 3	20	0.01
75. LS87?	22	1+ 3	22	0.02
76. LS100	9	1+ 3	7	0.01
77. LS103	14	1+11	1,826	0.65
78. LS105	14	1+ 4	35	0.03
79. LS106	14	1+ 4	34	0.03
81. LS111	14	1+ 4	26	0.03
83. LS115	21	1+ 6	109	0.08
84. LS116	16	1+11	7,897	3.28
Chang&Lee 1	5	1+3	5	0.00
Chang&Lee 2	7	1+9	1,589	0.55
Chang&Lee 3	5	1+9	206	0.07
Chang&Lee 4	5	1+6	26	0.01
Chang&Lee 5	9	1+3	4	0.00
Chang&Lee 6	9	1+3	26	0.01
Chang&Lee 7	7	1+5	24	0.01
Chang&Lee 8	9	2+11	3,104	1.01
Chang&Lee 9	8	3+7	163	0.04

3 Conclusion

A Prolog technology theorem prover has numerous advantages. For problems that are not too difficult, i.e., if the proof is not too deep or the branching factor too large, a PTTP can rapidly explore the search space and return an answer quickly. Given the use of depth-first search, memory requirements are almost neglible. It can be a very useful reasoning utility. It is exceptionally easy to use, since its inference system and search strategy are essentially fixed. Prolog computations can be easily embedded in the theorem-proving process because it is implemented as an extension of Prolog itself. Because of its simplicity, it is comparatively easy to implement correctly and its behavior is comprehensible.

It is not, however, a panacea for the problem of theorem proving in general. Solutions to really hard problems will always require human assistance in specifying strategies and to determine where to search for a solution. This is strong motivation for the related effort at Argonne National Laboratory to use Prolog technology (i.e., an implementation of the Warren abstract machine) to speed up the existing implementation of the powerful interactive theorem-proving system

LMA+ITP [11,12,13].

Our choice has been to try to extend Prolog to general-purpose theorem proving in such a way as to get the highest possible inference rate. We are willing to perform more inferences if their individual cost can be kept low. The Argonne approach, on the other hand, is to use Prolog technology to modify their system to make exactly the same inferences as before, only faster. Model elimination is probably the closest match of an inference system for the full first-order predicate calculus to Prolog, so that using Prolog technology to speed up the other inference operations in LMA+ITP will not yield the same high inference rate as for a Prolog technology theorem prover. Thus, there is a trade-off between the higher inference rate of a PTTP and the greater flexibility of LMA+ITP that often allows much smaller search spaces.

For handling the hard problems that are currently beyond its reach, a Prolog technology theorem prover can be improved by incorporating some of the refinements suggested above. In addition, it may be speeded up by improvements in Prolog machine technology. It should always be possible to build a PTTP that runs at a respectable fraction of Prolog's speed. Projected machines with execution measured in megalips would make possible a PTTP orders of magnitude faster than the current one, which is already quite fast. Finally, a PTTP can be used as a reasoning component in a larger theorem-proving system that decomposes problems into subproblems which are then dealt with successively by a PTTP.

References

[1] Boyer, R.S. and J S. Moore. The sharing of structure in theorem-proving programs. In B. Meltzer and D. Michie (eds.). *Machine Intelligence 7*. Edinburgh University Press, Edinburgh, Scotland, 1972.

[2] Chang, C.L. and R.C.T. Lee. *Symbolic Logic and Mechanical Theorem Proving*. Academic Press, New York, New York, 1973.

[3] Cohen, J. Describing Prolog by its interpretation and compilation. *Communications of the ACM 28*, 12 (December 1985), 1311–1324.

[4] Colmerauer, A. Prolog and infinite trees. In Clark, K.L. and S.A. Tarnlund (eds.). *Logic Programming*. Academic Press, New York, New York, 1982.

[5] Fleisig, S., D. Loveland, A.K. Smiley III, and D.L. Yarmush. An implementation of the model elimination proof procedure. *Journal of the ACM 21*, 1 (January 1974), 124–139.

[6] Korf, R.E. Depth-first iterative-deepening: an optimal admissible tree search. *Artificial Intelligence 27*, 1 (September 1985), 97–109.

[7] Lawrence, J.D. and J.D. Starkey. Experimental tests of resolution based theorem-proving strategies. Technical Report, Computer Science Department, Washington State University, Pullman, Washington, April 1974.

[8] Loveland, D.W. A simplified format for the model elimination procedure. *J. ACM 16*, 3 (July 1969), 349–363.

[9] Loveland, D.W. *Automated Theorem Proving: A Logical Basis*. North-Holland, Amsterdam, the Netherlands, 1978.

[10] Loveland, D.W. and M.E. Stickel. The hole in goal trees: some guidance from resolution theory. *IEEE Transactions on Computers C-25*, 4 (April 1976), 335–341.

[11] Lusk, E.L., W.W. McCune, and R.A. Overbeek. Logic Machine Architecture: kernel functions. *Proceedings of the 6th Conference on Automated Deduction*, New York, New York, June 1982, 70–84.

[12] Lusk, E.L., W.W. McCune, and R.A. Overbeek. Logic Machine Architecture: inference mechanisms. *Proceedings of the 6th Conference on Automated Deduction*, New York, New York, June 1982, 85–108.

[13] Lusk, E.L. and R.A. Overbeek. A portable environment for research in automated reasoning. *Proceedings of the 7th Conference on Automated Deduction*, Napa, California, May 1984, 43–52.

[14] Michie, D., R. Ross, and G.J. Shannan. G-deduction. In B. Meltzer and D. Michie (eds.). *Machine Intelligence 7*. John Wiley and Sons, New York, New York, 1972, pp. 141–165.

[15] Nilsson, N.J. *Principles of Artificial Intelligence*. Tioga Publishing Co., Palo Alto, California, 1980.

[16] Plaisted, D.A. The occur-check problem in Prolog. *New Generation Computing 2*, 4 (1984), 309–322.

[17] Reboh, R., B. Raphael, R.A. Yates, R.E. Kling, and C. Velarde. Study of automatic theorem-proving programs. Technical Note 72, Artificial Intelligence Center, SRI International, Menlo Park, California, November 1972.

[18] Shostak, R.E. Refutation graphs. *Artificial Intelligence 7*, 1 (Spring 1976), 51–64.

[19] Stickel, M.E. A Prolog technology theorem prover. *New Generation Computing 2*, 4 (1984), 371–383.

[20] Stickel, M.E. and W.M. Tyson. An analysis of consecutively bounded depth-first search with applications in automated deduction. *Proceedings of the Ninth International Joint Conference on Artificial Intelligence*, Los Angeles, California, August 1985, 1073–1075.

[21] Warren, D.H.D. An abstract Prolog instruction set. Technical Note 309, Artificial Intelligence Center, SRI International, Menlo Park, California, October 1983.

[22] Wilson, G.A. and J. Minker. Resolution, refinements, and search strategies: a comparative study. *IEEE Transactions on Computers C-25*, 8 (August 1976), 782–801.

[23] Wos, L.T. Unpublished notes, Argonne National Laboratory, about 1965.

Paths to High-Performance Automated Theorem Proving

Ralph Butler
Ewing Lusk
William McCune
Ross Overbeek

Mathematics and Computer Science Division
Argonne National Laboratory
Argonne, Illinois 60439

ABSTRACT

We present four components of a strategy for the implementation of high-performance automated theorem proving systems. These are 1) clause compilation, 2) multiprocessing, 3) database indexing, and 4) clause-set "compaction". We describe each of these techniques and show how they can be integrated into a coherent system with significantly higher performance than traditional systems. The strategy described has been implemented, and we present results showing the effects of these techniques on Sam's Lemma, a relatively difficult theorem-proving problem.

1. Introduction

This is a report on the continuing evolution of the series of powerful, general-purpose theorem-proving programs begun at Northern Illinois University and now continuing at Argonne National Laboratory. After several preliminary efforts in the late sixties and early seventies, the first heavily-used system in the series was AURA, described in [9] and [15]. This was a very fast system, written in IBM assembler language and PL/I, which sacrificed portability and some measure of flexibility for speed. Nor was it particularly convenient to use. In the late seventies and early eighties, realizing how long-lived a full-featured theorem-proving system is likely to be, we developed the LMA subroutine package[4, 5] and the ITP automated reasoning system[7]. LMA/ITP emphasized portability (it was written in Pascal), ease of use (interactive, convenient input, etc.) and flexibility (highly structured, easy to add new features), but was not as fast as AURA (by a factor of roughly ten when the same features were being used in each).

We are now beginning to implement the next-generation system, and are returning to performance as a primary design criterion, but with no intention of sacrificing either portability or functionality. In this paper we describe the combination of new technologies underlying the design of the new system, and present some experimental results demonstrating the effects of these ideas. It is the intention of this paper to convey the essence of these ideas and their significance for theorem proving; the implementation details will be published elsewhere.

The new technologies described are:

1. Clause compilation. This idea is a direct descendant of David Warren's idea of compiling Prolog clauses[11, 12]. We show here what this means in the context of theorem proving, and briefly describe the implementation of the target machine for the compilation.

2. Multiprocessing. Both the most cost-effective computers and the very highest performance computers of the near future will be multiprocessors. We describe opportunities for parallelism in theorem proving, our implementation strategy, and give some preliminary results from runs on an Encore Multimax multiprocessor.

3. Database indexing. The database component of our previous systems has always been a major factor in their effectiveness. Clause compilation modifies the methods used in our earlier systems. We describe this and reiterate the necessity of separating the database and deduction components in automated reasoning systems.

* This work was supported by the Applied Mathematical Sciences subprogram of the Office of Energy Research, U.S. Department of Energy, under contract nr. W-31-109-Eng-38.

4. Clause-set compaction. This is a technique for preprocessing a set of related, non-unit clauses in such a way that certain duplicate computations occurring in their use for deduction are eliminated, thus providing increased efficiency.

None of these techniques has any effect on the deductions and other theorem-proving operations (e. g., subsumption) that occur during a given run. They affect only speed.

2. Compiling Clauses for Theorem-Proving Operations

The idea of compiling logic is due to D. H. D. Warren[13], who developed it the context of Prolog. We describe here the extension of his idea to the theorem-proving context. A theorem prover typically manipulates objects such as terms, literals, and clauses. Each object may play many different roles during the course of theorem-proving run. For example, a clause may be used in the formation of a resolvent, may be tested to determine whether it subsumes another clause or is subsumed by it, or be used as a rewrite rule. The standard approach to theorem prover implementation is to represent these objects as data structures and operate upon them with a variety of general-purpose algorithms. It is possible to replace the operation of a general-purpose algorithm on an object (representing a role in which the object is used) by a special-purpose algorithm (particularized to the object and the role). If all of the roles are replaced by special-purpose algorithms, the need for a representation of the object as a data structure disappears entirely. To make this clear, we consider a simple example.

2.1. An Example

Consider the unit clause `max(1,x,1)`, where x represents a variable. (This is an axiom of Sam's Lemma, described in reference [9].) We anticipate that it may be necessary to unify other literals with this one, and so we compile a special-purpose algorithm for unifying others literals with this one. We make the assumption that this algorithm will only be invoked when we know that the literal to be tested is of the form $\mathtt{max}(a_1, a_2, a_3)$, where a_1, a_2, and a_3 are unknown. Then the algorithm to determine whether the literal to be tested is indeed unifiable with `max(1,x,1)` is:

```
if (a1 is 1 or a variable and
    a3 is 1 or a variable) then

    return success

else

    return failure
```

(We ignore for the time being how one keeps track of the unifying substitution.) Note that a change of role for the literal requires the compilation of a different algorithm. The algorithm to determine whether a literal $\mathtt{max}(a_1, a_2, a_3)$ is more general than `max(1,x,1)`, as would be required during back subsumption, is:

```
if (a1 is 1 or a variable and
    a2 is a variable different from a1 and from a3 and
    a3 is 1 or a variable) then

    return success

else

    return failure
```

The test to determine whether `max(a1,a2,a3)` is an instance of `max(1,x,1)`, as would be required during forward subsumption, is:

```
if (a1 and a3 are both 1) then

    return success

else

    return failure
```

2.2. Choice of Object Languages

Warren's original Prolog compiler compiled Prolog into the machine language of the DEC-10. For convenience and portability, many current Prolog systems use as a target language the machine (or assembler) language of an abstract machine based on a design by Warren[12], which is then executed on an emulator. In our application of compilation to theorem proving, we use an extension to Warren's machine, but it is important to realize that the notion of compilation is separate from any particular choice of object language. Extensions to Warren's original design are necessary in two areas: unification instructions must be included to handle the occurs check and one-way matching (not crucial in Prolog but very much so in theorem proving), and data structures to support parallelism must be provided. (see next section). (The extensions for the occurs check and one-way matching, together with a tutorial on the Warren abstract machine in general are given in [1]).

2.3. Compiling Non-Units

Let us now consider a non-unit clause such as one of the associativity axioms from Sam's Lemma:

```
¬max(x,y,xy)   ¬max(y,z,yz)   ¬max(x,yz,xyz)   max(xy,z,xyz)
```

One of its roles (in our attack on the problem) is that of hyperresolution nucleus[10]. That is, during the run a positive clause will be chosen from which to make deductions; we call this clause the *given* clause. If the given clause unifies with one of the negative literals above, then other positive clauses must be found with which to remove the remaining negative literals, leaving a new, deduced positive clause which is processed (simplification, forward and backward subsumption) and possibly added to the clause space (perhaps compiled for some set of roles and added to data structures for others.) So the nucleus above can be compiled into an algorithm like:

```
if (the given clause unifies with max(x,y,xy)) then

    apply the resulting substitution to the remaining literals
    attempt to find a clause that unifies with max(y,z,yz)

    if success then

        apply the resulting substitution to the remaining literals
        attempt to find a clause that unifies with max(x,yz,xyz)

        if success then

            apply the resulting substitution to max(xy,z,xyz)
            process and add it (with all substitutions applied)
          as the new derived clause
```

Two additional parts of the algorithm are compiled, corresponding to matching the given clause with each of the second and third literals first, respectively. The three sections of the algorithm correspond to the three negative literals, which represent in turn three ways for the clause to be used (three roles) in conjunction with the given clause to produce a resolvent. The algorithm is perhaps more easily understood if expressed in Prolog. Suppose we view a given clause like `max(a,b,c)` as converted to `callmax(a,b,c)` in order to invoke this procedure. Then the procedure could be written as

```
callmax(X,Y,XY)  :-    max(Y,Z,YZ),
                       max(X,YZ,XYZ),
                       gen_clause(max(XY,Z,XYZ)).
callmax(Y,Z,YZ)  :-    max(X,Y,XY),
                       max(X,YZ,XYZ),
                       gen_clause(max(XY,Z,XYZ)).
callmax(X,YZ,XYZ)  :- max(X,Y,XY),
                       max(Y,Z,YZ),
                       gen_clause(max(XY,Z,XYZ)).
```

Here `gen_clause` is a built-in procedure for processing a new clause, including compilation. This is then an expression of the algorithm for hyperresolution, particularized for this nucleus. Compilation is similar to Prolog compilation of the above procedure. The occurs check will be included. The unification tests for the individual literals will be compiled as described above. Finding a clause to unify with a given literal will be done via the indexing method described in Section 4.

All the ideas from the field of optimizing compilers apply. The simple examples given here do not adequately indicate the savings in execution time that may be gained by taking advantage of information about the clause available at compile time. As in compiling other languages, compilers that do careful register allocation and sophisticated optimization can make a big difference.

3. Multiprocessing

We are now concluding the period in computer history in which most multiprocessors have been experimental, one-of-a-kind machines. Shared-memory machines with as many as twenty processors are now commercially available, as are non-shared-memory machines with over one thousand processors. The largest, fastest computers, such as those at the top end of the CRAY family, are multiprocessors. Hence application programs requiring intensive computation (here theorem proving certainly fills the bill) must be parallelized in order take advantage of state-of-the-art hardware.

3.1. Parallel Theorem Proving

Automated theorem proving provides abundant opportunities for parallelism. Here are only a few:

1. The pipeline:

   select given clause →
   generate resolvents →
   simplify resolvents →
   forward subsumption →
   backward subsumption →
   integrate into knowledge base

   can be run as a collection of concurrently executing processes, given some restraints imposed by completeness.
2. When a given clause generates multiple resolvents, they can be processed in parallel.
3. The literals of a clause can in some cases be attacked in parallel.
4. The application of rewrite rules to the literals of a clause and to the arguments of a term may proceed in parallel.

Problems associated with parallel programs are:

1. The design of parallel algorithms (the next step after the identification of potential). This is (merely?) a programming problem.
2. The expression of the parallel algorithm. This involves the choice of a parallel programming language or extensions to an existing programming language. The design of such languages and extensions is currently an active research area.
3. Choice of an execution environment. (This is not independent of problem 2.) In order to defer this choice as long as possible, portability must be a major factor in the choice of language.

3.2. Our Approach

We are implementing our new theorem prover in a mixture of Prolog and PARLOG[3], which is a dialect of logic programming designed to support multiprocessing. The PARLOG and Prolog code will both be compiled into the machine language for WAM[2], our parallel version of an extended Warren abstract machine. The primary change to the Warren machine to allow parallelism is that the machine stack becomes a tree, and the synchronization problems associated with multiple processes manipulating this tree must be solved. This implementation will be described in full in a forthcoming paper. WAM itself is implemented in C, with the parallelism expressed via monitors implemented as macros. The success of this environment for providing parallelism, efficiency, and portability across multiprocessors with differing synchronization primitives has been been described elsewhere[8, 6].

4. Database Indexing

A fundamental operation in theorem proving is the lookup by key of an object (or the code associated with one of its roles) in the current knowledge base. The use of a key based on unification properties to rapidly isolate the set of literals or terms which are likely to unify with a given term is a crucial ingredient in both AURA and LMA. Historically a major bottleneck, it is likely to become even more so as compilation reduces other sources of overhead.

In the long run, the most efficient solution to this problem may come from associative memory hardware. Until that solution becomes feasible, however, it is necessary to isolate the theorem prover's interface to its database, so that it can adapt to technological advances emerging from the database industry.

4.1. The Prolog Approach

The speed of DEC-10 Prolog is due not only to Warren's compilation technique but also to his notion of "indexing" the compiled clauses to allow rapid access to sections of the compiled code, bypassing a sequential search through the clauses of a Prolog procedure. The idea is to hash on the first argument of the literal invoking a particular procedure to find the first clause head which is likely to unify with it. This avoids repeated attempted unifications which are doomed to fail.

Indexing on the first argument works well for most Prolog programs, in which the clauses for a procedure (those with the same predicate in the head literal) may be easily distinguished by the first argument and terms are often not particularly complex. In theorem proving, however, our experience has been that it is usually necessary to discriminate more finely among literals and terms in the clause space. There may be very many literals with the same predicate symbol, and the first argument may not be sufficient to distinguish them adequately. Furthermore, there may be deeply nested terms. The methods we used to deal with this problem in our earlier systems are described in [10] and [4]. Another interesting approach to this problem in the Prolog context can be found in [14].

4.2. Our Approach

Once roles of objects are compiled to code, the database contains code fragments. The role of the indexing scheme is then to retrieve the address of the code, rather than a data structure representing an object. The index is implemented as a more-or-less standard hash mechanism. What is unusual is the way keys are created. The values associated with the keys can be any type of data item the user wishes; in our case they are pointers to code of the type described in Section 2. The keys are designed to help in determining which literals or terms *might* unify with a given term; the code pointed to by the value carries out the actual unification test. The goal is to avoid executing code which we can tell in advance will result in unification failure.

A given literal or term may have more than one key-value pair in the index, and the keys generated for term when it is first added to the index are not necessarily the same as the keys used to retrieve it, which depend on role. We illustrate this situation with a few examples.

4.3. Example

Consider adding the literal `p(a)` to the index. Two keys will be generated:

```
p a
p gc
```

where `gc` represents a "generic constant" symbol, whose use will be apparent once we look at retrieval. If we add the literal `p(x)` to the index, where x is a variable, the two keys generated are

```
p gv
p gc
```

where `gv` represents a "generic variable". Now suppose we want to retrieve literals that fully unify with `p(a)`. (This means that we want to execute the code representing literals that might unify with `p(a)`, of which there may be many.) The keys for full unification retrieval are:

```
p a
p gv
```

The first will find p(a) and the second will find p(x), each of which unifies with p(a). To retrieve literals that are instances of p(a), we use only the one key

```
p a
```

To retrieve literals that are generalizations of p(x) we use the key

```
p gv
```

This key is sufficient because only variables are generalizations of a variable. The keys necessary to retrieve instances of p(x) are

```
p gv
p gc
```

since both variables and constants are instances of a variable. Here the use of the "generic constant" is illustrated, since we want to retrieve p(a) when looking for instances of p(x), but cannot generate all the possible keys for various constants.

In this example, we have shown the keys as comprised of 2-tuples. In our implementation, we use 4-tuples, thus allowing indexing on the functor and first three arguments of a literal or term. The key

```
p gv
```

would actually be represented as

```
p gv undef undef
```

The procedures to maintain indices have been added as built-in functions for WAM, so that they are available as Prolog or PARLOG intrinsics. A related intrinsic is that to invoke a procedure by address. Thus the normal sequence of events during, say, forward subsumption testing of a newly generated literal, is to use the index to find appropriate code and then invoke it. That is, the Prolog-like code for the first clause in Section 2 becomes

```
callmax(X,Y,XY,Database,Generated_clause) :-
        retrieve(Database,max(Y,Z,YZ),Code_1),
        invoke(Code_1,[Y,Z,YZ]),
        retrieve(Database,max(X,YZ,XYZ),Code_2)
        invoke(Code_2,[X,YZ,XYZ]),
        Generated_clause = max(XY,Z,XYZ).
```

Here retrieve is a foreign subroutine that takes a database and a term and returns (on successive backtracks) a sequence of pointers to code for terms that might unify with the current instantiation of the given term, and `invoke` is a foreign subroutine that invokes the unification code with a list of arguments. The call to `invoke` succeeds if the unification is successful. It is assumed in this case that the code contains the appropriate occurs-check instructions. The details of this implementation will be presented in a later paper.

5. Clause-set Compaction

Motivation for the notion of clause-set compaction comes from looking at what happens during the use of a set of hyperresolution nuclei. Let us consider first the two axioms representing associativity of the join operation in Sam's Lemma:

```
¬max(x,y,xy)  ¬max(y,z,yz)  ¬max(x,yz,xyz)  max(xy,z,xyz)
¬max(x,y,xy)  ¬max(y,z,yz)  ¬max(xy,z,xyz)  max(x,yz,xyz)
```

Clause-set compaction is a technique for representing this set of clauses in such a way that fewer unifications are required for their use.

At the outer level, the six negative literals, considered individually, are all the same, in that a literal, say `max(1,x,1)`, that unifies with any one of them unifies with all of them. Thus a single unification in essence produces six (binary) resolvents. It is possible to then find collections of negative literals in these six clauses which are essentially the same, so that again a single unification does the work of many.

It is interesting to note that whereas indexing is an attack on the number of unsuccessful unifications, this approach is an attempt to reduce the number of successful unifications required for the generation of a set of resolvents from a given set of clauses.

We use a Prolog program to process theorem prover input clause into a "compact" form prior to the compilation process described in Section 2.

6. Experimental Results

The theorem prover based on these approaches is still in a preliminary stage, and does not yet provide a convenient theorem-proving environment. It is far from bing optimized at this point. Nonetheless, we have been able to illustrate, at least partially, the effect of all of these ideas on Sam's Lemma, a classic problem in lattice theory from the theorem-proving literature. We compare the results of running this problem on ITP and the as-yet-unnamed new WAM-based theorem prover that incorporates the ideas presented in this paper. In addition we give preliminary results from running it on an Encore Multimax.

In these runs the set of support consisted of all the unit clauses. See [9] for the precise statement of the problem. The first set of experiments was run on a VAX 11/780:

theorem prover	*time in minutes*
ITP	153.8
New theorem prover, without compaction	13.2
New theorem prover, with compaction	12.5

Thus the new approach is providing a speedup of a factor of eleven, largely due to clause compilation. The results from compaction are somewhat disappointing, showing about a 5% effect. We attribute this to the still primitive state of our clause compiler, which currently undoes some of the optimizations made by the compactor.

The Encore Multimax is a shared-memory machine with twenty National Semiconductor 32032's, each of which is roughly the speed of a VAX 11/750. The multiprocessing aspect of the new system is in a *very* early stage, with several parts that are still unnecessarily sequential. Even so, encouraging speedups were observed on the Encore. All of these runs are with the new system, without compaction.

# of processes	time in minutes	speedup
1	19.3	1.00
2	10.7	1.81
4	7.3	2.66
8	5.6	3.47

The nature of the basic theorem-proving algorithms is such that we would expect to achieve substantially greater degrees of parallelism as the system matures. At this point, we believe that speedups of at least twelve to fifteen should be possible on a twenty-processor system.

7. Conclusion

We believe that the concepts of clause-compilation, multiprocessing, indexing, and compaction (more or less in decreasing order of significance) can lead to a new generation of high-performance theorem provers. In fact, special hardware may compound the speed increases in two ways.

1. Interest in Prolog has created several efforts at hardware implementations of the Warren abstract machine. At least one of these efforts is incorporating the extensions to the Warren machine necessary to support theorem proving.
2. The commercial database industry provides a large market for devices enabling fast access to large databases. Associative memory hardware may be adaptable to the indexing methods of Section 4.

If all of these trends come together, we may see within the next few years theorem provers from two to four orders of magnitude faster than today's systems This will radically increase the significance of the entire field of automated theorem proving.

References

1. J. Gabriel, T. Lindholm, E. Lusk, and R. A. Overbeek, "A tutorial on the Warren abstract machine," Technical Report ANL-84-84, Argonne National Laboratory, Argonne, Illinois (October, 1984).
2. J. Gabriel, T. Lindholm, E. L. Lusk, and R. A. Overbeek, "Logic Programming on the HEP," in *Parallel MIMD Computation: The HEP Supercomputer and its Applications*, ed. J. S. Kowalik, The MIT Press (1985).
3. Steven Gregory, *Design, application and implementation of a parallel logic programming language*, Doctoral Thesis, Imperial College of Science and Technology, London, September 1985.
4. E. Lusk, William McCune, and R. Overbeek, "Logic machine architecture: kernel functions," pp. 70-84 in *Proceedings of the Sixth Conference on Automated Deduction, Springer-Verlag Lecture Notes in Computer Science, Vol. 138*, ed. D. W. Loveland, Springer-Verlag, New York (1982).
5. E. Lusk, William McCune, and R. Overbeek, "Logic machine architecture: inference mechanisms," pp. 85-108 in *Proceedings of the Sixth Conference on Automated Deduction, Springer-Verlag Lecture Notes in Computer Science, Vol. 138*, ed. D. W. Loveland, Springer-Verlag, New York (1982).
6. Ewing L. Lusk and Ross A. Overbeek, "Implementation of Monitors with Macros: A Programming Aid for the HEP and Other Parallel Processors," Technical Report ANL-83-97, Argonne National Laboratory, Argonne, Illinois (December 1983).
7. Ewing L. Lusk and Ross A. Overbeek, "A Portable Environment for Research in Automated Reasoning," pp. 43-52 in *Proceedings of the 7th International Conference on Automated Deduction, Springer-Verlag Lecture Notes in Computer Science, Vol. 170*, ed. R. E. Shostak, Springer-Verlag, New York (1984).
8. E. L. Lusk, R. L. Stevens, and R. A. Overbeek, "A tutorial on the use of monitors in C: writing portable code for multiprocessors," ANL-85-2, Argonne National Laboratory (January 1985).
9. J. McCharen, R. Overbeek, and L. Wos, "Problems and experiments for and with automated theorem-

proving programs,'' *IEEE Transactions on Computers* **C-25**(8), pp. 773-782 (1976).

10. R. Overbeek, ''An implementation of hyper-resolution,'' *Computers and Mathematics with Applications* **1**, pp. 201-214 (1975).

11. D. H. D. Warren, ''Implementing Prolog - compiling predicate logic programs,'' DAI Research Reports 39 and 40, University of Edinburgh (May 1977).

12. D. H. D. Warren, ''An Abstract Prolog Instruction Set,'' SRI Technical Note 309, SRI International (October 1983).

13. D. H. D. Warren, ''Applied logic - its use and implementation as a programming tool,'' SRI International Technical Note 290 (June 1983). University of Edinburgh Ph.D. thesis, 1977

14. M. J. Wise and D. M. W. Powers, ''Indexing Prolog Clauses via Superimposed Code Words and Field Encoded Words,'' *Proceedings of the 1984 International Symposium on Logic Programming, Atlantic City*, pp. 203-211 (February, 1984).

15. L. Wos, S. Winker, and E. Lusk, ''An automated reasoning system,'' *Proceedings of the AFIPS National Computer Conference*, pp. 697-702

Purely Functional Implementation of a Logic

F K Hanna and N Daeche
University of Kent
Canterbury, Kent, UK

A new approach to the computational implementation of a theorem prover is outlined. Its main feature is that the logic is defined as a data type within a purely functional programming language. Its principle advantages are the brevity and simplicity of an implementation. As an example, the program to implement a polymorphic, higher-order logic is only half a dozen pages long. Such a program can very effectively serve as a formal specification of a logic.

1. Introduction

The aim of this paper is to describe a new technique for the computational implementation of a logic. The technique has two distinctive features. The first is that the logic is implemented as an 'algebraic' data type within a *purely* functional programming language (ie, one without *any* imperative features). The main way in which this is achieved is by treating *signatures* as a 'first-class' data type.

The second distinctive feature of the technique is the emphasis it places on *derivations* (or 'proofs') rather than on theorems, and in the way that it employs the *'propositions as types'* analogy. Derivation construction (or 'inferencing') exactly parallels term construction. This leads to a considerable conceptual (and actual) simplification.

The net effect of these two features is that the computational implementation of a logic can be *very compact indeed*. For example, it is possible to implement a polymorphic, higher-order logic in *only about half a dozen pages*. This brevity (along with the mathematical 'purity' associated with functional programming languages) brings several important advantages. The main ones are that it does not take very long to implement a new logic, and that such an implementation can very effectively serve as a *formal definition* of the logic. Further, such a technique leads to an implementation which appears to be reasonably efficient, both spacewise and timewise.

We will illustrate this technique by discussing an implementation of a particular logic (a polymorphic, higher-order predicate one) but we emphasise that the principles are generally applicable.

1.1 Background

The concepts underlying the approach to be described come from many sources. The two main ones are, however, the ML/LCF system and the CLEAR specification language. We thus briefly review the principal features of these two formalisms, and also of two other ones, VERITAS and MIRANDA, that will be referenced later.

The ML/LCF theorem prover In the ML/LCF approach [GMW79, P85] to theorem proving, the *object-logic* is embedded in a strongly-typed, executable *meta-language*. The meta-language provides abstract types for the *term*, *formula* and *theorem* syntactic categories of the object-logic. The only way in which a 'user' can construct values of these types is by using the constructor functions provided. The implementation of these functions includes checks to guarantee the well-formedness and soundness of all values constructed.

The (object-logic) theory, with respect to which terms are defined and inferencing is carried out, is implicitly defined by the *state* of the meta-language interpreter. By means of procedure calls, the user can alter this state; in particular new (object-logic) symbols and axioms can be added to the state, or previous theory definitions can be re-loaded.

One of the important advantages gained by this approach to embedding a logic in an executable meta-language is that *proof strategies* may be represented as (functional) values. For example, functions (called '*tactics*') may be written for implementing a *goal-directed* style of inferencing, and higher-order functions (called '*tacticals*') may be written for composing tactics together.

The CLEAR language CLEAR [BG78] is a language intended for writing large specifications in a structured way. Values in CLEAR denote *sorted theories* in an equational logic. The language provided operators for constructing theories (defined in terms of constants and axioms) and for manipulating theory values. For example, theories may be *enriched*, *combined*, *parameterised* and *abstracted*. The '*sig*' data type that we introduce below shares many features with the CLEAR 'theory presentation' type.

The VERITAS logic The particular logic whose implementation will be described in this paper is one named VERITAS. It is a polymorphic, higher-order predicate logic, originally devised for use as a *specification language* for describing and formally verifying the correctness of digital systems. (A case study (of significant size and complexity) illustrating this use of the logic may be found in [HD85]).

For descriptive purposes it is convenient at first to ignore the *inferential* aspects of the logic, and focus attention on just the structure of the *terms*. We shall refer to this as the 'VERITAS language'. We remark that the VERITAS language is effectively the typed λ-calculus, augmented with some of the more expressive types present in Martin-Löf's 'Intuitionistic Type Theory' [ML84]. We shall use an ordinary mathematical typeface when describing VERITAS entities (for example, $\forall n : nat.\ n = n$.)

The MIRANDA language The meta-language that we use in this paper is MIRANDA [T84]. It is a *purely* functional language, that is, it has no imperative features (eg, assignment) at all. The approach we shall be describing exploits the '*algebraic*' data type mechanism of MIRANDA. An *algebraic* type comprises two (user-defined) components:

- a tagged union of a set of Cartesian product types (the tags being known as *constructors*), and
- a set of *laws*. These laws may be used to *restrict* the domain of applicability of the constructors (essentially by mapping unwanted elements to $\perp$).

Such types correspond to partial, free, Σ-algebras.

We shall be using MIRANDA to define an *executable specification* of the VERITAS logic. In order to distinguish MIRANDA programs (or '*scripts*' as they are termed) from object-level entities, we use a typewriter-like typeface for the former (as, for example, in `tm = Sym sg 2`).

2. Term construction

We begin by focussing attention on the VERITAS language. Terms in VERITAS are, to a first approximation, the same as those of the typed λ-calculus. That is, a *term* is either a *symbol*, an *application* or a λ-*abstraction*. Each term has a *sort*, which is recursively defined in terms of the sorts of its subterms according to the usual rules.

Described meta-linguistically, a VERITAS term is simply a MIRANDA value of type `term`. Such a value can be constructed by (and *only* by) using one of a small collection of *term constructors*. For example, the `App` constructor ('*application*') takes two MIRANDA values of type `term` and (provided they satisfy the *laws* associated with the constructor) yields as a MIRANDA value their *application*. Since the MIRANDA interpreter has no implicit 'state' to rely upon, the representation of values of type `term` must (unlike the corresponding values in ML/LCF) be *context-independent* entities.

The structure of the `term`, `dec` and `sig` data types is defined in MIRANDA by means of the following type definition

```
term  ::=  Sym sig int      |    — symbol
           App term term    |    — application
           Lam term         |    — λ-abstraction
           Univ sig int     |    — universe
           Pi term               — exponential

dec   ::=  Dec string term       — declaration

sig   ::=  Empty            |    — the null signature
           Decsig sig dec        — 'declaration' signature
```

We shall, in the following sections, be extending this type definition to include 'derivations' as well as terms.

Signatures We start our description of this algebraic type by considering the **sig** data type. The purpose of a VERITAS *signature* is to define the *symbols* (and their *attributes*) that may be used in building terms. A signature is essentially a list of *declarations*.

It is useful to define a simple signature to serve as an example. We (arbitrarily) choose to base this example on the language of elementary number theory:

$bool : V0$	— *the sort 'bool' (of truth values)*	*(7)*
$equal : (\Pi s : V0.\ s \rightarrow s \rightarrow bool)$	— *a polymorphic equality predicate*	*(6)*
$nat : V0$	— *the sort 'nat' (of natural numbers)*	*(5)*
$zero : nat$		*(4)*
$suc : nat \rightarrow nat$	— *the 'successor' function*	*(3)*
$plus : nat \rightarrow nat \rightarrow nat$		*(2)*
$m : nat$	— *a typical 'variable'*	*(1)*

Let us assume that this signature (as a MIRANDA value of type **sig**) is bound to a MIRANDA identifier **sg_nats**. (Later, we will describe how a **sig** value such as this may be constructed, but, for the moment we will just think of it as if it were a list of declarations.) Now, with this signature available as an example, we consider in turn each of the constructors defined in the above type definition.

Constructing symbols The constructor Sym a signature and a number, and forms a *symbol* (ie, an atomic term) as its value. For example, the MIRANDA expressions

```
m_sym = Sym sg_nats 1      and      suc_sym = Sym sg_nats 3
```

build the symbols *m* and *suc*. Because Sym is a *constructor*, such values may be 'destructed' to yield their components (ie, a signature and a number). We call the signature component the *support signature* of the symbol, and the number component its *offset*. By reference to the *offset*-numbered component of the signature, the *sort* and the (alphameric) *name* of the symbol are defined. The reader familiar with Martin-Löf's ITT may find it helpful to identify the notion of the 'support signature' of a term with what in ITT are termed the 'assumptions of a hypothetical judgment'.

In passing, we note that, in the definition of the data type, the *name* of the symbol plays absolutely no role. In particular, there is no notion whatsoever of 'capture'; the provision for allowing names to be associated with symbols is purely for the user's convenience.

Constructing applications The constructor App takes two terms, and yields an *application*. For example

```
tm_1 = App suc_sym m_sym
```

builds the term *suc m*.

The *law* associated with the algebraic data type (or, alternatively phrased, the *definition* of the VERITAS language) requires that the application be *well-sorted*, and further, that both components are defined on the same support signature (which is then taken as the support signature of the overall term).

Constructing universes The sort structure of the VERITAS language is based on Martin-Löf's ITT. In particular, sorts are themselves terms. For example, the sort *nat* or the sort $nat \rightarrow nat$ are both terms. These terms themselves have a sort; the sort of any 'small' sort is the symbol $V0$, a 'small universe'. In turn, the sort of $V0$ is $V1$, and of $V1$ is $V2$, *etc, ad inf.* The constructor function `Univ` constructs elements of this infinite family. For example `Univ sg_nats 1` builds the 'large universe' symbol $V1$.

Constructing declarations Now consider *declarations.* A declaration, as we have already noted, serves to define the *attributes* of a symbol, namely its *sort* and its *name.* For example, the MIRANDA expression `n_dec = Dec "n" (Sym sg_nats 5)` builds the declaration $n : nat$. Often, it will be useful to provide infixed variants of function symbol names (eg, '+' as well as '*plus*').

Constructing signatures A signature is essentially a sequence of declarations, built by using the constructor `Sig` to extend an existing signature. The value `Empty` is a null signature. As an example, let us build a new signature and bind it to the MIRANDA identifier `sg_nats1`) by *extending* the signature `sg_nats` with the declaration $n : nat$, thus

```
sg_nats1  =  Decsig sg_nats n_dec
```

This is the signature

$$
\begin{array}{ll}
bool : V0 & \text{— (8)} \\
equal : (\Pi s : V0.\ s \rightarrow s \rightarrow bool) & \text{— (7)} \\
nat : V0 & \text{— (6)} \\
zero : nat & \text{— (5)} \\
suc : nat \rightarrow nat & \text{— (4)} \\
plus : nat \rightarrow nat \rightarrow nat & \text{— (3)} \\
m : nat & \text{— (2)} \\
n : nat & \text{— (1)}
\end{array}
$$

Constructing λ-abstractions A central feature of the approach we are advocating is the way in which terms containing 'bound variables' are build. It is best explained with an example. Suppose it is desired to build the term $\lambda n : nat.\ plus\ n\ zero$ on the support signature `sg_nats`. (Note carefully that the symbol n is *not* defined in this particular signature.)

In order to build this term, the *body* of the abstraction (ie, the term *plus n zero*) is first constructed on the signature `sg_nats1` (ie, the required signature extended with the declaration $n : nat$), and then the `Lam` ('make λ-abstraction') constructor is used to form the abstraction:

```
Lam body
  where body = App (App plus_sym n_sym) zero_sym
               where plus_sym  =  Sym sg_nats1 3
                     n_sym     =  Sym sg_nats1 1
                     zero_sym  =  Sym sg_nats1 5
```

This yields the desired term, $\lambda n : nat.\ plus\ n\ zero$, on the desired signature, `sg_nats`. In effect, the `Lam` constructor strips the outermost declaration off the support signature of a term, and incorporates it as the λ-binding in the term.

Polymorphic sorts Some symbols, '=' being a good example, are inherently *polymorphic.* In order to be able to express their sorts we use the Martin-Löf 'dependent exponential' binding operator, Π. The Π operator can be regarded as a way of forming *'parameterised'* types. As an example of the use of the Π operator, consider the 'Π' sort that occurs in the declaration of the symbol *equal* (as shown in the above signature). The 'dependent' sort specifies that the symbol *equal* expects an argument s of sort $V0$ (ie, a 'small' sort) and then two arguments of sort s, yielding a term of sort *bool*. Thus, the *sort* of the second and third arguments depend upon the *'value'* of the first argument. For example, the sort of the term *equal nat* is $nat \rightarrow nat \rightarrow bool$. Thus, the term $n = m$ would be expressed as *equal nat n m* Dependent exponentials are formed using the term constructor

$$\texttt{Pi} : \texttt{term} \rightarrow \texttt{term}$$

(note that it takes only a *single* argument). This constructor behaves in an identical way to the **Lam** ('make λ-abstraction) constructor. That is, the 'declaration' component for the Π binding operator is taken from the front of the support signature of the term to which the constructor is applied. It is convenient to *abbreviate* 'non-dependent' instances of 'Π' using '→', as for example, $nat \rightarrow nat$ in place of $\Pi x : nat.\ nat$.

Other polymorphic terms We note that, in higher-order logic, polymorphic terms tend to occur fairly frequently. An important example is the universal quantification combinator of Curry and Feys

$$forall : (\Pi s : V0.\ (s \rightarrow bool) \rightarrow bool)$$

Using this combinator, a term like $\forall n : nat.\ n = n$ can be expressed as $forall\ nat\ (\lambda n : nat.\ n = n)$.

Other examples of polymorphism are provided by lists, numbers modulo-n, rings, etc. In fact, this mechanism turns out to be capable of expressing *very* polymorphic terms, as, for example when a polymorphic function that is (in effect) passed as an argument is 'simultaneously' used at different instances of its sort.

3. Observations

Although we will be going on to describe the inferential aspects of the logic, this will involve techniques which are essentially identical to those already described. The schemes that we have outlined can be seen to have inherited features from ML/LCF (*'term'* as a secure data type), from CLEAR (signatures as 'first-class' values), from de Bruijn's scheme [dB72] (symbols defined by numerical offset), and from ITT (sorts as terms). Terms are abstract entities, *inseparable* from their support signatures, and the *laws* associated with the data-type ensures that only well-sorted terms can be formed. The 'abstract' nature of symbols means that there is no concept of 'capture', and that determining α-equivalence of terms simply involves testing their component parts (less the 'name' components in signatures) for identity.

The fact that terms are inseparable from their support signatures means that they are *context-independent* pure values, and may therefore be constructed and manipulated within a *purely* functional language.

4. Inferencing

So far, in this account, we have been solely concerned with *terms* and with *term construction*. We now widen the discussion to include *derivations* and *derivation construction*. In the approach we are describing, we place the emphasis on *derivations* (ie, proofs of theorems) rather than, as is usually done, on *theorems*. We have several reasons for this choice; one of the main ones is that derivations turn out to be formally *very similar* in their nature to terms. There is, in fact, an identification (at first sight, a somewhat unlikely one) between terms and derivations known as the 'Propositions as Types' principle. This principle was first noted by Howard, and subsequently exploited by Martin-Löf in his Intuitionistic Type Theory [ML84].

The key idea in this principle is to introduce the notion of the *sort* of a derivation, and define it as being the *theorem* that the derivation establishes. Atomic derivations are *axioms*, and derivation constructor functions are the *rules of inference* of the logic.

In order to incorporate this principle within the framework we have so far described, we need to generalise the notion of a signature, so that it can include what we term *proclamations* of axioms, as well as declarations of symbols. The role of a *proclamation* is to introduce, and define the attributes of, an axiom. Thus, a signature now describes a collection of symbols and of axioms, and thus it defines an *axiomatic theory*. (Technically, it corresponds to what in CLEAR is called a *theory presentation*.)

Derivation Constructors The extension required to the MIRANDA algebraic data type which defined the 'VERITAS language' (as described earlier) so as to define the 'VERITAS logic' is a relatively simple one: the only changes are are addition of

- An algebraic data type `proc` (for '*proclamation*') with a constructor `Proc` for building values of this type.
- An algebraic data type `deriv` (for '*derivation*') and a small collection of constructors (one corresponding to each rule of inference of the logic) for building values of this type.

Unlike first-order logic, higher-order logic is known to be essentially *incomplete*. For this reason, the choice of the particular set of inference rules to provide is somewhat arbitrary. The set that we have used is

```
deriv  ::=  Axiom sig int            |   — axiom
            Mp deriv deriv           |   — modus ponen
            Spec deriv term          |   — specialisation
            Gen deriv                |   — generalisation
            Disch deriv              |   — discharge
            Beta term                |   — β-reduce
            Taut term                |   — tautology
            Rw deriv deriv index         — rewrite

proc   ::=  Proc name term               — proclamation

sig    ::=  Empty                    |   — the null signature
            Decsig sig dec           |   — 'declaration' signature
            Procsig sig proc             — 'proclamation' signature
```

The 'logical' signature It is convenient to begin the description of the user's view of the above data type with the `sg_logic` constant. This is of type `sig`, and its purpose is to provide a signature defining the 'logical constants' and 'logical axioms' associated with the VERITAS logic. This signature is

$bool : V0$	— *declaration of the sort 'bool'*	*(11)*
$T : bool$	— *declaration of 'true'*	*(10)*
$F : bool$	— *declaration of 'false'*	*(9)*
$forall : (\Pi s : V0.\ (s \rightarrow bool) \rightarrow bool)$	— *the* $\forall$ *combinator*	*(8)*
$\rightarrow : bool \rightarrow bool \rightarrow bool$	— *implication*	*(7)*
$\wedge : bool \rightarrow bool \rightarrow bool$	— *conjunction*	*(6)*
$\vee : bool \rightarrow bool \rightarrow bool$	— *disjunction*	*(5)*
$\neg : bool \rightarrow bool$	— *negation*	*(4)*
$= : (\Pi s : V0.\ s \rightarrow s \rightarrow bool)$	— *polymorphic equal*	*(3)*
$reflex : (\forall s : V0.\ \forall x : s.\ x \ = \ x)$	— *reflexivity of equality*	*(2)*
$exten :$ $\forall s, t : V0.\ \forall f, g : s \rightarrow t.$ $(\forall x : s.\ (f \ \ x) \ = \ (g \ \ x)) \rightarrow (f \ = \ g)$	— *extensionality*	*(1)*

The signature contains *declarations* for the sort *bool* (ie, propositional truth values), for the logical constants T and F, for the *forall* combinator. for the logical connectives $\rightarrow$, $\wedge$, $\vee$, $\neg$, and for (polymorphic) equals. It also contains two *proclamations*, introducing the axioms *reflex* and *exten*. The first of these *logical axioms* asserts the reflexivity of equality, the second asserts the extensionality of functions.

4.1 Inference rules

In this subsection, we briefly consider the *derivation constructor* functions defined in the above MIRANDA data abstraction. Our aim will be not to describe these in detail, but rather to illustrate the general principles involved and, in particular, to indicate their close relationship with the *term constructor* functions described earlier. In addition, since derivations are now 'first-class' values, we propose an *object-language* notation for derivations (essentially a simple extension of the ordinary notation used for terms).

Axiom The `Axiom` function is used to construct an *axiom* (or, alternatively expressed, atomic derivation). It takes as arguments a signature (which becomes the *support signature* of the derivation) and a number (the *offset* of the relevant proclamation in the signature). For example, the value of the MIRANDA expression

```
Axiom sg_logic 2
```

is an axiom, which is proclaimed to be named *reflex* and to be of sort (or, alternatively expressed, to assert the theorem)

$$\forall s : V0.\ (\forall x : s.\ x \ = \ x)$$

We note that the `Axiom` derivation constructor function plays an exactly analogous role to the `Sym` term constructor function. Because of this, the object-language notation we use for expressing an axiom is simply the *name* of the axiom (as defined in its proclamation). For example, the above atomic derivation is expressed as *reflex*.

Specialise The rule of inference known as 'specialisation' is provided by the `Spec` derivation constructor function. This function takes as arguments a derivation (of the theorem it is desired to specialise) and a term (of appropriate sort), and yields as its value a derivation of the specialised theorem. For example, the value of the MIRANDA expression

```
Spec dv tm
 where dv  =  Axiom sg_logic 2
       tm  =  Sym   sg_logic 11
```

is $\forall x : bool.\ x \ = \ x$. That is, the axiom *reflex* has been specialised by the (sort) symbol *bool*.

We note that the `Spec` derivation constructor function may be identified with the `App` term constructor (recollect that the 'function' component of an application has, in general, a dependent sort). Because of this, the object-language notation we propose for expressing a 'specialisation' is the juxtaposition of the derivation and the term. For example, the derivation *reflex bool F* is a proof of the theorem $\vdash F = F$.

Generalise The rule of inference known as 'generalise' is provided by the `Gen` derivation constructor function. This function closely parallels the `Lam` (λ-abstraction) term constructor discussed earlier. Informally described, it takes a *single* argument, a derivation of sort *tm*, assumed to be on a support signature of the form $d\smile sg$ where d is a declaration, and yields as its value a derivation of sort $\forall d.\ tm$, on a support signature sg.

Noting the parallel between generalisations and λ-abstractions, the object-language notation that we use for expressing a generalisation is of the form $\gamma\ \langle dec\rangle.\ \langle deriv\rangle$. As a very simple example, the derivation

$$\gamma\ p : bool.\quad reflex\ bool\ (\neg p)$$

is a proof of the theorem

$$\vdash \forall p : bool.\quad \neg p \ = \ \neg p$$

Discharge The rule of inference known as 'discharge' is provided by the `Disch` derivation constructor. This function is very similar to the `Gen` function (above), and thus also closely parallels the **Lam** function. Informally described, it takes a single argument, a derivation of sort *tm*, assumed to be on a support signature of the form $p \smile sg$, where *p* is a proclamation of an axiom of sort *tm'*, and yields as its value a derivation of sort $tm' \rightarrow tm$ on a support signature *sg*. The notation we use for expressing a discharge is of the form $\delta\langle proc\rangle.\ \langle deriv\rangle$. As a very simple example, the derivation $\gamma\ b : bool.\ (\delta p : b.\quad p)$ is a proof of the theorem $\vdash \forall b : bool.\ b \rightarrow b$.

Rewrite One of the most frequently used inferencing methods is rewriting. The **Rw** derivation constructor takes as arguments two derivations and an *index*. An index is simply a list of *selectors* that serves to identify (or 'point to') a subterm within a term. Of the two derivations given as arguments, one is an arbitrary one (call its sort *tm*) whilst the sort of the other one has the form of a 'rewrite rule', (ie, $tm_1 = tm_2$). Subject to various constraints being satisfied, the sort of the overall derivation is *tm'*, where *tm'* is *tm* with the identified subterm (required to be tm_1) replaced with tm_2.

The notation we use to express it is of the form $[RW\quad \langle deriv\rangle\ \langle deriv\rangle\ \langle pattern\rangle]$ where *pattern* is a template with $*$ indicating the chosen subterm.

Other derivation constructors There are another three derivation constructors defined in the 'VERITAS logic' data type. It would not be of interest to describe them in any detail here. We simply note that they provide for modus ponens (**Mp**), β-reduction (**Beta**), and establishing propositional tautologies (**Taut**).

Signature constructors This is a convenient point in the text to note that the **Decsig** constructor in the 'VERITAS language' data type (described earlier) has now, in the 'VERITAS logic' data type been joined by a further constructor, **Procsig**. This (corresponding exactly to **Decsig**) extends a signature with a a proclamation. The two functions are virtually identical in their definition.

We also remark, although we have no space to discuss the topic here, that using these signature constructors (and their corresponding destructors), a variety of high-level operations on signatures can easily be implemented, corresponding loosely to the theory-manipulation functions of CLEAR (ie, *combine*, *enrich*, etc).

5. Example

As we noted earlier, one of the features of the approach described here is its emphasis on *derivations* rather than on *theorems*. As an illustration of this, and also as a non-trivial illustration of the use of the concrete notation for derivations introduced above, we present a simple proof.

The fact that the (polymorphic) equality relation is *symmetric* can be expressed as

$$\vdash \forall s : V0.\ \forall a : s.\ \forall b : s.\ (a \ = \ b \ \rightarrow \ b \ = \ a)$$

[ie, for all sorts *s*, for all *a*, *b* of sort *s*, if $a = b$ then $b = a$.]
One derivation that serves to establish this theorem is

$$\begin{array}{l}\gamma\ s : VO. \\ \quad \gamma\ a : s. \\ \quad\quad \gamma\ b : s. \\ \quad\quad\quad \delta\ hyp : a = b. \\ \quad\quad\quad\quad [RW\quad (reflex\quad s\quad a)\quad hyp\quad ''\_\ =\ *'']\end{array}$$

This derivation can be read as:

Introduce a (sort) symbol *s*.
Introduce a symbol *a* of sort *s*.
Introduce a symbol *b* of sort *s*.
Assume an axiom *hyp* of sort $a \ = \ b$.
Rewrite (using the axiom *hyp* as a rewrite rule) the right-hand

operand of the result of specialising the axiom *reflex* first
with the sort s and then with the symbol a.
Discharge (ie, δ) the axiom *hyp*
Generalise (ie, γ) over the symbol b
Generalise over the symbol a
Generalise over the symbol s.

6. Observations

Overall, the definition of the entire VERITAS logic (ie, the type definitions for **term**, **deriv**, **sig**, **dec** and **proc**, and their associated laws) as an executable MIRANDA script *is no more than half* a *dozen pages in length*. (This includes error-reporting facilities and short comments.) The extreme brevity of this definition comes about partly because the higher-order features of the meta-language allow powerful, compact coding, partly because of the uniformity of the VERITAS logic, and partly because of the 'signature-based' approach to the handling of bound variables.

For most practical applications, it is desirable to define *parsing* functions and *unparsing* functions, to translate between the abstract meta-linguistic types and a concrete object-language. Again, the higher-order features of the meta-language allow very compact definitions of such functions. We remark that a set of parsing functions for terms, derivations and signatures, able to handle the full richness of a conventional, infixed logical notation (including facilities to automatically generate the required sorts with which to specialise polymorphic terms, such as *equal* and *forall*) are only a few pages long. The corresponding set of unparsing functions (including 'pretty-printed' layout and multi-line formatting) are likewise only a few pages long.

We note that there are great advantages which result from the decision to treat derivations as 'first-class' values. For example, we have found it invaluable to regard the process of creating a proof as being one of constructing a derivation using an interactive derivation editor, backed up with a *graphical* representation of the partially constructed proof tree. A *derivation editor* is a function that takes a *goal* and a stream (from the user) of *tactics* (as introduced in [GMW79]) and yields a stream of *responses* and (finally) a *derivation* that establishes the stated goal. A case-study (involving some 1,600 tactic applications) using this approach is described in [HD85].

Efficient implementation As we have seen, the object-logic has been defined (with complete rigour and transparent clarity) as an algebraic data type in MIRANDA. This means that, viewed meta-linguistically, *the object-logic is simply a partial, free, Σ-algebra.*

Whilst this is ideal, in all respects, as a *definition* of the logic, it is computationally inefficient. This inefficiency arises from the fact that the sorts of terms and derivations have to be computed afresh each time they are required. An *efficient* implementation may, however, easily be obtained, simply by replacing the 'algebraic' types by corresponding 'abstract' types in which the ('memo-ised') sorts are stored alongside each term or derivation. This does not add greatly to the complexity of the definition of the logic.

We note that, in this form, there is no reliance on the 'algebraic' type structure unique to MIRANDA; that is, this approach can be used to embed a logic in *any* programming language, provided only that the latter has provision for the introduction of user-defined abstract data types.

7. Conclusions

We can summarise the important features of the approach we have described, as follows:

- Our overall aim has been to embed, as a data type, a logic in a computational meta-language [as in ML/LCF].
- We choose to use a *purely* functional meta-language [MIRANDA].
- Following CLEAR, we treat signatures (or 'theory presentations') as a 'first-class' data type. Signatures are built up from declarations and proclamations, and serve to define the attributes of the symbols and axioms of the logic.
- We exploit the 'propositions as types' principle; that is, derivations (or 'proofs') are analogous to terms, and derivation construction (or 'inferencing') is analogous to term construction. In

particular, the sort of a derivation is the theorem it establishes.

- We follow Martin-Löf's type theory in treating sorts as being terms in their own right, and in using the Π operator to build 'dependent' exponential sorts. These two features allow polymorphism to be incorporated in a very natural way.
- We treat terms and derivations as being inseparable from their support signature. This means that they are context-independent, pure values. In turn, this allows the implementation to be a functional one.
- The same feature enables 'bound variables' (as found in λ-abstraction, Π, generalisation, discharge) to be treated in a very pure way. In particular, there is no notion of capture.

The main advantages which derive from this overall approach are generality, simplicity and rigour.

- As we have demonstrated, the approach is sufficiently *general* to allow (for instance) a polymorphic, higher-order predicate logic to be implemented. Likewise (although we have not yet had an opportunity to confirm this), we believe that logics like PPλ or ITT may be implemented along similar lines.
- The approach is sufficiently *simple* to enable a new logic to be implemented in a relatively short time (typically in weeks rather than months). The conceptual simplicity of the approach is also of benefit in limiting the complexity of the 'interface' presented to the user.
- Finally, the approach is a completely *rigorous* one; this method of *implementing* a logic may equally be regarded as providing a *formal specification* of it.

Although the approach offers these desirable features, it does not do so at the expense of computational efficiency. As was noted earlier, we have been using an implementation of the VERITAS logic along the general lines described here for 'serious' (ie, relatively large scale) work in connection with the specification and verification of digital systems.

8. Acknowledgments

Our decision to exploit the 'propositions as types' principle owes much to the illumination cast on this subject by colleagues of the authors during a series of Theoretical Computer Science seminars at Kent.

The VERITAS project is currently supported under grant GR/C/2287.5 from the UK Science and Engineering Research Council. We are grateful to David Turner for allowing us access to an Orion machine (supplied under SERC grant GR/D/26825) and a (pre-release) version of MIRANDA.

9. References

[BG78] Burstall, R.M., and Goguen, J.A., "Putting theories together to make specifications", Proc 5th IJCAI, pp1045-58, 1978

[dB72] de Bruijn, N.G., "Lambda Calculus Notation with Nameless Dummies", pp381-392, Koninkl. Nederl. Akademie van Wetenschappen, 1972.

[GMW79] Gordon, M., Milner, R. and Wadsworth, C., "Edinburgh LCF", Springer-Verlag, 1979.

[HD85] Hanna, F.K. and Daeche, N., "Specification and Verification using Higher-Order Logic: A Case Study", Technical Report, University of Kent, 1985.

[ML84] Martin-Löf, P. "Constructive mathematics and computer programming", pp501-518, Phil Trans R. Soc. London, A 312, 1984.

[T84] Turner, D.A. "Functional programs as executable specifications", *in* "Mathematical Logic and Programming Languages" *ed* Hoare and Shepherdson, Prentice Hall, 1984.

Note 'MIRANDA' is a trademark of Research Software Limited.

Causes for Events :
Their Computation and Applications

P.T. Cox; T. Pietrzykowski
Technical University of Nova Scotia
Canada·

Abstract

A formal definition is presented for the cause for an event with a given knowledge base. An event is understood to be any closed formula; a knowledge base is a set of closed formulae; and a cause is a prenex formula, the matrix of which is a conjunction of literals. The properties of minimality, basicness, consistency and nontriviality are defined to characterise causes that are useful and interesting. An algorithm for computing basic, nontrivial and minimal causes is presented, and its soundness is proved. An extensive example is provided to illustrate the application of causes. In this example is discussed the interaction between the knowledge base, event and causes, and in particular, the property of consistency is used to eliminate ambiguity, and to discover deficiencies of the knowledge base.

1. Introduction

The problem of performing deductions of new facts from sets of axioms is well-studied and understood. An equally important but far less explored problem is the derivation of hypotheses to explain observed events. In formal terms this involves finding an assumption that, together with some axioms, implies a given formula. Clearly, certain assumptions trivally have the desired property: for example, the formula itself, or the negation of any axiom. It is necessary therefore to apply some restrictions that will guarantee that the assumption is in some sense interesting.

Motivations for the study of this type of inductive reasoning range from purely theoretical interest, to practical applications such as expert systems. Related theoretical results to date are concerned with finding an appropriate theoretical foundation for cycles in unification [2, 11], and for all types of nonunifiability [4]. The importance of

This research was supported by NSERC Grants A0124 and A2493

hypothesis formation to expert systems is well known, but surprisingly little has been done to provide it with a firm logical foundation [8]. The very popular MYCIN-like systems [1, 10], widely used in practice, pay much attention to the probabalistic aspect [12] of inductive reasoning, but little to the strictly logical issues. Inductive reasoning in MYCIN-like systems is simulated by deduction.

Another area of related research is non-monotonic logic pioneered by Reiter [9], and further investigated by others [6,7].

An important issue related to hypothesis formation is the reintroduction of quantifiers into formulae containing Skolem functions. This has been successfully and thoroughly investigated [3], producing results essential to the method we describe below.

In the following we provide a precise definition of the cause of an event with a given knowledge base. We also define several restrictions on causes that guarantee them to be interesting in the sense that they are neither too general (minimal) nor trivial, have no further causes (basic), and do not trivially imply the event by producing inconsistency with the knowledge base. The issue of finiteness of such a set of causes is discussed, and an open problem is posed. We provide a method for computing causes that are basic and minimal, using linear resolution [5] and reverse Skolemization [3]. An extensive example is presented to illustrate the definition of cause, and the effects that the event and elements of the knowledge base have on the structure of the set of fundamental causes.

2. Causes and their computation

A conjunct is a closed prenex well-formed formula (wff) with a conjunction of literals as its matrix. If B is a set of wffs, we will also use B to denote the conjunction of the wffs in the set B when the meaning is clear from the context.

Definition: Let K be a set of wffs (the knowledge base, each element of which is called a descriptor), and e a wff (an event) such that $K \supset e$ does not hold. A conjunct c is called a cause of e with K iff $(c \wedge K) \supset e$ holds. When the knowledge base is understood in the context we will omit the phrase "with K". A cause c of e is:

(i) minimal iff for all causes x of e, $c \supset x$ implies $c \equiv x$.

(ii) consistent iff $c \wedge K$ is satisfiable.

(iii) <u>nontrivial</u> iff $c \supset e$ does not hold.

(iv) <u>basic</u> iff every consistent cause of c is trivial.

2.1: Examples and comments

Let K be the set of descriptors { $p \supset q$, $q \supset r$ } and e the event r, then p, q, r and $p \wedge s$ are all causes for e, where s is any literal. Clearly q is not basic, r is trivial, $p \wedge s$ is not minimal, and if s is the literal $\neg q$, $p \wedge s$ is also not consistent.

We will now discuss each of the restrictions of causes defined above and provide the motivation for each by considering examples.

Minimality: Let K be the set of descriptors { $\exists x P(x) \supset q$ } and e the event q. Then $\forall x P(x)$, $\exists x P(x) \wedge r$, and $\exists x P(x)$ are causes, but only the last one is minimal. This property eliminates causes that are unnecessarily general in terms of universal quantification, as well as those that contain literals which are not required to deduce the event. Note that there is always an infinite number of nonminimal causes for any event because of the presence of such irrelevant literals. There exist examples, however, for which there is an infinite number of minimal causes. This issue will be further discussed below.

Consistency: Let K be the set of descriptors { $p \supset q$ } and e the event r, then $p \wedge \neg q$ is a minimal, basic and nontrivial cause, however it is not consistent. The consistency property eliminates causes that are unrelated to the event, but because they produce unsatisfiability with the knowledge base are causes nevertheless.

Triviality: Let K be the set of descriptors { $p \supset q$ } and e the event r, then every cause is trivial since the event is unrelated to the knowledge base. Note that if there exists a nontrivial cause for an event then no trivial cause can also be basic. Nontriviality eliminates causes which are unrelated to the knowledge base, but are causes only because they directly imply the event.

Basicness: Let K be the set of descriptors { $p \supset q$, $q \supset r$, $p \supset t$ } and e the event $r \wedge t$, then $q \wedge t$, $r \wedge p$ and p are all causes of e, but only p is basic. Note that $r \wedge p$ is neither basic nor minimal. The property of basicness eliminates intermediate causes, that is, causes which themselves have causes.

Fundamental: A cause is called fundamental iff it is minimal, consistent, nontrivial and basic.

It is worthwhile noting the intimate relationship between the above properties, namely that a cause c is basic iff every consistent, nontrivial cause of c is not minimal.

2.2: Infiniteness of sets of causes

Let K be the set of descriptors { $\forall x(P(f(x)) \supset P(x))$ } and e the event P(a). Each of the wffs $P(f(a)), \ldots, P(f^i(a)), \ldots$ is a minimal cause; however, none of these causes are basic since each is a cause of the preceding formula in the sequence. If we extend the knowledge base to { $\forall x(R(x) \supset P(x))$, $\forall x(P(f(x)) \supset P(x))$ }, then the set of minimal and basic causes for the above event is { $R(f(a)), \ldots, R(f^i(a)), \ldots$ } which is clearly infinite. Note that in both cases all the above causes are nontrivial and consistent. It is interesting to note that the presence of the function symbol in the above example is essential in producing an infinite set of minimal and basic causes. We conjecture that if no function symbols occur in the knowledge base the number of minimal and basic causes is finite. We pose as an open problem finding a proof or counterexample to this conjecture.

3. Computing Causes

We will now briefly describe a method for computing causes for an event e with a given knowledge base K. First we convert K and negated e into clausal form, obtaining K' and ~e, then perform linear resolution with some element of ~e as top clause and input clauses from K'. Because of our assumption that K does not imply e, such a deduction cannot be a refutation, therefore there are two possibilities: either the deduction does not terminate, or it terminates in some clause d which we will call a dead end. Let the set of dead ends be D. We then negate each $d \subset D$ obtaining a conjunction of literals $p_1 \wedge \ldots \wedge p_n$ and apply the reverse Skolemization algorithm described in [3] to the expression $\Lambda(p_1, \ldots, p_n)$ obtaining a closed literal $Q_1 x_1 \ldots Q_k x_k \; \Lambda(q_1, \ldots, q_n)$ where $Q_1, \ldots Q_k$ are quantifiers and $q_1, \ldots, q_n$ are obtained from $p_1, \ldots, p_n$ by replacing Skolem terms by the universally quantified variables. Finally, we create the conjunct $c = Q_1 x_1 \ldots Q_k x_k (q_1 \wedge \ldots \wedge q_n)$.

Lemma: If c, e and K are as above then c is a basic cause of e with K.

Proof: Let us suppose the contrary, then there exists a consistent, nontrivial cause x of c with K, so that $(x \wedge K) \supset c$ and therefore $x \wedge K \wedge \neg c$ is unsatisfiable. Since x is a consistent cause, $x \wedge K$ is satisfiable, so there is a linear refutation of $x' \cup K' \cup \{\sim c\}$ with $\sim c$ as top clause, where x', K' and $\sim c$ are the clausal forms of x, K and $\neg c$ respectively. Since x is nontrivial $x \supset c$ does not hold, and therefore this linear refutation must involve clauses from K'. Let k be the first clause from K' used in this refutation, then since each clause in x' consists of a single literal, there must exist a literal in $\sim c$ which can be resolved with k. Hence by the soundness of the reverse Skolemization algorithm [3], k can also be resolved with d contradicting the fact that d is a dead end. □

We will illustrate the above using an example from [3] which illustrates the role of reverse Skolemization.

Let $K = \{ \forall x \forall y \forall u \forall v (P(x,y,u,v) \supset (R(x,y) \vee Q(u,v))),$

$\forall x \exists y (R(x,y) \supset T(y))),$

$\forall u \exists v (Q(u,v) \supset T(v))) \}$

and $e = \exists x T(x)$

Then $K' = \{ \{ \neg P(x,y,u,v), R(x,y), Q(u,v) \},$

$\{ \neg R(x, \varphi(x)), T(\varphi(x)) \},$

$\{ \neg Q(u, \alpha(s)), T(\alpha(s)) \} \}$

and $\sim e = \{ \{ \neg T(x) \} \}$

In these clauses as in the rest of this paper, lower case Greek letters are used for Skolem functions. In the above we have only one dead end which is $\neg P(x,\varphi(x),u,\alpha(u))$. After applying the reverse Skolemization algorithm we obtain the two distinct causes $\exists x \forall y \exists u \forall v M$ and $\exists u \forall v \exists x \forall y M$ where M is $P(x,y,u,v)$.

We now give a further example in which the event is a universally quantified disjunction.

Let K = { $a \supset \forall x(Q(x) \vee R(x))$, $b \supset \forall x(R(x) \vee T(x)))$ }

and e = $\forall x(R(x) \vee Q(x))$

Then K' = { { $\neg a$, $Q(x)$, $R(x)$ }, { $\neg b$, $R(x)$, $T(x)$ } }

and $\sim e$ = { { $\neg R(\alpha)$ }, { $\neg Q(\alpha)$ } }

We obtain two dead ends which are { $\neg a$, } and { $\neg b$, $T(\alpha)$ }. After applying the reverse Skolemization algorithm we obtain the two distinct causes a and $\forall x(b \vee \neg T(x))$ which are fundamental.

3.1: Comments on the algorithm

Let K be the knowledge base { $P(a) \supset Q(a)$, $Q(a) \supset R$ } and e the event R. Our algorithm will return the single basic cause $P(a)$. There exist, however, other basic causes such as $\forall x Q(x)$, but none of them are minimal. It is interesting to note that $Q(a)$ itself is a minimal but not basic cause. Fortunately our algorithm will not produce the cause $\forall x Q(x)$ or other similar basic but nonminimal causes.

To illustrate the power of the algorithm we consider the knowledge base { $\forall x(R(x) \supset P(x))$, $\forall x(\exists y P(y) \supset P(x))$ } and the event $P(a)$. If we apply our algorithm, we first obtain the sets of clauses { { $\neg R(x)$, $P(x)$ }, { $\neg P(\alpha(x))$, $P(x)$ } } from the knowledge base and { $\neg P(a)$ } from the event. From these the following infinite set of dead ends will be produced: { { $\neg R(\alpha(a))$ }, . . ., { $\neg R(\alpha^i(a))$ }, . . . }. Interestingly, after applying reverse Skolemization to each of these we obtain the same cause $\forall x R(x)$ which is the only minimal and basic cause for this event.

There are some basic and minimal causes of a pathological nature which our algorithm will not generate. For example, consider the knowledge base { $\forall x(\exists y P(y) \supset P(x))$ } and the event $P(a)$ which has a minimal and basic cause $\forall x P(x)$.

Applying our algorithm, we first obtain the sets of clauses { { $\neg P(\alpha(x))$, $P(x)$ } } from the knowledge base and { $\neg P(a)$ } from the event. Clearly we can generate no dead ends from these sets and consequently will obtain no causes.

Finally we note that causes generated by the algorithm are not always minimal. For example, let **K** be { $q \wedge p \supset r$, $p \supset r$ } and **e** be r. Our algorithm will produce two causes $q \wedge p$ and p of which only the latter is minimal. This situation can easily be prevented by removing dead ends which are subsumed by other dead ends. This is the only kind of nonminimality that can occur since the algorithm does not introduce any redundant information into the dead ends, and the minimality of the reverse Skolemization algorithm [3] guarantees that no unnecessary universal quantification is performed.

4. Application to the identification problem

In the previous sections we provided the basic definitions of cause and an algorithm for computing causes for an event. The abstract character of the definitions, however, makes it somewhat difficult for the reader to see how to apply them in practical situations such as in the hypothesis formation component of an expert system. We will therefore present an extensive example which applies these concepts to the problem of identifying a pattern on the basis of certain features. The example used is rather specific, but the techniques applied have a far more general character.

4.1: Description of the problem

The problem involves recognising graphical objects according to features they possess: in particular we will investigate the recognition of uppercase letters of the Roman alphabet. The data which is gathered about the pattern to be identified consists of a set of points together with their neighbourhoods, classified according to their geometrical properties. Points are classifed into the following three categories:

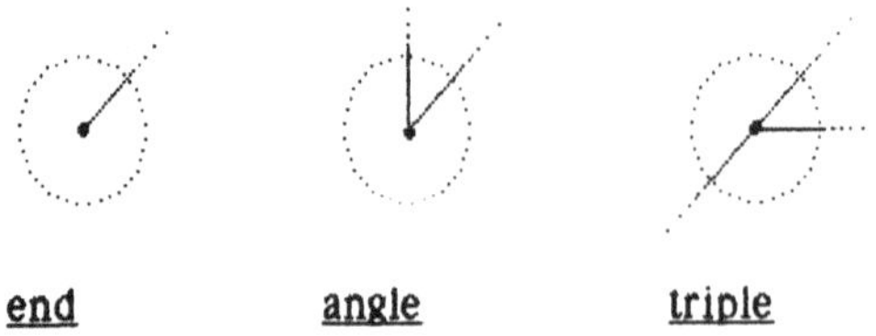

For example, the letter E has three distinct ends, two distinct angles and one triple:

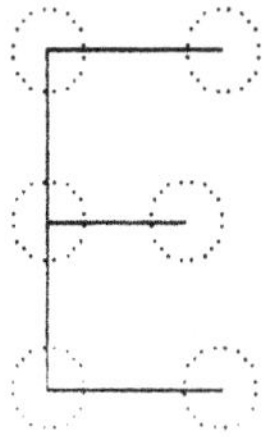

From the observed data, an event is constructed that captures the information using <u>characteristic predicates</u> corresponding to the types of points: **end**, **ang** and **tri**. In the following we will discuss how events should be constructed from the observed data.

Suppose we discover that the observed pattern contains some ends and triple points. This information can be expressed as the event $\exists x\,end(x) \wedge \exists y\,tri(y)$ which is rather weak. If we observe, however, that there are definitely two distinct end points, we may construct the stronger event $\exists x \exists z(end(x) \wedge end(z) \wedge x \neq z) \wedge \exists y\,tri(y)$. Clearly this type of construction may be extended arbitrarily. An even stronger event can be constructed if the observer is positive that the pattern contains only certain types of points. For example, if the pattern is the letter B and the observer is correctly convinced that he has observed all the points, he may construct the event $\forall x(ang(x) \vee tri(x))$ which states that the **only** types of points in the pattern are angles and triples. We will further discuss the use of universal quantifiers in events below.

We will now discuss the knowledge base for this example by considering the kinds of descriptors it can contain. For the purposes of this discussion we will assume that there are only five possible letters:

A B C D E

<u>Existential descriptors</u>. These specify that a letter must contain points of certain types. For example, the letter B must contain at least one angle and at least one triple. This information is expressed by the formula:

$$B \supset \exists x \exists y(ang(x) \wedge tri(y))$$

In this formula, the predicate B corresponds to the assertion that the letter in question is B.

Quantitative descriptors. A quantitative descriptor specifies that a letter contains no less than a specified number of points of a particular type. For example the fact that the letter E contains no less than two angles is expressed by:

$$E \supset \exists x \exists y(ang(x) \wedge ang(y) \wedge x \neq y)$$

Negative descriptors. These specify the types of points which do not occur in a letter. For example, the letter C has no angles or triples:

$$C \supset \forall x(\neg ang(x) \wedge \neg tri(x))$$

Positive descriptors. A positive descriptor specifies that a letter contains only points of certain types. For example:

$$B \supset \forall x(ang(x) \vee tri(x))$$

Note that negative and positive descriptors play dual roles. A negative descriptor is useful when a letter possesses most of the possible types of points, while a positive descriptor is convenient in the opposite situation.

Distinguishing descriptors. These specify that certain types are mutually exclusive. For example $\forall x(end(x) \supset (\neg ang(x) \wedge \neg tri(x)))$

4.2: Example of application of descriptors

Suppose e is the event $\exists x \exists y(ang(y) \wedge end(x))$. Let us assume that the knowledge base K_1 consists of the following existential descriptors:

(1) $A \supset \exists x \exists y \exists z(ang(x) \wedge end(y) \wedge tri(z))$

(2) $B \supset \exists x \exists y(ang(x) \wedge tri(y))$

(3) $C \supset \exists x\ end(x)$

(4) $D \supset \exists x\ ang(x)$

(5) $E \supset \exists x \exists y \exists z(ang(x) \wedge end(y) \wedge tri(z))$

The set of fundamental causes for e with K_1 is $\{ A, B \wedge C, C \wedge D, E \}$. Hence the observed letter is either A, or E, or both B and C, or both C and D. Obviously a letter cannot be simultaneously B and C, or both C and D. This, however, is a characteristic of our example, and in other application areas such as medical diagnosis, such combinations are legitimate since two illnesses could occur simultaneously. Therefore we will not resolve the problem here by adding descriptors that will directly exclude such possibilities, since there are more subtle and generally applicable methods, as follows.

Let us now consider the knowledge base K_2 obtained by adding the following negative descriptors to K_1.

(6) $B \supset \forall x \neg end(x)$

(7) $C \supset \forall x(\neg ang(x) \wedge \neg tri(x))$

(8) $D \supset \forall x(\neg end(x) \wedge \neg tri(x))$

The two causes $B \wedge C$ and $C \wedge D$ produced above are now inconsistent with the knowledge base K_2. We will show this in the case of the cause $B \wedge C$. After converting K_2 and $B \wedge C$ into clausal form we obtain the following clauses, among others:

$\{ \neg B, \neg end(x) \}$ from (6)

$\{ \neg C, end(\alpha) \}$ from (3)

$\{ B \}$

$\{ C \}$

Obviously the above set of clauses is unsatisfiable so that the cause $B \wedge C$ is inconsistent. Therefore the set of fundamental causes is now reduced to $\{ A, E \}$.

The above ambiguity can also be eliminated by the use of positive descriptors as we will now show. Suppose the knowledge base K_3 is obtained by adding the following positive and distinguishing descriptors to K_1.

(9) $A \supset \forall x(end(x) \vee ang(x) \vee tri(x))$

(10) $B \supset \forall x(ang(x) \vee tri(x))$

(11) $C \supset \forall x\ end(x)$

(12) $D \supset \forall x\ ang(x)$

(13) $E \supset \forall x(end(x) \vee ang(x) \vee tri(x))$

(14) $\forall x(end(x) \supset (\neg ang(x) \wedge \neg tri(x)))$

(15) $\forall x(ang(x) \supset \neg tri(x))$

Again the two causes $B \wedge C$ and $C \wedge D$ of e with K_1 are inconsistent with the knowledge base K_3. We will show this in the case of the cause $B \wedge C$. After converting K_3 and $B \wedge C$ into clausal form we obtain the following clauses, among others:

$\{ \neg B, ang(x), tri(x) \}$	from (10)
$\{ \neg end(x), \neg ang(x) \}$	from (14)
$\{ \neg end(x), \neg tri(x) \}$	from (14)
$\{ \neg C, end(\alpha) \}$	from (3)
$\{ B \}$	
$\{ C \}$	

Again the above set of clauses is unsatisfiable so that the cause $B \wedge C$ is inconsistent. Therefore the set of fundamental causes is again reduced to $\{ A, E \}$.

Let us note that it is impossible to distinguish between these two causes for the above event. However, by adding to the event information concerning the number of observed points of particular types, we can use quantitative descriptors to further distinguish between the possible causes. Let the new event e' be $e \wedge \exists x \exists y(tri(x) \wedge tri(y) \wedge x \neq y)$. With either K_2 or K_3 as the knowledge base we will obtain $\{ A \wedge \forall x(x \neq x), E \wedge \forall x(x \neq x) \}$ as the set of minimal, nontrivial and basic causes; these are, however, inconsistent. Now we will add quantitative descriptors to K_2 to obtain K_4 Among these new descriptors is the following:

(16) $A \supset \exists x \exists y(tri(x) \wedge tri(y) \wedge y \neq x))$

The set of fundamental causes now consists of the single conjunct **A**.

As promised above, our final example deals with the use of universal quantifiers in events. Suppose observation of a picture reveals that there are two angles: on this basis we cannot decide whether the letter is B, D or E. If, however, we also observe that the picture contains **only** angles, we should be able to conclude that the letter is in fact D. The event which captures this situation is $\forall x\ ang(x) \wedge \exists z \exists y(ang(z) \wedge ang(y) \wedge z \neq y)$. Since A and C both contain less then two angles, they will be ruled out as causes by methods discussed above. The basic causes involving B, D and E are as follows:

$B \wedge \forall x\ ang(x)$, D, and $E \wedge \forall x\ ang(x)$

The only consistent one, however, is D, because the others produce unsatisfiability with the knowledge base. $B \wedge \forall x\ ang(x)$ together with descriptors (2) and (15) is unsatisfiable; and $E \wedge \forall x\ ang(x)$ together with (5) and (15) is unsatisfiable.

5. Application of fundamental cause to expert systems

As mentioned earlier, an essential part of an expert system is its inductive mechanism for producing plausible explanations for observed phenomena. Such an inductive mechanism consists of three components: a logical structure for the knowledge base, a language for expressing observations, and a procedure for producing hypotheses from them.

The logical structure of the knowledge base should be such that information about the domain can be expressed naturally and directly without a bias toward a particular inference system. For example, if the knowledge base is structured as in MYCIN [1,10] using rules of the deductive form:

$$\text{symptom}_1 \wedge \ldots \wedge \text{symptom}_n \supset \text{disease}$$

then collecting information becomes awkward since there is an assumption that all "symptoms" of a "disease" are known a priori. In reality, however, information is usually more readily available in the inductive form:

$$\text{disease} \supset \text{symptom}_1$$
$$\vdots$$
$$\text{disease} \supset \text{symptom}_n$$

Although it at first appears that the above two forms are merely duals of each other, the deductive form is severely inadequate for expressing information involving quantification and negation. For instance in the example of the previous section, the negative, positive and quantitative descriptors related to the letter A can be independently added to or deleted from a knowledge base in inductive form, but would all have to be included in one large and unwieldy rule in a base in deductive form.

The language for expressing observations should have the full power of predicate calculus.

The computational mechanism should produce causes which are: pertinent to the observations without any extraneous conditions or unnecessary generality (minimal), which do not themselves have any apparent cause (basic), and are not simply an augmentation of the event unrelated to the knowledge base (nontrivial). It is well known that there is no decision procedure for inconsistency. In our system when inconsistency between a cause and the base is discovered, this information can be used in several ways. First, it can be used to eliminate a cause consisting of a conjunction of incompatible causes: for example, the cause $B \wedge C$ together with the base K_2 or K_3 in the last section. A second application of inconsistency is illustrated in the last section when, in the absence of the descriptor (16), the event $e \wedge \exists x \exists y(\mathrm{tri}(x) \wedge \mathrm{tri}(y) \wedge x \neq y)$ which contains two distinct triple points has only inconsistent causes, showing that the base lacks some information.

One important function of an expert system is to provide guidance for making further observations. For example, in a medical diagnosis system, some observed symptoms may lead to several hypotheses, each accompanied by some suggestions for further tests to be performed to reach the correct diagnosis. For example, the set of fundamental causes for the event $\exists x \exists y(\mathrm{ang}(x) \wedge \mathrm{ang}(y) \wedge x \neq y)$ with the knowledge base K_4 of the last section is { B, D, E }. For each cause one can identify features missing from the observations. In the case of cause B for example, we search for dead ends using the set of clauses $\{B\} \cup K_4 \cup \{\sim e\}$ by linear deduction with {B} as top clause. Among others we obtain the dead end $\mathrm{tri}(\alpha)$, which indicates that if B is a cause we should be able to observe a triple point.

6. Concluding remarks

The results reported above are of a preliminary character but appear to provide an interesting area for further investigation. In particular, it should be worthwhile to examine the case when the descriptors produce Horn clauses and events are conjunctions of positive literals. Another topic worth further research is the last issue mentioned in section 5, that is the search for missing observations that would confirm a hypothesis.

References

[1] Buchanan, B.G.; Shortliffe, E.H. (eds), **Rule-Based Expert Systems**, Addison-Wesley (1984).

[2] Colmerauer, A. et al., **Prolog II: Reference Manual and Theoretical Model**, Groupe d'Intelligence Artificielle, Faculte des Sciences de Luminy, Marseille (1982).

[3] Cox, P.T.; Pietrzykowski, T., A Complete Nonredundant Algorithm for Reversed Skolemization, **Theoretical Comp. Science 28**, (1984), 239-261.

[4] Cox, P.T.; Pietrzykowski, T., Surface Deduction: a uniform mechanism for logic programming, **Proc. 1985 Logic Prog. Symp.**, Boston (1985), 220-227.

[5] Loveland, D.W., A linear format for resolution, **Lectures in Mathematics 1-5 (Symp. on Autom. Demonstration)**, Springer-Verlag, Berlin (1970), 147-162.

[6] McCarthy, J., Circumscription - A Form of Non-monotonic Reasoning, **Artificial Intelligence Journal** v.3, no.s 1&2 (1980), 27-40.

[7] McDermott, D.; Doyle, J. Non-monotonic Logic I, **Artificial Intelligence Journal** v.3, no.s 1&2 (1980), 41-72.

[8] Pietrzykowski, T., **Mechanical Hypothesis Formation**, Res. Rept. CS-78-33, Department of Computer Science, University of Waterloo (1978).

[9] Reiter, R., A Logic for Default Reasoning, **Artificial Intelligence Journal** v.3, no.s 1&2 (1980), 81-132.

[10] Shortliffe, E.H., **Computer-Based Medical Consultation: MYCIN**, American Elsevier, New York (1976).

[11] van Emden, M.H.; Lloyd, J.W., A Logical Reconstruction of Prolog II, **Proc. 2nd Int'l. Conf. on Logic Prog.** Uppsala (1984), 35-40..

[12] Zadeh, L.A., Approximate Reasoning Based on Fuzzy Logic, **Proc. IJCAI-79**, Tokyo (1979), 1004-1010.

How to Clear a Block:
Plan Formation in Situational Logic

Zohar Manna
Stanford University
Weizmann Institute

Richard Waldinger
SRI International

Abstract

Problems in robot planning are approached by proving theorems in a new formulation of situational logic. A machine-oriented deductive-tableau inference system is adapted to this logic, with special attention being paid to the derivation of conditional and recursive plans. Equations and equivalences of the situational logic have been built into a unification algorithm for the system. Inductive proofs of theorems for even the simplest planning problems have been found to require challenging generalizations.

Program Synthesis

For many years, the authors have been working on *program synthesis*, the automated derivation of a computer program to meet a given specification. We have settled on a deductive approach to this problem, in which program derivation is regarded as a task in theorem proving (Manna and Waldinger [80]). To construct a program, we prove a theorem that establishes the existence of an output meeting the specified conditions. The proof is restricted to be constructive, in that it must describe a computational method for finding the output. This method becomes the basis for the program we extract from the proof.

For the most part, we have focused on the synthesis of *applicative* programs, which yield an output but produce no side effects. We are now interested in adapting our deductive approach to the synthesis of *imperative* programs, which may alter data structures or produce other side effects.

Plans are closely analogous to imperative programs, in that actions are like computer instructions, tests are like conditional branches, and the world is like a huge data structure. For this reason, we may hope that techniques for the synthesis of imperative programs may carry over into the planning domain. Conversely, we may hope that insights we develop by looking at a relatively simple planning domain, such as the blocks world, will then carry over to program synthesis in a more complex domain, involving array assignments, destructive list operations, and other alteration of data structures.

Clearing a Block

Consider the problem of clearing a given block, when we are not told whether the block is already clear or, if not, how many blocks are on top of it. Assume we are in a blocks world in which blocks are all the same size, so that only one block can fit directly on top of another, and in which the robot arm may lift only one block at a time. Then we might expect a planning system to produce the following program:

$$makeclear(a) \;\Leftarrow\; \begin{cases} \textit{if } clear(a) \\ \textit{then } \Lambda \\ \textit{else } makeclear(hat(a)); \\ \qquad put(hat(a), table). \end{cases}$$

This research was supported in part by the National Science Foundation under grants MCS-82-14523 and MCS-81-05565, by the Defense Advanced Research Projects Agency under Contract N00039-84-C-0211, by the United States Air Force Office of Scientific Research under Contract AFOSR-81-0014, by the Office of Naval Research under Contract N00014-84-C-0706, by United States Army Research under Contract DAJA-45-84-C-0040, and by a contract from the International Business Machines Corporation.

In other words, to clear a given block a, first test if it is already clear. If not, clear the block that is on top of block a, and then put that block on the table. Here Λ is the empty sequence of instructions, corresponding to no action at all, and $hat(a)$ is the block directly on top of a, if one exists.

A fundamental difficulty in applying a theorem-proving approach to plan construction is that the meaning of an expression in a plan depends on the situation, whereas in ordinary logic the meaning of an expression does not change. Thus, the block designated by $hat(a)$ or the truth value designated by $clear(a)$ may change from one state to the next. The traditional approach to circumventing this difficulty relies on a *situational logic*, i.e., one in which we can refer explicitly to situations or states of the world.

Elements of Situational Logic

Situational logic is actually a theory in first-order predicate logic. Our version of situational logic admits several sorts of terms.

The *static* terms denote the same element regardless of state. They include

- *state* terms, which (under any model) denote states
- *plan* terms, which denote plans
- *object* terms, which denote particular objects, such as blocks or boxes.

The *fluent* terms designate different elements from one state to the next. They include

- *object-fluent* terms, which, in given states, designate objects
- *propositional-fluent* terms, which, in given states, designate truth values.

For example, $hat(a)$ is an object fluent term and $clear(a)$ is a propositional-fluent term.

We may think of object and propositional fluents as functions mapping states into objects and truth values, respectively. Syntactically, however, they are denoted by terms, not function symbols. To determine what elements these terms designate with respect to a given state, we resort to the *in* function @ and the *in* predicate symbol $\tilde{@}$.

Definition (in function). If e is an object fluent term and s is a state term, $e@s$ is an object term denoting the object *designated by e in state s.*

For example, $hat(a)@s_0$ denotes the block designated by the object fluent term $hat(a)$ in state s_0.

In general, we shall introduce pairs of function symbols $f(u_1, \ldots, u_n)$ and $f'(x_1, \ldots, x_n, w)$ together, with the convention that f takes object fluents $u_1, \ldots, u_n$ as arguments and yields an object fluent, and f' takes objects $x_1, \ldots, x_n$ and a state w as arguments and yields an object. The two symbols are linked in each case by the *function linkage* axiom

$$f(u_1, \ldots, u_n)@w = f'(u_1@w,\ u_2@w,\ \ldots, u_n@w,\ w).$$

(Implicitly, variables in axioms are universally quantified.)

For example, corresponding to the function *hat*, which yields a block fluent, we have a function *hat′*, which yields a fixed block. Here, $hat'(x, w)$ denotes the block on top of block x in state w. The appropriate instance of the function linkage axiom is

$$hat(u)@w = hat'(u@w,\ w).$$

Thus $hat(a)@s$ denotes the block on top of block $a@s$ in state s. (This is not necessarily the same as the block on top of $a@s$ in some other state s'.)

It is important for us to have both entities *hat* and *hat′*. While explicit states and functions that yield fixed objects are useful specification constructs, we shall only admit fluents into our plans. This means that the plan can only access the present state, not hypothetical or future states. (In fact, the plan cannot even recall facts about past states unless it has explicitly stored, or "remembered," them.)

The *in* predicate symbol $\tilde{@}$ is analogous to the *in* function symbol @.

Definition (*in relation*). If e is a propositional-fluent term and s is a state term, $e\tilde{@}s$ is a proposition denoting the truth-value *designated by e in state s*.

For example, $clear(a)\tilde{@}s_0$ denotes the truth-value designated by the propositional-fluent term $clear(a)$ in state s.

As for object fluents, we shall introduce pairs of function symbols $r(u_1, \ldots, u_n)$ and predicate symbols $R(x_1, \ldots, x_n, w)$ together, with the convention that r takes object fluents $u_1, \ldots, u_n$ as arguments and yields a propositional fluent, and R takes objects $x_1, \ldots, x_n$ and a state w as arguments. The two symbols are linked in each case by the *predicate linkage* axiom

$$r(u_1, \ldots, u_n)\tilde{@}w \equiv R(u_1@w, \ldots, u_n@w, w).$$

For example, corresponding to the function *clear*, which yields a propositional fluent, we have an actual relation *Clear*. The instance of the predicate linkage axiom that relates them is

$$clear(u)\tilde{@}w \equiv Clear(u@w, w).$$

Thus $clear(a)\tilde{@}s$ is true if the block $a@s$ is clear in state s. Normally we shall drop the tilde from $\tilde{@}$, writing @, relying on context to make the meaning clear.

Certain fluent constants are to denote the same object regardless of state. For example, we may require that our input parameter a always denotes the same block and the constant *table* always denote the same piece of furniture. In this case, we will identify the object fluent with the corresponding fixed object.

Definition (*rigid designator*). An object fluent constant u is a *rigid designator* if

$$u@w = u$$

for all states w.

For example, the fact that *table* is a rigid designator is expressed by the axiom

$$table@w = table$$

for all states w. On the other hand, some constant fluents, such as *the-highest-block* or *the-president*, are likely not to be rigid designators.

Executing and Composing Plans

We need to describe the effects of executing a plan in a given state.

Definition (*execution*). If s is a state term and p is a plan term, $s;_s p$ is a state term denoting the state obtained by executing plan p in state s.

For example, $s_0;_s put(hat(a), table)$ denotes the state obtained if we are in state s_0 and put the hat of block a on the table.

Two plans are regarded as equal if they have the same effect on all states: this is expressed by the *equality axiom*

$$z_1 = z_2 \equiv (\forall w)[w;_s z_1 = w;_s z_2]$$

for all plans z_1 and z_2. Here w ranges over states.

The composition function $;_p$ allows us to combine two plans.

Definition (*composition*). If p_1 and p_2 are plan terms, $p_1;_p p_2$ is a plan term denoting the *composition* of p_1 and p_2. Executing $p_1; p_2$ is the same as executing p_1 followed by p_2, that is

$$w;_s (z_1;_p z_2) = (w;_s z_1);_s z_2,$$

for all states w and plans z_1 and z_2.

Normally we shall drop the subscripts from $;_s$ and $;_p$ and write ";", relying on context to make the meaning clear.

Composition of plans can be shown to be *associative*, i.e.,

$$(z_1; z_2); z_3 = z_1; (z_2; z_3),$$

for all plans z_1, z_2, and z_3. We shall often write $z_1; z_2; z_3$, dropping the parentheses.

Definition (*empty plan*). The empty plan Λ corresponds to no action at all; in other words

$$w; \Lambda = w$$

for all states w.

The empty plan behaves as an identity under composition; that is, we can show

$$\Lambda; z = z; \Lambda = z$$

for all plans z.

Specifying Facts and Actions

Facts about the world may be expressed as situational-logic axioms. For example, the principle property of the *hat* function is expressed by the *hat* axiom

$$\begin{array}{l} \textit{if not } (\textit{clear}(v)@w) \\ \textit{then } \textit{on}(\textit{hat}(v),\ v)@w. \end{array}$$

for all blocks v and states w. (As usual, we omit sort conditions such as $\textit{state}(w)$ from the axiom, for simplicity). In other words, if block v is not clear, then its hat is directly on top of it. Here v is an object (block) fluent variable, and w is a state variable. (If v is not clear, its hat is a "nonexistent" object, not a block.)

The effects of actions may also be described by situational-logic axioms. For example, the primary effect of putting a block on the table may be expressed by the *put-table* axiom

$$\begin{array}{l} \textit{if } \textit{clear}(u)@w \\ \textit{then } \textit{On}(u@w,\ \textit{table},\ w; \textit{put}(u,\ \textit{table})), \end{array}$$

for all blocks u and states w. Here u is a block fluent, w is a state variable, and the constant *table* is a rigid designator. The axiom expresses that after putting a block on the table, the block will indeed be on the table, provided it was clear beforehand. (The effects of attempting to move an unclear block are not specified and are hence unpredictable.)

Note that we do not conclude that $\textit{on}(u, \textit{table})@[w; \textit{put}(u, \textit{table})]$ in the consequent of the above axiom, because we have no way of knowing that u will designate the same block in state $w; \textit{put}(u, \textit{table})$ that it did in state w. For example, if u is taken to be $\textit{hat}(a)$, the axiom allows us to conclude that if $\textit{clear}(\textit{hat}(a))@s_0$, then $\textit{On}(\textit{hat}(a)@s_0,\ \textit{table},\ s_0; \textit{put}(\textit{hat}(a), \textit{table}))$.

In other words, the block that was on block a initially is on the table after executing the plan step. We cannot conclude that $on(hat(a), table)@[s_0; put(hat(a), table)]$, that is, that $hat(a)$ is on the table after executing the plan step. In fact, in this state, a is clear and $hat(a)$ no longer designates a block.

Plan Formation

To construct a plan for achieving a condition $Q[a, s_0, w_f]$, where a is the input fluent, s_0 the initial state, and w_f the final state, we prove the theorem

$$(\forall a)(\forall s_0)(\exists z)Q[a, s_0, s_0; z].$$

Here z is a plan variable. In other words, we show for any input a and initial state s_0 the existence of a plan z such that, if we are in state s_0 and execute plan z, we produce a state in which Q is true.

For example, the plan for clearing a block is constructed by proving a theorem

$$(\forall a)(\forall s_0)(\exists z)[Clear(a@s_0,\ s_0; z)].$$

In other words, the block initially designated by a is to be clear after execution of the desired plan z.

In the balance of this paper, we present a machine-oriented deductive system for situational logic in which we can prove such theorems and derive the corresponding plans at the same time. We shall use the proof of the above theorem, and the concomitant derivation of the *makeclear* plan, as a continuing example.

The Situational-Logic Deductive System

To support the synthesis of applicative programs, we developed a *deductive-tableau* theorem proving system (Manna and Waldinger [80], [85a]), which combines nonclausal resolution, well-founded induction, and conditional term rewriting within a single framework. In this paper, we carry the system over into situational logic. Although a full introduction to the deductive-tableau system is not possible here, we describe just enough to make this paper self-contained.

Deductive Tableaus

The fundamental structure of the system, the *deductive tableau*, is a set of rows, each of which contains a situational-logic sentence, either an *assertion* or a *goal*, and an optional plan term, the *plan entry*. We can assume the sentences are quantifier-free. For the time being, let us forget about the plan entry.

Under a given interpretation, a tableau is *true* whenever the following condition holds:

> If all instances of each of the assertions are true,
> then some instance of at least one of the goals is true.

Thus, variables in assertions have tacit universal quantification, while variables in goals have tacit existential quantification. In a given theory, a tableau is *valid* if it is true under all models for the theory.

To prove a given sentence valid, we remove its quantifiers (by skolemization) and enter it as the initial goal in a tableau. Any other valid sentences of the theory we are willing to assume may be entered into the tableau as assertions. The resulting tableau is valid if and only if the given sentence is valid. The deduction rules of the system attempt to establish validity.

The deduction rules add new rows to the tableau without altering its validity; in particular, if the new tableau is valid, the original tableau is also valid. The deductive process continues until we derive as a goal the propositional constant *true*, which is always true; or until we derive as an assertion the propositional constant *false*, which is always false. The tableau is then automatically valid, and hence the original sentence is also valid.

In deriving a plan $f(a)$, we prove a theorem of form

$$(\forall a)(\forall s_0)(\exists z)Q[a,\ s_0,\ s_0; z].$$

In skolemizing this, we obtain the sentence

$$Q[a,\ s_0,\ s_0;z],$$

where a and s_0 are constants and z is a variable. (Since this sentence is a theorem or goal to be proved, its existentially quantified variables become variables, while its universally quantified variables become skolem constants or functions. The situation is precisely the opposite for axioms or assertions.)

To prove the theorem, we establish the validity of the initial tableau

assertions	goals	plan: $f(a)$
	$Q[a,\ s_0,\ s_0;z]$	z

Certain valid sentences of the blocks-world theory, such as the axioms for blocks-world actions, would be included as assertions.

Note that the initial tableau includes a plan entry z. The plan entry is the mechanism for extracting a program from a proof of the given theorem. Throughout the derivation, we maintain the following *correctness* property:

For any model of the theory, and for any goal [or assertion] in the tableau,
if some instance of one of the goals is true [assertions is false],
then the corresponding instance t of the plan entry (if any)
will satisfy the desired condition $Q[a,\ s_0,\ s_0;t]$.

In other words, executing the plan segment t produces a state $s_0;t$ that satisfies the desired condition. The initial goal already satisfies the property in a trivial way, since it is the same as the desired condition. Each of the deduction rules of our system preserves these correctness properties.

Basic Properties

It may be evident that there is a duality between assertions and goals; namely, in a given theory,

a tableau that contains an assertion $\mathcal{A}$ is valid
if and only if
the tableau that contains instead the goal (*not* $\mathcal{A}$), with the same plan entry, is valid.

On the other hand,

a tableau that contains a goal $\mathcal{G}$ is valid
if and only if
the tableau that contains instead the assertion (*not* $\mathcal{G}$), with the same plan entry, is valid.

This means that we could push all the goals over into the assertion column simply by negating them, obtaining a refutation procedure; the plan entries would be unchanged, and the correctness properties are also unchanged. (This is done in conventional resolution systems.) Or we could push all the assertions into the goal column. Nevertheless, the distinction between assertions and goals has intuitive significance, and we retain it in our exposition.

Two other properties of tableaus are useful. First, the variables of any row in the tableau are dummies and may be renamed systematically without changing its validity or its correctness properties. Second, we may add to a tableau any instance of any of its rows, preserving validity and correctness properties.

Primitive Plans

Not all the symbols in our situational-logic theory are permitted in the plans we produce. This is because some constructs are valuable for specification but cannot be executed. Symbols that may appear in plans are said to be *primitive*. For example, we do not regard states, such as s_0, or the *in* function @ as primitive, because we do not want to admit then into our plans. On the other hand, we do regard the input constant a and the *hat* and *clear* function symbols as primitive.

In deriving a plan, it is not enough to end a derivation merely because we have derived the final goal *true* or assertion *false*; we also require that the associated plan entry t be primitive, i.e., that it contain only primitive symbols. If so, we can discontinue the process and return

$$f(a) \Leftarrow t.$$

This is because we have maintained the correctness condition that the plan entry of any goal [or assertion] will satisfy the desired condition in the case in which that goal [or assertion] is true [or false]. Since the truth symbol true is always true and the truth symbol *false* is always *false*, the plan entry t will always satisfy the condition.

In the next section, we begin to introduce the deduction rules of our system, emphasizing those rules that need to be adapted for situational logic and those that play a major role in the *makeclear* derivation.

Propositional Simplification

We assume that all the sentences in a tableau have been subjected to full propositional simplification. Rules such as

$$\mathcal{P} \textit{ and true} \rightarrow \mathcal{P}$$
$$\mathcal{P} \textit{ and } \mathcal{P} \rightarrow \mathcal{P}$$
$$\textit{not}\,(\textit{not}\,\mathcal{P}) \rightarrow \mathcal{P}$$

have been applied as much as possible before the assertion or goal is entered. Thus propositional simplification is not in itself a deduction rule but rather a part of other deduction rules.

Formation of Conditionals

The resolution rule accounts for the introduction of conditional expressions, or tests, into the derived plan and also plays a major role in ordinary reasoning. Because a special adaptation of the rule is necessary to form conditionals in situational logic without introducing states into the plan, we first consider applications of the rule that do not form conditionals.

The Resolution Rule: Ground Version

We begin by disregarding the plan entries and consider the ground version, in which there are no variables. We describe the rule in a tableau notation.

assertions	goals
$\mathcal{F}[\mathcal{P}]$	
$\mathcal{G}[\mathcal{P}]$	
$\mathcal{F}[\textit{true}]$ *or* $\mathcal{G}[\textit{false}]$.	

More precisely, if our tableau contains two assertions, $\mathcal{F}[\mathcal{P}]$ and $\mathcal{G}[\mathcal{P}]$, which share a common subsentence $\mathcal{P}$, we may replace all occurrences of $\mathcal{P}$ in $\mathcal{F}[\mathcal{P}]$ with *true*, replace all occurrences of $\mathcal{P}$ in $\mathcal{G}[\mathcal{P}]$ with *false*, take the disjunction of the results, and (after propositional simplification) add it to the tableau as a new assertion.

The rationale for this rule is as follows: We suppose that $\mathcal{F}[\mathcal{P}]$ and $\mathcal{G}[\mathcal{P}]$ are true under a given model, and show that then ($\mathcal{F}[\textit{true}]$ *or* $\mathcal{G}[\textit{false}]$) is also true.

We distinguish between two cases. In the case in which $\mathcal{P}$ is true, because $\mathcal{F}[\mathcal{P}]$ is true, its equivalent (in this case) $\mathcal{F}[true]$ is true. On the other hand, in the case in which $\mathcal{P}$ is false, because $\mathcal{G}[\mathcal{P}]$ is true, its equivalent $\mathcal{G}[false]$ is true. In either case, the disjunction ($\mathcal{F}[true]$ *or* $\mathcal{G}[false]$) is true.

Note that the rule is asymmetric in its treatment of $\mathcal{F}[\mathcal{P}]$ and $\mathcal{G}[\mathcal{P}]$. In fact, it can be restricted according to the "polarity" of the occurrences of $\mathcal{P}$, the common subsentence. We may require that some occurrence of $\mathcal{P}$ in $\mathcal{F}[\mathcal{P}]$ have ***negative polarity*** (i.e., it must be within the scope of an odd number of implicit or explicit negations) and that some occurrence of $\mathcal{P}$ in $\mathcal{G}[\mathcal{P}]$ be of ***positive polarity*** (i.e., it must be within the scope of an even number of implicit or explicit negations). Thus, in applying the rule between assertions $((not\ P)\ or\ Q)$ and $(P\ or\ R)$, the role of $\mathcal{F}[\mathcal{P}]$ must be played by $((not\ P^-)\ or\ Q)$, in which P has negative polarity, and the role of $\mathcal{G}[\mathcal{P}]$ by $(P^+\ or\ R)$, in which P has positive polarity, yielding the new assertion $(Q\ or\ \mathcal{R})$, after propositional simplification. Reversing the roles of the two assertions yields the trivial assertion *true*, which is of no value in the proof. This strategy has been shown by Murray [82] to retain completeness for first-order logic.

We have applied the rule between two assertions but, by duality, the rule can just as well be applied between two goals or between an assertion and a goal. In the goal-goal case, the new goal is a conjunction rather than a disjunction. In applying the polarity strategy, each goal must be considered to be within the scope of an implicit negation.

The Resolution Rule: General Version

We have up to now been considering the ground case, in which the sentences have no variables. In the general case, the rule may be expressed as follows:

assertions	goals
$\mathcal{F}[\mathcal{P}]$	
$\mathcal{G}[\mathcal{P}']$	
$\mathcal{F}\theta[true]$ *or* $\mathcal{G}\theta[false]$	

More precisely, suppose our tableau contains two assertions $\mathcal{F}[\mathcal{P}]$ and $\mathcal{G}[\mathcal{P}']$, which have been renamed so that they have no variables in common. The subsentences $\mathcal{P}$ and $\mathcal{P}'$ are not necessarily identical, but they are unifiable, with most-general unifier θ; thus $\mathcal{P}\theta = \mathcal{P}'\theta$. Then we may apply θ to $\mathcal{F}[\mathcal{P}]$ and $\mathcal{G}[\mathcal{P}']$, replace all occurrences of $\mathcal{P}\theta$ in $(\mathcal{G}[\mathcal{P}])\theta$ with *true*, replace all occurrences of $\mathcal{P}'\theta$ in $(\mathcal{G}[\mathcal{P}'])\theta$ with *false*, take the disjunction of the results, and (after propositional simplification) add it to our tableau as a new assertion. In other words, after applying the most-general unifier θ, we use the ground version of the rule.

Now let us consider the plan entries of the rows. If each of the given rows has no plan entry, neither does the newly derived row. If exactly one of the rows has a plan entry, the appropriate instance of that entry is inherited by the new row. The case in which both rows have plan entries is treated by the situational-logic extension of the resolution rule.

Although most of the ideas of this paper have not been incorporated into our implemented deductive-tableau system, Dag Mellgren [private communication] has implemented a version of Fay's [79] unification algorithm (see also Hullot [80]) that incorporates properties of situational logic and other theories. This allows us to omit these properties from the assertions of the tableau and gives us shorter proofs. In the blocks-world theory, the following properties are included:

$$clear(u)@w = Clear(u@w, w)$$

(i.e., the predicate linkage axiom for *clear*)

$$a@w = a$$

(i.e., the input is a rigid designator)

$$w; (z_1; z_2) = (w; z_1); z_2$$

(i.e., the composition axiom), for all states w, block fluents u, and plans z_1 and z_2.

In an equational theory, there is not necessarily a most-general unifier. The algorithm may be pulsed to produce more and more unifiers. For example, when we unify

$$Clear(a@s_0, s_0; z) \quad \text{and} \quad Clear(v@w,\ w; put(u, table)),$$

we obtain the substitutions

$$\{v \leftarrow a,\ w \leftarrow s_0,\ z \leftarrow put(u, table)\},$$

$$\{v \leftarrow a,\ w \leftarrow s_0; z_1,\ z \leftarrow z_1; put(u, table)\},$$

and others. In applying the resolution rule, we must consider each of these unifiers.

Examples

Let us illustrate the resolution rule with an example from the *makeclear* derivation.

Example (Resolution). Suppose our tableau contains the initial goal

	1. $\boxed{Clear(a@s_0,\ s_0; z)}^{-}$	z

and the axiom

if $on(u,v)@w$ *and* $clear(u)@w$ *then* $\boxed{Clear(v@w,\ w; put(u, table))}^{+}$		

The axiom asserts that after putting a block on the table, the block underneath it is clear. As we have seen above, the two boxed subsentences are unifiable in the blocks-world theory. One of the unifiers is

$$\{v \leftarrow a,\ w \leftarrow s_0; z_1,\ z \leftarrow z_1;\ put(u, table)\}.$$

The polarity of the boxed subsentences is indicated by their annotation. (The goal is negative because goals are within the scope of an implicit negation.) Let us apply the resolution rule, taking $\mathcal{P}$ and $\mathcal{P}'$ to be the boxed subsentences and θ to be the above unifier. Recall that by duality we can push the assertion into the goal column by negating it. We obtain

	true *and* *not* $\begin{pmatrix} \textit{if } on(u,a) @ s_0; z_1 \textit{ and } clear(u) @ s_0; z_1 \\ \textit{then false} \end{pmatrix}$	$z_1;\ put(u, table)$

which simplifies propositionally to

	2. $on(u,a) @ s_0; z_1$ *and* $clear(u) @ s_0; z_1$	$z_1;\ put(u, table)$

In other words, if after executing some plan z_1, some block u is on block a but is itself clear, we can achieve our desired condition by first executing plan z_1 and then putting block u onto the table. ∎

To show another step of the *makeclear* derivation, we give a further example of branch-free resolution.

Example (resolution). The boxed subsentence of the new goal,

	2. $\boxed{on(u,a) @ s_0; z_1}^{-}$ *and* $clear(u) @ s_0; z_1$	$z_1;\ put(u, table)$

unifies with the boxed subsentence of the *hat* axiom,

if not $(clear(v) @ w)$ *then* $\boxed{on(hat(v), v)@w}^{+}$		

with most-general unifier

$$\{v \leftarrow a,\ u \leftarrow hat(a),\ w \leftarrow s_0; z_1\}.$$

Applying the resolution rule, we obtain (after propositional simplification)

	3. $clear(hat(a))@s_0; z_1$ *and* *not* $(clear(a)@s_0; z_1)$	$z_1;\ put(hat(a), table)$

In other words, if, after executing some plan step z_1, the block a is not clear but the block $hat(a)$ is, we can achieve our desired condition by first executing plan z_1 and then putting $hat(a)$ on the table. ∎

Resolution with Conditional Formation

In applying the resolution rule between two rows, both of which have output entries, we must generate a conditional. If we applied the ordinary resolution rule in such a case, we would be forced to introduce tests which contained the $\tilde{@}$ predicate symbol and explicit references to states. We would have no way of executing the resulting nonprimitive plans. To avoid introducing nonprimitives into the plan entry, we employ the following (ground) version of the resolution rule:

assertions	goals	plan: $f(a)$
	$\mathcal{F}[p @ s_0; e]$	$e; e_1$
	$\mathcal{G}[p @ s_0; e]$	$e; e_2$
	$\mathcal{F}[true]$ *and* $\mathcal{G}[false]$	$e;$ *if* p *then* e_1 *else* e_2

In other words, suppose our tableau contains two goals, both of which refer to the truth of the same propositional fluent p after the execution of a common plan e. Suppose further that e is an initial segment of the plan entries for each of the two goals. Then we can introduce the same new goal as the branch-free version of the rule. The plan entry associated with this goal has as its initial segment the initial segment e of the given plan. The final segment is a conditional whose test is the common propositional fluent p and whose *then*-clause and *else*-clause are the final segments e_1 and e_2 of the given plans.

The rationale for this plan entry is as follows: We suppose the new goal ($\mathcal{F}[true]$ *and* $\mathcal{G}[false]$) is true and show that the associated plan entry satisfies the desired condition.

We distinguish between two cases. In the case in which $p @ s_0; e$ is true, the conjunct $\mathcal{F}[true]$ is true, hence the given goal $\mathcal{F}[p @ s_0; e]$ is also true, and hence the associated plan entry $e; e_1$ satisfies the desired condition. In this case, the conditional plan e; (*if* p *then* e_1 *else* e_2) will also satisfy the condition, because, when executed in state $s_0; e$, the result of the test of p will be true.

Similarly, in the case in which $p @ s_0; e$ is false, the given goal $\mathcal{G}[p @ s_0; e]$ is true, the associated plan entry $e; e_2$ satisfies the desired condition, and the conditional plan e; (*if* p *then* e_1 *else* e_2) will also satisfy the condition. Thus, in either case the conditional plan satisfies the desired condition.

Of course, the rule applies to assertions as well as to goals. The polarity strategy may be imposed as before. We have given the ground version of the rule; in the general version, in which the rows may have variables, we first apply the most-general unifier of the subsentences $p \,@\, s_0; e$ and $p' \,@\, s_0; e'$; then we use the ground version of the rule.

We illustrate with an example.

Example (*resolution with conditional formation*). Suppose our tableau contains the two goals

	$\boxed{clear(a)@s_0; z}^{-}$	$\Lambda; z$
	$not\ \boxed{clear(a)@s_0; \Lambda}^{+}$	$\Lambda;$ $makeclear(hat(a));$ $put(hat(a), table)$

The boxed subsentences are unifiable, with most-general unifier $\{z \leftarrow \Lambda\}$. The unified subsentences both refer to the truth of the same propositional fluent *clear*(a) after the execution of a common plan, the empty step Λ. The plan Λ is an initial segment for the plan entries of each of the given goals. Therefore we can apply the resolution rule to obtain (after propositional simplification)

	true	$\Lambda; \left(\begin{array}{l} if\ clear(a) \\ then\ \Lambda \\ else\ makeclear(hat(a)); \\ \quad put(hat(a), table) \end{array}\right)$

Note that because of our use of Fay's algorithm to build in properties of the blocks-world theory, the first goal could have been $Clear(a@s_0,\ s_0; z)$, and the empty plan step Λ could have been dropped from the second goal and from both given plan entries. ▪

Resolution with Equality Matching

Sometimes in an attempt to apply the resolution rule, two subsentences will fail to unify completely but will "nearly" unify; that is, all but certain pairs of subterms will unify. In such cases, instead of abandoning the attempt altogether, it may be advantageous to go ahead and apply the rule, but impose certain conditions upon the conclusion. This is the effect of applying the resolution rule with equality matching. In its simplest ground version, the rule may be expressed as follows:

assertions	goals
	$\mathcal{F}[\mathcal{P}\langle s\rangle]$
	$\mathcal{G}[\mathcal{P}\langle t\rangle]$
	$s = t\ and\ \mathcal{F}[true]\ and\ \mathcal{G}[false]$

Here $\mathcal{P}\langle s\rangle$ and $\mathcal{P}\langle t\rangle$ are identical except that certain occurrences of s in $\mathcal{P}\langle s\rangle$ are replaced by t in $\mathcal{P}\langle t\rangle$. If they were really identical, we could apply the ordinary resolution rule to obtain the new goal $(\mathcal{F}[true]\ and\ \mathcal{G}[false])$. Instead, we obtain this goal with the additional conjunct $s = t$. The treatment of the output entry is analogous to that for the original resolution rule.

This rule is a nonclausal version of the E-resolution rule (Morris [69]) or the RUE-resolution rule (Digricoli [86]). In Manna and Waldinger [86], we generalize the rule to allow more than one pair of mismatched terms and to employ binary relations weaker than equality, but we shall not require these extensions here.

In the nonground version, in which the sentences may contain variables, we apply a substitution to the given rows and then apply the ground version of the rule to the results. The substitution is the outcome of a failed attempt to unify the two sentences. We shall see that for a given pair of sentences, the substitution we employ and the pair of mismatched subterms we obtain are not necessarily unique. Some of the strategic aspects of choosing the substitution and term pair are discussed in Digricoli [86].

Example (*resolution with equality matching*). Suppose our tableau contains the goal

	$\boxed{Clear\big(hat(a)@s_0;z_1,\ s_0;z_1\big)}^{-}$ *and* $q(z_1)$	$z_1;$ $put\big(hat(a, table\big)$

and the assertion

if $r(u,a)$ *then* $\boxed{Clear\big(u@w,\ w; makeclear(u)\big)}^{+}$		

The two boxed subsentences are not unifiable. However, if we apply the substitution

$$\{u \leftarrow hat(a),\ w \leftarrow s_0;z_1\},$$

we obtain the sentences

$$Clear\big(hat(a)@s_0;z_1,\ s_0;z_1\big) \quad \text{and} \quad Clear\big(hat(a)@s_0;z_1,\ (s_0;z_1);makeclear\big(hat(a)\big)\big)$$

Our mismatched terms are then

$$s_0;z_1 \quad \text{and} \quad (s_0;z_1);makeclear\big(hat(a)\big).$$

The conclusion of the rule is then (before simplification)

	$s_0;z_1 \;=\; (s_0;z_1);makeclear\big(hat(a)\big)$ *and* *true and* $q(z_1)$ *and* *not* $\big($*if* $r\big(hat(a),a\big)$ *then false*$\big)$	$z_1;\ put\big(hat(a), table\big)$

On the other hand, if we apply the substitution

$$\{w \leftarrow s_0,\ z_1 \leftarrow makeclear(u)\}$$

the boxed subsentences become

$$Clear\big(hat(a)@s_0;makeclear(u),\ s_0;makeclear(u)\big)$$

and

$$Clear\big(u@s_0,\ s_0;makeclear(u)\big).$$

Our mismatched terms are then

$$hat(a)@s_0;makeclear(u) \quad \text{and} \quad u@s_0,$$

and the conclusion of the rule (after simplification this time) is then

	$hat(a)@s_0;makeclear(u) \;=\; u@s_0$ *and* $q\big(makeclear(u)\big)$ *and* $r(u,a)$	$makeclear(u);$ $put(u, table)$

Other substitutions are possible if we alter Fay's unification algorithm to discover mismatched terms. ▟

Formation of Recursion

The mathematical induction rule accounts for the introduction of the basic repetitive construct, recursion, into the plan being derived. We employ well-founded induction, i.e., induction over a well-founded relation; this is a single very general rule that applies to many subject domains.

The Mathematical Induction Rule

A well-founded relation $\prec_\omega$ is one that admits no infinite decreasing sequences, i.e., sequences $x_1, x_2, x_3, \ldots$ such that

$$x_1 \succ_\omega x_2 \ \textit{and}\ x_2 \succ_\omega x_3 \ \textit{and}\ \ldots.$$

For instance the less-than relation $<$ is well-founded in the theory of nonnegative integers but not in the theory of real numbers. A well-founded relation need not be transitive.

The instance of the *well-founded induction rule* we require can be expressed as follows (the general rule is more complex, notationally):

Suppose that our initial tableau is

assertions	goals	plan: $f(a)$
	$Q[a, s_0, z]$	z

In other words, we are attempting to construct a program f that, for an arbitrary input a and initial state s_0, yields a plan z satisfying the condition $Q[a, s_0, z]$. According to the well-founded induction principle, we may prove this assuming inductively that the program f will yield a plan $f(u)$ satisfying the condition $Q[u, w, f(u)]$ for any input u and initial state w, provided that the pair $\langle u, w\rangle$ is less than the pair $\langle a, s_0\rangle$ with respect to some well-founded relation $\prec_\alpha$, that is, $\langle u, w\rangle \prec_\alpha \langle a, s_0\rangle$. In other words, we may add to our tableau the new assertion

if $\langle u, w\rangle \prec_\alpha \langle a, s_0\rangle$ *then* $Q[u,\ w,\ f(u)]$		

Here u and w are both variables. The well-founded relation $\prec_\alpha$ is arbitrary; its selection may be deferred until later in the proof.

Example (induction principle). The initial tableau in the *makeclear* derivation is

assertions	goals	plan: *makeclear*(a)
	1. $Clear(a@s_0,\ s_0; z)$	z

By application of the well-founded induction rule, we may add to our tableau the new assertion

if $\langle u,\ w\rangle \prec_\alpha \langle a,\ s_0\rangle$ *then* $Clear(u@w,\ w; makeclear(u))$		

In other words, we may assume inductively that the *makeclear* program will produce a plan *makeclear*(u) that satisfies the desired condition for any input u in any initial state w, provided that the block-state pair $\langle u,\ w\rangle$ is less than the pair $\langle a,\ s_0\rangle$ with respect to some well-founded relation $\prec_\alpha$. ▟

Use of the induction hypothesis in the proof may account for the introduction of a recursive call into the derived program.

Example (*formation of recursive calls*). In the *makeclear* derivation, we have obtained the goal

	3. $\boxed{clear(hat(a))@s_0; z_1}^-$ *and* $not\ (clear(a)@s_0; z_1)$	$z_1;\ put(hat(a), table)$

In our situational logic, the boxed subsentence "nearly" unifies with the boxed subsentence of our induction hypothesis,

if $\langle u,\ w\rangle \prec_\alpha \langle a,\ s_0\rangle$ *then* $\boxed{Clear(u@w,\ w; makeclear(u))}^+$		

The Fay unification algorithm makes use of the equivalence

$$clear(hat(a))@s_0; z_1 \quad \equiv \quad Clear(hat(a)@s_0; z_1,\ s_0; z_1),$$

an instance of the predicate linkage axiom. The resulting substitution is

$$\{w \leftarrow s_0,\ z_1 \leftarrow makeclear(u)\}.$$

(The algorithm will also yield more complex substitutions, but this one is general enough to allow the proof to go through.) The mismatched subterms are

$$hat(a)@s_0; makeclear(u) \quad \text{and} \quad u@s_0.$$

We obtain the new goal

	4. $hat(a)@s_0; makeclear(u) \ = \ u@s_0$ *and* $not\ (clear(a)@s_0; makeclear(u))$ *and* $\langle u,\ s_0\rangle \prec_\alpha \langle a,\ s_0\rangle$	$makeclear(u);$ $put(hat(a), table)$

▟

Note that at this stage of the derivation a recursive call $makeclear(u)$ has been introduced into the plan entry. The condition $\langle u,\ s_0\rangle \prec_\alpha \langle a,\ s_0\rangle$ in the goal ensures that this recursive call will not lead to nontermination. For any nonterminating computation, there corresponds an infinite decreasing sequence of pairs $\langle s_0, makeclear(a)\rangle$, $\langle s_0,\ makeclear(u)\rangle$, $\langle s_0,\ makeclear(u')\rangle$, ..., contrary to the well-foundedness of $\prec_\alpha$.

The Choice of a Well-founded Relation

Although the well-founded induction principle is the same from one theory to the next, each theory has its own well-founded relations. For the blocks-world theory, one relation of particular importance is the *on* relation, which holds if one block is immediately on top of another. In a given state, this relation is well-founded because we assume towers of blocks cannot be infinite.

More precisely, we define the well-founded relation $\prec_{on_w}$ by the following axiom:

Definition (*on relation*).

$$u \prec_{on_w} v \ \equiv \ on(u, v)@w$$

(Note that for each state w we obtain a different relation $\prec_{on_w}$). This relation has the property

$$\begin{array}{l} \textit{if not } (clear(v)@w) \\ \textit{then } hat(v) \prec_{on_w} v. \end{array}$$

We actually take well-founded relations to be objects in our theory and regard the expression $u \prec_\alpha v$ as a notation for a three-place relation $\prec(\alpha, u, v)$, where α is a variable that ranges over well-founded relations.

The *on* relation applies to blocks, but the desired relation $\prec_\alpha$ applies to block-state pairs. However, for any well-founded relation $\prec_\beta$, there exists a corresponding well-founded *first-projection* relation $\prec_{\pi_1(\beta)}$ on pairs, defined by the following axiom.

Definition (*first-projection relation*).

$$\begin{array}{l} \langle u_1,\ u_2\rangle \prec_{\pi_1(\beta)} \langle v_1,\ v_2\rangle \\ \quad \equiv \\ u_1 \prec_\beta v_1. \end{array}$$

In other words, two pairs are related by the first-projection relation $\prec_{\pi_1(\beta)}$ if their first components are related by $\prec_\beta$. As usual we omit the sort conditions, but here β is a variable that ranges over well-founded relations.

By application of rules of the system to the above properties, we may reduce our most recent goal

	4. $hat(a)@s_0; makeclear(u) = u@s_0$ *and* $not\ (clear(a)@s_0; makeclear(u))$ *and* $\langle u,\ s_0\rangle \prec_\alpha \langle a,\ s_0\rangle$	$makeclear(u);$ $put(hat(a), table)$

to obtain

	5. $hat(a)@s_0; makeclear(hat(a)) = hat(a)@s_0$ *and* $not\ (clear(a)@s_0; makeclear(hat(a)))$	$makeclear(hat(a));$ $put(hat(a), table)$

Through these steps, the well-founded relation $\prec_\alpha$ on block-state pairs is chosen to be $\prec_{\pi_1(on_{s_0})}$, the first projection of the *on* relation in the initial state s_0.

At this stage, we have completed the derivation of the entire *else*-branch of the *makeclear* program.

The Need for Generalization

One might hope that the derivation is nearly complete; all that remains is to dispense with the remaining conjuncts of our goal 5,

$$hat(a)@s_0; makeclear(hat(a)) = hat(a)@s_0$$

and

$$not\ (clear(a)@s_0; makeclear(hat(a)))$$

and introduce the conditional. In fact, closer examination of the above two conditions indicates that they are not so straightforward.

The first condition, $hat(a)@s_0; makeclear(hat(a)) = hat(a)@s_0$, requires that after clearing $hat(a)$, the value of $hat(a)$ is the same as it was before. In other words, we must show that the *makeclear* program we are constructing will not move $hat(a)$ in the process of clearing it. In fact, nothing in the specification of *makeclear* forces it to be so well behaved. In fact, if *makeclear* were trying to be economical with table space, it might clear *hat*(a) by putting underneath it all the blocks that were previously on top of it, as illustrated below:

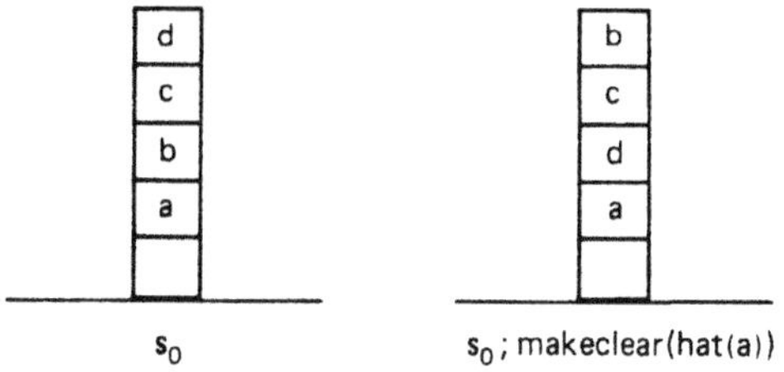

Here a hypothetical *makeclear* program has cleared *hat*(a), that is, b, by putting c and d underneath b. The subsequent value of *hat*(a) is d, not b, contrary to the condition. Attempting to put *hat*(a) on the table will lead to unpredictable results, because d is not clear.

The second remaining condition of the goal, *not* $(clear(a) @ s_0; makeclear(hat(a)))$, requires that, in the process of clearing *hat*(a), we do not inadvertently clear a. Again, there is nothing in the specification that prevents an over-ambitious *makeclear* program from clearing a or any other block when it was only asked to clear *hat*(a). Attempting to move *hat*(a) will then lead to unpredictable results, because *hat*(a) is not a block.

The only knowledge we have about *makeclear* is that given in our induction hypothesis, which depends in turn on what is required by our specification. We have not specified what *makeclear*(a) does to blocks underneath its input parameter a or elsewhere on the table. Thus it is actually impossible to prove the two conditions.

In proving a given theorem by induction, it is often necessary to prove a stronger, more general theorem, so as to have the benefit of a stronger induction hypothesis. This strengthening is mentioned by Polya [57] (see also Manna and Waldinger [85b]) and is done automatically by the system of Boyer and Moore [79]. By analogy, in constructing a program to meet a given specification, it is often necessary to impose a stronger specification, so as to have the benefit of more powerful recursive calls.

This turns out to be the case with the *makeclear* problem: rather than constructing the program to meet the given specification, it is necessary to construct the program to meet the following stronger specification:

$$(\forall a)(\forall s_0)(\exists z)\begin{bmatrix} Clear(a@s_0,\ s_0;z)\ \mathit{and} \\ (\forall g)(\forall h)\begin{bmatrix} \mathit{if}\ Overeq(a@s_0,\ g,\ s_0) \\ \mathit{then}\ \mathit{if}\ On(g,\ h,\ s_0) \\ \qquad \mathit{then}\ On(g,\ h,\ s_0;z) \end{bmatrix} \end{bmatrix}$$

In other words, executing the desired plan z in state z_0 should clear block a without moving a or any of the blocks that are directly or indirectly underneath a. (Here *Overeq*(x, y, w) holds if block x is the same as block y or rests directly or indirectly on top of it.) This theorem is strong enough to enable us to show that, in clearing *hat*(a), or $hat(hat(a))$, or $hat(hat(hat(a)))$, or ..., we do not move *hat*(a) itself. The induction hypothesis is also strong enough to enable us to prove the new condition in the theorem.

With human intuition, it may not be difficult to formulate such strengthened theorems. But the strengthening required by this problem seems to be beyond the capabilities of the Boyer-Moore system or other current theorem provers.

Discussion

In this section we touch on some matters we have been unable to treat in this paper.

Historical Remarks

The HACKER system of Sussman [73] was applied to the problem of clearing a block and obtained a plan similar to ours. The system relied on somewhat higher-level knowledge. Furthermore, it employed a technique of successive debugging and did not attempt to establish the correctness of the plans it constructed.

Situational logic was introduced into the artificial-intelligence literature by McCarthy [63], but that logic was not used for planning. A variant of the logic was used in the QA3 system (Green [69]) and in the PROW system (Waldinger and Lee [69]). In this variant, predicate symbols had explicit state arguments; for this reason, the QA3 and PROW logic admitted the derivation of nonexecutable conditional plans. The present logic is closer to McCarthy's original.

The Frame Problem

One obstacle to employing a situational logic is the so-called *frame problem* (see McCarthy and Hayes [69]). In addition to saying what relations are changed by a given action, it is also necessary to provide *frame axioms* which state explicitly what relations are left unchanged. The resulting large axiom set for each action has a disastrous effect on the search space. Furthermore, it seems fundamentally wrong for a logical system to force us to spell out each frame axiom individually.

The present paper does not address the frame problem itself. We assume that the appropriate frame axioms are present, introduced either individually or by some mechanism such as McCarthy's circumscription (see Lifschitz [85]). We believe that the strategic aspects of the frame problem can be dealt with in part by Stickel's [85] theory resolution rule.

The Problem of Control

Many people believe that a theorem-proving approach is inadequate for planning because a general-purpose theorem prover will never be able to compete with a system with strategies designed especially for problem solving. Although we have not yet engaged the strategic question, we propose to overlay a general-purpose theorem prover with a special strategic component for planning. For example, the WARPLAN system (Warren [74]) might be regarded as a situational-logic theorem prover provided with a strategy that enables it to imitate the STRIPS planning system (Fikes and Nilsson [71]). We speculate that, in the same way, a situational-logic theorem prover could be induced to mimic any dedicated planning system, given the appropriate strategic component.

Acknowledgments

The authors would like to thank Martin Abadi, Peter Ladkin, Vladimir Lifschitz, John McCarthy, and Jonathan Traugott for reading, suggestions, and discussion.

References

Boyer and Moore [79]
R. S. Boyer and J S. Moore, *A Computational Logic*, Academic Press, Orlando, Fla., 1979.

Digricoli [86]
V. J. Digricoli, Equality-based binary resolution, *Journal of the ACM*, Vol. 33, No. 2, April 1986, to appear.

Fay [79]
M. Fay, First-order unification in an equational theory, *Proceedings of the Fourth Workshop on Automated Deduction*, Austin, Texas, Feb. 1979, pp. 161–167.

Fikes and Nilsson [71]
R. E. Fikes and N. J. Nilsson, STRIPS: A new approach to the application of theorem proving to problem solving, *Artificial Intelligence*, Vol. 2, No. 3–4, Winter 1971, pp. 189–208.

Green [69]
C. C. Green, Application of theorem proving to problem solving, *Proceedings of the International Joint Conference on Artificial Intelligence*, Washington, D.C., May 7–9, 1969, pp. 219–239.

Hullot [80]
J.-M. Hullot, Canonical forms and unification, *Proceedings of the Fifth Conference on Automated Deduction*, Les Arcs, France, July 8–11, 1980, pp. 318–334.

Kowalski [79]
R. Kowalski, *Logic for Problem Solving*, North-Holland, New York, N.Y., 1979.

Lifschitz [85]
V. Lifschitz, Circumscription in the blocks world, Unpublished Report, Stanford University, Stanford, Calif., Dec. 1985.

McCarthy [63]
J. McCarthy, Situations, actions, and causal laws, Technical Report, Stanford University, Stanford, Calif., 1963. Reprinted in *Semantic Information Processing*, Marvin Minsky, editor, MIT Press, Cambridge, Mass., 1968, pp. 410–417.

McCarthy and Hayes [69]
J. McCarthy and P. Hayes, Some philosophical problems from the standpoint of artificial intelligence, *Machine Intelligence 4*, B. Meltzer and D. Michie, editors, American Elsevier, New York, N. Y., 1969, pp. 463–502.

Manna and Waldinger [80]
Z. Manna and R. Waldinger, A deductive approach to program synthesis, *ACM Transactions on Programming Languages and Systems*, Vol. 2, No. 1, Jan. 1980, pp. 90–121.

Manna and Waldinger [85a]
Z. Manna and R. Waldinger, The origin of the binary-search paradigm, *Proceedings of the Ninth International Joint Conference on Artificial Intelligence*, Los Angeles, Calif., Aug. 18–23, 1985, pp. 222–224.

Manna and Waldinger [85b]
Z. Manna and R. Waldinger, *The Logical Basis for Computer Programming*, Vol. 1: *Deductive Reasoning*, Addison-Wesley, Reading, Mass., 1985.

Manna and Waldinger [86]
Z. Manna and R. Waldinger, Special relations in automated deduction, *Journal of the ACM*, Vol. 33, No. 1, Jan. 1986, pp. 1–60.

Martelli and Rossi [86]
A. Martelli and G. Rossi, An algorithm for unification in equational theories, *Proceedings of the Third Symposium on Logic Programming*, Salt Lake City, Utah, Sept. 21–25, 1986 (to appear).

Morris [69]
J. B. Morris, E-resolution: Extension of resolution to include the equality relation, *International Joint Conference on Artificial Intelligence*, Washington, D.C., May 7–9, 1969, pp. 287–294.

Murray [82]
N. V. Murray, Completely nonclausal theorem proving, *Artificial Intelligence*, Vol. 18, No. 1, 1982, pp. 67–85.

Polya [57]
G. Polya, *How to Solve It*, Doubleday and Company, Garden City, N.Y., 1957.

Sacerdoti [77]
E. D. Sacerdoti, *A Structure for Plans and Behavior*, Elsevier North-Holland, New York, N. Y., 1977.

Stickel [85]
M. E. Stickel, Automated deduction by theory resolution, *Journal of Automated Reasoning*, Vol. 1, No. 4, 1985, pp. 333–355.

Sussman [73]
G. J. Sussman, *A Computational Model of Skill Acquisition*, Ph. D. Thesis, MIT, Cambridge, Mass., 1973.

Waldinger and Lee [69]
R. J. Waldinger and R. C. T. Lee, PROW: A step toward automatic program writing, Proceedings of the International Joint Conference on Artificial Intelligence, Washington, D. C., May 7–9, 1969, pp. 241–252.

Warren [74]
D. H. D. Warren, WARPLAN: A system for generating plans, Technical Report, University of Edinburgh, Edinburgh, Scotland, 1974.

DEDUCTIVE SYNTHESIS OF SORTING PROGRAMS

Jonathan Traugott
Computer Science Department
Stanford University

INTRODUCTION

Sorting is a useful domain in which to judge the attributes of a program synthesis system. By discovering a sorting algorithm such as *quicksort*, an automated system can demonstrate both intelligence and practical utility; by failing to generate a full gamut of other sorting algorithms the system can expose its limitations. In practice, automated systems have been quite restrictive in the types of program they can produce. For example in [La] the formalism limits one to tail-recursive sorting programs and in [Sm] to programs that conform to a preset divide-and-conquer schema. Those frameworks with greater generality tend to sacrifice the formality necessary for automation; in [Da] and [Br] manual derivations are presented that generate a good cross-section of the sorting algorithms but also contain many steps that require unmechanical ingenuity. It is not clear how the derivations in either of these papers could be automated.

In this paper we will manually synthesize several sorting programs using the deductive synthesis framework proposed by Manna and Waldinger ([MW1],[MW3]). These derivations illustrate a combination of generality and formality inherent in the deductive approach and lacking in the systems mentioned above; the variety of algorithms we obtain compares favorably with [Br] and [Da], yet our derivations are just as formal as in automated systems such as [La] and [Sm]. We ultimately hope for a computer system to generate similar derivations automatically.

The initial section of this paper includes a summary of the tableau framework and its various deduction rules. For the sake of brevity we give only a bare definition of the the system. For a more complete definition the reader is refered to [MW3]. We have augmented Manna and Waldinger's original formalism by the use of many-sorted logic. This addition is not necessary for the derivations to go through but does significantly reduce their length and complexity.

Section 2 includes the formal derivation of a recursive *quicksort* program. In Section 3 we develope the ideas presented in [MW1] for the generation of auxiliary procedures and show how the auxiliary procedures for *quicksort* might be discovered. Section 4 sketches alternative paths in the synthesis tree that lead to other sorting programs, including *mergesort*, *selectsort*, *insertsort* and *bubblesort*. Section 5 discusses related work in program synthesis.

1. DEDUCTIVE SYNTHESIS

In the deductive approach, the synthesis of programs is reduced to a problem in theorem proving. The program specification is represented by a sentence in first-order logic and the target program is extracted from a proof of the sentence. The program specification is typically of the form

$$F[b] \supset \exists z G[b, z].$$

This represents the claim that, given an input b satisfying $F[b]$, there exists some output z satisfying the output condition $G[b, z]$. A constructive proof of this sentence will involve actually finding z, and thus producing an acceptable output.

Recursive calls in the target program are obtained by the use of mathematical induction. Let f be our target program. Then inductively we may assume that f satisfies the above sentence for all inputs x

This research was supported in part by the National Science Foundation under grants MCS-82-14523 and MCS-81-05565, by DARPA under Contract N00039-84-C-0211, by the United States Air Force Office of Scientific Research under Contract AFOSR-81-0014, by the Office of Naval Research under Contract N00014-84-C-0706, and by a contract from the International Business Machines Corporation.

that are less than b with respect to some founded relation $\prec$ (where a well founded relation is one which admits no infinite decreasing sequences). In other words we may assume that $x \prec b \supset (F[x] \supset G[x, f(x)])$. By using this induction hypothesis within a proof, the original output z may become instantiated to a term containing f. In this way a recursive call is introduced.

DEDUCTIVE TABLEAUX

Let us now consider the precise mechanism by which proofs are carried out within the tableau framework. We wish to prove a sentence of the form $F[b] \supset \exists z G[b, z]$. Typically, this sentence will be proved within a particular theory T, characterized by the set of axioms and properties $T_1, \ldots, T_n$. Proving the above sentence in T is equivalent to showing the validity of $\exists z(T_1 \wedge T_2 \wedge, \ldots, \wedge T_n \wedge F[b] \quad \supset \quad G[b, z])$. In the tableau framework we treat the properties $T_1, \ldots, T_n$ and the input condition $F[b]$ as assertions. The output condition $G[b, z]$ is the initial goal. Each of these sentences occupies a separate row of the tableau.

Formally, the tableau corresponds to a first-order-logic sentence in which the conjunction of the assertions implies the disjunction of the goals. Free variables in the tableau represent existentially quantified variables in the associated sentence. Each goal in the tableau is associated with an output entry consisting of a single term (possibly containing the *if-then-else* constructor). The term is a *suitable* program output with respect to that goal; if the goal is true then the term will satisfy the original program specification.

Deduction rules are used to add new rows to the tableau (generally goals). This is done in such a way as to preserve validity of the associated sentence. Generally we will show that the new goal implies the disjunction of two or more existing goals in the tableau, a sufficient condition for preservation of validity. The proof is complete when a new goal is obtained consisting of the propositional symbol *true* . In this case the associated sentence also reduces to *true* and is therefore valid. The output entry for this goal will contain our final program.

Notation

Since much of what follows is presented in symbolic form, it is useful to fix the meaning of certain symbols from the outset. These symbols are listed below along with their meaning.

$\prec$: an infix binary operator.

F,G,H: sentences.

P,Q,R: subsentences.

r,s,t: terms.

I,J,K: expressions (terms or sentences).

$O(I)$: an operator O containing I in one of its argument places.

Replacement

The deduction rules described below all utilize the syntactic procedure of subexpression replacement. We are given an expression I containing occurrences of a subexpression J; we produce a new expression which is identical to I except that the occurrences of J have been replaced with some new expression K. We express the replacement operation informally as follows:

If we are considering an expression denoted by $I[J]$ and later write $I[K]$, we refer to the expression obtained by replacing all occurrences of J in I with K. If we are considering an expression denoted by $I\langle J\rangle$ and later write $I\langle K\rangle$, we refer to the expression obtained by replacing particular occurrences of J in I (those denoted by $\langle J\rangle$) with K. If we are considering an expression denoted by $O(I)$ and later write $O(J)$, we refer to the expression obtained by replacing the operand I of O with J.

Monotonicity

Central to many of our deduction rules is the concept of *monotonicity*. Roughly speaking, an expression is *monotonically increasing* (or *increasing* for short) in some subexpression if, by replacing the subexpression with something bigger, one obtains a bigger expression. The expression is *monotonically decreasing* if performing the same replacement results in a smaller expression. "Bigger" and "smaller" will always be relative to particular operators. For example the expression $x \geq (3 - b)$ is increasing in b with respect to the operators $<$ and $\supset$ since increasing b (in the sense of $<$) will result in a bigger sentence (in the sense of $\supset$). A formal definition of monotonicity is given in [MW3]. From this definition we may prove the following theorem:

Theorem (Montonicity)

Let $G[I]$ be increasing in I with respect to $(\prec_1, \prec_2)$; then $I_1 \prec_1 I_2 \supset G[I_1] \prec_2 G[I_2]$.

Let $G[I]$ be decreasing in I with respect to $(\prec_1, \prec_2)$; then $I_1 \prec_1 I_2 \supset G[I_2] \prec_2 G[I_1]$.

When refering to the monotonicity of sentences we will often speak in terms of *polarity*. If $G[I]$ is increasing in I with respect to $(\prec, \supset)$, then we say that I is of positive polarity in $G[I]$ with respect to $\prec$. We denote this fact symbolically by $G[I^{+\prec}]$. Similarly if $G[I]$ is decreasing in I with respect to $(\prec, \supset)$ then we say that I is negative in $G[I]$ with respect to $\prec$ and write $G[I^{-\prec}]$. If I is a subsentence then will usually assume that $\prec$ is the implication connective. In this case we write simply $G[I^+]$ or $G[I^-]$.

DEDUCTION RULES

In this section we will give a brief overview of the deduction rules used in this paper. A more complete treatment may be found in [MW3].

1. Resolution

Nonclausal resolution may be used to add a new row to the tableau using two existing rows. This rule is an extension of Robinson's original resolution principle. It was developed by Manna and Waldinger [MW1] and independently by Murray [Mu]. The principal advantage over standard resolution is that expressions do not need to put into conjunctive normal form. First we define the ground version where resolved sentences contain no variables.

Definition (Ground Resolution)

Suppose we have the following two rows of a tableau (where F and G are ground sentences):

Assertions	Goals	Outputs
1	$F[P]$	t_1
2	$G[P]$	t_2

Then we may add the new row:

3	$F[true] \;\wedge\; G[false]$	if P then t_1 else t_2

(We assume that the occurrences of *true* and *false* are then removed using transfomation rules such as $true \wedge Q \Longrightarrow Q$.)

We need to show that the addition of goal 3 preserves validity of the tableau. As noted earlier it is sufficient to show that the new goal implies the disjunction of goals 1 and 2. To see this, assume that goal 3 is true and consider the case in which P is true. From $F[true]$ we may infer goal 1, $F[P]$. On the other hand, if P is false then from $G[false]$ we may infer goal 2, $G[P]$.

We must also show that the new output entry is suitable: Suppose that goal 3 is true. If P is true, then, as shown above, so is goal 1. In this case t_1 is a suitable output. On the other hand, if P is false then goal 2 is true and t_2 is a suitable output. In either case, the conditional expression (if P then t_1 else t_2) is a suitable output.

The resolution rule may also be used between two assertions, between an assertion and a goal, or between two assertions. For example, from the assertion $F[P]$ and the goal $G[P]$ we may deduce the new goal $\neg F[true] \wedge G[false]$. This follows from the above rule since the assertion F is equivalent to the goal $\neg F$.

In the general version of the resolution rule, resolved subexpressions need not be identical but only unifiable. An application of the general rule consists in first applying the unifying substitution to the relevant tableau entries and then proceeding exactly as in the ground version. For the remaining deduction rules, the general versions may be inferred in a like manner.

2. Relation Replacement

Another rule that allows us to produce a new row from two given rows is *relation replacement.* Roughly speaking this rule allows us to use relational information contained in one goal in order to make replacements in another. Relation replacement is a generalization of paramodulation.

Definition (Relation replacement)

If we have the two goals:

	$F[I \prec J]$	t_1
	$G\langle I^{-\prec}\rangle$	t_2

then we may add the new goal

	$G\langle J\rangle \ \wedge \ F[false]$	if $I \prec J$ then t_2 else t_1

By $G\langle I^{-\prec}\rangle$ we denote the fact that I has negative polarity in G with respect to $\prec$.

In order to justify this rule we must show that the new goal implies the disjunction of the previous two and that the new output is suitable. Suppose that the new goal is true. Consider the case in which $I \prec J$ is true. Then the monotonicity theorem together with $G\langle J^{-\prec}\rangle$ allows us to infer $G\langle I\rangle$. In this case t_2 is a suitable output. On the other hand if $I \prec J$ is false then from $F[false]$ we may infer $F[I \prec J]$. In this case t_1 is a suitable output. In either case one of the two given goals is true and the conditional (if $I \prec J$ then t_2 else t_1) is a suitable output.

A dual version of the rule allows for the replaced subexpression to occur with positive rather than negative polarity. In this case from the goals $F[I \prec J]$ and $G\langle J^{+\prec}\rangle$ we obtain the new goal $G\langle I\rangle \wedge F[false]$. As usual, by the duality property, we may also apply the rule to assertions.

3. Relation Matching

Sometimes a deduction rule will not go through because unification fails on a single term. For example we may try to resolve the goals $F[x, a]$ and $\neg F[c, b]$ but fail because we are unable to unify a and b. A solution is to assume that $a = b$; this allows us to replace a by b in $F[x, a]$ and successfully unify the two goals. In this section we will show how the same approach may be used to extend an arbitrary deduction rule so as to handle mismatched subexpressions.

Definition (Relation Matching)

Let D be an arbitrary deduction rule that operates on goals $G_1\langle J\rangle, \ldots, G_n\langle J\rangle$ and produces the new goal $D(G_1\langle J\rangle, \ldots, G_n\langle J\rangle)$. Then, for some i, given the goals

	$G_1\langle J\rangle$

⋮

	$G_{i-1}\langle J\rangle$
	$G_i\langle I^{-\prec}\rangle$
	$G_{i+1}\langle J\rangle$

⋮

	$G_n\langle J\rangle$

D with ≺ matching may be applied to obtain

	$I \prec J \quad \wedge \quad D(G_1\langle J\rangle, \ldots, G_i\langle J\rangle, \ldots, G_n\langle J\rangle)$

The justification for matching is straightforward. Assume that the new goal is true; we wish to show that one of the original goals is also true. By the soundness of D we know that $G_k\langle J\rangle$ is true for some $k \in [1, n]$. If k is different than i then we are done. Otherwise $G_i\langle J^{-\prec}\rangle$ is true. From the monotonicity theorem and the fact that $I \prec J$ we may then infer the goal $G_i\langle I\rangle$.

We have defined the ground version of relation matching in which no unification is required. In the general version we suppose that a potential unifier is applied to $G_1\langle J_1\rangle$ through $G_n\langle J_n\rangle$ in order to unify the relevant subexpressions, and in particular J_1 through J_n. The unifier succeeds on all subexpressions except for J_i which is instantiated to I rather than to J. At this stage, the ground version of the rule applies.

The relation-matching rule is quite useful because it allows us to take advantage of semantic information in a goal (i.e., the polarity of subexpressions) to make inferences that would otherwise be blocked by failed unification. [MW3] defines separate matching rules for resolution and relation replacement. Our rule combines and generalizes these rules.

SORTED LOGIC

We adopt an approach similar to [Wal] in which each symbol of the theory is associated with a domain sort and/or a range sort. (For example the *length* function over strings would have a domain sort of string and a range sort of integer). Using this information the system can make sure that expressions in the theory are well sorted (i.e., that operators have operands of the appropriate sort), and that only terms of equal of "conformable" sorts be unifiable. In general the notions of well-sortedness and unifiablity can be quite problematic for a sorted logic. In our case we are fortunate to have a sort structure that allows for simple definitions of these concepts. Formally we proceed as follows:

• Assign sorts to each of the symbols in our theory: A variable or constant symbol is assigned a unique sort. A function or predicate symbol is assigned one domain sort for each of its operands. The function symbol is also assigned a unique range sort.

We define

• Well-sortedness: Variables and constants are well sorted. For an *n-ary* operator O and expressions t_1 through t_n, $O(t_1, \ldots, t_n)$ is well sorted if, for each i in $[1, n]$, the sort of t_i is a subsort of or equals the domain sort of O in its ith argument.

• Unifiability: Two expressions are unifiable if they are unifiable in the normal sense with the restriction that a variable x may only be instantiated to a term t on condition that the sort of t is a subsort of or is equal to the sort of x. We assume the use of commutative unification and associative pattern matching.

In our theory there are two sorts: *integer* and *string*; *integer* is a subsort of *string* (i.e., strings of length one are identified with integers.)

The Theory

In this section we present the axioms of our theory. For the sake of brevity we include only those axioms that are used in the subsequent derivations. For completeness an actual system would include a full set of Peano-style axioms for all operators in the theory.

Sort declarations

We list functions according to their range sorts. Domain sorts are defined by the sorts of the variables given as arguments.

integer variables: u , v.

string variables: w , x , y , z.

string constants: b , c , d , λ

integer functions: $f(x,y)$, $g(x,y)$, $h(u,x)$.
string functions: $x{\cdot}y$, $head(x)$, $tail(x)$, $left(x)$, $right(x)$.
predicates: $u \in x$, $ord(x)$, $perm(x,y)$, $x \prec_{bag} y$, $x \prec_{tail} y$, $x \prec_{string} y$.

The above functions have the following meanings: $x{\cdot}y$ denotes the result of appending y to x; $head(x)$ and $tail(x)$ denote respectively the first element of x and the string of remaining elements; $left(x)$ and $right(x)$ denote the left and right halves of x. The functions f, g, and h are skolem functions that result from removing quantifiers from the axioms in which they appear. Informally one may think of these functions as representing arbitrary integers. To encourage this interpretation we will omit the arguments of f, g and h in what follows (while of course keeping track of the arguments logically).

The predicates may be interpreted as follows: $u \in x$ says that u is a member of x; $ord(x)$ says that the elements of x are in increasing order from left to right; $perm(x,y)$ says that x and y are permutations of one another (i.e. that they contain the same bag of elements); $x \prec_{bag} y$ says that the elements of x form a proper subbag of those in y; $x \prec_{tail} y$ says that x equals $tail(y)$ for nonempty y; $x \prec_{string} y$ says that x is a proper substring of y.

Axioms

The following properties will be included as assertions in the initial tableau.

- For nonempty x, x is equal to $tail(x)$ appended to $head(x)$.

$$x \neq \lambda \supset x = head(x){\cdot}tail(x)$$ *head-tail* axiom

- For nonempty x, if $tail(x)$ is empty then the head of x is x itself.

$$x \neq \lambda \wedge tail(x) = \lambda \supset x = head(x)$$ single element property

- For nonempty, nonatomic x, x is equal to $right(x)$ appened to $left(x)$, and both $left(x)$ and $right(x)$ are different from (i.e. shorter than) x itself.

$$\begin{Bmatrix} x \neq \lambda \;\wedge \\ tail(x) \neq \lambda \end{Bmatrix} \supset \begin{Bmatrix} x = left(x){\cdot}right(x) \;\wedge \\ left(x) \neq x \;\wedge \\ right(x) \neq x \end{Bmatrix}$$ *left-right* axiom

- u is member of $v{\cdot}x$ when either u equals v or u is a member of x.

$$u \in v{\cdot}x \equiv u = v \vee u \in x$$ $\in$-*insert* axiom

- Every string is permutation of itself.

$$perm(x,x)$$ *perm-reflexive* axiom

- Two strings will remain permutations after adding or removing a common element.

$$perm(u{\cdot}x, y{\cdot}u{\cdot}z) \equiv perm(x, y{\cdot}z)$$ *perm-insert* axiom

- The empty string is ordered.

$$ord(\lambda)$$ *ord-empty* axiom

- Every atomic string is ordered.

$$ord(u)$$ *ord-atom* property

- If x is ordered and u is less than or equal to all the elements of x then $u{\cdot}x$ is also ordered.

$$\left\{\begin{array}{l} ord(x) \ \land \\ h \in x \supset u \leq h \end{array}\right\} \supset ord(u{\cdot}x) \qquad \textit{ord-insert} \text{ axiom}$$

• If x and y are ordered and all the elements of x are less than or equal to all the elements of y then $x{\cdot}y$ is ordered.

$$\left\{\begin{array}{l} ord(x) \ \land \\ ord(y) \ \land \\ f \in x \land g \in y \ \supset \ f \leq g) \end{array}\right\} \supset ord(x{\cdot}y) \qquad \textit{ord-append} \text{ property}$$

• If x is a proper substring of y then x is less than y with respect to the well-founded $\prec_{string}$ relation.

$$y = w{\cdot}x{\cdot}z \land x \neq y \ \supset \ x \prec_{string} y \qquad \prec_{string} \text{ axiom}$$

• If appending a nonempty string to x produces a permutation of y then x is less than y with respect to the well-founded $\prec_{bag}$ relation.

$$perm(w{\cdot}u{\cdot}x, y) \ \supset \ x \prec_{bag} y \qquad \prec_{bag} \text{ axiom}$$

Specification for Sort

Given a string b we wish to produce a string z containing the same elements as b but arranged in increasing order. Formally, z must satisfy the output condition $perm(b,z) \land ord(z)$. Our initial goal is therefore

	$perm(b,z) \ \land \ ord(z)$	z

Accordingly our induction hypothesis is

$x \prec_{\omega} b \ \supset \ (perm(x, sort(x)) \land ord(sort(x)))$	

Intuitively this assertion says that for all x less than b, $sort(x)$ satisfies the original output condition. The sense in which x is less than b will be defined by the well-founded relation $\prec_{\omega}$. The previous section contained axioms defining three different well-founded relations: the $\prec_{tail}$ relation, the $\prec_{string}$ relation and the $\prec_{bag}$ relation; $\prec_{\omega}$ may be instantiated to any one of these. (Formally $x \prec_{\omega} y$ represents a three-place predicate $\prec(x,y,\omega)$. By instantiating ω we obtain a particular well-founded relation.)

We are now ready to begin our derivation!

2. DERIVATION OF QUICKSORT

In this section we will give a detailed derivation for the program *quicksort*. In the next section we will see how auxiliary procedures for this program can be derived.

Quicksort

We will now derive a recursive version of Hoare's original *quicksort* algorithm. In *quicksort* we partition a string b into three parts, a discriminating element u, those elements of b less than or equal to u and the remaining elements greater than u. By sorting the last two parts and appending to either side of u we obtain the final output.

THE BASE CASE

Recall that our initial goal is

Assertions	Goals	Outputs
1	$perm(b,z) \ \wedge\ \boxed{ord(z)}$	z

By applying the substitution θ: $\{z \leftarrow \lambda\}$, we can unify the boxed expression with the *ord-empty* axiom below

$\boxed{ord(\lambda)}$		

Applying resolution then produces the new goal

2	$perm(b,\lambda)$	λ

At this stage we would like to resolve with the *perm-reflexive* axiom

$perm(x,x)$		

Taking θ to be $\{x \leftarrow b\}$ will unify the first pair of arguments but will fail to unify the second. In this case resolution with equality matching may be used to obtain

3	$b = \lambda$	λ

Goal 3 represents the base case for our sorting program. It states that for an empty input string, a suitable output is the empty string itself.

THE INDUCTIVE CASE

Let us go back to our initial goal

1	$\boxed{perm(b,z)}\ \wedge\ ord(z)$	z

We now work on the *perm* conjunct using the *perm-insert* axiom

$\boxed{perm(u{\cdot}w, x{\cdot}u{\cdot}y)}\ \equiv\ perm(w, x{\cdot}y)$	

We would like to unify the boxed expressions. Taking θ to be $\{z \leftarrow x{\cdot}u{\cdot}y\}$ unifies the second arguments of *perm* but fails to unify b and $u{\cdot}w$. We will therefore apply resolution with equality matching. The resulting goal is

4	$perm(w, x{\cdot}y)\ \wedge$ $ord(x{\cdot}u{\cdot}y)\ \wedge$ $\boxed{b = u{\cdot}w}$	$x{\cdot}u{\cdot}y$

To prove the last conjunct we may use the *head-tail* axiom

$x \neq \lambda\ \supset\ \boxed{x = head(x){\cdot}tail(x)}$	

To unify the two boxed expressions we take θ to be $\{x \leftarrow b,\ u \leftarrow head(b),\ w \leftarrow tail(b)\}$. Applying resolution then gives

5	$perm(tail(b), x{\cdot}y)\ \wedge$ $\boxed{ord(x{\cdot}head(b){\cdot}y)}\ \wedge$ $b \neq \lambda$	$x{\cdot}head(b){\cdot}y$

The result of the last two steps is to divide the input b into $head(b){\cdot}tail(b)$ and the output z into $x{\cdot}head(b){\cdot}y$. The original requirement $perm(b, z)$ then becomes $perm(head(b){\cdot}tail(b), x{\cdot}head(b){\cdot}y)$. By the *perm-insert* axiom this is equivalent to $perm(tail(b), x{\cdot}y)$.

Notice that goal 5 already contains the basic principle for *quicksort*: we must partition $tail(b)$ into two parts x and y such that all the elements of x come before $head(b)$ and all those of y come after. Our task now is to find x and y in terms of b.

The second conjunct requires that $x{\cdot}head(b){\cdot}y$ be ordered. We can break this into three separate conditions: that x be ordered, that $head(b){\cdot}y$ be ordered, and that all the elements of x be less than or equal to all the elements of $head(b){\cdot}y$. Formally, this step is accomplished by resolving with the ord-append property below

$\left\{\begin{array}{l} ord(x)\ \wedge \\ ord(w)\ \wedge \\ f \in x \wedge g \in w\ \supset\ f \leq g) \end{array}\right\} \supset \boxed{ord(x{\cdot}w)}$	

Taking θ to be $\{w \leftarrow head(b){\cdot}y\}$ we can resolve on the boxed expressions. The resulting goal is

6	$perm(tail(b), x{\cdot}y)\ \wedge$ $ord(x)\ \wedge\ \boxed{ord(head(b){\cdot}y)}\ \wedge$ $f \in x \wedge g \in head(b){\cdot}y\ \supset\ f \leq g\ \wedge$ $b \neq \lambda$	$x{\cdot}head(b){\cdot}y$

Let us similarly break up $ord(head(b){\cdot}y)$. Since $head(b)$ is of sort *integer* we can use the *ord-insert* property

$\left\{\begin{array}{l} ord(y)\ \wedge \\ h \in y \supset u \leq h \end{array}\right\} \supset \boxed{ord(u{\cdot}y)}$	

Taking $u \leftarrow head(b)$ we resolve to obtain

7	$perm(tail(b), x{\cdot}y)\ \wedge$ $ord(x)\ \wedge\ ord(y)\ \wedge$ $f \in x \wedge \boxed{g \in head(b){\cdot}y}\ \supset\ f \leq g\ \wedge$ $h \in y\ \supset\ head(b) \leq h\ \wedge$ $b \neq \lambda$	$x{\cdot}head(b){\cdot}y$

Now recall the $\in$-*insert* property

$\boxed{u \in v{\cdot}y}\ \equiv\ (u = v \vee u \in y)$	

Taking θ to be $\{u \leftarrow g\ ,\ v \leftarrow head(b)\}$ we can apply equivalence replacement to replace $g \in head(b){\cdot}y$ by the disjunction $g = head(b) \vee g \in y$. The resulting goal is

8	$perm(tail(b), x{\cdot}y)\ \wedge$ $ord(x)\ \wedge\ ord(y)\ \wedge$ $(f \in x \wedge (g = head(b) \vee g \in y))\ \supset\ f \leq g\ \wedge$ $h \in y\ \supset\ head(b) \leq h\ \wedge$ $b \neq \lambda$	$x{\cdot}head(b){\cdot}y$

Notice that the above goal contains an equality. Performing a case analysis on this equality allows us to break up the fourth conjunct: If g equals $head(b)$ then we can replace $g = head(b)$ with $true$ and all occurrences of g with $head(b)$. This yields $(f \in x \wedge (true \vee head(b) \in y)) \supset f \leq head(b)$, which simplifies to $f \in x \supset f \leq head(b)$. On the other hand if $g = head(b)$ is false then from the same conjunct we obtain $f \in x \wedge g \in y \supset f \leq g$. In either case, the fourth conjunct is implied by the conjunction

$$(f \in x \supset f \leq head(b)) \wedge (f \in x \wedge g \in y \supset f \leq g).$$

Thus it is sound to add the new goal

9	$perm(tail(b), x \cdot y) \wedge$ $\boxed{ord(x)} \wedge \boxed{ord(y)} \wedge$ $f \in x \supset f \leq head(b) \wedge$ $f \in x \wedge g \in y \supset f \leq g \wedge$ $h \in y \supset head(b) \leq h \wedge$ $b \neq \lambda$	$x \cdot head(b) \cdot y$

Formally goal 9 is obtained by a single application of equality replacement between goal 8 and itself. Let $F[g = head(b)]$ and $G[g]$ both represent goal 8. Then according to the relation replacement rule we can add the new goal $F[false] \wedge G[head(b)]$. After simplifying (including equality removal) this yields goal 9 above.

Now recall the induction hypothesis

$w \prec_\omega b \supset (perm(w, sort(w)) \wedge \boxed{ord(sort(w))})$	

We may resolve successively with each *ord* conjunct in goal 9 to obtain

10	$perm(tail(b), \boxed{sort(w_1)^{+perm}} \cdot sort(w_2)) \wedge$ $f \in \boxed{sort(w_1)^{+perm}} \supset f \leq head(b) \wedge$ $g \in \boxed{sort(w_1)^{+perm}} \wedge h \in sort(w_2) \supset g \leq h \wedge$ $h \in sort(w_2) \supset head(b) \leq h \wedge$ $w_1 \prec_\omega b \wedge w_2 \prec_\omega b \wedge$ $b \neq \lambda$	$sort(w_1) \cdot$ $head(b) \cdot$ $sort(w_2))$

We still haven't used some information supplied by the induction hypothesis: that $sort(w)$ is a permutation of w. This fact together with the positive polarity of $sort(w_1)$ (with respect to *perm*) allows us to greatly simplify the above goal. Formally let $F[perm(w, sort(w))]$ and $G[sort(w_1)^{perm+}]$ represent the induction hypothesis and goal 10 respectively. Then according to the relation replacement rule we may derive $\neg F[false] \wedge G[w_1]$. An analogous step may be applied to w_2. After simplification these two steps yield

12	$perm(tail(b), w_1 \cdot w_2) \wedge$ $f \in w_1 \supset f \leq head(b) \wedge$ $g \in w_1 \wedge h \in w_2 \supset g \leq h \wedge$ $h \in w_2 \supset head(b) \leq h \wedge$ $\boxed{w_1 \prec_\omega b} \wedge \boxed{w_2 \prec_\omega b} \wedge$ $b \neq \lambda$	$sort(w_1) \cdot$ $head(b) \cdot$ $sort(w_2))$

These replacements are intuitively motivated by the fact that $\in$ and *perm* only care about the elements in a string, not their order; for these relations $sort(w)$ is effectively indistinguishable from w. We have now made use of the induction hypothesis. But has it been applied correctly? The test is whether we can prove the two conjuncts $w_1 \prec_\omega b$ and $w_2 \prec_\omega b$. These conditions ensure that we are not trying to derive a nonterminating program. Let us start by resolving each conjunct with the $\prec_{bag}$ axiom $perm(x \cdot u \cdot w, y) \supset w \prec_{bag} y$. We obtain

13	$perm(tail(b), w_1 \cdot w_2)\ \wedge$ $f \in w_1\ \supset\ f \leq head(b)\ \wedge$ $g \in w_1 \wedge h \in w_2\ \supset\ g \leq h\ \wedge$ $h \in w_2\ \supset\ head(b) \leq h\ \wedge$ $\boxed{perm(x_1 \cdot u_1 \cdot w_1, b)}\ \wedge\ perm(x_2 \cdot u_2 \cdot w_2, b)\ \wedge$ $b \neq \lambda$	$sort(w_1)\cdot$ $head(b)\cdot$ $sort(w_2)$

Now recall the *perm-insert* axiom

$\boxed{perm(v \cdot y, x \cdot u \cdot w)}\ \equiv\ perm(y, x \cdot w)$	

We wish to resolve the two boxed expressions. We first attempt to unify using commutative unification and taking the substitution θ to be $\{x \leftarrow x_1, u \leftarrow u_1, w \leftarrow w_1\}$. A mismatch persists between b and $v \cdot y$, leading us to use resolution with equality matching, resulting in the new goal

14	$perm(tail(b), w_1 \cdot w_2)\ \wedge$ $f \in w_1\ \supset\ f \leq head(b)\ \wedge$ $g \in w_1 \wedge h \in w_2\ \supset\ g \leq h\ \wedge$ $h \in w_2\ \supset\ head(b) \leq h\ \wedge$ $perm(y, x_1 \cdot w_1)\ \wedge\ \boxed{perm(x_2 \cdot u_2 \cdot w_2, b)}\ \wedge$ $b = v \cdot y\ \wedge\ b \neq \lambda$	$sort(w_1)\cdot$ $head(b)\cdot$ $sort(w_2)$

An analogous step using the last *perm* conjunct gives

15	$perm(tail(b), w_1 \cdot w_2)\ \wedge$ $f \in w_1\ \supset\ f \leq head(b)\ \wedge$ $g \in w_1 \wedge h \in w_2\ \supset\ g \leq h\ \wedge$ $h \in w_2\ \supset\ head(b) \leq h\ \wedge$ $perm(y, x_1 \cdot w_1)\ \wedge\ perm(y', x_2 \cdot w_2)\ \wedge$ $\boxed{b = v \cdot y}\ \wedge\ \boxed{b = v' \cdot y'}\ \wedge\ b \neq \lambda$	$sort(w_1)\cdot$ $head(b)\cdot$ $sort(w_2)$

By taking θ to be $\{x \leftarrow b, v \leftarrow head(b), v' \leftarrow head(b), y \leftarrow tail(b), y' \leftarrow tail(b)\}$ we may unify the boxed expressions with the *head-tail* axiom $x \neq \lambda \supset x = head(x) \cdot tail(x)$ and resolve to obtain

16	$perm(tail(b), w_1 \cdot w_2)\ \wedge$ $f \in w_1\ \supset\ f \leq head(b)\ \wedge$ $g \in w_1 \wedge h \in w_2\ \supset\ g \leq h\ \wedge$ $h \in w_2\ \supset\ head(b) \leq h\ \wedge$ $perm(tail(b), x_1 \cdot w_1)\ \wedge\ perm(tail(b), x_2 \cdot w_2)\ \wedge$ $\boxed{b \neq \lambda}$	$sort(w_1)\cdot$ $head(b)\cdot$ $sort(w_2)$

Notice that the last two *perm* conjuncts are redundant given the first conjunct: once we find w_1 and w_2 to satisfy $perm(tail(b), w_1, w_2)$ then the expressions $perm(tail(b), x_1 \cdot w_1)$ and $perm(tail(b), x_2 \cdot w_2)$ will also be satisfied by taking x_1 and x_2 to be w_2 and w_1 respectively. In other words the original conditions that w_1 and w_2 be smaller than b are already implicit in the first conjunct. Thus, in a sense, we have dispensed with the ordering conditions that we originally set out to prove in goal 12.

Earlier we derived the base case

3	$\boxed{b = \lambda}$	λ

Since $b \neq \lambda$ is shorthand for $\neg(b = \lambda)$ we may resolve with the boxed expression in goal 16 to obtain

17	$perm(tail(b), w_1 \cdot w_2) \wedge$ $f \in w_1 \supset f \leq head(b) \wedge$ $g \in w_1 \wedge h \in w_2 \supset g \leq h \wedge$ $h \in w_2 \supset head(b) \leq h \wedge$ $perm(tail(b), x_1 \cdot w_1) \wedge perm(tail(b), x_2 \cdot w_2)$	if $b = \lambda$ then λ else $\begin{pmatrix} sort(w_1) \cdot \\ head(b) \cdot \\ sort(w_2) \end{pmatrix}$

Informally the above goal requires that we partition the elements of $tail(b)$ into w_1 and w_2, so that w_1 contain only those elements less than or equal to $head(b)$, and w_2 contain only elements greater than or equal to $head(b)$. At this point the system would need to derive two auxiliary functions to perform this partitioning. The two functions would be defined by the following assertion:

$perm(x, les(x) \cdot gre(x)) \wedge$ $v_1 \in les(x) \supset v_1 \leq head(b) \wedge$ $v_2 \in les(x) \wedge v_3 \in gre(x) \supset v_2 \leq v_3 \wedge$ $v_4 \in gre(x) \supset head(b) \leq v_4 \wedge$ $perm(x, sk_1(b) \cdot les(b)) \wedge perm(x, sk_2(b) \cdot gre(b))$	

In a later section we will see how this assertion could be automatically derived as a lemma during the proof (along with programs to compute *les* and *gre*).

Taking θ to be

$$\{v_1 \leftarrow f,\ v_2 \leftarrow g,\ v_3 \leftarrow h,\ v_4 \leftarrow h,\ x \leftarrow tail(b),$$
$$w_1 \leftarrow les(tail(b)),\ w_2 \leftarrow gre(tail(b)),\ x_1 \leftarrow sk_1(b),\ x_2 \leftarrow sk_2(b)\}$$

we may resolve the above assertion with goal 17 to obtain

18	*true*	if $b = \lambda$ then λ else $\begin{pmatrix} sort(les(head(b), tail(b))) \cdot \\ head(b) \cdot \\ sort(gre(head(b), tail(b))) \end{pmatrix}$

The final program for *quicksort* is therefore

$$sort(x) \Longleftarrow \left\{ \begin{array}{l} \text{if } x = \lambda \text{ then } \lambda \\ \text{else } \begin{pmatrix} sort(les(head(x), tail(x))) \cdot \\ head(x) \cdot \\ sort(gre(head(x), tail(x))) \end{pmatrix} \end{array} \right\}$$

3. GENERATION OF AUXILIARY PROCEDURES

In the previous derivation we included an assertion that defined the functions *les* and *gre* corresponding to auxiliary procedures called by the main program. But this assertion is of a different character from the original theory axioms; in practice it would not initially be known to the system. In this section we will see how the system could automatically discover and prove assertions of this type, and in the process generate programs for the functions they define. The method presented here is a generalization of [MW1].

Auxiliary Tableaux

Suppose that in the course of a derivation we obtain a goal of the form

		$f(b)$
i	$G[t[b], x]$	$r[b, x]$

and subsequently the new goal

j	$F[G[t'[b], x']$	$r'[b, x']$

Notice that goal j contains a replica of the original goal $G[t[b], x]$. This suggests that it may be possible to prove G by induction, provided that $t'[b]$ is smaller than $t[b]$ with respect to some well-founded relation. Accordingly we will introduce a new tableau with G as the initial goal,

		$fnew(c)$
1	$G[c, x]$	x

To allow the use of induction, we have generalized the original goal, replacing the term $t[b]$ with an arbitrary constant c. For this goal the corresponding induction hypothesis is

$y \prec_\omega c \supset G[y, fnew(y)]$		

We call the new tableau an *auxiliary tableau* since it may be used to derive an auxiliary procedure to compute the function *fnew*. By applying similar deduction steps to those that lead from goal i to goal j, we hope to obtain the new goal

k	$F'[G[t''[c], x']]$	$r''[c, x']$

The subsentence G may now be "proved" by resolving with the above induction hypothesis, resulting in the new goal

$k+1$	$F'[true] \wedge t'' \prec_\omega c$	$r''(c, fnew(t''(c)))$

Suppose that the proof goes through with the final goal

	true	$s[c]$

The resulting program for *fnew* is then

$$fnew(y) \Leftarrow s[y]$$

By completing the auxiliary tableau we have proved $G[c, x]$ by taking x to be $fnew(c)$. Since c was arbitrary we have more generally shown that $\forall y G[y, fnew(y)]$. We may therefore return to the main tableau and add this sentence as a derived lemma.

$G[y, fnew(y)]$		

Note that this assertion unifies with goal i

i	$G[t[b], x]$	$r[b, x]$

By resolving we obtain the new goal

	true	$r[b, fnew(t[b])]$

so our final program is

$$f(x) \Leftarrow r[x, fnew(t[x])]$$
$$fnew(y) \Leftarrow s[y]$$

The introduction of auxiliary procedures was illustrated for the case in which G is of the form $G[t, x]$. In general we allow any number of terms and free variables to be included. In the auxiliary tableau, each term will correspond to a separate input constant, and each free variable to the output for a distinct auxiliary procedure.

The above procedure for generating an auxiliary tableau was purely syntactic and would therefore be natural to automate.

Auxiliary Procedures for Quicksort

Let us now illustrate the above procedure by showing how the system would derive auxiliary procedures for the program *quicksort*. Recall that toward the end of the *quicksort* derivation we obtained the goal

A	$perm(tail(b), w_1 \cdot w_2) \ \wedge$ $f \in w_1 \ \supset \ f \leq head(b) \ \wedge$ $g \in w_1 \wedge h \in w_2 \ \supset \ g \leq h \ \wedge$ $h \in w_2 \ \supset \ head(b) \leq h \ \wedge$ $perm(tail(b), x_1 \cdot w_1) \ \wedge \ perm(tail(b), x_2 \cdot w_2)$	t_A

where t_A stands for some complicated term. This goal was resolved with the assertion

$perm(x, les(x) \cdot gre(x)) \ \wedge$ $v_1 \in les(x) \ \supset \ v_1 \leq head(b) \ \wedge$ $v_2 \in les(x) \wedge v_3 \in gre(x) \ \supset \ v_2 \leq v_3 \ \wedge$ $v_4 \in gre(x) \ \supset \ head(b) \leq v_4 \ \wedge$ $perm(tail(b), sk_1(b) \cdot les(b)) \ \wedge \ perm(tail(b), sk_2(b) \cdot gre(b))$	

to obtain the final program for *quicksort* . But in general such specialized assertions are not initially available to the system; let us see how the proof might proceed without the above assertion.

As we did earlier in the *quicksort* derivation, we might begin by resolving with the *perm-insert* axiom

$\boxed{perm(v \cdot y, w \cdot v \cdot w_2')} \ \equiv \ perm(y, w \cdot w_2')$	

By taking θ to be $\{w \leftarrow w_1, w_2 \leftarrow v \cdot w_2'\}$ we can resolve (by equality-matching $tail(b)$ and $v \cdot y$) the boxed expression with the first conjunct of goal A to get

B	$perm(y, w_1 \cdot w_2')) \ \wedge$ $f \in w_1 \ \supset \ f \leq head(b) \ \wedge$ $g \in w_1 \wedge h \in v \cdot w_2' \ \supset \ g \leq h \ \wedge$ $h \in v \cdot w_2' \ \supset \ head(b) \leq h \ \wedge$ $perm(tail(b), x_1 \cdot w_1) \ \wedge$ $perm(tail(b), x_2 \cdot v \cdot w_2') \ \wedge$ $tail(b) = v \cdot y$	t_B

A similar step applied to the last *perm* conjunct gives

C	$perm(y, w_1 \cdot w_2')) \ \wedge$ $f \in w_1 \ \supset \ f \leq head(b) \ \wedge$ $g \in w_1 \wedge h \in v \cdot w_2' \ \supset \ g \leq h \ \wedge$ $h \in v \cdot w_2' \ \supset \ head(b) \leq h \ \wedge$ $perm(tail(b), x_1 \cdot w_1) \ \wedge$ $perm(y', x_2 \cdot w_2') \ \wedge$ $\boxed{tail(b) = v \cdot y} \ \wedge \ \boxed{tail(b) = v \cdot y'}$	t_C

By use of the *head-tail*, $\in$-*insert* and *perm-insert* axioms we eventually obtain

D	$perm(tail(tail(b)), w_1 \cdot w_2') \ \wedge$ $f \in w_1 \ \supset \ f \leq head(b) \ \wedge$ $g \in w_1 \wedge h \in w_2' \ \supset \ g \leq h \ \wedge$ $h \in w_2' \ \supset \ head(b) \leq h \ \wedge$ $perm(tail(tail(b)), x_1' \cdot w_1) \ \wedge$ $perm(tail(tail(b)), x_2 \cdot w_2') \ \wedge$ $g \in w_1 \ \supset \ g \leq head(b) \ \wedge$ $tail(tail(b)) \neq \lambda \ \wedge \ head(b) \leq head(tail(b))$	t_D

Notice that goal D contains a replica of goal A. Specifically, let

$$G[tail(b), w_1, w_2, x_1, x_2]$$

represent goal A, then goal D is of the form

$$F[G[tail(tail(b)), w_1, w_2', x_1', x_2]].$$

Accordingly we may form an auxiliary tableau containing the initial goal

		$\langle les(d), \ \ gre(d) \rangle$
1	$G[d, w_1, w_2, x_1, x_2]$	$\langle w_1, \ \ w_2 \rangle$

This expands to

		$\langle les(d), \ \ gre(d) \rangle$
1	$perm(d, w_1, w_2) \ \wedge$ $f \in w_1 \ \supset \ f \leq head(b) \ \wedge$ $g \in w_1 \wedge h \in w_2 \ \supset \ g \leq h \ \wedge$ $h \in w_2 \ \supset \ head(b) \leq h \ \wedge$ $perm(d, x_1 \cdot w_1) \ \wedge \ perm(d, x_2 \cdot w_2)$	$\langle w_1, \ \ w_2 \rangle$

Notice that we have ignored the output columns for x_1 and for x_2; we do not need to know how to compute x_1 and x_2 since they do not occur in the output for *sort* (as do w_1 and w_2). A proof of the above sentence within the tableau framework requires appoximately 20 steps. The resulting programs for *les* and *gre* are

$$\langle les(x), \ \ gre(x) \rangle \Leftarrow \left\{ \begin{array}{l} \text{if } x = \lambda \text{ then } \langle \lambda, \ \ \lambda \rangle \\ \text{else} \left\{ \begin{array}{l} \text{if } head(x) \leq head(b) \\ \text{then } \langle head(x) \cdot les(tail(x)), \ \ gre(tail(x)) \rangle \\ \text{else } \langle les(tail(x)), \ \ head(x) \cdot gre(tail(x)) \rangle \end{array} \right\} \end{array} \right\}$$

The b occurring in the procedures for *les* and *gre* is a global constant, identical to the the argument b initially given to *quicksort*.

Goal 1 corresponds to the quantified sentence

$$(\forall x)\ (\exists w_1, w_2, x_1, x_2)\ (\forall v_1, v_2, v_3, v_4) \left\{ \begin{array}{l} perm(x, w_1, w_2) \\ v_1 \in w_1 \ \ \supset \ \ v_1 \leq head(b) \\ v_2 \in w_1 \wedge v_3 \in w_2 \ \ \supset \ \ v_2 \leq v_3 \\ v_4 \in w_2 \ \ \supset \ \ head(b) \leq v_4 \\ perm(x, x_1 \cdot w_1) \ \ \wedge \ \ perm(x, x_2 \cdot w_2) \end{array} \right\}$$

which we have proved by taking w_1 to be $les(x)$ and w_2 to be $gre(x)$. Accordingly, we may add the following assertion to the main tableau:

$perm(x, les(x)\cdot gre(x)$ $\wedge$ $v_1 \in les(x) \supset v_1 \le head(b)$ $\wedge$ $v_2 \in les(x) \wedge v_3 \in gre(x) \supset v_2 \le v_3$ $\wedge$ $v_4 \in gre(x) \supset v_4 \ge head(b)$ $\wedge$ $perm(x, sk_1(x)\cdot gre(x))$ $\wedge$ $perm(x, sk_2(x)\cdot les(x))$	

allowing us the proceed with the original derivation

The method used to derive auxiliary procedures for *quicksort* may be applied analogously to obtain auxiliary procedures for the other sorting programs.

4. OTHER SORTING PROGRAMS

In this section we will see how alternative paths in the derivation tree lead to other sorting programs, including *mergesort, selectsort, insertsort, bubblesort* and two other novel sorting programs.

Mergesort

Let us now see how the Manna-Waldinger formalism may be used to derive another common sorting program: *mergesort*. In *mergesort* we first split b roughly in half, sort the two halves and then interleave to obtain the final output. The split requires that b contain at least two elements; we must handle separately the case in which b is empty and the case in which b contains a single element. If b is empty a suitable output is the empty string itself. This was represented earlier by goal 3. If b contains a single element then a suitable output is b itself. This is represented by

4	$b \neq \lambda \wedge tail(b) = \lambda$	b

Goal 4 may be derived from the initial goal in three steps using the *perm-reflexive* axiom, the *ord-atom* property and the single element property.

Now let us see how the recursive branch of *mergesort* is derived. We start by using the *left-right* axiom

$\begin{Bmatrix} y \neq \lambda \ \wedge \\ tail(y) \neq \lambda \end{Bmatrix} \supset \begin{Bmatrix} \boxed{y} = left(y)\cdot right(y) \ \wedge \\ left(y) \neq y \ \wedge \\ right(y) \neq y \end{Bmatrix}$	

By taking y to be b we may use equality replacement between this assertion and goal 1 to obtain

5	$perm(left(b)^{-perm}\cdot right(b)^{-perm}, z)$ $\wedge$ $ord(z)$ $\wedge$ $b \neq \lambda \wedge tail(b) \neq \lambda$	z

The induction hypothesis tells us that $left(b)$ and $right(b)$ are permutations of b. Thus using the relation replacement rule we can replace $left(b)$ and $right(b)$ by $sort(left(b))$ and $sortright(b))$ respectively. Eventually this leads to the goal

9	$perm(sort(left(b))\cdot sort(right(b)), z)$ $\wedge$ $ord(z)$ $\wedge$ $b \neq \lambda \wedge tail(b) \neq \lambda$	z

Informally, this goal requires that we produce a ordered string z from two given ordered strings. At this stage the auxiliary function *merge* would be derived to accomplish this task. The defining assertion for *merge* is

$ord(x) \wedge ord(y) \supset$ $perm(x\cdot y, merge(x, y))$ $\wedge$ $ord(merge(x, y))$	

Intuitively, this assertion states that given two ordered strings x and y, *merge* will form a new ordered string by merging their elements. Resolving the assertion with goal 9 gives

10	$b \neq \lambda \wedge tail(b) \neq \lambda \;\wedge$ $\boxed{ord(sort(left(b)))} \;\wedge$ $\boxed{ord(sort(right(b))}$	$merge(sort(left(b)), sort(right(b)))$

By again using the induction hypothesis we eventually obtain

14	$b \neq \lambda \wedge \boxed{tail(b) \neq \lambda}$	$merge(sort(left(b)), sort(right(b)))$

Resolving with the two base cases completes the proof and results in the final program

$$sort(x) \Leftarrow \left\{\begin{array}{l}\text{if } x = \lambda \text{ then } \lambda \\ \text{else } \left\{\begin{array}{l}\text{if } tail(x) = \lambda \text{ then } x \\ \text{else } merge\begin{pmatrix}sort(left(x)), \\ sort(right(x))\end{pmatrix}\end{array}\right\}\end{array}\right\}$$

The auxiliary procedure *merge* could be discovered in a manner analogous to the auxiliary procedures for *quicksort* . Our derivation of *merge* required 25 steps.

Selectsort

Let us see how a slightly different choice at the beginning of first derivation could have led to a final program of *selectsort* rather than *quicksort* . In the *quicksort* derivation we used the *perm-insert* axiom to derive

4a	$perm(w, x{\cdot}y) \;\wedge$ $ord(x{\cdot}u{\cdot}y) \;\wedge$ $\boxed{b = u{\cdot}w}$	$x{\cdot}u{\cdot}y$

Goal 4a was obtained via resolution with equality matching: we unified the second arguments of each *perm* conjunct and matched the first arguments, b and $u{\cdot}w$. But since *perm* is symmetric, and our unification algorithm is commutative, we could have tried instead to unify *opposite* arguments. In this case we would derive the following goal:

4b	$perm(w, x{\cdot}y) \;\wedge$ $ord(u{\cdot}w) \;\wedge$ $\boxed{b = x{\cdot}u{\cdot}y}$	$u{\cdot}w$

While goals 4a and 4b were obtained in a similar manner and exhibit a similar appearence they in fact lead to two quite different sorting programs. That 4a leads to *quicksort* was seen earlier. Let us now see how 4b leads to a *selectsort* algorithm.

The condition $b = x{\cdot}u{\cdot}y$ requires that u be an element of b and that the string $x{\cdot}y$ contain the remaining elements of b; $perm(w, x{\cdot}y)$ requires that the elements of w be the same as those in $x{\cdot}y$; finally $ord(u{\cdot}w)$ requires that u be less than or equal to all the elements of w, and that w itself be ordered. Taken together, the three conditions then require that u be the minimum element of b and that w contain the remaining elements of b arranged in increasing order. If these conditions are fulfilled, then the output entry, $u{\cdot}w$ effectively determines the recursive branch of a *selectsort* program: select the minimum element of b, that is u, sort the remaining elements in w and append. In fact from goal 4b we eventually derive the following *selectsort* program:

$$selectsort(x) \Leftarrow \left\{\begin{array}{l}\text{if } x = \lambda \text{ then } \lambda \\ \text{else } min(x){\cdot}selectsort(lem(x){\cdot}rem(x))\end{array}\right\}$$

The functions *min*, *lem* and *rem* correspond to auxiliary procedures derived in the course of the proof: *min*(*b*) returns the minimum element of *b*; *lem*(*b*) and *rem*(*b*) together return the remaining elements of *b* (specifically, the substrings of *b* to the left and right of *min*(*b*), respectively). A complete derivation of the above program required 17 steps. An additional 22 steps were required to derive programs for the auxiliary functions. We have also derived the following iterative version of *selectsort*

$$selectnew(x) \Longleftarrow snew2(\lambda, x)$$

$$snew2(y,z) \Longleftarrow \left\{ \begin{array}{l} \text{if } z = \lambda \text{ then } y \\ \text{else } snew2\begin{pmatrix} y \cdot min(z), \\ lem(z) \cdot rem(z) \end{pmatrix} \end{array} \right\}$$

Insertsort

In deriving the *mergesort* program we started by using the *left* and *right* functions in order to decompose the input *b* into the two substrings *left*(*b*) and *right*(*b*). But *left* and *right* are not the only decomposition operators that are primitive in our theory; we also included the functions *head* and *tail*. Instead of splitting *b* into left and right halves, we could have used these functions to split *b* into *head*(*b*) (the first element of *b*) and *tail*(*b*) (the remaining elements of *b*). By proceeding in this fashion we obtain the following *insertsort* program:

$$insertsort(x) \Longleftarrow \left\{ \begin{array}{l} \text{if } x = \lambda \text{ then } \lambda \\ \text{else } insert\begin{pmatrix} head(x), \\ insertsort(tail(x)) \end{pmatrix} \end{array} \right\}$$

A complete derivation of the above program required 12 steps. The derivation of *insert* required an additional 16 steps. We have also derived an iterative version of *insertsort* analogous to the iterative version of *selectsort* shown above.

Sorting Programs without Auxiliary Procedures

It is interesting to consider how the system might proceed if it did not include a mechanism for the generation of auxiliary procedures. Surprisingly the derivations could still be completed. We have derived the following two programs:

$$slowsort(x) \Longleftarrow \left\{ \begin{array}{l} \text{if } x = \lambda \text{ then } \lambda \\ \text{else if } tail(x) = \lambda \text{ then } \lambda \\ \text{else } \left\{ \begin{array}{l} \text{if } head(x) \leq head(slowsort(tail(x))) \\ \text{then } head(x) \cdot slowsort(tail(x)) \\ \text{else } \left\{ \begin{array}{l} head(slowsort(tail(x))) \cdot \\ slowsort(head(x) \cdot tail(slowsort(tail(x)))) \end{array} \right\} \end{array} \right\} \end{array} \right\}$$

$$sleepsort(x) \Longleftarrow \left\{ \begin{array}{l} \text{if } x = \lambda \text{ the } \lambda \\ \text{else if } tail(x) = \lambda \text{ then } x \\ \text{else } \left\{ \begin{array}{l} \text{if } head(sleepsort(left(x))) \leq head(sleepsort(right(x))) \\ \text{then } \left\{ \begin{array}{l} head(sleepsort(left(x))) \cdot \\ sleepsort(tail(sleepsort(left(x))) \cdot sleepsort(right(x))) \end{array} \right\} \\ \text{else } \left\{ \begin{array}{l} head(sleepsort(right(x))) \cdot \\ sleepsort((sleepsort(left(x))) \cdot tail(sleepsort(right(x)))) \end{array} \right\} \end{array} \right\} \end{array} \right\}$$

While not particuarly efficient (!), these programs are interesting because they were not known in advance but discovered inadvertantly while trying to derive other programs: We discovered the *slowsort* program by following a natural branch in the *insertsort* derivation; the *sleepsort* program was discovered by intentionally following a similar branch in the *mergesort* derivation. Both programs took us by surprise since we had not considered the possibility a sorting algorithm without auxiliary procedures. This experience suggests that by systematically exploring the derivation space an automated system might discover other novel sorting programs.

Bubblesort

In our original theory we included two pairs of decomposition operators: $\{head, tail\}$ and $\{left, right\}$. Let us now add a third pair of functions: $\{front, last\}$. The function $last(x)$ returns the last element of x and $front(x)$ returns the substring of x containing all but the last element. Using these functions it is possible to derive the following bubblesort program:

$$bubblesort(x) \Longleftarrow \begin{Bmatrix} \text{if } x = \lambda \text{ then } \lambda \\ \text{else } bubblesort(front(sift(x))) \cdot last(sift(x)) \end{Bmatrix}$$

The auxiliary function *sift* moves the maximum element of its argument to the end of the string, leaving the other elements in place. The derivation of *bubblesort* took 26 steps. The derivation of *sift* took and extra 17 steps.

5. RELATED WORK

Broy [Br] offers perhaps the most extensive treatment of sorting, deriving 12 different algorithms from a common specification. Here the emphasis is on human program development rather than computer-based synthesis. The derivations are paradigms for the kind of systematic programming advocated by Dijkstra, in which program construction is broken into small formal steps, which gradually transform a specification into algorithmic code. The choice of which tranformation to apply is seen as a design decision regarding the target program.

Clark and Darlington [CD],[Da] use a predicate logic framework similar to that of [MW1]. By applying Peano-style theory axioms a first-order specification is decomposed into subproblems. Those parts that replicate the original specification are satisfied by recursion. The remaining portion becomes a specification for new auxiliary procedures. In [CD] the system is used to manually derive four applicative sorting programs over lists and in [Da] to derive six array sorting programs. The focus of both papers is on program analysis and specifically on how the relationship between algorithms is brought out by tracing common origins in the derivation tree. Our analysis of program structure in Section 5 is a similar exercise.

Laaser [La] presents a fully automated system capable of synthesizing recursive programs such as *mergesort* and *insertsort*. The system is based on systematic generation and testing of possible outputs. Each output candidate is substituted into the program specification; this yields a test condition under which the output is acceptable. A theorem prover is then used in an attempt to simplify the condition. If the condition reduces to false then the output candidate is rejected; otherwise the simplified expression is used as a guard in the derived program. Recursive output is restricted to one-level-down decomposition of the input structure (e.g., decomposing a binary tree into left and right branches). This excludes such programs as *quicksort* and *selectsort* where an auxiliary procedure is needed for decomposition.

Follet [Fo] performs manual derivations for in-place versions of *quicksort* and *insertsort*. The system initially uses domain knowledge to split a top-level goal into subgoals. Techniques borrowed from planning [Wa],[MW2] are then used to achieve the various subgoals simultaneously. One technique involves *passing back* a goal over parts of the code already synthesized. This produces a new goal which takes into account side-effects of the subsequent program segment. Follet's system includes a computer implementation which can synthesize certain simple programs such as one for finding the maximum element of an array, but fails to synthesize the more complex sorting programs. The problem of passing back goals appears to be one of the stumbling blocks in the latter case.

In Green and Barstow [GB] an informal method is given for transforming certain recursive sorts into more efficient iterative ones. The recursive forms require production of all elements before any can be consumed. The transformation involves interleaving the two operations so that partial results may be obtained without use of a recursion stack. We have accomplished similar transformations formally using the deductive framework. The resulting programs are listed in Section 4.

In Smith [Sm] a system is presented that can synthesize four of the recursive sorts in this paper but with more human assistance than in Laaser's system. At the top level, a knowledge-based approach is used, including program schemas and certain design strategies for filling in unknown operators. Once particular operators have beem chosen, a theorem prover is used to deduce specifications for those remaining. At this stage primitive operators may be matched with the new specifications, or else auxiliary procedures may be synthesized by returning to the top level. One interesting aspect of Smith's system is that the main programs for all four sorts are derived using a single program schema, based on the divide-and-conquer principle. The commonality of these programs is clearly brought out by the unifying schema. Unfortunately many of the low-level auxiliary procedures required do not conform to the general schema and must be synthesized by some other method.

A principal advantage of the deductive approach over the frameworks used in [La] and [Sm] is its generality. In these systems, automation was achieved in part by specializing the synthesis mechanism. In Smith, program structure was circumscribed by a set of pre-existing schemas; in Laaser recursion was fixed by the method if input decomposition. Such constraints tend to limit the derivable programs to forms already conceived by the system's designer. The derivations given here suggest that automation might be achieved within a more general framework. In deductive synthesis, program structure is determined by proof structure, which is unconstrained by the system. Similarly there are no limits on recursion since a general induction rule will handle any type of input decomposition. This flexibility allowed us to derive a wider class of algorithms than in either [La] or [Sm].

While the programs given in this paper were derived manually, we ultimately hope that a computer implementation of the system (of which none currently exists) will find similar programs automatically. Our derivations are intended to suggest this possibility; the proofs are short in relation to program complexity (on the order of 40 steps per program) and individual derivation steps are uncontrived. We have also shown how an automated system might discover auxiliary procedures, a common "eureka" step in program construction.

Acknowledgments

The author is indebted to Richard Waldinger for many valuable comments and suggestions throughout the synthesis of this paper.

BIBLIOGRAPHY

[Br]: Broy (1978). A case study in program developement: sorting. Technical Report TUM-INFO-11-78-31-350, Technisische Universität München.

[CD]: Clark, Darlington (1979). Algorithm classification through synthesis. *The Computer Journal*, Vol. 23, Number 1.

[Da]: Darlington (1978) A synthesis of several sort programs. *Acta Infomatica*, Vol 11 No. 1.

[Fo]: Follet (1981). Automatic program synthesis. Ph.D. Dissertation. Computer Science Department, University of New South Wales.

[GB]: Green, Barstow (1977). On program synthesis knowledge. Stanford AI memo. November 1977.

[La]: Laaser (1979). Synthesis of recursive programs. Ph.D. Dissertation. Department of Computer Science, Stanford University.

[Mu]: Murray (1978). A proof procedure for non-clausal first-order-logic. Technical Report, Syracuse University.

[MW1]: Manna, Waldinger(1980). A deductive approach to program synthesis. *ACM Transactions on Programming Languages and Systems*, Vol. 2, No 1.

[MW2]: Manna, Waldinger (1977). Studies in Automatic Programming Logic. New York: North Holland.

[MW3]: Manna, Waldinger (1986). Special relations in automated deduction. *Journal of the ACM* January 1986.

[Sm]: Smith (1985). Top-down synthesis of simple divide-and-conquer algorithms. *Artificial Intelligence*, 1985.

[Wa]: Waldinger (1975). Achieving several goals simultaneously. Technical Note 107, SRI project 2245. July 1975.

[Wal]: Walther (1984). A mechanical solution of Schubert's Steamroller by many-sorted resolution. Technical Note, Institut für Informatik, Universität Karlsruhe.

Extended Abstracts

of

Current Deduction Systems

The TPS Theorem Proving System

Peter B. Andrews, Frank Pfenning,
Sunil Issar, C. P. Klapper
Mathematics Department, Carnegie Mellon University,
Pittsburgh, Pa. 15213, U.S.A.

TPS is a theorem proving environment for first- and higher-order logic. It provides for automatic, semi-automatic, and interactive modes of proof, and contains various facilities useful for research on theorem proving. As its logical language TPS uses Church's formulation of type theory with λ-notation, in which most theorems of mathematics can be expressed very directly.

A new version of TPS, called TPS3, is under development. It has grown naturally out of the system TPS1 which was discussed in [5] and [2]. In automatic mode TPS1 combined theorem-proving methods for first-order logic with Huet's unification algorithm for typed λ-calculus, found acceptable general matings [1], and used these to construct proofs in natural deduction style. Among the theorems proved completely automatically by TPS1 is Cantor's Theorem that a set has more subsets than members.

TPS3 is being designed as a more general and flexible system for dealing with higher-order logic. The notion of a general mating has been replaced by that of an *expansion tree proof* [6]. Primitive substitutions (which play a role similar to the splitting rules of [4]) as well as higher-order unification are used to generate substitution terms during the search for an expansion tree proof.

TPS3 combines essential ideas of Miller [7] with tactics [3] to provide a flexible and extensible way of translating expansion tree proofs into a variety of logical calculi, including Gentzen's sequent calculus and natural deductions. TPS3 will also have facilities (based on ideas in [8] and [9]) for translating such proofs into expansion trees. This will facilitate constructing examples of expansion trees for research purposes, and will permit awkward natural deduction proofs to be automatically restructured by translating them into expansion tree proofs and then back into well structured natural deduction proofs.

As an interactive tool TPS has many features which facilitate constructing, displaying, and manipulating logical formulas, formal proofs, and expansion trees. A completely interactive version of TPS called ETPS (Educational Theorem Proving System) has been developed for use by students in logic courses, and has been used for several years at Carnegie Mellon University. ETPS permits students to concentrate on basic decisions about applying rules of inference while constructing formal proofs, and relieves them of many of the details of writing the proofs. ETPS figures out intelligent default arguments for commands and the types of many variables. It

---

This work is supported by NSF grant DCR-8402532.

permits students to work from the bottom up as well as from the top down, reorganize their proofs in various ways, and readily construct formulas from others which already occur somewhere in the proof. Instructors are not limited to a fixed logical language or rules of inference. They can specify language and rules in a simple format. TPS uses this information to automatically produce commands which apply the rules and fast matching functions which test their applicability. There are facilities for automatically recording information about how ETPS is being used (and misused) and for processing grades. When the instructor allows it, perplexed students can obtain automatically generated advice about what to try next. At present this advice is based on obvious features of the available rules, but eventually the full power of the automatic theorem-proving facilities will be used (invisibly) to generate this advice.

TPS3 is being written in Common Lisp. Considerable effort has been devoted to designing TPS as a system which is in a perpetual process of development. Much documentation is produced automatically and is available on-line. It always reflects the current state of the program. TPS is interfaced with Scribe and TeX so that output can contain mathematical and logical symbols.

References

1. Peter B. Andrews, *Theorem Proving via General Matings*, Journal of the Association for Computing Machinery **28** (1981), 193-214.

2. Peter B. Andrews, Dale A. Miller, Eve Longini Cohen, Frank Pfenning, "Automating Higher-Order Logic," in *Automated Theorem Proving: After 25 Years*, edited by W. W. Bledsoe and D. W. Loveland, Contemporary Mathematics series, vol. 29, American Mathematical Society, 1984, 169-192.

3. Michael J. Gordon, Arthur J. Milner, Christopher P. Wadsworth. *Edinburgh LCF*, Lecture Notes in Computer Science 78, Springer Verlag, 1979.

4. Gérard P. Huet, "A Mechanization of Type Theory," in *Proceedings of the Third International Joint Conference on Artificial Intelligence*, IJCAI, 1973, 139-146.

5. Dale A. Miller, Eve Longini Cohen, Peter B. Andrews, "A Look at TPS," in *6th Conference on Automated Deduction, New York*, edited by Donald W. Loveland, Lecture Notes in Computer Science 138, Springer-Verlag, June 1982, 50-69.

6. Dale A. Miller. *Proofs in Higher-Order Logic*, Ph.D. Thesis, Carnegie-Mellon University, October, 1983. 81 pp.

7. Dale A. Miller, "Expansion Tree Proofs and Their Conversion to Natural Deduction Proofs," in *7th International Conference on Automated Deduction, Napa, California, USA*, edited by R. E. Shostak, Lecture Notes in Computer Science 170, Springer-Verlag, May 14-16, 1984, 375-393.

8. Frank Pfenning, "Analytic and Non-analytic Proofs," in *7th International Conference on Automated Deduction, Napa, California, USA*, edited by R. E. Shostak, Lecture Notes in Computer Science 170, Springer-Verlag, May 14-16, 1984, 394-413.

9. Frank Pfenning. *Proof Transformations in Higher-Order Logic*, Ph.D. Thesis, Carnegie-Mellon University, 1986.

TRSPEC: A Term Rewriting Based System for Algebraic Specifications

J. Avenhaus, B. Benninghofen, R. Göbel, K. Madlener

Universität Kaiserslautern
Fachbereich Informatik
6750 Kaiserslautern
West-Germany

Term rewriting systems can be used both, as an algebraic programming language and as a tool for proving inductive properties of these programs (e.g.: [HH 82],[MU 80]). The TRSPEC system allows to specify functions by term rewriting systems and to prove assertions (equations) of these functions. The proofs are done by completion algorithms ala Knuth-Bendix (inductionless induction). The system is implemented in Common-Lisp on an Apollo-Workstation under Aegis. It contains the following components:

- A parser for transforming hierarchical structured specifications of functions (rewrite programs) and assertions (equations) into an internal representation used by the other tools of the system.

- A checker to check basic properties of rewrite programs:
  - **Consistency:** Rewrite rules of a higher level in the specification must not destroy the algebra on a lower level. This property is guaranteed by syntactical checks on the rewrite rules.
  - **Sufficient Completeness:** All functions have to be totally defined by the rewrite program.
  - **Unique Termination:** The rewrite program is confluent and terminating.

- A compiler for transforming functions specified by rewrite programs into executable lisp functions for computing the normal forms of terms.

- A <u>prover</u> for proving the assertions to hold in the initial algebra specified by the rewrite programs. The prover bases on a specialized Knuth-Bendix algorithm for inductive proofs, where the confluence is only required for ground terms but not for arbitrary terms. This restriction allows to consider fewer critical pairs, which increases the chance to get a finite term rewriting system [GO 85a]. Also it allows to handle cyclic rules like the commutativity by an extension of the Knuth-Bendix algorithm to globally finite term rewriting systems [GO 85b].
  Inductive lemmata are often necessary, where the Knuth-Bendix algorithm would diverge without these lemmata. Therefore, the prover generates hypotheses (potential lemmata) by generalizing rules, and completes them independently and simultaneously with the initial rules and equations.

In future we plan to extend the system to an even more realistic tool for using rewriting systems as a functional programming language. The topics to be treated include:

- **User Interface**: Extending the system to a comfortable programming environment for a functional programming language based on rewriting.

- **Term Orderings:** Considering new term orderings, especially orderings which are compatible to cyclic rules (associativity and commutatitvity) and implementing them as a part of the checker and the prover.

- **Efficiency**: Reducing the number of critical pairs for inductive completion, recognizing divergency and creating lemmata.

References

[A 84] Avenhaus, J.:
On the Termination of the Knuth-Bendix Completion Algorithm.
Int. Report 120/84, Universität Kaiserslautern, 1984.

[A 86] Avenhaus, J.:
On the Descriptive Power of Term Rewriting Systems.
To appear in J. Symb. Comp. 2, 1986.

[GO 83a] Göbel, R.:
A Completion Procedure for Globally Finite Term Rewriting Systems.
Proceedings of an NSF Workshop on the Rewrite Rule Laboratory, General Electric, Schenectady, 1983.

[GO 83b] Göbel, R.:
Rewrite Rules with Conditions for Algebraic Specificiations.
Proc. of 2nd Workshop on Theory and Applications of Abstract Data Types (Abstract), Passau, West-Germany, 1983.

[GO 85a] Göbel, R.:
Completion of Globally Finite TRS for Inductive Proofs.
Proc. GWAI 85, Springer Lecture Notes, 1985.

[GO 85b] Göbel, R.:
A Specialized Knuth-Bendix Algorithm for Inductive Proofs.
Proc. Combinatorial Algorithms in Algebraic Structures, Universität Kaiserslautern, 1985.

[HH 82] Huet, G., Hullot, J.:
Proofs by Induction in Equational Theories with Constructors.
JACM 25(2), pp. 239-266, 1982.

[MU 80] Musser, D.:
On Proving Inductive Properties of Abstract Data Types.
7th Annual ACM Symp. POPL, Las Vegas, 1980.

Highly Parallel Inference Machine

The highly parallel inference machine will be a VSLI-implementation of the **Logical Connection Machine** (LCM), which is a theorem prover for full first order logic.

The starting point of this work was a simple version of a theorem prover for full first-order logic (not restricted to clausal logic). The prover has been implemented in TLC-LISP on a 64KByte desk top computer and consists of two separate parts. The first part is a preprocessing module for non-normal form formulas transforming the input formula to its definitional form, a formula in clausal form which is different from the normal form used in other approaches. The second part of the system is a theorem prover for normal form formulas which is based on the connection method. It incorporates factorisation, avoidance of circuits, elimination of useless clauses and weak unification to improve efficiency.

Presently a more advanced *sequential theorem prover* is implemented on a Targon 35 (Nixdorf) and, independently, a *parallel version* of this theorem prover is under construction:

The theorem prover is divided in four parts.

Input:	Several logical notations (not necessarily normal form).
Preprocessing:	The partial evaluation part which contains well-known processes like subsumption, pure literal reduction, weak unification, etc. as well as particular reductions for the Connection Method.
Main proof procedure:	based on the Connection Method, a complete and sound calculus for first order predicate logic. Some of the essential features incorporated within the procedure: Factorisation (applying the mgu of two literals to all subgoals); top-down strategy with bottom-up support and intelligent boundary handling.
Output:	Some intelligent support available, e.g. the interface which clarifies the proof structure and gives hints for improving the input formula in case of nonprovability.

Area of application

All Prolog applications and various problems which cannot be solved using Prolog (e.g. Negation in heads, incompleteness and inconsistency of Prolog).

Implementation

All procedures of all parts are designed in a purely modular way. The sequential version of this theorem prover is implemented in FRANZ-Lisp, while the parallel version will be written in FP2, a functional programming language for parallel processes.

State of Development

There exist preliminary sequential versions for the main part. The preprocessing part of the PROLOG-like theorem prover has been implemented. The global structure of the parallel proof process has already been designed.

This project is part of the ESPRIT Project 415 and is supported by the CEC and Nixdorf Comp. AG. .

The author's address: M. Bayerl
Institut für Informatik
der Technischen Universität München
Arcisstraße 21
Postfach 20 24 20
D-8000 München 2

Automatic Theorem Proving in the ISDV System

Christoph Beierle, Walter Olthoff, Angelika Voss
University of Kaiserslautern, Department of Computer Science
P.O. Box 30 49, 6750 Kaiserslautern, West Germany
UUCP: ..!mcvax!unido!uklirb!beierle

The Integrated Software Development and Verification (ISDV)-project ([1], [5]) aimed at the integration of well-established software engineering techniques like stepwise-refinement, of powerful mechanisms for formal program specification (abstract data type theory) and of realizing the 'verify-while-develop' paradigm (proof tasks are processed as early as possible). In the ISDV system each development step is either carried out by mechanical tools guaranteeing correctness, or verified by appropriate theorem proving systems.

Currently, the ISDV environment employs two automatic theorem provers: the resolution-based Markgraf Karl Refutation Procedure for proofs in first order predicate logic and its extension to handle induction (MKRP, [2]), and the Rewrite Rule laboratory (RRLab, [3]) which is based on the Knuth-Bendix algorithm and which is used for proofs in equational theories (including "inductionless induction").

Software development and formal program specification are carried out within the algebraic specification development language ASPIK and its support environment SPESY. ASPIK specifications are arbitrary mixtures of loose algebraic specifications containing first order formulae and algorithmic algebraic specifications consisting of constructive carrier and operation definitions ([4]). Beside ASPIK the imperative programming language ModPascal ([6]) is used in SPESY. It extends Pascal and provides module constructs and related features such that ModPascal can be used as an imperative counterpart of ASPIK. Both, ASPIK and ModPascal are object-oriented and allow hierarchical design. Both languages are supplied with a formal algebraic semantics. This allows to express verification issues uniformly in algebraic terms, and therefore we e. g. do not employ Hoare-style verification for ModPascal.

Our realization of the 'verify-while-develop' paradigm distinguishes between object-related and inter-object-related verification tasks. In the first case we consider the axiom set A(s) associated to an ASPIK specification s: It may consist of arbitrary first order predicate calculus formulae built on the specification's signature. If the specification is loose questions of interest are: Is A(s) consistent? Is A(s) ∪ {new_axiom} still consistent? Is (formula) derivable from A(s)? For the answer an appropriate proof task is generated by SPESY and passed over to MKRP or RRLab.

If the specification s is algorithmic we have additional information available: s is connected to a canonical term algebra CTA(s), its carriers are term-generated by a set of constructors, and its operations are closed on the s carriers. This allows to use (structural) induction proofs for the above questions of interest. Additionally we may ask: Is A(s) inductively derivable from the algorithmic operation definitions of s? We use either the induction module of MKRP in order to find a proof or the 'induction-less induction' component of RRLab.

Inter-object-related verification tasks arise if a refinement or implementation step is performed (i. e. s is refined to or implemented by s'). Since both objects are semantically described by algebras (this holds for ASPIK as well as for ModPascal), there are algebraic correctness notions for these relations. E. g. s is implemented by s' correctly if there is an abstraction homomorphism from s' to s. Checking the homomorphism property of such a user supplied abstraction function amounts to an examination of certain equations analogous to the axioms in the object-related case above. Therefore we pass over the prooftasks to MKRP or its induction module or to RRLab. - Another application area for MKRP and RRLab is the program transformation component of SPESY. It applies recursion removal rules to algorithmic operation definitions if certain semantical conditions of the operations involved are satisfied; here again SPESY generates appropriate proof tasks for MKRP and/or RRLab.

Verified objects and object relations are marked in the SPESY data base. Together with the object-oriented software development philosophy this allows to limit the proof effort for a specific goal only to those objects (or object-relations) that currently are not marked 'verified'. Beside support of incremental verification the data base management uses sophisticated revision algorithms to update and maintain the proof-theoretic state of objects (e. g. after editing a non-toplevel object).

Accomplishments: Our work has shown that notions and formalisms of abstract data type theory, of modern software engineering paradigms, of algebraic verification techniques and automated theorem proving tools can be appropriately integrated into a single comprehensive system offering a convenient and effective support environment for the development of reliable software. SPESY has been used in several academic and real-world applications, including the specification and verification of a compiler's lexical analyser, a software package for code analysis, and a financial accounting problem. However, our experiences with the two automatic theorem provers we used showed that, while they are powerful enough to handle the arising proof tasks involving no (or only standard) induction schemes, the bottleneck were the induction proofs. Since in our data type applications induction is often necessary, a promising approach to handle also the more complicated cases in the near future seems to be a moderate interactive approach where the user is able to indicate some induction scheme to be tried by the theorem proving system.

References:

[1] Beierle, C., Olthoff, W., Voss, A.: Software Development Environments Integrating Specification and Programming Languages. Proc. German Chapter of the ACM Workshop on Software Architecture and Modular Programming, Teubner Verlag, Stuttgart, 1986.

[2] Karl Mark G. Raph: The Markgraf Karl Refutation Procedure. University of Kaiserslautern, Memo SEKI-MK-84-01, 1984.

[3] Thomas, C.: The Rewrite Rule Laboratory (in German). University of Kaiserslautern, Memo SEKI-84-01, 1984.

[4] Beierle, C., Voss, A.: Algebraic Specifications and Implementations in an Integrated Software Development and Verification System. University of Kaiserslautern, Memo SEKI-85-12, 1985 (joint SEKI-Memo containing the Ph.D. thesis by Ch. Beierle and the Ph.D. thesis by A. Voss).

[5] Beierle, C., Gerlach, M., Goebel, R., Olthoff, W., Raulefs, P., Voss, A.: Integrated Program Development and Verification. In: H.-L. Hausen (ed.): Symposium on Software Validation. North-Holland Publ. Co., Amsterdam, 1983.

[6] Olthoff, W.: An Overview of ModPascal. SIGPLAN Notices, 20 (10), Oct. 1985.

THE KARLSRUHE INDUCTION THEOREM PROVING SYSTEM

S.Biundo, B.Hummel, D.Hutter, C.Walther
Institut für Informatik I, Universität Karlsruhe, D-7500 Karlsruhe

Mathematical induction is the basic technique to prove properties of recursively defined functions. Hence proofs by induction play a central role in several subfields of mathematics such as arithmetic, formal logic, formal languages, algebra etc. In computer science induction proofs are indispensable when properties of loops and recursive procedures have to be verified.

Based on the works of Aubin [1] and Boyer and Moore [3], we develop the induction theorem proving system **INKA**, which shall extend the well known techniques of automated induction provers in the following research topics:
(1) Development of techniques to guide a many-sorted resolution and paramodulation based theorem prover to simplify formulas goal-directed. (2) Development of strong heuristics to generalize formulas about recursively defined functions to make them amenable for proofs by induction. (3) Development of techniques to prove formulas containing existential quantifiers using methods of deductive program synthesis. (4) Development of methods to mechanize termination proofs for the recursively defined functions submitted to the system.

This research project (started in 1984) is part of the Sonderforschungsbereich 314 "Künstliche Intelligenz" supported by the Deutsche Forschungsgemeinschaft and performed in close cooperation with the MKRP theorem proving project [4]. The INKA system is implemented in INTERLISP (consisting of approx. 3 MB source code) and currently runs on a SIEMENS 7561 machine. It is presently being reimplemented on SYMBOLICS 3640 and µVAX machines in COMMONLISP.

The INKA System Given a finite set of axioms AX and a formula φ an induction theorem prover attempts to find a proof of φ from AX using some sound first-order inference rules as well as some induction rules. In our system the axiom set AX consists of first-order formulas representing the various domains under consideration and also algorithms computing functions defined on these domains. The domain of natural numbers, for example, can be defined by an expression like **structure 0 succ(nat):nat**. Given this expression as input, the system extends AX by a set of formulas which corresponds to the first-order part of Peano's axioms for natural numbers (or to similar axioms for the other domains). The algorithm for the addition of natural numbers,for example, can be defined by an expression like **function plus(x,y:nat):nat = [if x=0 then y, if ex u:nat x=succ(u) then succ(plus(u y))]**. In order to guarantee that an algorithm submitted to the system computes a total function, the system provides a so-called admissibility test, which consists in proving the following properties (formulated as first-order formulas): (1) at least one of the cases in the algorithm's definition is always satisfied (completeness property), (2) at most one of the cases in the algorithm's definition is always satisfied (uniqueness property), and (3) the algorithm terminates according to some well-founded order relation (termination property), where certain order relations are assumed to be well-founded of course. If the admissibility test succeeds, a set of first-order formulas representing the algorithm is added to the set of axioms.

Given the set of axioms AX and a formula φ the system attempts to prove φ from AX in the following way: Firstly φ is simplified by symbolic evaluation of terms and by simple logical transformations. If the simplified formula φ' contains an existential quantification, it is given to the synthesis module which tries to find a solving term for the existentially quantified variable. Else φ' is negated, transformed into clausal form and then the basic prover is used to modify the formula in a logically equivalent way. If the prover is not able to deduce the empty clause (i.e. to deduce φ' from AX without induction), it comes up with a set of clauses which are transformed by negation and re-skolemization yielding a formula φ'' which is equivalent (but usually 'simpler') to φ'. φ'' is then passed to the induction module, which generates (eventually after a generalization step) an induction formula from φ'' by means of some induction rule, where we presently use Aubin's method [1] for the generation of induction formulas. Each induction formula is seperately proved by the system, i.e. it is treated like the original formula φ.

The Basic Prover Induction provers depend on a basic (non inductive) inference machine, which either proves or at least simplifies the formulas obtained by application of some induction rule. Since the domains of discourse in induction theorem proving are many-sorted, we use a many-sorted version of a resolution calculus with paramodulation [6]. We have investigated techniques of how to guide a theorem prover based on this calculus to simplify formulas goal-directed [5]: For symbolic evaluation of terms we use function definitions to open up function calls in the given formula by performing certain resolution and paramodulation steps. These deductions are driven by information about the cases and the termination of the recursive functions involved. We use also lemmata (i.e. formulas already proved) to enable further symbolic evaluations. If further simplifying deductions are not possible, the resulting clause set is stored to be returned to the induction module in case the prover fails to generate the empty clause.

Generalization It often happens that a given formula is not directly provable by induction, i.e. the induction formula obtained from the initial one provides an induction hypothesis which is too weak to be useable in the induction step. In such cases it is necessary to find a more general formula which is sufficient for the initial one such that the induction hypothesis obtained 'carries' the induction. Several techniques have been proposed in the literature to generalize a given formula. We have analyzed and compared these proposals. The results of this study (which will be published in a forthcoming report) shall serve as the basis for the generalization module of the INKA-system.

The Synthesis Module The synthesis module tries to find existence proofs using techniques of deductive program synthesis [2]. An existentially quantified variable in a formula is replaced by a skolem term. Using this new formula as a specification an algorithmic definition for the skolem function will be derived. A successful synthesis also represents an induction proof of the original formula. The synthesis module presently works interactively and performs a user-guided synthesis of functions like addition and difference of natural numbers and also the 'falsify' algorithm of [3]. Additionally several non-recursive algorithms have been derived automatically. This module is an experimental system used to develop strategies and heuristics which shall be implemented by a control component in order to synthesize function definitions with as little user guidance as possible.

Termination Proofs For recursively defined algorithms, the system has to prove that the algorithm terminates according to some well-founded order relation. Presently the system recognizes only termination according to the immediate structural predecessor order and lexicographic versions thereof, c.f. [1]. For more complex orders we intend to develop the approach of termination proofs using induction lemmata [3]: we have developed a technique to synthesize induction lemmata from algorithms, which is currently under implementation. With this technique it is possible in many cases to recognize automatically the conditions under which an algorithm, like for instance 'half', 'remove' e.t.c, returns a result which is 'count-smaller' [3] than one of its arguments - thus providing a means to prove the termination of certain algorithms without user guidance. A description of this technique and experimental results obtained with it will be published in a forthcoming report.

REFERENCES

1. **AUBIN, R.** Mechanizing Structural Induction. Part I: Formal System, Part II: Strategies. Theoretical Computer Science 9 (1979).

2. **BIUNDO, S.** A Synthesis System Mechanizing Proofs by Induction. Proceedings Seventh ECAI (Brighton 1986).

3. **BOYER, R.S. and MOORE, J S.** A Computational Logic. Academic Press (1979).

4. **EISINGER, N. and OHLBACH, H.J.** The Markgraph Karl Refutation Procedure (MKRP). Prooceedings Eighth CADE (Oxford 1986), this volume.

5. **HUTTER, D.** Using resolution and paramodulation for induction proofs. Interner Bericht 6/86, Universität Karlsruhe, Inst. f. Informatik I (1986).

6. **WALTHER, C.** A many-sorted calculus based on resolution and paramodulation. Proceedings Eighth IJCAI (Karlsruhe 1983).

Overview of a Theorem-Prover for A Computational Logic

Robert S. Boyer[1] and J Strother Moore
Institute for Computing Science and Computer Applications
University of Texas at Austin

The Logic and the Implementation

A brief description of our theorem-proving system may be found in [6]. The theorem prover and its logic, as of 1979, are described completely in [1]. A slight revision of the logic is presented in [2], where also a "metafunction" facility is described which permits the user to define new proof procedures in the logic, prove them correct mechanically, and have them used efficiently in subsequent proof attempts. During the period 1980-1985 a linear arithmetic decision procedure was integrated into the rule-driven simplifier. The problems of integrating a decision procedure into a heuristic theorem prover for a richer theory are discussed in [8].

Applications of the System

The theorem prover has been used in a wide variety of applications by many researchers. Below is an annotated list of selected references.

Elementary List Processing: Many elementary theorems about list processing are discussed among the examples in [1]. The appendix includes theorems proved about such concepts as concatenation, membership, permuting (including reversing and sorting) and tree exploration.

Elementary Number Theory: Euclid's Theorem and the existence and uniqueness of prime factorizations are proved in [1]. A version of the pigeon hole principle and Fermat's theorem are proved in [6]. Wilson's Theorem is proved in [16]. Finally, Gauss' Law of Quadratic Reciprocity has been checked; the theorem, its definitions, and the lemmas suggested by Russinoff are included among the examples in the standard distribution of the theorem-proving system.

Metamathematics: The soundness and completeness of a decision procedure for propositional calculus, similar to the Wang algorithm, is proved in [1]. The soundness of an arithmetic simplifier for the logic is proved in [2]. The Turing completeness of Pure LISP is proved in [9]. The recursive unsolvability of the halting problem for Pure LISP is proved in [7]. The Tautology Theorem, i.e., that every tautology has a proof in Shoenfield's formal system, is proved in [17]. The Church-Rosser theorem is proved in [18]. Goedel's incompleteness theorem is proved in [19].

Communications Protocols: Safety properties of two transport protocols, the Stenning protocol and

---

[1]Current address: MCC, AI Program, 9430 Research Blvd., Austin, Texas 78759

the "NanoTCP" protocol, are proved in [11].

Concurrent Algorithms: A mechanized theory of "simple" sorting networks and a proof of the equivalence of sequential and parallel executions of an insertion sort program are described in [14]. A more general treatment of sorting networks and an equivalence proof for a bitonic sort are given in [12]. A proof of the optimality of a given transformation for introducing concurrency into sorting networks is described in [15].

Fortran Programs: A verification condition generator for a subset of ANSI Fortran 66 and 77 is presented in [3]. The same paper describes the correctness proof of a Fortran implementation of the Boyer-Moore fast string searching algorithm. A correctness proof for a Fortran implementation of an integer square root algorithm based on Newton's method is described in [4]. The proof of a linear time majority vote algorithm in Fortran is given in [5].

Real Time Control: A simple real time control problem is considered in [10]. The paper presents a recursive definition of a "simulator" for a simple physical system -- a vehicle attempting to navigate a straightline course in a varying cross-wind. Two theorems are proved about the simulated vehicle: the vehicle does not wander outside of a certain corridor if the wind changes "smoothly" and the vehicle homes to the proper course if the wind stays steady for a certain amount of time.

Assembly Language: A simple assembly language for a stack machine is formalized in [1]. The book also gives a correctness proof for a function that compiles expressions into that assembly language. In our standard benchmark of definitions and theorems is a collection that defines another simple assembly language, including "jump" and "move to memory" instructions, and proves the correctness of a program that iteratively computes the sum of the integers from 0 to n. The correctness proof is complicated by the fact that the program instructions are fetched from the same memory being modified by the execution of the program. The list of events is included in the standard distribution of our theorem-proving system.

Hardware Verification: The correctness of a ripple carry adder is given in [13]. The adder is a recursively defined function which maps a pair of bit vectors and an input carry bit to a bit vector and an output carry bit. The theorem establishes that the natural number interpretation of the output is the Peano sum of the natural number interpretations of the inputs, with appropriate consideration of the carry flags. An analogous result is proved for twos-complement integer arithmetic. The recursive description of the circuit can be used to generate an adder of arbitrary width. A 16-bit wide version is shown. Propagate-generate and conditional-sum adders have also been proved correct. Also in [13] is the correctness proof of the combinational logic for a 16-bit wide arithmetic logical unit providing the standard operations on bit vectors, natural numbers, and integers. The dissertation then presents a recursively described microcoded cpu comparable in complexity to a PDP-11 and proves that the device correctly implements an instruction set defined by a high-level interpreter.

Obtaining a Copy of the System

The system is available free of license and copyright. Our theorem-prover can be got from Arpanet host Utexas-20, a DEC 2060 running Tops-20, in the directory aux:<cl.thm>. The 2060 and Multics Maclisp version, the Symbolics 3600 Zetalisp version (Release 6.0), and the Vax and Sun Franz Lisp version all work from the same source code. First get and read the file "aux:<cl.thm>-read-.-this-", which describes the files that contain the sources, the documentation, and our standard benchmark of some 800 theorems, including many of the theorems mentioned above. Login as anonymous, with any password. The theorem-prover is occasionally distributed on magnetic tape, but we strongly encourage people to obtain the system via the tcp/ip file transfer protocol (ftp) if at all possible. There is no fee for the system if you obtain it by tcp/ip, but we may charge for tape copying. Electronic correspondence may be sent to cl.boyer@utexas-20.arpa or cl.moore@utexas-20.arpa. We will attend immediately to any report of the derivation of an inconsistency; but there is otherwise very little user support currently available.

References

[1] R. S. Boyer and J S. Moore.
A Computational Logic.
Academic Press, New York, 1979.

[2] R. S. Boyer and J S. Moore.
Metafunctions: Proving Them Correct and Using Them Efficiently as New Proof Procedures.
In R. S. Boyer and J S. Moore (editors), *The Correctness Problem in Computer Science.* Academic Press, London, 1981.

[3] R. S. Boyer and J S. Moore.
A Verification Condition Generator for FORTRAN.
In R. S. Boyer and J S. Moore (editors), *The Correctness Problem in Computer Science.* Academic Press, London, 1981.

[4] R. S. Boyer and J S. Moore.
The Mechanical Verification of a FORTRAN Square Root Program.
CSL Report, SRI International, 1981.

[5] R. S. Boyer and J S. Moore.
MJRTY - A Fast Majority Vote Algorithm.
Technical Report ICSCA-CMP-32, Institute for Computing Science and Computer Applications, University of Texas at Austin, 1982.

[6] R. S. Boyer and J S. Moore.
Proof Checking the RSA Public Key Encryption Algorithm.
American Mathematical Monthly 91(3):181-189, 1984.

[7] R. S. Boyer and J S. Moore.
A Mechanical Proof of the Unsolvability of the Halting Problem.
JACM 31(3):441-458, 1984.

[8] R. S. Boyer and J S. Moore.
Integrating Decision Procedures into Heuristic Theorem Provers: A Case Study with Linear Arithmetic.
In *Machine Intelligence.* Oxford University Press, (to appear, 1986).

[9] R. S. Boyer and J S. Moore.
A Mechanical Proof of the Turing Completeness of Pure Lisp.
In W.W. Bledsoe and D.W. Loveland (editors), *Automated Theorem Proving: After 25 Years*, pages 133-167. American Mathematical Society, Providence, R.I., 1984.

[10] R. S. Boyer, M. W. Green and J S. Moore.
The Use of a Formal Simulator to Verify a Simple Real Time Control Program.
Technical Report ICSA-CMP-29, University of Texas at Austin, 1982.

[11] Benedetto Lorenzo Di Vito.
Verification of Communications Protcols and Abstract Process Models.
PhD Thesis ICSCA-CMP-25, Institute for Computing Science and Computer Applications, University of Texas at Austin, 1982.

[12] Huang, C.-H., and Lengauer, C.
The Automated Proof of a Trace Transformation for a Bitonic Sort.
Technical Report TR-84-30, Department of Computer Sciences, The University of Texas at Austin, Oct., 1984.

[13] Warren A. Hunt, Jr.
FM8501: A Verified Microprocessor.
Technical Report 47, University of Texas at Austin, December, 1985.

[14] Lengauer, C.
On the Role of Automated Theorem Proving in the Compile-Time Derivation of Concurrency.
Journal of Automated Reasoning 1(1):75-101, 1985.

[15] Lengauer, C., and Huang, C.-H.
A Mechanically Certified Theorem about Optimal Concurrency of Sorting Networks, and Its Proof.
Technical Report TR-85-23, Department of Computer Sciences, The University of Texas at Austin, Oct., 1985.

[16] David M. Russinoff.
A Mechanical Proof of Wilson's Theorem.
Masters Thesis, Department of Computer Sciences, University of Texas at Austin, 1983.

[17] N. Shankar.
Towards Mechanical Metamathematics.
Journal of Automated Reasoning 1(1), 1985.

[18] N. Shankar.
A Mechanical Proof of the Church-Rosser Theorem.
Technical Report ICSCA-CMP-45, Institute for Computing Science, University of Texas at Austin, 1985.

[19] N. Shankar.
Checking the proof of Godel's incompleteness theorem.
Technical Report, Institute for Computing Science, University of Texas at Austin, 1986.

GEO-Prover – A Geometry Theorem Prover Developed at UT*

SHANG-CHING CHOU

ICSCA, University of Texas at Austin, Austin, Texas 78712, USA

1 Introduction

The GEO-Prover is based on an algebraic method introduced by Wu Wen-Tsün for proving geometry theorems involving only equality [6], [7]. It is the further development of a prover (referred to as GEO-Prover0) described in the first two sections of [2]. The project started in the spring of 1983 after the GEO-Prover0 succeeded in proving about 100 nontrivial geometry theorems. The implementation was completed in early 1985. Since then, the system has been extended to include more facilities, such as proving theorems in geometries other than Euclid's (with the same inputs), mechanically deriving formulas in geometry [3], and proving geometry theorems using Gröbner bases [5].

2 Principles for Implementation

GEO-Prover0 was not satisfactory in several aspects. This led us to study the related problems more closely and develop the GEO-Prover.

2.1 Selection of Independent Variables. GEO-Prover0 needs the user to specify independent variables or parameters. Some leading representatives in the community of automated theorem proving repeatedly asked the question, "How does using different choices of parameters affect the soundness?"

In my opinion, specifying parameters is a important and natural step in the algebraic reformulation of geometry problems. The specification can only come from concrete geometry statements. It reflects an assumption, which is often implicit in geometry statements, that the figures to be considered in the statements are in *general positions* or "generic positions". A precise geometric statement usually gives the clear indication of the choice of parameters.

Some researchers overlooked this key issue, thus causing some kinds of unsoundness in their algebraic formulations of geometry problems or methods. For example, their methods can prove "AC is congruent to altitude CF of a triangle ABC" by adding inappropriate subsidiary conditions. Provers which prove an invalid "theorem" are almost useless, even if the likelihood that this will happen is small.

Incorrect specification of parameters can cause the above described unsoundness. Our provers require the user to choose parameters correctly. However, to make that responsibility at a higher level than in GEO-Prover0, we classify a large class of geometry statements – statements of constructive type. For such kinds of geometry statements, the choice of parameters can be made by the program.

Example. (Simson's Theorem) Let D be a point on the circumscribed circle (O) of a triangle ABC. From D three perpendiculars are drawn to the three sides BC, CA and AB of ABC. Let E, F and G be the three feet respectively. Show that E, F and G are collinear.

One of the possible inputs to our theorem prover is

```
((point-order A B C O D E F G) (eqdistance O A O B) (eqdistance O A O C)
 (eqdistance O A O D) (collinear G A B) (collinear E B C) (collinear F A C)
 (perpendicular D E B C) (perpendicular D F C A) (perpendicular D G A B) (collinear E F G))
```

Lists 2–9 are the geometric conditions (hypotheses) of the problem; the last one is the conclusion. The most important is the first, which specifies the order of points to be constructed. For the above order, we first have three arbitrary points A, B and C. Then the center O is determined (constructed) by the conditions corresponding to lists 2 and 3, etc. Our program assigns the coordinates exactly the same as choice 3 of example 2 in [2].

The specification of order of points is not unique. It is the user's responsibility to arrange the order of points correctly so that the *meaning* of the original geometry statement is not changed. The above input is most convenient for the user. The program, however, also provides for other input in case the problems cannot be fitted into the above format or the user wishes to control the choice of coordinates in a way that is more efficient for proofs.

2.2 The Completeness. Wu's complete method needs decomposition or factorization of polynomials over algebraic extension fields. From a practical point of view, the factorization for quadratic cases is sufficient for almost all cases. We have found a practical algorithm for factorization of quadratic polynomials over successive

---

* This is an extended abstract. The work reported here was supported by NSF Grant DCR-8503498.

quadratic extension fields. Thus, our prover is complete for quadratic cases in the metric geometry introduced by Wu. However, it is still not complete in Euclidean geometry because deciding the existence of real solutions of systems of polynomial equations is beyond the scope of Wu's method. For linear cases, however, Wu's method is *complete for all* geometries. In some current research, even this elementary fact has been overlooked.

2.3 Adding Nondegenerate Conditions Adding nondegenerate conditions is intimately connected with the selection of parameters. One of the advantages of Wu's method is that the user need not to worry about nondegenerate conditions – the program can produce these conditions during the proof. The nondegenerate conditions produced are in polynomial inequation forms $c_1 \neq 0, \ldots, c_r \neq 0$. Now several questions arise:

(2.3.1) *Will the theorem to be proved be weakened by introducing these new additional conditions? Especially, will the hypotheses become inconsistent by adding these new nondegenerate conditions?*

(2.3.2) *Are all these conditions necessary? Can those inequations which are necessary for mechanical proofs be translated back into their geometric forms by the program?*

The answers to the above questions can be fully found in practice only if we have a practical algorithm for decomposition and checking irreducibility. For the quadratic cases, we have satisfactory answers in practice.

For example, for the above input of Simson's theorem, the program proves the theorem by adding the following nondegenerate conditions in geometric forms: A, B and C are not collinear; AB, BC, and CA are non-isotropic. Isotropic lines do not exist in Euclidean geometry. However, they do exist in the metric geometry which is more general than Euclidean geometry.

3 Experimental Results

More than 200 theorems in various elementary geometries have been proved by our prover. We proved several extremely hard theorems in Euclidean geometry; e.g., a theorem of V. Thèbault, which was open for more than 40 years and was solved only quite recently [1]. Our mechanical solution is more general than the original one. We also have used the prover to prove and find theorems in projective geometry.

In Bolyai-Lobachevskian geometry, our prover has confirmed about 7 possibly new theorems such as the 9-point circle theorem and the Butterfly theorem (As a consequence of a well-known result, however, certain theorems with a projective nature are valid in this geometry if they are valid in projective geometry). We proved more than 40 theorems in Minkowskian plane geometry and found an unexpected meta-theorem inspired by our experimental results.

Quite recently, the prover has been extended to prove theorems involving inequality [4]. About 30 of such theorems have been proved mechanically up to the last step, which requires the decision of the definiteness or semi-definiteness of some polynomials produced by the program. An unexpected fact is that in most cases the definiteness can be easily confirmed just by inspecting the polynomials produced. For example, a hard theorem –"In a triangle, the internal bisector on the smaller side is larger than the internal bisector on the larger side"– can be reduced to deciding the definiteness of the expression

$$\frac{((y_3^2+1)y_2^2+y_3^2)((y_3^4+y_3^2)y_2^4+(y_3^4+3y_3^2+1)y_2^2+y_3^2)}{y_1^2(y_2^2+y_3^2)(y_2^2+1)^2(y_3^2+1)^2}.$$

For most other theorems, the expressions are simpler than the above, e.g., y_1^2/y_2^2 for Pasch Axiom. As we know, deciding the definiteness of polynomials is a hard problem, which is closely related to Hilbert's 17th problem.

References

[1] K.B. Taylor, "Three Circles with Collinear Centers", *American Mathematical Monthly* **90** (1983), 486-487.

[2] S.C. Chou, "Proving Elementary Geometry Theorems Using Wu's Algorithm", in *Automated Theorem Proving: After 25 years*, American Mathematical Society, Contemporary Mathematics **29** (1984), 243-286.

[3] S.C. Chou, "A Method for Mechanical Derivation of Formulas in Elementary Geometry", February 1986, submitted to *Journal of Automated Reasoning*.

[4] S.C. Chou, "A Step toward Mechanical Proving Geometry Theorems Involving Inequality – Experimental Results", Preprint, March 1986, Institute for Computing Science, University of Texas at Austin.

[5] S.C. Chou and W.F. Schelter, "Proving Geometry Theorems with Rewrite Rules", Preprint, December 1985.

[6] Wu Wen-tsün, "On the Decision Problem and the Mechanization of Theorem Proving in Elementary Geometry", *Scientia Sinica* **21** (1978), 157-179.

[7] Wu Wen-tsün, *Basic Principles of Mechanical Theorem Proving In Geometries*, (in Chinese) Peking 1984.

The Markgraf Karl Refutation Procedure (MKRP)

N. Eisinger, H. J. Ohlbach
FB Informatik, Universität Kaiserslautern
D-6750 Kaiserslautern, West Germany

System Summary

MKRP is a general purpose theorem prover for many sorted first order logic [13]. It accepts type specifications and arbitrary first order formulae, simplifies them, and translates them into clauses. If possible, the problem is split into several independent subproblems. The clause part of the system uses an extension of Kowalski's connection graph procedure [5,4]. The clause set is represented as a graph, where various types of edges (links) stand for various relations between literals, such as resolvability or unifiability. These links provide the basis for resolution, factoring, subsumption, tautology detection, and similar operations. Equality is handled with "paramodulated connection graphs" based on so called P-links. A P-link connects one side of an equation with a unifiable subterm and thus indicates a potential paramodulation operation [11].The basic algorithm is a loop:

WHILE empty clause not found and graph not collapsed:

(1) Select a chain of resolution, factoring, or paramodulation links.
(2) Derive the final (and some intermediate) clauses generated by the corresponding sequence of operations.
(3) Insert these clauses into the graph and generate the links connecting the new clauses with the old graph using link inheritance [7].
(4) Remove the links operated upon.
(5) Perform any possible reduction on the graph.

There are various heuristics for the selection of an appropriate link chain in step (1), ranging from the valuation of the expected length and term depth of the derived clauses, over the detection of complex resolution, factoring, and subsumption sequences with the total effect to reduce the clause graph, up to the extraction of subgraphs corresponding to complicated unit resolution chains [1]. Step (5) includes the following operations:

- Elimination of tautologies, subsumed clauses, and pure clauses (a pure clause contains a linkless literal).
- Elimination of links that would produce a tautology, a subsumed clause, or a pure clause.
- Elimination of redundant literals. A literal is redundant, if there is an operation chain terminating with a clause that does not contain this literal, but is a variant of a subset of the parent clause. This is a generalization of Robinson's "replacement principle". A powerful look ahead mechanism based on the link structure of the graph has been implemented, which finds most of such chains.
- Generalization of clauses: a limited look ahead mechanism finds operation chains starting from a clause C and terminating with a clause containing less instantiated literals than C.

To some extent the data-driven mechanisms used in steps (1) and (5) are the same. They override the traditional strategies and in their combination account for most of the system's strength. Only a small number of mostly boolean options is available to influence the algorithms, thus the system works largely automatically rather than interactively. A detailed description of the current system is given in [6].

Implementation

Altogether the MKRP theorem prover consists of about 70 000 lines of Interlisp code corresponding to about 3 MB of compiled code. It is currently running on the Siemens 7500 series. There have been portations to IBM computers and to Symbolics Lisp Machines using the Interlisp compatibility package. The system has been used as the basis to develop the inference machine of the induction theorem prover of Walther et al. at Karlsruhe [3].

Current Activities and Future Plans

Two modules have been implemented, but not yet integrated into the system:

- A module that automatically translates unsorted clause sets into many sorted clause sets [9].
- A universal equality procedure that unifies two given terms under arbitrary equational theories [2]. This module will be used to replace paramodulation by E-resolution as the basic equality handling mechanism.

The whole system is currently being redesigned and reimplemented in COMMON LISP (on Symbolics, Microvax and Apollo workstations). It is based on a further theoretical generalization of the connection graph procedure, with links suitable for representing theory resolution operations involving arbitrarily many clauses at once [12, 8]. Frequently occurring axioms are handled by special purpose unification algorithms, thus relieving the user from specifying these axioms and the selection component from considering deduction steps whose effect can also be achieved by simple computation. The output of a run will be transformed into a so called refutation graph [10], which is an abstract and compact representation of not only one, but a whole class of proofs. This graph is to be manipulated in order to optimize the proof or to generate resolution proofs with reversed direction, e.g. to translate a refutation into a positive proof. It will also serve as a basis for translating the proof into the Gentzen calculus and finally into natural language.

Related activities include the investigation of languages like LCF or PRL to provide a more adequate representation level for mathematical problems and for the relevant control knowledge, and the design of knowledge bases for particular mathematical fields.

Acknowledgement: This work was partially supported by the Deutsche Forschungsgemeinschaft, SFB 314, "Künstliche Intelligenz".

References

1. Antoniou, G., Ohlbach, H.J., *Terminator*, Proc. 8th IJCAI, Karlsruhe (1983), 916-919.
2. Bläsius, K.H., *Equality Reasoning with Equality-Paths*, Proc. GWAI-85, Springer (1985).
3. Biundo, S., Hummel, B., Hutter, D., Walther, C., *The Karlsruhe Induction Theorem Proving System*, Proc. CADE-8 (this volume), Oxford (1986).
4. Eisinger, N., *Completeness, Confluence, and Related Properties of Clause Graph Resolution*, Dissertation, FB Informatik, Univ. Kaiserslautern (forthcoming 1986).
5. Kowalski, R., *A Proof Procedure Using Connection Graphs*, JACM, Vol .22 No.4 (1975), 424-436.
6. Karl Mark G Raph: *The Markgraf Karl Refutation Procedure*, Seki-84-08-KI, FB Inf., Univ.Kaiserslautern.
7. Ohlbach, H.J., *Link Inheritance in Abstract Clause Graphs*, JAR (forthcoming 1986).
8. Ohlbach, H.J., *The Semantic Clause Graph Procedure - A First Overview*, (submitted to GWAI-86).
9. Schmidt-Schauss, M. *Mechanical Generation of Sorts in Clause Sets*, Memo-Seki-85-6-KI, FB Informatik, Univ. Kaiserslautern (1985).
10. Shostak, R.E., *Refutation Graphs*, Artificial Intelligence 7 (1976), 51-64.
11. Siekmann, J., Wrightson, G., *Paramodulated Connection Graphs*, Acta Informatica (1979)
12. Stickel, M.E., *Automated Deduction by Theory Resolution*, JAR, Vol.1, No.4 (1985), 333-356.
13. Walther, C., *A Many-Sorted Calculus based on Resolution and Paramodulation*, Proc. 8th IJCAI, Karlsruhe (1983), 882-891.

The J-Machine: Functional Programming with Combinators

Jacek Gibert

Department of Computer Science, University of Melbourne, Parkville, Victoria, 3052, Australia

Functional programming languages, such as Turner's KRC (and its predecessor SASL) or Backus' FP, are inspired by the combinatory logic approach to the treatment of computable functions. The main advantage of such programming languages lies in their clear mathematical basis, which allows one to describe algorithms in terms of functions, thus exhibiting the algebraic properties on which algorithms depend, and providing powerful verification methods.

In Backus' FP one needs only to use elementary algebraic laws for operators to do proofs of equivalence and correctness for functional programs. However, the complete algebraic treatment of functional programs is made complicated because an additional domain is needed for reasoning about higher order functions. This is because Backus' FP is an incomplete version of the combinatory theory. Backus uses only a restricted set of very high level combinators, called *program forming operators*, that can be expressed by FP Algebra of Programs, e.g. he does not allow combinators of higher types, therefore general higher order functions or new *program forming operators* cannot be defined by the FP algebra.

Turner allows any higher order functional program to be defined in his SASL by closely emulating the combinatory logic approach to the treatment of higher order computations. He uses a small number of primitive combinators to translate functional programs written in SASL into corresponding combinator expressions which are then executed by a *reduction machine*. However, compilation of a program in SASL is slow and the resulting combinatory code is far removed from the source program, therefore hard to interpret and debug. The computation of such code progresses in very small steps and is deeply nested.

In [Gibert84a] [Gibert86] we further advance the idea of using combinators for a combinatory based implementation of functional programs by improving on weak points of Turner's and Backus' approaches. We aim to construct an optimal algebraic programming system which has the power to accommodate any functional programming language inspired by the lambda calculus. That is, our system should provide a functional language with a natural computational model and an associated elementary verification system. At the same time, it should lay a basis for an efficient machine architecture to implement this functional language.

For this purpose, the *J* reduction language is designed in accordance with the algebraic construction of our combinatory system, the *J* algebra of functions. Having a programming and a theory language combined in one *J* language avoids an additional metalevel for proving properties of programs and allows our combinatory reduction machine, the **J-Machine**, to assist a programmer in carrying out the proofs and program transformations. For example, the equivalences between functional programs can be simply proven by analysing the symbolic computations of the programs since the evaluation of a functional program closely follows the equations of the J algebra. The basic proof tool used by the J-Machine to prove functional equivalences between programs is the derivation rule called ***extensionality***.

A programmer can express in the J language all computable functions of any order and of any number of arguments in a simple and efficient way using recurrence relations (simple iterations) over the domain of streams (arbitrarily long tuples). They are grouped into classes indexed by variables ranging over natural numbers. Therefore, a property that is satisfied by a whole class of functions can be proven by the J-Machine using ***simple induction*** on an index variable.

The efficient implementation can be achieved by utilizing such properties of the J language as distributivity and freedom from bound variables in parallel executions of functional programs. These properties imply that subexpressions of most language expressions contain all necessary information to be evaluated separately and possibly in parallel.

The author has made a preliminary implementation of a *sequential* J-Machine in the C language on a Vax 11/780 at the University of Melbourne which is described in [Gibert84b]. A preliminary performance analysis of this J-Machine interpreter has been very encouraging. The recent advent of hardware concepts like data flow or reduction architecture makes a VLSI implementation of a parallel J-Machine feasible and appealing, since the J-Machine is mainly oriented towards symbol manipulation and it aims at exposing flows of data and parallelism in ordinary functional programs.

References

[Gibert84a]

J. Gibert, Functional Programming with Combinators, in *Proceedings of the Logic and Computation Conference*, vol. 1, J. N. Crossley and J. Jaffar, (eds.), Monash University, Australia, JAN 1984.

[Gibert84b]

J. Gibert, J-Machine Users' Manual, Technical Report 84/10, University of Melbourne, Parkville, Australia, SEP 1984.

[Gibert86]

J. Gibert, Functional Programming with Combinators, *to appear in Journal of Symbolic Computation*, 1986.

The Illinois Prover: A General Purpose Resolution Theorem Prover

Steven Greenbaum
DCL 240
Department of Computer Science
University of Illinois
1304 W. Springfield Avenue
Urbana, Illinois 61801

David A. Plaisted
Department of Computer Science
New West Hall 035A
University of North Carolina at Chapel Hill
Chapel Hill, North Carolina 27514

The Illinois prover is a resolution theorem prover for first-order logic with equality, implemented as part of the first author's Ph.D. thesis at the University of Illinois at Urbana-Champaign. This prover was originally intended as a testbed for the abstraction strategies given in Plaisted[81]. We needed a resolution theorem prover mainly for comparison purposes to know how good the abstraction strategy was. However, the resolution theorem prover improved so dramatically that our focus shifted to it instead. Abstraction may be useful on problems with many predicate and function symbols and short proofs, unlike the examples we tried.

The purpose of this prover is to be uniformly good on problems of moderate difficulty without any user guidance at all. The prover is intended for users who know nothing about mechanical theorem proving. This differs from the ITP prover of Argonne National Laboratory (Lusk and Overbeek[84]) which is intended for very hard problems and makes use of extended human interaction. We feel that a prover such as ours will have applications to program verification and other areas where the proofs are not extremely hard, just tedious for humans. We have run our prover on a wide variety of problems from a number of sources, and the default strategy does seem to be uniformly good. The prover also has a number of switches which an advanced user can set to improve performance on specific domains if desired.

The prover is implemented in Franz lisp and the compiled prover takes about a meg of memory. The basic strategy is a refinement of locking resolution in which negative literals resolve before positive literals, together with a refinement of unit resolution. The former strategy simulates forward chaining and the latter strategy simulates a restricted back chaining, so the overall strategy of the prover is a combination of forward and backward chaining. Much attention has been given to design of algorithms and data structures for the sake of time and storage efficiency. We feel that the success of the prover is due largely to the priority system. Each potential resolution is given a priority, based on the number of literals in the clauses, the sizes of the clauses, and the effort expended to generate the clauses. At each step the resolution of lowest numerical priority is performed. The prover also has a version of "set-of-support" that uses the priority system to prefer supported clauses. The features in the prover evolved gradually with extended use. Many problems that were difficult or impossible for early versions of the prover are now trivial.

The prover also has a number of input transformations. One of these permits it to simulate paramodulation by resolution. This has been extended to permit an implementation of a variant of the Knuth-Bendix completion procedure for equational systems. Also, the structure-preserving clause form translation of Plaisted and Greenbaum[86] has been built in. This non-standard clause form translation avoids many of the problems with the usual translation to clause form, overcoming some of the objections to resolution based on problems with translation to clause form. Finally, a method of rewriting before Skolemizing is built in, which has led to a dramatic improvement in one problem. The idea is to replace certain predicates by their definitions before translating to clause form.

We give sample times on various problems; garbage collection time is included. These times are for a VAX 750, and are all for the default strategy. That is, the user simply types in the theorem, and the prover finds the proof with no additional guidance. In some published works it is difficult to evaluate the results presented because the degree of user guidance is not given, nor is the general utility of the strategy given. The nine examples at the end of Chang and Lee[73] can all be obtained in a few seconds, most in about one second. Three of the examples from Plaisted[82] can also be obtained in similar times, as can the examples from Reboh et al[72]. The "latin squares" problem from Robinson[63] takes 46 seconds. The "schubert steamroller" problem from Walther[84] takes 56 seconds in the form given. Completion of the three axioms of group theory to obtain a canonical term rewriting system takes about 20 seconds. The problem of showing that $P(x) \cap P(y) = P(x \cap y)$ where P(x) is the powerset of x, takes 5 or 10 seconds, using input transformations given above. Search spaces are also fairly small. Of course, coding in C could substantially reduce the times, but we have no plans to do so.

We feel that these results show that we have achieved our goal of a good uniform proof procedure based on resolution. We know of no other prover that is as good on this class of problems with this little guidance. We have found that good engineering techniques such as careful use of priorities are much more important than mathematical completeness proofs of new strategies. This suggests that much of the emphasis in the sixties on new strategies may have been misguided, and also suggests that the belief that resolution has "failed" in some sense, is premature. Also, we strongly disagree with some published work that suggests that all strategies are about equally good. We have found marked and consistent differences between different versions of our prover on many examples.

We do not mean to emphasize this prover as much as the lessons we have learned, which we feel should enable almost anyone to write a small, fast, good resolution theorem prover. We also feel that such provers may justify a renewed application of uniform proof procedures to certain sub-problems in artificial intelligence.

This prover has been used extensively by Sam Kamin's group at the University of Illinois for verifying properties of abstract program specifications. It has been distributed to several sites. For information about obtaining a copy, contact one of the authors at the addresses given above. A finished version should be ready shortly.

References

Chang, C. and Lee, R., Symbolic Logic and Mechanical Theorem Proving (Academic Press, New York, 1973).

Lusk, E. and Overbeek. R., A portable environment for research in automated reasoning, Proceedings of the 7th Conference in Automated Deduction, R. Shostak, ed., Lecture Notes in Computer Science 170, G. Goos and J. Hartmanis, eds. (Springer-Verlag, New York, 1984), pp. 43-52.

Plaisted, D., Theorem proving with abstraction, Artificial Intelligence 16 (1981) 47 - 108.

Plaisted, D., A simplified problem reduction format, Artificial Intelligence 18 (1982) 227-261.

Plaisted, D., and Greenbaum, S., A structure-preserving clause form translation, Journal of Symbolic Computation, 1986, to appear.

Reboh, R, Raphael, B., Yates, R., Kling, R., and Velarde, C., Study of automatic theorem-proving programs, Technical note 75, Artificial Intelligence Center, SRI International, Menlo Park, California, 1972.

Robinson, J., Theorem proving on the computer, J. ACM 10 (1963) 163-174.

Walther, C., Schubert's Steamroller - a case study in many sorted resolution, technical report, Institut for Informatik, May, 1984.

Theorem proving systems of the Formel project

Gérard Huet

INRIA
Domaine de Voluceau
78150 Rocquencourt
FRANCE

1 On overview of the Formel Project

The Formel project, at Inria Rocquencourt, has for long-range goal the mechanization of reasoning, and its application to the design of programming environments for verifiably correct software. Over the years various research prototypes of reasoning systems have been implemented for experimental purposes. The main two areas are equational reasoning by rewrite rules techniques, and higher-order constructive type theory.

2 Equational and Rewriting Reasoning

2.1 The KB system

Designed and implemented originally by J. M. Hullot [11], the KB system implements the Knuth-Bendix completion algorithm, including a full treatment of commutative-associative operators. A subsystem implements the Huet-Hullot [10] system for proofs in initial algebras. It is also posiible to experiment with propositional logic tautology checking, à la Hsiang-Fages [8]. A trace system permits to extract equational proofs from completions. The system exists in Maclisp on Multics, in Zeta Lisp on Symbolics 3600, and in Franz Lisp on Vax-Unix. A French manual is available [9].

2.2 The Al-Zebra system

This system, designed and implemented by Ph. Le Chenadec, solves word problems in finitely presented algebras. It deals with specialized data structures for representing the associative and commutative-associative structures of groups and rings. Extensive experimentations with group theory have led to a monography on the topic [13]. This system has been implemented in Franz Lisp on Vax-Unix. A French manual is available [12].

3 Higher-Order Reasoning

The main efforts of the project are now directed towards higher order proof systems for programming environments. Two systems are under development: a new implementation of ML, and an implementation of the Calculus of Constructions.

3.1 ML

The programming langauge ML was originally the meta-language of the LCF proof assistant. The current Lisp implementation, done in collaboration with Cambridge University, is available under Maclisp, ZetaLisp, FranzLisp and Le_Lisp. It is fully documented in the ML Handbook [1]. A new implementation Le_ML is now under development, which will serve as meta-language of our future provers.

3.2 The Constructions System

The Calculus of Constructions is a higher-order formalism for constructive proofs in natural deduction style, inspired from Automath and related systems of Girard and Martin-Löf. It is documented in [2,4,6,7]. An implementation in ML is in progress. Numerous examples are reported in[5,14].

References

[1] "The ML Handbook, Version 6.1." Internal document, Projet Formel, Inria (July 1985).

[2] Th. Coquand. "Une théorie des constructions." Thèse de troisième cycle, Université Paris VII (Jan. 85).

[3] Th. Coquand. "An analysis of Girard's paradox." First Conference on Logic in Computer Science, Boston (June 1986).

[4] Th. Coquand, G. Huet. "A Theory of Constructions." Preliminary version, presented at the International Symposium on Semantics of Data Types, Sophia-Antipolis (June 84).

[5] Th. Coquand, G. Huet. "Constructions: A Higher Order Proof System for Mechanizing Mathematics." EUROCAL85, Linz, Springer-Verlag LNCS 203 (1985).

[6] Th. Coquand, G. Huet. "Concepts Mathématiques et Informatiques Formalisés dans le Calcul des Constructions." Colloque de Logique, Orsay (Juil. 1985). To appear, North-Holland.

[7] Th. Coquand, G. Huet. "The Calculus of Constructions." To appear, JCSS (1986).

[8] F. Fages. "Formes canoniques dans les algèbres booléennes et application à la démonstration automatique en logique de premier ordre." Thèse de 3ème cycle, Univ. de Paris VI (Juin 1983).

[9] F. Fages. "Le système KB: présentation et bibliographie, mise en œuvre." Rapport Interne Inria (1984).

[10] G. Huet, J.M. Hullot. "Proofs by Induction in Equational Theories With Constructors." JCSS **25,2** (1982) 239–266.

[11] J.M. Hullot "Compilation de Formes Canoniques dans les Théories Equationnelles." Thèse de 3ème cycle, U. de Paris Sud (Nov. 80).

[12] Ph. Le Chenadec. "Le système Al-Zebra pour algèbres finiment présentées, manuel de référence." Rapport Technique, Greco de Programmation (1985).

[13] Ph. Le Chenadec. "Canonical forms in finitely presented algebras". Lecture Notes in Theoretical Computer Science, Pitman-Wiley (1986).

[14] C. Mohring. "Algorithm Development in the Calculus of Constructions." IEEE Symposium on Logic in Computer Science, Cambridge, Mass. (June 1986).

The Passau RAP System: Prototyping Algebraic Specifications Using Conditional Narrowing

Heinrich Hussmann
Universität Passau
Fakultät für Mathematik und Informatik
Postfach 2540
D-8390 Passau
F. R. G.

Abstract

The Passau RAP system is an interpreter for algebraic specifications which contains a conditional narrowing algorithm as its central part. The system is in use since the end of 1984, a revised version is currently under development.

The aim of the system is to supply a most flexible prototyping facility for algebraic specifications.This means that the specification is given an operational semantics with a logic-programming flavour. The interpreter is not only able to evaluate specified functions, but also to solve systems of equations formulated over the specified functions. This comprises

- systematic test of the specification and hence a comparison with the informal requirements behind it
- automatic generation of test data for later implementations
- proof of simple algebraic properties of the specified functions.

The specification language used is that of hierarchical algebraic specifications ([Wirsing et al. 83]). This language contains a modularisation mechanism as well as a type concept. The axioms of specifications are in general of the form

$$t_1 = s_1 \ \& \ldots \& \ t_n = s_n \ \Rightarrow \ l = r,$$

interpreted by the system as conditional rewrite rules.

Given a set R of such conditional rewrite rules and a system of equations T

$$p_1 = q_1, \ldots , p_m = q_m,$$

the conditional narrowing algorithm enumerates solutions of T, i.e. substitutions σ such that

$$\sigma p_1 = \sigma q_1, \ldots , \ \sigma p_m = \sigma q_m$$

are derivable in the conditional-equational calculus from the axiom set R.

The conditional narrowing algorithm combines term rewriting and resolution techniques similar to [Fribourg 84]. It consists in a search in a tree of goals. A goal is a pair of a system of equations and a substitution. The start goal is given by $\langle T ; \emptyset \rangle$ where T are the equations to solve and $\emptyset$ the empty substitution. The algorithm performs steps of two kinds:

Unification step:
If in a goal ‹T ; σ› there is a most general unifier τ for all the left hand side / right hand side pairs in T , then the substitution τσ is a solution.

Narrowing step:
If in a goal ‹T ; σ› there is a subterm T/u of T which is unifiable with the left hand side l of a rule $t_1 = s_1$ & ... & $t_n = s_n \Rightarrow l = r$ using a most general unifier τ, then the new subgoal

$$\langle \tau t_1 = \tau s_1, \dots, \tau t_n = \tau s_n, \tau T[u \leftarrow r] \ ; \ \tau\sigma \rangle$$

has to be considered.

The narrowing steps define a tree of subgoals the root of which is the start goal. The enumeration of solutions is obtained by a fair search of this tree for subgoals where a unification step is applicable.
If the given set of rules is confluent, then this algorithm has been proven correct and complete w.r.t. conditional-equational logic ([Hussmann 85]).

The actual implementation in RAP contains a number of additional optimisations:
- simplification of subgoals by conditional term rewriting
- expansion of auxiliary variables
- use of irreducible function symbols for decomposition of subgoals and unsatisfiability detection
- removal of subgoals which are a special case of another subgoal or a known solution (subsumption).

Further useful properties of the system are:
- optional depth-first and breadth-first search of the proof tree
- an interactive debugger
- a facility for switching off the various optimizations for experimental purposes.

The system is written in PASCAL (approx. 8000 l.o.c), it is currently available for the VAX/VMS, UNIX 4.X BSD and VM/CMS operating systems. It has been installed and experimented at several external sites. The experimentation done in Passau with the system includes classical abstract data types as well as rather large and new specifications. The most complex specification tested with RAP up to now is a functional description of the INTEL 8085 microprocessor which contains about 250 axioms.

References:

[Fribourg 84]
L. Fribourg, Oriented equational clauses as a programming language. Proc. 11th ICALP, Lecture Notes in Computer Science 172, pp. 162-172, 1984.

[Hussmann 85]
H. Hussmann, Unification in conditional-equational theories. Proc. EUROCAL 85 Conf., Lecture Notes in Computer Science 204, pp. 543-553, 1985.

[Wirsing et al. 83]
M. Wirsing, P. Pepper, H. Partsch, W. Dosch, M. Broy, On hierarchies of abstract data types. Acta Informatica 20, pp. 1-33, 1983.

RRL: A Rewrite Rule Laboratory†

Deepak Kapur
Corporate Research & Development
General Electric Co.
Schenectady, NY, USA

G. Sivakumar
Dept. of Computer Science
University of Illinois
Urbana, IL, USA

Hantao Zhang
Dept. of Computer Science
Rensselaer Polytechnic Institute
Troy, NY, USA

The RRL (Rewrite Rule Laboratory) is a theorem prover based on rewriting techniques developed at General Electric Research and Development Center, and Rensselaer Polytechnic Institute. It is implemented in Franz Lisp and it runs on Vax computers, SUN workstations, as well as Symbolics Lisp machines. The work on this project is being conducted in cooperation with Prof. Guttag's group at MIT under a grant by the National Science Foundation.

The kernel of RRL is the extended Knuth-Bendix completion procedure which attempts to generate a complete (canonical) rewriting system from a finite set of formulae. An equational approach to theorem proving for first-order predicate calculus developed by Kapur and Narendran has also been implemented in RRL.

The input to RRL is a first-order theory specified by a finite collection of axioms - a mixture of equations, conditional equations and arbitrary first-order formulae, in which operators can be commutative or associative-commutative. The system attempts to generate a complete set of rules which serves as a decision procedure for the first-order theory given as the input. For proving a first-order formula, RRL uses the proof-by-contradiction method. The set of hypotheses and the negation of the conclusion are given as the input, and RRL attempts to generate a contradiction, which is a complete system including the rule true → false.

RRL uses recursive path orderings and lexicographic recursive path orderings to orient equations into terminating rewrite rules. This can be done either with the user's help in which case, whenever needed, the user is asked for precedence relation on function symbols appearing in the axioms; it is always possible to undo any wrong or premature choice made by backtracking to the appropriate point. Or, RRL can be run in an automatic mode in which the system itself determines an ordering for rules that guarantees termination of rewriting.

RRL supports different strategies for normalization and generation of critical pairs, crucial concepts in the completion procedure. These strategies play a role in the completion process similar to different strategies and heuristics available in resolution-based theorem provers.

---

† This project is partially supported by the National Science Foundation Grant nos. DCR-8211621 and DCR-8408461.

Complete rewriting systems generated by RRL can be used to prove theorems in an equational theory, check for consistency of a theory, prove a first-order formula using the refutational approach, as well as prove inductive properties using the *inductionless induction* method. RRL also has algorithms to prove the sufficient-completeness property of equational specifications of abstract data types.

RRL has been used (i) to generate decision procedures for many free algebraic structures including free groups, free l-r systems, free r-l systems, free rings, free boolean rings, free modules, etc; (ii) to show the equivalence of different axiomatizations of free groups, free boolean rings, etc; (iii) to prove many first-order formulae, including most examples from the AAR newsletters, such as Schubert's steamroller example, Lewis Carroll's puzzle, Smullyan's problem, Peter Andrews' problem involving equivalences, and the Truthtellers and Liars Puzzle, as well as (iv) for solving word problems for finitely presented abelian monoids and abelian groups. RRL has also been used to prove many inductive properties of abstract data types from their equational specifications, and to check the sufficient-completeness and consistency properties of abstract data type specifications.

Apart from being a theorem prover, RRL provides an environment to facilitate research into the term rewriting approach to automated deduction. By running experiments on RRL, we have been able to introduce a number of improvements in the algorithms in RRL. In particular, we have been led to study normalization strategies and complexity of matching and unification algorithms, identify redundant critical-pairs, and examine variations in the completion procedure to make it faster. Providing support for such activities and experiments is one of the main purposes of building RRL.

A Partial List of References

1. Dershowitz, N., "Termination," *Proc. of the First Intl. Conf. on Rewriting Techniques and Applications*, Dijon, France, May 1985, LNCS 202, Springer Verlag, pp. 180–224.

2. Kapur, D., and Sivakumar, G., "Architecture of and Experiments with RRL, a Rewrite Rule Laboratory," *Proc. of a NSF Workshop on the Rewrite Rule Laboratory*, Sept. 6–9, 1983, General Electric CRD Report 84GEN004, April 1984, pp.33–56.

3. Kapur, D., and Narendran, P., "An Equational Approach to Theorem Proving in First-order Predicate Calculus," *Proc. of the 9th IJCAI*, Los Angeles, August 1985, pp. 1146–1153.

4. Kapur, D., Narendran, P., and Zhang, H., "Proof by Induction using Test Sets," to appear in the *Proc. of 8th Intl. Conf. on Automated Deduction (CADE-8)*, Oxford, England, July 1986.

5. Kapur, D., and Zhang, H., *RRL: A Rewrite Rule Laboratory – User's Manual.* Unpublished Manuscript, General Electric Corporate Research and Development, Schenectady, NY, March 1986.

6. Knuth, D.E., and Bendix, P.B., "Simple Word Problems in Universal Algebras," *Computational Problems in Abstract Algebras* (ed. Leech), Pergamon Press, 1970, pp. 263–297.

7. Musser, D.R., "Proving Inductive Properties of Abstract Data Types," Proc. of 7th ACM Symposium on Principles of Programming Languages, Las Vegas, Nevada, Jan. 1980.

A Geometry Theorem Prover Based on Buchberger's Algorithm

B. Kutzler, S. Stifter
Working Group CAMP, Institut für Mathematik
Universität Linz, A-4040 Linz, Austria

The System we describe is based on Buchberger's algorithm for computing Gröbner bases /Buchberger 1965/85/ and aims in automatically proving geometry theorems that can be formulated as algebraic problems of a special kind. Problems of this kind first have been investigated by /Wu 1978/84/. Roughly, in many cases one can find an adequate algebraic formulation of a given geometrical problem by choosing some coordinate system and coordinates for the points of the geometrical entities involved and expressing the hypotheses and the conjecture in form of polynomial equations using elementary vector calculus.

Example:
"The midpoints of the sides of an arbitrary quadrangle form a parallelogram."
An algebraical formulation of this geometrical problem could be as follows:

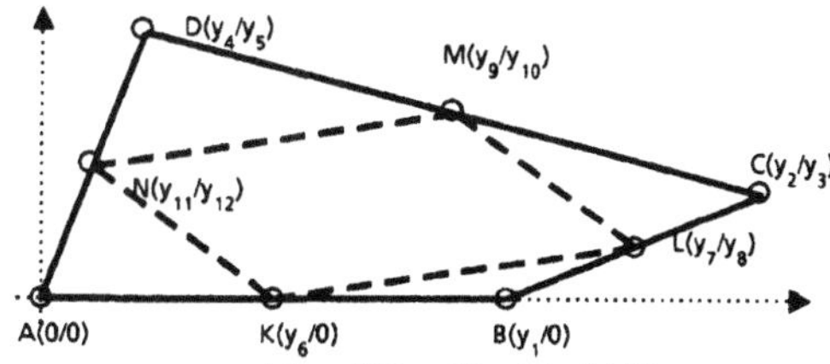

hypotheses:
$h_1 = 2y_6 - y_1 = 0$ (since K is midpoint of AB),
$h_2 = 2y_7 - y_1 - y_2 = 0$, $h_3 = 2y_8 - y_3 = 0$ (since L is midpoint of BC),
$h_4 = 2y_9 - y_2 - y_4 = 0$, $h_5 = 2y_{10} - y_3 - y_5 = 0$ (since M is midpoint of CD),
$h_6 = 2y_{11} - y_4 = 0$, $h_7 = 2y_{12} - y_5 = 0$ (since N is midpoint of AD),

conjecture:
$c = (y_7 - y_6)(y_{10} - y_{12}) - y_8(y_9 - y_{11}) = 0$ (i.e. KL is parallel to NM).

In general, the hypotheses and the conjecture are not sufficient for proving theorems. For most theorems one also has to exclude some cases (for instance "degenerate cases") by means of an inequality $s \neq 0$. Incorporating such inequalities, the problem is (following /Wu 1984/):

Problem:

Given: $h_1, \ldots, h_n, c \in \mathbf{Q}[y_1, \ldots, y_m]$

Find: $s \in \mathbf{Q}[y_1, \ldots, y_m]$, such that

$$\neg(\forall z_1, \ldots, z_m)(h_1(z_1, \ldots, z_m) = 0 \wedge \ldots \wedge h_n(z_1, \ldots, z_m) = 0 \Rightarrow s(z_1, \ldots, z_m) = 0) \wedge$$
$$(\forall z_1, \ldots, z_m)(h_1(z_1, \ldots, z_m) = 0 \wedge \ldots \wedge h_n(z_1, \ldots, z_m) = 0 \wedge s(z_1, \ldots, z_m) \neq 0 \Rightarrow c(z_1, \ldots, z_m) = 0)$$

Wu gave a complete decision procedure for this problem in algebraically closed fields using Ritt's bases (no full implementation exists). But complete procedures in algebraically closed fields are only of theoretical interest, since they argue about the complex numbers, whereas geometry problems are assertions about the real numbers. Therefore, for practical theorem proving only confirmations of geometry theorems can be made (since if c vanishes on all complex zeros of the hypothesis polynomials, it vanishes as well on all real zeros). Our prover uses Gröbner bases and Buchberger's algorithm and, as Wu's method, allows only to confirm theorems.

Our methods are based on a guess about the independence of some of the variables. This guess can be motivated by considering a possible "construction process" of the geometrical entities where "independent" variables are used for describing points that can be chosen arbitrarily, "dependent" variables are used for describing points that have to meet certain properties in relation to the points introduced earlier in the "construction". Our methods first give a complete decision whether or not the guess was correct and then construct a polynomial s that depends only on the independent variables such that $s.c \in \mathrm{Ideal}(h_1,\ldots,h_n)$ (i.e. $s.c = \Sigma_{i=1,\ldots,n} p_i h_i$) in case such an s exists. It is straightforward to see that such an s fulfills the two conditions required in the above problem.

The first method makes use of the fact that the question whether or not there exists such an s is equivalent to the "Main Problem of Ideal Theory" for the polynomial ring over the field of the rational functions in the independent variables, which is readily solved by a straightforward application of Buchberger's algorithm. The second method is based on a new notion of reduction, which we called "pseudo-reduction" and allows computation in the polynomial ring over the rationals. Both methods are equivalent but show quite a difference in efficiency. We give a rough sketch of the second procedure:

Algorithm:
input $h_1,\ldots,h_n,c,y_{i1},\ldots,y_{ir}$ /:comment $y_{i1},\ldots,y_{ir}$ denote the independent variables :/
$\{g_1,\ldots,g_t\}$: = Gröbner basis of $\{h_1,\ldots,h_t\}$ in $\mathbf{Q}[y_{i1},\ldots,y_{ir},y_1,\ldots,y_{i1-1},y_{i1+1},\ldots,y_{ir-1},y_{ir+1},\ldots,y_m]$
if $(\exists i)(g_i \in \mathbf{Q}[y_{i1},\ldots,y_{ir}])$ then output "The variables are not independent."; stop
(r,s): = the final pseudoremainder and the product of all necessary multipliers when
iteratively $(y_{i1},\ldots,y_{ir})$-pseudo-reducing c modulo $\{g_1,\ldots,g_t\}$
if $r=0$ then output "Theorem is confirmed assuming $s\neq 0$"

Using this algorithm based on an implementation of Buchberger's algorithm by R. Gebauer in the SAC-2 computer algebra system on an IBM 4341, the above theorem was proved in about 1 second considering $y_1,\ldots,y_5$ as independent variables. A detailed description as well as computing time statistics for approximately 50 non-trivial examples of geometry theorems and the correctness proofs of both methods can be found in /Kutzler,Stifter 1986a,b/. Other investigations on using Buchberger's algorithm for geometry theorem proving have been reported in /Kapur 1986/ and /Chou,Schelter 1986/.

References:

BUCHBERGER B., 1965/85: Gröbner Bases - An Algorithmic Method in Polynomial Ideal Theory. Chapter 6 in N.K. Bose (ed.): 'Multidimensional Systems Theory', R.Reidel Publ.Comp.

CHOU S.C., SCHELTER W.F., 1986: Proving Geometry Theorems with Rewrite Rules. Subm. to J. of Automated Reasoning.

KAPUR D., 1986: Geometry Theorem Proving Using Gröbner Bases. To appear as application letter in the J. of Symbolic Computation.

KUTZLER B., STIFTER S., 1986a: Automated Geometry Theorem Proving Using Buchberger's Algorithm. To appear in the Proc. of SYMSAC'86, Waterloo, Canada.

KUTZLER B., STIFTER S., 1986b: On the Application of Buchberger's Algorithm for Automated Geometry Theorem Proving. To appear as application letter in the J. of Symbolic Computation.

WU W.T., 1978/84: Basic Principles of Mechanical Theorem Proving in Elementary Geometries. Journal of System Sciences and Mathematical Sciences, vol. 4, no. 3, pp. 207-235.

REVE a Rewrite Rule Laboratory

Pierre LESCANNE

Centre de Recherche en Informatique de Nancy
Campus Scientifique BP 239
54506 Vandoeuvre Cedex France

1 — System Sumary

REVE is a rewrite rule laboratory intended to conduct experiments in equational theories based on rewriting techniques. It was built to provide both convenient tools for an experimented user and to make experiments using the most recent techniques in the field. Among others, it offers tools to compute convergent rewriting systems i.e., confluent and noetherian and to perform proofs by induction based on the method of proof by consistency.

REVE has four versions, REVE-1 was the prototype [LES,83], REVE-2 is a robust software for working on term rewriting systems, REVE-3 is a laboratory for equational rewriting based on Jouannaud and Kirchner's completion algorithm [J-K,84], [K-K,85], REVEUR-4 [R-Z,84] is a laboratory for conditional rewriting based on Remy and Zhang's ideas.

2 — Applications

REVE has its mains applications in equational algebra [LES-83] and abstract data type specifications. It has also applications in Petri nets properties [C-J,85] and data bases theory [C-K,85].

3 — Implementation

REVE is written in CLU and is available on VAX/UNIX and SUN workstation. Its architecture in described in [FOR-84]

4 — Developments

The main developments are currently on termination algorithms, proof by induction, introduction of new equational theories through unification algorithms, efficiency of equational rewriting, introduction of new strategies for conditional rewriting.

5 — List of participants

REVE-2 Alhem Ben Cherifa (CRIN), Dave Detlefs (MIT), Randy Forgaard (MIT), John Guttag (MIT), Azzedine Lazrek (CRIN), Pierre Lescanne (CRIN).

REVE-3 Claude Kirchner (CRIN), Hélène Kirchner (CRIN), Jalel Mzali (CRIN), Kathy Yelick (MIT).

REVEUR-4 Jean-Luc Remy (CRIN), Hantao Zhang (CRIN).

6 — References

[FOR,84] R. FORGAARD, "A program for generating and analyzing term rewriting system", Master of science in computer science MIT 1984. Jouannaud Kirchner

[C-J,85] C. CHOPPY, C. JOHNEN "Petrireve: Proving Petri Net Properties With Rewriting Systems", Proc. 1st Conference on Rewriting Techniques and Applications, Dijon (France), Lecture Notes in Computer Science vol. 202, Springer Verlag, pp. 271-286, 1985.

[C-K,85] S.S. COSMADAKIS, P.C. KANELLAKIS, "Two Applications of Equational Theories to Data base Theory", Proc. 1st Conference on Rewriting Techniques and Applications, Dijon (France), Lecture Notes in Computer Science vol. 202, Springer Verlag, pp. 107-123, 1985.

[J-K,84] J.P. JOUANNAUD, H. KIRCHNER, "Completion of a set of rules modulo a set of equations", Proceedings 11th ACM Conference of Principles of Programming Languages, Salt Lake City (Utah, USA), 1984.

[K-K,85] C. KIRCHNER & H. KIRCHNER, "Implementation of a general completion procedure parameterized by built-in theories and strategies", Proceedings of the EUROCAL 85 conference, 1985.

[LES,83] P. LESCANNE "Computer Experiments with the REVE Term Rewriting System Generator" 10th POPL, Austin, Texas, January 1983, pp 99-108

[R-Z,84] J.L. REMY, H. ZHANG, "REVEUR 4: a System for Validating Conditional Algebraic Specifications of Abstract Data Types" Proceedings of the 5th ECAI Pisa, 1984.

ITP at Argonne National Laboratory

Ewing Lusk, William McCune, and Ross Overbeek

Mathematics and Computer Science Division
Argonne National Laboratory
Argonne, Illinois 60439
U. S. A.

System Summary

ITP is a relatively easy-to-use, full-featured classical theorem prover. It can be used interactively or in batch mode (with its command stream coming from a file). The input language for control consists of about thirty commands, and the input languages for data consist of variations on the format of clauses.

The system's overall operation can be summarized as follows, in simplified form. There are four main lists: the axioms, set of support, have-been-given list (usually empty on input), and demodulators (unit equalities to be used as rewrite rules). The basic algorithm is a loop:

```
while (null clause not found and set of support not empty)
    (1) select a clause from the set of support (the given clause)
    (2) derive clauses from it, the axioms, and the have-been-given list
    for each newly generated clause,
        (3) apply all existing demodulators
        (4) check for subsumption by any existing clause
        (5) add the clause to the clause space (deleting clauses it
            subsumes and applying it to all existing clauses if it
            is a new demodulator
    move the given clause to the have-been-given list
```

This algorithm is modifiable using a host of options. In step (2) many inference rules can be selected, including hyper-resolution, Unit-resulting(UR)-resolution, paramodulation, unit resolution, binary resolution, linked UR-resolution[10], and others. A general pattern-matching language is available to encode heuristics governing selection in step (1), keeping of new clauses in step (5), and choosing the rewrite direction of new demodulators.

User interface features include interactive, prompting commands, menus for the options, and limited on-line help. Miscellaneous other features include a primitive case analysis mechanism, an interface to Prolog (described in [5]), and a proof examination subsystem. There is no special provision for induction.

Applications

ITP has been used for research in mathematics and logic, and (duplicating work originally done with AURA, its predecessor), has solved open questions in these areas. Other researchers have used it to study circuit design and validation and to verify properties of Petri nets representing fault-tolerant hardware configurations. It has been used to solve innumerable logic puzzles and mathematical exercises (primarily in algebra).

Architecture and Implementation

LMA (Logic Machine Architecture) is a layered set of subroutines supporting a formula database and an extensive set of inference operations. ITP (Interactive Theorem Prover) is a theorem prover built on the LMA package. The entire system is composed of three layers:

1. An indexed database for items of the abstract data type "object", which is used (in layer 2) to implement clauses, literals, and terms. Objects have subobjects, but only one copy of any particular object is kept. Objects are indexed by unification properties, making it possible to rapidly access the set of terms which are likely to be unifiable with a given object. Once an object is found via the index, it is possible to merely follow pointers to find the superterms, literals, and clauses containing the object. This mechanism is described in [8] and in [3].
2. Implementation of the abstract data types term, literal, clause, and list in terms of objects, together with clause-based inference rules, demodulation, and subsumption. Inference rules are implemented as a common abstract algorithm

with a special set of "exits" to define actions to be taken as specific points in the algorithm. This technique is described in [4].

3. This layer is ITP. It contains the interactive command loop, the implementation of the basic theorem-proving algorithm described above, the menu system for control of options, the proof examination facility, and various useful utilities. It is described in [5].

The system in total contains about 80,000 lines of Pascal, and is portable to any computing environment with a robust Pascal compiler. It has been run on VAX/UNIX, VAX/VMS, IBM/CMS, the Cray X-MP, and Sun/3, Ridge/32, Apollo, and Perq workstations. The UNIX, VMS, and CMS versions are distributed at nominal cost by

Numerical Algorithms Group, Inc.
1101 31st Street
Downers Grove, IL 60515
U.S.A.

It has been distributed to approximately two hundred sites. Documentation[1, 6, 7] includes reference manuals for Layers 1 and 2 and a tutorial-style manual for Layer 3 (ITP). The reference manuals for Layers 1 and 2 are of use to those using the LMA subroutine package as a foundation for special-purpose inference-based systems.

History and Future Plans

The ITP automated reasoning system was written at Argonne National Laboratory by Ross Overbeek, Ewing Lusk, and Bill McCune. It is the successor to the AURA system[9] developed by Ross Overbeek and others at Northern Illinois University and Argonne. It is currently being re-implemented using the ideas described in [2]. The new implementation is expected to look much the same to the user, but be approximately one hundred times faster.

References

1. Charles H. Applebaum, *Logic Machine Architecture Database Support System: Layer 1 User's Manual,* The MITRE Corporation, Bedford, MA (November, 1985).
2. Ralph Butler, Ewing Lusk, William McCune, and Ross Overbeek, "Paths to High Performance Automated Theorem Proving," *CADE-8* (July, 1986).
3. E. Lusk, William McCune, and R. Overbeek, "Logic machine architecture: kernel functions," pp. 70-84 in *Proceedings of the Sixth Conference on Automated Deduction, Springer-Verlag Lecture Notes in Computer Science, Vol. 138,* ed. D. W. Loveland, Springer-Verlag, New York (1982).
4. E. Lusk, William McCune, and R. Overbeek, "Logic machine architecture: inference mechanisms," pp. 85-108 in *Proceedings of the Sixth Conference on Automated Deduction, Springer-Verlag Lecture Notes in Computer Science, Vol. 138,* ed. D. W. Loveland, Springer-Verlag, New York (1982).
5. Ewing L. Lusk and Ross A. Overbeek, "A Portable Environment for Research in Automated Reasoning," pp. 43-52 in *Proceedings of the 7th International Conference on Automated Deduction, Springer-Verlag Lecture Notes in Computer Science, Vol. 170,* ed. R. E. Shostak, Springer-Verlag, New York (1984).
6. Ewing L. Lusk and Ross A. Overbeek, "Logic Machine Architecture inference mechanisms - layer 2 user reference manual- Release 2.0," ANL-82-84, Argonne National Laboratory (April, 1984).
7. Ewing L. Lusk and Ross A. Overbeek, "The automated reasoning system ITP," ANL-84-27, Argonne National Laboratory (April, 1984).
8. R. Overbeek, "An implementation of hyper-resolution," *Computers and Mathematics with Applications* **1**, pp. 201-214 (1975).
9. L. Wos, S. Winker, and E. Lusk, "An automated reasoning system," *Proceedings of the AFIPS National Computer Conference,* pp. 697-702 (1981).
10. L. Wos, R. Veroff, B. Smith, and W. McCune, "The Linked Inference Principle, II: The User's Viewpoint," in *Proceedings of the 7th International Conference on Automated Deduction, Springer-Verlag Lecture Notes in Computer Science, Vol. 170,* ed. R. E. Shostak, Springer-Verlag, New York

AUTOLOGIC at University of Victoria

Charles G. Morgan
Department of Philosophy
University of Victoria
Victoria, British Columbia
Canada V8W 2Y2

The AUTOLOGIC system was originally designed to find proofs of proposed theorems in arbitrarily specified (usually non-classical) propositional logics. As will become apparent below, the present implementation will handle either propositional or first order logics whose quantifier is "classical" in the sense that from E(x) one may infer E(t), where E is any well formed expression with free variable x and t is any term. We make no assumptions about the underlying motivation or semantic theory for the permitted logics. Our theorem prover is strictly syntactically driven and is therefore suitable for application to logics for which no semantic theory is known. It is applicable to logics both weaker and stronger than classical logic, as well as to classical systems. For complete details and proofs of appropriate theorems, the interested reader may consult [1].

For all logics we consider, the object language is presumed to contain propositional constants (and perhaps predicate constants) and arbitrary sentential connectives. The meta-language consists of the object language extended by the addition propositional variables (and perhaps term variables as well).

A logic is specified by (i) a finite list of axioms (meta-linguistic expressions containing no variables) or axiom schemes (meta-linguistic expressions containing free variables), and (ii) a finite list of inference rules or inference rule schemes. Each inference rule must consist of a finite list of "antecedent" expressions and a single "consequent" expression.

Basically, AUTOLOGIC consists of a simple implementation of a backward proof tree search. In the most elementary case, the root node of the tree consists of the proposed theorem. Other nodes on the tree consist of sequences of zero or more meta-linguistic expressions. Intuitively, the expressions on a given node are a list of "lemmas" which, if proved, would permit the proof of the desired theorem. Generation of the empty node indicates that a proof of the desired theorem has been found.

Since AUTOLOGIC manipulates sequences of meta-linguistic expressions, it may be used for the determination of some elementary meta-theoretical facts about a logic, in addition to theoremhood. Any constant which appears on the root node but which does not appear in the axioms or inference rules functions as a "Skolemized" universal quantifier, since such a constant may be replaced throughout a proof by any other appropriately typed expression that has no variables in common with the original proof. Similarly, any variable which appears on the root node functions as a "Skolemized" existential quantifier. A conjunctive query may be posed by using a sequence on the root node that contains more than one expression.

In its simplest form, AUTOLOGIC generates new nodes from an old node by first designating one member of the old node as a "target" formula and then applying the following:

(N.1) The target formula is checked to see if it can be unified with any of the axioms. For each match found, a new node is generated by deleting the target formula from the old node sequence and making the unification substitutions in the remaining formulas of the sequence.

(N.2) The target formula is checked to see if it can be unified with the consequent of any rule of inference. For each match found, a new node is generated by replacing the target formula in the old node sequence by the sequence of antecedent formulas from the rule in question and making the unification substitutions in the resulting sequence.

Both correctness and completeness can be proved for the system as described thus far; see [1] for details.

It should be noted that as a consequence of its syntactic orientation, in most cases AUTOLOGIC will not find disproofs of proposed theorems, since it will almost always be possible to generate at least one new node from an old node. For example, consider the rule of modus ponens. Using > as the symbol for the conditional, the rule permits the inference of y from the two formulas x > y and x. When searching for a proof of A, a node containing x and x > A will be generated. Next a node containing z and z > x and x > A would be generated. If the empty node were never generated (i.e., if A were not a theorem of the logic), then the generation of additional nodes would obviously continue in this fashion forever.

For most cases of serious interest, the number of nodes generated increases much too quickly for AUTOLOGIC to be of much use for any but the simplest of problems. Much research needs to be devoted to speed-up techniques. Even in its simplest implementation, the system permits the addition of derived inference rules and known theorems. The simple technique of comparing any new unit node with axioms and known theorems results in a substantial improvement in performance. We have also incorporated a "non-theorem" check, which allows the user to specify a list of non-theorems; any node containing a non-theorem is pruned. A special conditional chaining technique has been implemented. Variations on subsumption pruning and node ordering have been studied as well.

All of the speed-up techniques so far mentioned preserve completeness. We are currently investigating techniques which do not preserve completeness but which result in vast speed improvements for certain classes of problems. We are also investigating methods of including non-classical quantifiers permitting the user to specify arbitrary quantifier inference rules.

References

1. C. G. Morgan, "Autologic," Logique et Analyse, no. 110-111, pp. 257-282 (1985).

THINKER

Francis Jeffry Pelletier
Departments of Philosophy and Computing Science
University of Alberta
Edmonton, Alberta T6G 2E1
Canada

THINKER is an automatic theorem proving system which generates proofs of theorems (including arguments with premises) in the natural deduction format of Kalish & Montague (1964). The logic in which THINKER operates is first-order predicate logic with identity (this is an update from the report, Pelletier 1982). The theorem to be proved, and any premises for the argument, are entered in a fully-parenthesized form, and can have any wff composed of & (and), v (or), > (if-then), ~ (not), = (if and only if), A (universal quantifer) E (existential quantifier), - (identity), variables, constants, and predicates of any adicity. THINKER performs no pre-processing of formulas (eg., into clause form) but instead operates directly on the natural form.

The natural deduction system allows for assumptions to be made at various places, and subproofs to be generated from these assumptions. The proof which is generated is fully formatted (indented, with subproofs displayed) and fully annotated (each line of the proof mentions which previous line it comes from). Handling of the subproofs is the trickiest part of the program, since once a subproof is complete, the lines generated in the course of constructing the subproof are no longer available for use in the rest of the proof.

The user interface for THINKER includes interactive prompting commands, various debugging facilities, a statistics collector (to see, for example, how many times certain ultimately unsuccessful strategies were tried), a "unfinished proof" examination mechanism, and a help facility that allows the user to add new lines to a partial proof or to suggest new subproofs to try. A post-processor is available to print out very tidy proofs on the Xerox 9700 printer.

THINKER has performed quite well on a wide variety of elementary problems (see Pelletier 1986 for a sample). Its success is mostly due to its use of a natural deduction format (which allows one to partition antecedent lines into different types, depending upon what its main connective is) and a heavy use of memory to "remember" (by means of hash tables) all formulas of all types that have already occurred in the proof. Thus, for a trivial example, if the current subgoal is the formula P, a check will be made to see whether (among other things) a formula (Q > P) occurs among the antecedent lines, and if so whether also the formula Q does. THINKER does this simply by storing the string ⓐ > P) in a hash table (if <u>any</u> formula of that form is an antecedent line). A pointer from this template tells what the actual values of ⓐ these are - that is, for each formula of that form that is among the antecedent lines, what is the left side of the > . Since this too will be a string, it can be quickly looked up to determine whether it occurs in the antecedent lines. Templates are used for all connectives and free variables. Identity is handled by keeping track of equivalence classes of variables. (All these continually change due to new antecedent lines and completion of subproofs).

The reason that the management of formulas is so efficient has to do with the use of SPITBOL as the programming language. One merely needs to define each check as a SPITBOL pattern and use the SPITBOL TABLE facility. The use of strings, rather than lists or trees, enables us to keep the advantages of the natural form (the advantages in selecting a proof strategy) while allowing fast access to the "form" of various formulas (since we need not process any tricky data structures).

THINKER is currently available on VAX 780/UNIX and AMDAHL 5680/ MTS. Future extensions include the introduction of arbitrary functions, the use of (pseudo) parallelism to generate subproofs, and addiction of the lambda-calculus. This last is important since the ultimate goal of the system is to be attached to a natural language system to compute inferences and presuppositions. The NLU system under development makes heavy use of the lambda calculus in its representation of the logical form.

Pelletier, F.J. (1982) Completely Non-Clausal, Completely Heuristically Driven, Automatic Theorem Proving. Dept. Computing Science, University of Alberta, Tech. Report TR82-7.

Pelletier, F.J. (1986) "Seventy-Five Problems for Testing Automatic Theorem Provers" Jour, Automated Reasoning (forthcoming).

The KLAUS Automated Deduction System

Mark E. Stickel
Artificial Intelligence Center
SRI International
Menlo Park, California 94025

The KLAUS Automated Deduction System (KADS) is under development as part of the KLAUS project for research on the interactive acquisition and use of knowledge through natural language.[1] KADS was previously informally named CG5.

The principal inference operation is nonclausal resolution with possible resolution operations encoded in a connection graph [2]. The nonclausal representation eliminates redundancy introduced by translating formulas to clause form and improves readability. Special control connectives can be used to restrict use of the formulas. For example, $A \supset B$ specifies standard implication and allows either A or B to be resolved with either an axiom or a goal; $A \rightarrow B$ specifies forward chaining, i.e., A must be resolved with an axiom; $B \leftarrow A$ specifies backward chaining, i.e., B must be resolved with a goal.

Evaluation functions determine the sequence of inference operations in KADS. At each step, KADS resolves on the highest rated link. The resolvent is then evaluated for retention and links to the new formula are evaluated for retention and priority.

KADS supports incorporating theories for more efficient deduction, including by demodulation, associative and commutative unification, many-sorted unification, and theory resolution [4].

Theory resolution operations can be specified by rules. For example, the rule (as opposed to the assertion) $(x < y) \wedge (y < z) \supset (x < z)$ is used to specify that theory-resolution links are to be created between formulas. For example, using this rule, $(a < b)$ and $(b < c)$ in the formulas $(a < b) \vee P$ and $(b < c) \vee Q$ are linked and the single-step theory resolvent of the two formulas is $P \vee Q \vee (a < c)$.

KADS is the current assertional component (ABox) of the experimental Krypton knowledge-representation system [1]. Theory resolution is used to make the connections between information in the ABox and the TBox (the terminological component). In particular, connection-graph links

[1]This research was supported by the Defense Advanced Research Projects Agency under Contract N00039-84-K-0078 with the Naval Electronic Systems Command. The views and conclusions contained herein are those of the author and should not be interpreted as necessarily representing the official policies, either expressed or implied, of the Defense Advanced Research Projects Agency or the United States government.

can link literals that are not syntactically complementary, but are inconsistent according to the terminological definitions in the TBox. Thus, $Man(John)$ and $\neg Person(John)$ can be directly resolved on to derive the empty clause by use of appropriate TBox definitions. Also, connection-graph links can have a residue as well as a pair of literals to resolve on. The residue includes negated conditions for the inconsistency of the pair of literals if they are not directly inconsistent. For example, if $Coed(x)$ is defined in the Tbox as x is a woman and x is a student, $\neg Woman(John)$ can by obtained by resolving $Student(John)$ and $\neg Coed(John)$. Theory resolution links are computed directly from TBox definitions in Krypton and do not use the previously mentioned rule facility.

Besides the resolution theorem prover, KADS includes two other theorem provers: an interpreter-based Prolog technology theorem prover (a faster compiler-based Prolog technology theorem prover is described in these proceedings [5]) and an implementation of the Knuth-Bendix method with associative-commutative completion [3]. All three theorem provers are constructed using many of the same lower-level functions, but are not integrated at the top level.

KADS is not yet a polished system suitable for widespread use, but is a research tool under continuous development that still lacks some functionality and user amenities. It runs on the Symbolics 3600 Lisp Machine.

References

[1] Brachman, R.J., V. Pigman Gilbert, and H.J. Levesque. An essential hybrid reasoning system: knowledge and symbol level accounts of Krypton. *Proceedings of the Ninth International Joint Conference on Artificial Intelligence,* Los Angeles, California, August 1985, 532–539.

[2] Stickel, M.E. A nonclausal connection-graph resolution theorem-proving program. *Proceedings of the AAAI-82 National Conference on Artificial Intelligence,* Pittsburgh, Pennsylvania, August 1982, 229–233.

[3] Stickel, M.E. A case study of theorem proving by the Knuth-Bendix method: discovering that $x^3 = x$ implies ring commutativity. *Proceedings of the 7th International Conference on Automated Deduction,* Napa, California, May 1984, 248–258.

[4] Stickel, M.E. Automated deduction by theory resolution. *Journal of Automated Reasoning 1,* 4 (1985), 333–355.

[5] Stickel, M.E. A Prolog technology theorem prover: implementation by an extended Prolog compiler. *Proceedings of the 8th International Conference on Automated Deduction,* Oxford, England, July 1986.

The KRIPKE Automated Theorem Proving System

Paul B. Thistlewaite, Michael A. McRobbie and Robert K. Meyer
Automated Reasoning Project
Research School of Social Sciences
Australian National University
GPO Box 4, Canberra, ACT, 2601
Australia

System Summary

KRIPKE is a special purpose automated theorem proving system for a wide variety of non-classical logics, in particular relevant logics (see [1]), though it can be straightforwardly adapted to become an automated theorem proving system for a number of related modal logics.

This system implements a decision procedure for a class of non-classical logics originally due to Saul Kripke in [3] and then successively refined or extended to a wider class of such logics by Meyer in [7], McRobbie in [5] and Thistlewaite in [8]. This decision procedure is based on an elaborate Gentzen-style proof-theoretic analysis of these logics, the most sophisticated and efficient version of which is described in [10]. Put very roughly this decision procedure may be thought of as being something like the *specialization* of the decision procedure for intuitionistic logic based on, for example, the *set-based* intuitionistic Gentzen system G3 given in [2], to a class of similar Gentzen systems though ones that are characteristically *multiset-based* and unlike G3, lack the structural inference rule of weakening (also known as monotonicity or thinning). This specialization introduces complexities that makes the Kripke decision procedure many orders of magnitude more difficult than that for intuitionistic logic.

The implementation of the Kripke decision procedure involves a heuristically based but essentially breadth-first search of the proof search tree constructed from an extensively normal-formed version of the input. The space/time efficiency of this implementation is greatly enhanced by a wide variety of node-pruning techniques, the most important of which is the use of algebraic models to prune the proof search tree. Briefly this technique involves evaluating an appropriate representation of the multiset at a node in a proof search tree against an algebraic model of the logic in question. Those that (essentially) evaluate as true are passed back to the theorem prover, while those that evaluate as false are discarded. Note that these logics do *not* have *finite* characteristic models of this type, and so this technique, though extremely efficient, is still only approximate. Classes of such models can be specified at run time and the theorem prover is able to dynamically alter this class of models with respect to the properties of the multisets being generated in the proof search tree. (It should be noted that this technique is perfectly generalizable to an unlimited number of non-classical logics, e.g. intuitionistic and modal logics - see [4]). A range of other node-pruning techniques can be specified at run time as can many other features. If necessary, many of these features can also be altered interactively while the theorem prover is running. The best and most detailed description of KRIPKE can be found in [9].

Applications

Though a special purpose automatic theorem proving system, KRIPKE has been used extensively in a number of logical investigations. The principal one of these is sketched in [6]. Here KRIPKE was able to solve a large number of difficult and recondite open problems in the course of this investigation. It has also been routinely used to verify the provability or otherwise of a number of theorems in [1].

Implementation, History and Future Plans

KRIPKE is written in PASCAL and has been implemented on VAXs under UNIX and VMS, and on DEC KL-10s, Pyramids and SUN-3 computer workstations at the Australian National University and a number of other Universities. A partial history of the development of this system can be found in [6]. At present there are no plans to make this system available for distribution. However it is expected that it will, at least in part, be incorporated into some of the software that the Automated Reasoning Project will be developing.

References

1. A.R. Anderson and N.D. Belnap Jr., *Entailment: The Logic of Relevance and Necessity*, Vol.1, Princeton University Press, New Jersey, 1975.

2. S.C. Kleene, *Introduction to Metamathematics*, North-Holland, Amsterdam, 1952.

3. S.A. Kripke, "The Problem of Entailment", (Abstract), *Journal of Symbolic Logic*, 24 (1959), 324.

4. G.J. McGovern, *Automatic Theorem-Proving for Some Non-Classical Logics*, M.A. Thesis, La Trobe University, 1984.

5. M.A. McRobbie, *A Proof Theoretic Investigation of Relevant and Modal Logics*, Ph.D. Thesis, Australian National University, 1979.

6. M.A. McRobbie, R.K. Meyer and P.B. Thistlewaite, "Computer Aided Investigations into the Decision Problem for Relevant Logics: The Search for a Free Associative Connective", pp. 236-267 in *Proceedings of the Sixth Australian Computer Science Conference*, ed. L. Goldschlager, Basser Department of Computer Science, University of Sydney, 1983.

7. R.K. Meyer, *Topics in Modal and Many-Valued Logics*, Ph.D. Thesis, University of Pittsburgh, 1966.

8. P.B. Thistlewaite, *Automated Theorem-Proving in Non-Classical Logics*, Ph.D. Thesis, Australian National University, 1984.

9. P.B. Thistlewaite, M.A. McRobbie and R.K. Meyer, *Automated Theorem-Proving in Non-Classical Logics*, Manuscript, 1985.

10. P.B. Thistlewaite, M.A. McRobbie and R.K. Meyer, "Advanced Theorem-Proving Techniques for Relevant Logics", *Logique et Analyse*, 28 (1985), 233-256.

SHD-PROVER AT UNIVERSITY OF TEXAS AT AUSTIN

Tie Cheng Wang
Automatic Theorem Proving Project
The University of Texas at Austin
RLM 13.150
Austin, Texas 78712

Outline of the System

SHD-prover (Semantically Guided Hierarchical Deduction Theorem Prover) is a general-purpose theorem prover for proving theorems in first order logic. The system was developed at the University of Texas at Austin by the ATP group directed by W. W. Bledsoe. Besides the main proof procedure, the system contains a user interface and a data file generator.

For a simplest use of the system, the user needs only to supply the theorem to be proved and call the SHD-prover. The prover will run automatically until a proof is found or until the process is abandoned. The input theorem can be represented in the form of a well formed formula or in the form of set of clauses.

The user interface contains interactive commands and on-line help message, which allow the user to examine the deduction process, trace the history of the deduction, change control parameters, discard unpromising deduction paths, select interesting subgoals, etc.

The data file generator is designed for users to convey domain dependent knowledge and other human-like heuristics to help the prover to solve difficult problems. Instead of providing the main prover with the theorem to be proved straightforwardly, the user can provide it with a data file, which is produced originally by the data file generator from the input theorem, then modified by the user by doing the follows:

- Adding a model (example) of the input theorem;
- Applying a partial set of support strategy;
- Modifying certain control parameters.

The Basic Algorithm

The main proof procedure of SHD-prover is called hierarchical deduction, which is a particular goal-oriented and resolution-based proof procedure. This procedure proves a theorem by traversing a tree of nodes. Each node contains a different set of clauses and other information. All candidate goal clauses are contained in a goal-list. Each literal of each goal clause is indexed by a node name, through which a set of nodes can be located to obtain rule clauses for the resolution upon that literal.

At the beginning of a deduction, there is only a root node which contains all input clauses, while the goal-list contains only the input goal clauses. The literals of the input goal clauses are all indexed by 1, which corresponds to the name of the root node. The general proof process is as follows:

1. The first goal clause G of the goal-list is taken. Along with the index of the left-most literal of G, a set of rule clauses is obtained by retrieving the node indicated by this index and all parent nodes of this node. This index will be the parent name of the new node being produced in this round of deduction.

2. All "legal" resolvents are produced by resolving the goal clause against each clause of the set of rules upon the left-most literal of the goal clause. For each of the resolvents produced, the indices of the literals inherited from the goal clause are retained, but the indices of the literals inherited from the rule clause are replaced by a larger integer which is just the name of a node being produced.

3. If $\Box$ (the empty clause) is obtained, then the procedure returns "proved". Otherwise, the resolvents will be stored into the new node and also be inserted into the goal-list according to their priorities. Then the above process is repeated.

The "legal" resolvents produced by the prover are called hierarchical resolvents. They are produced under a number of completeness-preserving constraints. Mainly, they are local subsumption constraint, constraints on common tails, proper reduction refinement, global subsumption constraint, and a level subgoal reordering refinement.

Some domain dependent knowledge and human-like heuristics can be used to help the SHD-prover by applying a partial set of support strategy and/or by providing the prover with a well defined model of the input theorem. If a model is supplied, then only those subgoals that are false in the given model may be developed, those that are true in the model are allowed to be used only as rules, and those whose values are "unkown" in the model are discarded.

Implementation and Applications

SHD-prover is currently written with Zetalisp and running on SYMBOLICS LISP machines. This prover has been used to prove (without interaction) a series of difficult theorems in mathematics and logic.

The user can refer to [1] and [2] for a description of the SHD-prover. The method of designing models for SHD-prover is presented in [2]. A user's guide is available on request. For using our system, please contact us by the address shown on the title page.

References

[1] Wang, T.-C. Hierarchical deduction. Tech. Report ATP-78, Univ. of Texas at Austin (March 1984).

[2] Wang, T.-C. Designing examples for semantically guided hierarchical deduction. Proc. IJCAI-9, 1201-1207 (1985).

This series reports new developments in computer science research and teaching – quickly, informally and at a high level. The type of material considered for publication includes preliminary drafts of original papers and monographs, technical reports of high quality and broad interest, advanced level lectures, reports of meetings, provided they are of exceptional interest and focused on a single topic. The timeliness of a manuscript is more important than its form which may be unfinished or tentative. If possible, a subject index should be included. Publication of Lecture Notes is intended as a service to the international computer science community, in that a commercial publisher, Springer-Verlag, can offer a wide distribution of documents which would otherwise have a restricted readership. Once published and copyrighted, they can be documented in the scientific literature.

Manuscripts

Manuscripts should be no less than 100 and preferably no more than 500 pages in length.
They are reproduced by a photographic process and therefore must be typed with extreme care. Symbols not on the typewriter should be inserted by hand in indelible black ink. Corrections to the typescript should be made by pasting in the new text or painting out errors with white correction fluid. Authors receive 75 free copies and are free to use the material in other publications. The typescript is reduced slightly in size during reproduction; best results will not be obtained unless the text on any one page is kept within the overall limit of 18 x 26.5 cm (7 x 10½ inches). On request, the publisher will supply special paper with the typing area outlined.
Manuscripts should be sent to Prof. G. Goos, GMD Forschungsstelle an der Universität Karlsruhe, Haid- und Neu-Str. 7, 7500 Karlsruhe 1, Germany, Prof. J. Hartmanis, Cornell University, Dept. of Computer-Science, Ithaca, NY/USA 14850, or directly to Springer-Verlag Heidelberg.

Springer-Verlag, Heidelberger Platz 3, D-1000 Berlin 33
Springer-Verlag, Tiergartenstraße 17, D-6900 Heidelberg 1
Springer-Verlag, 175 Fifth Avenue, New York, NY 10010/USA
Springer-Verlag, 37-3, Hongo 3-chome, Bunkyo-ku, Tokyo 113, Japan

ISBN 3-540-16780-3
ISBN 0-387-16780-3